Iron Control: Practice and Research

Held at the 46th Annual Hydrometallurgy Meeting

Quebec City, Canada
11-15 September 2016

Editors:

E.M.L. Peek
M. Buarzaiga
J.E. Dutrizac

ISBN: 978-1-5108-5937-1

Printed from e-media with permission by:

Curran Associates, Inc.
57 Morehouse Lane
Red Hook, NY 12571

Printed by Curran Associates, Inc. (2018)

Metallurgical Society of CIM (MetSoc)
Suite 1250
3500 de Maisonneuve West
Westmount, Quebec H3Z 3C1
Canada

Phone: (514) 939-2710, EXT. 1329

metsoc@cim.org

Additional copies of this publication are available from:

Curran Associates, Inc.
57 Morehouse Lane
Red Hook, NY 12571 USA
Phone: 845-758-0400
Fax: 845-758-2633
Email: curran@proceedings.com
Web: www.proceedings.com

Iron Control – Practice and Research

Editors:
Edgar Peek
Mohamed Buarzaiga
and John Dutrizac

MET SOC
CIM ICM

Iron Control: Practice and Research

XXVIII | Québec City, Canada
IMPC 2016 Hosting COM 2016 CONFERENCE OF METALLURGISTS MET SOC

PROCEEDINGS OF THE 46th Annual Hydrometallurgy Meeting
SEPTEMBER 11 TO 15, 2016, QUÉBEC CITY, QUÉBEC, CANADA

Iron Control: Practice and Research

4th International Symposium on Iron Control in Hydrometallurgy

46th Annual Hydrometallurgy Meeting

Editors

E.M.L. Peek, M. Buarzaiga and J.E. Dutrizac
Quebec City, Canada

Symposium Organized by the Hydrometallurgy Section of the Metallurgy and Materials Society of the Canadian Institute of Mining, Metallurgy and Petroleum

MET SOC
CIM ICM

A Publication of The
Canadian Institute of Mining, Metallurgy and Petroleum
1250 - 3500 de Maisonneuve Blvd. West
Westmount, Québec, Canada H3Z 3C1
http://www.metsoc.org

MET SOC
CIM
ICM

Foreword

Iron control in hydrometallurgical processes remains a challenging area affecting both operational stability and project economics. Iron is often the predominant impurity in non-ferrous ore processing and its handling and control has a major impact on the design and economics of the process flowsheet. For example, the high cost of Ni recovery via hydrometallurgy in the current laterite ore flowsheets is dictated largely by the need to control iron.

The Fourth International Symposium on Iron Control in Hydrometallurgy aims to bring together international experts from around the world to share knowledge and lessons learned in order to help improve iron control in our industry. This truly international symposium builds on the foundations established by the first symposium, held in Toronto in 1986, the second symposium in Ottawa in 1996, and the third symposium in Montreal in 2006. Although great technical progress has been made in the thirty years since the first meeting, the iron control issue persists in most areas of hydrometallurgical processing. John Monhemius, one of the inaugural editors and organizers of the first Symposium, will reflect upon this challenge through a plenary lecture sponsored and organized by the Historical Metallurgy Section of MetSoc.

The Iron Control: Practice and Research volume contains papers which cover all aspects of the broad iron control problem in hydrometallurgy and illustrate that innovation is very much alive in the industry. The editors pondered long and hard on how to group these papers: by unit process operation, by commodity (cation), or by anion. In the end we decided to do all three. Hence, we have a session on how iron, and also arsenic, impacts solvent extraction, ion exchange and electrowinning; a chapter on how iron behaves in chloride systems, which addresses more than just acid regeneration; and several traditional sessions on iron management in Ni (laterites), Zn (jarosite) and Cu (chalcopyrite) flow sheets. Finally, we decided to group some papers that focus purely on iron itself and some on arsenic handling. Throughout all these sessions you will encounter flow sheet development, engineering equipment requirements, science and environmental compliance topics such as disposal concerns. There are also a few glimpses of iron becoming a commercial saleable by-product from non-ferrous metal production. It is hoped that this volume will become a valuable standard reference for iron control in hydrometallurgy.

This symposium proceedings volume is the result of over two years of effort that included the formulation of the program by the Organizing Committee and the preparation of manuscripts by the authors. Once the manuscripts were received, the tasks of refereeing and editing the various submissions were assumed by the editors, who are indebted to the authors for their efforts in preparing the papers and their understanding of the need for a consistently formatted volume.

We wish to thank all participants and their respective employers for their valued contributions and support. The indexing and formatting of the proceedings volume was done by the CIM-MetSoc staff. We especially recognize the contributions made by Ronona Saunders, Brigitte Farah and Janice Burke (CIM-MetSoc), the Iron Control 2016 organizing committee, and its designated referees for paper review.

The organization of the present Iron Control in Hydrometallurgy Symposium has been greatly facilitated by the active support provided by numerous co-sponsoring organizations. Special reference is made to the members of the Mining and Materials Processing Institute of Japan (MMIJ), who contributed greatly to the contents of this Symposium.

Québec City, Québec, Canada
June 2016

Mohamed Buarzaiga
John E. Dutrizac
Edgar M.L Peek
Editors

Editors' Biographies

Edgar Peek received a M.Sc. (1991) and Ph.D (1996) degree in Raw Materials Technology from the Department of Mining and Petroleum Engineering of the Delft University of Technology in The Netherlands. In 1995 he joined the Research and Development group of Falconbridge Limited in Sudbury, Ontario to work on the laboratory and pilot plant aspects of nickel laterite flow sheet development. In 1999 Edgar joined the Falconbridge Projects & Engineering group in Toronto to work on the joint hydrometallurgical and pyrometallurgical scoping study for the Koniambo laterite ore deposit. From 2002 to 2006 he was seconded to the Technical Services group of the Falconbridge Nikkelverk AS refinery in Kristiansand, Norway. In 2006 Xstrata Plc. acquired Falconbridge Ltd. and Edgar moved back to Sudbury, Ontario to become the Manager of Extractive Metallurgy at the newly created Xstrata Process Support (XPS). In 2011 Edgar became the Manager of Process Development, Rare Metals Division, Neo Performance Materials based in Peterborough, Ontario. In 2012 this company was acquired by Molycorp Inc and due to deteriorating metal markets Edgar became the interim plant manager of a Ga/In recycling facility and led a number of major operational restructuring efforts. Currently, Edgar is the Director of Technology and Operational Excellence in the Molycorp Rare Metals Division which includes significant commercial analysis responsibilities. Edgar is the former Treasurer and Chair of the Hydrometallurgy Section of the Metallurgical Society of CIM, the former co-chair of Chloride Metallurgy 2002 and the 2014 recipient of the Sherritt Hydrometallurgy award.

Mohamed Buarzaiga is a hydrometallurgist with Vale Base Metals. He graduated from the University of Toronto with B.A.Sc. and M.A.Sc. degrees in metallurgy and materials science, then received a Ph.D. degree in metals and materials engineering from the University of British Columbia, Canada. Upon graduation in 1999, he joined the Hydrometallurgy group at the Falconbridge Technology Center in Sudbury, Ontario and worked there for six years conducting R&D activities related to various aspects of Ni and Co metallurgy. In 2005, Dr. Buarzaiga joined the Hydrometallurgy group at the Vale Technical Excellence Center in Mississauga, Ontario as a principal research engineer. He is currently working for Vale New Caledonia as a Manager of Operational Capacity Improvement.

John Dutrizac attended the University of Toronto where he obtained his B.A.Sc. degree in metallurgical engineering in 1963 and his Ph.D. in 1967. He worked briefly in industry before joining CANMET as a research scientist in 1968. At CANMET, his principal research interests have been iron precipitation, the characterization of both iron-rich precipitates and copper electrorefining circuits, zinc processing and the leaching of sulphide minerals. He has published over 300 papers on these and related topics and has made numerous presentations to technical audiences. Dr. Dutrizac has been involved with many professional societies, and notably, co-chaired the first three Iron Control in Hydrometallurgy Symposia. John Dutrizac is a past chairman of the Hydrometallurgy Section and a past president of the Metallurgical Society of CIM; he has received many national and international awards which recognize his contributions to the hydrometallurgy community. Dr. Dutrizac retired from CANMET in 2008, but continued his research involvement with that organization, initially as an Emeritus Scientist and most recently as a part-time scientist examining iron precipitation in the presence of the rare earth elements.

Symposium Organizing Committee

Chair and Co-Chair
Edgar Peek, Molycorp Minerals Canada ULC, Peterborough, Ontario
Mohamed Buarzaiga, Vale, Noumea, New Caledonia

Editors
John Dutrizac, CANMET, Ottawa, Canada
Mohamed Buarzaiga, Vale, Noumea, New Caledonia
Edgar Peek, Molycorp Minerals Canada ULC, Peterborough, Ontario

Organizing Committee
Edgar Peek, Molycorp Mineral Canada ULC, Peterborogh, Ontario
Mohamed Buarzaiga, Vale, Noumea, New Caledonia
Levente Becze, Molcorp Minerals Canada ULC, Peterborough, Ontario, Canada (**Short Course Organizer**)
Dimitrios Filippou, Rio Tinto (TiO_2 division), Sorel, Quebec, Canada (**Technical Tour Organizer**)
John Dutrizac, CANMET, Ottawa, Canada
Mayooran Somanathan, Westpro Machinery, Toronto, Ontario, Canada (**Publicity**)
Ahmad Ghahremaninezhad, Queens University, Kingston, Ontario, Canada
Takashi Nakamura, Tohoku University, Sendai, Japan
Jo vandenBroucke, Nyrstar, Budel, The Netherlands
Mariam Melashvili, SGS Minerals, Peterborough, Ontario, Canada
Etsuro Shibata, Tohoku University, Sendai, Japan

Reviewers
Edgar Peek, Molycorp Mineral Canada ULC, Peterborough, Ontario, Canada
Mohamed Buarzaiga, Vale, Noumea, New Caledonia
Levente Becze, Molcorp Minerals Canada ULC, Peterborough, Ontario, Canada
Dimitrios Filippou, Rio Tinto (TiO2 division), Sorel, Quebec, Canada
John Dutrizac, CANMET, Ottawa, Canada
Takashi Nakamura, Tohoku University, Sendai, Japan
Jo vandenBroucke, Nyrstar, Budel, The Netherlands
Mike Collins, Sherritt International, Fort Sasketchewan, Alberta, Canada
Etsuro Shibata, Tohoku University, Sendai, Japan

Table of Contents

Iron Control –Practice and Research

(4th International Symposium on Iron Control in Hydrometallurgy)
(46th Annual Hydrometallurgy Meeting)

Plenary Lecture

THE IRON ELEPHANT

A BRIEF HISTORY OF HYDROMETALLURGISTS' STRUGGLES WITH ELEMENT NO. 26

A.J. Monhemius*

Emeritus Professor
Department of Earth Science and Engineering
Royal School of Mines
Imperial College London SW7 2BP UK
**Corresponding author: j.monhemius@ic.ac.uk*

ABSTRACT

The paper reviews the history of modern iron control processes in hydrometallurgy. Particular attention is paid to the processes developed in the 1960s by the electrolytic zinc industry for the removal of iron from zinc sulphate leach liquors - namely the Jarosite, Goethite and Hematite processes. Problems faced by the industry in the safe disposal of the iron residues produced by these processes are discussed. The lessons learnt over the past half century by the zinc industry are projected on to the current efforts to develop hydrometallurgical processes for the treatment of copper sulphide concentrates. It is argued that any attempt to create a "hydrometallurgical copper smelter" is doomed to ultimate failure, unless and until methods are developed that can recover iron from solution in a form that can be directly utilised by the steel or pigment industries.

KEYWORDS

History of iron control; jarosite process; goethite process; hematite process; copper sulphide concentrate leaching.

INTRODUCTION

You suddenly realise that age is catching up with you and that technology is moving on apace when you are asked to talk about the history of a process or a technology that you've grown up with. Such was my reaction when, out of the blue, I received an email last year from the chairman of the CIM's Historical Metallurgy Committee, asking me if I would give a talk on the early developments of iron control in hydrometallurgy. Once I'd got over the initial shock of realising that I had now become living history, I was of course flattered that the invitation was to give a talk at the prestigious IMPC, a conference series that has been going for as long as I have been in the business. That this particular IMPC was to include the 4th Symposium on Iron Control in Hydrometallurgy, made the invitation even more appealing and appropriate, because I was the co-chairman of the First Symposium on Iron Control, together with my old friend, John Dutrizac (known by many as Dr Jarosite), which was held 30 years ago in Toronto.

I first became involved with iron at the very start of my career in hydrometallurgy – the research topic for my Masters degree thesis at UBC, back in the mid 1960s, was "Reductive leaching of Goethite with SO_2". My thesis supervisor was the late Professor Ian Warren, who himself was a protégé of Professor Frank Forward, one of the founding fathers of modern hydrometallurgy. Little did I realise at that time, now 50 years ago, that the problem of iron and how to deal with it in hydrometallurgical processes would be a theme that followed me throughout much of my 40-year career in academia.

IRON MINERALOGY AND CHEMISTRY

So what is it about iron in hydrometallurgy that has merited four decennial, international conferences – 1986 in Toronto, 1996 in Ottawa, 2006 in Montreal and now 2016 in Quebec, not to mention the thousands of scientific and technical research papers and patents that have been published worldwide outside these conferences?

As we all know, iron is a pretty common element and it is one that we certainly couldn't live without. It's an essential part of our biology, where it plays a critical role in the transference of oxygen from the air into our bloodstreams, and it is indispensable for our modern way of life in the form of metallic iron and steels, from which most of the things around us are manufactured.

Let's start with a few facts about the geological occurrences of iron, element number 26 in the Periodic Table. To source these, I turned to the very first paper in the Proceedings of the First Iron Control Symposium, which was by Chen and Cabri (1986) from CANMET, entitled "Mineralogical overview of iron control in hydrometallurgical processing". In this excellent review, we read that iron is the second most abundant element in planet earth, after aluminium, and it is the fourth most abundant element in the earth's solid crust. Iron is an essential component of the crystal structure of over 600 minerals, most of which are classed as oxide minerals (i.e. oxides, silicates, phosphates, arsenates, sulphates, hydroxides and so on) and relatively few are sulphide minerals. Some of this large number of minerals are considered to be ore minerals, but most are gangue minerals, as far as the mineral processor is concerned. Chen and Cabri listed the most important of the iron-bearing minerals in a series of tables in their paper, giving either their chemical formulae, or, where minerals form solid solutions, their compositional ranges. They classified the occurrence of iron in non-ferrous ores into three different modes: i. Ore minerals; ii. Gangue minerals; and iii. Solid solution minerals.

i. Ore minerals: where iron is an essential constituent of a mineral used as a source of another metal, e.g. chalcopyrite, $CuFeS_2$, a copper ore. Other examples are pentlandite, $(Fe,Ni)_9S_8$, a nickel ore, and ilmenite, $FeTiO_3$, a titanium ore.

ii. Gangue minerals: where iron is an essential constituent of a mineral that is undesirable in the ore, e.g. pyrite, FeS_2, often the most abundant species in sulphide ores. Iron oxides, carbonates and silicates are common gangue minerals in many oxidised non-ferrous ore bodies.

iii. Solid solution minerals: this is the substitution of iron for an essential element in the crystal structure of either an ore or a gangue mineral. The most common example of a solid solution ore mineral is sphalerite, (Zn,Fe)S, the principal source of zinc, where up to 17% of the zinc can be substituted by iron. Much of the early work on iron removal from hydrometallurgical solutions was centred on the treatment of zinc leach liquors and we shall consider this work in some detail shortly. Other examples of solid solution ore minerals include tetrahedrite, $(Cu,Fe,Zn)_{12}(Sb,As)_4S_{13}$, and cobaltite, (CoFe)AsS.

Given that non-ferrous metals have been produced from ores for hundreds of years, or in the cases of copper and lead, for thousands of years, one might be forgiven for asking why, suddenly, has iron become the focus of so much attention by non-ferrous metallurgists? The simple answer is that it is due to the rise in importance of hydrometallurgy compared with pyrometallurgy for the production of non-ferrous metals. In the traditional high temperature smelting processes used for the production of copper, lead, zinc and other non-ferrous metals up to the beginning of the 20th century, iron in the smelter feeds ended up in the slags, together with many other deleterious impurity elements. Iron in silicate slags is present mainly as fayalite, ferrous silicate, Fe_2SiO_4. This is a chemically and environmentally inert material, which can be safely discarded without detrimental effects to the local surroundings. This benign method of discarding unwanted iron is one of the main advantages that pyrometallurgy has over hydrometallurgy.

To understand the problems faced by hydrometallurgists who want to discard iron, we need to remind ourselves of the chemical behaviour of dissolved iron. The most concise way of doing this is to use an Eh-pH, or Pourbaix-type diagram, as shown in Figure 1.

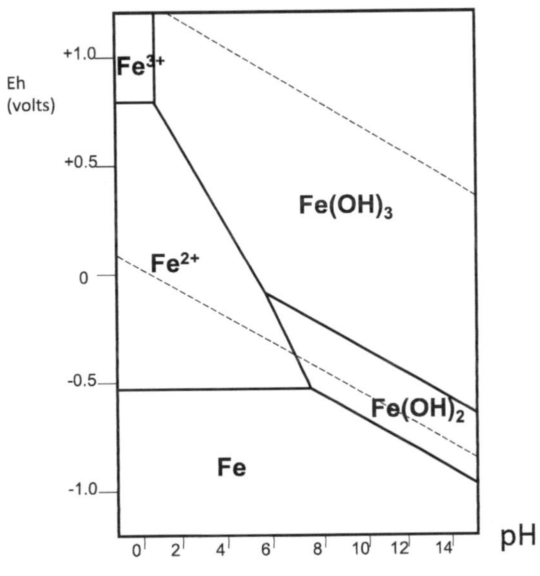

Figure 1 - Eh – pH diagram for the Fe – H_2O system at 298°K

This diagram, which illustrates the behaviour of the simple Fe-H_2O system, shows that iron in the ferrous oxidation state (Fe^{2+}) is soluble across a broad pH range from highly acidic to mildly alkaline, whereas ferric iron (Fe^{3+}) is soluble only in acid solutions (pH<3) and oxidising conditions (Eh>0.77V). Outside these conditions, iron is insoluble and, in the absence of other anions, at ambient temperatures exists as ferric hydroxide, $Fe(OH)_3$, or in alkaline reducing conditions, as ferrous hydroxide, $Fe(OH)_2$.

The relatively small area of solubility of ferric iron, up in the top left hand corner of the Pourbaix diagram, shows us that, theoretically, it is quite easy to get iron out of solution by oxidising it to the ferric state and making sure the pH is greater than about 3. This will swiftly cause iron to precipitate from solution as ferric hydroxide. Most other common base metals (Cu, Zn, Pb, etc) remain soluble under these conditions and so hydroxide precipitation provides in principle a ready method of separating dissolved iron from other more valuable metals in leach liquors.

The precipitation of so-called ferric hydroxide is a very complex process, which even today is not fully understood and it was even less so, back in the 1960s. Ferric iron is a powerful complexant, which will form complex ions with many anions, such as chloride, sulphate, fluoride, or phosphate, and the presence of different anions affects the form and composition of the hydrolysis product of ferric iron. Kinetic factors are also important, thus temperature, rate of change of pH and even the type of neutralising agent used, e.g. lime or sodium hydroxide, can all affect the outcome of hydrolysis. The usual result of rapid hydrolysis at ambient temperatures is the formation of a ferric hydroxide gel, with an open network structure, producing a soft, low density, material with a very high internal surface area. From a practical, process point of view, the physico-chemical properties of freshly precipitated ferric hydroxide have two main consequences: firstly, the large surface area of the hydroxide gel makes it an excellent absorbent material for dissolved metal and semi-metal ions and so it can be used to scavenge deleterious impurity ions from process solutions; and secondly, the open structure of the gel occludes a lot of mother liquor, which makes filtration and washing of the gel extremely difficult. In practice, this latter property means that in general it is not practically or economically feasible to use hydroxide precipitation to remove more than about 1 to 2 g/L of ferric iron from process solutions.

Such was the situation with regard to iron precipitation up to the early 1960s and it manifested itself most clearly in the electrolytic zinc industry, and so, in looking at the history of iron control, we should start with the zinc industry, which is where most of the iron control processes that we are familiar with today were pioneered. It also happens that, in 1980, I published a review of the electrolytic zinc process, which provides a convenient snap-shot of the state of development of the iron removal processes at that time, a decade or so after their first introduction into the zinc industry (Monhemius, 1980).

THE ELECTROLYTIC ZINC PROCESS

Until the beginning of the 20th century, zinc was produced exclusively by pyrometallurgical processes, mainly horizontal or vertical retorts, in which zinc oxide ores were mixed with carbon and heated in the absence of air. The carbon reduced the zinc oxide to zinc metal, which at the reaction temperatures used was formed as a vapour that was condensed to liquid metallic zinc in the cooler parts of the reactor.

The advent of the First World War in 1914 brought about an increased demand for zinc for military purposes and this provided the impetus for the industrial application of a completely new method of zinc production, which had been experimented with and developed only at bench scale during the late 19th century. This new technology was of course what we now know as the electrolytic zinc process, namely, the use of electrolysis to plate pure zinc on to an electrode from zinc sulphate solution. The first industrial electrolytic zinc plants were both started up in 1915, one by Cominco, in Trail, British Columbia, and the other by the Anaconda Copper Company in Anaconda, Montana. The process proved to be successful and its use grew during the first half of the 20th century, until by the 1950s, just under half the world's zinc was being produced electrolytically. The growth in the use of the electrolytic process then stalled for a decade or so until, in the 1970s, it began to grow again and its growth has continued steadily until today, when it now accounts for more than 90% of the world's zinc production. The reasons behind this two-stage historical growth pattern can be related directly to the behaviour of iron in the electrolytic zinc process. To understand this relationship, we need to look at the process in some detail.

In principle, the electrolytic zinc process is very simple and can be described by three chemical equations:

Step 1.	Roasting	$ZnS + 1.5O_2 = ZnO + SO_2$
Step 2.	Leaching	$ZnO + H_2SO_4 = ZnSO_4 + H_2O$
Step 3.	Electrowinning	$ZnSO_4 + H_2O = Zn^0 + H_2SO_4 + 0.5O_2$

It can be seen from these three equations why the electrolytic zinc process is often referred to as the "Roast-Leach-Electrowin", or "RLE", process. Thus the first step involves roasting zinc sulphide (sphalerite) mineral flotation concentrate in air to convert it to zinc oxide, forming gaseous sulphur dioxide, which is converted to sulphuric acid as an essential co-product of zinc production. Roasting is carried out at about 900°C in multiple-hearth roasters, or fluidised-bed reactors, the latter being more usual in modern plants. The hot roasted ore, known as "calcine", is transferred to the leaching tanks for the second step, where it is dissolved in sulphuric acid, contained in the acidic spent electrolyte returned from the electrowinning cells. After leaching, the neutral zinc sulphate solution is treated to remove impurities and is then sent to the electrowinning tankhouse for the third and final step, where zinc is plated on to aluminium cathodes and oxygen is released at the surface of lead anodes in the electowinning cells. A simplified flowsheet of this process is shown in Figure 2.

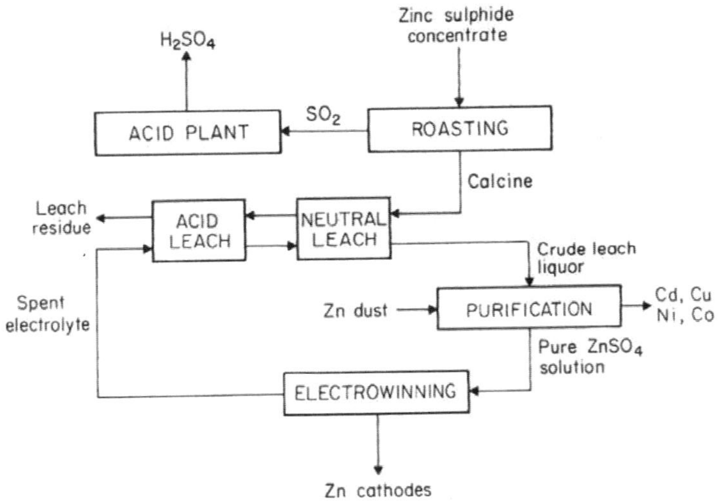

Figure 2 - RLE Process - Simple Flow Diagram

During the roasting process, any iron in the zinc concentrate, which typically can run as high as 12 wt%, will combine chemically with zinc to form a mixed oxide, known as zinc ferrite, $ZnFe_2O_4$. Zinc ferrite is insoluble under the conditions of acidity and temperature used to dissolve the major zinc oxide part of the roaster calcine. Thus in the traditional two-stage leaching process, ferritic zinc reported in the leach residue and this loss of zinc resulted in overall zinc recoveries in the range of 85-93%, depending on the iron content of the zinc concentrate. These relatively low recoveries were the most serious drawback of the electrolytic process and restricted its application to low-iron zinc concentrates. This was the situation until the early 1960s, at which time 40-50% of the world's zinc was produced by the electrolytic process, with the remainder coming from various pyrometallurgical processes.

It was well known even then that zinc ferrites will dissolve readily in sulphuric acid, provided that strong acid solutions are used and the temperature is kept close to boiling. However, under these conditions, most of the iron in the calcine will dissolve, in addition to the zinc and the insuperable problem at the time was how to get the dissolved iron out of the leach solution before sending it to electrolysis for recovery of the zinc. The only known way of removing iron was to neuralise the acidity of the leach solution to raise its pH to about 3 to cause the precipitation of iron as ferric hydroxide. As already described, this is a bulky, gelatinous, material that occludes a great deal of solution and the difficulties of filtering and washing residues that contain large amounts of ferric hydroxide were insuperable. Thus the only practical approach at that time was to use only low-iron zinc concentrates for the electrolytic process, while high-iron concentrates were treated pyrometallurgically, where the iron could be readily handled by causing it to enter the slag phase, which could be safely discarded.

A step-change in the history of zinc ore processing occurred in the mid-1960s, when several processes were developed in which iron could be precipitated from solution as crystalline, easily filterable materials, which could be readily washed to remove leach solution. Such materials include iron oxides, both hydrated and anhydrous, and importantly, basic iron sulphates known as jarosites. These new processes were incorporated into the traditional two-stage zinc leaching process, so that not only oxide zinc, but also ferritic zinc could be recovered from the calcine feeds, so that typical recoveries in zinc leaching plants rose to 95-97%. The introduction of these iron-removal processes rang the death knell for zinc pyrometallurgy, as the electrolytic zinc process began its second growth phase, steadily replacing the old-fashioned pyrometallurgical processes, until today it accounts for over 90% of the world's zinc production.

The Jarosite Process

The most important of the iron removal processes developed in the 1960s is the Jarosite process, in which iron is precipitated from the acidic zinc sulphate leach liquors as one of a group of basic ferric sulphates, known as jarosites. The composition of jarosite is $MFe_3(SO_4)_2(OH)_6$, where M represents a monovalent cation from the group comprising Na^+, NH_4^+, K^+, Ag^+, Rb^+, $0.5Pb^{2+}$ and H_3O^+. Precipitation is brought about by adjusting the pH to about 1.5, at a temperature of about 95°C, and adding a source of the monovalent cation, which is usually NH_4^+ or Na^+ in industrial practice, leading to the formation of ammonium or sodium jarosite, respectively. The jarosite forming reaction can be represented in simplified form as follows:

$$3Fe^{3+} + 2SO_4^{2-} + M^+ + 6H_2O = MFe_3(SO_4)_2(OH)_6 + 6H^+ \qquad (1)$$

In order to ensure complete reaction, the hydrogen ions formed during the reaction have to be neutralised to maintain the pH at about 1.5. The precipitated jarosite is in a crystalline form that can be readily filtered and washed free of the mother liquor.

As occasionally happens with technological advances, very similar versions of this process were developed independently, and virtually simultaneously, in different parts of the world. In this case, patents on variants of the jarosite process were lodged by zinc companies in Spain, Australia and Norway within months of each other in the mid-1960s, as detailed in the table below. These three companies later formed a consortium to license jarosite technology to the rest of the world's zinc industry. The jarosite process has been described extensively in the literature. It is very flexible, capable of being readily integrated into existing as well as new zinc leach plants and it is adaptable enough to cope with different plant practices (Gordon & Pickering, 1975).

Country	Company	Patent Number	Date
Spain	Asturiana de Zinc SA	Spanish Pat. 304601	Oct. 12, 1964
Australia	Electrolytic Zinc Company of Australia Ltd.	Australian Pat. 401724	Mar. 31, 1965
Norway	Det Norske Zinkkompani A/S (Norzink)	Norwegian Pat. 108047	Apr. 30, 1965

The Integrated Jarosite process developed by Norzink incorporated most of the various process options and a simplified flowsheet of the Norwegian process is shown in Figure 3 (Steintveit, 1970). Comparing this flowsheet with that of the basic electrolytic process shown in Figure 2, it can be seen that three extra unit operations have been added, namely, pre-neutralisation, jarosite precipitation and jarosite leaching. The solid residues from the neutral leaching tanks, which contain the undissolved zinc ferrites from the calcine feed, are transferred to the so-called "hot acid leach" tanks. Here, leaching is carried out at 95°C in a very acidic solution comprising a mixture of return electrolyte from the tankhouse, plus extra sulphuric acid. These conditions break down the zinc ferrite structure and most of the zinc and iron dissolves, together with much of the copper and cadmium contained in the neutral leach residues. The hot acid leach solution leaving the stage is still very acidic, typically containing about 40 g/L free acid. This is partially neutralised in the pre-neutralisation stage by the addition of fresh zinc calcine to reduce the acidity to about 10 g/L (pH ~1). Only oxide zinc in this calcine dissolves and so the ferritic residue from pre-neutralisation is returned to hot acid leaching. The hot pre-neutralised solution then passes to the next stage where iron is precipitated as ammonium jarosite, $NH_4Fe_3(SO_4)_2(OH)_6$, by the addition of ammonia to the solution. The temperature is maintained at 95°C in these tanks and the pH is controlled at 1.5 by the addition of fresh zinc calcine to consume the acid produced by the hydrolytic jarosite reaction (Equation 1).

Jarosite precipitation decreases the iron concentration over a period of several hours from an initial value of typically around 20 g/L in the pre-neutralised solution down to about 1 g/L. This amount of iron is deliberately left in the solution passing to the neutral leach tanks in order to form ferric hydroxide, which has the beneficial effect of absorbing certain deleterious impurities from the leach solution. The jarosite precipitated at this stage also contains undissolved ferritic zinc from the calcine added as a neutralising agent during the precipitation reaction. In order to recover this undissolved zinc, the final step in the process is jarosite leaching, which is carried out under conditions similar to those used in the hot acid leaching stage. This treatment dissolves the zinc ferrite in the mixture, while the jarosite itself remains virtually unattacked. The solution from jarosite leaching is returned to the pre-neutralisation stage, while the acid-washed jarosite is discarded. The overall recovery of zinc by this process from typical zinc calcines containing 56-57% Zn and 10-11% Fe is 98%, with similar high recoveries of other valuable metals in the calcine, such as Cd, Cu, Pb and Ag.

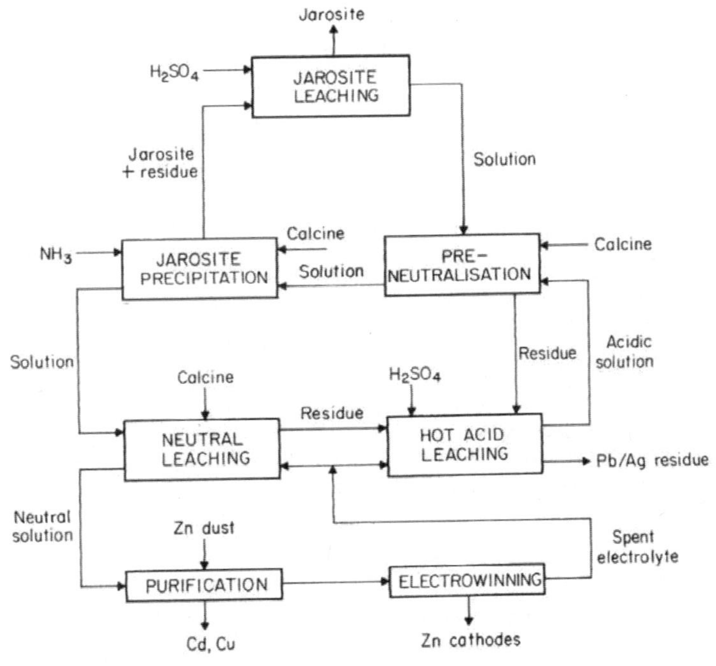

Figure 3. Integrated Jarosite Process

The main advantages of the Jarosite Process, compared with the other iron removal processes discussed subsequently, arise principally from the nature of jarosite itself. Being a basic sulphate, it is precipitated from relatively acid solutions, thus allowing efficient usage of calcine added as neutralising agent. It also enables excess sulphate to be removed from the closed leaching cycle. This is a distinct advantage in many zinc plants where excess sulphate originating from the calcine creates sulphate balance problems in the leach circuit. Furthermore once formed, jarosite is very resistant to acid attack and therefore undissolved calcine mixed with jarosite can be readily recovered by acid washing. These properties confer a considerable degree of flexibility on the operation of the process and allow for optimisation to meet particular process requirements.

The principal disadvantages are two-fold. Firstly the necessity to add a precipitation reagent, normally ammonia, to cause the formation of jarosite is an on-going operational expense. The theoretical consumption of ammonia is 37 kg/ 1,000 kg of jarosite, but in practice, the consumption is usually less than 30 kg/1,000 kg, because ammonium jarosite always contains some H_3O^+ in solid solution. Secondly, the low iron content of jarosite, theoretically 35 %, means that considerable quantities are produced for disposal. For a calcine of the composition quoted above, i.e. 56-57% Zn and 10-11 % Fe, of the order of 0.53 ton of jarosite will be produced per ton of zinc produced.

The Conversion Process

The Conversion process, developed by Outokumpu in Finland in the early 1970s (Huggare et al., 1974), is a modification of the Jarosite process in which zinc ferrite leaching and jarosite precipitation take place simultaneously in the same reactor. The overall reaction may be represented in simplified form as follows:

$$3ZnFe_2O_4 + 6H_2SO_4 + (NH_4)_2SO_4 = 2NH_4Fe_3(SO_4)_2(OH)_6 + 3ZnSO_4 \quad (2)$$

The process depends upon the fact that zinc ferrites are more soluble in sulphuric acid than jarosites. Thus, by controlling the acidity at an optimum value, conditions can be created where ferrites will dissolve while jarosites will precipitate. As shown by the above equation, the reaction consumes sulphuric acid and so the process is controlled by the feed rate of acid to the Conversion reactor. This is the key difference between the Conversion process and the Jarosite process. In the latter, the starting solution contains excess acid, which is neutralised by the addition of calcine, and there is the danger that improper control can lead to over-neutralisation, which in turn can lead to precipitates with difficult solid-liquid separation characteristics. In contrast, because the Conversion reaction is controlled by adding acid, the problems of over-neutralisation are eliminated and the process is very stable in operation, producing very consistent precipitates. The major drawback of this process is that long residence times in the Conversion reactor are necessary.

The Goethite Process.

In the Goethite process, iron is precipitated from solution as hydrated ferric oxide, FeOOH. The process used commercially was developed a few years after the Jarosite process by the Société de la Vieille Montagne in Belgium (SVM, 1968) and involves reduction of iron to the ferrous state, followed by oxidation with air at a temperature of about 90°C and at a pH controlled at about 3.0. The reaction involved is:

$$4Fe^{2+} + O_2 + 6H_2O = 4FeOOH + 8H^+ \quad (3)$$

The oxidation of ferrous iron by oxygen is catalysed by copper present in the leach liquor, and it is vital that the rate of oxidation is balanced by the rate of precipitation, so that the concentration of ferric iron in solution does not rise above 1 g/L. There is no reagent requirement to supply monovalent cations, as in the Jarosite process and, theoretically, no sulphate is removed from the process stream in the iron product. However, in practice, sulphate contamination of the goethite is quite heavy, due to adsorption and the formation of some basic sulphates, and the iron product usually contains 2-5 wt% sulphur.

The main differences between the Goethite and Jarosite processes occur after the hot acid leaching of the zinc ferrite residues. In the Goethite process, the liquor from hot acid leaching, containing 100 g/L Zn, 25-30 g/L Fe^{3+} and 50-60 g/L H_2SO_4, is first subjected to a reduction stage, where ferric iron is reduced to the ferrous state by reaction with unroasted zinc sulphide concentrate at 90°C:

$$2Fe^{3+} + ZnS = 2Fe^{2+} + Zn^{2+} + S^0 \tag{4}$$

After reduction of iron is completed, unreacted zinc sulphide, together with the elemental sulphur formed by the reaction, is separated and returned to the roaster. The solution is then pre-neutralised to 3-5 g/L H_2SO_4 with calcine. The ferritic residue from pre-neutralisation is separated and returned to hot acid leaching and the solution is passed to the precipitation reactor. Air is injected to oxidise the ferrous iron, which hydrolyses and precipitates as crystalline goethite. Calcine is added during precipitation to consume the acid produced by hydrolysis and thus to control the pH at the desired value. Following iron precipitation, a solid-liquid separation is made, with the solution returning to neutral leaching and the goethite precipitate, plus undissolved calcine, being discarded. It is not possible to use an acid wash to recover undissolved zinc in the discard material because goethite, unlike jarosite, would redissolve. A flowsheet of the Goethite Process is shown in Figure 4. A detailed description of Metallurgie Hoboken-Overpelt's (now Umicore) zinc electro-winning plant, which used Vieille Montagne (VM) technology, including the Goethite process, was given in a paper by Van Den Neste (1977).

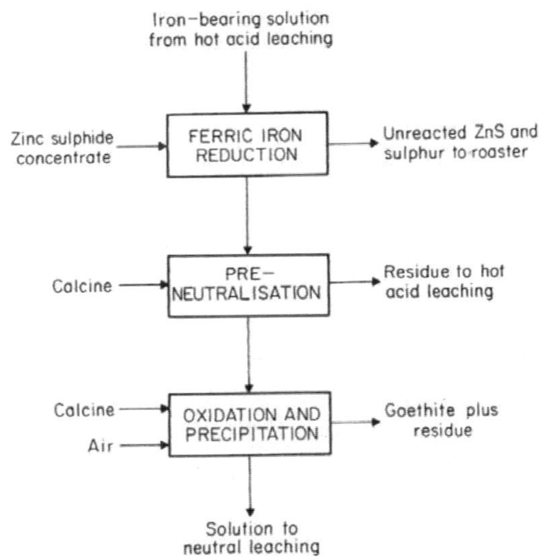

Figure 4. The Goethite Process

A rather similar process, now known as the paragoethite process, was developed in Australia a couple of years after the VM process by the Electrolytic Zinc Company of Australasia Ltd (now Pasminco EZ) (Allen et al., 1970), although this process version was not used commercially until 1985, when it was first used by the Italian company, ENIRISORCE, at its zinc refinery in Porto Vesme, Sardinia (Cubeddu et al., 1996). In the paragoethite process, iron in the ferric state is precipitated directly, without prior reduction, as a hydrated ferric oxide of undetermined nature. This is done by controlled addition of the iron-bearing liquor to a continuous precipitation reactor, where the ferric iron concentration is maintained at less than 1 g/L. Precipitation is carried out at 70-90°C and the pH is maintained at 2.8 by the addition of calcine as neutralising agent. The reaction involved is:

$$2Fe^{3+} + 4H_2O = Fe_2O_3.H_2O + 6H^+ \qquad (5)$$

Comparison of this reaction with that given above for the Vieille Montagne process (Equation 3) shows that more acid is produced by this process and therefore the neutralisation requirements are greater. Consequently, the loss of zinc, due to undissolved calcine in the iron product, will be greater in this process.

The Goethite process does not have the inherent flexibility of the Jarosite process. Very careful control of the conditions during precipitation, especially pH, is required for successful operation of the process. The relative solubility of goethite in sulphuric acid is a major disadvantage, because the iron precipitate cannot be acid-washed to recover undissolved zinc arising from the calcine added for pH control. This problem can be overcome by using selected calcine with a low ferrite content for neutralisation during goethite precipitation. However, this remedy is not generally available to plants tied to single major sources of raw material, where gross variations in the compositions of the calcines do not occur. In such plants, the use of the Goethite process would almost inevitably lead to lower overall zinc recoveries than could be achieved with the Jarosite process and this is the principal reason for the much wider usage of the latter process.

The Hematite Process

A process for the precipitation of iron as hematite, Fe_2O_3, from zinc leach liquors was first developed in Japan by the Akita Zinc Company and put into operation in 1972 in its Iijima zinc refinery (Onozaki et al., 1986). Ferritic zinc leach residue from the main leach circuit is re-leached with spent electrolyte and make-up acid in the presence of SO_2. The reaction is carried out at 95-100°C in autoclaves, lined with lead and acid-resistant brick, at a total pressure of 0.2MPa (30 psi). The ferrites dissolve readily in the presence of SO_2 and iron enters solution in the divalent state:

$$ZnFe_2O_4 + SO_2 + 2H_2SO_4 = ZnSO_4 + 2FeSO_4 + 2H_2O \qquad (6)$$

Excess SO_2 is stripped from the solution and copper is removed as sulphide by H_2S precipitation. The solution, containing about 90 g/L Zn, 60 g/L Fe and 20 g/L H_2SO_4 is neutralised with limestone, first at pH 2 to produce a marketable grade of gypsum and then at pH 4-5, with air oxidation to oxidise and precipitate some of the iron and other impurities. The precipitation of gypsum helps to maintain the sulphate balance in the process by removing sulphate formed during the oxidation of SO_2.

The neutralised solution, which contains about 45 g/L iron is then passed to titanium-clad autoclaves, where the iron is oxidised and precipitated as hematite by oxidation with oxygen at 200°C and a total pressure of 2MPa (300 psi). The reaction involved is:

$$2FeSO_4 + 2H_2O + 0.5O_2 = Fe_2O_3 + 2\ H_2SO_4 \qquad (7)$$

The residence time in the autoclaves is about 3 hours. The final solution, after hematite precipitation, contains 3-4 g/l Fe and this is returned to the main leach circuit. A flowsheet of the process is shown in Figure 5.

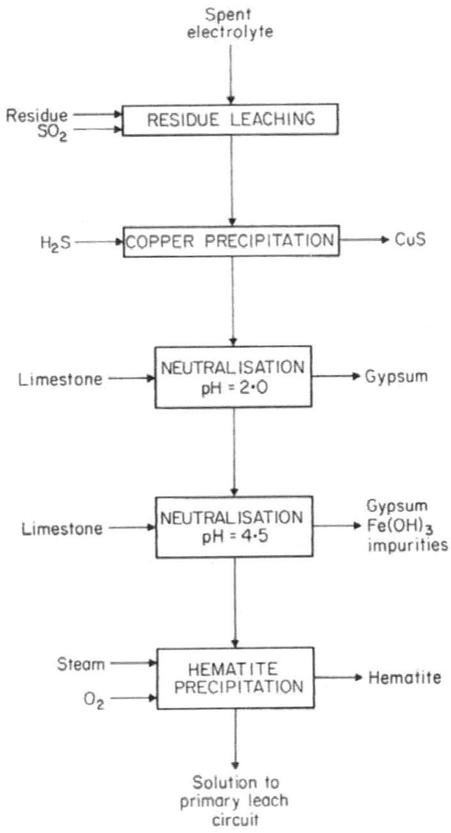

Figure 5. The Hematite Process

At the high temperatures used for iron precipitation in this process, hematite will continue to form even in relatively acidic conditions. Thus the necessity to add calcine to consume protons produced by the hydrolysis reaction is eliminated. This is a major advantage of this process and it means that theoretically no zinc should be lost with the iron residue. However, in practice, the iron residues contain 0-5-1.0% Zn, together with about 3% S, the sulphur arising presumably from the co-precipitation of basic iron sulphates. A second advantage of the process is the high iron content of the hematite residues, theoretically 70% Fe, but in practice closer to 60% Fe, which leads to much smaller quantities of iron residues for disposal. For example, a given quantity of iron precipitated as hematite will have less than half the weight that the same quantity of iron would have if precipitated as jarosite.

Ruhr Zink introduced a similar hematite process at its zinc plant in Datteln, Germany, in 1979 (Von Röpenack, 1986). RZ's objectives were to eliminate residues from the process that were suitable only for dumping and in particular to produce iron as hematite with a purity suitable for steel production. Unfortunately, RZ never achieved the latter objective as its hematite always contained too much zinc and sulphur for the steel-makers. Instead RZ was only ever able to sell its hematite to the cement and tile industries. Thus the economic justification for the process was always under strain and after only just over a decade of operation, the RZ hematite process ceased operation in 1993. The major disadvantage of the hematite process is, of course, the high capital and operating cost of the pressure equipment used and, at present, the process is only still in use in the original Akita zinc plant in Japan.

DISPOSAL OF IRON WASTES

For a typical sphalerite concentrate containing 53 wt% Zn and 7 wt% Fe, the compositions and quantities of the iron residues that would be produced by the three iron-removal processes are shown in Table 1. It can be seen that jarosite has the lowest concentration of contained iron and consequently there is more of it produced per ton of zinc compared with the other two processes. Thus for every ton of zinc metal produced, about half a ton of jarosite will arise. A typical zinc refinery producing 150,000 tons of zinc per year has to dispose of about 75,000 tons of jarosite per annum. A similar sized plant using one of the other two processes would produce about 50,000 tons of goethite, or 30,000 tons of hematite, yearly.

Process	Jarosite	Goethite	Hematite
Wt/wt. zinc metal	0.5	0.32	0.20
Fe wt%	25-30	40-45	50-60
Zn wt%	4-6	5-8	0.5-1.0
S wt%	10-12	2.5-5	3

Table 1: Compositions and Quantities of Iron Residues (Onozaki et al, 1986)

When these processes were first introduced in the late 60s and early 70s, disposal of the iron residues was not considered to be a particularly difficult or onerous task. Different disposal strategies were adopted depending on the local circumstances of each plant. Thus Norzink in Norway used to dump its jarosite into the deep fjord on which its plant was sited, whereas in Australia, EZ used to ship its jarosite in ocean barges to the edge of the continental shelf and dump it into the deep ocean. Asturiana de Zinc in northern Spain, like many other plants that followed it, disposed of its jarosite in lined tailings ponds in the vicinity of the plant, with recovery and recycling of process water from the ponds.

The initial success of these iron control processes, which resulted in much higher zinc recoveries from the electrolytic process and also widened the range of zinc concentrate compositions that could be economically treated by the process, led to an acceleration in the replacement of the older pyrometallurgical zinc processes by the more modern, "cleaner", electrolytic process. Thus the proportion of world zinc produced electrolytically jumped from about 45% in 1960 to 75% by 1980. However this period of growth coincided with birth and growth of the environmental movement and the introduction of environmental legislation and stricter regulation in much of the industrialised world. It wasn't long before the environmental activists and subsequently, the regulators, began to target the iron residues produced by the electrolytic zinc plants around the world, in particular those producing jarosites for discard, as being sources of long term pollution of the environment. Licenses for new jarosite ponds became harder to get, or in some countries, impossible. But the zinc industry, having sunk so much capital into building electrolytic zinc plants and having no alternative zinc production technology to turn to, spent much of the next two decades, i.e. the eighties and nineties, devising various "work-arounds", such as jarofix, jarochaux, etc., (Rosato & Agnew, 1996) to enable continuation of the use of the technically successful iron control processes devised a quarter of a century earlier.

EPILOGUE - THE IRON ELEPHANT

At the most recent decennial Hydrometallurgy conference, held in 2014 in Victoria, BC, the concept of the "hydrometallurgical copper smelter" came under discussion again. Thus, in spite of the vast amount research and development work carried out in the last quarter of the last century to devise hydrometallurgical processes to produce copper from sulphide concentrates, all of it ultimately commercially unsuccessful, the goal of a hydrometallurgical process that can compete head-on with modern copper smelters continues to tantalise up-coming generations of hydrometallurgists.

The "elephant in the room" that haunts and, thus far, has defeated the hydrometallurgical copper smelter concept is how to deal successfully with iron in the copper concentrates – hence my title for this paper: "the Iron Elephant". The size of this elephant for would-be copper makers is several times larger than that faced by today's zinc producers. A typical copper concentrate will contain around 30 wt% Fe and 25 wt% Cu, compared with a typical zinc concentrate with around 10 wt% Fe and more than 50 wt% Zn. Thus a ton of copper produced hydrometallurgically from copper concentrate would necessitate the disposal of around six times as much iron as that arising from the production of a ton of electrolytic zinc. Given the difficulties faced by today's zinc producers in disposing of their iron residues in ways that satisfy the demands of the regulators, the NGOs and the general public, the task of disposing of the vastly greater amounts iron from any copper concentrate leaching process has thus far proved insuperable.

For thousands of years, since the Copper Age began, copper smelters have been using an elegantly simple way of disposing of the iron that is an inevitable by-product of copper making processes – it ends up in the slag. As mentioned earlier, copper smelting slags are composed principally of ferrous silicate, or fayalite, which is an inert, dense, environmentally stable, glassy substance that immobilises iron and other deleterious impurity metals in its silicate matrix. Unlike jarosites and goethites of hydrometallurgical origin, copper smelting slags can be safely dumped without special precautions and with little or no danger of long-term adverse consequences for the environment. This is an enormous advantage for the smelting route and it accounts for the fact that, in spite of all the very significant advances that have been made in copper hydrometallurgy over the past half century, copper smelting still accounts for about three quarters of the world's copper production and, in my opinion, this situation will not change until hydrometallurgists can find a way to slay the Iron Elephant. I first aired this point of view publicly at an AIME annual conference in New Orleans in 1979 (Thorsen & Monhemius) and the situation has not changed much in the intervening nearly 40 years.

The best way that this formidable beast can be dispatched is to convert the iron in copper concentrates into iron oxide with a purity acceptable for steelmaking or pigments, the only industries with capacities big enough to absorb the quantities of iron oxide that would be produced by a commercially-sized hydrometallurgical copper smelter. To achieve the required purities, the best currently-available technical solution is to use solvent extraction to purify dissolved iron from all contaminants. The most likely solvent extraction reagents for this purpose are either carboxylic or alkyl phosphoric acids, both of which have a high affinity for ferric iron; then to make pure Fe_2O_3 powder for direct use as a pigment, or following pelletisation, sintering and reduction, to produce DRI pellets as feed for EAF steelmaking. Pure iron oxide can be produced by stripping the iron-loaded organic acids with hydrochloric acid and then using the well-established industrial process of pyrohydrolysis to convert the $FeCl_3$ strip solution to Fe_2O_3 and HCl (Jewell & Marcotte, 1986). An alternative, but industrially unproven method is hydrolytic stripping (Monhemius & Thorsen, 1980) in which the iron-loaded carboxylic acid is reacted with water at temperatures of 150-200°C to precipitate hematite directly from the organic phase and to regenerate the organic acid for recycle to extraction.

REFERENCES

Allen R.W., Haigh C.J. and Hamdorf C.J. (1970). An improved method of removing dissolved ferric iron from iron bearing solutions, Australian Patent No 424,095, 15 May.

Chen T.T. and Cabri L.J. (1986). Mineralogical overview of iron control in hydrometallurgical processing, in Iron Control in Hydrometallurgy, Dutrizac J.E. and Monhemius A.J. (Eds), Ellis Horwood Ltd, Chichester, 19-55.

Cubeddu F., Piasentin M., Reilly F., Meregalli l. and Tolomio M. (1996). The paragoethite process at the ENIRISORCE-Porto Vesme plant, in Iron Control and Disposal, Dutrizac J.E. and Harris G.B. (Eds), CIMMP, Montreal, 147-161.

Iron Control in Laterite Processing

BEHAVIOR OF IRON AT THE AMBATOVY NICKEL OPERATION IN MADAGASCAR

*M. Collins[1], B. Haugrud[1], and E. McConaghy[1], G. Smith[2]

[1]*Technologies*
Sherritt International Corporation
8301 – 113 Street
Fort Saskatchewan, Alberta, Canada T8L 4K7
*(*Corresponding author: mcollins@sherritt-tech.com)*

[2]*Ambatovy*
Fokontany d`Amboarikarivo
Commune d`Amboditandraho
Toamasina II – Region Atsinanana, Madagascar

ABSTRACT

Laterite ore processed by the Ambatovy Joint Venture in Madagascar is primarily limonitic in nature, containing about 40 to 45 wt.% Fe and 1.1 wt.% Ni. The process selected for treatment of this ore considered the behavior and deportment of iron in each step of the commercial flowsheet. Such considerations included nickel and cobalt recoveries, reagent consumptions, liquid-solid separation characteristics, and both intermediate and final product purities. In this paper, the results for the first three years of commercial operation at Ambatovy are compared with the design parameters, with emphasis on the behavior of iron.

KEYWORDS

Ambatovy, nickel, cobalt, laterite, high pressure acid leach, sulphide precipitation, Sherritt

INTRODUCTION

The main component of nickeliferous laterite ore is iron, and the Ambatovy deposit in Madagascar is no exception. The Ambatovy ore is primarily limonitic in nature, containing about 40 to 45 wt.% Fe and 1.1 wt.% Ni. Since the nickel content of the ore is roughly one fortieth that of iron, successful recovery of nickel depends on close control of the behavior of iron in the selected process.

Exploration of the Ambatovy area, which is located to the east of Antananarivo, the capital of Madagascar, dates back to 1960, when the Malagasy Service Géologique carried out a limited pitting and auger-drilling program. Phelps Dodge Corporation (now Freeport-McMoRan Copper & Gold, Inc.) was granted the mining rights to the property in 1995 and carried out significant resource definition, environmental assessment and engineering studies between 1995 and 2001 (Chou, Rooke, Williams, & Hanks, 2000). In 2003, Dynatec Corporation, which was formed in 1997 by the merger of Dynatec International Ltd. and Sherritt International Consultants Inc., signed a joint venture agreement with subsidiaries of Phelps Dodge to evaluate the Ambatovy Nickel Project. Dynatec then became extensively involved in the development of the project, including process selection and pilot plant testing in Fort Saskatchewan (Collins et al., 2004, 2005, 2009; Collins & Vardill, 2005). Dynatec acquired 100% of the project in 2005 and a joint venture between Dynatec, Sumitomo Corporation, Korea Resources Corporation (KORES), and SNC-Lavalin Inc. (SLI) was formed shortly thereafter to provide funding for further development. SLI was responsible for engineering, procurement and construction management of the commercial facilities. Sherritt International Corporation reacquired Dynatec Corporation, including 40% ownership in the Ambatovy Nickel Project, in 2007. The current Ambatovy Joint Venture consists of an integrated nickel and cobalt mining, processing, refining and marketing enterprise between subsidiaries of Sherritt (40% owner and mine and plant operator), Sumitomo Corporation (32.5% owner) and KORES (27.5% owner).

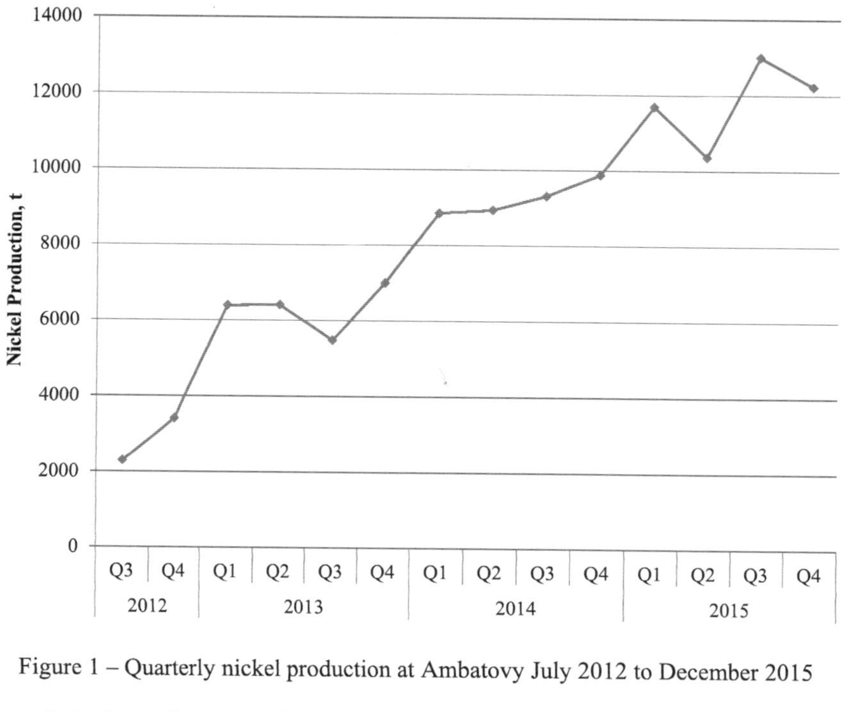

Figure 1 – Quarterly nickel production at Ambatovy July 2012 to December 2015

Commissioning and start-up of the Ambatovy mine and plant facilities were completed in 2012, with produced acid first introduced to the ore leach in March 2012 and test nickel briquettes first produced in the refinery on June 26, 2012. Ramp-up of the facilities continues to advance (Figure 1). Commercial production (defined as 70% of design capacity) was achieved in January 2014 and the production level required to achieve financial completion (90% of design capacity for 90 days) was achieved in March 2015 (Valle, Benz, Chalkley, Collins, Dobson, Holmwood, Malevich, Tuffrey, & Vrolson, 2016). The LME approved the good delivery listing of Ambatovy nickel briquettes in September 2015. Ambatovy has an annual design capacity of 60,000 tonnes of nickel and 5,600 tonnes of cobalt, which makes it the largest finished nickel and cobalt operation from laterite ore in the world.

Iron is a common component in base metal ores and concentrates, and development of a successful process for recovering the base metals requires an understanding of the behaviour of iron in these processes. Discussions of iron behaviour in several process flowsheets developed by Sherritt/Dynatec, including HPAL processing, have been provided previously (Buban, Collins, & Masters, 1999; Masters, Barta, Stiksma, & Collins, 2006).

PROCESS DESCRIPTION

The metals recovery process in practice at Ambatovy is shown in Figure 2 (Collins et al., 2004; Collins, Yuan, Sitter, & St. Jean, 2015).

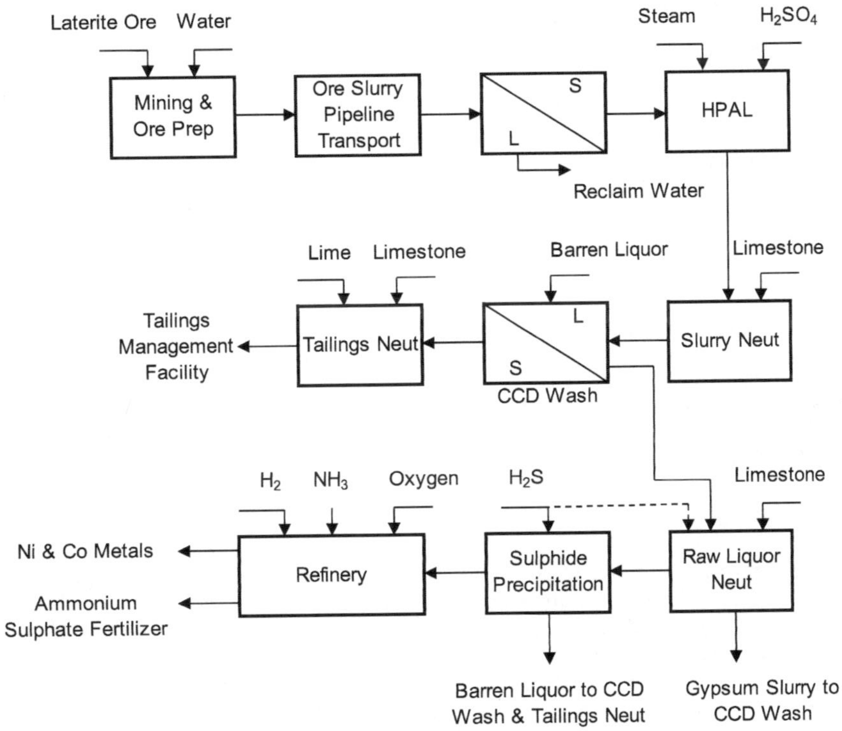

Figure 2 – Simplified Ambatovy overall process flowsheet

The leaching process is very similar to that in place at Moa Bay, Cuba (Chalkley & Toirac, 1997), and both the leaching and nickel refining processes are very similar to those practiced at Murrin Murrin, Australia (Hewitt & Belke, 2004). The laterite ore at Ambatovy (mostly limonite, with iron present primarily as goethite) is mined, scrubbed, screened, thickened and transferred to the pressure acid leach (PAL) plant via a pipeline. In the PAL plant, the slurry is thickened further and then fed along with sulphuric acid to the high pressure acid leach (HPAL) autoclaves. After HPAL, the leached slurry is partially neutralized with limestone (slurry neutralization) followed by a counter-current decantation (CCD) wash, which separates the nickel- and cobalt-bearing solution from the residue (mostly iron oxide – hematite). Sulphuric acid in the CCD product solution is further neutralized with limestone in the raw liquor neutralization circuit. The neutralized raw liquor is then contacted with hydrogen sulphide in the sulphide precipitation circuit, to produce a clean mixed sulphide intermediate containing the nickel and cobalt that was leached from the ore. The barren liquor from sulphide precipitation is partially neutralized with limestone before being used to wash the leached ore slurry in the CCD circuit. Excess partially neutralized barren liquor is combined with the washed leach residue, further neutralized with limestone and lime, and directed to the tailings management facility.

In the adjacent refinery, nickel and cobalt in the mixed sulphide intermediate produced in the PAL plant are re-leached into solution, the solution is purified, and nickel and cobalt are separated by solvent extraction before being precipitated as metal powders by hydrogen reduction, and formed into briquettes for sale. Ammonium sulphate fertilizer is also recovered in the refinery, as a by-product of the process.

Mining and Ore Preparation

The Ambatovy mine site is about 14 km northeast of the town of Moramanga and 80 km east of Antananarivo, the capital city of Madagascar. A 220 km long and 60 cm diameter pipeline delivers prepared laterite slurry from the mine site to the pressure acid leach plant, which is located at the port city of Toamasina on Madagascar's east coast.

Ore is surface mined by hydraulic excavators and delivered to the nearby Ore Preparation Plant (OPP) in trucks. At the OPP, the ore is prepared into a slurry with water and oversize material, enriched in acid-consuming components like magnesium and aluminium, is scrubbed before being separated by screening and rejected. The recovered fines are thickened to about 35 wt.% solids for pipeline delivery to the pressure acid leach plant. The average composition of the prepared ore delivered to the pressure acid leach plant in the past three years is provided in Table 1. The iron content of the prepared ore is about 40 times that of nickel.

Table 1 – Average composition of prepared ore (pipeline discharge)

	Ni	Co	Al	Cu	Fe	Mg	Mn	Zn	Ni:Fe Ratio
2013 Mean Analysis, wt.%	1.07	0.09	3.35	0.02	44.0	0.91	0.69	0.05	0.024
2014 Mean Analysis, wt.%	1.09	0.09	4.06	0.02	43.2	1.12	0.77	0.04	0.025
2015 Mean Analysis, wt.%	1.14	0.09	3.77	0.02	43.6	1.12	0.70	0.04	0.026

Ore Slurry Thickening

Ore slurry received from the mine via the pipeline is thickened further in the pressure acid leach plant to minimize water input with the ore to the HPAL autoclaves. Maximizing the solids content of the autoclave feed enables steam and acid consumptions in HPAL to be minimized, while maximizing nickel throughput. The extent of iron extraction in HPAL is also dependent on the solids content of the feed, since the concentration of ferric iron in the HPAL product liquor reaches a solubility limit under the autoclave operating conditions chosen. As such, increased water input per unit solids leads to increased solution production and increased iron extraction in HPAL.

The design solids content of the thickened ore is 42 wt.% solids. In practice, the solids content has been more typically in the range of 38 to 40 wt.% solids, leading to marginally higher iron extraction in HPAL than the design (see below). It is expected that operation of a second ore thickener, which was brought online in late 2014, will lead to increased solids content in the HPAL feed going forward.

High Pressure Acid Leaching

Nickel and cobalt are extracted from the ore in horizontal, multi-compartment autoclaves by addition of sulphuric acid to dissolve the host minerals, e.g. goethite (equation 1). The high temperature of the HPAL process, 260°C at Ambatovy, promotes hydrolysis of metal sulphates and precipitation of oxides from the solution, e.g., hydrolysis of ferric sulphate and precipitation of iron as hematite (equation 2), and hydrolysis of aluminum sulphate and precipitation of aluminum as alunite (equations 3 and 4), minimizing net acid consumption. The hematite and alunite precipitates have improved liquid-solid separation characteristics compared with the original ore (Papangelakis, Georgiou, & Rubisov, 1996).

$$2FeOOH + 3H_2SO_4 \rightarrow 2Fe_2(SO_4)_3 + 4H_2O \quad (1)$$

$$Fe_2(SO_4)_3 + 3H_2O \rightarrow Fe_2O_3 + 3H_2SO_4 \quad (2)$$

$$2Al(OH)_3 + 3H_2SO_4 \rightarrow Al_2(SO_4)_3 + 6H_2O \quad (3)$$

$$3Al_2(SO_4)_3 + 14H_2O \rightarrow 2(H_3O)Al_3(SO_4)_2(OH)_6 + 5H_2SO_4 \quad (4)$$

The autoclave discharge slurry, following three stages of flashing, is sampled regularly to monitor solution quality and metals extractions. Nickel and cobalt extractions are both 95% in the plant design. The actual nickel and cobalt extractions achieved in production typically range between about 93.5 and 96.5%, and have remained remarkably close to the design value since the earliest days of the operation. Monthly average values for nickel and cobalt extractions in HPAL are provided in Figure 3.

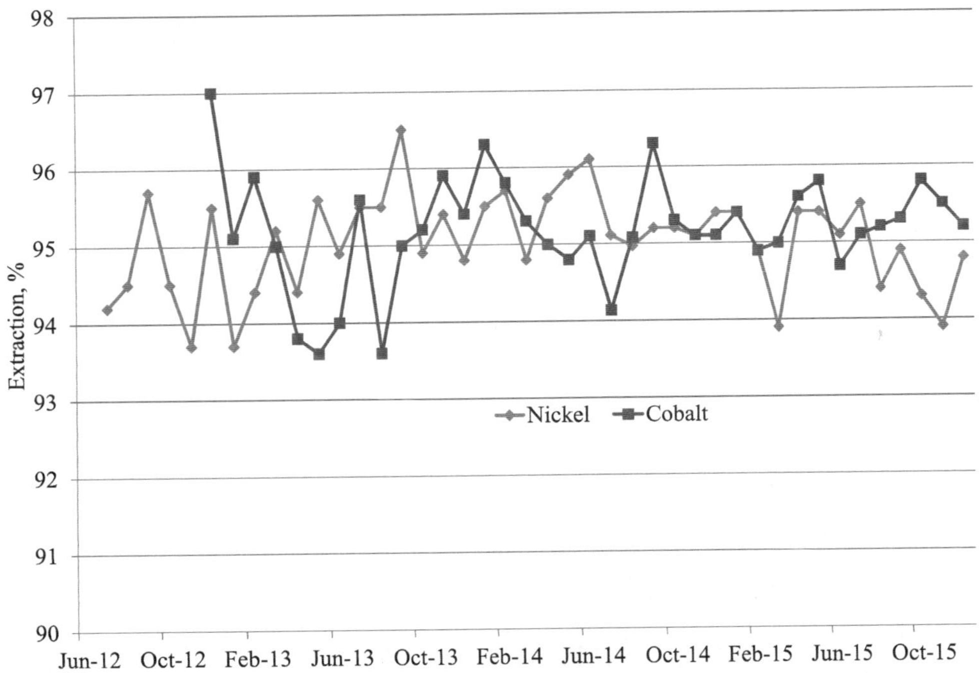

Figure 3 – Nickel and cobalt extractions in HPAL at Ambatovy July 2012 to December 2015

The extent of ferric iron hydrolysis and precipitation from solution in HPAL (equation 2) depends on the autoclave temperature and retention time, and the acid addition rate, since excess acid leads to less complete hydrolysis of ferric iron. The HPAL autoclave temperature is relatively constant at Ambatovy, as is the autoclave retention time, which varies according to the slurry feed rate. Although the autoclave feed pumps may be adjusted to cover a wide range of flow, and the flows are adjusted as required in response to the needs of the remainder of the operation, the typical operating range of the HPAL autoclave feed pumps is relatively narrow. Plant throughput instead varies primarily according to how many of the five HPAL autoclave trains are in operation at a time. Hence, rather than either autoclave temperature or retention time, the concentration of ferric iron in the HPAL discharge solution at Ambatovy primarily varies according to the acid addition rate.

In addition to ferric iron, the HPAL solution may contain ferrous iron. The concentration of ferrous iron in the solution is relatively independent of the operating conditions in HPAL, but instead depends on the nature of the ore, since reducing components in the ore can react with ferric iron in the autoclave to form ferrous (e.g., reaction with organic carbon in the ore, such as found in humic acids, as represented in equation 5). Ferrous iron does not hydrolyze and regenerate sulphuric acid under the autoclave conditions. As such, in addition to putting more iron into solution that then must be handled downstream, the presence of reducing components in the ore leads to increased acid consumption in HPAL.

$$C + 2Fe_2(SO_4)_3 + 2H_2O \rightarrow CO_2 + 4FeSO_4 + 2H_2SO_4 \qquad (5)$$

The total iron (ferric plus ferrous) content of the HPAL discharge solution is shown for the past three years of operation in Figure 4. Each point on the graph represents an average of up to 20 separate analyses, corresponding to four samples of discharge solution per day for each of the five autoclave trains. The acid addition rate tended to be on the high side of what was actually required to achieve target nickel extraction in 2013, resulting in relatively high iron concentration in solution (4.4 g/L on average in 2013). As plant operation matured and the ore feed rate increased throughout 2014, control of acid addition improved. The average concentration of iron in the autoclave discharge solution in 2014 and 2015 was about 2.2 g/L. The nickel to iron weight ratio in the autoclave discharge solution was about 2.3:1 at this time (see Table 2), compared with about 0.025:1 in the autoclave feed.

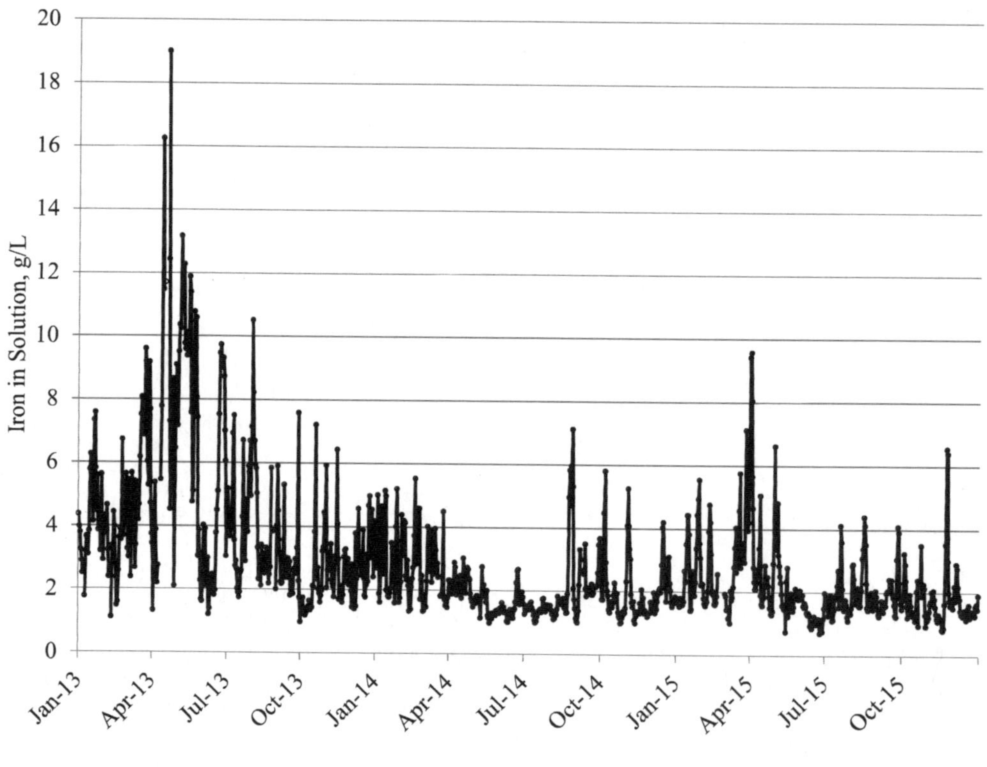

Figure 4 – Iron concentration in HPAL discharge solution in 2013, 2014 and 2015

Table 2 – Average solution compositions in the PAL plant in 2015

Process Stream	Solution Analysis, g/L								Ni:Fe
	Ni	Co	Al	Fe	Mg	Mn	H_2SO_4	pH	Ratio
HPAL Discharge	5.05	0.37	5.74	2.21	4.24	2.54	30.9		2.29
Neutralized Slurry	5.09	0.36	5.57	1.71	4.19	2.39	5.4		2.98
Raw Liquor (from CCD)	3.60	0.26	4.17	1.00	3.85	2.13	2.4	1.8	3.60
Neutralized Raw Liquor	3.42	0.25	1.80	0.358	3.79	2.07	--	3.7	9.55
Barren Liquor	0.10	0.01	1.65	0.308	3.38	1.85	3.1		

When the Ambatovy ore is net oxidizing in character (i.e., when there is no ferrous iron in the discharge solution), the total iron concentration is identical to the ferric iron concentration in solution. In 2014 and 2015, the discharge solution typically contained about 1 to 2 g/L ferric iron (the minimum points on the curve for the total iron analyses in Figure 4). This corresponds to a net extraction of about 0.5 to 1% of the iron in the ore. When the ore is net reducing in character, i.e., when the discharge solution contains ferrous iron, the discharge solution still contains 1 to 2 g/L ferric iron, but it also contains a highly variable concentration of ferrous iron. In 2014 and 2015, the total iron concentration was as high as about 10 g/L (Figure 4), corresponding to up to about 8 or 9 g/L ferrous iron. Net iron extraction in HPAL was about 5% at this time. The plant was designed to handle a discharge solution containing 1.5 g/L iron, with a net extraction of 0.7% iron in HPAL. Excursions in the iron content of the solution due to processing of reducing ore occurred approximately once per month from August 2014 to October 2015 (see Figure 4). More recently, Ambatovy ores have been more strongly oxidizing, and this trend is contingent on the areas where ore is being mined.

Slurry Neutralization and Residue Washing

The discharge slurry from the autoclave is partially neutralized with limestone, to about 5 g/L H_2SO_4, in the slurry neutralization circuit prior to advancing to the six-stage, counter-current decantation (CCD) wash circuit. About 0.5 g/L ferric iron hydrolyzes and precipitates from solution as a result of the lower free acid concentration in the slurry neutralization step, increasing the Ni:Fe weight ratio in the solution to about 3:1 (see Table 2). The recycle of iron-bearing residues from the refinery, which contain nickel in the entrained solution, to the PAL plant at this point in the flowsheet also contributes to the observed increase in the Ni:Fe ratio in solution.

The pH of the solution continues to increase in the CCD wash circuit, since solids from the raw liquor neutralization circuit, which have residual neutralizing power, are washed in the same circuit as the leach residue. As a result, ferric iron in solution continues to hydrolyze and precipitate in the CCD wash circuit. At the same time, partially neutralized barren liquor from sulphide precipitation, rather than fresh water, is used to wash the leach residue in the wash circuit, and the barren liquor contains ferrous sulphate (see below). The overall result is a net decrease of iron in solution relative to nickel and the nickel to iron ratio in the product liquor from the wash circuit (raw liquor) is marginally higher than it is in the solution following slurry neutralization, as shown in Table 2.

Raw Liquor Neutralization

Overflow solution from the lead thickener in the CCD wash circuit is transferred to the raw liquor neutralization circuit, where additional limestone is added to neutralize sulphuric acid and precipitate ferric iron and other impurities, to produce a solution suitable for recovery of nickel and cobalt as a clean mixed sulphide intermediate product in the sulphide precipitation step. The rate of limestone addition is selected to optimize intermediate product quality and to maximize recoveries of nickel and cobalt in the sulphide precipitation circuit, while at the same time minimizing precipitation of nickel and cobalt in the raw liquor neutralization circuit itself. Both the efficiency of limestone utilization and the thickening properties of the neutralized slurry are enhanced by recycle of raw liquor neutralization thickener underflow slurry (gypsum slurry) to the raw liquor neutralization circuit. The net production of gypsum slurry is directed to the CCD circuit, where it is washed along with the leach residue.

When the ore being processed is oxidizing in character (see above), the raw liquor contains up to about 1 g/L, more typically up to about 0.5 g/L, of hexavalent chromium. To prevent this Cr(VI) from reacting with hydrogen sulphide gas in sulphide precipitation, which would result in elemental sulphur reporting to the mixed sulphide intermediate (equation 6), Cr(VI) is pre-reduced to Cr(III) by addition of hydrogen sulphide, as required, immediately prior to the limestone addition in the raw liquor neutralization circuit. Elemental sulphur formed in raw liquor neutralization is then thickened along with the gypsum and metal hydroxides formed in this step and separated from the solution prior to sulphide precipitation. Keeping elemental sulphur out of the mixed sulphides is important for maximizing the efficiency of nickel recovery in the refinery. Feed selection and blending at the mine helps to minimize the Cr(VI) concentration in the HPAL solution, and hence minimizes hydrogen sulphide consumption in raw liquor neutralization.

$$2H_2CrO_4 + 3H_2S + 3H_2SO_4 \rightarrow Cr_2(SO_4)_3 + 3S^\circ + 8H_2O \quad (6)$$

Ferric iron in solution reacts with H_2S in a similar fashion to Cr(VI), once all of the Cr(VI) has been reduced, as shown in equation 7, hence ferric iron is also undesirable in the sulphide precipitation feed solution (neutralized raw liquor). At the pH range selected for raw liquor neutralization at Ambatovy, about pH 3.6 to 3.8, ferric iron is hydrolyzed and precipitates essentially quantitatively from solution, as shown in equation 8. Ferrous iron is untouched at this pH range, and most of the ferrous iron in the raw liquor remains in the feed to sulphide precipitation. However, since a small stoichiometric excess of H_2S is added in the pre-reduction step at the beginning of the raw liquor neutralization circuit to ensure complete reduction of Cr(VI), a small fraction of the ferric iron in the raw liquor reacts with this excess H_2S and is converted to ferrous iron prior to neutralization. At the same time, since air is sparged into the raw liquor neutralization tanks to strip carbon dioxide from the slurry, a small fraction of the ferrous iron in solution is oxidized to ferric and precipitates from solution in the raw liquor neutralization step. The net result is that the sulphide precipitation feed solution typically contains about 0.5 to 1 g/L less iron than the raw liquor, essentially all in the ferrous oxidation state (see Table 2).

$$Fe_2(SO_4)_3 + H_2S \rightarrow 2FeSO_4 + S^\circ + H_2SO_4 \quad (7)$$

$$Fe_2(SO_4)_3 + 3CaCO_3 + 9H_2O \rightarrow 2Fe(OH)_3 + 3CaSO_4{\cdot}2\,H_2O + 3CO_2 \quad (8)$$

Sulphide Precipitation

The neutralized raw liquor is combined with hydrogen sulphide gas in the sulphide precipitation vessels to recover nickel and cobalt as a mixed sulphide intermediate. Optimal metals recoveries in sulphide precipitation are achieved at about 95 to 105°C and with about 200 kPa or more hydrogen sulphide partial pressure in the reactors. The discharge slurry from the flash tanks is thickened and the mixed sulphides are washed in a two-stage CCD circuit. The mixed sulphide intermediate product composition is shown in Table 3.

Table 3 – Average composition of the mixed sulphide intermediate product in 2015

	Ni	Co	Al	Ca	Cu	Fe	Mg	Mn	Si	S	Zn	Ni:Fe
Assay, wt.%	53.8	3.90	0.08	<0.2	1.68	0.43	0.01	0.01	0.04	36.3	3.31	125

A key reason for selection of sulphide precipitation over other nickel recovery methods, e.g., hydroxide precipitation, is the favorable rejection of impurities that can be obtained by sulphide precipitation. As shown in Table 3, exceptionally good rejection of Al, Ca, Mg, Mn and Si is achieved at Ambatovy in the PAL plant. Rejection of iron is also good in sulphide precipitation; the nickel to iron weight ratio in the mixed sulphide intermediate product is about 125:1, compared with about 10:1 in the feed solution (see Tables 2 and 3).

As described below, the capacity of the refinery is limited by the iron content of the mixed sulphide intermediate, especially the capacity of the liquid-solid separation steps used in the refinery to remove iron-bearing residues. As such, even though the largest part of the iron in the ore is readily rejected prior to sulphide precipitation, the behavior of iron within sulphide precipitation is important to the overall success of the operation.

Although precipitation of the sulphides of the base metals Cu, Zn, Ni and Co is highly favored over that of iron, it is known that a small quantity of iron sulphide is produced under commercial sulphide precipitation conditions (e.g., at the Moa operation, see Kofluk & Freeman, 2006). Figure 5 compares the concentration of iron in the feed to sulphide precipitation (neutralized raw liquor) with the iron content of the mixed sulphide intermediate product. A relationship between the two is clear in the figure. Because of the large recycle of mixed sulphide that is used as seed material in sulphide precipitation, direct calculation of the instantaneous extent of iron precipitation in sulphide precipitation based on solids analyses is not readily achieved. At the same time, because the extent of iron precipitation is quite small, direct calculation of the instantaneous extent of precipitation based on solution analyses is also not reliable. However, the overall average extent of iron precipitation in sulphide precipitation in 2015 may be readily calculated based on average feed and discharge solution analyses for the year (see Table 2). Using magnesium or manganese as tie elements, the average extent of precipitation of iron in sulphide precipitation is calculated to be 3.5% (magnesium tie) or 4% (manganese tie) in 2015.

The annual average iron content of the mixed sulphide in 2015 (0.43 wt.%, see Table 3) corresponds to 7% of the iron in the feed solution, using nickel as a tie element between solids and solution. The difference between the values calculated for the extent of iron precipitation in sulphide precipitation based on product solids assays (7%) and feed and product solution analyses (3.5 to 4%) may be due to entrained hematite in the mixed sulphide intermediate product. There are two significant potential sources of entrained hematite in the mixed sulphide. One potential source is flashed steam from the HPAL circuit, since this steam is used to preheat the sulphide precipitation feed, and the flashed steam can contain entrained leach slurry. A similar source of iron in the mixed sulphide has been described for the Moa operation (Kofluk & Freeman, 2006). Iron-bearing solids carried over with the raw liquor neutralization thickener overflow feed to sulphide precipitation may also contribute to the iron content of the mixed sulphides. Such iron-bearing solids likely also originate as HPAL leach residue, in this case suspended in the CCD circuit lead thickener overflow solution that is the feed to raw liquor neutralization. Following the discussion above with respect to the calculated extent of precipitation of iron in sulphide precipitation based on solution analyses, iron entrainment as hematite (originating from either preheat steam or neutralized raw liquor suspended solids) could account for about 0.2 wt.% Fe in the mixed sulphide on average in 2015. The quantity of entrained hematite is clearly not uniform at Ambatovy, as the mixed sulphide intermediate product contained as little as 0.05 wt.% Fe in 2015, when the neutralized raw liquor contained 0.1 g/L Fe or less (see Figure 5). These values are consistent with precipitation of about 3 to 4% of the iron in solution and essentially no entrainment of hematite at that time.

The design value for the iron content of Ambatovy mixed sulphide is 0.53 wt.%. Although the design value is very close to that of the actual average iron content of the mixed sulphide in current operation, the range in the iron content of this intermediate product has been quite broad, e.g., from 0.05 to 2.1 wt.% in 2015. The high value of 2.1 wt.% Fe was obtained in early April 2015, when strongly reducing ore was being processed and the neutralized raw liquor contained 3 g/L Fe (see Figure 5). These values in early April are consistent with a range in the possible behavior of iron in the circuit, e.g., a) precipitation of 4% of the iron in the feed solution and entrained iron from hematite accounting for 0.1 to 0.2 wt.% Fe in the mixed sulphide, b) precipitation of 3.5% of the iron in the feed solution and entrained iron from hematite accounting for 0.4 wt.% Fe in the mixed sulphide, or c) precipitation of 3% of the iron in the feed solution and entrained iron from hematite accounting for 0.6 wt.% Fe in the mixed sulphide. In any case, excursions to 2 wt.% Fe in the mixed sulphide are undesirable, as this instantaneously puts an iron load on the refinery that is four times that of the design.

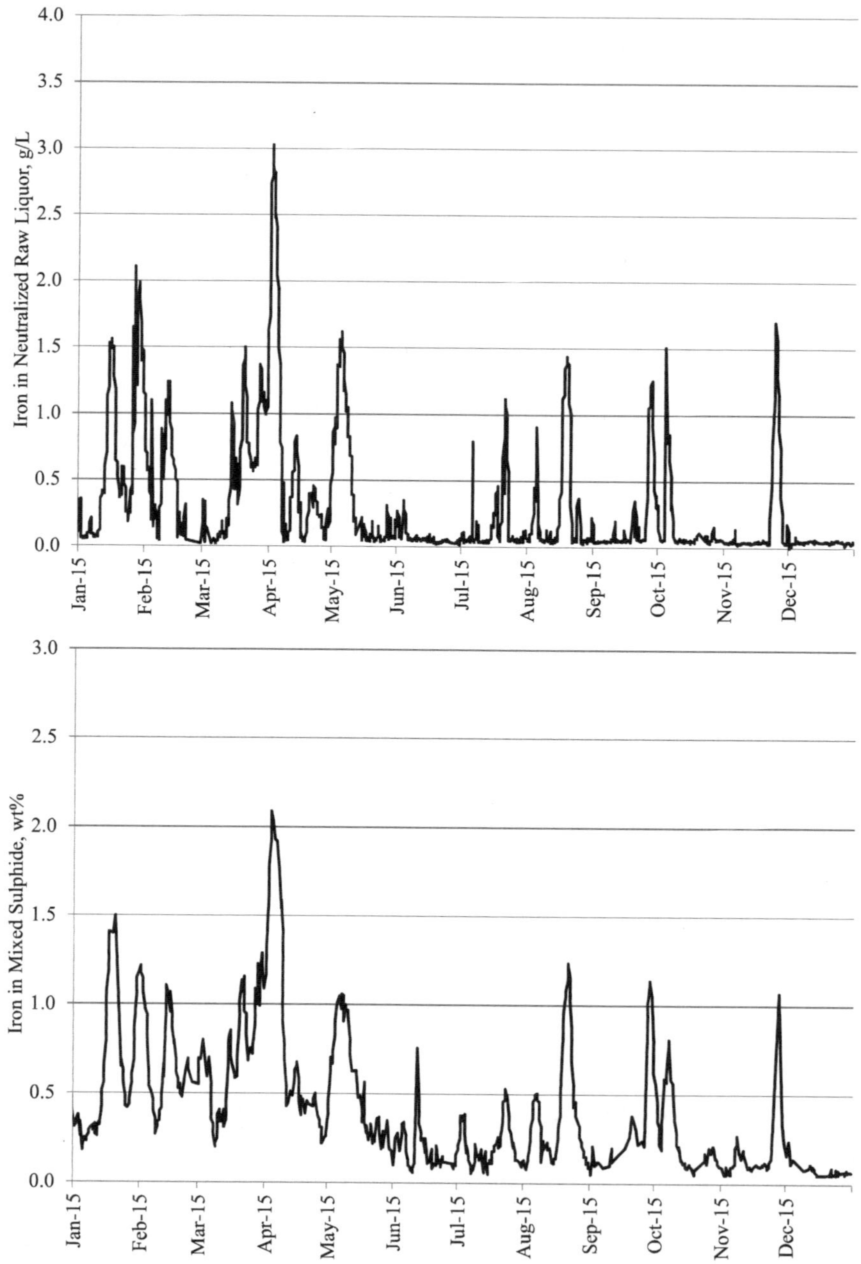

Figure 5 – Iron concentration in feed to sulphide precipitation (upper) and mixed sulphide (lower) in 2015

Refinery

In the refinery (Figure 6), the mixed sulphide intermediate, along with solids recovered from sulphide stripping of the reduction end solutions (the metals-depleted solutions after nickel reduction and cobalt reduction), is leached in horizontal autoclaves under oxygen pressure to produce a high strength nickel sulphate solution. The typical leaching conditions include 165°C with 500 kPa oxygen partial pressure. The resulting leach solution typically contains about 80 to 95 g/L Ni and 0.2 to 0.6 g/L Fe (average values were 89 g/L Ni and 0.34 g/L Fe in 2015, see Table 4). About one half of the iron in the mixed sulphide intermediate reported to solution in the sulphide leach in 2015, consistent with about half of the iron in the intermediate feed being present as iron sulphide (and dissolving in the sulphide leach), and the other half being present as entrained HPAL leach residue (and remaining essentially inert in the sulphide leach). The nickel to iron weight ratio in the sulphide leach discharge solution was 260:1 on average in 2015, compared with 125:1 in the mixed sulphide.

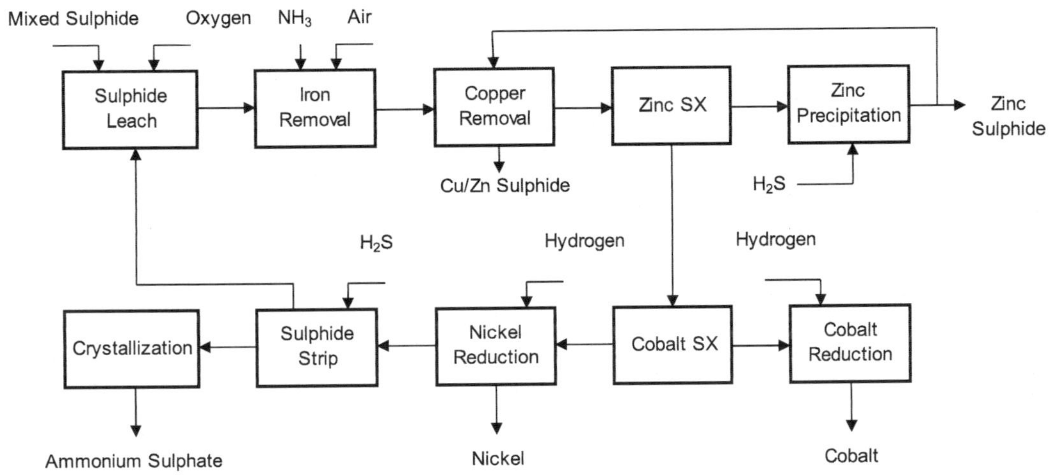

Figure 6 –Ambatovy refinery flowsheet

Table 4 – Average solution compositions in the refinery in 2015

Process Stream	Analysis, g/L						Ni:Fe
	Ni	Co	Cu	Fe	Zn	H_2SO_4	Ratio
Sulphide Leach Discharge	88.6	6.3	2.6	0.34	5.8	5.9	260
Iron Removal Filtrate	86.4	6.2	1.6	<0.001	5.7		86,000
Copper Removal Filtrate	81.9	5.9	0.001	<0.001	7.1		82,000
Nickel Reduction Feed	64.1	0.003	0.002	0.002	0.006		32,000
Nickel Reduction End Sol'n	0.9	0.003	0.001	0.004	0.002		

Ammonia is added to the leach solution in the iron removal circuit, which consists of a series of cascading stirred tank reactors, to increase the pH and promote precipitation of impurities. Air is sparged into these iron removal tanks to oxidize ferrous iron to ferric iron, which hydrolyzes and precipitates. The iron precipitate, along with the mixed sulphide leach residue, is directed to the pressure acid leach plant, where it joins the HPAL residue. Discharge from the iron removal circuit has been treated by various means to separate the iron precipitate since start-up of the plant. Initially, the entire slurry was filtered. However, it was soon apparent that the installed filters were undersized, and one of the sulphide leach thickeners was then repurposed to act as an iron removal thickener, with the thickener overflow solution directed to the original filters. Currently, the plan is to install duty and standby centrifuges in the iron removal circuit in early 2016, in place of the thickener, which can then be returned to service in the sulphide leach circuit. The centrifuges are also expected to provide significantly better dewatering of the iron precipitate, decreasing the recycle of nickel and cobalt in the entrained solution to the pressure acid leach plant and reducing the soluble metal losses in the CCD wash circuit in the PAL plant.

The iron-free solution is contacted with zinc sulphide in the copper removal circuit, which consists of a stirred tank, to precipitate copper as copper sulphide. Excess zinc sulphide is added to ensure complete removal of copper from solution. The zinc sulphide that is added to the copper removal tank is obtained from the zinc precipitation circuit; the resulting copper-zinc sulphide is sold. The iron-free solution routinely contains <1 mg/L Fe. There is no measurable pick-up of iron in the copper removal step; on average the copper-free solution contained 1 mg/L or less each of iron and copper in 2015.

After liquid-solid separation, the copper-free solution is forwarded to zinc solvent extraction, the first of two solvent extraction circuits. Zinc is stripped from the loaded organic with sulphuric acid and precipitated as zinc sulphide from the resulting concentrated zinc sulphate solution using hydrogen sulphide. Part of the zinc sulphide is recycled to the copper removal circuit, as described above, and the remaining zinc sulphide is bagged and sold.

Nickel and cobalt in the raffinate from the zinc solvent extraction circuit are separated in the cobalt solvent extraction circuit, where cobalt is extracted, stripped from the loaded organic with sulphuric acid to form a concentrated cobalt sulphate solution, and purified. Nickel in the cobalt extraction raffinate, following adjustment with ammonia and ammonium sulphate, is reduced to nickel metal powder with hydrogen gas in reduction autoclaves. After washing and drying, the nickel powder is compacted into briquettes for sale. Similarly, cobalt is reduced to cobalt metal powder with hydrogen gas in separate reduction autoclaves and the washed and dried cobalt metal powder is compacted into briquettes for sale.

The end solutions after hydrogen reduction of nickel and cobalt contain high concentrations of ammonium sulphate. Small portions of these solutions are recycled to the nickel and cobalt reduction circuits while the remainders are combined and then stripped with hydrogen sulphide to recover the residual nickel and cobalt. The resulting sulphide precipitate is recycled to the sulphide leach circuit, where it joins the mixed sulphide intermediate from the pressure acid leach plant, as described above. The purified ammonium sulphate solution is fed to crystallizers in the ammonium sulphate plant, where ammonium sulphate fertilizer is produced for sale.

The quality of the cobalt product consistently exceeds the LME specification. The nickel quality originally met LME specification for all elements with the exception of silicon; the silicon specification of less than 0.005 wt.%, as well as the remaining LME specifications, have been routinely achieved since the fourth quarter of 2014. The iron content of the nickel product is well below the LME specification of 0.02 wt.% max.

SUMMARY OF IRON DEPORTMENT

Iron deportment in the Ambatovy pressure acid leach plant and refinery is summarized in Table 5, based on average data obtained for the commercial operation in 2015.

Table 5 – Summary of iron deportment in PAL and Refinery in 2015

Process Step or Stream	Iron Distribution, %		Ni:Fe Ratio		Ni:Fe in Product Stream Relative to Feed
	Solids	Solution	Solids	Solution	
Prepared Ore	100	--	0.026	--	1
HPAL	98.9*	1.1	--	2.3	87
Slurry Neutralization	99.2*	0.8	--	3.0	114
CCD	99.3*	0.7	--	3.6	138
Raw Liquor Neutralization	99.7*	0.28	--	9.6	365
Sulphide Precipitation	0.02	0.26	125	--	4,780
Sulphide Leaching	0.01	0.01	--	260	9,970
Iron Removal	0.01	--	--	86,000	3,300,000
Copper Removal	--	--	--	82,000	3,130,000
Cobalt SX Raffinate			--	32,000	1,230,000
Nickel Reduction			22,000	--	850,000

* overall deportment within the pressure acid leach plant with combined solids reporting to CCD discharge

About 99% of the iron in the ore is rejected to the leach residue in the HPAL step, about three-quarters of the remainder is precipitated in the subsequent neutralization and washing steps, and the equivalent of about 7% of the remaining about one-quarter of 1% reports to the mixed sulphide intermediate. About half of the iron in the mixed sulphide intermediate reports to the solution in sulphide leaching, consistent with about half of the iron in the mixed sulphide intermediate being present as insoluble hematite (entrained HPAL leach residue). Virtually all of the iron in the sulphide leach solution is subsequently precipitated in the iron removal step, producing purified solution containing less than 1 mg/L Fe and with a nickel to iron enrichment factor of about three million to one compared with the ore. Trace iron contamination is evident in downstream circuits (including in sampling and preparation of nickel briquettes for analysis), so that the weight ratio of nickel to iron in the final nickel briquettes was about 22,000 to 1 on average in 2015, compared with over 80,000 to 1 following both iron removal and copper removal. Overall enrichment in the Ni:Fe ratio in the nickel briquettes is nearly one million to one compared with the ore.

ACKNOWLEDGEMENTS

The authors thank the management of Sherritt International Corporation and the Ambatovy Joint Venture for granting permission to publish and present this paper.

REFERENCES

Buban, K.R., Collins, M.J., & Masters, I.M. (1999). Iron control in zinc pressure leach processes. *JOM, Vol. 51, No. 12,* 23-25.

Campbell, F., & Vardill, W.D. (1998). PAL process energy saving opportunities since the construction of Moa Bay. In: *IBC Nickel Conference* (pp. 1-34). Sydney, Australia.

Chalkley, M.E., & Toirac, I.L. (1997). The acid pressure leach process for nickel and cobalt laterite. Part I: Review of operations at Moa. In: W.C. Cooper & I. Mihaylov (Eds.), *Hydrometallurgy and Refining of Nickel and Cobalt* (pp. 341-353). Montreal, QC: CIM.

Chou, E.C., Rooke, P.M., Williams, A.R., & Hanks, J.T. (2000). Nickel and cobalt recovery from Madagascar laterite. In: P.R. Taylor (Ed.), *EPD Congress 2000* (pp. 469-488). Warrendale, PA: TMS.

Collins, M.J., Barta, L.A., Buban, K.R., Kalanchey, R., Owusu, G., Raudsepp, R., Stiksma, J., & Masters, I.M. (2004). Process development studies carried out by Dynatec as part of the Ambatovy nickel project evaluation. In: M.J. Collins & V.G. Papangelakis (Eds.), *Pressure Hydrometallurgy 2004* (pp. 259-276). Montreal, QC: CIM.

Collins, M.J., Barta, L.A., Buban, K.R., Kalanchey, R., Owusu, G., Raudsepp, R., Stiksma, J., & Masters, I.M. (2005). Process development by Dynatec for the Ambatovy nickel project. *CIM Bulletin, 98* (No. 1089), 1-7.

Collins, M.J., & Vardill, W.D. (2005). Introduction to the Ambatovy nickel project. In: *ALTA Nickel/Cobalt 2005*. Perth, Australia: ALTA.

Collins, M.J., Buban, K.R., Holloway, P.C., Masters, I.M., & Raudsepp, R. (2009). Ambatovy laterite ore preparation plant and high pressure acid leach pilot plant operation. In: J.J. Budac, R. Fraser, & I. Mihaylov (Eds.), *Hydrometallurgy of Nickel and Cobalt 2009* (pp. 499-501). Montreal, QC: CIM.

Collins, M.J., Yuan, D., Sitter, M., & St. Jean, S. (2015). Sustainable hydrometallurgical processing at the Ambatovy nickel operation in Madagascar. In: M. Agnew, T. Feldmann, N. Ahern, L. Centomo, G. Houlachi, J. Dutrizac, & D. Yuan (Eds.), *Lucy Rosato Memorial Symposium on Sustainable Hydrometallurgical Processes and Technologies*. Westmount, QC: CIM.

Hewitt, B., & Belke, J. (2004). Murrin Murrin CCD1 rake mechanism modification: the decision and the result. In: M.J. Collins & V.G. Papangelakis (Eds.), *Pressure Hydrometallurgy 2004* (pp. 227-246). Montreal, QC: CIM.

Kofluk, R.P., & Freeman, G.K.W. (2006). Iron control in the Moa Bay laterite operation. In: J.E. Dutrizac & P.A. Riveros (Eds.), *Iron Control Technologies* (pp. 573-589). Montreal, QC: CIM.

Masters, I.M., Barta, L.A., Stiksma, J., & Collins, M.J. (2006). Iron control in processes developed at Dynatec. In: J.E. Dutrizac & P.A. Riveros (Eds.), *Iron Control Technologies* (pp. 653-671). Montreal, QC: CIM.

Papangelakis, V.G., Georgiou, D., & Rubisov, D.H. (1996). Control of iron during the sulphuric acid pressure leaching of limonitic laterites. In: J.E. Dutrizac & G.B. Harris (Eds.), *Iron Control and Disposal* (pp. 263-274). Montreal, QC: CIM.

Valle, L., Benz, M., Chalkley, M., Collins, M., Dobson, T., Holmwood, R., Malevich, A., Tuffrey, N., & Vrolson, R. (2016). Completing the Ambatovy ramp-up: the road to successful Financial Completion. In: *ALTA Nickel/Cobalt/Copper 2016*. Perth, Australia: ALTA.

THE BEHAVIOUR OF IRON IN THE HIGH PRESSURE ACID LEACH PROCESS

Masahiro Tanaka[1] and Takao Oishi[2], Go Ohara[3], Toru Kitazaki[4]

[1234]Coral Bay Nickel Corporation (CBNC)
Rio Tuba, Municipality of Bataraza, Palawan 5306, Philippines
*(*Corresponding author: masahiro.tanaka@smm.com.ph)*

ABSTRACT

The High Pressure Acid Leach (HPAL) plant for nickel-cobalt mixed sulphide production at Coral Bay Nickel Corporation (CBNC) in the Philippines has been in operation since 2005 with a capacity of 10,000 t/year of nickel. Subsequent to the commissioning of the second plant in 2008, the production capacity increased to 22,000 t/year of nickel (Llerin et al., 2011). CBNC's annual nickel production reached more than 24,000 tons of nickel in 2014. The concepts to achieve this capacity are the high availability and the reciprocally flexible operation of both plants. The availability depends on some factors, such as the control of impurities in the process and the equipment condition, to mention a few. In the HPAL process, it is not only the autoclave operating conditions that affect nickel extraction but also the impurities in the ore greatly influence it. The major element in the ore is iron and most of this iron is precipitated as Fe_2O_3, while the rest remains as Fe^{2+} or Fe^{3+} sulphate in solution. These ions in solution reduce the availability, so it is important to control the formation of iron ions by controlling the impurities in feed ore as well as the HPAL conditions. This paper presents the general HPAL process and the behaviour of impurities, focusing on iron behaviour in the HPAL process.

KEYWORDS

High Pressure Acid Leach, HPAL, Ni, Fe, Fe_2O_3, impurities, Coral Bay Nickel Corporation

INTRODUCTION

Coral Bay Nickel Corporation Profile

Coral Bay Nickel Corporation (CBNC) was founded on April 2002, as a joint venture of four companies; namely, Mitsui Corporation (18%), Sojitz Corporation (18%), Nickel Asia Corporation (10%) and Sumitomo Metal Mining Corporation (SMM) with a controlling stake of 54%. Aside from its capital share, Nickel Asia Corporation also supplies the ore and limestone for the High Pressure Acid Leach (HPAL) plant located in Rio Tuba, Philippines, as shown in Figure 1 (Tsuchida, 2008). The objective of the CBNC project is to produce 22,000 t/year (original design) of nickel and 1,500 t/year of cobalt, as sulphide, by the treatment of nickel laterite ore which is a low grade nickel oxide ore in the Philippines (Mines and Geosciences Bureau, 2014). The sulphide product has been exported as a raw material to the SMM Nickel Refinery in Japan for 30 years. The product of this plant, a mixed (Ni/Co) sulphide, is shipped to the SMM's Nickel Refinery in Niihama, Japan. SMM Corporation provides the technology and management personnel to the plant. SMM Corporation also supplies the sulphuric acid, which is the main reagent for the HPAL process.

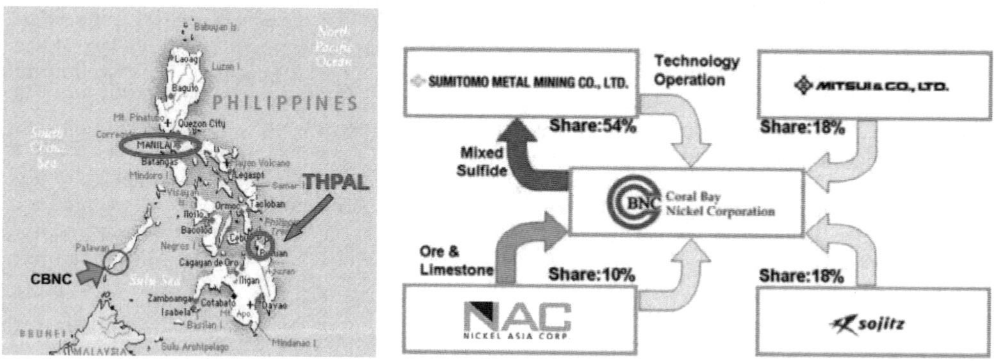

Figure 1 – CBNC project location and illustration of the relationship of stake holders

Nickel Asia Corporation (NAC), which has mined saprolitic ores for many years, was able to accumulate at least 20 million dry tons of laterite, a low grade nickel ore which is considered waste in the mining operations due to its only 1% nickel content. It is from the laterite ore that nickel is extracted and converted by the CBNC HPAL plant to an intermediate sulphide product having a 56-58 wt.% concentration of nickel. The Niihama Nickel Refinery (NNR) in Japan refines this intermediate product to 99.99 wt.% Ni metal using the proven Matte Chlorine Leach Electrowinning (MCLE) process of Sumitomo Metal Mining (SMM).

The plant takes into consideration the improvements, technical data and operational know-how gained from the successful operation of CBNC, the sister company of Taganito HPAL Nickel Corporation (THPAL). The objective of the Taganito HPAL project is to produce 30,000 t/year of nickel metal and 2,600 t/year of cobalt, as a mixed sulphide, by the treatment of nickel laterite ore in the Philippines (Shibayama, 2015). THPAL was founded in August 2008 and became a joint venture of three companies, namely Mitsui Corporation, Nickel Asia Corporation (NAC) and SMM with a controlling stake of 62.5%. Aside from its capital share, Taganito Mining Corporation (TMC), under the control of Nickel Asia Corporation, supplies the ore for the HPAL plant located in Taganito, Claver, Mindanao, Philippines, as shown in Figure 1. The product of this plant, a mixed (Ni/Co) sulphide, is also shipped to the SMM Nickel Refinery in Japan. SMM provides HPAL technology and dispatches management personnel to THPAL. Figure 2 below illustrates this relationship.

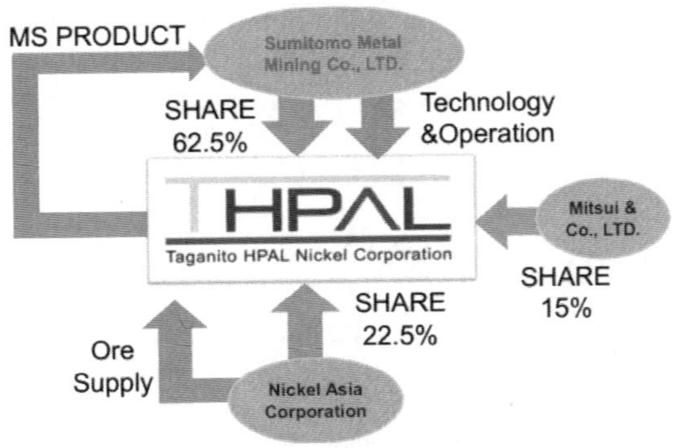

Figure 2 – Illustration of the relationship of the stake holders in the Taganito HPAL operation

Process Design

The process flowsheet shown in Figure 3 illustrates the current CBNC process (Salanga, et al., 2015). Each major processing circuit is briefly explained below.

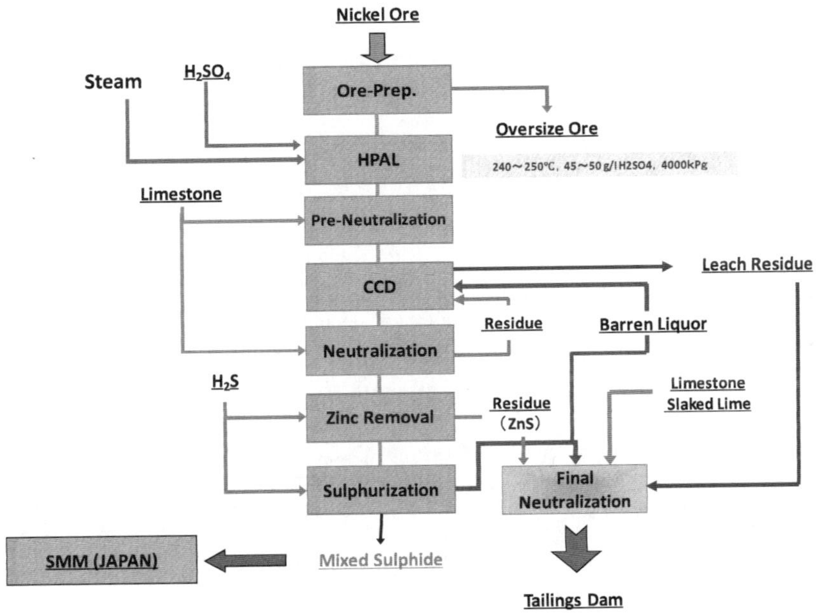

Figure 3 – Coral Bay Nickel Corporation process flow sheet, as explained below

Ore Preparation

The objective of the ore preparation process is to prepare the ore feed slurry for HPAL. Laterite ore is wet-screened in this circuit and transferred to a thickener, where the ore slurry is thickened to around 40 wt.% solids. Table 1 shows the typical ore slurry assay.

Table 1 – Ore slurry assay

Component	Concentration (wt.%)
Ni	1.0-1.3
Co	0.09-0.15
Mg	1.0-2.0
Si	5.0-8.0

High Pressure Acid Leach (HPAL)

The ore slurry is pre-heated to 200°C before being pumped to the autoclave using a positive displacement pump. 98 wt.% H_2SO_4 is injected to initiate the leaching reaction. The autoclave is controlled at 240-250°C and 4.0 MPa. Leached slurry with a free acid concentration of 45-50 g/L is de-pressurized in series of flash tanks and is sent to the pre-neutralization circuit. The recovered steam from the flash tanks is recycled back to the pre-heaters to heat the ore slurry prior to its injection into the autoclave.

$$NiO + H_2SO_4 \rightarrow NiSO_4 + H_2O \quad (1)$$

Pre-Neutralization

The leached slurry, containing about 45-50 g/L free acid concentration, is neutralized to pH 2-3 using limestone slurry. The leach slurry is then sent to the countercurrent decantation circuit for residue washing.

Countercurrent Decantation (CCD)

Leached slurry is washed in a series of thickeners using barren liquor from the sulfurization circuit. The pregnant liquor overflow is sent to the neutralization circuit. The thickened solids from the CCD circuit are sent to the final neutralization circuit for treatment prior to disposal in the tailings dam.

Neutralization

The pregnant liquor from the CCD circuit is neutralized with limestone slurry. The objective is to remove impurities such as iron, aluminum and chromium by precipitating them as hydroxides. The resulting solids are thickened and sent back to the CCD circuit; the clear neutralized overflow is sent to the zinc removal circuit.

Zinc Removal

Zinc is a major impurity in the final product. It will co-precipitate with nickel and cobalt in the sulfurization circuit; hence, zinc is removed from the pregnant liquor using H_2S gas in the zinc removal circuit. The precipitated ZnS solids are removed using polishing filters and are sent to the final neutralization circuit. The Zn-free pregnant liquor passes to the sulfurization circuit.

Sulfurization

Nickel and cobalt are precipitated from the Zn-free pregnant liquor using H_2S gas. Barren liquor is sent to the CCD circuit as wash water for the leach residue. The mixed sulphide product is bagged using flexible containers. Table 2 below shows a typical mixed sulphide product analysis.

Table 2 – Mixed sulphide product profile

Component	Concentration (wt.%)	Component	Concentration (wt.%)
Ni	56-58	S	30-35
Co	4-5	Moisture	10-15
Fe	0.2-0.7	Particle Size D50	40-50 μm

IMPURITIES CONTROL IN THE HPAL OPERATION

Operational Status of the Autoclave

In general, there are several impurities contained in the ore and these are dissolved in the HPAL process. As an example, iron in the laterite ore is dissolved as ferric ion and ferrous ion. In the autoclave, most of the ferric ions form hematite (Fe_2O_3) through forced hydrolysis at elevated temperatures of more than 200°C; the sulphuric acid consumed during leaching is regenerated by the hydrolysis reaction. It is for this reason that the HPAL process is preferred over atmospheric leaching (AL) for the extraction of nickel and cobalt from lateritic ores, which typically contain around 50 wt.% iron. Therefore, the HPAL process is a useful technology for the recovery of Ni from oxide ore of low grade. The chemical reactions describing the behavior of oxide ore in the leaching process using sulphuric acid (H_2SO_4) are shown below. M indicates Ni, Co, Cu, Fe, Mg, Mn and Zn.

$$MO + H_2SO_4 \rightarrow MSO_4 + H_2O \quad (2)$$

$$FeOOH + 3H^+ \rightarrow Fe^{3+} + 2H_2O \quad (3)$$

$$Fe^{3+} + 1.5H_2O \rightarrow 0.5Fe_2O_3 + 3H^+ \quad (4)$$

Various researchers have reported that there is positive effect of temperature, leaching duration and acid dosage on nickel extraction, on the enhancement of forced iron hydrolysis, etc. It is also clear that the presence of residual iron in the pregnant leach solution results in increased consumption of the neutralizing reagent and also has a negative effect in the downstream purification processes. This is the main reason that iron removal control in the HPAL process is an essential item for the operation of the whole plant. In this regard, iron behaviour in the HPAL process was investigated.

The Iron Behaviour in the HPAL Process

The oxidation-reduction potential (ORP) of the leached slurry has a great impact on the total iron concentration. Generally, the ferric/ferrous concentration ratio depends on the operating ORP in the autoclave under low pH conditions. It is apparent that the total iron concentration is significantly elevated at low ORP due to the steep increase in the Fe^{2+} ion concentration with decreasing ORP, as is shown in Figure 4. In order to reduce reagent costs and improve product quality, it is preferable to control the ORP in the autoclave to enhance iron removal.

Figure 4 shows that if the ORP is greater than 500 mV, the Fe^{2+} concentration in the slurry is reduced. It is therefore possible to have a reduction in the cost of the neutralizing reagent used in the process. Additionally, this does not cause an adverse effect in the downstream processes. One particular effect of the reduced Fe concentration is in the filtration circuit, where high overall operational load performance may be possible. The overall operational load (actual plant performance) compared to the installed/design plant capacity is a numerical value that measures the efficiency of the operation. It is clearly desirable to have a high operational load value. Figure 5 shows that a low Fe concentration leads to a higher overall operational load. In addition, there is no dependence of the operational load on the ORP and Fe^{3+} concentration; therefore, equation (4) is in a state close to equilibrium in the autoclaves. However, the Fe^{3+} concentration has been found to depend on the temperature and the free acid concentration in the autoclave; therefore, it is desirable to operate at the appropriate temperature and free acid concentration. This will be explained in more detail later in this paper. For now, we will focus on the ORP in the autoclave.

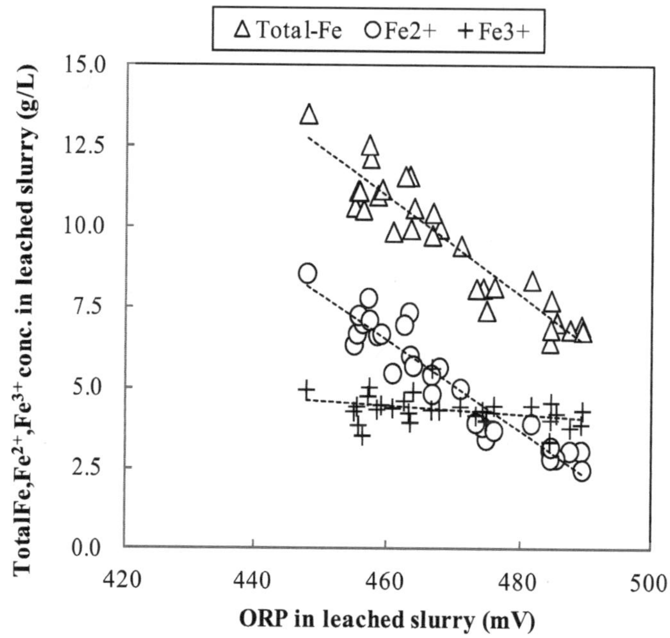

Figure 4 – Iron concentration vs. ORP in the leach slurry

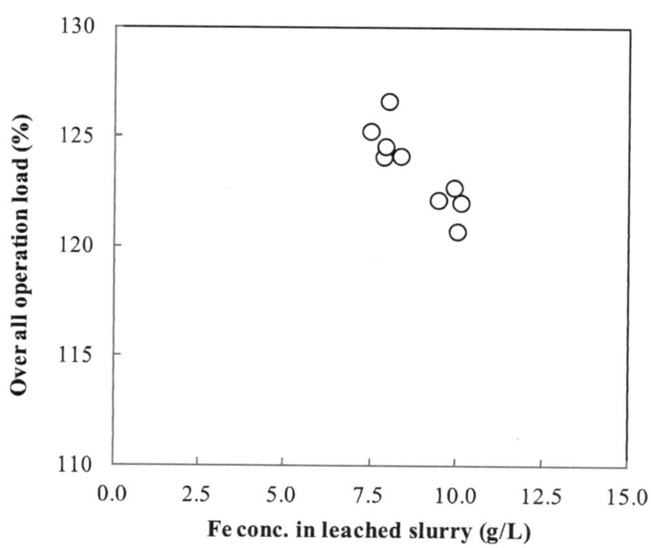

Figure 5 – Relationship between the operational load and the Fe concentration of the leach slurry

The ORP cannot be excessively increased because high ORP values lead to increased concentrations of harmful Cr^{6+}, as shown in Figure 6. The increased Cr^{6+} concentration creates an abnormal condition in the final neutralization circuit since the Cr^{6+} needs to be reduced before disposal to the tailings dam. The concentration of Fe^{2+} has been reduced with an increase of the ORP. On the other hand, harmful Cr^{6+} also increased with increasing ORP. However, Fe^{2+} oxidizes more easily than Cr^{3+}; thus, the presence of a significant Fe^{2+} concentration indicates that most of the chromium is present as Cr^{3+} (Figures 4 and 6). A high oxidation-reduction potential is required to facilitate the production of hematite; it is suitable for Fe removal from the HPAL solution. However, in such conditions, Cr^{6+} is formed and it is difficult to remove from the solution. Reducing the Cr^{6+} concentration will require added cost.

In contrast, a low ORP is also not desirable because the oxide film that protects the titanium alloy liner of the autoclave is reduced; the alloy is then dissolved in the solution. The ORP is adjusted to the desirable range of 500-600 mV. This controlling range makes it possible to achieve a high overall operational load. The original capacity of the facility was 22,000 t/year of nickel, but now 24,000 t/year of nickel is achieved.

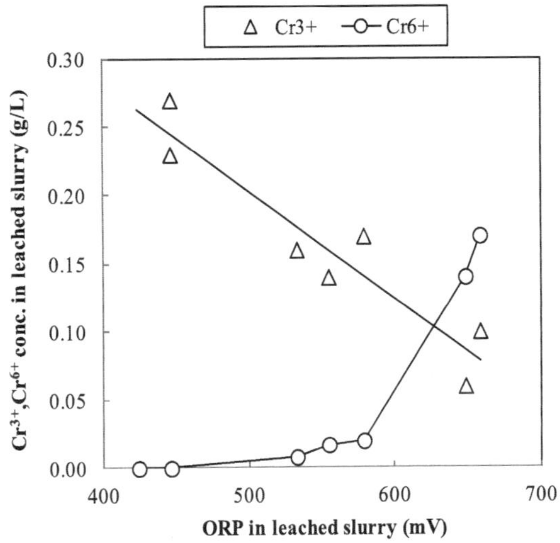

Figure – 6 – Chromium concentration vs. ORP in the leach slurry

The Iron Control in the HPAL Process

Ore Blend Control

There are three main types of the ore used in CBNC: Type A, Type B and Type C. These are classified according to their nickel grade. The main mineral of these ores is goethite, but the components are different for each type. These ores greatly influence the acid consumption and ORP for the HPAL process, and they also influence the physical characteristics of the slurry, such as the thickening behaviour and viscosity. Thus, defining the ore mix and blending the ore is carried out to supply a generally desirable ore feed. Table 3 shows the typical elemental analyses and features for each kind of ore feed at CBNC. The ORP and several impurities are monitored and taken into consideration for ore blending. By adjusting the ore blend, it is possible to keep a high ORP in the leach process. However, it is not only the ORP and other impurities that need to be monitored. For example, an increase in Mg content means an increase in sulphuric acid consumption and it also has an effect on the leaching reaction and the physical characteristics of the ore slurry. The achievement of a relatively low sulphuric acid consumption requires a lower magnesium containing ore feed for the leaching reaction.

Table 3 – Typical elemental analyses of the three ore types

Type	Ni(%)	Co(%)	Fe(%)	Mg(%)	S(%)	Carbon(%)	Remarks
A	<0.8	<0.1	>50	<1.0	0.05	0.25-0.30	Low-Ni,Mg
B	<1.2	0.1-0.2	40-50	<1.5	0.03	0.20-0.25	Middle-Ni,Mg
C	<1.5	0.1-0.2	40-50	<3.0	0.01	<0.2	High-Ni,Mg

Autoclave Temperature Control

Adjusting the ore feed to HPAL allows the control of the sulphuric acid addition for leaching, as well as the amount of base required for the neutralization of the residual acid in the downstream process. Furthermore, the hematite (Fe_2O_3) precipitation reaction in the autoclave is accelerated by an increase in the reaction temperature. Raising the temperature of the reactants speeds up the chemical reactions because the activation energy barrier is more easily overcome. The activation energy is the energy required in a system so that a chemical reaction can be initiated.

$$\text{Rate of reaction} = C \times e^{-Ea/RT} \tag{5}$$

Where, C = rate constant, Ea = activation energy, R = molar gas constant (8.314 J $mol^{-1}K^{-1}$), and T = absolute temperature.

The Fe^{3+} concentration in the leach slurry is reduced with increasing operating temperature because of the increase in the reaction rate and the lower solubility of hematite. As shown in Figure 7, the Fe^{3+} concentration drastically decreases with increasing temperature. This suggests that temperature is also very important for improving the equilibrium of the forced hydrolysis reaction of ferric sulphate, where the ferric concentration seems to be saturated at elevated temperatures.

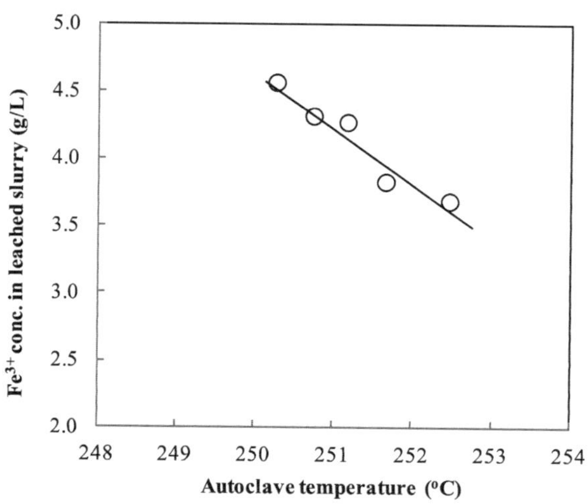

Figure 7 – Relationship between Fe^{3+} concentration and autoclave temperature

Free Acid Control

Free acid in the leach slurry is also an important parameter in the HPAL process. During leaching in the autoclave, impurities are leached in addition to the dissolution of nickel. Because of these reactions, excess addition of acid is needed. The adjustment of the sulphuric acid addition is based on the amount of free acid present in the leach solution. As shown in Figure 8, hematite formation is also affected by the concentration of free acid. When the free acid concentration of the leach solution is low, the Fe^{3+} concentration is also low. This means that the hematite reaction is proceeding. On the other hand, when the free acid concentration increases, the Fe^{3+} in the solution also increases; this means that the hematite precipitation reaction is not occurring to a sufficient degree. With these findings, hematite precipitation is best achieved at a low free acid concentration during leaching. However, this is not a desired condition because of the low leaching rate of Ni at low acid concentrations. Therefore, for optimum HPAL operation, it is necessary to determine the acid concentration required for both the optimum leaching rate for Ni and the effective precipitation of hematite.

$$Fe^{3+} + 3/2H_2O \rightarrow 1/2Fe_2O_3 + 3H^+ \quad (4)$$

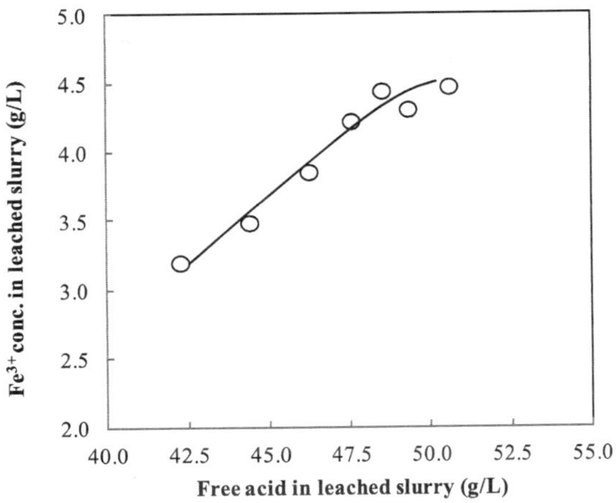

Figure 8 – Relationship between the Fe^{3+} concentration and free acid in the leach slurry

ADDITIONAL PROJECTS

Apart from nickel and cobalt recovery from the ore, CBNC investigated, at a pilot plant scale, the recovery of other elements present in the ore, such as chromite and scandium.

Chromite Pilot Plant

This pilot plant was designed by Sumitomo Metal Mining (SMM) with the objective of recovering chromite from the ore. The construction of the pilot plant started in March 2013 and was completed in July 2013; the commissioning period ran from August 2013 to July 2014. By August 2014, a chromite sample (Cr_2O_3>41wt%, Cr/Fe>1.2) had been produced.

Scandium Pilot Plant

Scandium is found in the ore used by CBNC to make the mixed sulphide. The scandium pilot plant was also designed by Sumitomo Metal Mining. The initial stage commenced in December 2012. Scandium recovery circuit commissioning began in October 2013 and was completed in November 2014; sample production (Sc_2O_3>99.9 wt.%) is in progress. The feed for the plant comes from the barren liquor of the mixed sulphide precipitation process.

CONCLUSIONS

In this paper, the effect of iron behaviour in the HPAL autoclave was reviewed. It was concluded that an increase in the operating ORP in the leaching reaction can be achieved by controlling the ore blend. The following points summarize the findings.

Adjusting the ore blend makes it is possible to control the ORP of the leach slurry. Therefore, it is possible to control the Fe concentration in the leach slurry. To suppress the generation of harmful chromium (Cr^{6+}), to decrease the amount of neutralizing reagents and to prevent equipment damage, the ORP should be adjusted and maintained in the range of 500 – 600 mV. Furthermore, Fe removal is promoted by hematite precipitation in the autoclave, and this reaction is achieved under high temperature conditions. Hematite formation is also best achieved at lower free acid concentrations. However, this condition decreases the nickel leaching rate. It is therefore necessary to determine the optimum condition for both Ni extraction and Fe precipitation in the autoclave. Other than Fe control, maintaining a desirably high Al/Mg ratio in the ore can give an effective leaching rate and a lower sulphuric acid consumption, while still achieving a high nickel extraction. By controlling all these operational parameters, CBNC is able to successfully maintain a high-load operation, equivalent to 120% of equipment design and to produce 24,000 t/year of nickel by optimizing the above mentioned parameters. Additionally, to maintain a world-class HPAL technology, CBNC is trying to recover metals other than nickel and cobalt from its ores.

REFERENCES

Llerin, J.E, Nishikawa, I., Kawata, M. (2011). Coral Bay Nickel HPAL plant expansion project. *ALTA 2011.*

Mines and Geosciences Bureau, Department of Environment and Natural Resources (2014). Philippine Metallic Minerals Production as of December 2014.

Salanga, L.G., Oishi, T.O. & Kitazaki, T. (2015). Current plant operation of Coral Bay Nickel Corporation. *ALTA 2015.*

Shibayama, K. (Ed.) (2015). Taganito HPAL plant project. *Nickel Processing 2015.*

Tsuchida, N. (2008). Process development and commissioning of HPAL plant for recovery of nickel and cobalt from low grade laterite nickel ore. *Journal of MMIJ* (124, 549-553).

IRON CONTROL STRATEGIES IN PRESSURE HYDROMETALLURGY

*R.G. McDonald and D.J. Robinson

CSIRO Mineral Resources
Waterford, WA, Australia 6152
*(*Corresponding author: Robbie.Mcdonald@csiro.au)*

ABSTRACT

High iron solution concentrations often result in the formation of basic ferric sulfate during the high temperature oxidation of sulfidic materials such as base metal and refractory gold concentrates. Consequently, the "Hot Cure Process" followed by limestone neutralization is required to facilitate subsequent processing and/or disposal of the leach residue. The present paper discusses strategies that can be employed to influence the composition and nature of the high-iron content leach residues produced during pressure leaching.

KEYWORDS

Iron removal, basic ferric sulphate, nickel, pressure oxidation, nitric acid

INTRODUCTION

Control of iron deportment in hydrometallurgical processing remains a key challenge for a range of reasons. Strategies generally relate to the control of conditions under which iron is optimally rejected (Demopoulos, 2009). Supersaturation provides the driving force for crystallization and this can be controlled by a number of factors that include (1) pH and/or Eh control, (2) the presence of matrix anions and impurities, (3) dilution, (4) complexation and dissociation and (5) manipulation of the rate at which the metal to be precipitated is released to solution. Each of these factors is impacted by temperature. Good control of crystallization is essential to the control of precipitate properties such as the degree of crystallinity, stability, particle size and quality which covers both the extent of impurity uptake and suitability of the precipitate for waste disposal. The quality and properties of iron precipitates in turn can impact downstream processing.

It is the aim of this paper to discuss the generation of iron-containing residues produced during the pressure oxidation of sulfidic ores and concentrates and how the nature of these can be controlled to potentially facilitate further processing. A brief overview of several related topics will now be given and include, where relevant, discussion of work undertaken in our laboratories. In addition, brief discussions of selected process options that enable good control of iron rejection will be provided.

BENCH SCALE LABORATORY RESULTS

Pressure oxidation tests were undertaken at batch scale using a Parr 1 gallon Grade 3 titanium autoclave fitted with serpentine cooling coil, dual pitched blade impeller and, gas injection and sampling ports. Pressure oxidation was undertaken using continuous injection of oxygen gas at the required partial pressure. For some tests the rate of oxygen injection was maintained at a constant value using a Hastings Teledyne Mass Flow Controller which in addition to temperature and pressure, were logged via LABVIEW-based custom designed software.

Samples taken during these experiments were processed using vacuum filtration, the filter cake re-pulped and washed several times with de-ionized water before drying overnight at 70°C. Analysis were conducted via Inductively Coupled Plasma-Optical Emission Spectroscopy (ICP-OES) with the solids first fluxed with a 12:22 lithium borate/metaborate mix. Sulfur analysis was conducted separately via carbon/sulfur analyzer.

Free acid values were determined using the calcium EDTA method with auto-detection of the titration endpoint. Ferrous iron was measured by titration by the dichromate method using barium diphenylamine-4-sulfonate indicator.

Quantitative X-Ray Diffraction (QXRD) analysis for the residue samples was undertaken using published crystallographic models and Brüker TOPAS software.

IRON CONTROL DURING PRESSURE LEACHING

Pressure Oxidation of Refractory Gold Ores

The encapsulation of sub-microscopic gold within iron sulfide minerals often requires hydrometallurgical pre-treatment to facilitate good gold recovery. Pressure oxidation is one approach that has been employed to achieve this and a summary of such plants operating at the time is given in the article of Adams (2005). Reviews on the treatment of refractory gold ores are also available (e.g. Haque, 1987; Iglesias & Carranza, 1994).

The pressure oxidation of iron sulfide minerals in refractory gold ores is normally conducted under conditions that generate ferric sulfate as the intermediate soluble species. Subsequent hydrolysis can produce several iron-containing materials that include hematite, jarosite and basic ferric sulfate via well-known reactions:

$$Fe_2(SO_4)_3 + 3H_2O \rightarrow Fe_2O_3 + 3H_2SO_4 \quad (1)$$

$$3Fe_2(SO_4)_3 + 14H_2O \rightarrow 2(H_3O)Fe_3(SO_4)_2(OH)_6 + 5H_2SO_4 \quad (2)$$

$$Fe_2(SO_4)_3 + 2H_2O \rightarrow 2FeOHSO_4 + H_2SO_4 \quad (3)$$

To enable the subsequent recovery of gold, and since both hydronium jarosite and basic ferric sulfate are unstable, neutralization with (slaked) lime prior to cyanidation is required:

$$(H_3O)Fe_3(SO_4)_2(OH)_6 + 2Ca(OH)_2 \rightarrow 3Fe(OH)_3 + 2CaSO_4.2H_2O \quad (4)$$

$$FeOHSO_4 + Ca(OH)_2 + 2H_2O \rightarrow Fe(OH)_3 + CaSO_4.2H_2O \quad (5)$$

In our work, a Telfer concentrate containing Sulfidization, Acidification, Recycle and Thickening (SART) material was leached with cyanide to remove the poorly crystalline copper sulfides and enable mineralogical characterization by QXRD. It was pressure oxidised under various conditions as described in Table 1 for the purposes of leaching the copper and releasing the refractory gold for subsequent extraction and recovery. The mineralogical compositions of the concentrate and leach residues are given in Table 2.

Table 1 – Experimental details for the pressure oxidation of a Telfer copper/gold concentrate* at 10% w/w

Experiment	Temperature (°C)	Sodium chloride concentration (g/L)
POX A	220	0
POX B	220	10
POX C	180	0
POX D	180	10
POX E	220	1

* Elemental content (% w/w): Cu (17.9), Fe (26.4), Al (0.63), Ni (0.08), Co (0.41), Mg (0.10), Ca (0.44), Na (0.13), K (0.12), Si (3.94), S (29.8).

The concentrate is clearly of sufficiently high iron sulfide content that basic ferric sulfate is readily formed during leaching, with non-saline process water, even without the up-front presence of sulfuric acid. While in a saline environment the formation of natrojarosite was more favoured and this was sufficiently stable for gold leaching to be undertaken without pre-neutralisation; as has been noted by others (e.g. King et al., 2011). Lower temperatures, while favouring hydronium jarosite formation over basic ferric sulfate, resulted in less complete oxidation of the sulfide minerals. The tests described were undertaken to enable comparison of cyanidation with copper catalyzed ammonium thiosulfate leaching of gold from the residues. It was shown, though the results will not be presented here, that leaching in the presence of sodium chloride produced residues readily leachable in the thiosulfate system but not when sodium chloride was absent (Breuer et al., 2014). This behaviour was sample specific and to date has not been fully explained by us. It was also recognised that leaching in the presence of sodium chloride will results in the solubilisation of some gold since the precipitation of natrojarosite at temperatures above 200°C has been shown not to result in the co-precipitation and/or adsorption of gold chloride (Blais, 2010). An approach to the recovery of this gold chloride from solution under pressure oxidation conditions has been proposed (Parisien-La Salle, 2012). The paper by Demopoulos et al. (2012) summarises preliminary studies to examine the behaviour of gold during pressure oxidation.

Table 2 – Quantitative mineralogy for Telfer concentrate calculated after cyanide leaching to remove SART material and the POX residues generated using conditions described in Table 1

Mineral Phase	Sample						
	CN tail	Calc. head	POX A	POX B	POX C	POX D	POX E
Quartz	14.4	12.7	15.3	9.1	13.3	11.8	13.1
Other silicates	7.5	6.6	4.4	3.3	5.7	3.6	3.6
Pyrite	73.0	64.2	0.0	0.0	19.5	4.3	0.0
Chalcopyrite	1.5	1.3	0.0	0.0	0.3	0.0	0.0
Cobaltite	1.2	1.0	0.0	0.0	0.0	0.0	0.0
Other copper sulfides	0.0	12.0	0.0	0.0	0.0	0.0	0.0
Sulfur	0.0	0.0	0.0	0.0	6.5	3.5	0.0
Basic ferric sulfate	0.0	0.0	27.1	0.0	3.0	0.0	31.8
Natrojarosite	0.0	0.0	0.8	87.6	8.6	76.8	13.9
Hematite	0.0	0.0	49.1	0.0	39.9	0.0	34.3
unaccounted	2.4	2.1	3.2	0.0	3.3	0.0	3.3

The development of the "Hot Cure Process" for the treatment of basic ferric sulfate present in gold-containing residues (e.g. Fleming, 2010; King et al., 2011) has enabled lime to be replaced by cheap limestone. The use of heat and acid generated during pressure oxidation means that only a small amount of additional energy is required to dissolve the basic ferric sulfate (and hydronium jarosite) prior to neutralization:

$$2FeOHSO_4 + H_2SO_4 \rightarrow Fe_2(SO_4)_3 \quad (6)$$

$$Fe_2(SO_4)_3 + 3CaCO_3 + 9H_2O \rightarrow 2Fe(OH)_3 + 3CaSO_4.2H_2O + 3CO_2 \quad (7)$$

Clearly, however, avoidance of these steps via the production of more stable pressure oxidation residues is desirable in gold processing. Conversely if a ferric sulfate lixiviant is being generated, as in the Sepon process (Baxter et al., 2004), the promotion of basic ferric sulfate formation is highly desirable.

Pressure Oxidation of Chalcopyrite

The leaching of chalcopyrite has been conducted over a wide range of conditions that range from ambient and atmospheric to high temperature and pressure, typically 220-225°C and 7-10 bar oxygen partial pressure. The success of these approaches is typically predicated upon optimising the Eh/corrosion current and to a lesser extent pH, thereby reducing the impact of surface passivation upon the rate at which copper is extracted.

At high temperatures and pressures oxidation in the system is near complete as represented by the following equations for chalcopyrite leaching:

$$2CuFeS_2 + 8.5O_2 + 2H_2O \rightarrow 2CuSO_4 + Fe_2O_3 + 2H_2SO_4 \quad (8)$$

$$6CuFeS_2 + M_2SO_4 + 25.5O_2 + 9H_2O \rightarrow 6CuSO_4 + 2MFe_3(SO_4)_2(OH)_6 + 3H_2SO_4 \; (M=H_3O, Na) \quad (9)$$

$$2CuFeS_2 + 5.5O_2 + 2H_2SO_4 \rightarrow 2CuSO_4 + 2FeOHSO_4 + H_2O \quad (10)$$

Around fifty batch tests for a single chalcopyrite concentrate (elemental content, %w/w: Cu 23.6, Fe 25.3, Al 0.34, Mg 1.46, Si 5.85, S 29.2) have been conducted in our laboratories employing a range of operating conditions and each of the iron-containing phases given in equations (8)-(10) have been generated in residues. Tests were either conducted at 10% or 20% w/w pulp density.

Basic ferric sulfate was detected in several tests when acid was not added up-front and when a 20% w/w pulp density was employed. The addition of a small amount of acid, 30 g/L, to simulate low strength copper liquor recycle enhanced the amount of basic ferric sulfate formed. A high acid addition, 100 g/L, was tested for comparison. The composition of the leach residues as a function of time are shown in Figure 1.

(A) (B) (C)

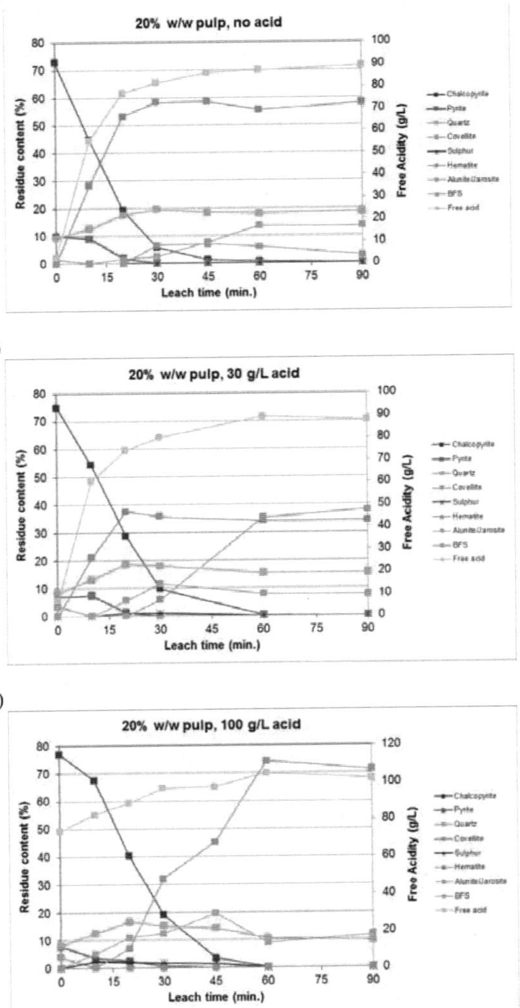

Figure 1 – QXRD data for leach residues obtained during the POX of a chalcopyrite concentrate (20% w/w) at 220°C with $p(O_2)$ 700 kPa: (A) No up-front acid addition, (B) 30 g/L sulfuric acid addition and (C) 100 g/L sulfuric acid addition

Examination of Figure 1 indicates that increasing acid addition slows the rate at which chalcopyrite reacts. Furthermore the formation of basic ferric sulfate does not commence until the free acidity is above 60 g/L. That this phase is thermodynamically unstable relative to hematite and jarosite is confirmed by its decreasing content (toward the back end of the reaction) in the residue formed without up-front acid addition (Figure 1(A)), although the free acidity is continuing to increase. The amount of basic ferric sulfate would be expected to decrease further with increasing time beyond the 90 minutes of these tests. When up-front acid is added the extent of basic ferric sulfate formation is greater (Figure 1(B)) and residue has a moderate sulfur content, 8.6%. For high up-front acid addition basic ferric sulfate and jarosite are dominant (Figure 1(C)) and generate a residue with high sulfur content, 17.2%. These observations are consistent with a series of studies (e.g. Umetsu et al., 1977; 1978) to examine the hydrolysis of ferric sulfate at elevated temperatures, 185°C and 200°C, and which determined the impact of various metal sulfate salt additions upon the precipitates that formed. Typically in our study the copper concentration was ~60 g/L which would enable a free acidity of ~85 g/L to be sustained before high sulfur-containing residues become more stable than hematite at a temperature of 200°C (Umetsu et al., 1979). The window of free acidity over which the residue changes from having low to high sulfur content is narrow and the acidity above which sulfur content starts to increase is often referred to as the breakpoint value. The breakpoint acidity at 200°C in the presence of copper sulfate does not change greatly with increasing copper concentration above 45 g/L and therefore controlling the formation of basic ferric sulfate during chalcopyrite concentration requires either the pulp density and/or the incoming acid concentration to be decreased.

Co-treatment of Nickel Ores

The in situ generation of sulfuric acid can be employed for leaching oxidic materials, for example through the combination of a nickel sulfide ore/concentrate with a nickel laterite ore with injection of oxygen under HPAL conditions, specifically at 250°C. The combination of sulfidic materials with nickel laterites in this manner appears to have been first reported by Ferron and Fleming (2004) though a patent was granted to INCO several decades ago relating to the combination of pyritic and nickeliferous oxide ores and leaching under pressure oxidation conditions (Opratko et al., 1974). Typically after blending mined laterite ore to achieve a consistent plant feed, HPAL requires acid additions between 250-400 kg/t dry ore given the TDS content of the process water is sufficiently low.

The work of Ferron and Fleming (2004) examined various sulfidic materials including elemental sulfur (with lignosol dispersant), pyrrhotite concentrate including in the presence of added chloride (i.e. Platsol™ conditions) and demonstrated the process to be technically feasible. The study did not extend to reporting the composition of the leach residues produced. More recently Quinn et al. (2009) reported the outcomes of batch and pilot scale test work dubbing the process Combined Pressure Acid Leaching (CPAL). The work, generally undertaken at 220-240°C and oxygen overpressure of 800 kPa employed a combination of sulfidic material blending and high iron content heap leach liquor to generate sufficient in situ acid though where a deficit occurred, this was made up by the addition of sulfuric acid. The benefit of blending sulfidic material with laterites upon rheological behaviour was also noted and this is the subject of the invention of Rodriguez (2009). Although demonstrating nickel extractions in excess of 90%, the work of Quinn et al. (2009) did not report on the mineralogy of the residues to confirm their composition and explain the mass balance data generated.

Table 3 – Elemental composition of nickel concentrate and laterite feeds employed for co-processing test work (% w/w)

Feed	Ni	Co	Cu	Mg	Mn	Fe	Al	Cr	Ca	Na	Si	S
Concentrate	8.11	0.138	1.28	0.651	0.037	45.0	0.318	0.175	0.421	0.082	2.48	30.3
Laterite	1.90	0.131	0.011	3.32	0.456	24.6	2.47	0.857	0.170	0.201	16.8	0.02

It has been demonstrated that the addition of a low grade nickel concentrate (Table 3) can provide sufficient in situ generated acid to leach the laterite component of the mixture (McDonald et al., 2012). It is not clear that such a process could be sustained commercially unless the low grade concentrate, potentially from a disseminated sulfide ore body, could be generated locally. This is certainly possible in Western Australia as companies such as Poseidon Nickel have nickel resources that include sulfide and oxide deposits as well as oxidised dump material from previously mined ores. That said, the sweetening of nickel laterite feedstock via the addition of a portion of higher grade oxidised sulfidic overburden has been practised at the Glencore, Murrin Operation for some time; this has both increased the grade of the autoclave feed material and enabled the pulp density of this feed to be increased.

(A)

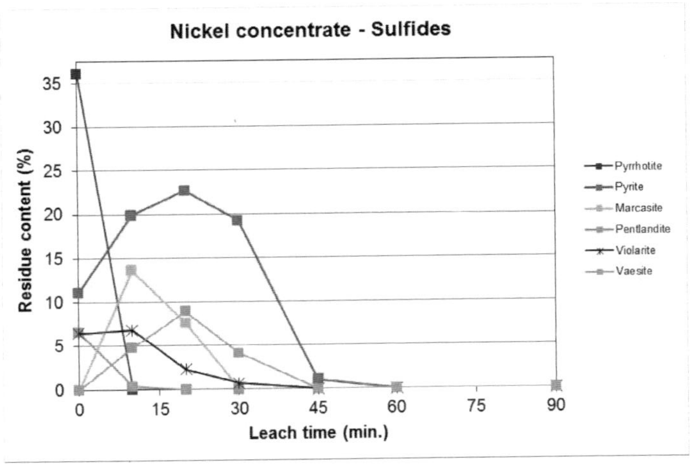

(B)

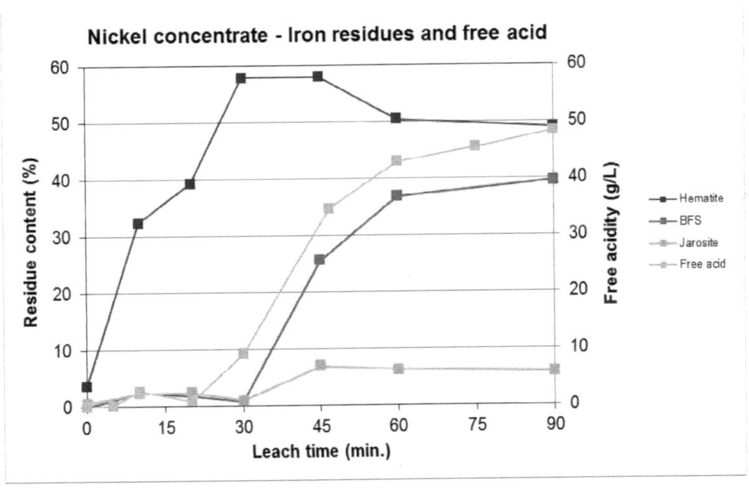

Figure 2 – QXRD data for leach residues obtained during the POX of a nickel concentrate (10% w/w) at 250°C with $p(O_2)$ of 100 kPa: (A) Sulfide minerals (B) Iron-containing products and free acidity

A close examination of the residue mineralogy in previously published work (McDonald et al., 2012) is provided by the data shown in Figures 2 and 3. Constant oxygen mass flow control was employed and enabled the nickel concentrate to leach more slowly than normally hence the sulfide minerals were only completely leach in just over 45 minutes (Figure 2A). Apart from the early hydrothermal conversion of pyrrhotite to pyrite and marcasite with co-generation of hematite, these reactions combined being acid neutral, most of the iron sulfides, viz. pyrite, leached after 30 minutes and this resulted in the rapid generation of sulfuric acid. This was accompanied by the formation of basic ferric sulfate and a small amount of jarosite (Figure 2B). In comparison the leaching of the sulfide minerals during co-leaching followed a similar pattern to the concentrate except that the acid generated was primarily required to complete leaching of the nontronite in the laterite ore. This drop in the amount of acid present in the system was sufficient to favour the formation of hematite over basic ferric sulfate. It was indicated from the unit cell parameters of the jarosite phase that this was rich in aluminium and that only around 5% of the iron in the residue reported to this phase. In summary the co-leaching of sulfidic and oxidic ores enables good control of the iron-containing phases in the leach residue while employing the in situ generated sulfuric acid to extract more metals values.

(A)

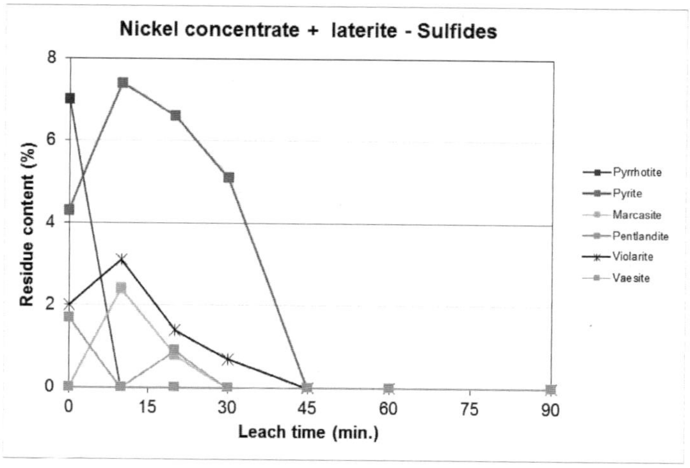

(B)

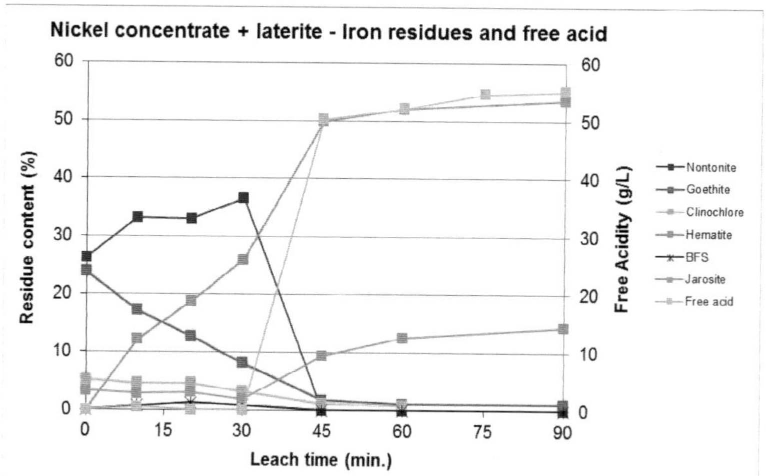

Figure 3 – QXRD data for leach residues obtained during the POX of a nickel concentrate/laterite blend (10.5/19.5% w/w) at 250°C with $p(O_2)$ of 100 kPa: (A) Sulfide minerals (B) Laterite and iron-containing minerals, and free acidity

Nitric Acid Leaching

Another paper in this conference discusses the iron control strategy employed in the Direct Nickel process (McCarthy et al., 2016). This approach involves the staged heating of PLS under atmospheric conditions to distil off water and nitric acid. The colligative properties of the PLS, which is highly concentrated in magnesium nitrate, enable the removal of high concentration nitric acid generated by iron hydrolysis to thereby sustain iron rejection. Hydrolysed liquor with iron content of 10 mg/L can be produced in this manner from which subsequent aluminium removal is undertaken prior to recovery of mixed hydroxide precipitate, the primary product of the process. The primary attraction of the Direct Nickel approach is the high recovery of nitric acid from the barren liquor stream via the decomposition of a magnesium nitrate melt. This acid is combined with that collected during iron hydrolysis. A potential by-product, MgO, that has been shown to have good neutralisation properties in hydrometallurgical applications, including back in the process itself, is generated by the decomposition of the melt.

BGRIMM has also developed a nitric acid-based process for nickel laterite ore that, among other differences, includes a high temperature (and pressure) leaching operation (Ma et al., 2013; 2015). While both nitric acid leaching-based approaches produce an environmentally acceptable hematite-containing residue, in the Direct Nickel process the high iron-content residue represents a potential by-product that would reduce the amount of waste material sent to tailings. In order to generate a high quality hematite product using a high temperature approach it is necessary to undertake pre-treatment of the ore, specifically desiliconization, as employed in the studies of Wang et al. (2012) and Zhang et al. (2014).

The use of nitric acid potentially represents an elegant approach to the treatment of low grade nickel sulfide ores with low extraction of iron and still enable capture of NOx liberated for regeneration of the lixiviant (Brock et al., 2014a; .2014b). Here oxidic components present or added to the reaction mixture ensure reaction of the nitric acid to drive iron hydrolysis forming goethite and/or hematite.

In comparison, processes such as the Arseno and Nitrox process also recover NOx emissions but rely on their adsorption by recycled nitrate-rich liquors (Monhemius, 1987). Shukla et al. (1978) also reported that nitric acid recovery of 60% from NOx gases generated during the processing of a nickel/copper concentrate, suggesting that in principle recovery of up to 80% was achievable.

A description of other approaches that generate hematite-containing residues such as those based upon hydrochloric acid will be left to others to provide. Fundamental studies of the hydrolysis of ferric sulfate, nitrate and chloride solutions to determine the purity of hematite that could be formed appear to have been first reported by Voigt and Göbler (1986).

Solubilities of Metal Sulfates as a Function of Acidity and Temperature

The impacts of temperature upon the solubility of metal sulfates as a function of free acidity and the salts formed when the metals precipitate are well known (e.g. Linke and Seidell, 1958; Bruhn et al., 1965; Marshall and Slusher, 1965). These data are supported by various speciation studies, high temperature pH measurements and modeling using software packages such as OLI (www.olisystems.com).

The solubility of selected divalent metal sulfates in water as a function of temperature is shown in Figure 4. Such metal sulfates are often generated, including at high concentrations, under pressure leaching conditions from feed materials that include saprolitic nickel laterite ore and, copper and nickel concentrates. Although the solubilities of magnesium and nickel sulfates are low at temperatures above 185°C, the co-presence of sulfuric acid raises the solubility of both metals; this is shown for magnesium sulfate in Figure 5.

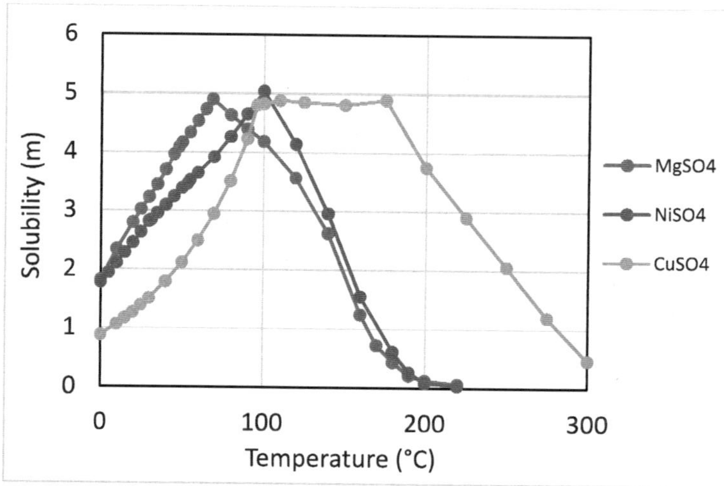

Figure 4 – Solubilities of magnesium, nickel and copper sulfates in water as a function of temperature

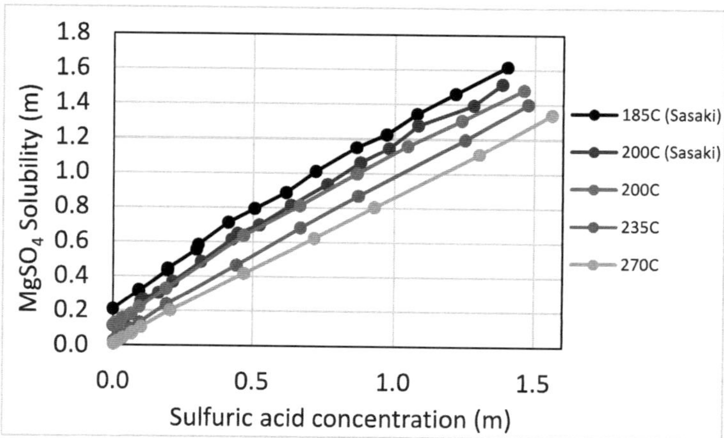

Figure 5 – Solubility of magnesium sulfate at temperatures of 185°C and above as a function of sulfuric acid concentration (Marshall & Slusher, 1965; Sasaki et al., 1994)

Modeling studies suggest the formation of soluble magnesium bisulfate species in the presence of sulfuric acid at high temperatures (Rubisov and Papangelakis, 2000):

$$5MgSO_4.H_2O(s) + 3H^+ + 3HSO_4^- \rightarrow Mg_5(HSO_4)_6(SO_4)_2 + 5H_2O \quad (11)$$

As sulfuric acid at high temperatures exists primarily as H^+ and HSO_4^-, equation (11) represents further buffering of the acid by the addition of magnesium sulfate (Jankovic et al., 2009). The associated decrease in pH enables control of the iron hydrolysis residue phases.

It is therefore expected that the addition of metal sulfates will buffer excess acid both enhancing metal solubility and reducing the "at temperature" pH. The latter is the primary driver for the type of iron-containing precipitate formed.

Basic Ferric Sulfate Mitigation Using Metal Sulfate Buffers

Recent work although focused upon the leaching mechanism of a low grade nickel concentrate under POX conditions also considered approaches to mitigate the formation of basic ferric sulfate. The buffering effect of metal sulfates in controlling "at temperature" pH as summarised by Tozawa et al. (1986), has rarely been considered in commercial leaching situations. The study of Cheng and Demopoulos (2004) examined the impact of zinc sulfate upon iron hydrolysis in order to establish the kinetics of the process with the particular aim of being able to shorten the Akita process retention time below 3 h. Even more recently Fleuriault (2016) examined mixtures of copper sulfate, ferrous or ferric sulfate and sulfuric to recreate the matrix compositions that arise from the pressure oxidation of copper sulfides. The breakpoint free acidity corresponding to the transition between low and high-sulfur containing iron hydrolysis residues was shown to increase with increasing copper sulfate concentration. Increasing initial acid concentration was often associated with a drop between the initial and final free acidity of the solution which reflects complete oxidation of the ferrous but incomplete hydrolysis to form residues dominated by basic ferric sulfate:

$$2FeSO_4 + 0.5O_2 + H_2SO_4 \rightarrow Fe_2(SO_4)_3 + H_2O \qquad (12)$$

$$Fe_2(SO_4)_3 + 2H_2O \rightarrow 2FeOHSO_4 + H_2SO_4 \qquad (13)$$

In the current study a simple comparison was undertaken to examine the pressure oxidation of nickel sulfide concentrate similar to that employed for co-processing of nickel sulfide and laterite both in the absence and presence of added magnesium sulfate buffer. Two tests were conducted at 250°C, at 9.25% w/w pulp density, with oxygen overpressure of ~100 kPa and oxygen gas flow rate of 1 L/min; for the second test magnesium sulfate was added the starting concentration of this salt was 0.9 m, i.e. 23.6 g/L Mg. While the primary goal was to study the hydrothermal conversion of metal sulfides the data obtained again indicated that various oxidation processes occurring proceeded at quite different rates; this will be reported elsewhere. Extraction and free acidity data are shown in Figure 6.

The extractions follow a similar pattern in both experiments except that the free acidity increased more rapidly and to a final much higher value when magnesium sulfate was present. This is consistent with a greater degree of hydrolysis to form hematite as shown in Figure 7(A). In fact, given the low aluminium content of the concentrate, hematite was exclusively detected as the iron hydrolysis product; as little alunite formed the incorporation of iron into this phase was minimal. The preferred formation of hematite is consistent with the results reported by Sasaki et al. (1994) who noted that the breakpoint acidity for iron hydrolysis in the presence of 27-33 g/L Mg was ~105 g/L; in contrast for 14-16 g/L Mg the breakpoint acidity was ~82 g/L. In comparison, when no magnesium sulfate was added, the final residue contained mainly basic ferric sulfate (Figure 7(B)). As a final comment here it should be noted that typically 85-90% of the iron was present in the ferrous state until after 70 minutes of leaching, thereafter decreasing to less than 10% of the total dissolved iron content.

(A)

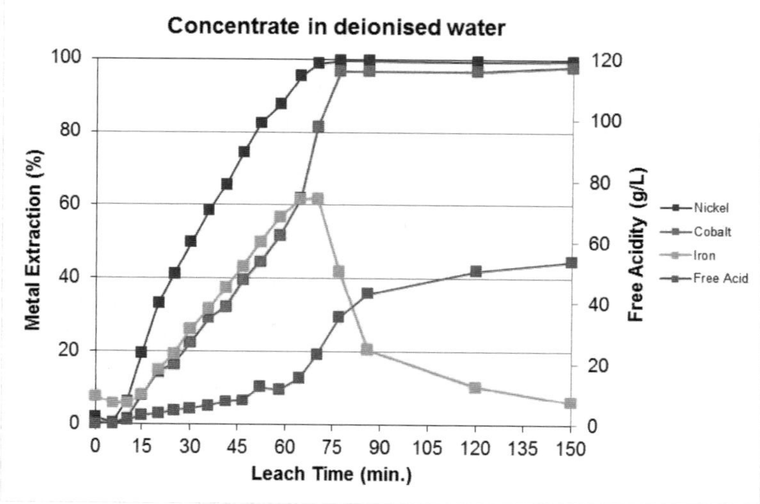

(B)

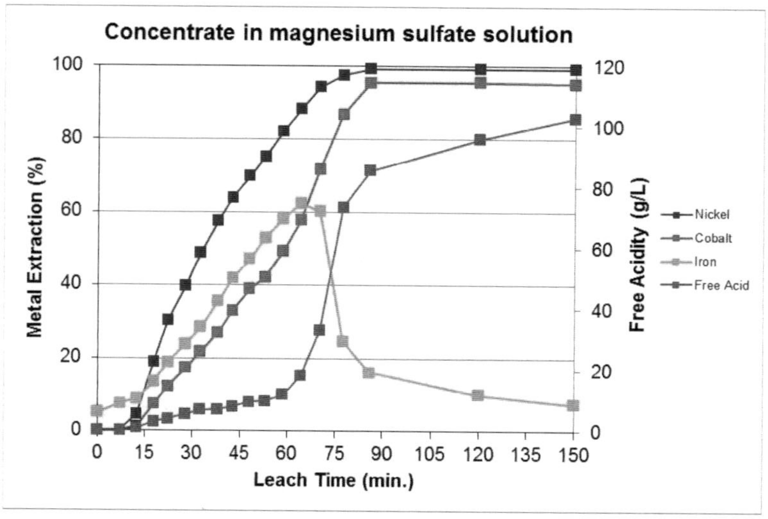

Figure 6 – Metal extraction and free acidity data for the pressure oxidation of nickel sulfide concentrate pulped in (A) deionised water and (B) deionised water plus magnesium sulfate

(A)

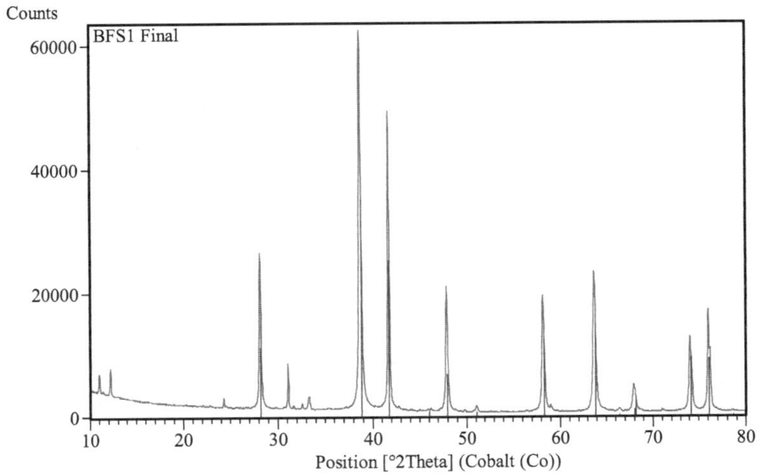

(B)

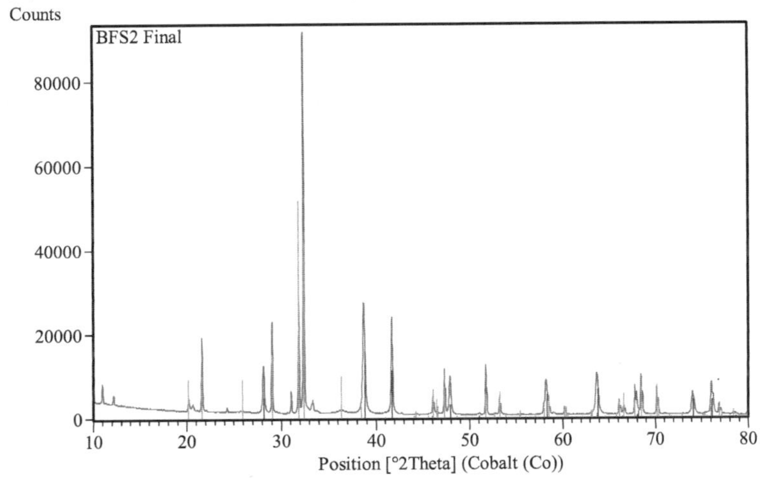

Figure 7 – XRD patterns showing for the final residues formed from (A) concentrate in deionised water (BFS1 Final) and (B) concentrate in magnesium sulfate solution (BFS2 Final) showing the peaks due to hematite (blue) and basic ferric sulfate (light green)

Using the analytical data from the second test it is possible to plot the dissolved $MgSO_4$ concentration against the H_2SO_4 concentration (which increases) as a function of reaction time and compare these with the solubility data of Figure 4; the additional data are shown in Figure 8. These clearly show that the solubility limit was normally exceeded, more so at the beginning of the reaction. It is therefore expected that kieserite, $MgSO_4.H_2O$, formed early during the reaction and this is consistent with the drop in concentration below the initial value of 0.9 m for the samples taken while the free acidity was low. The concentration drop probably arises due to scaling of the vessel notwithstanding that any kieserite sampled will re-dissolve as the temperature of the sampled material dropped. That the autoclave contents were not sampled homogeneously for the early samples was also consistent with the magnesium content of samples taken later in the reaction increasing above the initial value to almost 1.2 m at the conclusion of the reaction. Given, however, it is more pertinent to consider the total sulfate concentration (expressed as divalent metal sulfate), these data are shown for comparison in Figure 8. Clearly to avoid kieserite formation (and that of other divalent metal sulfates) with associated scaling a lower initial magnesium sulfate addition should be employed. Other strategies that could be considered include (1) decreasing the reaction temperature to enhance metal sulfate solubility although this will also increase the residence time, (2) increase the pulp density to enhance acid production provided this does not increase the free acidity above the breakpoint or negatively impact upon maintenance of the heat balance.

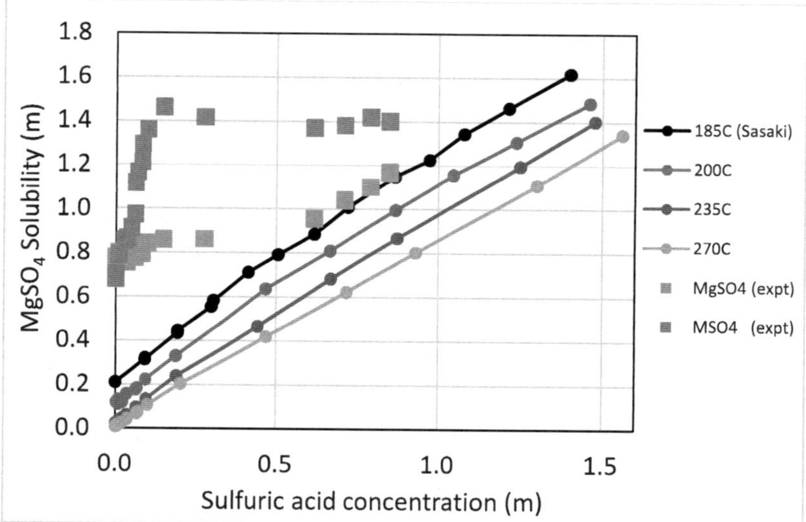

Figure 8 – Plots of magnesium sulfate solubility with superimposed data from the buffered pressure oxidation test showing timed data for (1) $MgSO_4$ versus H_2SO_4 and (2) total metal sulfate expressed as MSO_4 versus H_2SO_4

Kieserite scaling results in nickel losses, due to isomorphous substitution of nickel for magnesium, and can be both detrimental and potentially beneficial for other reasons. CSIRO examined samples of scale containing kieserite formed in the Bulong autoclave back in 2002. Its formation was attributed to the recycling of solvent extraction raffinate for use as leach process water combined with insufficient residual acid to keep the magnesium in solution that resulted in the autoclaves requiring shutdown for descaling within a matter of weeks. Conversely Queneau (1983) has suggested that the re-dissolution of this component of the scale during cooling should be beneficial for the removal of the scale during shutdown periods.

The co-presence of high magnesium concentrations with nickel that can be encountered, depending upon the type of feed, in HPAL operation is not considered to be an impediment during downstream processing. Solvent extraction reagents exist that enable the clean separation of nickel and magnesium (Cheng et al., 2010), and that would also allow the magnesium laden solution to be recycled to leaching. The main element of concern would likely be calcium which is liable to form scale in the pre-heaters.

In summary the addition of an appropriate divalent metal sulfate during the leaching of sulfidic materials with high iron sulfide content appears to be an appropriate strategy for controlling the iron hydrolysis products formed. In particular more environmentally acceptable (and stable) hematite can be preferentially formed though this will necessitate higher limestone requirement to neutralise the additional acid formed.

SUMMARY AND CONCLUSIONS

The control iron precipitation and more particularly the mineralogy of the iron residues during pressure leaching has been examined in this paper. Improved control and understanding of the implications of varying iron precipitation options may create opportunities to make changes to existing operations and contribute to the development of new processing opportunities.

The pressure oxidation of refractory gold concentrates is hampered by the production of basic ferric sulfate. As this material (and associated hydronium jarosite) is unstable under the conditions employed for cyanidation of gold, the residue is "hot cured" and the dissolved iron neutralised with limestone before cyanidation. Therefore any approach that enables more stable iron residues to be formed enables direct neutralisation of the excess acid with limestone and will potentially be beneficial. Such an approach could involve the addition of a divalent metal sulfate during pressure oxidation pre-treatment.

The use of lixiviants such as nitric acid and hydrochloric acid that promote the formation of environmentally acceptable, stable hematite residues was noted. The Direct Nickel process is able to generate a hematite by-product potentially enabling a greater fraction of the nickel laterite ore feed to be used and reducing the amount of waste material. This is in addition to the generation of a magnesia by-product that is suitable for hydrometallurgical use.

The pressure oxidation of high iron sulfide-containing copper and nickel concentrates can under certain conditions generate basic ferric sulfate. For high copper/nickel-containing concentrates the dissolved copper/nickel concentration may be sufficient to mitigate formation of this product and promote the formation of stable, high hematite content residues. However, where this is not the case the addition of a divalent metal sulfate buffer may be beneficial providing that, the divalent metal can be kept in solution, it does not impact downstream processing and can be recycled. Several candidate metals including the target metal could be considered here. Given that the free acidity breakpoint for each of the metal sulfate systems corresponds to a narrow range of acid concentrations over which the sulfur content of the residue can change dramatically, a high level of chemical control during leaching appears to be essential to optimise iron hydrolysis and to minimise the negative impacts of metal sulfate precipitation.

ACKNOWLEDGEMENTS

The authors would like to gratefully acknowledge the CSIRO analytical staff Milan Chovancek, Tuyen Pham and Sophia Surin for their assistance in providing the analytical and XRD data. Michael Jackson is also thanked for experimental measurements.

REFERENCES

Adams, M.D., 2005. Summary of gold plants and processes. In: Adams, M.D. (Ed.), Developments in Mineral Processing, Volume 15, Elsevier, Amsterdam, Chapter 41, pp 994-1012.

Baxter, K.G., Pavlides, A.G. and Dixon, D.G., 2004. Testing and modelling a novel iron control concept in the two-stage ferric leach/pressure oxidation process for the Sepon Copper Project. In: Collins, M.J. and Papangelakis, V.G. (Eds.), Pressure Hydrometallurgy 2004: Proceedings of the International Conference on the Use of Pressure Vessels for Metal Extraction and Recovery, Metallurgical Society of the Canadian Institute of Mining, Metallurgy and Petroleum, Montreal, pp. 57-76.

Blais, D., 2010. Behavior of gold(III) chloride during the precipitation of iron(III)/arsenic(V) in aqueous media under atmospheric and pressure conditions. M. Eng. Thesis, McGill University, 112pp.

Breuer, P., McDonald, R. and Zhang, H., 2014. Treatment of refractory gold ores. Case studies comparing cyanide and thiosulfate leaching to recover gold. ALTA 2014. Gold-Precious Metals Conference, 5th Annual Gold Event, ALTA Metallurgical Services, Melbourne, pp. 105-115.

Brock, G., Rodriguez, M. and McCarthy, F., 2014a. Method for the recovery of metals from sulphide ores. World Patent 2014/008540A1.

Brock, G., Rodriguez, M. and McCarthy, F., 2014b. Method for the recovery of metals from nickel bearing ores and concentrates. World Patent 2014/047672A1.

Bruhn, G., Gerlack, J. and Pawlek, F., 1965. Untersuchungen über die löslichkeiten von salzen und gasen in wasser und wäβrigen lösungen bei temperaturen oberhalb 100°C. Zeitschrift für anorganische und allgemeine Chemie 337, 68-79.

Cheng, C.Y., Boddy, G., Zhang, W., Godfrey, M., Robinson, D.J., Pranolo, Y., Zhu, Z. and Wang, W., 2010. Recovery of nickel and cobalt from laterite leach solutions using direct solvent extraction: Part 1. - selection of a synergistic SX system. Hydrometallurgy 104, 45-52.

Cheng, T.C. and Demopoulos, G.P., 2004. Hydrolysis of ferric sulfate in the presence of zinc sulfate at 200°C: Precipitation kinetics and product characterization. Industrial and Chemical Engineering Research 43(20), 6299-6308.

Demopoulos, G.P., Parisien-La Salle, J.-C. and Blais, D., 2012. The mobilization and potential "in-autoclave" recovery of gold during pressure oxidation and leaching of auriferous sulfide materials. In: Collins, M.J., Filippou, D., Harlamovs, J.R. and Peek, E. (Eds), Pressure Hydrometallurgy 2012, MetSoc, Montreal, pp. 15-29.

Fleming, C.A., 2010. Basic ferric sulfate - a potential killer in the processing of refractory gold concentrates by pressure oxidation. Minerals and Metallurgical Processing 27(2), 81-88.

Fleuriault, C., 2016. Iron phase control during pressure leaching at elevated temperature. M.Sc. Thesis, Colorado School of Mines, 142pp.

Haghtalab, A., Papangelakis, V.G. and Zhu, X., 2004, The electrolyte NRTL model and speciation approach as applied to multicomponent aqueous solutions of H_2SO_4, $Fe_2(SO_4)_3$, $MgSO_4$ and $Al_2(SO_4)_3$ at 230-270°C. Fluid Phase Equilibria 220, 199-209.

Haque, K.E., 1987. Gold leaching from refractory ores - Literature survey. Mineral Processing and Extractive Metallurgy Review 2, 235-253.

Iglesias, N. and Carranza, F., 1994. Refractory gold-bearing ores: a review of treatmetn methods and recent advances in biotechnological techniques. Hydrometallurgy 34, 383-395.

Jankovic, Z., Papangelakis, V.G. and Lvov, S.N., 2009. Effect of nickel sulphate and magnesium sulphate on pH of sulfuric acid at elevated temperatures. Journal of Applied Electrochemistry 39, 751-759.

King, J.A., Ji, J., Fleming, C.A. and Ferron, C.G., 2011. Reduction of lime consumption when treating refractory gold ores or concentrates. US Patent 8,029,751B2.

Linke, W.F. and Seidell, A., 1958. Solubilities. Inorganic and Metal-Organic Compounds. American Chemical Society, Washington D.C., Volumes I and II.

Ma, B., Wang, C., Yang, W., Yang, B. and Zhang, Y., 2013. Selected pressure leaching of Fe(II)--rich limonitic ores from Indonesia using nitric acid. Minerals Engineering 45, 151-158.

Ma, B., Yang, W., Yang, B., Wang, C., Chen, Y. and Zhang, Y., 2015. Pilot-scale plant study on the innovative nitric acid pressure leaching technology for nickel laterites. Hydrometallurgy 155, 88-94.

Marshall, W.L. and Slusher, R., 1965. Aqueous systems at hih temperature. XV. Solubility and hydrolytic instability of magnesium sulfate in sulfuric acid-water and deuterosulfuric acid-deuterium oxide solutions, 200°C to 350°C. Journal of Chemical and Engineering Data 10(4), 353-358.

McCarthy, F. and McDonald, R., 2016. Iron hydrolysis in the Direct Nickel Process. This conference.

McDonald, R., Rodriguez, M., Li, J., McDonald, R., Jackson, N. and Hosken, T., 2012. The co-processing of nickel sulphide and laterite materials using low oxygen pressures. In: Collins, M.J., Filippou, D., Harlamovs, J.R. and Peek, E. (Eds), Pressure Hydrometallurgy 2012, MetSoc, Montreal, pp. 211-225.

Monhemius, A.J., 1987. Recent advances in the treatment of refractory gold ores. II Meeting of the Southern Hemisphere on Mineral Technology, Rio de Janiero, pp 280-302.

Opratko, V., Bell, J.A.E. and Ferrajuolo, R.V, 1974. Acid leaching of lateritic ore. US Patent 3,809,549.

Parisien-La Salle, J.-C., 2012. In-situ adsorption of gold(III) chloride on activated carbon under pressure oxidation-leaching conditions. M. Eng. Thesis, McGill University, 88pp.

Quinn, J., Turner, J. and van der Meulen, D., 2009. The combined pressure acid leach (CPAL) process, batch and piloting testwork. In: Malhotra, D., Taylor, P., Spiller, E. and Lavier, M. (Eds), Recent Advances in Mineral Processing Plant Design, SME, Littleton, CO, pp 140-154.

Rodriguez, M., 2009. Rheological method for the hydrometallurgical recovery of base metals from ores. World Patent 2009/149522.

Rubisov, D.H. and Papangelakis, V.G., 2000. Sulphuric acid pressure leaching of laterites - speciation and prediciton of metal solubilities "at temperature". Hydrometallurgy 58, 13-26.

Queneau, P., Doane, R.E., Berggren, M.H. and Cooperrider, M.W., 1983. Controlling scale deposition during acid pressure leaching of laterite and garnierite ore. US Patent 4,415,542.

Sasaki, K., Ootsuka, K. and Tozawa, K., 1994. The effect of addition of magnesium sulphate on hydrolysis of ferric sulphate solutions at elevated temperatures - Hydrometallurgical studies on hydrolysis of ferric sulfate solutions at elevated temperatures (3rd report). Shigen-to-Sozai, 110, 643-652 (in Japanese).

Shukla, P.P., Mukherjee, T.K. and Gupta, C.K., 1978. A nitric acid route for processing a nickel/copper sulphide concentrate. Hydrometallurgy 3, 55-64.

Tozawa, K. and Sasaki, K., 1986. Effect of coexisting sulphates on precipitation of ferric oxide from ferric sulphate solutions at elevated temperatures. In: Dutrizac, J.E. and Monhemius, A.J. (Eds.), Iron Control in Hydrometallurgy, Ellis Horwood, Chichester, UK, pp 454-476.

Umetsu, Y., Tozawa, K. and Sasaki, K., 1977. The hydrolysis of ferric sulfate solutions at elevated temperatures. Canadian Metallurgical Quarterly, 16(1), 111-117.

Umetsu, Y., Sasaki, K. and Otsuka, K., 1978. Effects of the addition of zinc sulphate on the hydrolysis of ferric sulfate solutions at elevated temperature. Hydrometallurgical studies on hydrolysis of ferric sulfate solutions at elevated temperatures (2nd report). Journal of the Mining and Metallurgical Institute of Japan 94(1081), 883-887 (in Japanese).

Umetsu, Y., Sasaki, K. and Otsuka, K., 1979. Effects of the addition of copper sulphate on the hydrolysis of ferric sulfate solutions at elevated temperature. Hydrometallurgical studies on hydrolysis of ferric sulfate solutions at elevated temperatures (3rd report). Journal of the Mining and Metallurgical Institute of Japan 95(1102), 883-887 (in Japanese).

Voigt, B. and Göbler, A., 1986. Formation of pure haematite by hydrolysis of iron(III) salt solutions under hydrothermal conditions. Crystal Research and Technology 21(9), 1177-1183.

Wang, C. Zhang, Y. and Xu, Z., 2012. Pressure leaching of alkali pre-treated linonitic ore in nitric acid. In: Collins, M.J., Filippou, D., Harlamovs, J.R. and Peek, E. (Eds), Pressure Hydrometallurgy 2012, MetSoc, Montreal, pp. 175-185.

Zhang, Y.-L., Wang, C.-Y., Yang, Y.-Q., Yin, F. and Ma, B.-Z., 2015, Pressure nitric acid leaching of alkali pre-treated low-grade limonitic laterite ore. Rare Metals 34(1), 64-70.

IRON HYDROLYSIS IN THE DIRECT NICKEL PROCESS

*F. McCarthy[1], R. McDonald[2], G. Woodbridge[1], G. Brock[1] and D. Robinson[2]

[1]Direct Nickel Ltd
PO Box 7229, Waterford, WA 6152, Australia
*(*Corresponding author: fiona@directnickel.com)*

[2]CSIRO Mineral Resources
PO Box 7229, Waterford, WA 6152, Australia
Waterford, WA, Australia 6152

ABSTRACT

The Direct Nickel process is a hydrometallurgical technology being developed to treat nickel laterite ores by leaching at atmospheric pressure. The leaching process uses nitric acid in a relatively large excess, is thus non-selective, and much of the soluble iron goes into solution. The first stage of solution purification is iron hydrolysis and the Direct Nickel process utilises the colligative properties of the pregnant leach solution (PLS) to enable the removal nitric acid of varying concentrations via distillation. Once sufficient acid has been removed, the boiling point of the solution is increased allowing high temperatures (up to ~170°C) to be achieved under atmospheric pressure. The higher temperatures initiate the precipitation of the iron primarily as hematite and allow the removal of the acid generated by the hydrolysis reaction. The rejection of the iron is near complete with solution tenors typically falling from 65 g/L Fe, or more, to less than 50 mg/L Fe. The separation of clean nitric acid from the PLS at this stage allows for relatively simple acid recycling. A hematite product with iron content above 60% could also provide an extra revenue stream.

KEYWORDS

Hematite, nitric acid, hydrolysis, nickel, laterite

INTRODUCTION

The Direct Nickel (DNi) Process is a novel hydrometallurgical processing route designed to treat all types of nickel laterite ores in a single flow sheet using nitric acid under atmospheric conditions. A simplified flow sheet is shown in Figure 1.

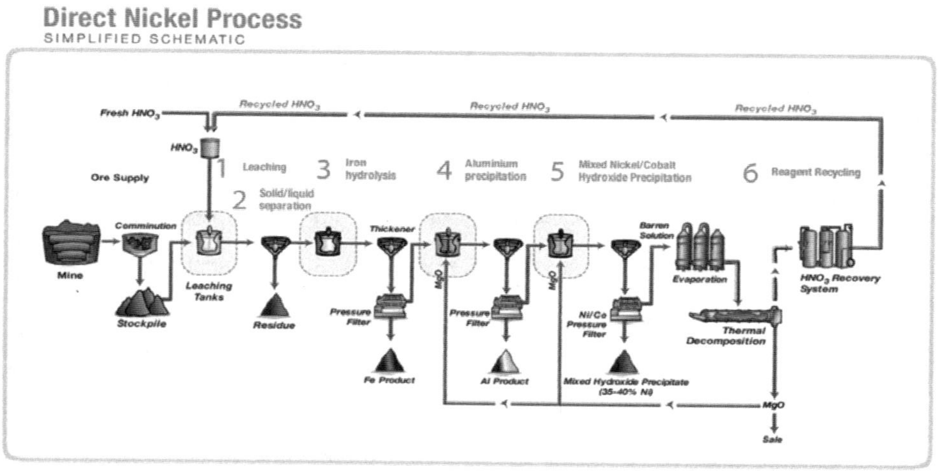

Figure 1 – Simplified flow sheet for the Direct Nickel Process

Between 2007 and 2012 the process was developed and refined at Direct Nickel's laboratory within the CSIRO Minerals' facilities in Perth, Australia, and in 2013 a series of pilot plant trials were run under continuous operating conditions. The process is designed to treat both limonite and saprolite in the same flow sheet, and after comminution, the blended ore is mixed with nitric acid and leached just below boiling point in agitated atmospheric leach tanks. The acid is added in excess, so typically for a blend of limonite and saprolite the acid addition can be in the order of 1.8-2.0 t acid/t dry ore. After a residence time of between 4-6 hours the majority of the soluble metals have leached into solution and the insoluble leach residue is separated from the pregnant leach solution (PLS) via counter-current decantation.

The PLS then is treated in the iron removal circuit; where more than 99% of the iron is precipitated as hematite, removed through filtration and, nitric acid is distilled and sent to the acid recycling circuit; note that this recycle stream is not shown in the simplified flow sheet. Following iron rejection the pH is adjusted using magnesium oxide to precipitate aluminium as a hydroxide. Once this is removed via filtration the pH is adjusted again with MgO to precipitate nickel and cobalt mixed hydroxide (MHP). This is filtered to produce a wet cake of around 35-40 wt.% nickel (on a dry basis), and approximately 1-2 wt.% cobalt. Following the nickel and cobalt precipitation, a barren solution containing primarily magnesium nitrate remains. This barren solution, which contains all of the nitrate ions added in leach minus those lost to products and waste streams, is evaporated to remove the free water and then decomposed to produce solid MgO and a gas stream containing NO_2, NO, H_2O and O_2. Part of the MgO produced is recycled as a neutralising agent, for use in the MHP and aluminium precipitation circuits as well as tails neutralisation, with the remainder available for sale. The gas stream is fed to the nitric acid regeneration circuit where 55 wt.% HNO_3 is recovered for reuse in the leaching module. It has been verified through modelling and the pilot plant performance that around 95% of the nitric acid can be recovered in this manner.

The pilot plant, located at CSIRO Minerals' Waterford site, has a capacity of 1 tonne per day of dry laterite ore and was operated throughout much of 2013. After full hot commissioning of each process module, the plant operated the full flow sheet from leaching through to acid recovery. All of the recycle loops (including nitric acid and MgO) were operated in continuous closed circuit for periods ranging from 10 days to 28 days. During these campaigns ores from two deposits in Indonesia and Brazil were tested, with blends ranging from 50:50 limonite:saprolite, to 25:75 limonite:saprolite, and 100% saprolite. This paper will discuss the performance of the iron hydrolysis reactions at bench-scale as well as some observations from the continuous trials conducted.

The Iron Hydrolysis Circuit

In the DNi process, the laterite ores are leached in nitric acid at ~110°C in a series of continuously stirred tank reactors. The main iron bearing mineral is goethite, which dissolves to form ferric nitrate:

$$FeO(OH)_{(s)} + 3HNO_{3(aq)} \rightarrow Fe(NO_3)_{3(aq)} + 2H_2O_{(l)} \quad (1)$$

The other group of important minerals are serpentines, from which magnesium dissolves to produce magnesium nitrate:

$$Mg_3Si_2O_5(OH)_{4(s)} + 6HNO_{3(aq)} \rightarrow 3Mg(NO_3)_{2(aq)} + 2SiO_{2(s)} + 5H_2O_{(l)} \quad (2)$$

The ore blend therefore has a strong influence on the amount of iron and magnesium in the post leach solution. The composition of typical post leach solutions from the three commonly classified ore types are shown in Table 1; note that the limonite and saprolite ores referred to are from different sources. The DNi process is designed to treat the whole ore body within a single flow sheet meaning that typical post leach solutions will be derived from a blend of ore types. An assay of the pilot plant leach solution produced from a 50% limonite/50% saprolite ore blend from an Indonesian deposit is also shown in Table 1.

Table 1 – Some typical post leach solution assays, mg/L

Ore Type	Ni	Co	Al	Fe	Mg	Cr	Mn	Ca
Limonite	2412	240	9206	116760	335	1324	1946	2412
Transition	5468	295	6490	103729	6291	1394	2559	5468
Saprolite	9297	53	1533	37912	40142	606	561	9297
Pilot Plant 50%/50% Limonite/Saprolite	7569	282	8597	125388	30304	1547	1989	576

The basis of the iron removal circuit involves heating the PLS to its boiling point and the subsequent distillation of nitric acid of varying compositions. This nitric acid is condensed and either sent to the nitric acid regeneration circuit or, if low in nitric acid concentration, neutralised and recycled to the process water circuit. As more of the residual nitric acid is distilled, the remaining solution increases in both magnesium nitrate and ferric nitrate concentrations. This raises the boiling point such that temperatures in excess of 170°C can be achieved under atmospheric pressures, although typical temperatures are nearer 160-165°C.

As pure nitric acid has an azeotropic concentration of 68.2 wt.% HNO_3, it should not be possible to remove 100% of the nitric acid by distillation, however the presence of both the magnesium and ferric nitrate salts alter the azeotropic behaviour and effectively allow almost all of the nitric acid to be removed by distillation. This phenomenon has previously been utilised to commercial effect in the production of super-azeotropic nitric acids (Jubin, Counce, Holland, Groenier, & North, 1987; Sloan, 1976). Clearly in the DNi process, the effect of ferric nitrate will decline once it has precipitated as a hematite solid, so sufficient volume of magnesium nitrate must be present to both act as an azeotrope modifier as well as to provide a liquid medium within which the hydrolysis reactions can occur. The exact mechanism for the thermal decomposition of ferric nitrate to hematite (Fe_2O_3) is still under discussion (Melnikov, Nascimento, Arkhangelsky, Zanoni Consolo, & de Oliveira, 2014), however it is clear that several intermediates can be formed. The overall reaction for the fully hydrated species is as follows:

$$2Fe(NO_3)_3.9H_2O_{(aq)} \rightarrow Fe_2O_{3(s)} + 6HNO_{3(aq)} + 6H_2O_{(l)} \quad (3)$$

Since at no stage of the iron hydrolysis reaction super-azeotropic nitric acid is produced (the stoichiometry of the above reaction (3) would suggest the formation of 78 wt.% HNO_3), it is clear that this reaction must either occur in the liquid phase prior to full removal of the free water in solution or during further dehydration of other metal nitrate salts.

BENCH SCALE LABORATORY RESULTS

A comprehensive series of bench scale tests have been conducted on a number of leach liquors from various ore sources. Much of the early bench work was focussed on understanding the possibilities and limitations surrounding the thermal iron hydrolysis step, including iron removal efficiency, precipitate quality, and material handling issues such as washing and filtration.

Iron hydrolysis bench scale tests were conducted in a 1 L glass vessel with a sealed glass lid, fitted with an agitator and a distillation trap (Dean and Stark apparatus) to allow the removal, via re-condensation, of the vapour products. The solutions were heated to temperatures of 160-170°C at the fastest rate available using an electric mantle and held for varying times under reflux conditions (distillation trap removed and condenser replaced on the vessel). Additional magnesium nitrate was added to solutions deficient in Mg prior to heating to ensure good fluidity at higher temperatures (except where indicated), and a seed sample of iron oxide was added to some of the tests to the solution before hydrolysis began. Typical seed additions of around 10 wt.% of the final precipitate amount were found to be sufficient when used. After hydrolysis, the slurry of molten salt (resulting from evaporation of the PLS) and the iron oxide particles is supersaturated due to the temperatures reached and would freeze if allowed to cool, so standard practice was to add a small amount of water to the vessel to allow for good solid/liquid separation (in the pilot plant a wash water off the filter was used so as not to lose the valuable pay metals within the wash). Filtration of the solutions and solids was completed using a standard laboratory vacuum filter array and assays of the samples analysed by Inductively Coupled Plasma-Optical Emission Spectroscopy (solids digested via metaborate/tetraborate fusion).

The Effect of Temperature and Holding Time

The first series of tests presented here gave some interesting results. The liquor was a combined solution, made up from several series of bench scale leach test liquors. The ores were primarily limonites or blends, and the liquor solution composition is shown in Table 2. No additional magnesium nitrate was added to this solution, despite it having a low Mg:Fe ratio slightly greater than 0.4.

Table 2 – Limonite Blend PLS assay

	Ni	Co	Al	Fe	Mg	Cr	Mn	Ca	Zn	Si	Density, g/cm³
	mg/L										
PLS	4393	297	6210	54876	24146	1396	1357	5045	46	2	1.506

The precipitate samples were produced at temperatures of 150, 155, 160 and 165°C after a hold time of 1 hour. The residues were filtered and then re-pulped twice in water. The residue was then dried at 110°C. The compositions of the residues after the first and second washes are given in Table 3. Washing was shown to be critical in removing value and gangue metals, as well as increasing the iron content of the residue.

Table 3 shows increasing amounts of aluminium are precipitated with the iron and examination of the XRD patterns for these solids shown in Figure 2 clearly indicates that for this set of residues increasing the hydrolysis temperature increased the amount of ferrihydrite in the solid. Given the extent of aluminium hydrolysis also increased this is consistent with the retardation of the ferrihydrite conversion to hematite by aluminium inclusion (Barrón, Michel, Liu, Koch, & Torrent, 2012). Previously, it has been seen that ferrihydrite precipitates formed in the presence of aluminium may contain aluminium hydroxide phases (Cismasu, Michel, Stebbins, Levard, & Brown, 2012). While no such phases were detected by XRD, these may be poorly crystalline and, being in low amounts, require other methods to detect. It was noted during these tests that the filterability of the final hydrolysis slurry was worse for the higher temperature residues, however it is not clear whether the increasing ferrihydrite content was responsible for this.

Table 3 – Residue compositions after first and second wash, wt.%

Temp, °C	Wash	Ni	Co	Al	Fe	Mg	Cr	Mn	Ca
150	1st	0.21	0.02	0.60	50.94	0.95	0.02	0.27	0.23
	2nd	0.04	0.00	0.35	55.07	0.06	0.00	0.25	0.02
155	1st	0.23	0.02	1.70	47.64	0.85	0.02	0.26	0.21
	2nd	0.10	0.01	1.47	51.61	0.09	0.01	0.25	0.03
160	1st	0.14	0.01	2.52	45.70	0.57	0.01	0.40	0.14
	2nd	0.05	0.00	2.25	48.76	0.14	0.00	0.41	0.03
165	1st	0.14	0.01	2.76	46.13	0.43	0.01	0.32	0.11
	2nd	0.10	0.01	2.77	49.75	0.22	0.01	0.33	0.06

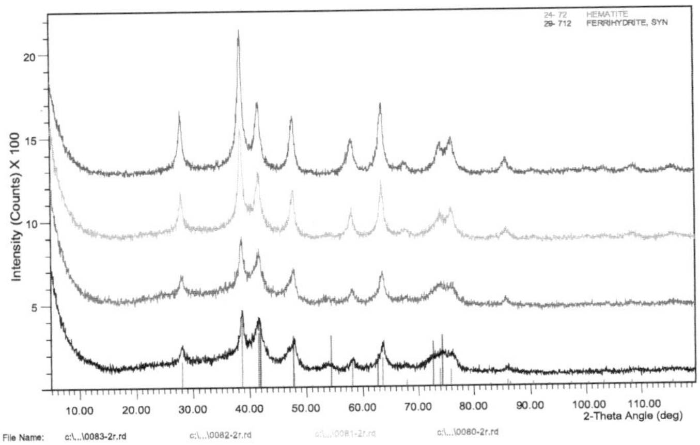

Figure 2 – XRD scans of iron precipitates formed at (1) 165°C (black, bottom), (2) 160°C (red), (3) 155°C (blue) and (4) 150°C (pink, top)

The second series of experiments presented here investigated the influence of reaction temperature and holding time at temperature for a synthetic solution based on saprolite leach liquor composition. The solution composition is shown in Table 4.

Table 4 – Synthetic Saprolite PLS assay

	Ni	Co	Al	Fe	Mg	Cr	Mn	Ca	Zn	Si	Density, g/cm³
	mg/L										
PLS	8592	195	6991	46174	64824	12	959	952	641	1	1.472

Three temperatures were tested; 160, 165 and 170°C, and hold times were varied; 1 minute, 1 hour and 2 hours. No seed material was added and no additional magnesium nitrate was added. Typically, saprolite leach solutions are high in magnesium; this solution had a Mg:Fe ratio of ~1.4 as produced. The residues were filtered and then re-pulped three times. The residue was then dried at 110°C. A portion of each dried residue was retained unground and submitted for particle size distribution (PSD) analysis.

As can be seen in Figure 3 the range of hold times and temperatures tested showed that increased temperature increased iron precipitation. Also the kinetics of iron precipitation were reasonably fast such that by 60 minutes hold time the maximum iron precipitation extent had been reached and further holding to 2 hours did not significantly enhance iron removal (it is possible that there is even a minor amount of re-dissolution). The optimum temperature for iron removal for this solution composition seems to be at or above 165°C. The co-precipitation of aluminium was similar at the lower temperatures (at around 10% of the aluminium precipitated), and increased significantly at hold times of 1 hour or longer at 170°C. Nickel co-precipitation followed a similar pattern as aluminium, with 0.16-0.34 wt.% of the nickel precipitating at the lower two temperatures and only small increases with increasing hold time. This jumped to 0.6 wt.% at 170°C for hold times of 1-2 hours.

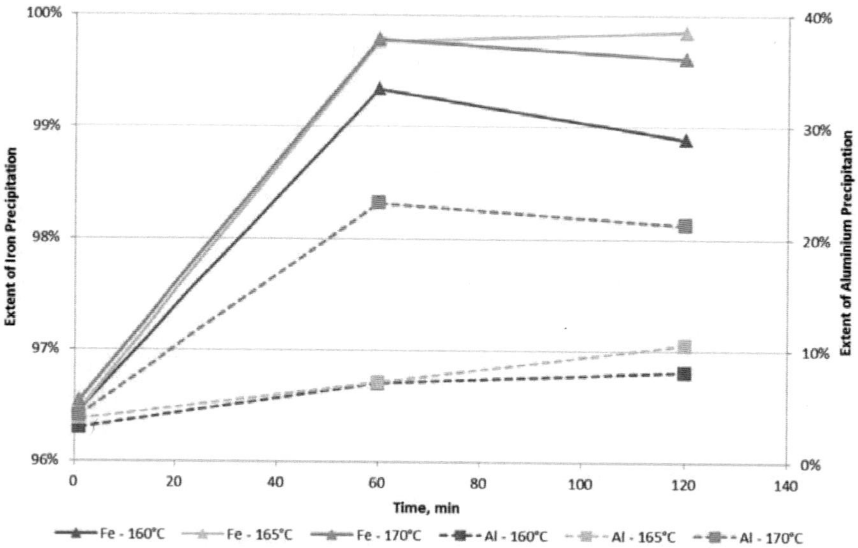

Figure 3 – Iron hydrolysis, effect of temperature and hold time on iron and aluminium precipitation extents

Filtration rates of the laboratory samples were again observed to deteriorate as the amount of aluminium precipitated increased. The particle size distribution of the 160 and 170°C samples was analysed using a Malvern Mastersizer laser diffraction particle size analyser to see if there was an observable trend. While there was a slight P_{80} increase for the 170°C samples the d(0.5) was not significantly different from the 160°C samples.

The liquor produced post iron hydrolysis for all the tests at or above 165°C should be suitable for the next stage of solution purification, aluminium removal. The post hydrolysis solution compositions for the 1 hour tests are shown in Table 5.

Table 5 – Post hydrolysis liquor compositions for the 1 hour tests, mg/L

Temp, °C	Ni	Co	Al	Fe	Mg	Cr	Mn	Ca
160	9798	226	7287	277	74838	0	1069	1125
165	10384	236	7755	38	78969	0	1048	1112
170	10335	237	5911	23	82438	0	755	1191

The Effect of Seed Composition and Magnesium to Iron Ratio in Feed Liquors

A third series of bench scale tests was conducted on a combined solution made up from several bench scale leach test liquors, very similar to that used above in the first series of tests. Again the ores were primarily limonites or blends, see Table 6.

Table 6 – Limonite blend PLS assay

	Ni	Co	Al	Fe	Mg	Cr	Mn	Ca	Zn	Si	Cl	Density,
	mg/L											g/cm³
PLS	4920	208	5573	57491	19662	1199	980	267	43	6	743	1.387

Three temperatures were tested: 167, 169, and 172°C (slightly higher than those used in the previous series of tests since the Fe was not completely hydrolysed in some of these tests, probably due to a higher initial free acid level). Two seed materials were selected from earlier tests (not presented here), demonstrating the wide variability that can be seen in the aluminium contents (Table 7).

Table 7 – Seed composition, wt%

Seed type	Ni	Co	Al	Fe	Mg	Cr	Mn	Ca	Zn	Si	Na
High Al	0.12	<0.02	5.27	37.02	0.76	1.59	0.04	<0.01	<0.01	<0.002	0
Low Al	0.05	<0.02	0.62	61.74	0.43	0.48	0.15	0.01	-	0.02	-

Magnesium nitrate (as a laboratory reagent hexahydrate salt) was added in all the tests, since it was shown in the first series of tests above (and in subsequent unpublished tests) that iron product quality is improved with increased PLS Mg to Fe ratios. The tests were conducted with a ratio of Mg to Fe of 1.2. All tests had a hold time at temperature of 60 minutes. Several tests were repeated with an increased PLS Mg:Fe ratio of 2.0.

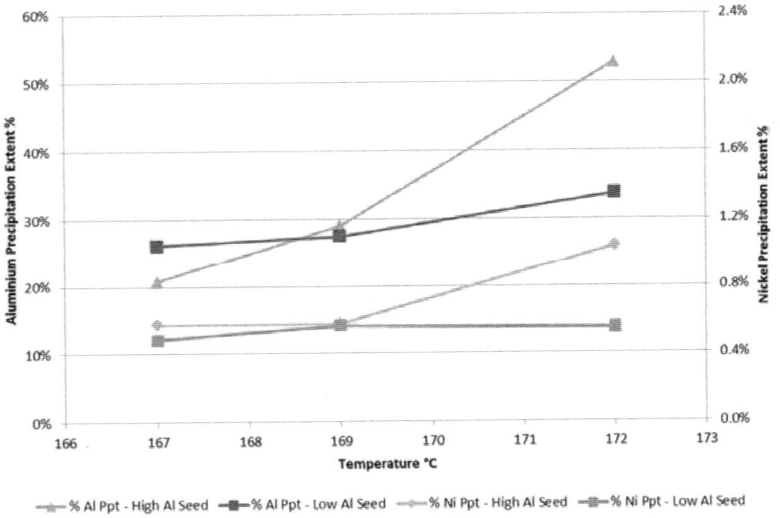

Figure 4 – Effect of seed composition on precipitation extents of aluminium and nickel

The effect of seed type was most strongly demonstrated by the extent of aluminium precipitated from the PLS, as shown in Figure 4. At lower temperatures the differences were minor, but at 172°C the amount of aluminium precipitated increased from 33% to 52% when the seed type was changed from low to high Al containing seed. A smaller effect was seen in the nickel precipitation extent; at 172°C the amount of nickel co-precipitated increased from 0.55% with the low Al containing seed to 1.04% with the high Al containing seed. Another observable effect was the drastically increased filtration times that correlate with increased Al contents in the iron precipitates, so at this stage it is unclear if the increased nickel co-precipitation is a result of actual precipitation or incomplete washing of the sample during sample filtration.

Increasing the Mg:Fe ratio from 1.2 to 2.0 did not appear to alter the precipitation extents in any significant way. The effect of the Mg:Fe ratio does have noteworthy implications when considered as part of the whole Direct Nickel process however, beyond just maintaining a sufficient reaction volume within the iron hydrolysis module. Changing the Mg:Fe ratios for ore blends that are deficient in Mg can be done in several ways; a recycle stream of barren solution can be blended with the incoming PLS, MgO can be added to pre-neutralise the free acid in the iron hydrolysis feed, or finally other nickel scavenge streams can be added (such as a second stage mixed hydroxide precipitate or second stage aluminium precipitate) which act in a similar manner to MgO. Each of these approaches alters the water balance of the process, primarily in changing the ratios of acid to water produced by distillation from each of the sequential iron hydrolysis tanks.

The justification for higher Mg:Fe ratios can be seen, however, in comparing the third series of tests with the results presented above for the first series. The closest two data points from each series of tests are compared in Table 8, in taking the 165°C test from the first series and the 167°C test from the third series with a low Al seed and a Mg:Fe ratio of 1.2. In the third series tests, magnesium nitrate hexahydrate was added to the PLS targeting a set Mg:Fe ratio, this data point was for Mg:Fe = 1.2, so enough magnesium nitrate was added to give effective Mg and Fe concentrations of around 55 and 46 g/L respectively. A distinct improvement in the iron product quality can be seen with the increased Mg:Fe ratio. The iron content increased significantly, likely as a direct result of a much lower quantity of Al precipitated (53% of the initial aluminium precipitated in the first series compared to 26% in the third). Nickel co-precipitation or entrainment is also reduced in the solid as a result of increasing Mg:Fe ratio. The mechanism for this is as yet undetermined.

Table 8 – First and third series feed PLS and iron precipitate compositions

Mg:Fe	Ni	Co	Al	Fe	Mg	Cr	Mn	Ca	Zn	Si
PLS					**mg/L**					
0.4	4393	297	6210	54876	24146	1396	1357	5045	46	2
1.2	4920	208	5573	57491	19662*	1199	980	267	43	6
Fe Solid					**wt.%**					
0.4	0.14	0.01	2.76	46.13	0.43	1.20	0.32	0.11	<0.002	0.04
1.2	0.02	<0.002	1.36	56.37	0.09	1.26	0.12	0.00	<0.002	0.03

* Prior to Mg:Fe ratio adjustment

Iron removal from the post iron hydrolysis liquors was essentially complete for the majority of the 169 and 172°C tests, with all of the samples recording iron and chromium assays below detection limits (i.e., <0.2 mg/L). A typical set of assays for these solutions is shown in Table 9.

Table 9 - Post hydrolysis liquor compositions for tests with high Al seed and Mg:Fe ratio of 1.2

Temp, °C	Ni	Co	Al	Fe	Mg	Cr	Mn	Ca	Zn	Si
					mg/L					
167	4630	202	3825	8	73517	0	856	904	46	4
169	4555	196	3799	0	75478	0	678	1031	47	4
172	4613	200	2344	0	76426	0	739	1065	48	1

Selected iron precipitates from these series of tests were analysed by XRD to determine if the changes in product handling properties as demonstrated in increased filtration times were a result of phase variations. These precipitates contained mainly hematite but varying amounts of ferrihydrite were also detected. Based on a limited number of scans, it was seen that the ratio of hematite to ferrihydrite was slightly increased at higher Mg:Fe ratio (see Figure 5 and Figure 6). The influence of seed type was not as clear, but it appeared that the high Al seed promoted ferrihydrite formation with the lower Mg:Fe ratio but not with the higher Mg:Fe ratio.

It has been demonstrated that hematite forms via ferrihydrite and subsequently goethite at high pH conditions (Vu, Shaw, & Benning, 2008). A similar pathway at low pH and at temperatures up to 75 °C has also been demonstrated (Das, Hendry, & Essilfie-Dughan, 2011), though with the kinetics of transformation being significantly slower. For the samples considered here, however, no goethite was detected in the XRD patterns which may indicate a direct (solid state) transformation mechanism from ferrihydrite to hematite is involved (Johnston and Lewis, 1983; Schwertmann, Friedl, & Stanjek, 1999; Liu, Cao, Li, Wei, & Hou, 2010).

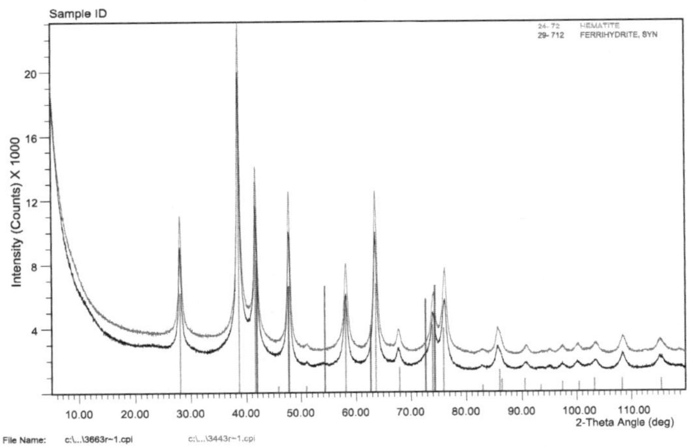

Figure 5 – XRD scans of iron precipitates produced at 172°C with Mg:Fe ratio of 2.0 and (1) low Al seed (black), (2) high Al seed (red)

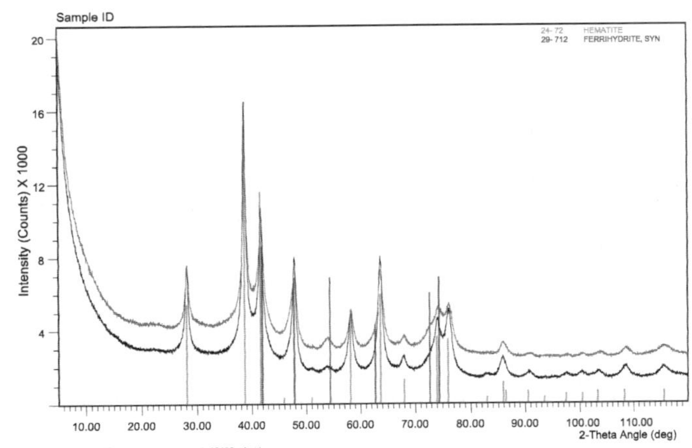

Figure 6 – XRD scans of iron precipitates produced at 172°C with Mg:Fe ratio of 1.2 and (1) low Al seed (black), (2) high Al seed (red)

The Effect of Magnesium to Iron Ratio in Feed PLS on Acid Generation

In the above series of tests, the effect of Mg:Fe ratio was examined in relation to the solid precipitate phases. As mentioned, a higher Mg:Fe ratio is preferred if higher Fe levels in the solid are important (i.e. if the hematite is to be sold as an iron ore), however the impact of these changing levels goes beyond the solid and also alters the water/acid balance being distilled from the iron hydrolysis circuit.

A fourth series of tests was conducted to assess how the distillate produced from iron hydrolysis alters within the Mg:Fe range 0.71-1.10 (as it has been previously shown that there was little to no difference between 1.2 and 2.0 in the third series of tests). This series of tests utilised solutions taken from the CCD1 overflow from one of the 2013 pilot plant campaigns which was treating a 50% limonite 50% saprolite blend from an Indonesian deposit; solution compositions are shown in Table 10. The free acid for this PLS is somewhat higher than was necessary for good extraction, as the sample was taken during one of the hot commissioning campaigns. The PLS had the base starting free acid of 252 g/L HNO_3.

Table 10 - Pilot Plant PLS and barren solution compositions

	Ni	Co	Al	Fe	Mg	Cr	Mn	HNO_3	Density
	mg/L							g/L	g/cm^3
PLS	4306	167	5234	73986	17504	889	1090	252	1.399
Barren	4	0	1	0	86747	0	1	-	1.375

The distillate produced in each of these tests was collected differently than the previous tests. In these experiments a separate distillate sample was taken for each 5°C temperature bracket while heating. Distillation typically starts between 115 and 120°C, and samples were collected for each 5°C interval until the final temperature of 170°C. In the previously reported series of tests the entire distillate was collected as a bulk sample.

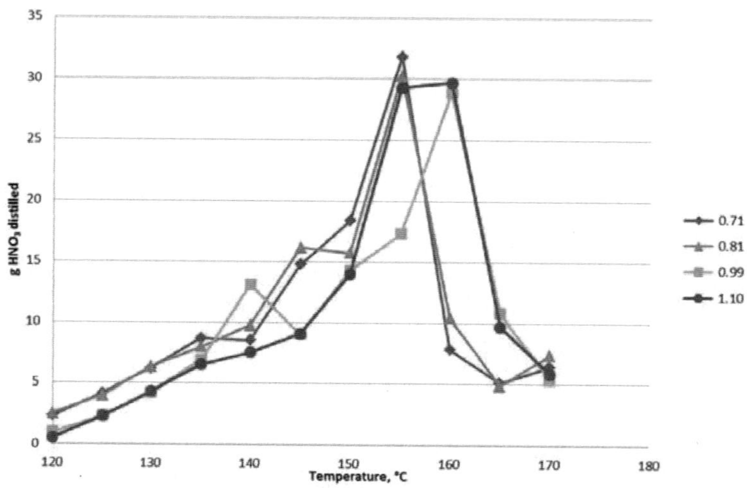

Figure 7 - Amount of nitric acid removed as distillate during the heating phase as a function of Mg:Fe ratio

The effect of changing the Mg:Fe ratio from 0.71 to 1.10 (by adding barren solution to PLS) was slight, and the distillate curves (amounts and strengths) are shown in Figure 7 and Figure 8. The iron precipitate compositions are shown in Table 11. XRD analysis again indicated that the hematite formed contains water as hydroxyl ions and the crystallite size is typically in the range of 10-15 nm (see below for discussion of crystallite size analysis).

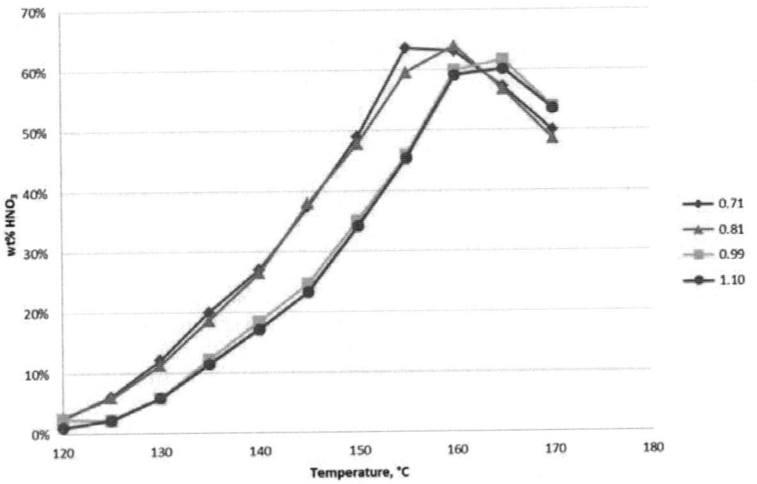

Figure 8 – Strength of nitric acid removed as distillate during the heating phase as a function of Mg:Fe ratio

Table 11 - Assays of iron precipitates formed with varying Mg:Fe ratio, wt.%

Mg:Fe	Ni	Co	Al	Fe	Mg	Cr	Mn
0.71	0.04	0.002	1.18	55.25	0.002	0.70	0.22
0.81	0.07	0.002	1.33	54.41	0.008	0.69	0.27
0.99	0.03	0.002	1.32	54.62	0.002	0.68	0.14
1.10	0.03	0.002	1.12	55.90	0.018	0.71	0.17

As is shown in Figure 7, the temperature of peak acid generation appears shifted to slightly higher temperatures with higher Mg:Fe ratio. The strength of the acid produced is also slightly lower, however this is likely due to the increased amounts of water that are associated with the increased Mg:Fe ratio. The composition of the iron precipitates are very similar, and certainly within the expected variance seen in these types of tests. The solution composition post iron hydrolysis was again similar to those presented previously with iron levels typically <200 mg/L.

The Effect of PLS Pre-Neutralisation

A fifth series of bench scale tests are presented here to demonstrate the effect of pre-neutralising the incoming PLS to the iron hydrolysis circuit with MgO. This series of tests again utilised the PLS solution taken from 2013 pilot plant campaigns (Table 10). The PLS was pre-neutralised with various amounts of MgO generated in the pilot plant having composition given in Table 12 to give a wide range of starting free acids, with the base case of PLS having a free acid level of 252 g/L. These are shown in Table 13.

Table 12 – Test Plant MgO composition

	Ni	Co	Al	Fe	Mg	Cr	Mn	Density
				Wt.%				g/cm³
MgO	0.1	0.01	0.03	0.2	56.0	0.01	0.4	3.394

Table 13 – Pre-neutralisation levels used in the comparative tests

Starting Free Acid, g/L	**252**	**207**	**190**	**136**	**87**	**20**
% Neutralised over base case	0%	18%	25%	46%	66%	92%

The distillate produced in each of these tests was collected similarly to those in the fourth series of tests, with a separate distillate sample taken for each 5°C temperature bracket while heating until the final temperature of 170°C.

The extent of neutralisation was determined to have no effect on the amount of acid released during hydrolysis beyond that which was removed as a result of the reduction in the starting free acid, i.e., the amount of acid released from the PLS as a result of hydrolysis of metals was the same regardless of the starting free acid. The distribution of the acid produced (Figure 9 and Figure 10), however, was significantly changed in the PLS samples with lower starting free acids (136 g/L sample and less) when compared with the higher free acid samples. The higher free acid samples (190-252 g/L) all have a peak acid generation point between 155-160°C which is lowered in the more pre-neutralised samples (150-155°C for the 136 g/L sample and 145-150°C for the 20 and 87 g/L samples). It is a reasonable assumption that the free acid stabilises the ferric nitrate, and higher free acid liquor requires a higher temperature to initiate iron hydrolysis (Equation 3). The distillate acid strength also reduces with decreasing starting free acid levels (Figure 10).

The composition of the solid precipitates produced in this series of tests varied and the analyses are shown in Table 14. The total extents of precipitation for each of the metals were similar within experimental error. It appears that a more complex series of phenomena is occurring across the range of pre-neutralisation levels tested. The iron content of the precipitate started at around 57 wt.% Fe and with increasing amounts of pre-neutralisation dropped to around 53-54 wt.%. At the very low free acid levels however, it increased again to 59-60 wt.% Fe. These samples were submitted for XRD to determine if phase differences could explain this variation. The main phase in each sample was again hematite; however the XRD patterns for the various samples indicated the presence of varying small amounts of ferrihydrite and in some instances goethite. These XRD data were analysed to determine unit cell parameters and crystallite sizes as calculated by the Brüker TOPAS software LVol-IB parameter. To enable estimates for the hematite crystallite size, contributions to the XRD pattern from the ferrihydrite were modelled using an uncalibrated hkl phase derived using a P3 space group and the starting a and c cell parameters of (Jansen, Kyek, Schäfer, & Schwertmann, 2002); a fixed ferrihydrite crystallite size of 10 nm was employed. It was subsequently noted that the refined ferrihydrite cell parameters varied little from the reported values of 2.955 Å and 9.37 Å. Further work is required to properly characterise this phase and to enable quantitative estimation of its content in the iron hydrolysis residues.

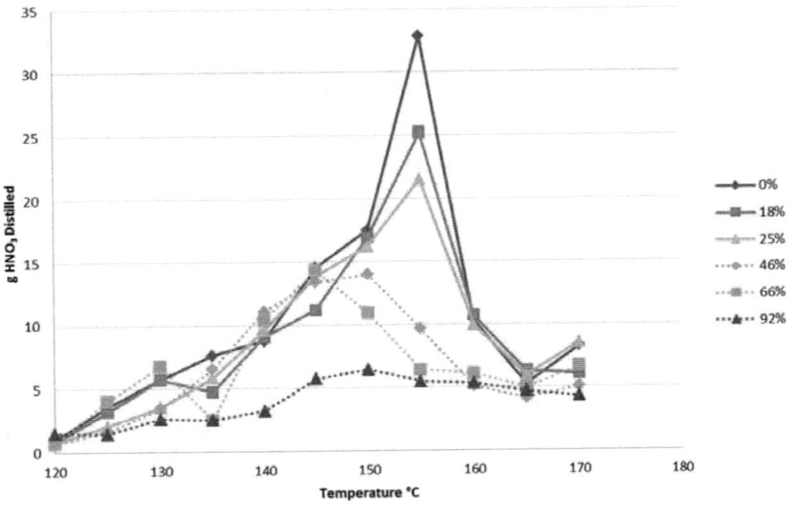

Figure 9 – Amount of nitric acid removed as distillate during the heating phase of iron hydrolysis as a function of pre-neutralisation

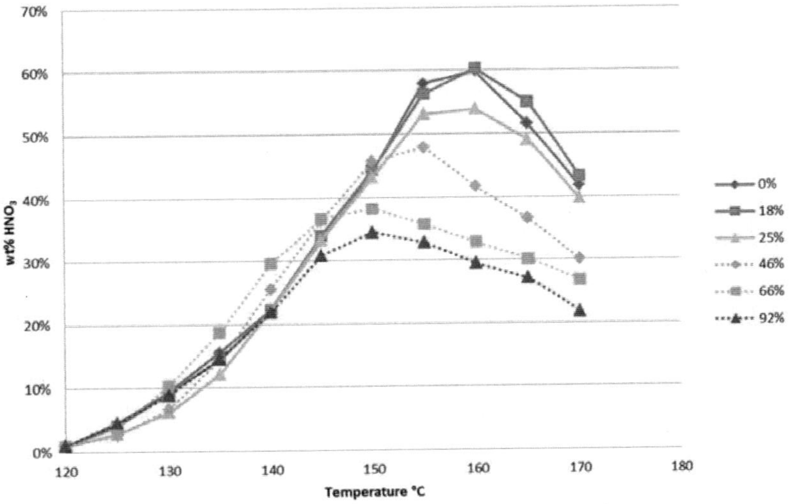

Figure 10 – Strength of nitric acid removed as distillate during the heating phase of iron hydrolysis as a function of pre-neutralisation

Table 14 – Assays of iron precipitates formed with varying amounts of pre-neutralisation, wt.%

% Neutralised	Ni	Co	Al	Fe	Mg	Cr	Mn
0	0.04	0.00	1.04	57.15	0.02	0.68	0.25
18	0.03	0.00	1.23	54.38	0.02	0.67	0.08
25	0.04	0.00	1.50	54.73	0.02	0.71	0.12
46	0.06	0.01	2.43	53.51	0.08	0.67	0.19
66	0.06	0.01	1.29	59.90	0.03	0.71	0.19
92	0.22	0.01	1.69	60.50	0.03	0.67	0.28

For hematite it has been previously noted (Wolska & Schwertmann, 1989) that the (hk0), (hk3) and (hk6) reflections for hematite can be sharper than the others suggesting an anisotropic crystallite size; this was modelled in the analysis of the XRD pattern. The results are given in Table 15.

Table 15 – XRD derived estimates for crystallite size (nm) and unit cell parameters (Å) of hematite in the iron precipitates produced with varying amounts of pre-neutralisation

% Neutralisation	CS (h k 3n)/(h k l)	Hem (a)	Hem (c)	CS (Goe)	Goe (abc)
0	12/11	5.028	13.802	ND	-
18	16/10	5.030	13.803	ND	-
25	15/12	5.026	13.803	ND	-
46*	13/15	5.024	13.822	14	4.611, 10.006, 3.012
66	16/13	5.023	13.819	ND	-
92	39/17	5.019	13.812	ND	-

* Relative amounts of hematite to goethite: 93% to 7%, Hem = hematite, Goe = goethite, CS = crystallite size (nm)

The data in Table 15 mostly indicate that the crystallite size of the hematite is close to isotropic in a most instances consistent with a spherical particle shape. In comparison the observed filterability of the samples increased from around 5 L/m^2/min for the base case to 8.7 and 14.7 L/m^2/min for the 18% and 25% neutralised samples, respectively. It is interesting to note that these samples appear to contain increasing amounts of ferrihydrite (Figure 11) and this is consistent with the decreasing Fe and increasing Al contents given also that the rate of transformation of ferrihydrite to hematite is known to be retarded by the presence of aluminium at high temperatures (Barrón et al., 2012). It would seem, however, that the ferrihydrite phase does not impact filterability in these samples.

The filterability of the subsequent samples then fell, with filter rates of 1.3, 0.5 and 0.2 L/m^2/min for the 46%, 66% and 92% samples, respectively, though the last two samples consist essentially of hematite and a trace amount of goethite. Clearly the macro behaviour of the samples in filtration conditions is not determined by the mineralogy. Instead, it seems to be primarily driven by the agglomeration of the hematite (and ferrihydrite) nanoparticles formed.

As a general observation, the filterability of all of the presented iron hydrolysis solids became worse as the temperature at which these were produced increased; the pH of the filtrates also increased with increasing hydrolysis temperature. Furthermore, additional washing with clean wash water led to de-agglomeration and breakthrough of ultrafine particles into the subsequent filtrate. While it is known that particle size, pH and ionic strength can all affect the rate and extent of aggregation for hematite nanoparticles (He, Wan, & Tokunaga, 2008), the current system represents a wider (and significantly lower) range of pH and ionic strength than normally considered in these studies.

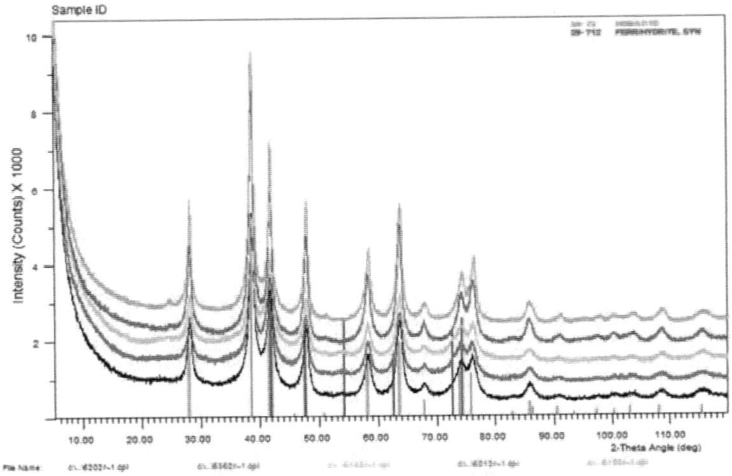

Figure 11 – XRD scans of iron precipitates produced after varying degrees of pre-neutralisation (1) base case (black, bottom), (2) 18% (red), (3) 25% (light blue), (4) 66% (pink) and (5) 92% (light green, top)

Figure 12 – SEM image of an iron hydrolysis product. The scale bar corresponds to a distance of 10 μm.

SEM imaging (Figure 12) indicates that the primary agglomerates are spherical and that these may agglomerate further to form larger agglomerates with non-spherical shape. The figure also shows loosely agglomerated material and separate nanometre-scale agglomerates. In comparison the controlled precipitation of iron as ferrihydrite and conversion to hematite can be employed to generate nanoparticles of various morphologies and having a much narrower size range (e.g. Supattarasakda, Petcharoen, Permpool, Sirivat & Lerdwijitjarud, 2013).

For the present suite of samples it was also notable that a significant amount of goethite was generated in the sample after 46% pre-neutralisation. The data in Figure 9 indicate that a larger amount of nitric acid was collected up to 130°C but much less between 130 and 135°C. Thus the free acid level was low which would favour the formation of goethite and the precipitation of aluminium.

Modelling suggests an iron deficient character for the hematite which has also been noted for hematite formed from goethite (Wolska & Schwertmann, 1989). This is consistent with the presence of hydroxyl anions instead of oxyanions in the structure. It has been noted elsewhere (Wolska, 1981) that hydrothermally formed hematite, hydrohematite, may contain 3.0-4.5% water in the form of hydroxyl anions that requires temperatures up to 1000°C to completely remove; in contrast to weakly held water that can be removed at temperatures of 160-200°C.

The hematite c parameter without substitutions is 13.7454 Å. According to (Neumann, Avelar, & da Costa, 2014), the extent of hydroxyl substitution can be calculated from the equation:

$$X_{OH-} = (c - 13.7454)/0.23725 \quad (4)$$

Using the range of c values found in the present samples from Table 15, between 13.802 and 13.822 Å, the calculated hydroxyl substitution for oxide (in mole fraction) for the hematite ranges between 0.239 and 0.323. This range of values corresponds to structural water contents between 4.4 and 6.1%.

An interesting feature of the XRD patterns obtained for the current samples is the existence of a broad peak centred at ~2.9 Å which has been ascribed to an ultradispersed (or highly dispersed) hematite phase (Drits, Sakharov, & Manceau, 1993). A different work (Drits, Sakharov, Salyn, & Manceau, 1993) indicated this ultradispersed phase could be intimately associated with ferrihydrite. The size of crystallites that can produce this broad peak, and associated weaker broad peaks, would be typically of order of magnitude, 10 Å or 1 nm. Further studies, including SEM or TEM imaging, are required to better understand the nature of the iron-containing materials formed during iron hydrolysis in the DNi process.

PILOT PLANT PERFORMANCE

The pilot plant, which was located at CSIRO Minerals' Waterford site in Western Australia, operated for 11 months in 2013. Much of the first 7 months of operation was dedicated to commissioning each stage of the circuit, trialling alternate configurations and optimising operating conditions and set points (e.g. flocculant types, temperature set points and residence times). Perhaps due to the large amount of bench scale test work completed prior to the pilot plant, operation of the iron hydrolysis circuit was smooth (outside of the expected mechanical issues typically seen with new equipment), and iron rejection was essentially complete. Likely due to the continuous nature of the pilot plant, the filterability of the iron precipitate was also very good.

The full flow sheet shown in Figure 1 (excluding ore preparation as this was outside the 2013 piloting scope) was run in a closed loop for a total of 4 months. Steady-state operation was achieved for approximately 50 days of continuous running for the iron hydrolysis circuit over three campaigns. Sufficient MgO had been produced by the pilot plant in previous campaigns that all the recycle loops were able to be implemented. Some additional operational data from the pilot plant campaigns have been presented elsewhere (Brock & McCarthy, 2014).

The iron hydrolysis tanks are shown in Figure 13, and consist of four heated tanks where the PLS feed from CCD 1 overflow was heated sequentially in each tank to a fixed set point. Here the PLS was heated from approximately 90°C to around 165-175°C. The distillates were collected and re-condensed from each tank (via plate and frame heat exchangers above each iron hydrolysis tank). This acid was sent back to either acid recovery, for the higher strength acid (tanks 2, 3 and 4) or combined with process water, for the weak acid distillate (<5 wt.% HNO_3).

Two ore blends were treated through iron hydrolysis to achieve steady state operation: a 50:50 limonite:saprolite blend and a 100% saprolite ore. The liquid compositions from both leach discharge and iron hydrolysis discharge are shown in Table 16 for the 50:50 blend. The corresponding solid compositions for the feed ore and the iron hydrolysis precipitate are shown in Table 17.

Figure 13 –Fe hydrolysis tanks (left) and Al precipitation tanks (right) during operation of the pilot plant

Table 16 – Average solution compositions of leach discharge and iron hydrolysis discharge during steady-state operation, treating a 50:50 blend of limonite and saprolite ores, mg/L

Solution assays	Ni	Co	Al	Fe	Mg	Cr	Mn	Ca	Si
Leach Tank 4 Liquor	7262	276	8342	121145	29161	1480	1941	553	10
Iron Hydrolysis Discharge Liquor	7376	301	5355	83	92714	9	1108	1994	5

As can be seen in Table 16 the liquor assays show a very high degree of iron and chromium removal, and a low degree of nickel and cobalt precipitation. Magnesium and calcium were both increased through iron hydrolysis in this campaign via the addition of barren solution. A second stage MHP thickener underflow was also produced during this period of operation and was also added to iron hydrolysis tank 1 for recovery of the contained nickel.

Table 17 – Average solids compositions for leach feed ore and iron hydrolysis precipitate during steady state operation, treating a 50:50 blend of limonite and saprolite ores, wt.%

Solids Assays	Ni	Co	Al	Fe	Mg	Cr	Mn	Ca	Si
Leach Feed Ore	1.6	0.09	2.8	30.6	6.6	1.4	0.6	0.2	9.0
Fe Precipitation Solids	0.16	0.01	1.4	58.4	0.1	0.8	0.7	0.01	0.1

In general the iron hydrolysis precipitate composition was reasonably consistent, with iron levels typically in the range 58-60 wt.% Fe, as seen in Table 17. Several operating temperatures for the final iron tank were tested throughout the pilot plant trials however, and the compositions of iron and aluminium varied as predicted in the previous bench scale tests; this can be seen in Figure 14. Samples were taken both from steady state and upset conditions as well as during ramp up, and all the compositions are included in the Figure 14, hence the wide scatter of data points. Based upon the bench scale work it is expected that higher iron content solids will be richer in hematite whereas lower iron content and higher aluminium content solids will contain higher amounts of ferrihydrite and/or goethite.

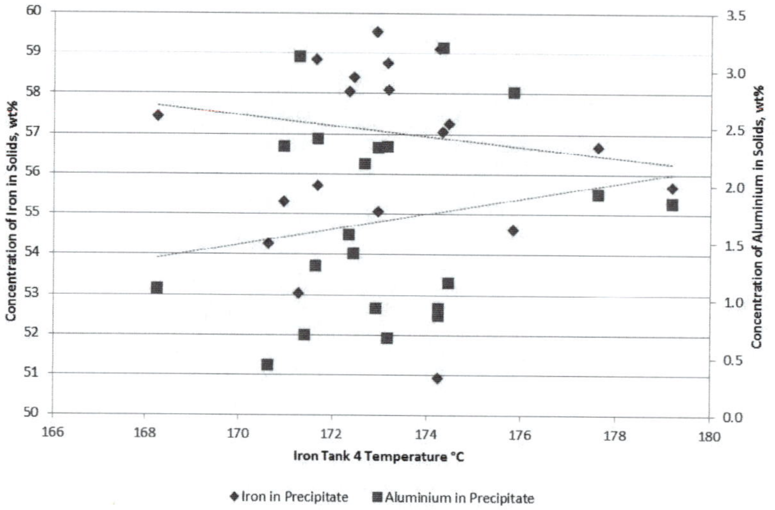

Figure 14 – Iron and aluminium levels in the iron precipitate as a function of temperature in the final iron hydrolysis tank

The thickener treating the iron solids averaged an underflow density of 37 wt.% solids, however long periods of operation over 40 wt.% were noted. The primary limitation to the underflow density was the pump selection for the thickener underflow stream. Due to the relatively small scale of the pilot plant (compared to commercial operations) a hose pump had to be chosen for its ability to pump small stream volumes, however hose pumps are unable to pump high solid containing slurries. As such the bed height in the thickener was not allowed to increase significantly and underflow densities were not allowed to develop fully. The iron product behaved well in the plate and frame pressure filter used in the pilot plant operation, and cake moistures of around 18 wt.% were typical. The issues with handling seen in the laboratory testing did not present as a problem in the pilot plant operation.

The main operational challenges for the iron circuit in the pilot plant were related to temperature control. Each of the tanks in the four tank heating circuit had a heating jacket through which hot thermal oil was circulated, and the rate of flow of the oil (and thus tank temperature) was controlled by automatic flow control valves. These valves did not perform as required, a combination of faulty equipment and poor suitability for operational conditions. Hence much of the pilot plant performance around iron hydrolysis was impacted negatively due to temperature set points not being maintained steadily. In practical terms this tended to reduce iron removal from around 99%+ to around 97%, or resulting in iron discharge liquors having iron tenors in the range of 800-2000 mg/L Fe.

A secondary challenge was some corrosion noted in the gas space in the distillate condensation area. The 304L stainless steel chosen for the heat exchangers was not suitable and resulted in several failures of the acid condensers. Corrosion coupons in both the liquid side and gas side of the iron tanks demonstrated that it was an issue where the nitric vapour is at its dew point, and no corrosion was detected in the liquid wetted areas of the iron tanks; this is consistent with a previous study (Balbaud, Sanchez, Fauvet, Santarini, & Picard, 2000). A redesign of the condensation system to PTFA lined spray condensers was made but the pilot plant was shut down prior to full installation. This system should be fully commissioned during the next pilot plant operation.

SUMMARY AND CONCLUSIONS

The iron hydrolysis circuit has proven to be a robust method for iron removal for the Direct Nickel process, both at laboratory scale as well as in continuous pilot operation. The unique nature of the nitrate environment allows for a near complete iron and chromium rejection from the leach PLS, which is important for the process due to the extremely high levels of iron that are possible in the leach liquors. The circuit utilises boiling point elevation to achieve the high temperatures required for both acid removal via distillation and hematite precipitation.

It has been demonstrated that a moderate amount of control of aluminium levels in the iron precipitate can be achieved through temperature control, but typically comes at the expense of iron grade. The process is relatively insensitive to residence time. Seed composition can enhance aluminium precipitation or suppress it within a moderate range of aluminium precipitation extents. The predominant phase in the iron precipitate is hematite, however small amounts of ferrihydrite can form, particularly when aluminium levels in the iron precipitate are increased or the degree of pre-neutralisation of the PLS is increased.

The pilot plant operated the full Direct Nickel flow sheet (except ore preparation) and the iron hydrolysis circuit treated PLS formed from two ore blends, a 50:50 limonite: saprolite blend as well as a 100% saprolite ore. The circuit generally operated well, producing an iron precipitate averaging around 58 to 60 wt.% Fe. When the circuit was operating at steady state and with the correct temperature set points, iron removal was greater than 99% efficient. The operation of the pilot plant highlighted one area of corrosion in the vapour space of the iron hydrolysis circuit. A redesign of the condensation system has been completed and will be tested in a subsequent operation of the plant.

ACKNOWLEDGEMENTS

The authors would like to acknowledge the substantial contributions made by staff of Direct Nickel and CSIRO Mineral Resources, Waterford, throughout the long collaboration between CSIRO and Direct Nickel.

REFERENCES

Balbaud, F., Sanchez, G., Fauvet, P., Santarini, G., & Picard, G. (2000). Mechanism of corrosion of AISI 304L stainless steel in the presence of nitric acid condensates. *Corrosion Science*, *42*(10), 1685–1707.

Barrón, V., Michel, F., Liu, Q., Koch, C., & Torrent, J. (2012). Effects of Al (III) on the ferrihydrite-ordered ferrimagnetic ferrihydrite-hematite transformation (Vol. 14, p. 2300). Presented at the EGU General Assembly Conference Abstracts.

Brock, G., & McCarthy, F. (2014). Direct Nickel Test Plant Program: 2013 in review. In *ALTA 2014 Nickel-Cobalt-Copper Conference* (pp. 80–99). Perth, Australia.

Cismasu, A. C., Michel, F. M., Stebbins, J. F., Levard, C., & Brown, G. E. (2012). Properties of impurity-bearing ferrihydrite I. Effects of Al content and precipitation rate on the structure of 2-line ferrihydrite. *Geochimica et Cosmochimica Acta*, *92*, 275–291.

Das, S., Hendry, M. J., & Essilfie-Dughan, J. (2011). Transformation of two-line ferrihydrite to goethite and hematite as a function of pH and temperature. *Environmental Science & Technology*, *45*(1), 268–275.

Drits, V., Sakharov, B., & Manceau, A. (1993). Structure of feroxyhite as determined by simulation of X-ray diffraction curves. *Clay Minerals*, *28*, 209–222.

Drits, V., Sakharov, B., Salyn, A., & Manceau, A. (1993). Structural model for ferrihydrite. *Clay Minerals*, *28*, 185–207.

Jansen, E., Kyek, A., Schäfer, W., & Schwertmann, U. (2002). The structure of six-line ferrihydrite. *Applied Physics A*, *74*(1), s1004–s1006.

Jubin, R. T., Counce, R. M., Holland, W. D., Groenier, W. S., & North, E. D. (1987). A simplified process for the extractive distillation of nitric acid. *Industrial & Engineering Chemistry Research*, *26*(5), 990–997.

Melnikov, P., Nascimento, V. A., Arkhangelsky, I. V., Zanoni Consolo, L. Z., & de Oliveira, L. C. S. (2014). Thermal decomposition mechanism of iron(III) nitrate and characterization of intermediate products by the technique of computerized modeling. *Journal of Thermal Analysis and Calorimetry*, *115*(1), 145–151.

Neumann, R., Avelar, A. N., & da Costa, G. M. (2014). Refinement of the isomorphic substitutions in goethite and hematite by the Rietveld method, and relevance to bauxite characterisation and processing. *Minerals Engineering*, *55*, 80–86.

Sloan, J. G. (1976). The Extractive Distillation Process for Nitric Acid Concentration Using Magnesium Nitrate. In *Thermodynamic Behavior of Electrolytes in Mixed Solvents* (Vol. 155, pp. 128–142). American Chemical Society.

Supattarasakda, K., Petcharoen, K., Permpool, T., Sirivat, A. & Lerdwijitjarud, W. (2013). Control of hematite nanoparticle size and shape by the chemical precipitation method. *Powder Technology*, *249*, 353-359.

Vu, H. P., Shaw, S., & Benning, L. G. (2008). Transformation of ferrihydrite to hematite: an in situ investigation on the kinetics and mechanisms. *Mineralogical Magazine*, *72*, 217–220.

Wolska, E. (1981). The structure of hydrohematite. *Zeitschrift Für Kristallographie - Crystalline Materials*, *154*(1-4), 69–76.

Wolska, E., & Schwertmann, U. (1989). Selective X-Ray Line Broadening in the goethite-derived hematite phase. *Physica Status Solidi (a)*, *114*, K11–K16.

SEQUENTIAL ACID LEACHING OF NICKEL LATERITES WITH IRON PRECIPITATION

*D.T. White[1] and J.D. Gillaspie[2]

[1]*Consultant*
Perth, Western Australia
*(*Corresponding author: david.t.white.mmt@gmail.com)*

[2]*Senior Project Manager*
Hazen Research Inc.
Golden, Colorado

ABSTRACT

Sulphuric acid atmospheric leaching of nickel saprolite / garnierite ores was proposed in the early twentieth century, prior to the development of the pressure acid leaching process for limonite ores. Sulphuric acid leaching with alkali addition for iron precipitation as jarosite has been considered periodically since the early 1970s, shortly after the development of the jarosite process for iron control in zinc refining. In the late 1990's a series of flowsheets were proposed and tested involving simultaneous atmospheric leaching, iron precipitation and acid regeneration. These were considered particularly suited to sequential leaching of limonite and saprolite ores. A number of laterite sulphuric acid leaching processes utilising this chemistry have been put into commercial operation. Similar sequential leaching flow sheets involving precipitation of iron as goethite were also developed and piloted. This paper will firstly describe some of the important historical developments of both the jarosite and goethite processes. It will then provide descriptions of the commercialised processes involving alkali jarosite/alunite precipitation, with special emphasis on the downstream impurity removal requirements necessary for efficient recovery of nickel and cobalt from solution.

KEYWORDS

Nickel, Laterite, Limonite, Saprolite, Sulphuric Acid Leaching, Atmospheric Leaching, Pressure Acid Leaching, Iron Precipitation, Jarosite, Goethite, Hematite, Alunite

INTRODUCTION

This paper is a summary and extension of previous works with which the authors have been involved, on the subject of sulphuric acid leaching of nickel laterites (Adams, et al., 2004), (Liu H. , Gillaspie, Lewis, Neudorf, & Barnett, 2004), (White, Miller, & Napier, 2006), (White, 2009), (White & Gillaspie, 2013), (White & Gillaspie, 2015).

Some of the historical developments of both the jarosite and goethite leaching-precipitation processes will be described, followed by a description of the commercialised sequential nickel laterite leaching processes, involving alkali jarosite/alunite precipitation and acid regeneration, with a comparison to the equivalent goethite based processes. These will include Jarosite Atmospheric Leaching (JAL), Goethite Atmospheric Leaching (GAL) and combined Pressure Acid Leaching – Atmospheric Leaching, termed EPAL®. There will also be an explanation of the downstream impurity removal requirements necessary for efficient recovery of nickel and cobalt from solution.

THE LATERITE PROFILE

An idealized nickel laterite profile is shown in Figure 1. The areas containing commercial nickel values are usually the goethite section of the limonite zone and the clay and saprolite zones. As indicated by its name, the goethite dominant section is high in iron and low in magnesium. Iron content can be in excess of 35%, with magnesium as low as 1%. Conversely, the saprolite zone is low in iron and high in magnesium. Iron grades of less than 10% and magnesium grades in excess of 15% are typical.

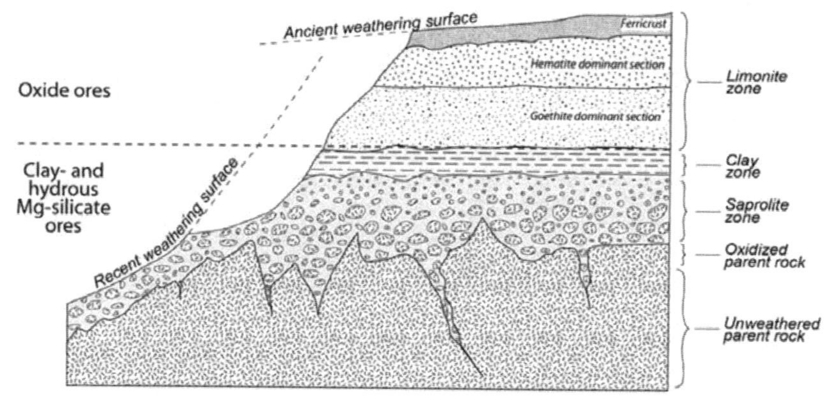

Figure 1 – Idealized cross section through a nickel laterite profile (Samama, 1986)

It is inevitable that in any ore body there will be pockets of different ore types within different zones and cross contamination will occur in mining, so a broader definition would apply in practice to each ore type, for the purpose of hydrometallurgical processing. In addition, the relative ratio of the limonite and saprolite ore types is variable among deposits and also varies as mining progresses through a deposit.

Enrichment of nickel by laterisation is typically considered an acidic leaching-deposition process. High iron goethite precipitates more readily, while the easier to leach magnesium silicate containing saprolite components migrating further before precipitating.

HISTORY OF LATERITE ACID LEACHING

The history of sulphuric acid leaching of nickel laterite ores dates back over a century. A patent specification in the name of Madagascar Minerals Syndicate discusses leaching of low iron, high magnesium "garnierite" ore, followed by solid-liquid separation and precipitation of a nickel-cobalt intermediate with magnesium hydroxide (or oxide) (Sulman & Picard, 1914). A further specification by the same group included leaching at greater than 90°C and sulphide precipitation using sodium sulphide (Na_2S), with subsequent recycling of the barren solution (Sulman & Picard , 1916).

Freeport considered atmospheric leaching for the Moa Bay project, but noted that iron dissolution was high (Carlson & Simons, 1961):

"Under normal laboratory conditions, the great bulk of this ore is soluble in mineral acids. No selectivity is apparent in the rates of solution of the various components up to the normal boiling point."

They therefore took advantage of the iron hydrolysis – precipitation and acid regeneration process that occurs con-currently with leaching at elevated temperatures in pressure acid leaching (PAL).

Goethite and Jarosite Precipitation

Development of the Jarosite process in the zinc industry in the 1960s led to an understanding that iron could be precipitated from nickel laterite atmospheric leach solutions, at moderate acid levels, by the addition of an alkali such as potassium, sodium, or ammonia, whilst maintaining a temperature in excess of 90 °C (Gordon & Pickering, 1975). Work in the late 1960s led to a number of patents being issued concerning the sulphuric acid leaching of nickel laterites with addition of an alkali to precipitate iron (Weston, 1974). The abstract of one of the patents summarizes the process intent well:

"A process for the leaching of nickel bearing laterites, beneficiated laterites or other nickel bearing products wherein the leach is conducted at a temperature below the atmospheric boiling point of the pulp and nickel and cobalt are effectively dissolved while iron is prevented from remaining in solution through the use of precipitating agents capable of introducing alkali metal ion or ammonia ion to the pulp."

Nickel extractions were in excess of 80%, at acid addition rates of around 500 kg/t. Terminal iron solution concentrations of less than 0.5 g/L were typically achieved. Both sodium and potassium additions were made to promote iron precipitation. The potential for using seawater to slurry the ore was noted.

Jarosite and goethite precipitation were discussed in several papers in the following two decades. At the CSIRO, it was noted in 1975 that:

"Processes requiring sulphation require the use of aqueous or concentrated sulphuric acid ... as the sulphating agent. In many cases the leach liquors produced are highly contaminated and recovery procedures play a vital part. Fortunately, it has recently become easier to remove dissolved iron using either the jarosite process or the goethite process." (Canterford, 1975)

Also in 1975, in a paper on the Goethite process it was stated that:

"The process has potential applications, outside the electrolytic zinc industry, in the purification of impure leach liquors from a wide variety of hydrometallurgical processes for the recovery of copper, nickel, cobalt, etc." (Davey & Scott, 1975)

In the 1980s and early 1990s, researchers at the National Technical University of Athens published a number of papers where precipitation of iron from nickel laterite acid leach solutions was considered. Initially goethite precipitation was investigated, (Agatzini & Burkin, 1985), (Agatzini, Kontopoulos, Marboutis, & Xenidis, 1986), with subsequent papers discussing jarosite precipitation. These complement each other, with leach solutions described in one paper (Panagiotopoulos & Kontopoulos, 1988), treated to precipitate iron as jarosite discussed in the second (Maraboutis & Kontopoulos, 1988). A further paper comparing atmospheric and pressure leaching was published in 1991 (Kontopoulos, 1991). They also made reference to the various techniques utilized for iron precipitation in the zinc industry and options considered for their own studies:

"Iron removal by precipitation has been extensively studied for the refining of solutions in the zinc industry, where the techniques of jarosite, goethite and hematite precipitation have found extensive industrial application.

"In our previous papers iron removal from sulphate solutions has been studied by precipitation as goethite ... In the present paper, jarosite precipitation has been studied."

AMAX Pressure Acid Leaching

AMAX researchers added an alkali to nickel laterite PAL slurry, primarily to improve settling properties (Duyvestyn, 1980). However it was noted that iron and aluminium precipitation occurs:

"...the alkali metal and ammonium cations react with solubilized iron and aluminum to form highly insoluble species of jarosites and alunites..."

AMAX separately considered atmospheric leaching of saprolite ore, after PAL of the limonite ore, with the aim of fully utilizing the laterite ore body in a single flowsheet (Duyvesteyn, Wicker, & Doane, 1979). One variant, consisted of leaching raw saprolite ore in PAL discharge liquor, followed by a second leach of roasted saprolite, with both atmospheric leach residues being subjected to PAL. It is apparent that this flowsheet served mainly to utilize (and neutralize) the free acid remaining after PAL to dissolve magnesium and associated nickel from saprolite, with the partially leached saprolite further treated by PAL to recover the remaining nickel values.

Australian Nickel Laterite Pressure Acid Leaching Operations

In the late 1990s, the Cawse and Bulong nickel laterite PAL operations utilized saline waters, resulting in similar chemistry to that of the AMAX alkali addition work. The Murrin Murrin operation subsequently added ammonium sulphate to achieve a similar result. A number of laboratory studies by the CSIRO identified that the precipitation products are mixed alunite-jarosite compounds, with the iron progressively converting from jarosite to hematite as the leach progresses (Johnson, McDonald, Whittington, Quan, & Muir, 2002), (Johnson & Whittington, 2004).

BHP Minerals

In the 1990s, BHP Minerals developed a number of hydrometallurgical technologies with the potential to improve on perceived shortcomings with PAL, to support development of the Gag Island nickel laterite project. The aims included being:

- Cheaper to build than a PAL facility;
- Simpler to operate than PAL in a remote environment;
- Able to process saprolite as well as limonite ore.

Options considered included sulphuric acid heap leaching (Duyvestyn, Liu, & Davis, 2001), agitated tank atmospheric leaching (Arroyo & Neudorf, 2001), (Arroyo & Neudorf, 2004), (BHP Minerals, 2001), (Liu H. , Gillaspie, Lewis, Neudorf, & Barnett, 2004), (Liu H. , Gillaspie, Lewis, Neudorf, & Barnett, 2008), and combinations of PAL and atmospheric leaching (Arroyo, Gillaspie, Neudorf, & Weenink, 2002), (Adams, et al., 2004), (White, Miller, & Napier, 2006), (Krebs & Liu, 2009).

Precipitation of iron with regeneration of (at least some of the) acid during the leach was identified as a desirable goal, for both chemistry control and minimizing acid consumption, emulating the same technique inherent in PAL (Liu H. , Gillaspie, Lewis, Neudorf, & Barnett, 2004). These processes became available to the Ravensthorpe nickel project, when BHP merged with Billiton in 2001. There was potential to use jarosite leaching technology because seawater was already proposed for the process and substantial saprolite resources existed as well as limonite.

MAJOR PROCESS OPTIONS DEVELOPED

Whilst a range of technologies were in the area of nickel laterite acid leaching, two technologies stand out in the area of atmospheric leaching with iron precipitation as either jarosite or goethite. These are:

- Sequential atmospheric leaching of limonite followed by leaching of saprolite with simultaneous iron precipitation and acid regeneration, with variants known as Jarosite Atmospheric Leaching (JAL) and Goethite Atmospheric Leaching (GAL).
- Pressure acid leaching of limonite, followed by atmospheric leaching of saprolite, which subsequently became known as Enhanced pressure acid leaching, or EPAL®, in both jarosite and goethite variants.

JAROSITE ATMOSPHERIC LEACHING

Flowsheet and Basic Leach Chemistry

A flowsheet for the jarosite atmospheric leach is shown in Figure 2. Both limonite and saprolite are slurried in an alkali containing water, such as seawater, to ensure there is enough for jarosite formation.

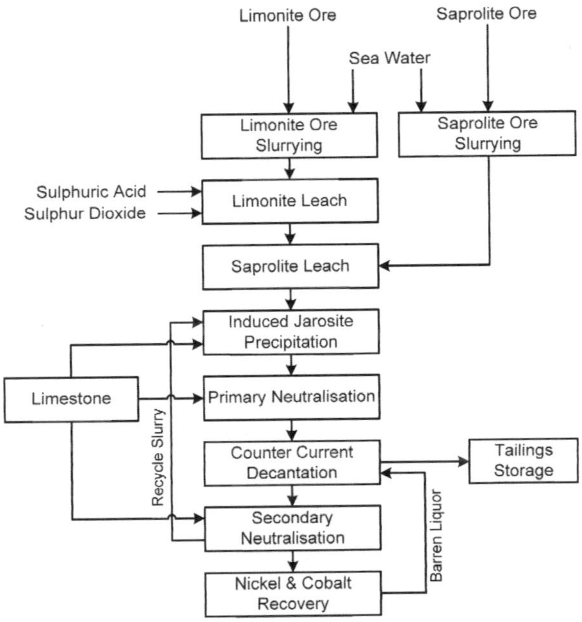

Figure 2 – Jarosite atmospheric leach flowsheet

Concentrated sulphuric acid is added directly to limonite ore slurry. This substantially dissolves the nickel containing goethite matrix leaching over 90% of the nickel and cobalt, along with substantial amounts of iron. Leach temperatures close to the boiling point of the liquor are attained, due to the heat of dilution of the acid. The goethite dissolution reaction is represented by equation 1.

$$2FeO(OH) + 3H_2SO_4 \rightarrow Fe_2(SO_4)_3 + 4H_2O \quad (1)$$

Sulphur dioxide or other suitable reductant may be added to assist with leaching of manganese minerals, which are generally higher in cobalt. This reaction, which is represented by equation 2 also suppresses chromium(VI) formation, ensuring downstream processes are not affected.

$$MnO_2 + SO_2 \rightarrow MnSO_4 \quad (2)$$

Saprolite slurry is then added. Much of the magnesium and nickel is leached, causing the free acid to drop to a level where jarosite precipitation begins, regenerating acid and furthering the leach process. Jarosite Precipitation and acid regeneration are represented by equation 3, with the saprolite leach by equation 4.

$$\tfrac{1}{3}Na_2SO_4 + Fe_2(SO_4)_3 + 4H_2O \rightarrow \tfrac{2}{3}NaFe_3(SO_4)_2(OH)_6 + 2H_2SO_4 \quad (3)$$

$$3MgO{*}2SiO_2{*}2H_2O + 3H_2SO_4 \rightarrow 3MgSO_4 + 2SiO_2 + 5H_2O \quad (4)$$

Overall acid addition rates are typically in the range of 700 to 900 kg per tonne of total ore leached, depending on the ratio of saprolite to limonite and the nickel extraction desired. The net reaction is as shown in equation 5, indicating jarosite precipitation can regenerate two thirds of the acid consumed in leaching iron.

$$2FeO(OH) + 3H_2SO_4 + \tfrac{1}{3}Na_2SO_4 \rightarrow \tfrac{2}{3}NaFe_3(SO_4)_2(OH)_6 + 2H_2SO_4 \quad (5)$$

The combined leach slurry passes to induced jarosite precipitation (IJP) and primary neutralization (PN). Recycle streams from downstream processes are leached initially, decreasing free acid and allowing more jarosite formation to occur. Limestone is then added to further neutralize free acid and precipitate iron.

Ore Preparation

Limonite and saprolite are typically mined and processed through slurry preparation plants separately, with some beneficiation typically achieved, based on coarse ore rejection. Some saprolite contamination of limonite is acceptable, as saprolite will leach in the high acid conditions of the limonite leach, but the same cannot necessarily be said of limonite in the saprolite leach.

Limonite Leach

The goethite component of limonite ore requires almost total dissolution for liberation of the nickel held in the matrix. Sulphuric acid is added at a rate of 1000 to 1500 kg per tonne of limonite ore. Limonite leach times are 3 to 4 hours at around 95 °C. Peak iron concentrations are up to 100 g/L. At the end of limonite leaching, free acid concentrations range from 10 to 30 g/L. Some jarosite may start to form in the limonite leach, especially where high alkali containing waters such as seawater are being used. Extractions of nickel, cobalt, iron and manganese are typically in excess of 90%

Saprolite Leach

Although saprolite is a relatively high acid consumer, the free acid driving force required for leaching the majority of the nickel is lower than for limonite. Sufficient saprolite ore slurry is mixed with the limonite leach discharge to reduce free acid levels below 30 g/L H_2SO_4, so that iron commences to precipitate as jarosite, simultaneously regenerating acid to maintain a driving force for continued nickel and cobalt leaching. Ferric iron in solution is reduced to less than 5 g/L, with free acid in the range of 20 to 25 g/L H_2SO_4. The free acid – iron interaction and leaching progress are illustrated in Figure 3. Saprolite leach times are in the range of 10 to 12 hours.

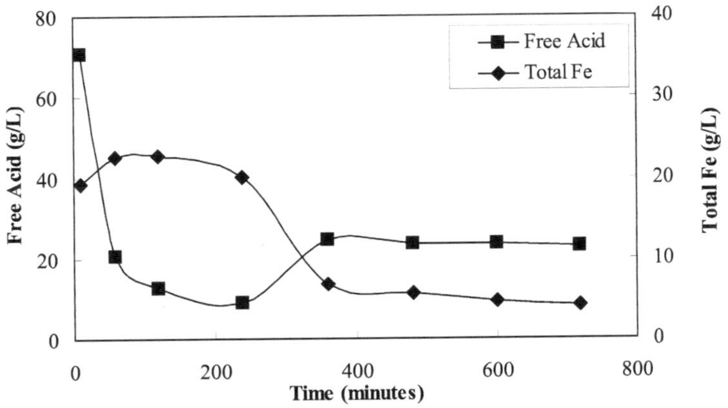

Figure 3 – Batch atmospheric leach free acid and iron in solution

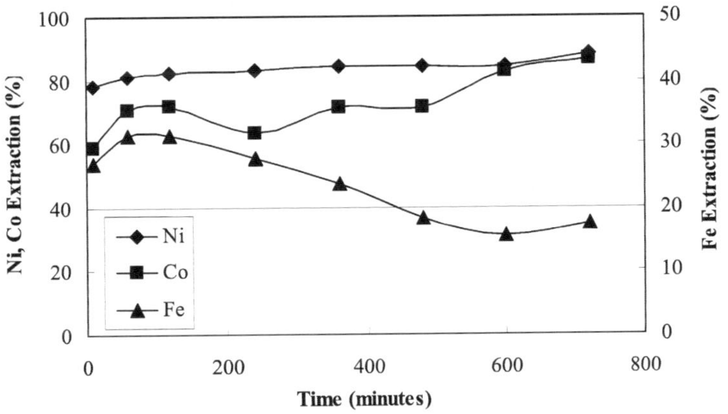

Figure 4 – Batch atmospheric leach nickel, cobalt and iron extractions

It has been found in practice that the reaction is self-seeding, with sufficient solids available in the agitated tanks to initiate and sustain jarosite precipitation.

Induced Jarosite Precipitation

The function of the IJP stage is to precipitate excess ferric iron from solution under hot, mildly acidic conditions. Some precipitation of other impurities, such as aluminium, also occurs at this stage.

Saprolite leach discharge is partially neutralized with recycled Secondary Neutralization Precipitation residue slurries, simultaneously recovering the metal values associated with those streams. The residual acid is then further reduced to approximately 5 g/L H_2SO_4 using ground limestone slurry. Temperature is maintained at 95 °C.

Primary Neutralization

In Primary Neutralization (PN), further limestone slurry is added in stages to precipitate the bulk of the remaining iron(III). Solution pH is controlled to avoid unrecoverable losses of nickel and cobalt.

Discharge from primary neutralization typically has a pH of 2 to 2.5 and has an iron(III) concentration of less than 0.5 g/L.

GOETHITE ATMOSPHERIC LEACHING

Goethite atmospheric leaching follows the same principles as the sequential Jarosite process, with limonite being leached first and saprolite addition used to neutralise acid and precipitate iron, with the basic goethite precipitation reaction shown in equation 6, being the reverse of the leaching reaction of equation 1.

$$Fe_2(SO_4)_3 + 4H_2O \rightarrow 2FeOOH + 3H_2SO_4 \quad (6)$$

Goethite precipitation has the potential to release all of the acid consumed in leaching iron, rather than the two thirds indicated for jarosite precipitation. This nominally leads to lower acid consumption than the jarosite process. However, goethite precipitation occurs at lower free acidity / higher pH than in the jarosite process. As a consequence, either more saprolite addition is required for iron precipitation, typically leading to lower saprolite nickel extractions, or alternatively limestone must be added. In any event, managing the final acidity is important when precipitating goethite. The lower free acid levels after saprolite leaching mean leaching of recycles from downstream processes requires careful consideration.

Options include:

- Retaining a similar flow sheet sequence to
- Figure **2**, with the addition of supplementary acid to aid recycle leaching;
- Altering the flow sheet so that the recycles are leached upstream of saprolite addition, to take advantage of the free acid available after limonite leaching.

Other differences include:

- Liquors used must clearly be low in alkali elements, sodium, potassium or ammonia, or some jarosite precipitation will occur. This has little impact on the practicalities of flowsheet operation, but may impact on tailings disposal. Naturally present alkali elements will participate in the reaction, but care should be taken to avoid or at least minimise further additions.
- Aluminium rejection may be more problematical.

ENHANCED PRESSURE ACID LEACHING

In EPAL®, much the same chemistry is applied as for the JAL or GAL processes, but the sequence of leaching is changed, as illustrated for the jarosite version in Figure 5.

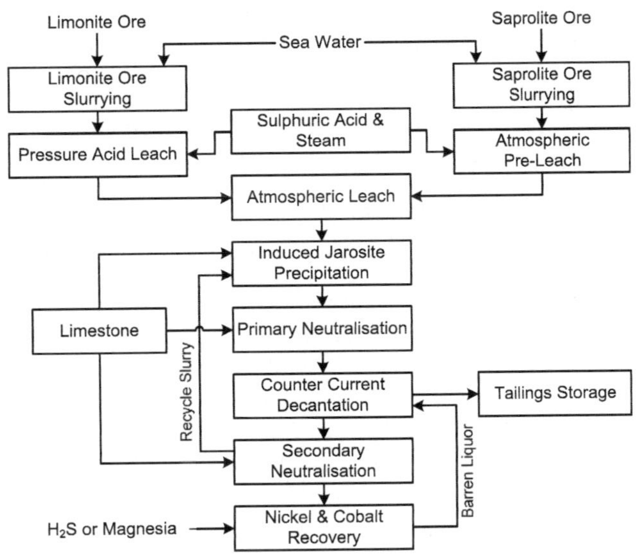

Figure 5 – Jarosite enhanced pressure acid leach flowsheet

Limonite Pressure Acid Leaching

Limonite ore is leached with sulphuric acid at around 250 °C. At this temperature, acid consumption is minimised because iron is hydrolysed to hematite, kinetics are rapid and nickel and cobalt extractions typically exceed 95%. Equation 1 remains relevant for leaching of goethite.

Hematite precipitation as shown in equation 7, results in nominally complete regeneration of acid, as shown in equation 8, although for practical purposes iron extractions of 1% to 2% are typical.

$$Fe_2(SO_4)_3 + 3H_2O \rightarrow Fe_2O_3 + 3H_2SO_4 \quad (7)$$

$$2FeO(OH) \rightarrow Fe_2O_3 + H_2O \quad (8)$$

A reductant may be added to aid leaching of manganese minerals, and suppress chromium(VI) formation, with elemental sulphur being quite effective at higher temperatures (Johnson J. , Whittington, McDonald, & Muir, 2004).

$$3MnO_2 + S + 2H_2SO_4 \rightarrow MnSO_4 + 2H_2O \quad (9)$$

In addition to the hematite precipitation – acid regeneration chemistry, sodium and potassium react with aluminium in solution to form alunite, resulting in low net extraction. Alunite precipitation also regenerates acid, helping to minimize acid consumption, analogous to jarosite precipitation in atmospheric leaching. Sodium alunite precipitation, as shown in equation 10, is an analogue of jarosite precipitation in equation 3.

$$\tfrac{1}{3}Na_2SO_4 + Al_2(SO_4)_3 + 4H_2O \rightarrow \tfrac{2}{3}NaAl_3(SO_4)_2(OH)_6 + 2H_2SO_4 \quad (10)$$

Aluminium to sulphur ratios in residues point to iron co-precipitation, such that an alunite-jarosite compound is formed, typically about two-thirds alunite and one-third jarosite. The sodium plus potassium to sulphur ratios indicate that 10 to 20% of the alunite-jarosite is in the hydronium form. A useful summary of these phenomena (Kyle, 2003) and detailed studies into the effects of alkali addition and leach liquor salinity on PAL have been made (Johnson, McDonald, Whittington, Quan, & Muir, 2002) (Johnson & Whittington, 2004). Some typical assays are shown in Table 1 and Table 2.

Table 1 – Pressure acid leach in saline waters - metal extractions **(White, Miller, & Napier, 2006)**

Stream	Acid Addition (kg/t)	Free Acid (g/L)	Metal Extractions (%)						
			Ni	Co	Fe	Al	Cr	Mg	Mn
Batch	320	34	97.4	97.6	1.6	0.7	1.3	87.6	98.0
Pilot Plant	325	26	95.7	95.8	2.3	2.2	1.7	88.7	96.1

Table 2 – Pressure acid leach in saline water - feed and residue assays **(White, Miller, & Napier, 2006)**

Stream	Elemental Analysis (wt %)							
	Ni	Co	Fe	Cr	Al	K	Na	S
Feed	2.2	0.088	27.6	0.28	2.0	0.02	0.04	0.05
Residue	0.11	0.004	30.0	0.30	2.3	0.09	0.80	3.0

Saprolite Atmospheric Leaching

There no particular advantage in leaching saprolite in an autoclave due to low iron, and the relative solubility limits of magnesium favour atmospheric leaching. If the saprolite to limonite ratio in the ore body is relatively low, the acid contained in the PAL discharge solution can be used for atmospheric leaching of saprolite. At higher saprolite to limonite ratios, additional acid is added in a Pre-leach step. This flexibility allows processing of the entire ore profile.

In the Pre-Leach step, concentrated sulphuric acid is added to saprolite slurry. This raises the slurry temperature to greater than 95 °C and leaches substantial amounts of nickel, cobalt, iron and other impurities. As with the limonite leaching in JAL, some jarosite may start to precipitate.

PAL and Pre-Leach discharge streams are combined and allowed to react to complete saprolite leaching in Atmospheric Leach. The functions of IJP and PN remain the same as those for the JAL process.

Goethite EPAL®

The main difference in the PAL stage of the Goethite EPAL process is that instead of formation of an alkali alunite, hydronium alunite is formed, as illustrated by equation 11. This is less complete than in alkali containing waters, typically resulting in 10% to 15% net aluminium extraction.

$$Al_2(SO_4)_3 + 7H_2O \rightarrow (H_3O)Al_3(SO_4)_2(OH)_6 + {}^{5}/_{2}H_2SO_4 \qquad (11)$$

Leaching of recycles may be carried out in PAL discharge, or added to PAL feed, depending on free acid availability and disposition of impurity elements. Whilst the overall process complexity is similar to the Jarosite EPAL process, there are different process drivers, including alkali element behaviour, impurities impact, and overall ore body mix, which determine which option is most appropriate.

OTHER ATMOSPHERIC LEACHING CONFIGURATIONS

It is feasible to carry out a saprolite only atmospheric leach. The main technical constraint of this configuration is the solubility limit of magnesium. It is also possible to pre-blend limonite and saprolite slurries. Recoveries are at best equivalent to the sequential limonite-saprolite leach (JAL), while producing end solutions of higher acidity and iron content. This has material implications for the quantity of limestone required for neutralization and equipment sizing for downstream solid-liquid separation and residue storage.

OPERATIONAL FLEXIBILITY – CONFIGURATION CHANGES

An aspect of plant design and operation that is apparent from the above description is that the atmospheric leach section of the EPAL® configuration can be used as an Atmospheric Leach plant, simply by piping limonite to feed the Pre-leach and adding saprolite directly to the atmospheric leach tanks. In the event that PAL becomes an operating constraint, operation as stand-alone atmospheric leach can be used to fully utilize the acid plant capacity and minimize production disruptions.

DISPOSITION OF OTHER ELEMENTS

Aluminium

Disposition of aluminium is of particular concern where the process calls for precipitation of a mixed hydroxide intermediate, or if a direct solvent extraction is utilized. Re-leaching of mixed hydroxide intermediates in ammonia solutions is impeded if aluminium is present (White, 2009). Crud formation occurs in cobalt-nickel separation with Cyanex 272 (Donegan, 2006) and aluminium can also adversely affect the performance of the Cyanex 301 reagent (Bacon, et al., 2005), (Okita, Singhal, & Perraud, 2006).

Pressure Acid Leaching. As discussed above, net aluminium extraction is around 1% when pressure acid leaching is carried out in saline waters and 10% to 15% in non-alkali containing solutions. Thus, aluminium removal in the PAL process step has the potential to be an "out" for aluminium, even if downstream removal steps prove to be ineffective. If necessary, nickel and cobalt containing recycle slurries from downstream precipitation processes can be recycled to PAL for metal recovery and partial aluminium precipitation.

Atmospheric Leach and Precipitation Processes. Aluminium is precipitated at the pH used for ferrous iron oxidation. The precipitated aluminium reports to the residue recycle stream and is re-leached in the Induced Jarosite Precipitation stage. There is therefore the risk of a recirculating load building up in the circuit between the Induced Jarosite Precipitation and Secondary Neutralization circuits. However, the pH of the intervening Primary Neutralization step is sufficiently high to precipitate excess aluminium and provide a safe bleed from the process.

A similar situation exists in the goethite version of the process, where at a sufficiently high temperature and pH, recycled aluminium is partially precipitated in Primary Neutralization. The control window for this is smaller than that for the jarosite process, necessitating close monitoring to balance aluminium precipitation with nickel and cobalt losses.

Magnesium

Magnesium exhibits reverse solubility at very high temperatures and has been known to precipitate in the PAL autoclave as kieserite (White, Miller, & Napier, 2006). A tolerable level is around 20 g/L in flash discharge, although this may be conservative. At the 95 °C temperature conditions of atmospheric leaching, magnesium solubility is around 50 g/L. A good working limit is around 40 g/L, as this allows for solution cooling without the risk of magnesium sulphate crystallization.

Calcium

When using seawater or recycle liquors, anhydrite scale formation in high temperature heaters in PAL requires careful management.

Calcium carbonate contained in limestone is added in the neutralization processes after nickel laterite leaching. The resulting calcium sulphate solids have the potential to form scale on equipment and piping surfaces. In the IJP and PN processes, this is typically done in-slurry, at temperatures in excess of 80°C. Few issues with scale formation occur, apparently due to the presence of a large surface area available from the leach residue solids. Adequate retention time for complete reaction of the limestone is required, to avoid reactions continuing to occur in downstream thickeners and pipework. Such reactions can lead to both process upsets and scale formation over the longer term.

The downstream secondary neutralization stage is susceptible to scaling because the feed liquor is effectively solids free, so more susceptible to precipitation of gypsum on plant equipment such as tank walls, pipes and agitators. Recirculation of gypsum containing precipitated solids is a viable control option, with operational improvements still occurring at a seed recycle rate of 500% of net precipitation. These include reduced scaling rates, higher settled slurry density and improved limestone consumption efficiency. Good design practice typically includes dedicated thickener underflow recirculation, and blending of seed slurry and feed liquor ahead of limestone addition to the first neutralization vessel. Deliberate addition of leach residue solids has also been shown to be effective (Tsuchida, Ozaki, Nakai, & Kobayashi, 2004).

Manganese

Manganese is in the ore in the manganese(IV) state, which is a strong oxidant, so will oxidize chromium to chromium(VI). Addition of a reducing agent such as sulphur dioxide, as discussed above avoids this, while also increasing manganese and associated cobalt dissolution. Manganese causes few problems with nickel recovery, being almost totally left in solution during sulphide precipitation and partially so during mixed hydroxide precipitation. Barren liquors may be treated to remove manganese by a combination of pH adjustment with lime and oxidation.

Chromium

Chromium(III) behaves in a similar manner to iron(III) so causes few problems in downstream processing. It precipitates after iron but before aluminium as the pH is raised. Chromium(VI) is quite persistent once formed and will cause contamination issues in downstream nickel and cobalt recovery. It can be eliminated by the addition of a reducing agent ahead of neutralization. Suitable reagents used include sulphur dioxide, hydrogen sulphide and their alkali equivalents.

Silicon

Silicon dissolves at temperatures prevailing in PAL operations. During pressure letdown, solubility decreases considerably, giving the potential for scale formation on vessel walls. AMAX devoted considerable effort to understanding and controlling the phenomena (Queneau, Berggren, Cooperrider, & Doane, 1983), with the draft tube crystallizer concept adapted by AMAX from the geothermal industry incorporated at the Cawse and Ravensthorpe operations.

Commercial operations seem to have had few major issues with downstream silica formation, especially if there is a secondary neutralization step, where levels are reduced to less than 30 mg/L. The magnesia based mixed hydroxide process further reduce this to less than 5 mg/L in processes analogous to lime softening (White, Miller, & Napier, 2006), (White, 2009).

LEACH RESIDUE STABILITY

The stability of jarosite containing residues from atmospheric leaching has been studied in depth (Reynolds, 2007). Residues are chemically stable if kept in contact with alkali containing liquors. Jarosite breakdown is also retarded in the presence of lime, which reacts with breakdown products and encapsulates the remaining jarosite.

Goethite Residues are considered to be more stable than those containing jarosite. Toxicity Characteristic Leaching Procedure tests have been carried out on pilot plant residues (Liu H. , Gillaspie, Lewis, Neudorf, & Barnett, 2004), with low values typically achieved.

CONCLUDING REMARKS

Sequential sulphuric acid leaching of laterites with iron precipitation is a versatile process, capable of treating both limonites and saprolites and variable ratios of these two ore types. The EPAL® variant is well suited to ore bodies which are predominantly limonite, and can also be operated in full atmospheric leach mode. Of the purely atmospheric leach options presented, sequential limonite-saprolite atmospheric leaching gives the best extractions for a given total acid addition and produces a leach slurry ready for further impurity precipitation, with a minimum of limestone addition.

In-slurry acid neutralization at elevated temperatures allows almost complete iron(III) precipitation, without risk of significant nickel or cobalt co-precipitation.

Jarosite chemistry lends itself well to managing aluminium in solution, with partial removal achieved in the neutralization step, providing a purge without risk of nickel and cobalt losses. Jarosite containing nickel laterite leach residues are chemically stable if kept in contact with alkali containing liquors. The Jarosite Atmospheric Leach process may be particularly suited to arid climates, where the only water source readily available is saline and it is possible to operate with zero liquid discharge.

The Goethite Atmospheric leach process has similar attributes to those of the Jarosite process, with a number of differences, including:

- Modestly lower overall nickel and cobalt extractions, with accompanying lower acid and limestone consumption.
- Recycle stream and aluminium management are less developed than for the jarosite based processes.
- Leach residue management is likely to be more straightforward.

REFERENCES

Adams, M., van der Meulen, D., Czerny, C., Adamini, P., Turner, J., Jayasekera, S., . . . Miller, G. (2004). Piloting of the Beneficiationand EPAL Circuits for Ravensthorpe Nickel Operations. *International Laterite Nickel Symposium* (pp. 193-202). TMS, Warrendale, PA.

Agatzini, S., & Burkin, A. R. (1985). A Statistical Approach to the Precipitation of Iron as Goethite. *Trans. IMM, 94*, C105-C114.

Agatzini, S., Kontopoulos, A., Marboutis, P., & Xenidis, A. (1986). Removal of Iron from Iron-Nickel-Cobalt Solutions by Precipitation and Solvent Extraction Techniques. *Iron Control in Hydrometallurgy* (pp. 353-373). Ellis Horwood Ltd.

Arroyo, J. C., & Neudorf, D. A. (2001). *U.S. Patent No. 6,261,527.*

Arroyo, J. C., & Neudorf, D. A. (2004). *U.S. Patent No. 6,680,035.*

Arroyo, J. C., Gillaspie, J. D., Neudorf, D. A., & Weenink, E. M. (2002). *U.S. Patent No. 6,379,636.*

Bacon, W. G., Colton, D. F., Krause, E., Mihaylov, I. O., Singhal, A., Vahed, A., & Duterque, J. P. (2005). Development of the Goro Nickel Process. *ALTA 2005 Nickel/Cobalt Conference* (pp. 1-20). ALTA Metallurgical Services, Melbourne, Australia.

BHP Minerals. (2001). Nickel Laterites … Taking the Pressure Off,. *New Caledonian Nickel Conference*, (pp. 1-19).

Canterford, J. H. (1975). Treatment of Nickeliferous Laterites. *Minerals Sci. Engng, vol. 7, No. 1, January*, 3-17.

Carlson, E. T., & Simons, C. S. (1961). Pressure Leaching of Nickeliferous Laterites with Sulfuric Acid. *Extractive Metallurgy of Copper, Nickel, and Cobalt* (pp. 363-396). Interscience Publishers, New York.

Davey, P. T., & Scott, T. R. (1975). Removal of Iron from Leach Liquors by the "Goethite" Process. *Hydrometallurgy, 2*, 25-33.

Donegan, S. (2006). Direct Solvent Extraction of Nickel at Bulong Operations. *Minerals Engineering 19*, 1234-1245.

Duyvesteyn, W. P., Wicker, G. R., & Doane, R. E. (1979). An Omnivorous Process for Laterite Deposits. *International Laterite Symposium 1979* (pp. Chapter 28, 553-570). AIME, New York, NY.

Duyvestyn, W. P. (1980). *U.S.Patent No. 4,195,065.*

Duyvestyn, W. P., Liu, H., & Davis, M. J. (2001). *U.S.Patent No. 6,312,500.*

Gordon, A. R., & Pickering, R. W. (1975). Improved Leaching Technologies in the Electrolytic Zinc Industry. *Metallurgical Transactions B, Vol 6B, March*, 43-53.

Johnson, J. A., & Whittington, B. I. (2004). Effect of Ammonium, Sodium and Potassium Sulphates and Chlorides in Pressure Acid Leaching of Western Australian Nickel Laterite Ore. *Pressure Hydrometallurgy 2004* (pp. 199-214). CIM, Montreal, QC.

Johnson, J. A., McDonald, R. G., Whittington, B. I., Quan, L. P., & Muir, D. M. (2002). Process Water Salinity Effects in the Pressure Acid Leaching of Cawse Nickel Laterite Orel. *Chloride Metallurgy 2002* (pp. 339-354). CIM, Montreal, QC.

Johnson, J., Whittington, B., McDonald, R., & Muir, D. (2004). Effect of Eh in the High Pressure Acid Leaching of Nickel Laterite Ores. *International Laterite Nickel Symposium 2004* (pp. 273-288). TMS, Warrendale, PA.

Kontopoulos, A. (1991). Sulfuric Acid Leaching of Laterites. *EPD Congress '91* (pp. 147-163). TMS, Warrendale, PA.

Krebs, D. G., & Liu, H. (2009). *U.S. Patent No. 7,559,972.*

Kyle, J. H. (2003). Jarosite/Alunite in Nickel Laterite Leaching - Friend or Foe? *ALTA 2003 Nickel/Cobalt-9* (pp. 1-28). ALTA Metallurgical Services, Castlemaine, Vic.

Liu, H., Gillaspie, J. D., Lewis, C. A., Neudorf, D. A., & Barnett, S. C. (2008). *U.S. Patent No. 7,416,711.*

Liu, H., Gillaspie, J., Lewis, C., Neudorf, D., & Barnett, S. (2004). Atmospheric Leaching of Laterites with Iron Precipitation as Goethite. *International Laterite Nickel Symposium* (pp. 347-368). TMS, Warrendale, PA.

Maraboutis, P., & Kontopoulos, A. (1988). Jarosite Precipitation from Iron – Nickel – Cobalt Sulfate Solutions. *Extractive Metallurgy of Nickel and Cobalt* (pp. 479-492). TMS, Warrendale, PA,.

Okita, Y., Singhal, A., & Perraud, J.-J. (2006). Iron Control in the Goro Nickel Process. *Iron Control Technologies* (pp. 635-651). CIM, Montreal, QC.

Panagiotopoulos, N., & Kontopoulos, A. (1988). Atmospheric Pressure Sulfuric Acid Leaching of Low-Grade Hematitic Laterites. *Extractive Metallurgy of Nickel and Cobalt* (pp. 447-459). TMS, Warrendale, PA.

Queneau, P. B., Berggren, M. H., Cooperrider, M. W., & Doane, R. E. (1983). Control of Silica Deposition during Pressure Let-Down of Acidic Leach Slurries. *Hydrometallurgy – Research, Development and Plant Practice* (pp. 121-137). TMS-AIME, Warrendale, PA.

Reynolds, G. A. (2007). The Jarosite Group of Compounds - Stability, Decomposition and Conversion. *MSc Thesis*. NSW, Australia: University of Newcastle.

Samama, J.-C. (1986). *Ore Fields and Continental Weathering*. New York: Van Nostrand Reinhold Co.

Sulman, H. L., & Picard , H. F. (1916). *U.S. Patent No. 1,193,734.*

Sulman, H. L., & Picard, H. F. (1914). *U.S. Patent No. 1,091,545.*

Tsuchida, N., Ozaki, Y., Nakai, O., & Kobayashi, H. (2004). Development of Process Design for Coral Bay Nickel Project. *International Laterite Nickel Symposium* (pp. 151-160). TMS, Warrendale, PA.

Weston, D. (1974). *U.S. Patent No. 3,793,430, U.S. Patent No. 3,793,432, U.S. Patent No. 3,798,304.*

White, D. T. (2009). Commercial Development of the Magnesia Mixed Hydroxide Process for Recovery of Nickel and Cobalt from Laterite Leach Solutions. *Hydrometallurgy of Nickel and Cobalt 2009* (pp. 351-367). CIM, Montreal, QC.

White, D. T., & Gillaspie, J. D. (2013). Acid Leaching of Nickel Laterites with Jarosite Precipitation. *Ni-Co 2013* (pp. 75-95). TMS, Warrendale PA.

White, D. T., & Gillaspie, J. D. (2015). Atmospheric Leaching of Nickel Laterites with Iron Precipitation. *ALTA 2015 Nickel-Cobalt-Copper Conference* (p. 20). Perth: ALTA Metallurgical Services, Melbourne, Vic.

White, D. T., Miller, M. J., & Napier, A. C. (2006). Impurity Disposition and Control in the Ravensthorpe Acid Leaching Process. *Iron Control Technologies* (pp. 591-610). CIM, Montreal, Quebec.

IRON BEHAVIOR IN NICKEL LATERITE INVERSE LEACHING PROCESS

*K. Jiang, H. Wang, S. Liu, K. Xie, and L. Su

Beijing General Research Institute of Mining and Metallurgy
Building 23, Zone 18 of ABP, No.188, South 4th Ring Road West
Beijing 100160, China
*(*Corresponding author: jiangkx@bgrimm.com)*

ABSTRACT

The reduction of iron dissolution and acid consumption is fairly important in the extraction of Ni and Co from laterite ores from both economic and efficiency perspectives. BGRIMM conducted detailed research to investigate the behavior of iron in a nickel laterite leaching process. The mineralogy of Ni and Co was studied. In limonite ore about 76% of nickel is present in goethite while in saprolite ore 79% of nickel is associated with magnesium and silicon minerals. Test work revealed that limonite can be decomposed completely with enough acid concentration during atmospheric leaching. More than 98% of Ni and Co were leached into solution while 85% of the iron was also dissolved with 130-140 g/L iron in solution. Iron precipitates as hematite in the temperature range 250-270°C without neutralization reagents. That's why limonite is treated by high pressure acid leaching (HPAL) technology to achieve nickel extraction and iron precipitation in the same autoclave. BGRIMM's test results indicated that the remaining acid in atmospheric leaching solution and acid released by iron precipitation at a lower temperature was enough to leach saprolite efficiently. The so-called inverse leaching technology leaches limonite at atmospheric condition and the leaching slurry combined with saprolite is pumped into an autoclave to leach saprolite at 150-170°C. In the autoclave iron precipitates as hematite without acid consumption. Total nickel and cobalt extractions in the test program were more than 90% with iron content in solution below 8 g/L. In this inverse process iron dissolves in the atmospheric leaching stage and precipitates as hematite in the low temperature leaching stage. Compared with HPAL the leaching temperature is reduced to 150-170°C from 250-270°C. Compared to atmospheric tank leaching acid consumption is reduced from 850-950 kg/t ore (atmospheric) to 550-650 kg/t ore (inverse).

KEYWORDS

Iron behaviour, nickel laterite, limonite, saprolite, inverse leaching, hematite

INTRODUCTION

Nickel laterite has become the main resource for nickel extraction. Laterite deposits typically contain an upper limonite ore and the lower saprolite ore layer. High pressure acid leaching (HPAL) with high metal recovery and low acid consumption has commonly been used to treat nickel limonite laterite (Georgiou & Papangelakis, 1998; Johnson, Cashmore, & Hockridge, 2005; Kaya & Topkaya, 2011; Önal & Topkaya, 2014; Rubisov & Papangelakis, 2000a, 2000b) while atmospheric tank leaching (ATL) with lower capital (CAPEX) and operating expenditures (OPEX) has been proposed to treat nickel saprolite laterite (Luo, Feng, Ou, Zhang, & Chen, 2010; MacCarthy, Addai-Mensah, & Nosrati, 2014; McDonald & Whittington, 2008; Thubakgale, Mbaya, & Kabongo, 2013; Whittington, Johnson, Quan, McDonald, & Muir, 2003). However, HPAL's CAPEX is as high as 25 USD/lbs. Ni due to the high temperature and pressure required for efficient leaching of nickel and simultaneous elimination of iron. ATL suffers from high acid consumption due to dissolution of iron, and its nickel recovery is obviously lower than that of HPAL. How to treat nickel laterite ore with both low CAPEX and OPEX is one of the most important issues for the extractive metallurgists. The control and stabilization of iron as well as reduction of acid consumption play a key role in the design and operation of hydrometallurgical processes for laterite ores.

Beijing General Research Institute of Mining and Metallurgy (BGRIMM) has studied iron dissolution and precipitation behaviour during laterite leaching and found that leaching of iron was mainly affected by solution acidity and acid generated from iron precipitation was sufficient to dissolve saprolite even though it was inadequate for leaching of limonite. Based on systematic thermodynamic analysis and experiments, a novel inverse leaching process for nickel laterite starting with a primary ATL stage and followed by a secondary autoclave leaching stage was developed. Limonite with high grade iron and low grade Ni and Mg was leached in the atmospheric tank (primary leaching) while saprolite with high grade Ni and Mg was leached in the autoclave (secondary leaching) with the limonite leached slurry. Sulphuric acid is added as lixiviant to the atmospheric tank leaching, while no acid is added to the autoclave leaching stage, since the acid will be auto-generated as Fe precipitates. BGRIMM finished detailed bench scale tests and pilot tests from 2012 to 2013. The process flow diagram is shown in Figure 1. The effects of acid dosage, pulp density, leaching temperature and retention time were examined in this test program. The speciation of iron in feed ores, leaching solution and residue were analysed. Herein, the iron behavior in the inverse leaching process of laterite is unveiled. The underlying mechanism and the relations between Ni and Co leaching and iron behaviour are discussed.

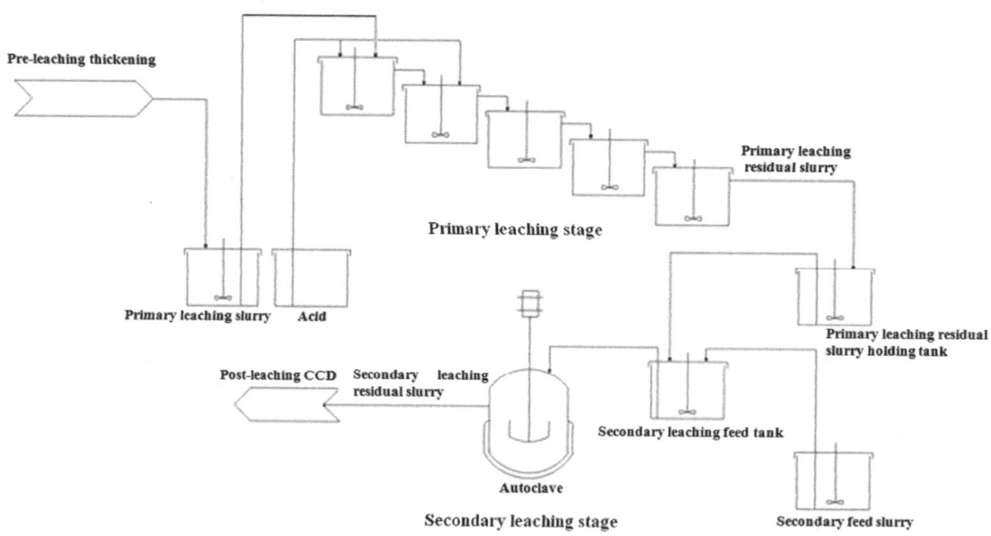

Figure 1 - The flow diagram of the inverse leaching process

After piloting a bankable feasibility study (BFS) was completed by MGRIMM for a nickel laterite project in the Philippines. The results showed that the inverse leaching process is very flexible to treat a wide proportion of limonite to saprolite laterite ores with high Ni and Co extraction, and the CAPEX and OPEX were estimated at less than 10 USD/lbs. Ni and 2.6 USD/lbs. Ni, respectively, just accounting for 40% of the HPAL's CAPEX and 65-70% of the ATL's OPEX. The process consumes at least 290 kg acid/t ore less than the ATL and its nickel extraction reached more than 92%, at least 10% higher than ATL.

EXPERIMENTAL

Ores

Limonitic and saprolitic laterite ore samples were obtained from a nickel laterite ore body. These two types of laterite ores were independently blended, crushed, scrubbed and screened into feeds for the two leaching stages. The head grade analysis of limonite and saprolite is shown in Table 1.

Table 1 - Head grade analysis of limonite and saprolite

Ore	Content, wt.%						
	Fe	Ni	Co	Mn	Al_2O_3	MgO	SiO_2
Limonite	40.15	1.14	0.11	0.83	4.12	2.47	8.39
Saprolite	5.64	1.56	0.40	0.09	0.30	29.85	47.38

Microscope, scanning electron microscopy (SEM) and X-ray diffraction (XRD) analysis showed that the predominant mineral phase of Fe was goethite. In limonite ore about 76.32% of the nickel was associated with goethite while in saprolite ore 79.49% of nickel existed with magnesium and silicon minerals. A detailed mineralogical evaluation is provided in Table 2 for limonite ore. It focuses on the distribution of Ni and Co between the various mineral phases. A nickel deportation study was also conducted for saprolite ore. The results are shown in Table 3.

Table 2 - Chemical phases and distribution of nickel and cobalt in limonite

Mineralogy	Ni wt.%	Ratio/%	Mineralogy	Co wt.%	Ratio/%
Absorption Ni	0.007	0.61	-	-	-
Free Ni oxide	0.039	3.42	Co oxide	0.043	39.09
Ni Sulfide	0.011	0.96	Co sulfide	0.0048	4.36
Limonite	0.87	76.32	Limonite	0.015	13.64
Ni in Mn oxide	0.065	5.70	Co in Mn oxide	0.031	28.81
Ni in silicate	0.12	10.53	Co in silicate	0.0064	5.82
Others	0.028	2.46	Others	0.0098	8.91
Total	1.14	100.00	Total	0.11	100.00

Table 3 - Chemical phases and distribution of nickel in saprolite

Mineralogy	Ni wt.%	Ratio/%
Absorption Ni	0.012	0.77
Free Ni oxide	0.092	5.90
Ni Sulfide	0.038	2.44
Limonite	0.17	10.90
Ni in talc	0.008	0.51
saprolite	1.24	79.49
Total	1.56	100.00

Primary Leaching

For each test, the feed limonite ore was first slurried and homogenized to the required pulp concentration (% solids). The pulp was then heated up to the desired temperature and immediately the acid addition commenced, marking the start of leaching. Acid addition was done by pumping the required amount of acid into the glass reactor with the aid of a pre-calibrated peristaltic pump. Samples were taken every 0.5 h for analysis until the end of the allotted leaching time. The replicate runs that produced slurries for the secondary leaching stage tests were performed under the determined optimal conditions without any sampling and filtration.

Secondary Leaching

Both the fresh primary leaching slurry and the prepared secondary feed saprolite slurry were poured into the autoclave. The autoclave was then shut, sealed and activated. The temperature controller was set to heat the autoclave to the desired temperature. Sampling was done at the end of the allotted total leaching time. Each sample from all runs was filtered immediately using a Buchner funnel. The filtrate, which is the PLS, was collected and analyzed for free acid and metal content. The filter cake was then thoroughly washed with acidified water to ensure that it was free from PLS. The washed filter cake was then weighed and prepared for assay.

RESULTS AND DISCUSSION

Iron Behavior in Primary Leaching

The Effect of the Acid Dosage

Acid concentration is an essential factor for leaching, especially for iron dissolution. Its presence ensures that the conditions remain favorable for leaching reactions to proceed. In this process, sulphuric acid (H_2SO_4) was used as a lixiviant. Iron and other main metal extractions were studied at the acid dosage range 300-1200 kg/t ore when the pulp concentration was 33% solids, the leaching temperature was 95°C, the leaching time was 6 h and the stirring speed was 400-450 rpm. The results are shown in Figure 2.

It can be seen in Figure 2 that the acid dosage significantly affected the iron dissolution and extractions of metals such as Ni, Co, Mg and Mn. Increasing the acid dosage from 300-1200 kg/t ore enhanced quite significantly the Fe and Ni extractions from 21.48% and 40.55% to 86.96% and 98.21%, respectively, which were more evident than incremental extractions of other metals. That is, the acid dosage had the largest impact on Fe and Ni extractions. At a fixed acid dosage, Ni extraction was always higher than Fe extraction for the reason that Ni not only existed in goethite but also existed in the forms of absorbate, oxide, sulphide and silicate Ni, which were much easier to leach. Because more than 76% Ni was evenly distributed in goethite, Ni would be leached with Fe during the dissolution of goethite (Equations 1-2), resulting in a linear dependence between Ni and Fe extractions, as shown in Figure 3.

$$FeOOH(s)+3H^{+}=Fe^{3+}+2H_2O(1) \quad (1)$$

$$NiO(s)+2H^{+}=Ni^{2+}+H_2O(1) \quad (2)$$

Notably, the increment in Ni extraction was 4.03% when the acid dosage increased from 1000-1100 kg/t ore and 0.3% when the acid dosage increased from 1100-1200 kg/t ore. Ni extraction increased little at the acid dosage above 1100 kg/t ore and more acid led to subsequent heavy neutralization needs and high costs, thus the acid dosage was selected to be 1100 kg/t ore, where Ni and Co extractions were 97.91% and 88.23%, respectively.

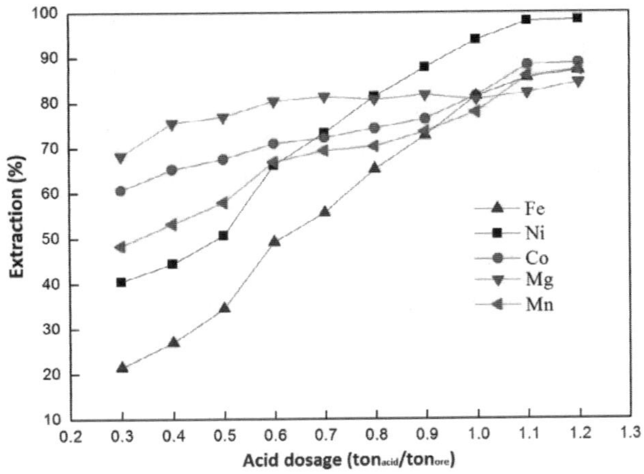

Figure 2 – Variation of iron and other metal extractions at different acid dosages in the leaching of limonite at 95°C for 6 hours

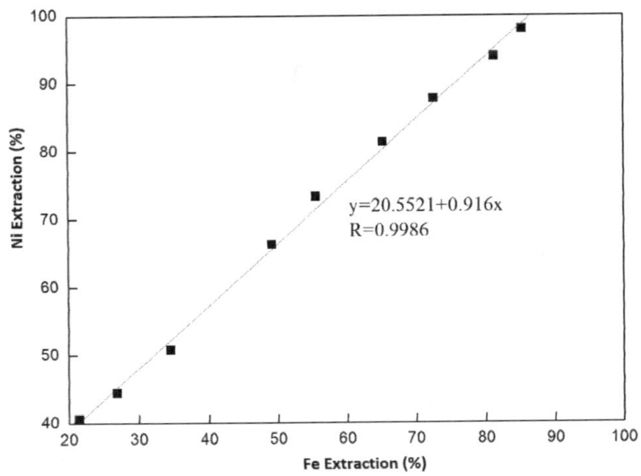

Figure 3 – Correlation between Fe extraction and Ni extraction in the leaching of limonite at 95°C for 6 hours

The Effect of the Leaching Temperature

Iron and other main metal extractions were studied at five different temperatures (75°C, 80°C, 85°C, 90°C and 95°C) when the pulp concentration was 33% solids, the acid dosage was 1100 kg/t ore, the leaching time was 6 h and the stirring speed was 400-450 rpm. The results are shown in Figure 4.

Figure 4 shows that extractions of iron, nickel, cobalt, magnesium and manganese as a function of increasing leaching temperature. The extractions of iron and nickel increased slowly, Mg extraction remained almost constant while the changes in Co and Mn extractions were relatively large. It indicated that the temperature has a greater affect on Co and Mn extractions. In order to improve nickel and cobalt extractions, accelerate the reaction rate and shorten the reaction time, 95°C was chosen as the optimum temperature, where the leaching yields of nickel and cobalt were 97.03% and 87.30%, respectively.

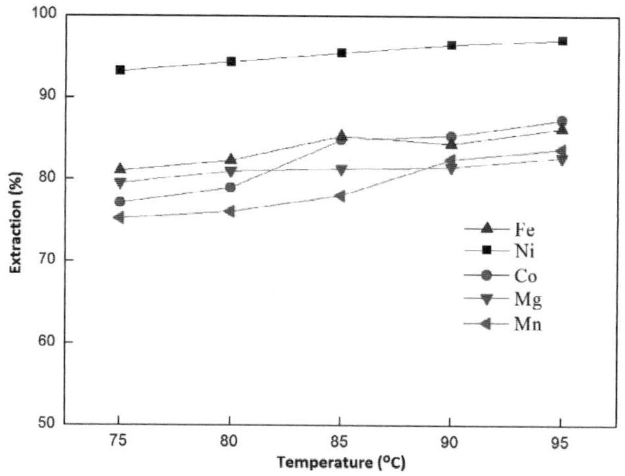

Figure 4 - Variation of iron and other metal extractions at different temperatures in the leaching of limonite with acid addition of 1100 kg/t ore for 6 hours

The Effect of the Pulp Density

The effect of varying pulp density (27%, 29%, 31%, 33% and 35% solids) on iron and other main metal extractions was studied when the leaching temperature was 95°C, the acid dosage was 1100 kg/t ore, the leaching time was 6 h and the stirring speed was 400-450 rpm. The results are shown in Figure 5.

Figure 5 shows that Fe, Ni, Co, Mg and Mn extractions increased as the pulp density increased from 27% to 35% solids, because more concentrated pulp density was accompanied by higher acid concentration, and the probability of effective collisions between minerals and acid would increase, leading to higher metal extraction in a certain reaction time. When the initial slurry concentration increased from 33% to 35%, cobalt, magnesium leaching yields decreased slightly, this is because higher pulp concentration increased the pulp viscosity, and in turn increased the loss of cobalt and magnesium into the leaching residue. However, Fe extraction increased a little bit, indicating more Fe was leached when compared to its loss in the liquid-solid separation procedure. Nickel and manganese extraction held almost constant, showing a balance between dissolution and entrainment loss. Considering maximizing cobalt and nickel extraction, reducing leaching solution volume and increasing concentration of valuable metal ions, 33% is selected as the optimal initial concentration of pulp density.

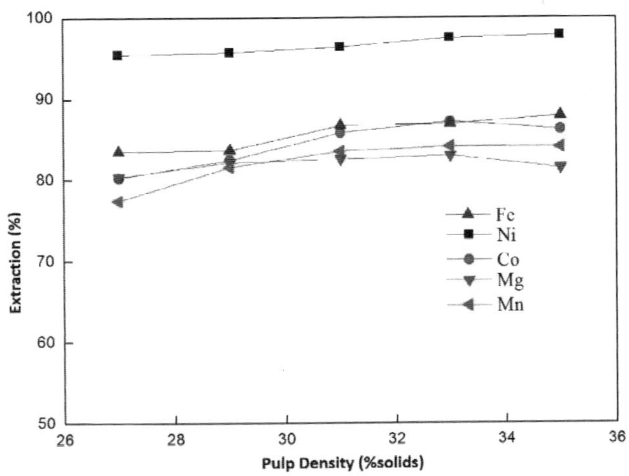

Figure 5 - Variation of iron and other metal extractions at different pulp density in the leaching of limonite with acid addition of 1100 kg/t ore at 95°C for 6 h

The Effect of the Leaching Time

The effect of varying the leaching time (2-6 h) on iron and other main metal extractions was studied when the leaching temperature was 95℃, the acid dosage was 1100 kg/t ore., the pulp density was 33% and the stirring speed was 400-450 rpm. The results are shown in Figure 6.

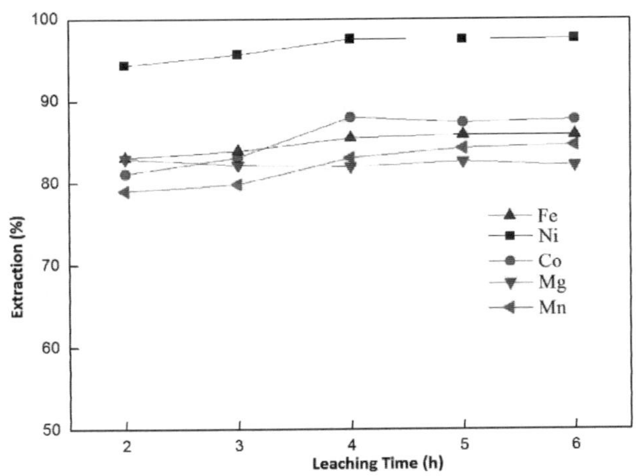

Figure 6 - Variation of iron and other metal extractions at different leaching times in the leaching of limonite with acid addition of 1.1 t/t ore at 95°C

As shown in Figure 6, when the leaching time was 2 hours, nickel extraction was 94.36%, which was acceptable and demonstrated the good reaction activity of nickel minerals and nice contacts between limonite and leaching agent. When the leaching time increased from 2 to 4 hours, nickel, cobalt, iron, and manganese extractions all increased, among them, cobalt extraction changed most distinctly, i.e. it increased from 81.12% to 87.96%. When the leaching time was further prolonged to 6 hours, no significant changes in extraction were observed. Based on these results, the optimum leaching time for the primary leaching stage was defined to be 4 hours.

Iron Behavior in Secondary Leaching

The Effect of the Leaching Temperature

Iron and other main metal extractions were studied at varying temperatures (110-160°C) when the saprolite pulp concentration was 33% solids, the saprolite addition amount was 0.9 t saprolite/t limonite, the leaching time was 3 h and the stirring speed was 600-650 rpm. The results are shown in Figure 7.

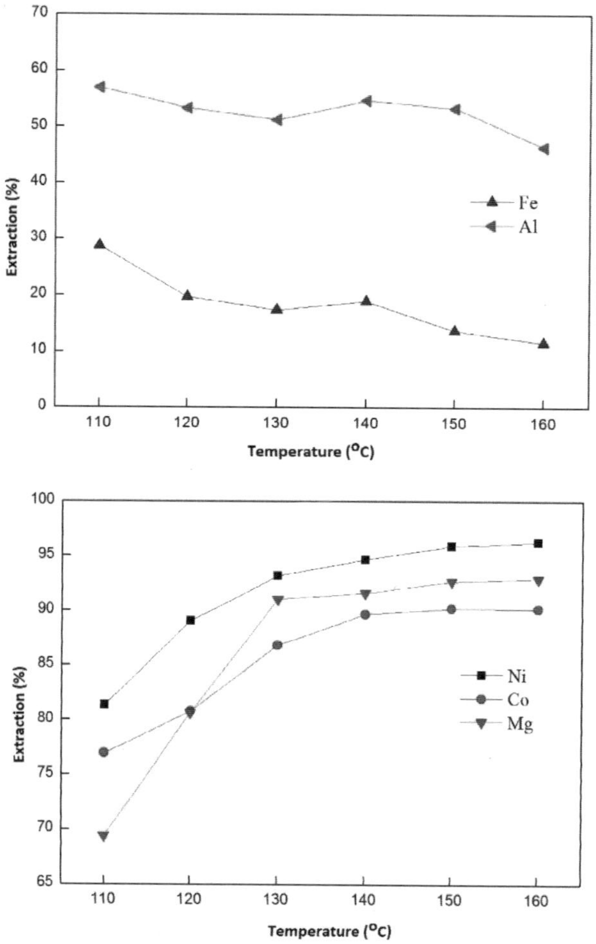

Figure 7 - Variation of iron and other metal extractions at different temperatures in the secondary leaching with a saprolite to limonite ratio of 0.9 t/t for 3 hours

From Figure 7, with the rise of temperature, nickel, cobalt and magnesium extractions were greatly increased, especially magnesium extraction. This is mainly because hydrolysis of Fe^{3+} increased with the increase of temperature and subsequently complemented free acid and speeded up the reaction, providing favourable conditions for leaching of the saprolite ores. The iron extraction was lower than 20% in the temperature range 120-160°C owing to its hydrolysis resulting from neutralization of free acid by saprolite. Iron extraction was lower than that of aluminium, indicating iron hydrolysis deposition rate was significantly higher than that of aluminium. Nickel and cobalt leaching yields reached more than 95% and 90%, respectively, at 150°C and were not significantly affected thereafter, so, 150°C was preferred as the optimum temperature for secondary autoclave leaching.

The Effect of the Saprolite to Limonite Ratio

The effect of varying the saprolite/limonite ore ratio 0.65-1.10 was studied when the leaching temperature was 150°C, the saprolite pulp concentration was 33% solids, the leaching time was 3 h and the stirring speed was 600-650 rpm. The results are shown in Figure 8.

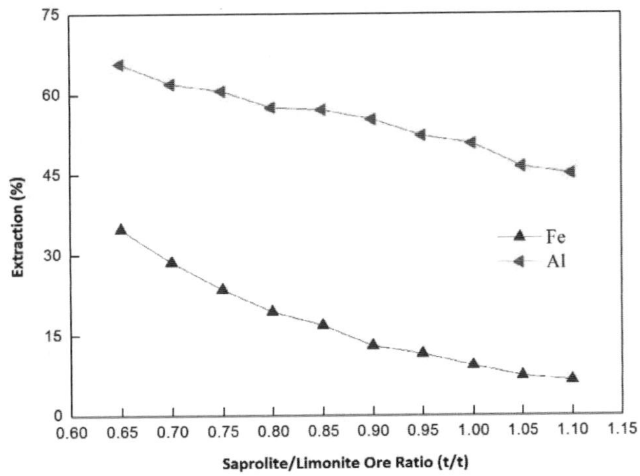

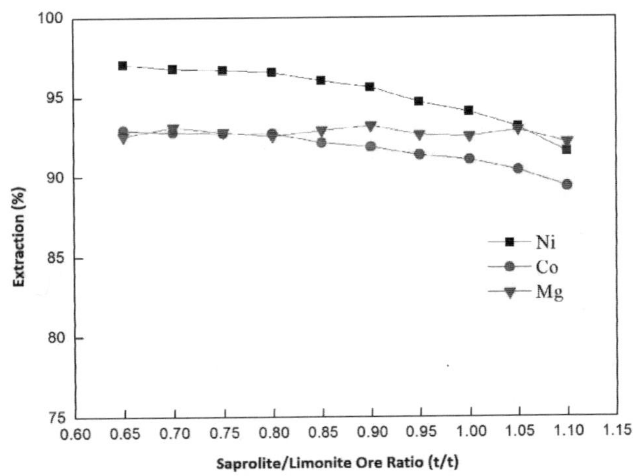

Figure 8 - Variation of iron and other metal extractions at different saprolite addition amounts in the secondary leaching at 150°C for 3 hours

Figure 8 shows that by increasing the addition amount of saprolite, the leaching yields of iron, nickel, cobalt, and aluminium declined to different extents while magnesium extraction only changed slightly. As the saprolite addition increased, the total acid/ore ratio decreased, that is, the same dosage of sulphuric acid leached more ore, thus free acid decreased, resulting in a decrease of the nickel and cobalt extractions. Nickel and cobalt extraction reached 94% and 91%, respectively, and iron extraction was

9.24% at the saprolite/limonite ore ratio 1.0 t/t. Iron extraction continued to decline while the decrease of nickel and cobalt extraction gradually expanded. Based on the results, the optimum saprolite/limonite ore ratio for the secondary leaching stage was defined to be 1.0 ton saprolite ore per ton limonite ore.

The Effect of the Leaching Time

The effect of varying leaching time (0.5-3 hours) on iron and other main metal extractions was studied when the leaching temperature was 150°C, the saprolite pulp concentration was 33% solids, saprolite/limonite ore ratio was 1.0 t/t and the stirring speed was 600-650 rpm. The results are shown in Figure 9.

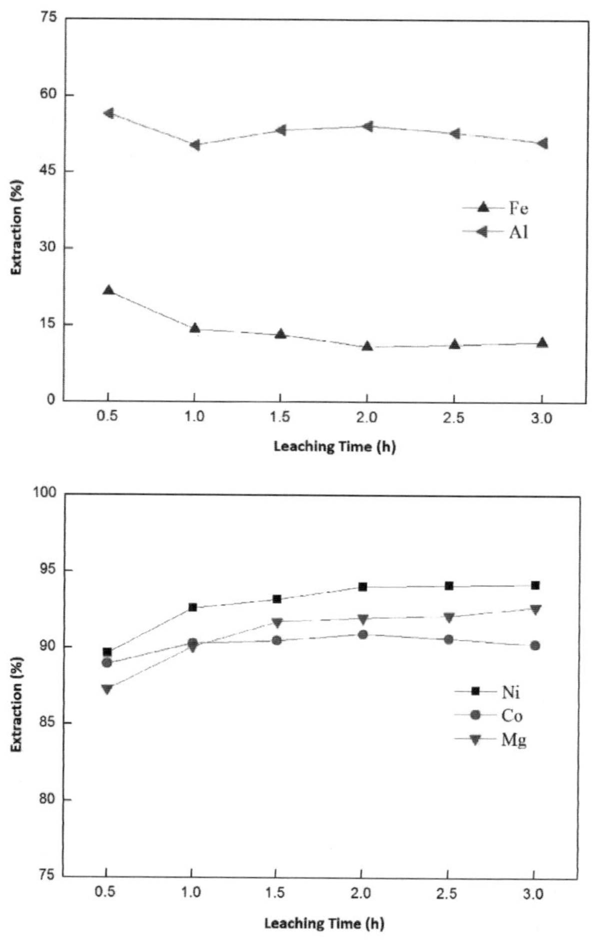

Figure 9 - Variation of iron and other metal extractions at different leaching time in the secondary leaching with a saprolite to limonite ratio of 1.0 t/t at 150°C

It can be seen from Figure 9 that iron extraction decreased to a minimum after 2 hours, and then remained constant. Minimum aluminum extraction was obtained after 1 hour, when the precipitation extents of iron and aluminum were 81.05% and 17.71%, respectively, again illustrating that the deposition rate of iron was significantly higher than that of aluminum. More than 80% nickel and cobalt was extracted from saprolite after 0.5 h leaching at 150°C, indicating a fast leaching reaction rate and a high reaction activity of minerals. Ni and Co extraction little changed when leaching time was more than 2 hours and could be improved through increasing leaching temperature or saprolite addition amount, namely, increase of the total acid/ore ratio. Based on these results, the optimum leaching time for the secondary leaching stage was defined to be 2 hours.

Inverse Leaching Comprehensive Tests and Residue Analysis

The results of these comprehensive tests indicated that the inverse leaching achieved an effective leaching yield of Ni and Co. The leaching conditions and results are listed in Table 4.

Table 4 - The inverse leaching conditions and results

Conditions	Results
Primary Leaching Stage: Feed ore: Limonite Acid dosage: 1100 t/t Initial pulp density: 33% solids Temperature: 95°C Leaching time: 4 h Stirring speed: 400-500 rpm	Primary Stage Extraction (%): Ni 97.34, Co 88.36, Fe 85.58, Mg 82.43, Mn 84.18, Al 65.43 Solution (g/L): Ni 4.50, Co 0.078, Fe 139.21 Mg 4.94, Mn 2.83, Al 5.78
Secondary Leaching Stage: Feed ore: Saprolite Saprolite addition amount: 1.0 t/t Saprolite pulp density: 33%solids Temperature: 150°C Leaching time: 2 h Stirring speed: 600-650 rpm	Total Acid consumption (kg/t): 550 Total extraction (%):Ni 94.18, Co 90.75, Fe 10.83, Mg 93.31, Mn 84.93, Al 58.64 Residue (wt.%): Ni 0.12, Co 0.013, Fe 28.55， Mg 0.91, Mn 0.097, Al 0.68 Solution (g/L): Ni 5.69, Co 0.39, Fe 7.62, Mg 40.55, Mn 1.76, Al 3.08

Ni and Co extractions were as high as 94.18% and 90.75%, respectively, while only 10.83% Fe was leached into the leaching solution. Ni, Co and Fe contents in the leaching residue were 0.12, 0.013 and 28.55 wt.%, respectively. Iron concentration in the PLS was less than 8 g/L. The acid consumption was 550 kg per ton ore, realizing reduction of acid consumption through control and precipitation of iron. Scanning electron microscope (SEM) and XRD analysis (Figure 10) of the leach residue showed that Ni and Co were almost completely leached and iron was precipitated into the residue in the form of hematite according to Equation 3. From Equation 3, the precipitation of ferric ions was accomplished and simultaneously the acid was generated to leach more laterite ore.

$$2Fe^{3+} + 3H_2O(1) = Fe_2O_3(s) + 6H^+ \tag{3}$$

From the primary and secondary leaching, the following results could be achieved.

- In the primary atmospheric leaching stage most of limonite was decomposed with Fe and Ni dissolving. The leaching extent of Fe and Ni were more than 85% and 97%, respectively.
- In the secondary pressure leaching stage the dissolution of Ni, Co, and Mg was done in an autoclave. With the decrease of the sulphuric acid concentration in solution iron began to precipitate as hematite above 150°C. Due to the higher reactivity of saprolite the extraction of Ni and Co was high at relatively lower leach temperature, compared to HPAL. In this combined process most of iron doesn't consume acid finally. Compared with traditional atmospheric leaching the acid consumption decreased from 850 kg/t ore to 550 kg/t ore.

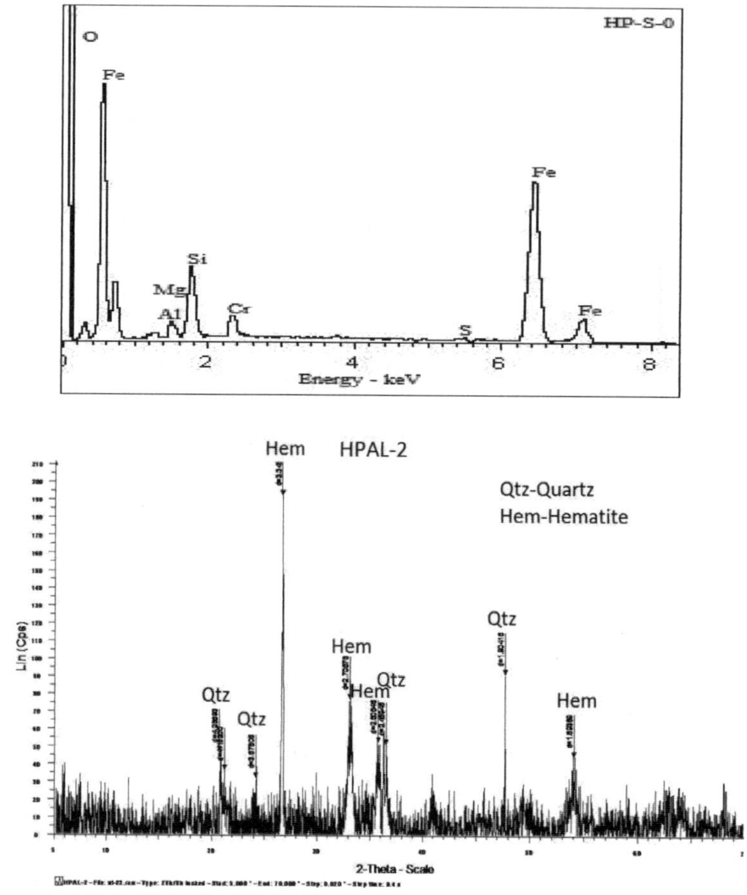

Figure 10 – The X-ray energy spectrum (top) and X- ray diffraction pattern (bottom) of the iron residue

CONCLUSIONS

The iron behavior in the inverse leaching process of limonite and saprolite laterite ores was unveiled. Ni was evenly distributed in goethite in limonite ores and leached with Fe during the atmospheric dissolution of goethite. Free acid from the atmospheric leach as well as acid generated from iron precipitation under mild autoclave conditions compared to HPAL was sufficient to dissolve saprolite. The precipitation of ferric ions was accomplished during neutralization by saprolite and iron was precipitated into the residue in the form of hematite.

Based on the control and stabilization of iron, an inverse leaching technology was developed. Limonite with high grade iron and low grade Ni and Mg was leached under atmospheric conditions (primary leaching) while saprolite with high grade Ni and Mg was leached in the autoclave (secondary leaching) with the limonite leached slurry. Ni and Co extractions were as high as 94.18% and 90.75%, respectively, while only 10.83% Fe was leached into the leaching solution. Compared to HPAL the leaching temperature was reduced to 150-170°C from 250-270°C. Compared to ATL the acid consumption was reduced to 550-650 kg/t ore from 850-950 kg/t ore.

ACKNOWLEDGEMENTS

This work was financially supported by the National Natural Science Foundation of China (Key Project No. 51434001).

REFERENCES

Georgiou, D. , & Papangelakis, V. G. (1998). Sulphuric acid pressure leaching of a limonitic laterite: Chemistry and kinetics. *Hydrometallurgy*, *49*(1-2), 23-46. doi: 10.1016/S0304-386X(98)00023-1

Johnson, J. A., Cashmore, B. C., & Hockridge, R. J. (2005). Optimisation of nickel extraction from laterite ores by high pressure acid leaching with addition of sodium sulphate. *Minerals Engineering*, *18*(13-14), 1297-1303. doi:10.1016/j.mineng.2005.05.013

Kaya, Ş., & Topkaya, Y. A. (2011). High pressure acid leaching of a refractory lateritic nickel ore. *Minerals Engineering*, *24*(11), 1188-1197. doi:10.1016/j.mineng.2011.05.004

Luo, W., Feng, Q., Ou, L., Zhang, G., & Chen, Y. (2010). Kinetics of saprolitic laterite leaching by sulphuric acid at atmospheric pressure. *Minerals Engineering*, *23*(6), 458-462. doi:10.1016/j.mineng.2011.05.004

MacCarthy, J., Addai-Mensah, J., & Nosrati, A. (2014). Atmospheric acid leaching of siliceous goethitic Ni laterite ore: Effect of solid loading and temperature. *Minerals Engineering*, *69*(8), 154-164. doi:10.1016/j.mineng.2014.08.005

McDonald, R. G., & Whittington, B. I. (2008). Atmospheric acid leaching of nickel laterites review: Part I. Sulphuric acid technologies. *Hydrometallurgy*, *91*(11), 35-55. doi:10.1016/j.hydromet.2007.11.009

Önal, M. A. R., & Topkaya, Y. A. (2014). Pressure acid leaching of Çaldağ lateritic nickel ore: An alternative to heap leaching. *Hydrometallurgy*, *142*(11), 98-107. doi:10.1016/j.hydromet.2013.11.011

Rubisov, D. H., & Papangelakis, V. G. (2000). Sulphuric acid pressure leaching of laterites — a comprehensive model of a continuous autoclave. *Hydrometallurgy*, *58*(2), 89-101. doi:10.1016/S0304-386X(00)00092-X

Rubisov, D. H., & Papangelakis, V. G. (2000). Sulphuric acid pressure leaching of laterites — speciation and prediction of metal solubilities "at temperature". *Hydrometallurgy*, *58*(1), 13-26. doi:10.1016/S0304-386X(00)00093-1

Thubakgale, C. K., Mbaya, R. K. K., & Kabongo, K. (2013). A study of atmospheric acid leaching of a South African nickel laterite. *Minerals Engineering*, *54*(4), 79-81. doi:10.1016/j.mineng.2013.04.006

Whittington, B. I., Johnson, J. A., Quan, L. P. McDonald, R. G., & Muir, D. M. (2003). Pressure acid leaching of arid-region nickel laterite ore: Part II. Effect of ore type. *Hydrometallurgy*, *70*(1-3), 47-62. doi:10.1016/S0304-386X(03)00044-6

Arsenic Management and the Role of Iron

ARSENIC REMOVAL BY CONTROLLED BIOLOGICAL IRON OXIDATION REACTIONS

*Paula Gonzalez Contreras[1], Henk Dijkman[1], Silvia Vega[2], Irene Sánchez-Andrea[3], Jan Weijma[2], Cees Buisman[2]

[1] PAQUES B.V.
T de Boerstraat 2
Balk, The Netherlands
*(*Corresponding author: p.gonzalezcontreras@paques.nl)*

[2] Sub Department of Environmental Technology
[3]Laboratory of Microbiology
Wageningen University
The Netherlands

ABSTRACT

More than 20 years ago, Paques B.V. introduced innovative biotechnologies to recover metals and to remove sulfate from aqueous streams. These technologies find their origin in the exploration of microorganisms involved in the global sulfur cycle. Currently, several sulfur cycle biotechnologies are applied successfully at full-scale. The sulfur cycle is closely linked with the iron cycle, and also the latter offers opportunities for application of innovative biotechnology for the mining industry. Microorganisms of the natural iron cycle carry out reactions that are not feasible by chemical methods such as ferrous iron oxidation with oxygen at pH below 4. Iron oxidation with oxygen can be conducted using microorganisms living at pH between 0.5 and 7 and at temperatures between 0 and 95°C. Remarkably these microorganisms can also carry out arsenite oxidation at similar acidic conditions and high temperatures. Making use of the extreme features of these microorganisms, Paques and Wageningen University have developed a biological process to precipitate arsenic as scorodite. This biological formation of scorodite is a novel combination of biological oxidation and biocrystallization. Depending on the level of saturation, biological oxidation rates and operational conditions, we could control the formation of the iron precipitates such as jarosite and scorodite. Currently new microorganisms have been harvested from hot springs and rock acid mine drainage to foster the growth of specialized microbial communities with potential high iron and arsenic oxidation capacities and higher resistance to other metals. In our paper we address the ongoing research and development of the bioscorodite process.

KEYWORDS

Biological oxidation, iron biological conversions, arsenic precipitation, jarosite, scorodite

THE USE OF IRON TO REMOVE ARSENIC

In 2015 several environmental disasters related to the release of mine wastewater and tailings occurred. On August 5th, a waste water spill occurred from the Gold King Mine near Silverton, Colorado. A dam holding the tailing pond was damaged and spilled three million US gallons (11 ML) of mine wastewater and tailings, including elements such as cadmium, lead and arsenic, into Cement Creek, a tributary of the Animas River in Colorado. "At their peak, arsenic levels were 300 times the normal level and lead was 3,500 times the normal level" (The Guardian, 2015). In early November, another collapse of a mining dam in the Brazilian state of Minas Gerais spilled between ten and sixteen thousand US gallons of water and sediment after a dam failed from an iron ore mine. Arsenic and mercury polluted the Doce River, arsenic levels were found more than ten times above the legal limit in some places of the river (The Guardian, 2015). Waste production is inevitable during Mining. However, a preventive waste management strategy could have minimised waste production, maximised waste reuse and classified waste by their hazardous degree.

Arsenic is present as a dissolved element in water or in acid bleeds together with other elements such as copper, which are valuable. In the current practise, it is also stored as arsenic trioxide powder, a very soluble and toxic form of arsenic (Edelstein, 2013). Arsenic trioxide powder is being stored until suitable technologies for its safe disposal are developed. Unfortunately, large amounts of arsenic cannot be reused and therefore the best option is to immobilise arsenic in its most stable solid phase to minimise contamination of other waste water or solids.

In the past century, calcium arsenate precipitation was the most used method to remove arsenic. Currently, calcium arsenate precipitation is discontinued because it has a high instability. Calcium arsenate sludge can leach up to 900-4,400 mg As L^{-1} under the TCLP test (maximum allowance test 5 mg As L^{-1}) (Henke, 2009). A better alternative to calcium precipitation is arsenic precipitation in ferrihydrite. Arsenical ferrihydrite precipitation has been chosen by the US EPA as the best demonstrated available technology (BDAT). Ferrihydrite has been widely used despite of the high iron consumption and neutralizing agent associated to its precipitation. Arsenical ferrihydrite precipitation requires high iron consumption with respect to arsenic, i.e. Fe/As higher than 4 and therefore large amounts of waste material are produced (Henke, 2009). Riveros *et al.* (2001) describe examples of the use of arsenical ferrihydrite in the metallurgical industry denoting high iron consumption. Not only Fe/As molar ratio higher than 4 are used for arsenical ferrihydrite precipitation, but also Fe/As molar ratio higher than 10 are used. The excessive use of iron leads to high sludge production. For example, around 40 ton per day of residue is produced by arsenical ferrihydrite precipitation, which only contains between 300-500 kg As per day (Riveros, 2001).

Arsenical ferrihydrite precipitates contain around 0.5-2 wt% of arsenic with a maximum solid content of 20-25 wt% (Henke 2009). The disposal and storage of these compounds is not entirely safe as they easily undergo physical and chemical changes with time, resulting in arsenic releases into the environment. Arsenic adsorption to ferrihydrite can also be affected by ions as phosphate, sulphate, carbonate, and dissolved organic species, affecting the removal of arsenic and the relative long term stability of ferrihydrite (Twidwell, 2011). Some authors suggest that arsenical ferrihydrite can be considered stable when an Fe/As molar ratio higher than 5 is provided, pH is kept slightly acidic and the contact with reducing substances such as reactive sulphides or reducing conditions is avoided (Riveros, 2001).

LOOKING FOR A SUITABLE ARSENIC STORAGE MEDIUM

Scorodite mineral ($FeAsO_4 \cdot 2H_2O$) is recognized by researchers and mining companies as an alternative to stabilize arsenic for long term storage (Figure 1) (Choi *et al.* 2014; Demopoulos, G. 2014; Radzinski *et al*, 2014). This is mainly because its low solubility, high arsenic content (30 wt%), high stability and compactness. Hydrothermal scorodite crystallisation employs acid pressure oxidation with temperatures above 150°C and oxygen overpressures up to 2000 kPa. Initially, hydrothermal scorodite was produced at pH 0.7 in ferric nitric medium and temperatures of 125°C (ideally 160°C) to ensure good crystallinity. Scorodite crystals have also been precipitated at atmospheric temperatures (~90°C) by chemical precipitation and gypsum addition, which has been extensively studied by Demopoulos and co-workers at McGill University (Filippou & Demopoulos, 1997). Their method based on the step wise neutralisation of saturated solution enabled the low temperature scorodite precipitation. Scorodite can also be precipitated from arsenite/ferric iron by controlling the oxidation of arsenic by peroxide or pure oxygen as demonstrated by Fujita et al (2008, 2009). This method is known today as the Dowa scorodite process.

Figure 1 - Mineral scorodite specimen from Hermerdon Bal Mine (Devon, England) (Obtained from a mineral supplier)

The high cost associated to chemical oxidation and high temperature precipitation processes have delayed the introduction in the market of scorodite technologies. Many companies have chosen for the arsenic precipitation at ambient temperature forming amorphous iron arsenate precipitates. The different arsenic precipitates formed and their stability are object of study and concern. A recent study comparing stability of scorodite with other arsenic containing phases was published by Madhumita *et al* (2013). In that study the authors compared leaching properties of scorodite to those of hydroxylapatites ($Ca_5(PO_4)_3(OH)$), ferrous arsenate (symplesite-type minerals - $Fe(AsO_4)_2 \cdot 8(H_2O)$), tooeleite ($Fe_6(AsO_3)_4(SO_4)(OH)_4 \cdot 4H_2O$) and arsenated-schwertmannite ($Fe_{16}O_{16}(OH)_{12}(SO_4)_2$). Scorodite and schwertmannite were the most stable arsenic bearing materials under aerobic leaching conditions. Under anoxic conditions the leaching of schwertmannite dramatically increased to 10 mg/L As.

Many researches are currently exploring the formation of scorodite. Emphasis is made on the characterisation and long-term stability of the "scorodite like" formed phases. Experts on scorodite precipitation have repeatedly highlighted in conferences that "not all scorodites are equal due to their differences in crystallinity". Their voices should be heard when developing a technology to produce scorodite, as the producing method could have significant effects on the stability of the formed scorodite. It has been determined, however, that scorodite releases arsenic in significant concentrations under alkaline pH or under anoxic conditions. Recently, encapsulation methods based on hydroxyapatite and fluoroapatite phases have been developed to protect scorodite against pH changes or exposure to reduced conditions at the disposal site (Katsarou, 2011; Demopoulos, 2014).

As an alternative to arsenic-iron precipitation, arsenic-sulfide can be also precipitated. Paques has demonstrated the biological production of H_2S for the purpose of copper recovery at a very large industrial scale using its THIOTEQ™ technology at the Pueblo Viejo gold mine located in the Dominican Republic (Figure 2). This plant both minimizes the amount of copper sent to the tailings pond and creates additional revenue. The required amount of H_2S is produced in a high rate bioreactor in which elemental sulfur is reduced by bacteria and a 10% H_2S containing gas stream is produced. The bioreactor can produce up to

20 ton H_2S per day for the recovery of copper. The copper concéntrate produced (covellite) has up to 60% copper content and is sold to a copper smelter (Mining Int., 2016). The same THIOTEQ™ technology can be used for arsenic sulfide precipitation. When arsenic is present in an acid bleed, arsenic oxide reacts with sulphide to form orpiment (As_2S_3), which has a very low solubility under anoxic conditions. Arsenate (AsO_4^{3-}) is first reduced by sulphide to arsenite (AsO_3^{3-}) and subsequently crystallised as orpiment at acidic conditions. Advantages of orpiment are its compact volume, its high arsenic content with 61% mass arsenic and no crystal water (Henke, 2009). Despite the low solubility of orpiment at anoxic conditions, it can release arsenite (As (III)) into solution if stored under aerobic conditions. Orpiment oxidation tends to increase at pH>8 and in the presence of oxidizing reagents such as manganese or ferric iron (Henke, 2009). Besides, under low oxygen conditions, carbonates may also dissolve orpiment. Therefore orpiment should be deposited in a (semi) closed landfill to avoid infiltration of oxygen and or carbonate rich water, or it should be converted to a more stable crystals such as scorodite.

Figure 2 - The largest biological H_2S "THIOTEQ™" is in operation at the new Pueblo Viejo mine at the Dominican Republic. This biological reactor has a design capacity of 20 tons H_2S per day

Although arsenic sulfide precipitation is technological feasible with Paques biosulfide reactors, Paques' ambitions are that all arsenic waste is immobilised into scorodite for long-term disposal. Paques has patented a bioprocess to crystallise arsenic into scorodite: the THIOTEQ™ scorodite process, also named the bioscorodite process. The bioscorodite process was developed in cooperation with Wageningen University. This process is based on biological iron and arsenic oxidation (Gonzalez *et al* 2010). Currently, Paques is enhancing the process by increasing the rates of the biological arsenite oxidation.

USING MICROORGANISMS FROM NATURE FOR NEW TECHNOLOGY DEVELOPMENTS

While microorganisms involved in the biological sulfur- and iron- cycles are responsible for the formation of acid mine waters, they can also offer a solution to the problem. When metal ores are exposed to both air and water, acidophilic iron- and sulfur-oxidizing bacteria and archaea transform insoluble metal sulfides to soluble metal sulfates with co-production of sulfuric acid. This results in strong acidification of the water body (Acid Mine Drainage). The low pH promotes the activities of these acidophiles and also increases the solubility of most (cationic) metals that occur as sulfide minerals. However, another group of bacteria which uses sulfate (rather than oxygen) as electron acceptor, the "sulfate-reducing bacteria" or SRB, can essentially reverse this process, regenerating metal sulfide minerals. Moreover, SRB as well as other types of microbes can directly reduce metals such as uranium, chromium, selenate or tellurate to insoluble forms. When acidophilic iron- and sulfur-oxidizing bacteria are used for the purpose of leaching, this process is known as "bioleaching". These microorganisms are characterised by their oxidative action on ferrous ions and/or forms of reduced sulphur species. Mesophile (40-45°C) and acidophile (pH 1.2-1.8) bioleaching bacteria have been used commercially in the BIOX® leaching plants (van Aswegen *et al*, 2006). Thermophilic bioleaching has also acquired a prominent role in the processing of refractory copper sulphides such as chalcopyrite ($CuFeS_2$). It has been demonstrated that bioleaching of copper from chalcopyrite concentrate is increased to 90% under thermophilic (60°C) compared to 12% under mesophilic (22°C) conditions (Plessis et al, 2006). Bioleaching processes are mainly profitable when dealing with low grade sulfide ores containing valuable metals at concentrations below 0.5% (wt/wt). The smelting and extraction of low grade sulfide ores often becomes non profitable because of their high gangue/metal ratio. Partial oxidation of sulfide minerals can be achieved by careful control of the bacterial oxidation process. The same is difficult to achieve in roasting and aqueous pressure oxidation. As a result, the bacterial process produces less acid. This will result in lower neutralisation requirements, contributing to reduce capital and operating costs. More information about biomining technologies can be found in Johnson (2014).

THE BIOSCORODITE EXTREME PERFORMERS

The microorganisms used in the biocrystallisation of scorodire are extremo-acidophiles, i.e. microorganisms living at pH between 0.5 and 4 and at temperatures between 20 and 90°C. These microorganisms have the largest potential to carry out the arsenic oxidation. Acidophiles can be found in mine drainage water or soils of low pH and also in thermal vents or hot springs (Konhauser, K., 2006).

These acidophilic microorganisms can obtain energy either from iron or sulphur oxidation. The carbon source is captured from the CO_2 present in the air and therefore addition of organic matter is not needed. When elemental sulfur is supplied as an energy source, the acidophiles oxidize sulfur to sulfate and protons, leading to sulfuric acid formation and therefore alkaline reagents are needed to neutralize the produced protons (Equation 1).

$$2S^0 + 3O_2 + 2H_2O \rightarrow 2H_2SO_4 \quad (1)$$

Iron as energy source for acidophiles leads in general to consumption of protons, minimizing the addition of neutralising reagents into the reactor. Biological oxidation equations for three different iron sources at acidophilic conditions (pH<4) are displayed in Equation 2, 3 and 4.

$$Fe^{2+} + \frac{1}{4}O_2 + H^+ \leftrightarrow Fe^{3+} + \frac{1}{2} H_2O \quad (2)$$

$$FeCO_3(s) + H^+ \leftrightarrow Fe^{2+} + HCO_3^- \quad (3)$$

$$Fe^0 + H_2O + \frac{1}{2}O_2 \leftrightarrow Fe^{2+} + 2OH^- \quad (4)$$

On the other hand, if the pH rises above 4, iron hydrolysis is expected to occur releasing protons into solution (Equation 5).

$$Fe^{3+} + 3H_2O \leftrightarrow Fe(OH)_{3(S)} + 3H^+ \quad (5)$$

Therefore, the iron source and the pH control are essential to form scorodite crystals. These are also important factors that govern the growth of the microorganisms.

Biological arsenite oxidation can reduce the cost related to oxidant reagent consumption. Biological oxidation mediated by microorganisms consumes four times less oxygen (from air) than chemical oxidation with peroxide (Equation 6).

$$H_3AsO_3^0 + \frac{1}{2} O_2 \leftrightarrow \frac{1}{2} H_2AsO_4^- + \frac{1}{2} HAsO_4^{2-} + 1.5H^+ \quad (6)$$

The arsenic oxidation also influences the saturation level and the quality of the crystals.

The THIOTEQ™ scorodite process has been tested at laboratory scale for simultaneous biological iron and arsenite oxidation at 70°C and pH 1.2 (Dinkla *et al*, 2013; Gonzalez *et al*, 2012; Gonzalez, 2014). Initially, the measured arsenite oxidation rates at 70°C were five times lower than the concomitant iron oxidation rates and therefore only diluted arsenite streams (< 1 g As^{3+}/l) could be treated.

In the biocrystallisation of scorodite, microorganisms carry out the oxidation of iron and arsenic using air. The saturation of the solution is controlled on microscale by controlling the rate of biological oxidation. To balance the rates of biological oxidation and crystallization, the parameters that affect these rates are of importance. The biological oxidation rate is influenced by the type of microorganism used, the retention of microorganisms in the reactor, metal concentrations, oxygen supply and environmental conditions such as temperature and pH. The crystallization rate is mainly influenced by the temperature, crystal (seeds) concentration in the reactor and the supersaturation level.

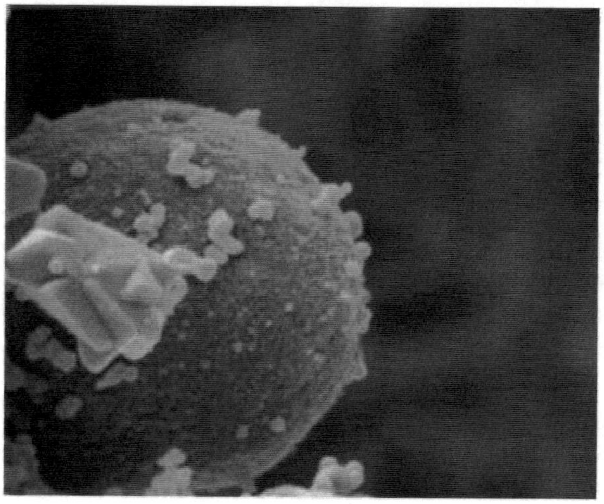

Figure 3 - Indirect biocrystallisation of bioscorodite on the surface of a microorganism

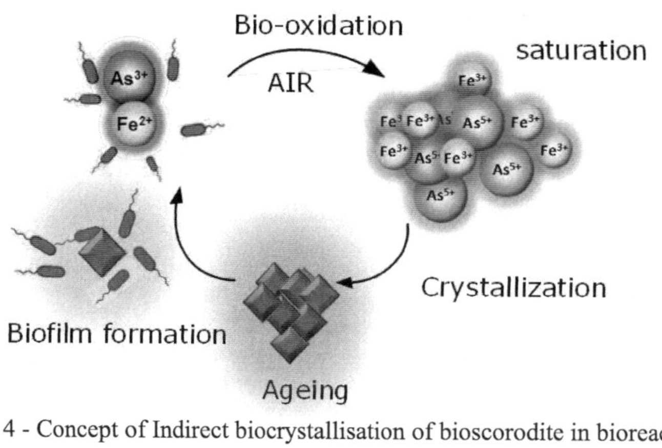

Figure 4 - Concept of Indirect biocrystallisation of bioscorodite in bioreactors

Figure 5 - Characteristic colours obtained in the different arsenic containing biocrystals at different temperatures

FINDING SUITABLE EXTREMOPHILES FOR NEW APPLICATIONS: BACK TO NATURE

In the same way that extremophilic microorganisms have been harvested for novel sulfate and sulfur reduction novel processes (Sanchez *et al.*, 2015), extremophilic microorganisms with potential arsenic resistance and arsenic oxidation capacities are currently being studied by Wageningen University and Paques.

For this purpose, extreme environments such as hot springs, volcanoes and rocks acid mine drainage were recognized as a potential source to foster the growth of thermoacidophilic microbial communities with potential capabilities to oxidise arsenite (As (III)). The first sampling was performed at El Tatio Geothermal field, located in the northern region of the arid Atacama Desert (Chili) at 4,300m of altitude (Figure 6). El Tatio is a natural habit of thermophilic microorganisms such as archaea and mat-forming microbial communities associated to the arsenic geochemical cycle. Furthermore, this geothermal is recognized as the principal natural source of arsenic to the Loa River, the main surface water resource in Atacama Desert. The geothermal has an average temperature above 50°C, high UV light radiation, a high concentration of metals and metalloids. The presence of arsenic in this geothermal has been confirmed by on-field measurements. The samples were taken from a maximum depth of 100 cm at the spring area of the middle basin called "pozo naranja", the orange pit. At the moment of sampling, a temperature between 55° and 60°C and redox values between 35 and 57 mV were measured. Samples were enriched on site and adapted to arsenic by batch cultivation at a local laboratory before they were transported to Wageningen University. The enrichment experiments yielded a population of a mixed culture of 10^6 cell/ml growing in the presence of 1 g/L As(V) and 0.5 g/L As(III).

Figure 6 - Map of the Sampling site and landscape of the Tatio geothermal basin and Lastarria Volcano. At the right picture, Silvia Vega (Wageningen University) is collecting one of the samples

Samples were also harvested from the Lastarria volcano, which is located along the border of Argentina and Chile. The volcano is part of the Lazufre volcanic system and is noted for the presence of molten sulfur lava flows. The isolation temperature of these samples was 100°C and pH of 1.2. The presence of arsenic was confirmed in these samples as well.

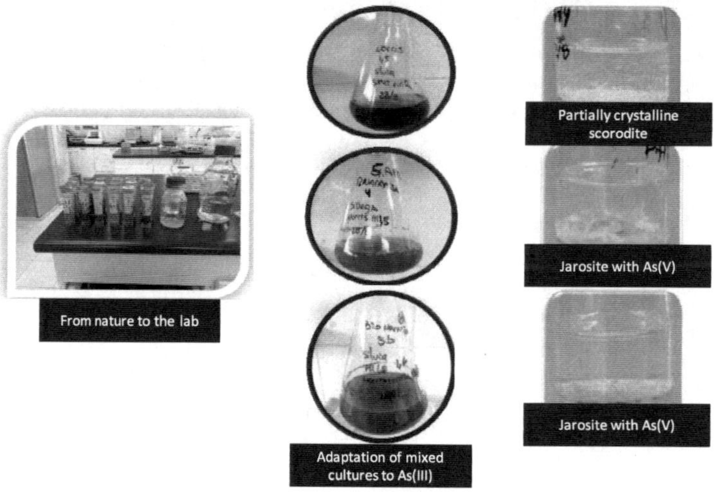

Figure 7 - Enrichment of the samples taken from Tatio Geothermal basin and first test to produce scorodite from biologically oxidised arsenic

EXPECTED RESULTS

Thermoacidophilic archaea such as *Sulfolobus* and *Acidianus* species have evolved genetic components to tolerate high levels of arsenic and conduct As(III) oxidation without losing their capability to oxidize ferrous iron. Based on the use of Paques' experience in developing biological processes, mixed cultures harvested from nature is more likely to lead to a successful technology than using a pure culture from a laboratory as the source of inoculum. The samples harvested from natural sources as described above are currently grown in the presence of high arsenic (III) concentrations (above 1 g/L As (III)). If biological arsenic and iron oxidation take place simultaneously at high rates, it is expected that the bioscorodite process can be applied to treat bleed streams containing As(III) and immobilise arsenic as scorodite in a single step process.

REFERENCES

Choi, Y., Ghahremaninezhad, A., & Ahern, N. (2014). Process for activated carbon-assisted oxidation of arsenic species in process solutions and waste waters. *COM 2014 - Conference of Metallurgists Proceedings*. Vancouver: Canadian Institute of Mining, Metallurgy and Petroleum.

Dinkla, I., Gonzalez-Contreras, P., Sekhar, C., Sandstrom, A., Weijma, J., & Buisman,C.J.N. (2012b). Quantifying microorganisms during biooxidation of arsenite and bioleaching of zinc sulfide. *Minerals Engineering*, 48, pp. 25-30.

Edelstein, D.L. (2013). Arsenic in mineral commodity summaries 2013. U.S. Geological Survey.

Demopoulos, G. (2014). Arsenic Immobilization Research Advances: Past, Present and Future. *COM 2014 - Conference of Metallurgists Proceedings*. Vancouver: Canadian Institute of Mining, Metallurgy and Petroleum.

Filippou, D. & Demopoulos, G. (1997). Arsenic immobilisation by controlled scorodite precipitation, *JOM,* 49, nr. 12, pp. 52-58.

Fujita, T., Taguchi, R., Abumiya, M., Matsumoto, M., Shibata, E., Nakamura, T., 2008. Novel atmospheric scorodite synthesis by oxidation of ferrous sulfate solution Part I. *Hydrometallurgy*, 90, (2-4), 92-102.

Fujita, T., R. Taguchi, E. Shibata, T. Nakamura, Preparation of an As(V) solution for scorodite synthesis and a proposal for an integrated As fixation process in a Zn refinery, *Hydrometallurgy*, Volume 96, Issue 4, May 2009, Pages 300-312.

Gonzalez-Contreras, P., Weijma, J., van der Weijden, R., & Buisman,C.J.N. (2010). Biogenic scorodite crystallization by *Acidianus sulfidivorans* for arsenic removal. *Environmental Science & Technology*, 44 (2), pp. 675-680.

Gonzalez-Contreras, P., Weijma, J., & Buisman,C.J.N. (2012). Continuous bioscorodite crystallization in CSTRs for arsenic removal and disposal. *Water Research,* 46 (18), pp. 5883-5892.

Gonzalez, P. (2014). The biological method to immobilise arsenic bearing waste as scorodite. *COM 2014 - Conference of Metallurgists Proceedings*. Vancouver: Canadian Institute of Mining, Metallurgy and Petroleum.

International Mining IM (2016). Copper recovery in: *Water management section*. Available in February number. Team Publishing Ltd.

Henke, K (2009). Arsenic in natural environments. In Henke, K. (Eds), *Arsenic Environmental Chemistry, Health Threats and Waste Treatment* (pp. 69-235). Wiley: United Kingdom).

Johnson, B. (2014). Biomining biotechnologies for extracting and recovering metals from ores and waste materials. *Current Opinion in Biotechnology*, 30, pp. 24-31.

Katsarou, L. (2011). Synthesis and applcation of hydroxyapatite and fluoroapatite to scorodite encapsulation. *Deparment of Mining and Material Engineering. McGill University (Montreal, Canada).*

Konhauser, K. (2006). Microbial weathering. In Wiley-Blackwell Publishing, *Introduction to geomicrobiology* (pp. 192-234). Maldon, Oxford, Carlton.

Madhumitha, R., Shan, J., Sáez, A.E., Wendell P. E. (2013). Scoping candidate minerals for stabilization of arsenic-bearing solid residuals. *Journal of Hazardous Materials,* 263, pp.525– 532.

Plessis, C., Batty, J, & Dew, B. (2006). Comercial Applications of Thermophile Bioleaching. In D. Rawlings & B.Johnson (Eds.), *Biomining* (pp. 57-80). Springer: Verlag Berlin.

Radzinski, R. & Ghahremaninezhad, A. (2015). Activated carbon assisted oxidation of arsenic species in process solutions and wastewasters: the oxidation reaction mechanism. *COM 2015 - Conference of Metallurgists Proceedings*. Toronto: Canadian Institute of Mining, Metallurgy and Petroleum.

Riveros, P.A., J.E. Dutrizac, & Spencer. P. Arsenic disposal practices in the metallurgy industry. *Canadian Metallurgical Quartely*, 2001. 40 (4): p. 395-420.

Sánchez-Andrea, I, Stams, A.J.M., Hedrich, S., Nancucheo, I. & Johnson, D.B. (2015). *Desulfosporosinus acididurans sp. nov.*: an acidophilic sulfate-reducing bacterium isolated from acidic sediments. *Environmental Science and Technology,* 49, (19), pp. 11746-11755.

Twidwell, L.G. & McCloskey, J.W. (2011). Removing from Aqueou solution and long term product storage. *JOM Journal of the Minerals Metals and Materials Society*, 2011. 63(8):p.94-100.

Van Aswegen, P.C., van Niekerk, J. & Olivier. W. (2006). The BIOX process for the treatment of refractory gold concentrates, in Bioming, D.E.Ralings and B.D. Johnson, Editors. 2006, Springer, p. 1-33.

Arsenic and mercury found in river days after Brazil dam burst. Retrieved on November 26, 2015, from **http://theguardian.com**.

Durango copes with 'orange nastiness' of toxic sludge river pollution. Retrieved on August 10, 2015, from **http://theguardian.com**.

REMOVAL OF ARSENIC AND DISSOLUTION OF COPPER FROM ARSENIC BEARING COPPER ORE BY HIGH PRESSURE OXIDATIVE LEACHING

*Atsushi Shibayama[1], Batnasan Altansukh[2], Ryuji Oinuma[2] and Kazutoshi Haga[2]

[1]*Graduate School of International Resource Science, Akita University*
1-1 Tegata-Gakuen Machi, Akita, Japan, 010-8502
*(*Corresponding author: sibayama@gipc.akita-u.ac.jp)*

[2]*Graduate School of Engineering and Resource Science, Akita University*
1-2 Tegata-Gakuen Machi, Akita, Japan, 010-8502

ABSTRACT

Recently, the arsenic grade in copper ores has been increasing year by year. Arsenic bearing minerals (enargite (Cu_3AsS_4) and tennantite ($Cu_{12}As_4S_{13}$)) coexist with typical copper minerals such as chalcopyrite ($CuFeS_2$). The removal of arsenic from copper ores that contain relatively high arsenic content is essential due to its adverse impact on the environment and the resulting economic consequences. In this study, arsenic removal and dissolution behaviour of Cu and As from enargite (Cu_3AsS_4) were investigated using ferric sulphate ($Fe_2(SO_4)_3$) solutions containing sodium chloride under high pressure oxidation conditions. The effects of numerous parameters including the concentration of $Fe_2(SO_4)_3$ and NaCl, temperature, total pressure, leaching time and particle size were studied while other variables such as pulp density and stirring speed were fixed at 100 g/L and 750 rpm. The results showed that more than 60% of the copper was dissolved in a 0.1 M ferric sulphate solution under the high pressure oxidative conditions, whereas the leaching rate of arsenic was less than 20%. SEM observation revealed that elemental sulphur was deposited on the surface of enargite forming a passivation layer and the residue was agglomerated. Those phenomena may cause a decrease in the copper leaching rate. It was found that more than 80% of copper was dissolved in the ferric sulphate solution in the presence of 0.34 M sodium chloride under the conditions optimized, whereas the dissolution of arsenic in the mixture did not exceed 2%. XRD analysis of the solid residues from the leaching experiments used ferric sulphate and mixtures of ferric sulphate and sodium chloride showed arsenic to be present as scorodite ($FeAsO_4$).

KEYWORDS

Arsenic bearing copper ore, high pressure leaching, ferric sulphate, ferric arsenate

INTRODUCTION

In recent years, there has been an increase of the arsenic (As) content of copper ores due to the gradual depletion of high grade copper ores. In some copper deposits, arsenic occurs in the form of minerals such as enargite and tennantite. The removal of arsenic in ores containing copper is critical to the production of pure copper and the management of the environmental challenges caused by the smelting process. Due to the environmental risk caused by arsenic emission and the deteriorated quality of the copper product, smelting of enargite concentrates is not an effective method for the extraction of copper (Padilla et al., 2010; Safarzadeh et al., 2014; Long et al., 2012). Therefore, a number of studies have been carried out to develop arsenic removal technologies from copper ores using various methods such as hydrometallurgy (Curreli et al., 2009; Tongamp et al. 2010), pyrometallurgy (Dalewski., 1999; Padilla. et al, 2012) and bio-oxidation (Ehrlich, 1964; Sasaki et al., 2010). Several hydrometallurgical leaching processes like sulphuric acid, alkaline sodium sulphide and sodium sulphate and ferric sulphate leaching have been developed to treat enargite ores and concentrates. The alkaline sodium sulphide leaching process has been developed for the removal of arsenic from enargite (Anderson, & Twidwell, 2008; Parada et al., 2014). The mechanism of the dissolution of arsenic from enargite in sodium sulphite solution is

$$2Cu_3AsS_4 + 3Na_2S \rightarrow 3Cu_2S + 2Na_3AsS_4 \quad (1)$$

Dutrizac & MacDonald (1972) proposed acidic ferric sulphate leaching process for the dissolution of copper and arsenic from natural and synthetic enargite and the mechanism of the enargite dissolution in that medium is illustrated in the following equation.

$$Cu_3AsS_4 + 11Fe^{3+} + 4H_2O \rightarrow 3Cu^{2+} + AsO_4^{3-} + 4S^0 + 8H^+ + 11Fe^{2+} \quad (2)$$

Sodium hydrosulphite leaching for the conversion of enargite to Cu_2S and Na_3AsS_4 in NaHS-NaOH solution has been suggested by Tongamp et al. (2009, 2010) and those researchers proposed the following mechanism for the dissolution of enargite:

$$2Cu_3AsS_4 + 3NaHS + 3NaOH \rightarrow 3Cu_2S + 2Na_3AsS_4 + 3H_2O \quad (3)$$

The advantage of these hydrometallurgical methods compared to the pyrometallurgical technique is the formation of a stable ferric arsenate, as scorodite or trisodium thioarsenide Na_3AsS_4. However, the dissolution efficiency of copper is restricted due to the formation of a sulphur passivation layer on the surface of enargite at atmospheric pressure (Rivera-Vasquez & Dixon, 2015; Sadegh Safarzadeh & Miller, 2014). Therefore other potential methods that could completely remove the arsenic and recover copper efficiently from enargite are vital. The high temperature pressure oxidation leaching (HTPOL) process is a promising method for treatment of enargite. Padilla et al. (2008) studied the dissolution of copper and arsenic from chalcopyrite-enargite concentrate using sulfidation and a high temperature-pressure oxidation leaching process (HTPOL) at 125-180 °C and a partial pressure of oxygen of 507-1520 kPa; they proposed the following leaching reaction for enargite.

$$Cu_3AsS_4 + 8.75O_2 + 2.5H_2O + 2H^+ \rightarrow 3Cu^{2+} + H_3AsO_4^- + 4HSO_4^- \quad (4)$$

The mechanism of the dissolution of enargite ore, with a high content of pyrite, in sulphuric acid solution at temperatures ranging between 160 - 200 °C and partial pressures of oxygen between 345-1034 kPa was advanced by Ruiz et al. (2011), as implied below:

$$Cu_3AsS_4 + 35Fe^{3+} + 20H_2O \rightarrow 3Cu^{2+} + H_3AsO_4^- + 4HSO_4^- + 33H^+ + 35Fe^{2+} \quad (5)$$

As a result of sulfidation and the pyrite content of the enargite, the dissolution kinetics of copper from enargite were faster and the copper recovery reached an acceptable value.

Therefore, we have studied the dissolution behaviour of enargite in ferric sulphate ($Fe_2(SO_4)_3$) and $Fe_2(SO_4)_3$-NaCl solutions using the high temperature-pressure oxidation leaching process. The objective of this work is to reveal the most important variables, namely the concentration of $Fe_2(SO_4)_3$ and NaCl, particle size, total pressure, leaching time and temperature in the medium and to understand the formation of ferric arsenate (scorodite) in the solid residue.

EXPERIMENTAL

Materials and Methods

An enargite concentrate from the Chinkuwa mine in Taiwan was used. The contents of copper, iron and arsenic in this sample were 25.7%, 12.3% and 9.3%, respectively. Leaching samples were prepared by crushing and grinding using a jaw crusher and a disc mill. The sample was ground to achieve different particle sizes from 150 µm to 45 µm and used in the experiments for studying the leaching behaviour of enargite under various concentrations of $Fe_2(SO_4)_3$, varying total pressures, different leaching times and different temperatures. Particle size distribution of the samples before and after leaching was studied using a particle size analyser. Chemical concentration of the aqueous leach samples were measured by Inductively Coupled Plasma-Optical Emission Spectrometer (ICP-OES), SPS-300, Seiko Instrument Inc. XRD analysis of the mineralogical composition of the enargite sample was done using the XRD-Rigaku, RINT-2200/PC system in the range between 10-80 ° 2Θ (deg) with a Cu Kα irradiation source at 40 kV and 50 mA. The XRD pattern of the sample is shown in Figure 1. Dominant mineral phases of the sample are enargite and pyrite. Reagent grade chemicals such as sulphuric acid, ferric sulphate and sodium chloride were used in the leaching studies.

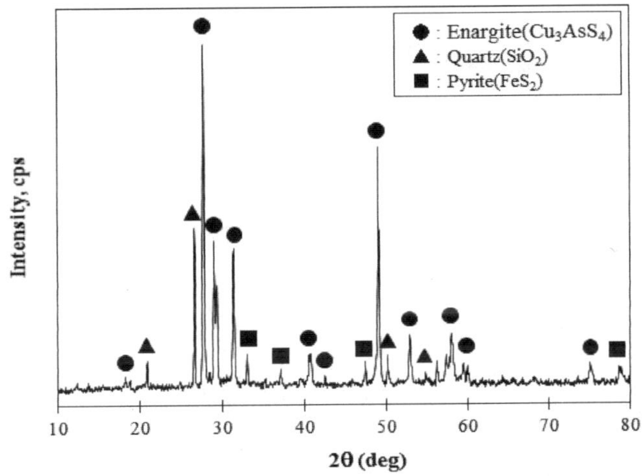

Figure 1 - XRD pattern of the enargite ore used in this study

Oxidation and Leaching of Enargite in an Autoclave

Since enargite leaching in acidic and alkaline media is inefficient at ambient temperature due to its slow leaching kinetics, the enargite was treated in an autoclave using H_2SO_4, $Fe_2(SO_4)_3$ and NaCl solutions at high temperature and high pressure under oxidation conditions supplied by oxygen gas. The autoclave used in this study is equipped with a 200 mL reaction vessel, impeller, inlet and outlet gas valves, pressure and temperature controller sensors, electrical heating system and oxygen gas cylinder. In all the leaching experiments, a 5 gram enargite sample was inserted into the reaction vessel together with 50 mL of the leaching solution (H_2SO_4, $Fe_2(SO_4)_3$ and $Fe_2(SO_4)_3$-NaCl mixture). Then the experimental parameters such as temperature, contact time, stirring speed and total pressure were set at the desired levels. The total pressure including the pressure of oxygen gas (P_{O2}) and the water vapour pressure (P_v) in the reaction vessel was controlled by a pressure gauge. The oxidized slurry samples from high temperature-pressure oxidative leaching (HTPOL) were cooled and filtered using a membrane filter for separation of the solution/filtrate and solid residue. Both phases were analysed by ICP-OES and XRD to determine the chemical and mineral compositions. The effects of varying parameters on enargite leaching were studied over a broad range; namely, concentration of H_2SO_4 (0.01-0.1 M), $Fe_2(SO_4)_3$ (0.01-0.1 M) and NaCl (0.17-0.68 M), particle size distribution (-150+106, -106+75, -75+45, -45 μm), leaching temperature (120-180 °C) and total pressure, $P_{total} = P_{O2} + P_v$, (0.5-2 MPa) at different leaching times (60 min, 120 min, 240 min).

RESULTS AND DISCUSSION

Enargite Leaching in Sulphuric Acid Solution in an Autoclave

The leaching behaviour of enargite in different concentrations of sulphuric acid (H_2SO_4), varying from 0.01 M to 0.1 M, was studied; the other experimental valuables such as average particle size, pulp density, stirring speed and total pressure, temperature were fixed at -150 μm, 100 g/L, 750 rpm and 1 MPa, 160 °C for 60 min. As shown in Figure 2, the different concentrations of sulphuric acid solution have only a minor effect on the dissolution of copper from enargite. The maximum leaching rate and grade of Cu are 25.4% and 36.5%, respectively, when the concentration of H_2SO_4 is 0.1 M. This may be related to the slow leaching kinetics of enargite and the formation of elemental sulphur in the sulphuric acid media.

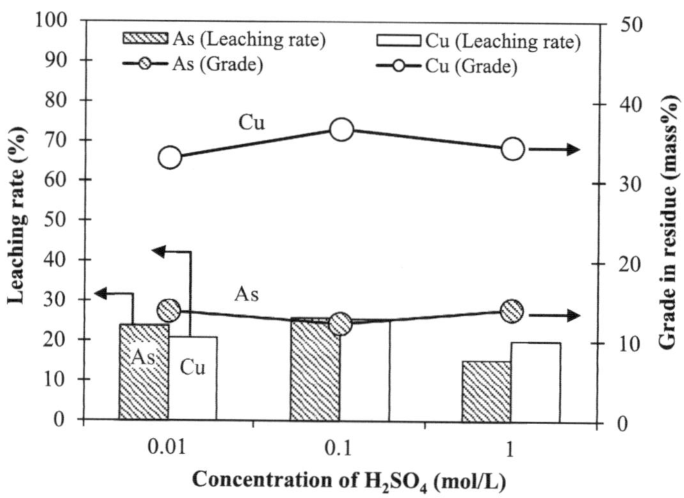

Figure 2 - Effect of sulphuric acid concentration on the dissolution of Cu and As from enargite (particle size: -150 μm, pulp density: 100 g/L, stirring speed: 750 rpm, P_{total}: 1 MPa, at 160 °C for 60 min

It was observed that sulphuric acid leaching of enargite is not a selective method for the removal of arsenic or for the complete dissolution of enargite because of the dissolution efficiencies of As and Cu are closely related in these media.

Enargite Leaching in Ferric Sulphate Solution in an Autoclave

The dissolution behaviour of copper and the selective removal of arsenic from enargite were investigated for various parameters including the concentration of ferric sulphate, NaCl concentration, particle size distribution, leaching time, temperature and total pressure in the autoclave. Some results obtained from the various leaching experiments are discussed in the following sections.

Effect of Ferric Sulphate Solution Concentration on Enargite Leaching

Varying concentrations of ferric sulphate ($Fe_2(SO_4)_3$) solutions were used for enargite leaching under the fixed conditions which were average particle size of -150 μm, pulp density of 100 g/L, stirring speed of 750 rpm, leaching time of 60 min, temperature of 160 °C and total pressure of 1 MPa. The effects of $Fe_2(SO_4)_3$ concentrations on the dissolution of Cu and As are summarized in Figure 3. The mechanism of dissolution of enargite in ferric sulphate can be represented by Equation 2. The dissolution rate of copper from enargite increases with increasing concentrations of ferric sulphate, from 0.01 to 0.1 M, and then decreases as the concentration of ferric sulphate increases further. In contrast, the dissolution rate of arsenic from enargite is nearly 6 % when the concentration of ferric sulphate increases from 0.01 to 0.1 M and the rate grows as the concentration of ferric sulphate increases from 0.25 to 1 M (Figure 3). Therefore, 0.1 M ferric sulphate solution was selected as the optimum concentration for the treatment of enargite due to the higher dissolution efficiency of copper and the lower dissolution of arsenic into the solution.

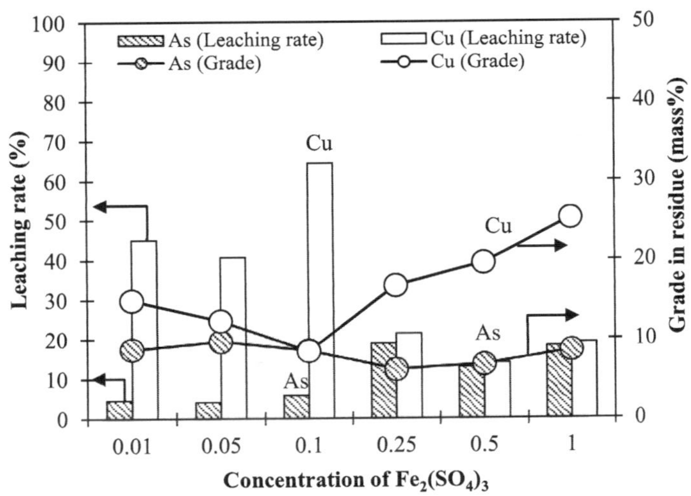

Figure 3 - Effect of the ferric sulphate concentration on the dissolution of Cu and As from enargite (particle size: -150 μm, pulp density: 100 g/L, stirring speed: 750 rpm, P_{total}: 1 MPa, at 160 °C for 60 min

These specific dissolution properties of copper and arsenic from enargite in 0.1 M $Fe_2(SO_4)_3$ solution offers the possibility to separate them and produce a leachate containing mostly copper. It appears that if the concentration of $Fe_2(SO_4)_3$ is higher than 0.1 M, extra SO_4^{2-} ions in the slurry promote the formation of elemental sulphur that deposits on the surface of enargite reducing the dissolution efficiency of copper.

Effect of Temperature on Enargite Leaching

The effect of temperature on the dissolution of copper from enargite was studied over the temperature range from 120 to 180 °C during high pressure oxidation using a 0.1 M solution of ferric sulphate. The experiments were carried out with an average particles size of -150 μm, a pulp density of 100 g/L, a steering speed of 750 rpm, a total pressure of 1 MPa and a leaching time of 60 min. The results obtained are summarized in Figure 4, where it can be seen that the leaching rate of copper from enargite increases significantly with increasing temperatures from 120 to 160 °C. A further increase in the temperature to 180 °C resulted in a small increase in the extraction of copper and has a slight reduction in arsenic dissolution. The dissolution of enargite in $Fe(SO_4)_{1.5}$ or $FeCl_3$ solution is slow at temperatures below 100 °C, and the complete dissolution of enargite is achieved only at temperatures above 170 °C (Riveros et al., 2001). In this study, the optimum temperature for the dissolution of enargite in 0.1 M $Fe_2(SO_4)_3$ solution was selected at 160 °C.

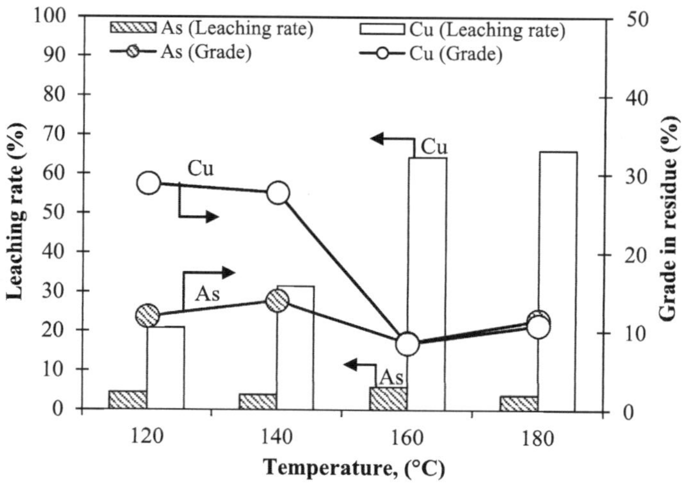

Figure 4 - Effect of temperature on the dissolution of Cu and As from enargite
(particle size: -150 μm, pulp density: 100 g/L, stirring speed: 750 rpm, P_{total}: 1 MPa, at 160 °C for 60 min

Effect of Particle Size on Enargite Leaching

The dissolution of enargite with various particles size, namely -150+106 μm, -106+75 μm, -75+45 μm, -45 μm, was studied using 0.1 M $Fe_2(SO_4)_3$ solutions at a temperature of 160 °C, a pulp density of 100 g/L, stirring speed of 750 rpm, total pressure of 1 MPa and a leaching time of 1 h (Figure 5).

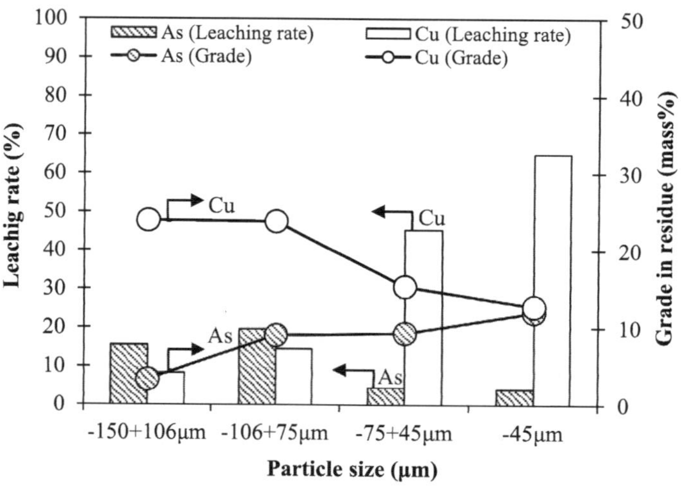

Figure 5 - Effect of particle size on the dissolution of Cu and As from enargite
($Fe_2(SO_4)_3$: 0.1 M, pulp density: 100 g/L, stirring speed: 750 rpm, P_{total}: 1 MPa, at 160 °C for 60 min

From the results it can be determined that, the leaching of the larger particles (-150+106 μm) of enargite was much slower than that of the small particles, and dissolution rate of copper from the larger particles was only 8.2% at these conditions (Figure 5). It can be seen that when the particle size decreases from the larger to the smaller, the dissolution rate of Cu significantly increases from 8.2% to 64.9%, whereas the dissolution of As decreases from 15.5% to 4.1%. The dissolution efficiencies of Cu and As from enargite with varying particle sizing have the opposite tendency. It can be concluded that particle size is an important variable in the leaching of enargite in ferric sulphate solution. It was observed that the particle size distribution (D_{50}) of enargite changed after the completion of the leaching. Further leaching experiments were conducted with -45 µm particles. The surfaces of the solid residues from the leaching experiment were examined by scanning electron microscopy with energy dispersive spectrometry (SEM-EDS) and the cross section SEM image of the residue after leaching in a 0.1 M $Fe_2(SO_4)_3$ solution is shown in Figure 6.

Figure 6 – SEM image of solid residue from enargite leaching in a 0.1 M $Fe_2(SO_4)_3$ solution

Effect of Sodium Chloride Additions on Enargite Leaching

Enargite with a particle size of -45 µm was dissolved in 0.1 M $Fe_2(SO_4)_3$ solutions in the presence of various concentrations of sodium chloride ranging from 0.17 M to 0.68 M. As shown in Figure 7, the leaching rate of copper increases remarkably from 64.9% to 81.9%. However, the rate decreases with further increases in the concentration of NaCl.

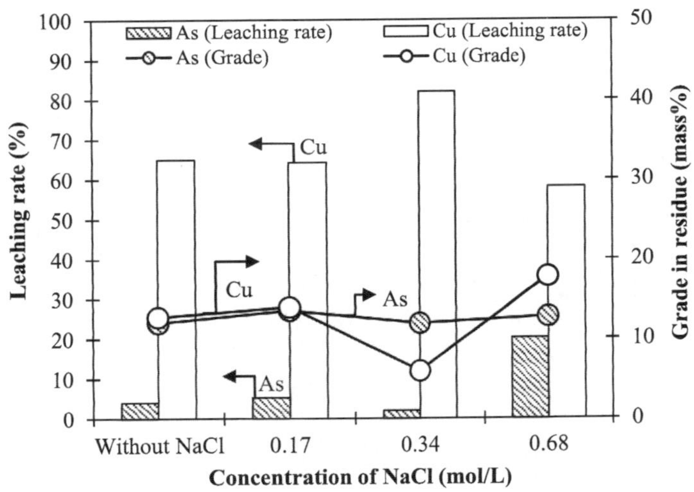

Figure 7 – Effect of NaCl additions on the dissolution of Cu and As from enargite
(Particle: -45µm, 0.1 M $Fe_2(SO_4)_3$, pulp density: 100 g/L, stirring: 750 rpm, 1 MPa P_{total}, 160 °C for 60 min

The leaching rate of arsenic decreases from 5.3% to 1.9% for the addition of 0.34 M NaCl, whereas it increases up to 20.2% when 0.68 M NaCl was added to the solution. The best concentration of NaCl was selected to be 0.34 M because of the higher copper extraction and the lower arsenic dissolution, thereby improving the separation of copper from arsenic. The other experimental variables, including pulp density, stirring speed, total pressure and temperature, leaching time, were fixed. The results presented in Figure 7 show that when 0.17 M NaCl is added to a 0.1 M $Fe_2(SO_4)_3$ solution, the leaching rate of copper does not change significantly compared to the 64.9% copper leaching rate in the absence of NaCl, in contrast to the behaviour when 0.34 M NaCl was added.

The leaching of copper from enargite in this medium at various times (1-4 h) and different total pressures, P_{total}, (0.5-2 MPa) were also investigated and the optimum leaching time and the pressure were selected as 1 h and 1 MPa, based on the experimental results which showed a higher dissolution of Cu from enargite while minimizing arsenic dissolution (data not shown). As a result of these experiments, the optimum conditions for enargite leaching were selected to be a pulp density of 100 g/L and 60 min retention time with 0.1 M $Fe_2(SO_4)_3$ solution in the presence of 0.34 M NaCl at the stirring speed of 750 rpm and at 160 °C. Under these conditions, 81.9% copper was dissolved from the enargite. The particle size distribution (D_{50}) in each residue obtained from ferric sulphate leaching in the absence and presence of 0.34 M NaCl solution reduced to 13.74 μm and 5.99 μm from 28.39 μm, respectively. It is obvious that the reduction of particle size occurs because of the more rapid dissolution of the smaller particles that influences the diffusion coefficients, surface charge and surface area, and the fragmentation of particles.

The XRD patterns of the solid residues obtained from enargite leaching with a 0.1 M $Fe_2(SO_4)_3$ solution in the presence and absence of 0.34 M NaCl are shown in Figure 8, from which it can be seen that the dissolution of enargite in the solution is nearly complete, thereby achieving 82% copper extraction from the sample under these conditions.

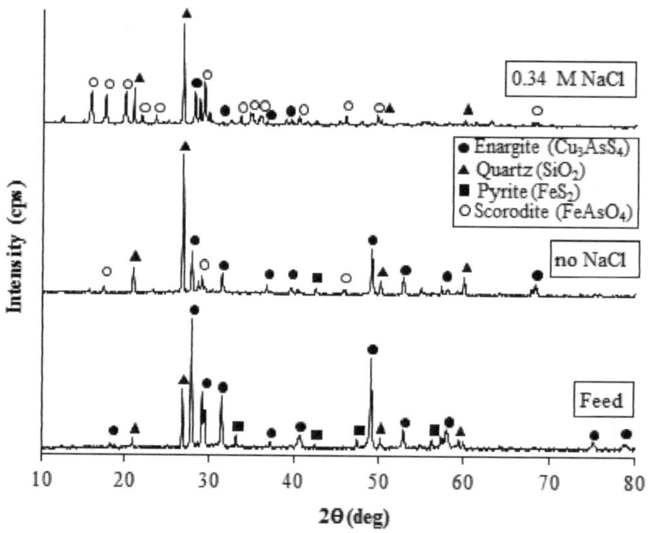

Figure 8 – XRD patterns of the enargite sample (feed) and solid residues after leaching with ferric sulphate in the absence and presence of NaCl

XRD and SEM observations of the solid residues after enargite leaching in a 0.1 M $Fe_2(SO_4)_3$ solution in the presence of 0.34 M NaCl identified a new ferric (iron) arsenate phase, such as scorodite ($FeAsSO_4$), which was formed by co-precipitation of ferric and arsenic ions in the solution under the oxidation conditions (Mayhew et al., 2010, 2013). It can be considered that ferric ion in the ferric sulphate could promote the dissolution of copper from enargite. The chloride ion in NaCl seems to have an influence on the particle size reduction and porosity of the enargite, resulting in an improvement of the leaching kinetics of copper and the reduction of the sulphur passivation layer on the enargite surface because of the formation of other ferric arsenates, such as scorodite.

Some researchers (Ruiz et al., 2011; Riveros et al., 2001) have studied enargite leaching in ferric sulphate solution with and without the addition of pyrite and sodium chloride. They concluded that the dissolution of enargite is a function of the ferric and chloride concentrations of the solution.

CONCLUSIONS

In this study, sulphuric acid, ferric sulphate and sodium chloride solutions were applied to the leaching of enargite under high pressure oxidative conditions using oxygen as an oxidant. A main objective of this study was to investigate the selective dissolution of copper and arsenic from enargite during leaching, in relation to various processing parameters including the concentration of leaching agents, particle size distribution, total pressure in the autoclave, temperature and leaching time. Based on the results, it was concluded that sulphuric acid could dissolve only about 25% copper from enargite, and it does not show any selectivity for arsenic under the studied conditions. Ferric sulphate leaching is a selective and efficient method for dissolution of copper from enargite, while minimizing arsenic leaching. Nearly 65% of the copper was dissolved from enargite in a 0.1 M ferric sulphate solution, whereas arsenic dissolution was only about 4% for the optimized conditions. It is suggested that ferric ion (Fe^{3+}) is an oxidizing agent which promotes copper extraction and diminishes arsenic dissolution from enargite. The dissolution rates of copper and arsenic were 82% and 2% during leaching with 0.1 M ferric sulphate and 0.34 M sodium chloride. It was observed that after leaching with ferric sulphate, the particle size distribution (D_{50}) of the samples with particle sizing greater than -75+45 μm increased causing a negative effect on the selectivity and dissolution efficiencies of copper and arsenic. In contrast, the particle size distribution (D_{50}) of finer particles (-45 μm) leached in $Fe_2(SO_4)_3$ and in $Fe_2(SO_4)_3$-NaCl solutions decreased, resulting in the higher dissolution of copper and the higher selectivity over arsenic. Chloride ions, added as sodium chloride, affect the morphology of the particles during the leaching and suppress sulphur passivation on the surface of the enargite.

It was suggested that high temperature-high pressure oxidative leaching with $Fe_2(SO_4)_3$-NaCl is an effective and selective method for the dissolution of copper and for the separation of arsenic from enargite.

ACKNOWLEDGEMENTS

The authors gratefully acknowledge the generous financial support from the Japan Society for the Promotion of Science (JSPS), through the New Frontier Leader Program for Rare-Metals and Resources in the Graduate School of Engineering and Resource Science, Akita University.

REFERENCES

Anderson, C. G., & Twidwell, L. G. (2008). The alkaline sulfide hydrometallurgical separation, recovery and fixation of tin, arsenic, antimony, mercury and gold. *The Southern African Institute of Mining and Metallurgy, Lead and Zinc*, 121-132.

Curreli, L., Garbarino, C., Ghiani, M., & Orrù, G. (2009). Arsenic leaching from a gold bearing enargite flotation concentrate. *Hydrometallurgy, 96*, 258–263. doi:10.1016/j.hydromet.2008.10.013

Dalewski, F. (1999). Removing arsenic from copper smelter gases. *The Journal of the Minerals, Metals & Materials Society, 51 (9)*, 24-26. doi: 10.1007/s11837-999-0153-0

Dutrizac, J. E. & MacDonald, R. J. C. (1972). The kinetics of dissolution of enargite in acidified ferric sulphate solutions. *Canadian Metallurgical Quarterly, 11(3)*, 469-476. doi: 10.1179/cmq.1972.11.3.469

Ehrlich, H. L. (1964). Bacterial oxidation of arsenopyrite and enargite. *Economic Geology, 59, 7*, 1306-1312. doi: 10.2113/gsecongeo.59.7.1306

Long, G., Peng, Yo., & Bradshaw. D. (2012). A review of copper–arsenic mineral removal from copper concentrates. *Minerals Engineering, 36–38*, 179–186. doi: 10.1016/j.mineng.2012.03.032

Mayhew, K., Parhar, P., & Salomon-de-Friedberg, H. (2010). CESL process as applied to enargite-rich copper concentrates. *Proceedings of Copper 2010, Book 5*, 1983-1998.

Mayhew, K., McCoy, T., & Mean, R. (2013). TECK'S CESL copper process: A commercial ready concentrate leach alternative. *Proceedings of Copper 2013, Book 2, 4*, 63-75.

Padilla, R., Aracena, A., & Ruiz, M. C. (2012). Reaction mechanism and kinetics of enargite oxidation at roasting temperatures. *Metallurgical and Materials Transactions B, 43 (5)*, 1119-1126. doi: 10.1007/s11663-012-9675-x

Padilla, R., Rivas, C.A., & Ruiz, M. C. (2008). Kinetics of pressure dissolution of enargite in sulphate-oxygen media. *Metallurgical and Materials Transactions B*, 39B, 399-407. doi: 10.1007/s11663-008-9151-9

Padilla, R., Rodríguez, G., & Ruiz, M.C. (2010). Copper and arsenic dissolution from chalcopyrite-enargite concentrate sulfidation and pressure leaching in H_2SO_4-O_2. *Hydrometallurgy, 100*, 152-156. doi:10.1016/j.hydromet.2009.11.006

Parada, F., Jeffrey, M. I., & Asselina, E. (2014). Leaching kinetics of enargite in alkaline sodium sulphide solutions. *Hydrometallurgy, 146*, 48-58. doi: 10.1016/j.hydromet.2014.03.003

Rivera-Vasquez, B. F., & Dixon, D. (2015). Rapid atmospheric leaching of enargite in acidic ferric sulfate media. *Hydrometallurgy, 152*, 149-158. doi: 10.1016/j.hydromet.2014.12.012

Riveros, P. A., Dutrizac, J. E., & Spencer, P. (2001). Arsenic disposal practices in the metallurgical industry. *Canadian Metallurgical Quarterly, 40(4)*, 395-420. doi: 10.1179/cmq.2001.40.4.395

Ruiz, M. C., Vera, M. V., & Padilla, R. (2011). Mechanism of enargite pressure leaching in the presence of pyrite. *Hydrometallurgy, 105*, 290–295. doi:10.1016/j.hydromet.2010.11.002

Sadegh Safarzadeh, M., & Miller, J. D. (2014). Reaction of enargite (Cu_3AsS_4) in hot concentrated sulfuric acid under an inert atmosphere. Part I: Enargite concentrate. *International Journal of Mineral Processing, 128*, 68-78. doi: 10.1016/j.minpro.2014.02.007

Sadegh Safarzadeh, M., & Miller, J. D. (2014). Reaction of enargite (Cu_3AsS_4) in hot concentrated sulphuric acid under an inert atmosphere. Part II: High-quality enargite. *International Journal of Mineral Processing, 128*, 79-85. doi: 10.1016/j.minpro.2014.02.006

Sasaki, K., Takatsugi, K., Kaneko, K., Kozai, N., Ohnuki, T., Tuovinen, O. H., Hirajima, T. (2010). Characterization of secondary arsenic-bearing precipitates formed in the bioleaching of enargite by *Acidithiobacillus ferrooxidans*. *Hydrometallurgy 104*, 424–431. doi:10.1016/j.hydromet.2009.12.012

Tongamp, W., Takasaki, Ya., & Shibayama, A. (2010). Precipitation of arsenic as Na_3AsS_4 from Cu_3AsS_4–NaHS–NaOH leach solutions. *Hydrometallurgy, 105*, 42–46. doi:10.1016/j.hydromet.2010.07.007

Tongamp, W., Takasaki, Ya., & Shibayama, A. (2009). Arsenic removal from copper ores and concentrates through alkaline leaching in NaHS media. *Hydrometallurgy, 98*, 213–218. doi:10.1016/j.hydromet.2009.04.020

FORMATION OF SCHWERTMANNITE-LIKE AND SCORODITE-LIKE COATINGS ON PYRITE AND ITS IMPLICATIONS IN ACID MINE DRAINAGE CONTROL

*C.B. Tabelin[1], R.D. Corpuz[1], S. Veerawattananun[1,2], M. Ito[1], N. Hiroyoshi[1] and T. Igarashi[1]

[1]*Faculty of Engineering, Hokkaido University*
Kita 8, Nishi 13, Kita-ku
Sapporo, Japan
*(*Corresponding author: carlito@eng.hokudai.ac.jp; carlito.tabelin@gmail.com)*

[2]*Faculty of Engineering, Chulalongkorn University*
254 Phayathai Road, Pathumwan
Bangkok, Thailand

ABSTRACT

Acid mine drainage (AMD), the very acidic and heavy metal-contaminated leachate generated when pyrite-bearing rocks are exposed to surface oxidizing conditions, is a serious environmental problem encountered after the closure of mines and mineral processing operations. In this study, we investigated the effects of soluble phases and metallic oxides, which are commonly found in strongly weathered pyritic wastes, on pyrite oxidation. Strongly weathered pyrite containing melanterite ($FeSO_4 \cdot 7H_2O$) as the main soluble phase was obtained from Furikusa mine, Japan. The sample was washed using two methods to vary the relative amounts of melanterite in the sample. The effect of metallic oxides on pyrite oxidation was evaluated using ultrapure hematite suspensions. Our results showed that under acidic conditions, the effect of soluble salts on the oxidation of pyrite was stronger than hematite. More thorough washing of the sample resulted in the formation of an arsenic-bearing iron-oxyhydroxysulfate coating that minimized pyrite dissolution. This oxidized layer became relatively thicker and mechanically stronger when hematite was introduced after the more thorough removal of soluble salts. Characterization of this "coating" coupled with geochemical modeling suggest that it is most probably composed of two materials: schwertmannite and scorodite. Based on these results, a new approach to suppress pyrite oxidation by directly using iron-oxyhydroxysulfate as coating material is introduced.

KEYWORDS

Acid mine drainage; Pyrite; Pyrite oxidation; Pyrite passivation

INTRODUCTION

Acid mine drainage (AMD), the very acidic and heavy metals-contaminated leachate produced in closed/abandoned mine sites (Evangelou, 1995), mineral processing and coal cleaning tailings (Evangelou, 1995), and tunnel construction waste rocks (Tabelin and Igarashi, 2009, Tabelin *et al.*, 2012), is one of the oldest and probably the most serious environmental problem of anthropogenic origin. When released to the environment, it rapidly pollutes river, soil and groundwater systems leading to the destruction of affected ecosystems. AMD is produced when pyrite (FeS_2), one of the primary gangue minerals in sulfide ores and coals, is exposed to oxidizing conditions as illustrated by the following reactions:

$$FeS_{2(s)} + 3.5O_{2(aq)} + H_2O \rightarrow Fe^{2+} + 2H^+ + 2SO_4^{2-} \quad (1)$$

$$FeS_{2(s)} + 14Fe^{3+} + 8H_2O \rightarrow 15Fe^{2+} + 16H^+ + 2SO_4^{2-} \quad (2)$$

$$Fe^{2+} + 0.25O_{2(aq)} + H^+ \rightarrow Fe^{3+} + 0.5H_2O \quad (3)$$

As shown above, one mole of pyrite could generate a lot of acidity (up to 16 moles of H^+) regardless of the type of oxidant. Further compounding this problem is the fact that natural pyrites incorporate heavy metals and toxic metalloids into their crystal structures (Abraitis *et al.*, 2004), so AMD generally contains toxic elements as well, which could threaten the health and well-being of individuals living around contaminated sites. Toxic elements found in AMD include cadmium (Cd), arsenic (As), lead (Pb) and mercury (Hg) that affect the cardio-vascular organs, nervous system, bladder, skin, kidneys and bones.

Pyrite is relatively resistant to both acidic and alkaline attacks, but dissolves quite easily in the presence of oxidants like O_2 and Fe^{3+} (Lowson, 1982). In addition, microorganisms in the natural environment like iron-oxidizing bacteria enhance this process by utilizing Fe^{2+}, one of the products of pyrite oxidation (Equations 1 and 2), as their energy source. The action of such microorganisms rapidly converts Fe^{2+} to Fe^{3+}, which further oxidizes pyrite leading to the generation of more AMD (Lowson, 1982). This catalytic effect of microorganisms further exacerbates the AMD problem in the natural environment. Although pyrite oxidation has been extensively studied because of its significance in AMD generation and treatment, photoelectrochemical cells, and extraction of gold and cobalt from pyritic ores, the majority of these studies were done either with freshly cleaved or thoroughly washed pyrite surfaces rather than the weathered pyrite typically found in the environment. In addition, the effects of secondary oxidation products like soluble salts and metallic oxyhydroxides on the overall pyrite oxidation dynamics are still not well understood. The latter is of particular importance because during AMD treatment by neutralization, it is common practice to dispose of the generated sludge back into the tailings dams. This means that particulate/colloidal metallic oxyhydroxides/oxides in these sludge could interact with the original mine tailings and lead to significant changes in pyrite oxidation dynamics.

In this study, we investigated the effects of soluble salts and metallic oxides on the oxidation of strongly weathered pyrite and examined the resulting oxidation product/s formed on the pyrite surface. By modifying the washing technique developed by McKibben and Barnes (1986), we created experimental conditions using the same sample that evaluated the effects of intimately associated soluble phases on pyrite oxidation. Several time-dependent batch-reactor type leaching experiments were also conducted in the presence of hematite as a representative metallic oxide. A combination of light and scanning electron microscopy with energy dispersive spectroscopy (SEM-EDX) was utilized to qualitatively characterize the evolution of the oxidized layer in the batch-reactor experiments. Powder X-ray diffraction (XRD), diffuse reflectance infrared Fourier transform spectroscopy (DRIFTS) and field emission electron probe microanalysis (FE-SEM) were also used to identify the morphology and chemical nature of the oxidized layer formed. Finally, geochemical modeling calculations based on actual solute activities of the leachates were performed to give insights into the factors and processes responsible for the observed results.

MATERIALS AND METHODS

Characterization of Pyrite and Hematite Samples

The pyrite sample used in this study came from Furikusa Mine in Aichi Prefecture, Japan. The sample had an as-delivered particle size distribution of < 2 mm and was directly screened without crushing to obtain several size fractions. Those with particle sizes between 500 and 710 μm were chosen because they were large enough to be seen without difficulty under an ordinary light microscope. Chemical and mineralogical compositions of both the bulk and 500-710 μm samples were determined by X-ray fluorescence (XRF) (NEXCG, Rigaku Corporation, Japan) and XRD (MultiFlex, Rigaku Corporation, Japan), respectively. Both of these analyses were done on samples ground to <50 μm in diameter. Ultrapure hematite powder (99.9% α-Fe_2O_3) used in the batch-reactor experiments was obtained from Wako Pure Chemical Industries Ltd., Japan. Its particle size distribution was determined by LASER diffraction (Microtrac® MT3300SX, Nikkiso Co., Ltd., Japan), and was found to be in the range of 0.6-50 μm with ca. 80% of the particles having diameters between 1 and 20 μm.

Batch-reactor Type Experiments

The amounts of soluble phases were varied by washing of the sample based on the method developed by McKibben and Barnes (1986). This method involves ultrasonic cleaning and desliming in ethanol, washing with 1 M HNO_3, rinsing with deionized (DI) water (18 MΩ·cm), and dewatering with acetone under vacuum. Because soluble salts readily dissolve in aqueous solutions, we varied their amounts in the sample by changing the liquid-to-solid (L/S) ratio during washing as detailed in Table 1. For the batch-reactor experiments, 1 gram of washed pyrite was mixed with 10 ml of DI water or hematite suspension in 50 ml Erlenmeyer flasks. These mixtures were shaken in a constant temperature water bath (25°C) at 120 rpm for up to 29 days. Some of the experiments were performed in duplicates to ascertain that the differences and trends observed were statistically significant. Four kinds of experiments were conducted in this study:

- Case 1A: The pyrite sample was washed using Method A and the leachant used was DI water.
- Case 1B: The pyrite sample was washed using Method B and the leachant used was DI water.
- Case 2A: The pyrite sample was washed using Method A and the leachant used was 200 mg/L hematite suspension.
- Case 2B: The pyrite sample was washed using Method B and the leachant used was 200 mg/L hematite suspension.

Selected samples were taken out of the shaker after 1, 3, 7, 17, 21 and 29 days, and their pH's were measured followed by filtration of the leachates through 0.2 μm syringe-driven membrane filters (Sartorius AG, Germany) for the chemical analyses. On the other hand, the solid fractions were washed thoroughly with DI water and dried in a vacuum drying oven at 40°C for 24 h.

Table 1 – Modifications in the washing method of McKibben and Barnes (1986) used in this study

Name	Step 1		Step 2		Step 3		Step 4	
	L/S ratio	Solvent	L/S ratio	Solvent	L/S ratio	Solvent	L/S ratio	Solvent
Method A	3	Ethanol	1	1 M HNO_3	1	DI water	1	Acetone
Method B	3	Ethanol	5	1 M HNO_3	10	DI water	1	Acetone

Note: L/S: liquid-to-solid ratio, Step 1: Sonicated for 30 s, Step 2: rotary shaken for 1 min, Step 3: washed three times, Step 4: rinsed four times.

Chemical Analysis, Geochemical Modeling and Microscopic Observations

Dissolved ions in the leachates were measured using an inductively coupled plasma atomic emission spectrometer (ICP-AES) (ICPE-9800, Shimadzu Corporation, Japan). Dissolution and precipitation of mineral phases that could affect the oxidation of pyrite were modeled using PHREEQC (Parkhurst and Apello, 1999) and the Geochemist's Workbench® (Bethke, 1992).

Coated pyrite crystals from the dried solid fractions were hand-picked and observed with a Keyence VHX-1000 digital microscope (Keyence Corporation, Japan), SEM-EDX (SSX-550, Shimadzu Corporation, Japan) and FE-SEM (JXA-8530F, JEOL Ltd., Japan). Elemental maps of the solid surfaces were also taken with SEM-EDX using an accelerating voltage of 15 KeV, and taken at 2,000 cps with 1000 ms time constant. Finally, coated pyrites were ground manually and analyzed by DRIFTS (FT/IR-6200HFV with DR PR0410-M attachment, Jasco Analytical Instruments, USA) and XRD.

RESULTS AND DISCUSSION

Properties of the Weathered Pyrite Sample

The bulk sample (< 2 mm) is mainly composed of pyrite with minor amounts of other minerals like arsenopyrite (FeAsS), pyrophillite ($Al_2Si_4O_{10}(OH)_2$), muscovite ($KAl_2(AlSi_3O_{10})(F,OH)_2$), melanterite ($FeSO_4 \cdot 7H_2O$) and quartz ($SiO_2$) (Figure 1). These identified minerals were consistent with the chemical composition of the sample (Table 2), which shows that Fe and S are its main components with lesser quantities of Si, Al, As, K and Ca. After sieving, most of the fine particles (*e.g.*, pyrophillite and muscovite) were removed (Figure 1 and Table 2), but the peaks of melanterite were still clearly visible in the XRD pattern. This is because melanterite was present in the sample not as separate particles but rather as relatively thick coatings on the surfaces of pyrite and arsenopyrite (Figure 2).

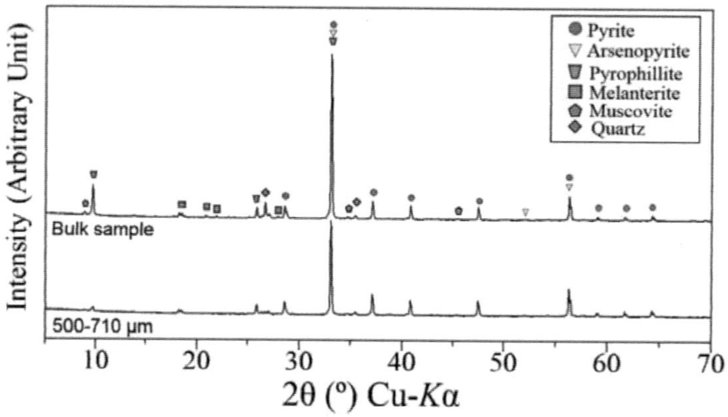

Figure 1 – XRD patterns of the weathered pyrite sample from Furikusa mine

Table 2 – Chemical composition of the pyrite sample from Furikusa mine

Size distribution	Elemental composition (wt.%)							
	Fe	S	Si	As	Al	Mg	K	Ca
Bulk (< 2 mm)	61.2	31.8	3.14	1.33	1.27	0.53	0.2	0.09
500-710 μm	66.4	29.9	1.42	1.09	0.35	0.42	0.04	0.06

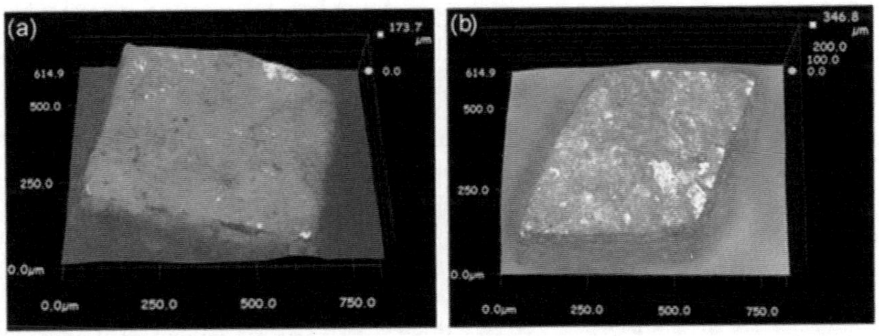

Figure 2 – 3D photomicrographs of pyrite (a) and arsenopyrite (b) with thick soluble salt coatings

<u>Effects of Soluble Phases on Pyrite Oxidation</u>

The washing methods produced samples with different amounts of soluble salts as illustrated in Figure 3. Between the two, Method B was more effective in the removal of soluble salts. These results also show that the main components of the salt-like coating on the sample were Fe and S consistent with the peaks of melanterite in the XRD pattern.

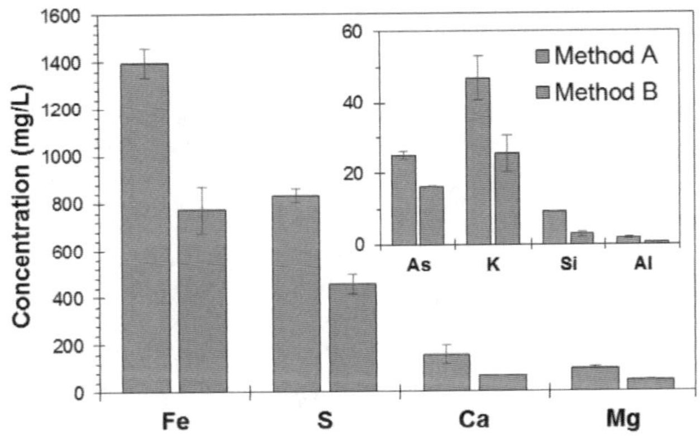

Figure 3 – Concentrations of Fe, S, Ca, Mg, K, As, Si and Al after washing with Methods A and B. These values were obtained after shaking mixtures of washed sample (1 g) and DI water (10 ml) for 30 minutes

These soluble salts strongly influenced not only the pH and solute concentrations of the leachates but also the overall oxidation of pyrite as illustrated in Figure 4. More soluble phases in the sample (Case 1A) resulted in more acidic pH and higher initial concentrations of Fe and S. The more acidic pH in Case 1A could be attributed to residual H_2SO_4 associated with the soluble phases (Schaufuss *et al.*, 1998), which was less thoroughly removed by Method A. After 1 day, the pH in both cases rapidly decreased, followed by a more gradual decline. This sudden pH drop coincided with dramatic increases in the concentrations of Fe, S and As in both cases. This means that further dissolution of the remaining soluble salts occurred at the same time as the oxidation of pyrite and arsenopyrite, both of which are capable of generating substantial amounts of H^+. Hydrolysis of Fe^{3+} due to the oxidation of the Fe^{2+} from melanterite could also have contributed to this rapid drop in pH (Stefansson, 2007). After 3 days, the concentrations of Fe, S and As in both cases suddenly decreased, but normalization of the results to background levels indicates that changes were more dramatic in Case 1B. Surprisingly, these relatively low concentrations of Fe, As and S persisted until the end of the experiment but only in Case 1B.

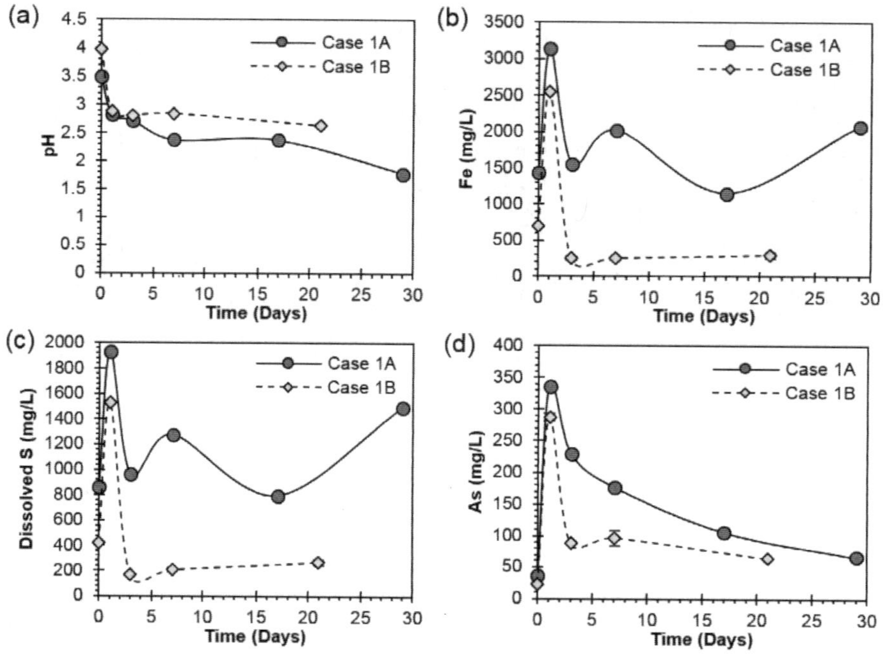

Figure 4 – Evolution of pH and the concentrations of Fe, S and As in Cases 1A and 1B; (a) pH vs time, (b) dissolved Fe vs time, (c) dissolved S vs time, and (d) dissolved As vs time

Figure 5 is an equilibrium activity-pH diagram created with the Geochemist's Workbench® based on the actual solute activities of the leachates calculated by PHREEQC. The stability line of schwermannite ($Fe_8O_8(OH)_6SO_4$) was drawn based on the reaction and Log K reported by Bigham *et al.* (1996). Colored regions superimposed in the diagram represented the ranges of Fe^{2+} and Fe^{3+} activities modeled by PHREEQC based on the experimental conditions of this study. Three mineral phases are predicted to precipitate in the experiments: K-jarosite ($KFe_3(SO_4)_2(OH)_6$), schwertmannite and scorodite ($FeAsO_4 \cdot 2H_2O$). Among the three, K-jarosite is the most soluble while scorodite is the most stable. Because the pH dropped below 3 in all cases after just one day, this model predicts that K-jarosite most likely re-dissolved leaving only scorodite and schwertmannite as the stable minerals at longer leaching times.

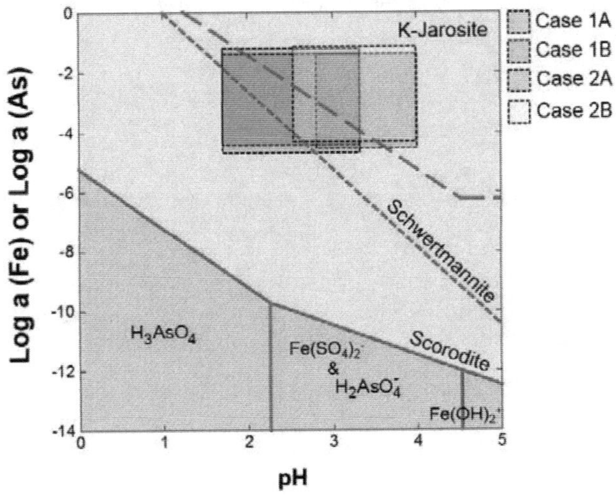

Figure 5 – Equilibrium activity-pH diagram of As and Fe at 25°C, 1.013 bars, activity of SO_4^{2-} = 10^{-2} and activity of K^+ = 10^{-3}

Microscopic observations of pyrite crystals after 3 and 7 days showed that distinct oxidized layers were formed on their surfaces, which were of differing colors depending on the washing method used (see Figure 6). In Case 1A, the color was grayish-green that turned bright-green after 7 days while in Case 1B, it was more of a combination of grayish-green and reddish-brown. SEM-EDX analysis of the coating containing both of these colorations showed that K signals were negligible and the grayish-green areas contained relatively more As and O than the reddish-brown ones (Figure 7). These results indicate that K-jarosite was most likely not part of the oxidation layer and the two colors corresponded to two kinds of As-bearing iron-oxyhydroxysulfates.

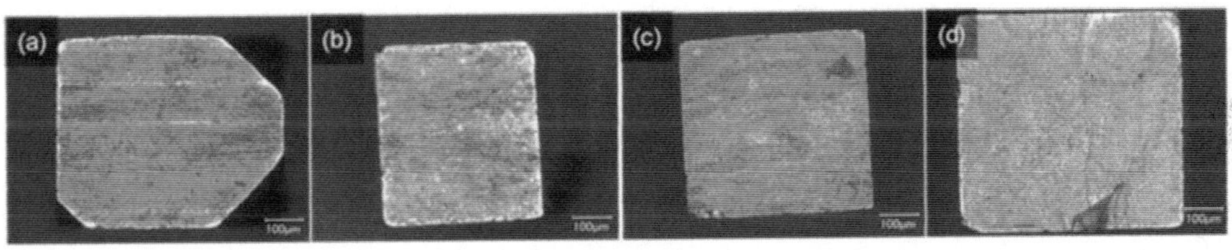

Figure 6 – Photomicrographs of the pyrite surfaces in Case 1A after (a) 3 days and (b) 7 days as well as those in Case 1B after (c) 3 days and (d) 7 days

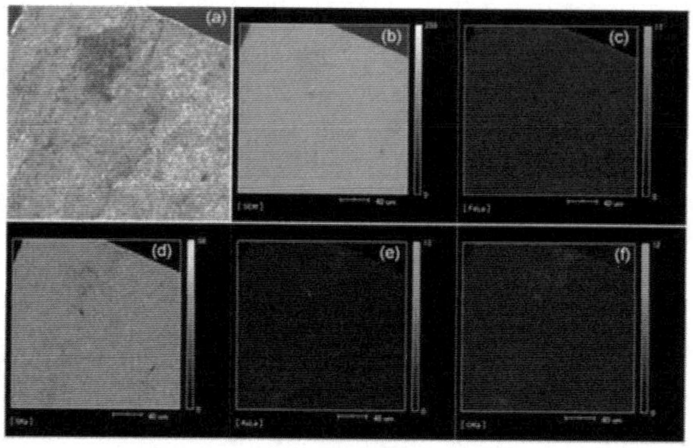

Figure 7 – Photomicrographs (a and b) and elemental maps of Fe (c), S (d), As (e) and O (f) of the oxidized layer in areas with two coloration (Case 1B after 3 days)

Effects of Hematite on the Oxidation of Weathered Pyrite

Figure 8 illustrates the effects of hematite on the evolution of the leachate pH and concentrations of Fe, As and S. Initially, the effect of hematite was negligible but after 1 day, it decreased the leachate pH and at the same time increased the concentrations of Fe, As and S. This effect could be attributed to the dissolution of hematite, which released Fe^{3+} into solution thereby enhancing pyrite oxidation (Equation 3). After 3 days, the concentrations of these ions became statistically the same compared with those of Case 1B, which persisted until the end of the experiment.

As illustrated in Figures 9 (a) and (c), the oxidized layer in Case 2A was very similar to the "cracked" and "flaky" morphology observed in Case 1A. The grayish-green oxidized layer formed in both of these cases grew with time as illustrated by the stronger signals of As and O obtained in our SEM-EDX analysis (data not shown), but they were mechanically weak as evident from their "cracked" and "flaky" morphology after 3 weeks. In comparison, oxidized layers in Cases 1B and 2B (Figures 9(b) and (d)), which are mixtures of the two colors, were less prone to cracking and flaking that resulted in the continued passivation of pyrite.

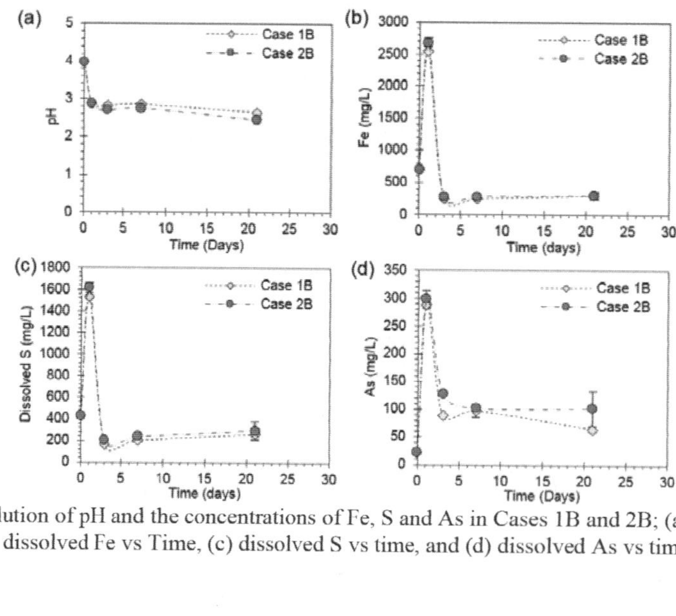

Figure 8 – Evolution of pH and the concentrations of Fe, S and As in Cases 1B and 2B; (a) pH vs time, (b) dissolved Fe vs Time, (c) dissolved S vs time, and (d) dissolved As vs time

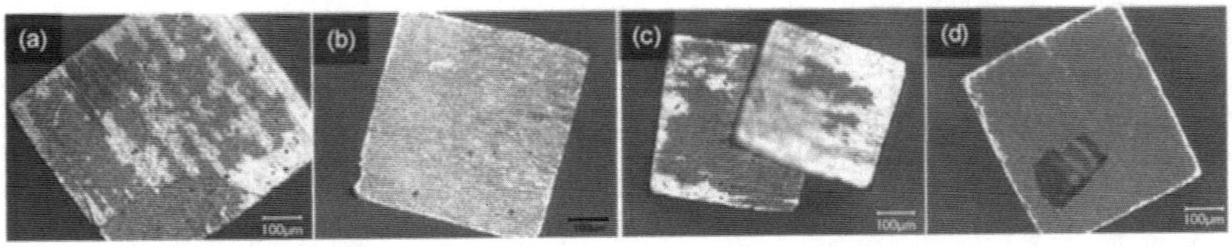

Figure 9 – Photomicrographs of pyrite surfaces in (a) Case 1A after 29 days, (b) Case 1B after 21 days, (c) Case 2A after 29 days, and (d) Case 2B after 21 days

Although the effects of hematite on the evolution of pH, and concentrations of Fe, S and As were less substantial than those of the soluble salts (Figures 4 and 8), it promoted the formation of a thicker and more mechanically stable oxidized layer on the pyrite surface. After 3 days, reddish-brown colored precipitates started accumulating on the pyrite surface, which grew with time until it almost completely covered the entire crystal (Figure 9(d)). Even with the constant shaking motion during the experiments (120 rpm), the oxidized layer remained adhered onto the pyrite surface almost entirely encapsulating the crystal (except around the crystal edges where attrition and shear forces were excessively high). FE-SEM and SEM-EDX results show that there are two types of materials in this reddish-brown coating: (1) clustered nano-sized spherical particles with relatively weak As signals, and (2) plate-like particles rich in As (Figure 10). The semi-quantitative elemental composition, distinct morphologies and geochemical model gave important information for their identification. The nano-spheres have very similar morphology to the rapidly synthesized schwertmannite reported by Regenspurg *et al.* (2004) while the plate-like particles resembled biogenic scorodite in the study of Gonzalez-Contreras *et al.* (2010).

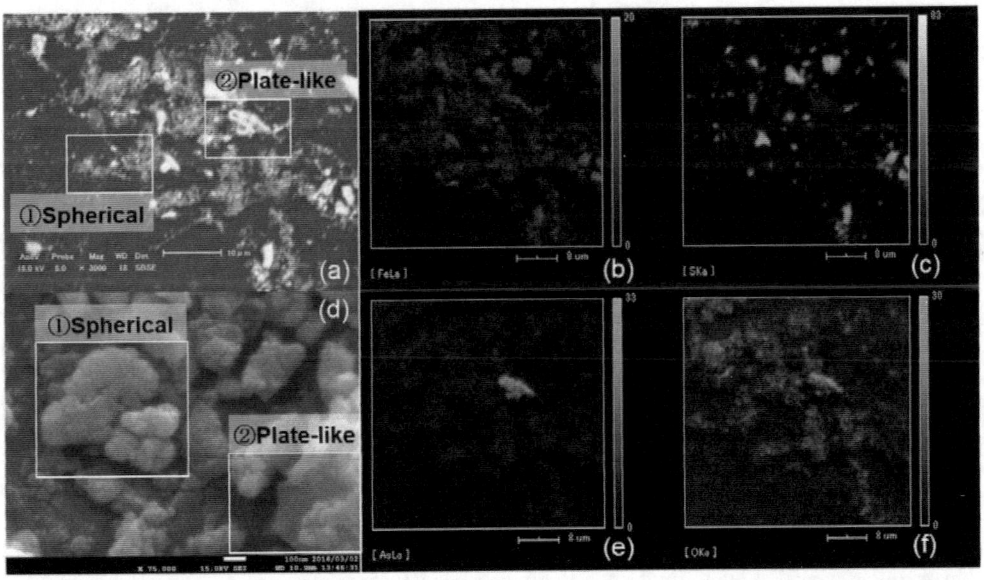

Figure 10 – SEM (a) and FE-SEM (d) photomicrographs as well as the elemental maps of Fe (b), S (c), As (e), and O (f) of the reddish-brown coating scraped from coated pyrites of Case 2B

To check whether these candidate minerals are present or not, DRIFTS spectra of hand-picked pyrite in Cases 1A and 2B after 7 days were measured (Figure 11(a)). The highest peak at 443 cm^{-1} that was detected in both samples was due to the S-S bonding of pyrite (Evangelou, 1995). Peaks at 424, 828, 864 and 920

cm^{-1} are assigned to arsenate vibrations in scorodite. The observed shifting of peaks between 1000 and 800 cm^{-1} in comparison to that of pure scorodite synthesized by Gonzalez-Contreras *et al.* (2010) could be attributed to S incorporation in our experiments. Sulfur in the structure of scorodite is fairly common especially when SO_4^{2-} concentration in the solution where it was formed is very high (Singhania *et al.*, 2006). Vibrational peaks of SO_4^{2-} (611 and 1140 cm^{-2}) and Fe-O bond (720 cm^{-1}), which are characteristic of schwertmannite, were also observed in the DRIFTS spectra (Regenspurg *et al.*, 2004). These peaks also shifted compared to pure schwertmannite most likely due to the adsorption/incorporation of As in its crystal structure (Figure 10). The XRD pattern of coated pyrite in Case 2B after 21 days shows small peaks of scorodite beneath the strong peaks of pyrite (Figure 10 (b)). In addition, the broad peaks of schwertmannite especially between 30-40 and 50-70 were visible in the XRD pattern. Based on these results and those presented earlier, the reddish-brown oxidized layer was most probably a mixture of scorodite (plate-like) and schwertmannite (clustered nano-sized spheres).

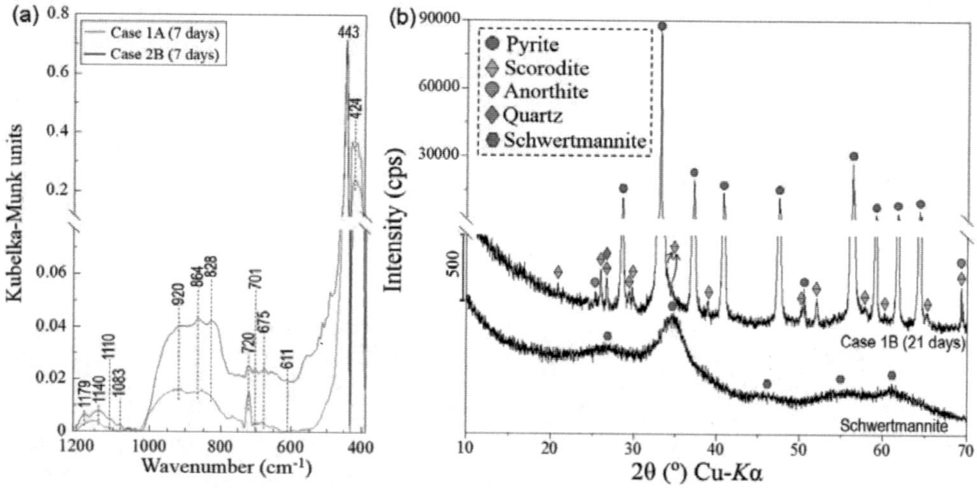

Figure 11 – (a) DRIFTS spectra of pyrite leached with and without hematite after 7 days, and (b) XRD pattern of coated pyrite after 21 days (Case 2B). The reference schwertmannite pattern in (b) was synthesized at 25°C using the method of HoungAloune *et al.*, (2014)

Implications to AMD Generation and Control

Based on our results, it is possible to encapsulate pyrite with a mechanically strong As-bearing iron-oxyhydroxysulfate coating under acidic conditions. The coating minimized not only the oxidation of pyrite but also the release of As. This means that iron-oxyhydroxysulfates could be used to control AMD generation in existing problem areas, which are already acidic, like tailings dams and abandoned mine workings by pyrite passivation.

Iron-oxyhydroxysulfates are also known precursor minerals of the more stable iron-oxyhydroxides like goethite and hematite, so their transformation, at least in theory, could facilitate the formation of iron-oxyhydroxide coatings on pyrite similar to the limonite-coated pyrite analog reported by Huminicki and Rimstidt (2009). Iron oxyhydroxides are the ideal coating material for pyrite passivation because dissolved Fe, its main component, is abundant in mine sites. Pyrite encapsulation using these minerals, however, had been difficult especially under stirred/shaken batch reactor conditions (Huminicki and Rimstidt, 2009).

As a potential solution to this dilemma, we propose using iron-oxyhydroxysulfates as precursor mineral instead of directly precipitating iron-oxyhydroxides at alkaline pH to passivate pyrite. These precursor minerals, could then be "aged" (*e.g.*, by raising the pH) to transform them into the more stable iron oxyhydroxides like goethite or hematite. Possible applications of this approach are in the treatment of tailings of porphyry copper mines where Fe, As, and SO_4^{2-} are known to exist in abundance as well as in AMD and As control of heap leaching waste rocks. This new concept might also provide a new perspective for the development of more economical and sustainable solutions to the AMD problem. However, more researches are still required to understand how the coating behaves with time or with changes in pH as well as whether it could continue protecting pyrite from oxidation when the precursor minerals start transforming into their thermodynamically more stable forms.

CONCLUSIONS

Weathered pyritic wastes typically contain soluble salts and metallic oxides that affect the overall dynamics of pyrite oxidation because of their strong influence on the leachate pH as well as in the formation of secondary mineral phases like scorodite, jarosite or schwertmannite. When the amounts of soluble salts decreased, mechanically strong oxidized layers were formed on the pyrite surface that suppressed pyrite oxidation. It was mostly composed of As-bearing iron-oxyhydroxysulfates/arsenates, which could potentially be used as precursor materials to ultimately passivate pyrite with iron-oxyhydroxides that are very stable under surface oxidizing conditions.

REFERENCES

Abraitis, P.K., Pattrick, R.A.D. & Vaughan, D.J. (2004). Variations in the compositional, textural and electrical properties of natural pyrite: A review. *International Journal of Mineral Processing* 74, 41-59. doi:10.1016/j.minpro.2003.09.002

Bethke, C.M. (1992). The Geochemist's Workbench – A user's guide to Rxn, Act2, Tact, React and Gtplot. Illinois: University of Illinois-Urbana.

Bigham, J.M., Schwertmann, U., Traina, S.J., Windland, R.L. & Wolf, M. (1996). Schwertmannite and the chemical modeling of iron in acid sulfate waters. *Geochim. Cosmochim. Acta* 60, 2111-2121. doi:10.1016/0016-7037(96)00091-9

Cornell, R.M. & Schwertmann, U. (2003). The iron oxides: Structures, properties, reactions, occurrences and uses, 2nd edition. Weinheim: Wiley-VCH Verlag GmbH & Co.

Evangelou, V.P. (1995). Pyrite oxidation and its control. New York: CRC Press.

HoungAloune, S., Kawaai, T., Hiroyoshi, N. & Ito, M. (2014). Study on schwertmannite production from copper heap leach solutions and its efficiency in arsenic removal from acidic sulfate solutions. *Hydrometallurgy* 147-148, 30-40. doi: 10.1016/j.hydromet.2014.04.001

Gonzalez-Contreras, P., Weijma, J., Van der Weijden, R. & Buisman, C.J.N. (2010). Biogenic scorodite crystallization by *Acidianus sulfidivorans* for arsenic removal. *Environ. Sci. Technol.* 44, 675-680. doi: 10.1021/es902063t

Huminicki, D.M.C. & Rimstidt, J.D. (2009). Iron oxyhydroxide coating of pyrite for acid mine drainage control. *Applied Geochemistry* 24, 1626-1634. doi:10.1016/j.apgeochem.2009.04.032

Lowson, R.T. (1982). Aqueous oxidation of pyrite by molecular oxygen. *Chemical Reviews* 82, 481-497. doi: 10.1021/cr00051a001

McKibben, M.A. & Barnes, H.L. (1986). Oxidation of pyrite in low temperature acidic solutions: Rate laws and surface textures. *Geochim. Cosmochim. Acta* 50, 1509-1520. doi:10.1016/0016-7037(86)90325-X

Parkhurst, D.L. & Appelo, C.A.J. (1999). User's Guide to PHREEQC (Version 2) – A computer program for speciation, batch-reactions, one-dimensional transport, and inverse geochemical calculations. U.S. Department of the Interior and U.S. Geological Survey, Water Resour. Invest. Rep., 99-4259.

Regenspurg, S., Brand, A. & Peiffer, S. (2004). Formation and stability of schwertmannite in acidic mining lakes. *Geochim. Cosmochim. Acta* 68, 1185-1197. doi:10.1016/j.gca.2003.07.015

Schaufuss, A.G., Nesbitt, H.W., Kartio, I., Laajalehto, K., Bancroft, G.M. & Szargan, R. (1998). Reactivity of surface chemical states on fractured pyrite. *Surface Science* 411, 321-328. doi:10.1016/S0039-6028(98)00355-0

Stefansson, A. (2007). Iron(III) hydrolysis and solubility at 25°C. Environ. Sci. Technol. 41, 6117-6123. doi: 10.1021/es070174h

Singhania, S., Wang, Q., Filippou, D., & Demopoulos, G.P. (2006). Acidity, valency and third-ion effects on the precipitation of scorodite from mixed sulfate solutions under atmospheric-pressure conditions. *Metallurgical and Materials Transaction B* 37B, 189-197. doi:10.1007/BF02693148

Tabelin, C.B. & Igarashi, T. (2009). Mechanisms of arsenic and lead release from hydrothermally altered rock. *Journal of Hazardous Materials* 169, 980-990. doi:10.1016/j.jhazmat.2009.04.049

Tabelin, C.B., Igarashi, T. & Takahashi, R. (2012). The roles of pyrite and calcite in the mobilization of arsenic and lead from hydrothermally altered rocks excavated in Hokkaido, Japan. *Journal of Geochemical Exploration* 119, 17-31. doi:10.1016/j.gexplo.2012.06.003

IMMOBILIZATION OF ARSENATE IN SOILS BY THE ADDITION OF MAGNESIUM OXIDE: AN EXPERIMENTAL AND MODELING INVESTIGATION

*T. Suzuki [1], F. Nakahara [1], T. Kawamoto [2], M. Niinae [1]

[1] *Department of Environmental Science and Engineering, Yamaguchi University, Ube, Japan*
*(*Corresponding author: tsuzuki@yamaguchi-u.ac.jp)*

[2] *Department of Sustainable Environmental Engineering, Yamaguchi University, Ube, Japan*

* The content of this paper has been published in *Journal of Hazardous Materials* 2015, 300, 680-687

ABSTRACT

MgO was chosen as an As(V) immobilization agent and a series of immobilization experiments was performed to obtain insights into the behavior of As(V) and MgO during leaching tests. Our experimental and modeling results demonstrated that As(V) immobilization by MgO consists of the following steps: (i) an increase in soil pH, (ii) desorption of As(V) from the soil, and (iii) the re-immobilization of As(V) by MgO/$Mg(OH)_2$ particles. Regarding the behavior of MgO, the modeling results showed that when the MgO dosage was 25 mg MgO/4g-drysoil or less, the majority of MgO was used to increase pH, and less than 1% of MgO was used to sorb As(V), which was consistent with the result of leaching tests showing that a high level of As(V) was leached at the MgO dosages. On the other hand, when the MgO dosage was above 25 mg MgO/4g-drysoil, the percentage of MgO used for As(V) sorption increased up to 35%, and correspondingly, the As(V) leaching level decreased to below 0.01 mg/L As at an MgO dosage of 75 mg MgO/4g-drysoil.

KEYWORDS

Kaolinite, Arsenate, Immobilization, Magnesium oxide

INTRODUCTION

One of the major arsenic transport pathways causing groundwater contamination is the migration of arsenic from artificially or naturally contaminated surface soils. Therefore, in this study, the immobilization of arsenate (As(V)) in soils was investigated using commercial-grade magnesium oxide (MgO) as a potential cost-effective measurement to prevent the occurrence of groundwater contamination with As(V). With respect to the use of MgO for remediating As(V)-contaminated environmental media, MgO has been used in several studies to remove As(V) from water (Yu et al., 2011; Jia et al., 2013; Tresintsi et al., 2014). Although these studies demonstrated that MgO effectively removes As(V) from water in water-MgO systems, the behavior of As(V) in water-soil-MgO systems has not been evaluated in the literature, and several questions remain to be answered to comprehensively evaluate the use of MgO as an As(V) immobilization agent in soils. For instance, as is well-documented in the literature (Ford, 2002; Lenoble et al., 2002; Zeng et al., 2008), iron oxy-hydroxide (FeOOH) is one of the important adsorbents for As(V) in both natural and engineered systems. Therefore, it is anticipated that the presence of FeOOH in soils significantly influences the behavior of As(V) in water-soil-MgO systems and consequently, the immobilization performance by MgO. Another important question to be answered is the behavior of MgO during the immobilization treatment. In our previous studies (Suzuki et al., 2013a, 2013b), commercial-grade MgO was demonstrated to effectively immobilize lead and fluoride, and the immobilization effectiveness was indicated to last more than 100 years against the exposure to acid rain (pH 4.0). However, these studies focused on the behavior of contaminants. As a result, little is known about the behavior of MgO during the immobilization treatment. The lack of understanding of the behavior of MgO during immobilization treatment is one of the main concerns leading to relatively low public confidence in immobilization treatments.

Accordingly, the objective of this study was to perform a series of experiments to better understand the behavior of both As(V) and MgO in water-soil-MgO systems using soils made of kaolinite or mixture of kaolinite and goethite, with the ultimate objective of evaluating the use of MgO as an As(V) immobilization agent. The insights obtained from this study are useful for increasing public confidence in As(V) immobilization by the addition of MgO as a cost-effective strategy for preventing the occurrence of groundwater contamination.

MATERIAL AND METHODS

Materials

ASP® 170 Georgia kaolinite was purchased from BASF Japan Ltd. (Tokyo, Japan). Preliminary experiments showed that the measured Brunauer, Emmett and Teller (BET) surface area of the kaolinite was 20.1 m^2/g, and the water content under water-saturated conditions was 28.6 wt.%. Goethite (α-FeOOH), with a minimum purity of 99%, was obtained from Alfa Aesar (Ward Hill, MA), and the measured BET surface area was 18.5 m^2/g. The immobilization material used in this study was commercial-grade MgO obtained from Ube Industries, Ltd. (Ube, Japan). The MgO is a light-burn MgO produced by calcining $Mg(OH)_2$, which is precipitated from seawater under high pH conditions. The average particle size is approximately 3 μm, and the BET surface area ranges from 25 to 35 m^2/g. The chemical composition is 91.7 wt.% MgO and 0.97 wt.% CaO, with the remainder including 3.22 wt.% ignition loss.

Batch As(V) Adsorption Experiment onto Kaolinite and Goethite

Batch adsorption experiments of kaolinite and goethite were performed in 50-mL polypropylene bottles at 23 ± 1 °C. Three stock solutions [As(V) solution (200 mg/L As provided as $Na_2HAsO_4 \cdot 7H_2O$), back electrolyte $NaNO_3$ solution (0.1 mol/L), and DDI water], which were pre-adjusted to the desired pH by HNO_3 or NaOH, were mixed in polypropylene bottles to prepare 40 mL of solution containing predetermined concentrations of As(V) and 0.01 mol/L of $NaNO_3$. After 8.0 g of kaolinite or 0.5 g of goethite was added to the bottle, the bottle was shaken using a horizontal shaker. After 48 hours (equilibrium was confirmed by preliminary experiments), the suspension was filtered through a 0.2-μm mixed cellulose ester type membrane filter and the final pH was measured. The filtrate was then analyzed for As(V) to calculate the As(V) sorbed on the adsorbents.

Batch As(V) Sorption Experiment with MgO

The batch sorption experiments were performed in 250-mL perfluoxy copolymer (PFA) bottles at 23 ± 1 °C. 200 ml of As(V) solution (100 mg/L As) was mixed with 1.0 g of MgO, and the pH was adjusted by HNO_3 to achieve a final pH value in the range of 9.5 to 10.4. After the bottle was shaken for 48 hours (equilibrium was confirmed by preliminary experiments), the suspension was filtered through a 0.1-μm glass fiber filter and the final pH was measured. The residue remaining on the filter was rinsed with 20 L of acetone to remove the aqueous phase without disturbing the chemical equilibrium between the aqueous and solid phases. The molar elemental ratio of As/Mg at the surface of the residue was then measured using a K-Alpha™+ X-ray photoelectron spectrometer (XPS) system (Thermo Fisher Scientific Inc., Waltham, MA).

Preparation of Artificially Contaminated Soils, Immobilization Treatment, and Leaching Tests

Artificially contaminated soils were prepared by mixing 200 g of soils made of kaolinite or mixture of kaolinite and goethite (Table 1) and 1 L of As(V) aqueous solution having a predetermined As(V) concentration for 24 hours. The aqueous pH was held constant at 6.8 during mixing by delivering a NaOH solution. After 24 hours of mixing, the mixture was centrifuged at 2,000×g for 15 minutes. The supernatant was filtered using a 0.2-μm mixed cellulose ester type membrane filter, and the filtrate was analyzed for As(V). The water-saturated contaminated soils obtained after removing the supernatant was dried at 105 °C for 4 hours and stored under naturally dry conditions until use. The As(V) concentrations in the artificially contaminated soils were calculated based on the mass balance of As(V) and are summarized in Table 1.

Table 1 – Soil composition and As(V) content in artificially contaminated soils used in this study

	Soil composition	As(V) content	
		(mg As/kg)	Cs-ini (μmol/g)
Soil A	Georgia kaolinite 100 wt.%	48.6	0.65
Soil B	Georgia kaolinite 100 wt.%	144	1.93
Soil C	Georgia kaolinite 98 wt.% + goethite 2 wt.%	245	3.27

MgO treatment for As(V) immobilization under water-saturated conditions was performed by thoroughly mixing 10 g of dried contaminated soils and a predetermined amount of MgO with a spatula in a 50-mL polypropylene centrifuge tube. 4.0 ml of DDI water was added to saturate the soils and the centrifuge tubes were vigorously shaken for 60 minutes using a horizontal shaker. The tubes were then left undisturbed for a week before subsequent leaching tests. After a week of immobilization treatment, the soils were naturally dried and the leaching tests were performed by mixing 4.0 g of the dried soils with 40 ml of leaching solution. The leaching solution consisted of DDI water adjusted to a pH in the range of 5.8 to 6.3 by NaOH or H_2SO_4 solutions. The mixture was then shaken for 6 hours using a horizontal shaker. This is a standardized method in compliance with the soil contamination countermeasures law in Japan. Additionally, two leaching tests with a leaching time of 24 hours and 48 hours were also performed in this study to investigate the influence of leaching time on the leached As(V) level. After the predetermined leaching time, the suspension was filtered through a 0.2-μm mixed cellulose ester type membrane filter and the filtrate was analyzed for pH, Mg^{2+} and As(V).

MODELING THE As(V) IMMOBILIZATION PERFORMANCE BY MgO TREATMENT

Under the leaching condition of this study (4.0 g of soils and 40 mL of leaching solution), the mass balance of As(V) in the leaching tests is expressed by Equation 1:

$$4.0\ C_{ini\text{-}s} = 4.0\ C_{equ\text{-}s} + 0.04\ C_{equ\text{-}w} + 0.917/40.3\times10^{6}\ m\ \alpha\ C_{equ\text{-}MgO} \quad (1)$$

in which $C_{ini\text{-}s}$ (μmol/g) is the initial As(V) concentration in soils (Table 1), $C_{equ\text{-}w}$ (μmol/L) is the equilibrium As(V) concentration in water, $C_{equ\text{-}s}$ (μmol/g) is the equilibrium As(V) concentration in soils, $C_{equ\text{-}MgO}$ (atom As/atom Mg) is the molar elemental ratio of As/Mg at the surface of MgO determined by XPS, 0.917 is the purity of MgO, 40.3 (g/mol) is the molar mass of MgO, 10^6 is a unit conversion factor to convert mol to μmol, m (g/4g-drysoil) is the amount of MgO added in 4 g of soils, and α is the fraction of the immobilizing agent used for As(V) sorption. $C_{equ\text{-}s}$ is further expressed by considering two components of soils; i.e., kaolinite and goethite:

$$C_{equ\text{-}s} = f_{Fe}\ C_{equ\text{-}Fe} + (1 - f_{Fe})\ C_{equ\text{-}ka} \quad (2)$$

in which f_{Fe} (-) is the weight fraction of goethite in soils, and $C_{equ\text{-}Fe}$ and $C_{equ\text{-}ka}$ (μmol/g) are the equilibrium As(V) concentration in goethite and kaolinite, respectively.

Our modeling process started with the analysis of (ad)sorption isotherms of kaolinite, goethite, and MgO obtained from batch experiments. The objective of this analysis was to express $C_{equ\text{-}ka}$, $C_{equ\text{-}Fe}$, and $C_{equ\text{-}MgO}$ in Equation 1 as a function of $C_{equ\text{-}w}$. For kaolinite and goethite, which adsorb As(V) onto the surface, if we use kaolinite as an example, the relationship between $C_{equ\text{-}ka}$ and $C_{equ\text{-}w}$ can be represented by the Langmuir equation $C_{equ\text{-}ka} = C_{max\text{-}ka}K_{L\text{-}ka}C_{equ\text{-}w}/(1+K_{L\text{-}ka}C_{equ\text{-}w})$, where $C_{max\text{-}ka}$ (μmol/g) is the maximum adsorption per unit mass kaolinite and $K_{L\text{-}ka}$ (L/μmol) is the Langmuir constant. On the other hand, based on XRD and TEM-EDS analysis, we assumed that the dominant mechanism of As(V) sorption by the immobilizing agent is the uptake by co-precipitation with $Mg(OH)_2$. With this assumption, $C_{equ\text{-}MgO}$ can be represented using $C_{equ\text{-}w}$ by the linear relationship of $C_{equ\text{-}MgO} = K_{MgO}C_{equ\text{-}w}$, where K_{MgO} ((atom As/atom Mg)/(μmol/L)) is an As(V) sorption coefficient to $Mg(OH)_2$.

After determining the Langmuir parameters (C_{max} and K_L) and K_{MgO}, the unknown values in Equation 1 were $C_{equ\text{-}w}$ and α. Therefore, in the second modeling process, we fitted α using the $C_{equ\text{-}w}$ values obtained from leaching tests for Soil B. Finally, we predicted the $C_{equ\text{-}w}$ values for Soils A and C using α with the assumption that α is independent of soil composition and As(V) content. The rationale for this assumption will be discussed later where relevant.

RESULTS AND DISCUSSION

Batch As(V) Adsorption Experiments onto Kaolinite and Goethite

The percentage of As(V) adsorbed onto kaolinite or goethite is shown in Figure 1 as a function of the equilibrium aqueous pH for different As(V) initial concentrations. The adsorption of As(V) on these adsorbents both strongly depended on the aqueous pH and was favored at a lower pH. This result indicated that these adsorbents play a significant role as As(V) adsorbents in acidic to weak alkaline pH conditions. However, the increase in the pH of the solution diminished the capability to adsorb As(V), and especially for kaolinite, practically no adsorption occurred at a pH above 10.0. To obtain the Langmuir parameters (C_{max} and K_L) for kaolinite and goethite as a function of the aqueous pH, the adsorption data shown in Figure 1 were divided at 0.5 pH intervals, and each dataset was analyzed using a linear form of the Langmuir adsorption equation (i.e., $C_{equ\text{-}w}/C_{equ\text{-}s} = 1/K_LC_{max} + C_{equ\text{-}w}/C_{max}$) by plotting $C_{equ\text{-}w}/C_{equ\text{-}s}$ vs. $C_{equ\text{-}w}$. The obtained Langmuir parameters are summarized in Table 2. These results clearly demonstrated that goethite has significantly higher C_{max} than kaolinite, especially at higher pH conditions, indicating the behavior of

As(V) in soils and the As(V) immobilization performance by MgO is strongly affected by the presence of goethite.

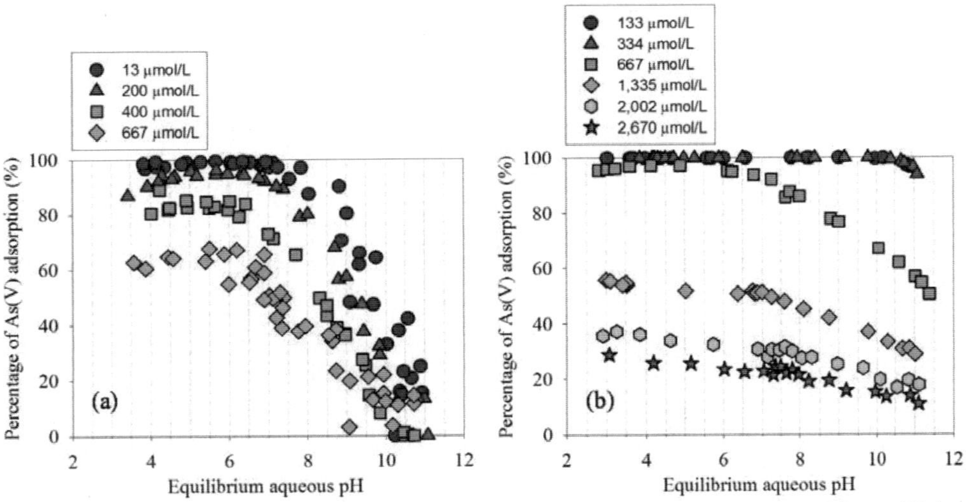

Figure 1 – Percentage of adsorbed As(V) onto (a) kaolinite and (b) goethite as a function of the equilibrium aqueous pH for different As(V) initial concentrations

Table 2 – Parameters in the Langmuir adsorption equation onto kaolinite and goethite, and a sorption coefficient to the MgO immobilizer

pH	Kaolinite			Goethite		
	$C_{max\text{-}ka}$ (µmol/g)	$K_{L\text{-}ka}$ (L/µmol)	Corr.coeff.	$C_{max\text{-}Fe}$ (µmol/g)	$K_{L\text{-}Fe}$ (L/µmol)	Corr.coeff.
6.5-6.9	1.92	0.121	0.979	63.3	0.0224	0.993
7.0-7.4	1.61	0.117	0.984	61.7	0.0162	0.985
7.5-7.9	1.33	0.077	0.995	63.7	0.0280	0.995
8.0-8.4	1.25	0.0251	0.958	56.8	0.0420	0.996
8.5-8.9	1.33	0.0101	0.881	48.5	0.0616	0.991
9.0-9.4	0.875	0.00834	0.917	51.8	0.0770	1.000
9.5-9.9	0.0560	0.0109	0.789	50.5	0.0204	0.986
10.0-10.4	No As(V) adsorption			46.5	0.0158	0.989
10.5-10.9	No As(V) adsorption			44.1	0.00357	0.850

Batch As(V) Sorption Experiment with MgO

The relationship between $C_{equ\text{-}MgO}$ and $C_{equ\text{-}w}$ obtained from the batch sorption experiments is shown in Figure 2. Taking into consideration the error in the data, it is reasonable to conclude that $C_{equ\text{-}MgO}$ can be expressed as a linear function of $C_{equ\text{-}w}$; $C_{equ\text{-}MgO} = K_{MgO}C_{equ\text{-}w}$, and that the aqueous pH does not affect the $C_{equ\text{-}MgO}/C_{equ\text{-}w}$ ratio in the pH range investigated in this study. Therefore, in the pH range of 9.5 to 10.4, which is relevant to soil pH after MgO treatment, K_{MgO} is determined to be 0.164 based on the slope of the linear regression line.

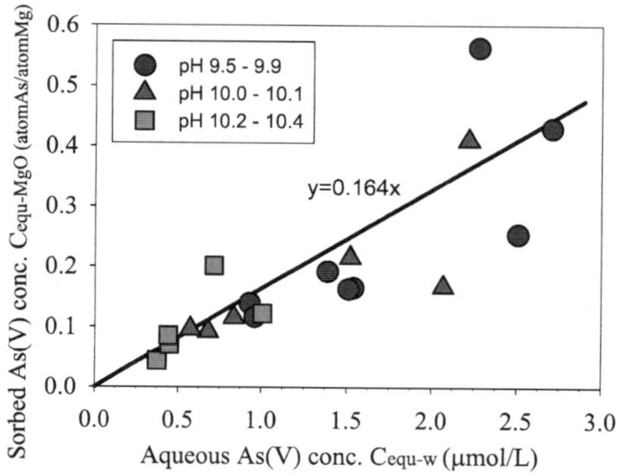

Figure 2 – Relationship between sorbed As(V) concentration in the very surface of $MgO/Mg(OH)_2$ particles and aqueous As(V) concentration in three different pH ranges

Modeling the Leached As(V) Concentration for Soil B

The final pH and $C_{equ\text{-}w}$ for Soil B treated with different dosages of MgO are shown in Figure 3 (left). The increase in MgO dosage gradually increased the soil pH, but the pH was maintained at approximately 10.4 at an MgO dosage of 50 mg/4g-drysoil or more due to limited $Mg(OH)_2$ water solubility. Taking into account the results shown in Figure 1 illustrating that As(V) adsorption onto kaolinite and goethite is not favored at high pH conditions, this result indicated that As(V) immobilization by MgO consists of the following steps: (i) an increase in soil pH, (ii) desorption of As(V) from soils, and (iii) the re-immobilization of As(V) by $MgO/Mg(OH)_2$ particles. In fact, the leached As(V) concentration was decreased by increasing the MgO dosage, and adding a sufficient amount of MgO (75 mg/4g-drysoil and more) effectively reduced the As(V) leaching level to below 0.01 mg/L or 0.13 μmol/L, the permissible level for leaching tests in Japan.

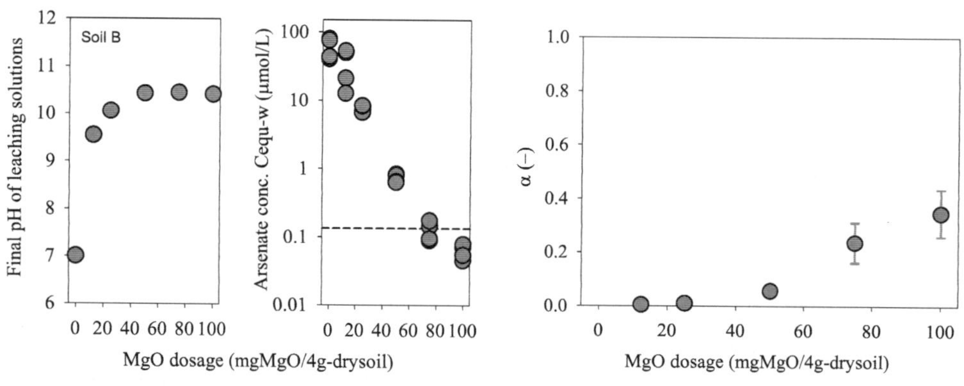

Figure 3 – (left) Experimental results of leaching tests for Soil B. The dashed line represents the environmental water quality standard for a leaching test in Japan (0.01 mg/L or 0.13 μmol/L). (right) The fraction of the MgO involved into As(V) sorption (α) as a function of dosage for Soil B

Figure 3 (right) shows the fitted α values calculated according to Equation 1 using Langmuir parameters corresponding to the pH of the leaching solution (Table 2), K_{MgO} (i.e., 0.164), and $C_{equ\text{-}w}$ shown in Figure 2. As shown in Figure 3 (right), when the MgO dosage was 25 mg MgO/4g-drysoil or less, α was less than 0.01, indicating that less than 1% of MgO was used to immobilize As(V). This result is reasonable because when the MgO dosage was small, the majority of MgO was supposed to be dissolved to provide OH^-, which is required to overcome the soil buffer capacity and to increase the pH of the system. On the other hand, when the MgO dosage was 50 mg MgO/4g-drysoil or more, only a minor fraction of MgO was consumed to increase the pH, and therefore, the remaining MgO could potentially be used to sorb As(V).

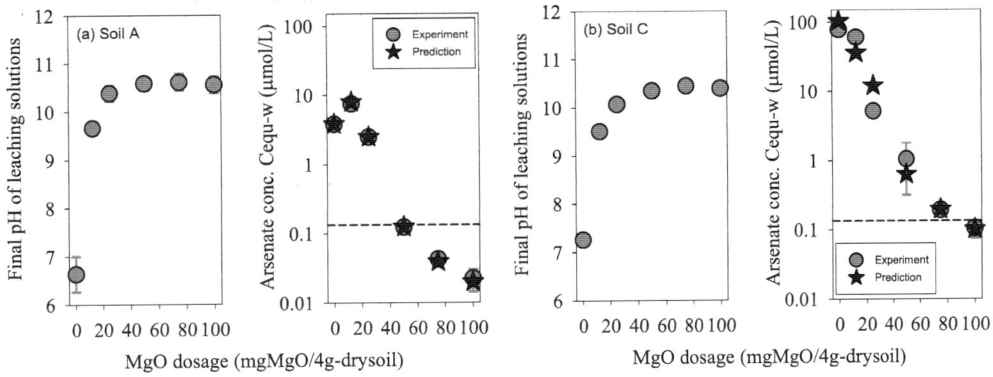

Figure 4 – Experimental and predicted results of leaching tests for Soils A and C. (left) experimentally measured final pH of the leaching solutions and (right) comparison of leached As(V) concentration between experimental results and prediction. The dashed line in the right figure represents the environmental water quality standard for a leaching test in Japan (0.01 mg/L or 0.13 μmol/L)

REFERENCES

Ford, R.G. (2002). Rates of hydrous ferric oxide crystallization and the influence on co-precipitated arsenate. *Environmental Science & Technology*, *36*, 2459–2463.

Jia, Y., Luo, T., Yu, X., Sun, B., Liu, H., & Huang, J. (2013). A facile template free solution approach for the synthesis of dypingite nanowires and subsequent decomposition to nanoporous MgO nanowires with excellent arsenate adsorption properties. *RSC Advances*, *16*, 5430–5437.

Lenoble, V., Bouras, O., Deluchat, V., Serpaud, B., & Bollinger, J.-C. (2002). Arsenic adsorption onto pillared clays and iron oxides. *Journal of Colloid and Interface Science*, *255*, 52–58.

Suzuki, T., Nakamura, A., Niinae, M., Nakata, H., Fujii, H., & Tasaka, Y. (2013). Lead immobilization in artificially contaminated kaolinite using magnesium oxide-based materials: Immobilization mechanisms and long-term evaluation. *Chemical Engineering Journal*, *232*, 380–387.

Suzuki, T., Nakamura, A., Niinae, M., Nakata, H., Fujii, H., & Tasaka, Y. (2013). Immobilization of fluoride in artificially contaminated kaolinite by the addition of commercial-grade magnesium oxide. *Chemical Engineering Journal*, *233*, 176–184.

Tresintsi, S., Simeonidis, K., Katsikini, M., Paloura, E.C., Bantsis, G., & Mitrakas, M. (2014). A novel approach for arsenic adsorbents regeneration using MgO., *Journal of Hazardous Materials*, 265, 217–225.

Yu, X., Luo, T., Jia, Y., Zhang, Y., Liu, J., & Huang, J. (2011). Porous hierarchically micro- / nanostructured MgO : Morphology control and their excellent performance in As(III) and As(V) removal. *The Journal of Physical Chemistry C*, *115*, 22242–22250.

Zheng, H., Arashiro, M., & Giammar, D. E. (2008). Effects of water chemistry and flow rate on arsenate removal by adsorption to an iron oxide-based sorbent. *Water Research*, *42*, 4629–4636.

THE TRANSFORMATION OF TWO-LINE FERRIHYDRITE INTO CRYSTALLINE PRODUCTS: INFLUENCE OF THE MEDIA (SULFATE VS. NITRATE)

Danni Zhang[1] Shaofeng Wang[1] and *Yongfeng Jia[1]

[1]*Institute of Applied Ecology, Chinese Academy of Sciences*
No. 72 Wenhua Road
Shenyang, China
*(*Corresponding author: yongfeng.jia@iae.ac.cn)*

ABSTRACT

Ferrihydrite is ubiquitous in soils, groundwater and aquatic sediments and is usually classified as "2-line" or "6-line" according to the number of diffuse XRD peaks. It has high surface area and reactivity; therefore, it may serve as a sink for sequestering trace metals, metalloids and organic matters via adsorption and co-precipitation. Although considerable attention was paid to the transformation process of this thermodynamically metastable solid, there are still some controversies on the mechanisms of the formation of goethite or hematite from the ferrihydrite precursor. This work systematically investigated the transformation of 2-line ferrihydrite produced by using different neutralization regent (CaO vs. NaOH) into crystalline products at different pHs (4, 7 and 8), three temperatures (25, 40, 75 °C) and in different media (sulfate vs. nitrate). X-ray diffraction (XRD), Raman, Fourier transform infrared spectroscopy (FTIR) and chemical extraction were employed to characterize the transformed solids as well as the redistribution of Fe between the amorphous and more crystalline phases. The results showed that the transformation product in nitrate media was hematite (major) and goethite (minor), which was independent of pH, temperature or neutralization reagent. In sulfate media, goethite was the dominant solid under slightly acidic conditions (pH 4) with 6-line ferrihydrite as the intermediate product, whereas neutral to slightly alkaline conditions still favor the formation of hematite. Chemical extraction results indicated that there was only ~3 % of amorphous phase detectable after aging at 25 °C for 280 days at slight acidic pH. In contrast, ~70 % of the iron remained as an amorphous phase under neutral to slightly alkaline conditions. The results suggested that the pH as well as the reaction media (sulfate ion) plays an important role in the 2-line ferrihydrite transformation process.

KEYWORDS

Two-line ferrihydrite, co-precipitation, transformation

INTRODUCTION

Ferrihydrite (Fh) is a poorly crystalline, disordered iron(III) oxyhydroxide commonly occurring in the environment in soils, surface and groundwater sediments (Jambor & Dutrizac, 1998). There are two types of Fh ("2-line" and "6-line") on the basis of the number of diffuse XRD peaks. Due to its high surface area and reactivity, it usually shows large adsorptive capacities for some trace elements such as As, Pb, and Cd (Das, Hendry, & Essilfie-Dugha, 2011a; Landry, Koretsky, Lund, Schaller, & Das, 2009), hence playing an important role in the sequestration of contaminants in the environment. In hydrometallurgical operations, Fh is considered as the dominant arsenic-bearing phase in the Fe^{III}-As^{V} co-precipitates generated during several arsenic removal and immobilization processes (Waychunas, Rea, Fuller, & Davis, 1993).

Although considerable attention has been paid to the chemical and physical properties of Fh, there still remains controversy regarding its crystal structure due to the difficulty in its characterization (Jambor & Dutrizac, 1998; Michel et al., 2007a). Drits et al. (1993) proposed the three-phase structural model for Fh suggested that Fh was a mixture of three components and the main structural difference between the two types of Fh was the size of their coherent scattering domains (CSD). However, a new single-phase structure model of Fh was proposed by Michel et al. (2007a) based on pair distribution function (PDF) analysis of high-energy X-ray total scattering data. The results indicated that the structure of Fh contains 20 % tetrahedral and 80 % octahedral Fe which is closely related to the Baker–Figgis δ–Keggin cluster. Furthermore, the authors suggested that the two types of Fh had similarities in atomic arrangement and the primary difference between them was the limited size of the CSD imposed by differences in the average primary crystallite size (Michel et al., 2007b). However, other authors raised the question about the correctness of the proposed single-phase structural model due to the lack of XRD and EXAFS results (Rancourt, & Meunier, 2008; Manceau, 2009), whereas, a recently published paper supported the single-phase structure by using the deuterated Fh as the sample to perform the PDF analysis (Harrington et al., 2011).

Ferrihydrite is a thermodynamically metastable solid and may transform to more crystalline and stable forms of iron oxyhydroxide. However, there is still no consensus on the mechanisms of formation of the more crystalline iron oxyhydroxide from the precursor Fh (Cudennec & Lecerf, 2006; Hansel, Benner, Nico, & Fendorf, 2004). Some published papers reported that low and high pH values favor the formation of goethite, while pH values near neutral favor the formation of hematite (Cudennec & Lecerf, 2006; Schwertmann, Stanjek & Becher, 2004). The transformation of 2-line Fh (precipitated in nitrate media) under various pHs and temperatures were systematically investigated recently by Das et al. (2011b). The results indicated a two-stage crystallization process, with goethite as an intermediate and hematite being the major product. In addition, the effects of some common cations (e.g. nickel and lead) and anions (e.g. phosphate and arsenic) on the transformation process of Fh were also widely studied (Das, Hendry, & Essilfie-Dugha, 2011a; Ford, Kemner, & Bertsch, 1999; Gálvez, Barrón, & Torrent, 1999).

Thus, the objectives of this work were (1) to investigate the influence of the reaction media (nitrate vs. sulfate) on the transformation process of Fh as well as the end product at various pHs and temperatures; (2) to study the redistribution of Fe between the amorphous and more crystalline phases during the transformation process.

EXPERIMENTAL

All chemicals were of analytical grade and deionized (DI) water was used for all experiments. All glassware was cleaned by soaking in dilute HNO_3 (5 %) overnight and rinsed three times with DI-water. Ferrihydrite used in this investigation refers to 2-line Fh unless otherwise specified.

Preparation and Aging Treatment of Ferrihydrite

Ferrihydrite was prepared via fast hydrolysis of ferric iron. The Fe^{III} stock solutions were prepared by dissolving $Fe_2(SO_4)_3 \cdot 5H_2O$ or $Fe(NO_3)_3 \cdot 9H_2O$ in DI-water. The acidic Fe^{III} solutions were neutralized to pH 4, 7 or 8 with 1 M NaOH solution or slaked lime while the mixtures were stirred mechanically. After stabilization for 1 h at room temperature, the slurries were transferred to tightly capped conical flasks: the total volume of each neutralized solution was 500 mL with an iron concentration of 3000 mg/L. The slurries were allowed to age in a water bath maintained at 25, 40, or 75 ± 2 °C by an immersion circulator.

The aging process lasted up to 280 days and, at prescribed time intervals, the solid phase was separated by filtration, washed three times with DI-water and vacuum-dried at 60 °C. All solid samples were manually ground into a fine powder before spectroscopic characterization and chemical extraction. Aliquots of weighed solid samples were dissolved in 1 M HCl solution and diluted with DI-water for the measurement of the total concentrations of Fe in the solid phase. Some samples especially those aged for a long time at higher temperatures were dissolved in 0.2 M, pH 3 NH_4-oxalate acid buffer + 0.1 M ascorbic acid solution in order to completely dissolve the transformed solids (Keon, Swartz, Brabander, Harvey, & Hemond, 2001).

The reference materials (goethite and hematite) were synthesized according to the methods described by Cornell and Schwertmann (2000) and the data for these samples were also obtained.

Chemical Extraction Procedure

Single-step 0.4 M HCl extraction was conducted for some selected samples. The solid solution ratio was 1:1000 (0.02 g to 20 mL). The suspensions were shaken for 1 hour at ambient temperature and then filtered through a 0.22-μm membrane for the analysis of the concentrations of Fe in the filtrates. Amorphous iron oxyhydroxide in the aged solid was the target phase for this extractant (Ford, Kemner, & Bertsch, 1999; Keon, Swartz, Brabander, Harvey, & Hemond, 2001).

Determination of the Iron Concentrations

The iron concentration of the solutions was determined on an atomic absorption spectrometer (AA240, Varian, Inc. Palo Alto, CA). The detection limit of the instrument for iron was 0.5 mg/L.

XRD Analysis

The mineralogical characteristics of the ferrihydrite precipitated at different pHs and aged at different temperatures were analyzed on a Rigaku D/max 2400 X-ray diffractometer (Rigaku Corporation, Japan) equipped with a copper target ($CuK_{\alpha 1}$ radiation, $\lambda = 1.5418$ Å), a crystal graphite monochromator and a scintillation detector. The equipment was run at 56 kV and 182 mA by step scanning from 10° to 80°2θ with increments of 0.02°2θ.

Attenuated Total Reflectance Fourier Transform Infrared (ATR-FTIR) Analysis

The infrared spectra of the powdered samples were collected on a Thermo Nicolet 6700 Fourier transform infrared spectrometer. A horizontal ATR accessory with a diamond crystal as reflection element was used. The scans ranged from 650 to 4000 cm^{-1} with the resolution of 4 cm^{-1} and the spectra were obtained as an average of 100 scans.

Raman Analysis

Raman spectra were collected on a Thermo DXR Raman microscope (Thermo Fisher Scientific, Madison, WI, USA) equipped with a solid laser diode operating at 780 nm and a 400 lines/mm grating. Approximately 0.1 g of the ground sample was put on a glass slide using a metal spatula. The data were obtained from 60 to 1800 cm^{-1} with an energy resolution of 4 cm^{-1}. The laser beam produced a spot size of approximately 3 μm in diameter using the 10×short distance objective. The scans were collected at 0.033 % (0.8 mW) of the laser output at the microscope exit to avoid phase transformation.

RESULTS AND DISCUSSION

XRD Characterization of the Solids

The XRD patterns of the fresh and aged iron(III) oxyhydroxides precipitated at different temperatures and pHs from sulfate media are compared in Figure 1. The freshly prepared solids made at different pH values showed characteristic 2-line Fh XRD peaks at ~34° and ~61°, and the results were consistent with previous reports (Jia, Xu, Fang, & Demopoulos, 2006). No measurable transformation of Fh was observed from XRD analysis even after aging for 280 days at 25 °C and pH 8. In comparison, the XRD patterns indicated that when aging at pH 4 and pH 7, at 25 °C, Fh partially transformed to goethite and hematite after 75 and 280 days, respectively. The transformation process was substantially shortened when higher temperatures were applied. For example, when the samples were aged at 75 °C, Fh transformed to hematite at neutral to slightly alkaline pH values after less than 7 days. Goethite was the dominant solid when aging under slightly acidic conditions whereas hematite was the major product under neutral to slightly alkaline conditions. This result was different from the observation by Das et al. (2011b) that the Fh precipitated in nitrate media transformed to hematite in both acidic and alkaline solutions.

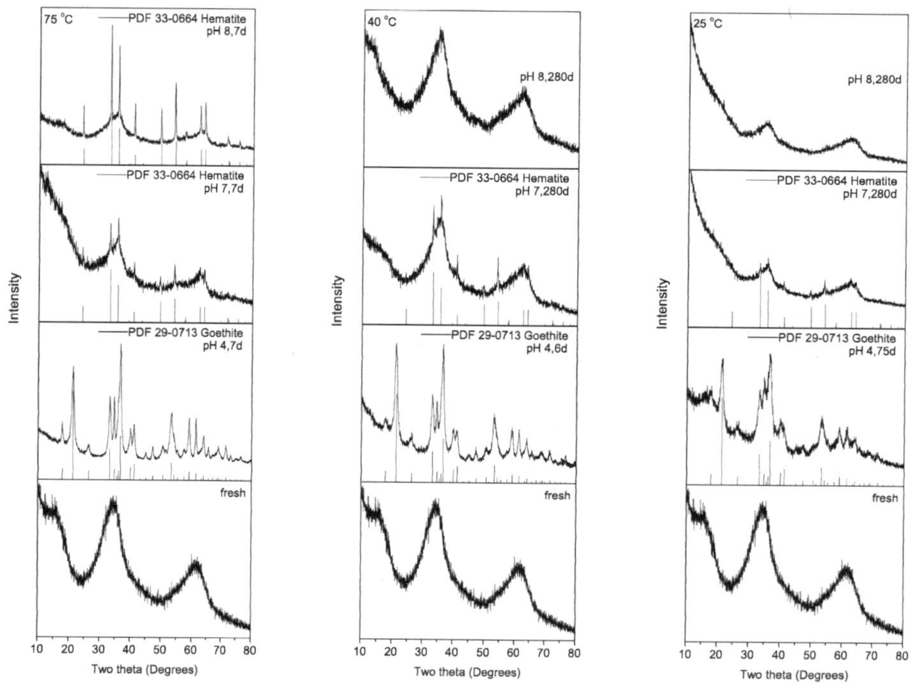

Figure 1 – XRD patterns of the iron(III) oxyhydroxides precipitated from sulfate media at pH 4, 7, and 8 (neutralized by CaO), aged at 25 °C, 40 °C, and 75 °C

The FTIR spectra of the freshly precipitated Fh and the aged samples were compared with those of the reference materials in Figure 2. The band position at ~791 and ~885 cm^{-1} are assigned to O-H deformation-2 and -1 vibrations of goethite, respectively, and are comparable with some published data (Ruan, Frost, Kloprogge, & Duong, 2002). The results showed that goethite had already formed after aging of the Fh at pH 4, 75 °C for 1 d in sulfate media. The band located at ~895 cm^{-1} was ascribed to the Fe-OH stretching vibration and hematite was the end product for those samples precipitated and aged at pH 8 from the same reaction media (Cagnasso, Boero, Franchini, & Chorover, 2010). However, for the characterization of the iron oxyhydroxides, especially for the heterogeneous mixtures, Raman spectroscopy is a more suitable method due to the broad overlapping bands in FTIR spectra which make solid phase identification more difficult (Das & Hendry, 2011). Fh shows characteristic Raman bands at ~362, ~490, and ~719 cm^{-1} which are ascribed to Fe-O in plane, out-of-plane as well as symmetric stretching vibration, respectively (Figure 3); this is consistent with the studies by Mazzetti and Thistlethwaite (2002). The strong bands located at ~302, ~385, and ~483 cm^{-1} was due to the Fe-OH symmetric bending, Fe-O-Fe/-OH symmetric stretching and Fe-OH asymmetric stretching vibrations of goethite, respectively. The characteristic hematite bands appeared at ~225, ~294 and ~407 cm^{-1} representing Fe-O symmetric stretching and bending vibrations (Legodi & de Waal, 2007).

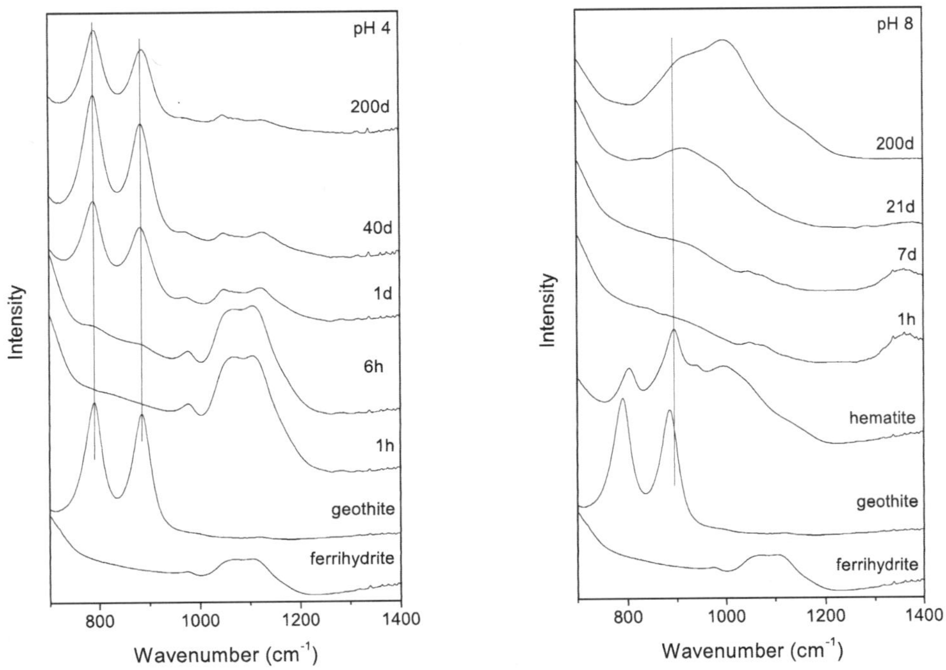

Figure 2 – FTIR spectra of the iron(III) oxyhydroxides precipitated from sulfate media at pH 4 and 8 (neutralized by CaO), and aged at 75 °C

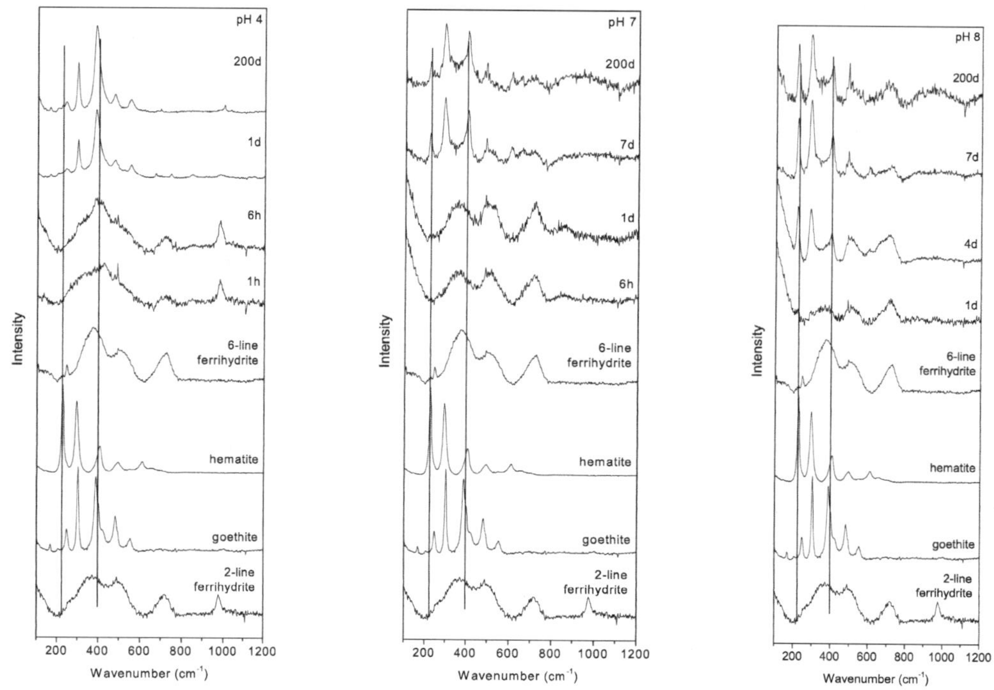

Figure 3 – Raman spectra of the iron(III) oxyhydroxides precipitated from sulfate media at pH 4, 7 and 8 (neutralized by CaO), and aged at 75 °C

Partitioning of Iron in the Solid Phase

There exists a range of methods including X-ray absorption spectroscopy (XAS), XRD and chemical extraction to investigate the solid phases during the crystallization process of Fh. XRD analysis was more widely accessible than XAS and is a conventional technology to identify solid phases. However, when it comes to the quantitative determination of the transformed solid phases, the extraction procedure was recommended and it is a more accurate method since XRD results indicated that most of the solid phase during the transformation process was a heterogeneous mixture consisting of the untransformed Fh and a more crystalline product (Figure 1).

The percentage of Fe extracted by 0.4 M HCl from the aged Fh at various pHs and temperatures was plotted as a function of aging time (Figure 4). For the pH 7 and 8 samples, the amount of Fe extracted by 0.4 M HCl solution was >95 % after aging for 1 day at 25 °C, and <70 % after aging for 280 days, suggesting that the untransformed very amorphous form of Fh was the dominant phase. For the pH 4 samples, the extractable Fe after aging for 280 days drastically decreased from >80 % to <3 % suggesting that almost all of the Fh converted to a more crystalline iron oxyhydroxide (i.e., goethite in this work). The results showed that <7 % of Fe can be extracted from the pH 7 and 8 samples and <0.5 % from pH 4 samples after aging for 280 days at 75 °C. The results suggested that the transformation rate of Fh is faster at slightly acidic conditions and this phenomenon is independent of the aging temperature. Previous studies observed that the rate of crystallization of Fh was faster with increasing pH when Fh was precipitated from nitrate media (Das, Hendry, & Essilfie-Dughan, 2011b; Schwertmann & Murad, 1983).

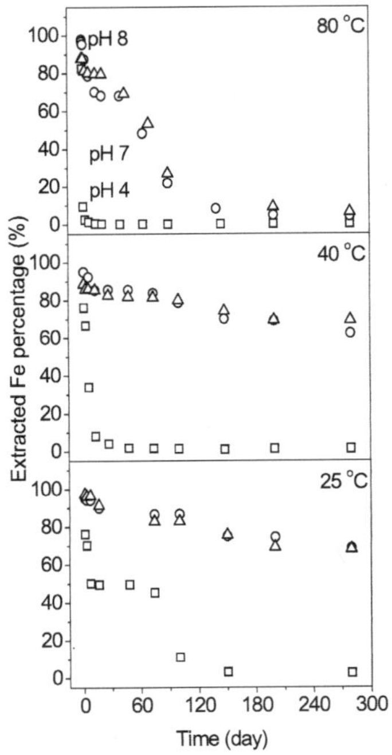

Figure 4 – The percentages of extracted Fe from different aged iron(III) oxyhydroxides with 0.4 M HCl solution at room temperature

Transformation Mechanisms

The solid phase within the first few hours of aging was characterized using XRD and the results were compared in Figure 5. In neutral to slightly alkaline conditions, hematite was the major end product no matter which neutralizing reagent (NaOH vs. CaO) or reaction media was used (data not shown). Therefore, only the results from the pH 4 samples are presented here. It can be seen from the XRD analysis that in sulfate media, Fh was first transformed to 6-line Fh after being aged for 4 hours, followed by conversion to goethite as the end product. However, in nitrate media, no intermediate product was observed. The characteristic hematite peak appeared after the sample was aged for 24 hours and then some weak peaks belonging to goethite were observed. These results are in agreement with previously reports (Das, Hendry, & Essilfie-Dughan, 2011b). In sulfate media, especially under acidic conditions, sulfate can be incorporated into the solid phase (Jia & Demopoulos, 2008), and it is a strong Fe^{3+} binding anion which may change the Fh transformation route (Zhu, Legg, Zhang, Gilbert, Ren, Banfield, & Waychunas, 2012). The overall results suggested that the reaction media as well as the pH values can exert a significant influence on the Fh crystallization process and the end product.

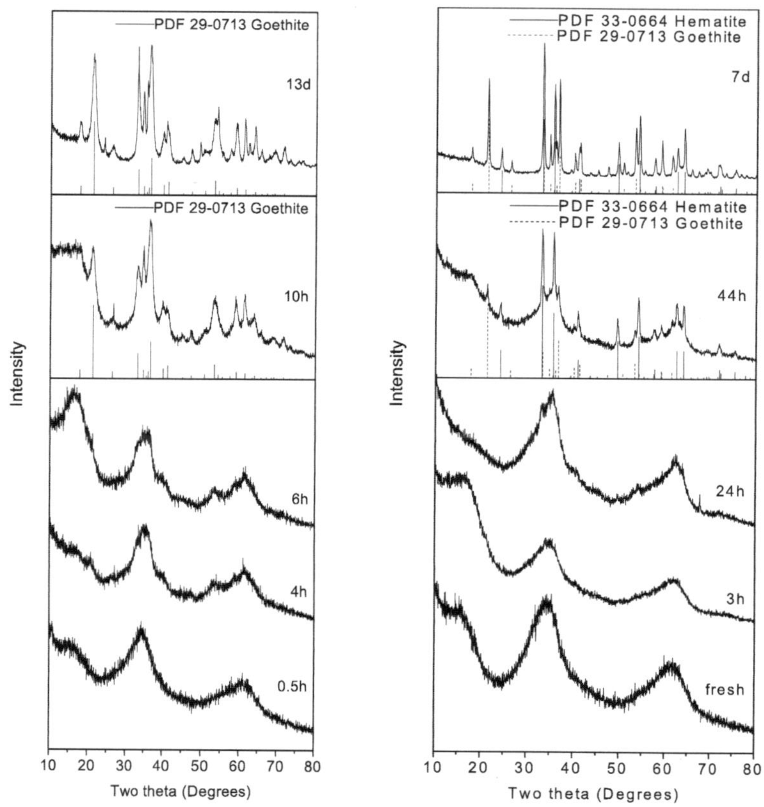

Figure 5 – XRD patterns of the fresh and 75 °C-aged Fh precipitated from sulfate media (left) and nitrate media (right) at pH 4 (neutralized by NaOH)

ACKNOWLEDGEMENTS

The authors thank the Natural Science Foundation of China for their financial support (41530643, 41303088).

REFERENCES

Cagnasso, M., Boero, V., Franchini, M. A., & Chorover, J. (2010). ATR-FTIR studies of phospholipid vesicle interactions with α-FeOOH and α-Fe_2O_3 surfaces. *Colloids and Surfaces B: Biointerfaces, 76*(2), 456–467. doi: 10.1016/j.colsurfb.2009.12.005

Cornell, R. M., & Schwertmann, U. (2000). *Iron oxides in the laboratory: preparation and characterization (2nd ed)*. doi: 10.1002/9783527613229

Cudennec, Y., & Lecerf, A. (2006). The transformation of ferrihydrite into geothite or hematite, revisited. *Journal of Solid State Chemistry, 179*(3), 716–722. doi: 10.1016/j.jssc.2005.11.030

Das, S., & Hendry, M. J. (2011). Application of Raman spectroscopy to identify iron minerals commonly found in mine wastes. *Chemical Geology, 290*(3–4), 101–108. doi:

10.1016/j.chemgeo.2011.09.001

Das, S., Hendry, M. J., & Essilfie-Dughan, J. (2011a). Effects of adsorbed arsenate on the rate of transformation of 2-line ferrihydrite at pH 10. *Environmental Science and Technology, 45*(13), 5557–5563. doi: 10.1021/es200107m

Das, S., Hendry, J. M., & Essilfie-Dughan, J. (2011b). Transformation of two-line ferrihydrite to goethite and hematite as a function of pH and temperature. *Environmental Science and Technology, 45*(1), 268–275. doi: 10.1021/es101903y

Drits, V. A., Sakharov, B. A., Salyn, A. L., & Manceau, A. (1993). Structural model for ferrihydrite. *Clay Minerals, 28*(2), 185–207. doi: 10.1180/claymin.1993.028.2.02

Ford, R. G. (2002). Rates of hydrous ferric oxide crystallization and the influence on co-precipitated arsenate. *Environmental Science and Technology, 36*(11), 2459–2463. doi: 10.1021/es015768d

Ford, R. G., Kemner, K. M., & Bertsch, P. M. (1999). Influence of sorbate-sorbent interactions on the crystallization kinetics of nickel-and lead-ferrihydrite co-precipitates. *Geochimica et Cosmochimica Acta, 63*(1), 39–48. doi: 10.1016/s0016-7037(99)00010-1

Gálvez, N., Barrón, V., & Torrent, J. (1999). Effect of phosphate on the crystallization of hematite, goethite, and lepidocrocite from ferrihydrite. *Clays and Clay Minerals, 47*(3), 304–311. doi: 10.1316/CCMN.1999.0470306

Hansel, C. M., Benner, S. G., Nico, P., & Fendorf, S. (2004). Structural constraints of ferric (hydr)oxides on dissimilatory iron reduction and the fate of Fe(II). *Geochimica et Cosmochimica Acta, 68*(15), 3217–3229. doi: 10.1016/j.gca.2003.10.041

Harrington, R., Hauser, D.B., Xu, W., Bhandari, N., Michel, F.M., Brown, G.E., Parise, J.B. (2011). Neutron pair distribution function study of two-line ferrihydrite. *Environmental Science and Technology, 45*(23), 9883–9890. doi: 10.1021/es2020633

Jambor, J. L., & Dutrizac, J. E. (1998). Occurrence and constitution of natural and synthetic ferrihydrite, a widespread iron oxyhydroxide. *Chemical Reviews, 98*(7), 2549–2586. doi: 10.1021/cr970105t

Jia, Y. F., & Demopoulos, G. P. (2008). Co-precipitation of arsenate with iron(III) in aqueous sulfate media: effect of time, lime as base and co-ions on arsenic retention. *Water Research, 42*(3), 661–668. doi: 10.1016/j.watres.2007.08.017

Jia, Y., Xu, L., Fang, Z., & Demopoulos, G. P. (2006). Observation of surface precipitation of arsenate on ferrihydrite. *Environmental Science and Technology, 40*(10), 3248–3253. doi: 10.1021/es051872+

Keon, N. E., Swartz, C. H., Brabander, D. J., Harvey, C., & Hemond, H. F. (2001). Validation of an arsenic sequential extraction method for evaluating mobility in sediments. *Environmental Science and Technology, 35*(13), 2778–2784. doi: 10.1021/es001511o

Landry, C. J., Koretsky, C. M., Lund, T. J., Schaller, M., & Das, S. (2009). Surface complexation modeling of Co(II) adsorption on mixtures of hydrous ferric oxide, quartz and kaolinite. *Geochimica et Cosmochimica Acta, 73*(13), 3723–3737. doi: 10.1016/j.gca.2009.03.028

Legodi, M. A., & de Waal, D. (2007). The preparation of magnetite, goethite, hematite and maghemite of pigment quality from mill scale iron waste. *Dyes and pigments, 74*(1), 161–168. doi:

10.1016/j.dyepig.2006.01.038

Manceau, A. (2009). Evaluation of the structural model for ferrihydrite derived from real-space modelling of high-energy X-ray diffraction data. *American Mineralogist, 44*(1), 19–34. doi: 10.1180/claymin.2009.044.1.19

Mazzetti, L., & Thistlethwaite, P. J. (2002). Raman spectra and thermal transformations of ferrihydrite and schwertmannite. *Journal of Raman Spectroscopy, 33*(2), 104–111. doi: 10.1002/jrs.830

Michel, F. M., Ehm, L., Antao, S. M., Lee, P. L., Chupas, P. J., Liu, G., Parise, J. B. (2007a). The structure of ferrihydrite, a nanocrystalline material. *Science, 316* (5832), 1726–1729. Doi:0.1126/science.1142525

Michel, F. M., Ehm, L., Liu, G., Han, W. Q., Antao, S. M., Chupas, P. J., Gillow, J. (2007b). Similarities in 2-line and 6-line ferrihydrite based on pair distribution function analysis of X-ray total scattering. *Chemistry of Materials, 19*(6), 1489–1496. doi: 10.1021/cm062585n

Rancourt, D. G., & Meunier, J. L. (2008). Constraints on structural models of ferrihydrite as a nanocrystalline material. *American Mineralogist, 93*(8–9), 1412–1417. doi: 10.2138/am.2008.2782

Ruan, H. D., Frost, R. L., Kloprogge, J. T., & Duong, L. (2002). Infrared spectroscopy of goethite dehydroxylation: III. FT-IR microscopy of in situ study of the thermal transformation of goethite to hematite. *Spectrochimica Acta Part A: Molecular and Biomolecular Spectroscopy, 58*(5), 967–981. doi: 10.1016/S1386-1425(01)00574-1

Schwertmann, U., & Murad, E. (1983). Effect of pH on the formation of goethite and hematite from ferrihydrite. *Clays and Clay Minerals, 31*(4) 277–284. doi: 10.1346/CCMN.1983.0310405

Schwertmann, U., Stanjek, H., & Becher, H. H. (2004). Long-term in vitro transformation of 2-line ferrihydrite to goethite/hematite at 4, 10, 15 and 25 °C. *Clay Minerals, 39*(4), 433–438. doi: 10.1180/0009855043940145

Waychunas, G. A., Rea, B. A., Fuller, C. C., & Davis, J. A. (1993). Surface chemistry of ferrihydrite: part 1. EXAFS studies of the geometry of co-precipitated and adsorbed arsenate. *Geochimica et Cosmochimica Acta, 57*(10), 2251–2269. doi: 10.1016/0016-7037(93)90567-G

Zhu, M., Legg, B., Zhang, H., Gilbert, B., Ren, Y., Banfield, J. F., & Waychunas, G. A. (2012). Early stage formation of iron oxyhydroxides during neutralization of simulated acid mine drainage solutions. *Environmental Science and Technology, 46*(15), 8140–8147. doi: 10.1021/es301268g

EFFECT OF SYNTHESIS CONDITIONS ON PARTICLE SHAPE OF SCORODITE

*E. Shibata and T. Nakamura

[1]*Institute of Multidisciplinary Research for Advanced Materials, Tohoku University*
2-1-1, Katahira, Aoba-ku
Sendai 980-8577, Japan
*(*Corresponding author: etsuro@tagen.tohoku.ac.jp)*

ABSTRACT

An atmospheric process for the synthesis of scorodite (DMSP®) has been developed, in which ferrous ions were oxidized by oxygen gas in the presence of pentavalent arsenic ions. Under optimum conditions, the synthesized scorodite ($FeAsO_4 \cdot 2H_2O$) was well crystallized and the crystals were faceted; the particle sizes were as large as 15 to 20 μm. In this study, iron (II) sulfate heptahydrate ($FeSO_4 \cdot 7H_2O$), arsenic (V) acid (60 mass% H_3AsO_4) solution, and distilled water were used to prepare the original experimental solution (700 ml), in which the contents of Fe(II) and As(V) were 55.9 g/L and 50.0 g/L, respectively. The effects of various synthesis conditions, such as the type of oxidizing gas (oxygen or air), gas injection rate, temperature, and the initial pH of the solution, on the size and shape of the scorodite particles were investigated. From the experimental results, the changes in the shape of the scorodite particles were clarified for each synthesis condition.

KEYWORDS

DMSP®, Arsenic, Stabilization, Scorodite, Ferrous oxidation

INTRODUCTION

DOWA has developed a process for synthesizing well-crystallized scorodite ($FeAsO_4{\bullet}2H_2O$). The process is named and trademarked as DMSP® (Fujita et al., 2008; Fujita et al., 2008; Fujita et al., 2009; Fujita et al., 2009; Fujita et al., 2009, Kubo et al., 2010). In this process, oxygen gas is injected into a sulfuric acid solution containing large amounts of As(V) and Fe(II). Faceted scorodite crystals as large as 15–20 µm were generated through the pre-formation of a gel-like precursor. However, it is known that the synthesized scorodite particles become smaller and more aggregated when the synthesis conditions deviate from the optimum conditions.

In this study, the effects of various synthesis conditions, such as the type of oxidizing gas (oxygen or air), gas flow rate, temperature, and initial pH of the solution, on the size and shape of the scorodite particles were investigated. The experimental results clarified the changes in the shape of the scorodite particles for each synthesis condition.

EXPERIMENTAL METHOD

Details of the experimental method have been reported in previous papers (Shibata et al., 2012; Shibata et al., 2014). A solution containing 55.9 g/L of Fe(II) and 50 g/L of As(V) was prepared from iron(III) sulfate n-hydrate ($Fe_2(SO_4)_3{\bullet}nH_2O$), arsenic(V) acid (60 mass% H_3AsO_4), and distilled water. The initial pH of the solution (pH: 0.4–1.6) was adjusted by adding sulfuric acid solution. In the experiments, the prepared solution (700 ml) was heated to the desired temperature (95, 70, and 50°C) with the injection of Ar gas. After the desired temperature was reached, the reaction was started by the injection of oxygen gas or air, instead of Ar gas, into the solution at a constant flow rate (100, 200, 400, 700, and 1000 mL/min). During the reaction, the solution was stirred using a shaft stirrer at 1000 rpm. The pH and ORP of the solution were measured during the reaction by using pH and ORP electrodes. The total reaction time was set at 420 min, and the sampling of the formed slurry was carried out at 10, 60, and 180 min. Table 1 summarizes the experimental conditions of this study.

Table 1 – Experimental conditions for the synthesis of scorodite particles

No.	Temp., °C	Oxidation gas	Initial pH	Flow rate, mL/min
1	95	O_2	1.47	100
2	95	O_2	1.56	200
3	95	O_2	1.55	400
4	95	O_2	1.59	700
5	95	O_2	1.57	1000
6	70	O_2	1.32	700
7	50	O_2	1.25	700
8	95	O_2	0.72	700
9	70	O_2	0.72	700
10	50	O_2	0.67	700
11	95	O_2	0.34	700
12	70	O_2	0.42	700
13	50	O_2	0.46	700
14	95	Air	1.57	700
15	70	Air	1.43	700
16	50	Air	1.17	700

The Fe(II) concentration of the solution was analyzed by redox titration using potassium permanganate. Total Fe and total As in the solution were analyzed using ICP-AES. The crystallinity and structures of the synthesized scorodite samples were determined using an X-ray diffractometer (XRD). In addition, the morphology of the particles was observed using scanning electron microscopy (SEM).

EXPERIMENTAL RESULTS AND DISCUSSION

Effect of the Oxygen Gas Injection Rate

Fe(II) in the solution was oxidized to Fe(III) by oxygen gas injection according to the following equation (1).

$$Fe^{2+}(aq) + H^{+}(aq) + 1/4O_2(g) = Fe^{3+}(aq) + 1/2H_2O(l) \quad (1)$$

The scorodite was formed by the reaction between Fe(III) and $H_3AsO_4^0$ because the arsenic existed as $H_3AsO_4^0$ in the solution at pH values lower than 2.

$$Fe^{3+}(aq) + H_3AsO_4^0(aq) + 2H_2O(l) = FeAsO_4{\cdot}2H_2O(s) + 3H^{+}(aq) \quad (2)$$

The total reaction can be expressed as equation (3). The pH of the solution decreased during the reaction owing to the formation of H^+. The ORP of the solution increased with an increase of the ratio between Fe(III) and Fe(II).

$$Fe^{2+}(aq) + H_3AsO_4^0(aq) + 1/4O_2(g) + 3/2H_2O(l) = FeAsO_4{\cdot}2H_2O(s) + 2H^{+}(aq) \quad (3)$$

Figures 1 and 2 show the change in the total As concentration and the pH of the solution with reaction time for each injection rate of oxygen gas (100, 200, 400, 700, and 1000 mL/min). The effect of the injection rate on scorodite formation was not significant. The reaction was restricted during the initial 60 min only for an injection rate of 700 mL/min. This disparity in the results was considered to be the result of an increase in the viscosity of solution owing to the formation of gel-like precursors during the initial stage of reaction. The structure and size of the formed scorodite particles were similar for each injection rate. The size of the scorodite particles was around 20 μm for a reaction time of 420 min.

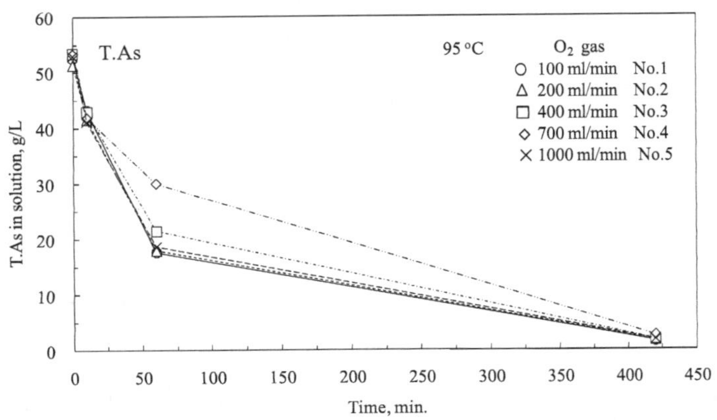

Figure 1 – Changes in total As concentrations of the solution plotted as a function of reaction time for each injection rate of oxygen gas

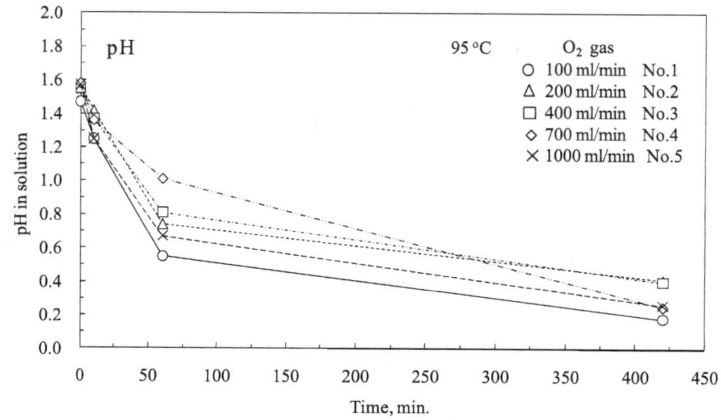

Figure 2 – Changes in the pH of the solution plotted as a function of reaction time for each injection rate of oxygen gas

Comparison Between Oxygen Gas and Air

Figure 3 shows the changes in the total As, Fe(II), and Fe(III) concentrations as function of the reaction time for oxygen gas injection at a rate of 100 mL/min and 95°C. Figure 4 shows the changes in the total As, Fe(II), and Fe(III) concentrations as a function of reaction time for air injection at 700 mL/min and 95°C. The formation of scorodite was obviously restrained by the rate of air injection; however, the total amount of injected oxygen was higher in the case of air injection (700 mL/min versus 100 mL/Min) compared to the case of oxygen gas injection. The oxidation reaction of Fe(II) to Fe(III) was estimated to decrease remarkably in the case of air injection. From these results, it is clear that the oxygen partial pressure is more important than the total amount of oxygen to promote the formation of scorodite.

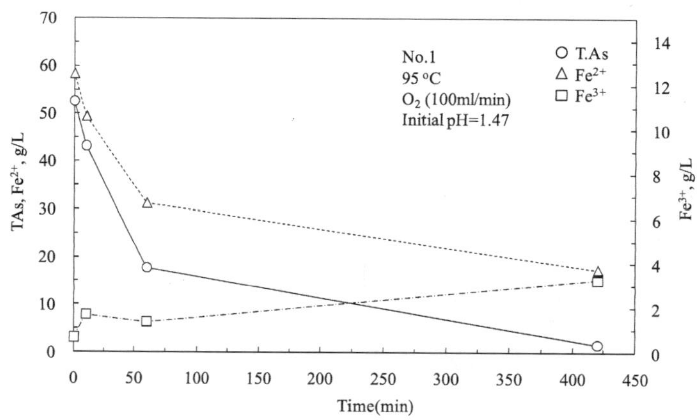

Figure 3 – Changes in the total As, Fe(II), and Fe(III) concentrations in the solution plotted as a function of reaction time for oxygen gas injection (100 mL/min, 95°C)

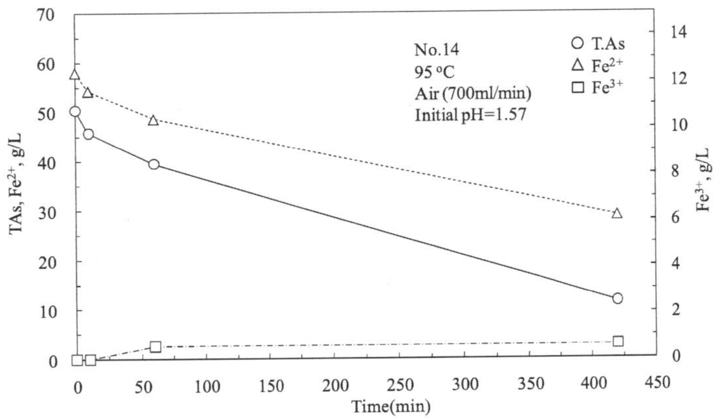

Figure 4 – Changes in the total As, Fe(II), and Fe(III) concentrations in the solution plotted as a function of reaction time in the case of air injection (700 ml/min, 95°C)

Figure 5 shows SEM images of the scorodite particles synthesized with oxygen gas injection (100 mL/min, 95°C) and with air injection (700 mL/min, 95°C). Within 10 minutes of reaction, the scorodite particles were around 10 μm in size under both conditions; however, the formation of gel-like precursors was more pronounced in the case of oxygen gas injection. The scorodite particles grew to sizes of 10–20 μm after a reaction time of 60 minutes. Finally, at 420 minutes of reaction time, the size of the scorodite particles was about 20 μm, and the surfaces of the crystals were covered with small particles in the case of oxygen gas injection. The formation of small scorodite particles at the end of reaction was estimated to be caused by nucleation due to an increase in the Fe(III) concentration of the solution. In contrast, in the case of air injection, the smaller scorodite particles aggregated instead of growing into faceted scorodite crystals.

Figure 5 – SEM images of the scorodite precipitated in the case of oxygen gas injection (100 mL/min, 95°C) and air injection (700 mL/min, 95°C)

Effect of the Reaction Temperature

Figures 6 to 8 show the changes in the total As, Fe(II), and Fe(III) concentrations of the solution as a function of the reaction time for each reaction temperature (95, 70, and 50°C) in the case of oxygen gas injection (700 mL/min). The decrease in the total As and Fe(II) concentrations at 70°C was less than that at 95°C. Further, the increase in the Fe(III) concentration was restrained at 70°C, although scorodite formation was slightly promoted at that temperature. The decrease in total As and Fe(II) concentrations was restrained at 50°C, but the Fe(III) concentration increased significantly. The scorodite formation reaction, shown in equation (2), was remarkably restrained at 50°C, and this is in contrast to the proceeding reaction for the oxidation of Fe(II) to Fe(III), shown in equation (1).

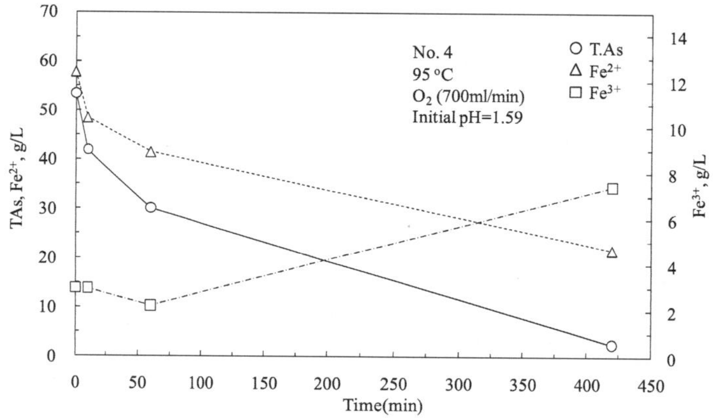

Figure 6 – Changes in the total As, Fe(II), and Fe(III) concentrations in the solution plotted as a function of reaction time in the case of oxygen gas injection (700 mL/min) at 95°C

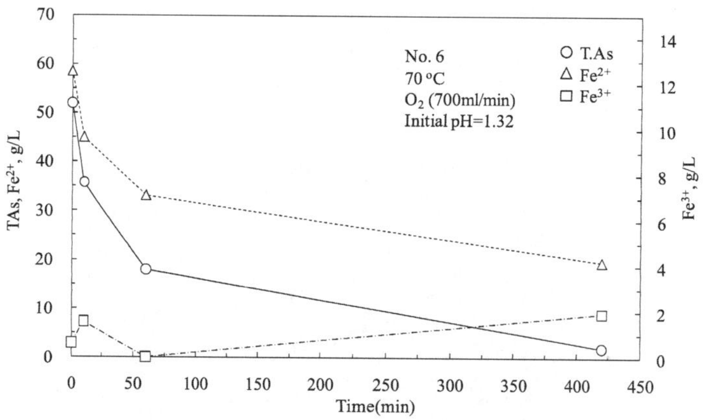

Figure 7 – Changes in the total As, Fe(II), and Fe(III) concentrations in the solution plotted as a function of reaction time in the case of oxygen gas injection (700 mL/min) at 70°C

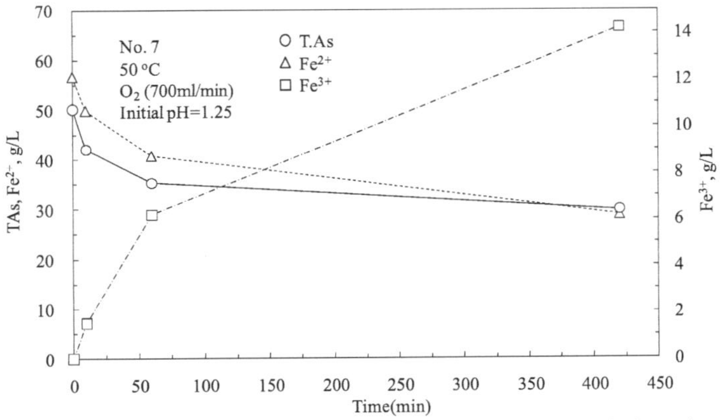

Figure 8 – Changes in the total As, Fe(II), and Fe(III) concentrations of the solution plotted as a function of reaction time in the case of oxygen gas injection (700 mL/min) at 50°C

Figure 9 shows the SEM images of the scorodite particles formed in the case of oxygen gas injection (700 mL/min) after 420 min at 95, 70, and 50°C. The morphology of the scorodite particles formed at the different reaction temperatures varies significantly. Faceted scorodite crystals about 20 μm in size formed at 95°C. Further, smaller particles with sizes of several μm are attached to the larger crystals. In contrast, at 70°C, aggregates of almond-shaped scorodite particles several μm in size formed. The scorodite formed at 50°C has a porous structure consisting of fibers; its crystallinity is lower, as confirmed by the broad peaks of the XRD patterns.

Figure 9 – SEM images of the scorodite particles precipitated during oxygen gas injection (700 mL/min) after 420 min of reaction at temperatures of 95, 70, and 50°C

Effect of the Initial pH of the Solution

The initial pH of the solution in experiment No.4 shown in Figure 6 was 1.59 for oxygen gas injection at 700 mL/min at 95°C. For comparison, Figures 10 and 11 show the changes in the total As, Fe(II), and Fe(III) concentrations as a function of the reaction time for different initial pH values (0.72 and 0.34). The final concentration of total As was higher in the lower initial pH experiment, and this suggests that scorodite formation was restrained at pH 0.34. In contrast, the rate of oxidation of Fe(II) to Fe(III) is about the same at both pH values as shown by the similar Fe(II) concentrations in both experiments. Therefore, the Fe(III) concentration of the solution increased when the initial pH of the solution was lower, presumably because less scorodite precipitated.

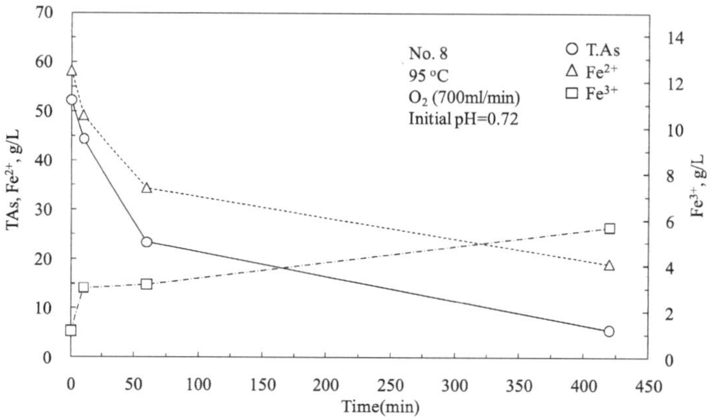

Figure 10 – Changes in the total As, Fe(II), and Fe(III) concentrations of the solution plotted as a function of reaction time in the case of oxygen gas injection (700 mL/min, 95°C) at an initial pH of 0.72

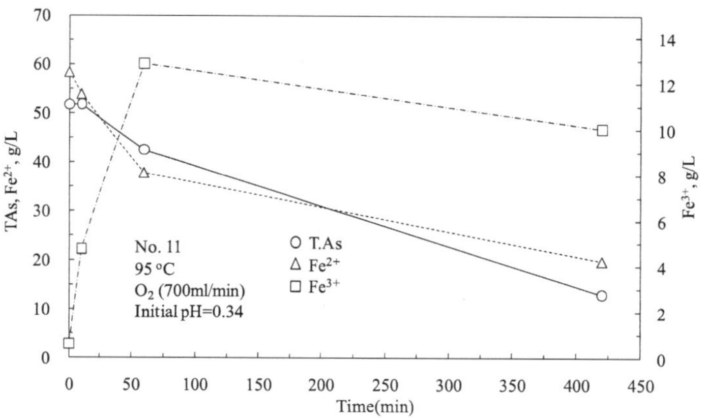

Figure 11 – Changes in the total As, Fe(II), and Fe(III) concentrations of the solution plotted as a function of reaction time in the case of oxygen gas injection (700 mL/min, 95°C) at an initial pH of 0.34

Figure 12 shows the SEM images of the scorodite particles formed during oxygen gas injection (700 mL/min, 95°C) under different initial pH conditions (1.59, 0.72, and 0.34). Larger faceted crystals formed after 420 minutes for an initial pH of 1.59. For initial pH values of 0.72 and 0.34, the scorodite developed a cauliflower-like structure, in which the initially formed particles became smaller at the lower initial pH values. The needle-like particles precipitated in the first 10 minutes of reaction at an initial pH of 0.34. After 10 minutes, these particles aggregated to form a cauliflower-like structure. In contrast, at an initial pH of 1.59, gel-like precursors formed at 10 and 60 minutes, but these precursors disappeared by the end of the reaction period. This suggests that the gel-like precursors transformed to the larger faceted scorodite crystals.

Figure 12 – SEM photos of the scorodite particles precipitated during oxygen gas injection (700 mL/min) at 95°C and different initial pH values of 1.59, 0.72, and 0.34

CONCLUSIONS

The shape and structure of the scorodite particles were investigated under different synthesis conditions, such as oxidizing gas (oxygen or air), rate of gas injection (100, 200, 400, 700, and 1000 mL/min), temperature (95, 70, and 50°C), and the initial pH of the solution (1.59, 0.72, and 0.34). Larger faceted scorodite crystals, which have a lower solubility in the environment, are preferred for disposal. For the synthesis of such large faceted crystals in this process, the synthesis conditions must be as follows: temperature of 95°C, oxygen oxidizing gas, and initial pH of the solution of about 1.5.

REFERENCES

Fujita, T. et al. (2008). Novel atmospheric scorodite synthesis by oxidation of ferrous sulfate solution. Part I. *Hydrometallurgy*, Vol.90, 92-102.

Fujita, T. et al. (2008). Novel atmospheric scorodite synthesis by oxidation of ferrous sulfate solution. Part II Effect of temperature and air. *Hydrometallurgy*, Vol. 90, 85-91.

Fujita, T. et al. (2009). Effect of pH on atmospheric scorodite synthesis by oxidation of ferrous ions: Physical properties and stability of the scorodite. *Hydrometallurgy*, Vol. 96, 189-198.

Fujita, T. et al. (2009). Preparation of an As(V) solution for scorodite synthesis and a proposal for an integrated As fixation process in a Zn refinery. *Hydrometallurgy,* Vol. 96, 300-312.

Fujita, T. et al. (2009). Immobilization of arsenic from novel synthesized scorodite-Analysis on solubility and stability. *Materials Transactions*, Vol. 50, 321-331.

Kubo, H., Abumiya, M., & Matsumoto, M. (2010). Dowa Mining Scorodite Process® -Application to copper hydrometallurgy. *Copper 2010*, GDMB, Germany, 2947-2958.

Shibata, E. et al. (2012). Impurity Effects on the Synthesis of Crystalline Scorodite. *Proc. 51st Annual Conference of Metallurgists of CIM* (COM 2012), 241-252.

Shibata, E. et al. (2014). Crystalline Scorodite Synthesized in Fe(II) Solution Containing As(V). *Proc. 7th International Symposium Hydrometallurgy (Hydro 2014)*, Vancouver, Canada, Vol. II, 533-573.

QUANTITATIVE MODELING OF A NEUTRALIZATION AND CO-PRECIPITATION PROCESS WITH FERRIHYDRITE FOR THE SIMULTANEOUS REMOVAL OF As(V) AND Hg(II)

T. Kato, T. Uchida, M. Yagisawa, S. Hobo and *C. Tokoro

Department of Science and Engineering, Waseda University
3-4-1 Okubo, Shinjuku-ku
Tokyo, Japan 169-8555
*(*Corresponding author: tokoro@waseda.jp)*

ABSTRACT

In this study, quantitative modeling was carried out for the treatment of simulated waste water containing As(V), Zn(II) and Hg(II) as toxic elements, and Fe(III), Na(I) and sulfate as coaxing ions. The objective was to predict the pH and residual concentrations of all the elements in solution after the neutralization/co-precipitation treatment. To accomplish this, the removal mechanism for As(V) and Hg(II) by ferrihydrite precipitation was first investigated for simulated waste water containing only As(V) or Hg(II). From the experimental results, it was confirmed that the removal mechanism involves not only surface complexation but also surface precipitation. However, surface complexation is the main mechanism for As(V) and Hg(II) removal to meet the target levels regulated in Japan. Next, the surface complexation model was constructed for the removal of As(V) and Hg(II) by ferrihydrite. The specific surface area and site density were determined from reference values proposed by Dzombak and Morel (1990). By comparing the experimental and calculated results, it was confirmed that our constructed model could successfully represent the experimental results. Furthermore, the surface complexation model was coupled with chemical equilibrium calculations and applied to the neutralization of simulated waste water containing As(V), Hg(II), Zn(II), Na(I), Fe(III) and sulfate. The constructed model could successfully represent the pH edge for As(V), Hg(II), Zn(II), Fe(III) and SO_4^{2-} removal. In addition, the constructed model could represent the ion balance. Thus, our constructed model is useful to estimate the amount of neutralizer that is needed to achieve a target pH.

KEYWORDS

Ferrihydrite, surface complexation, modeling, mercury, arsenic, sulfate

INTRODUCTION

Waste water generated in mining, such as acid mine drainage and mineral processing waste water, often contains several inorganic elements and other pollutants. Acid mine drainage (AMD) is a global environmental problem (Dold, 2008). In Japan, there are about 7,000 abandoned mines of which 450 generate AMD (JOGMEC, 2003). Moreover, the waste water produced in mineral processing contains significant quantities of suspended solid particles, heavy metal ions, flotation reagents, organics and other pollutants, etc. that result in a waste of water resources and environmental pollution. For example, waste water discharged from mines constitutes 1/10 of the total amount of the industrial waste water discharged in China (Chen et al., 2009).

These waste waters contain several inorganic toxic elements. Among these toxic elements, arsenic (As(V)) and mercury (Hg(II)) have long been known as toxic elements, and are regulated by Japanese effluent standards to less than 0.01 and 0.005 mg dm^{-3} in discharge water, respectively. Therefore, a highly efficient method is required for the treatment of inorganic toxic elements in waste water. Additionally, more efficient treatment methods to reduce the amount of sludge generated after waste water treatment is desirable because final disposal sites are limited. Co-precipitation with ferrihydrite is widely used as a robust and effective process for the simultaneous removal of several kinds of inorganic and organic species. However, quantitative knowledge for the removal process is limited.

Therefore, the objective of this study is to construct a quantitative model for the treatment of simulated waste water containing As(V), Zn(II) and Hg(II) as toxic elements, and Fe(III), Na(I) and sulfate as coaxing ions. This model was constructed based on chemical equilibrium calculations considering surface complexation with ferrihydrite. The model outputs would make it possible to calculate adequate neutralizer inputs and to predict pH and the residual concentrations of all the elements in solution after the neutralization/co-precipitation treatment. In a previous study, we constructed a model based on chemical equilibrium calculations, considering surface complexation with ferrihydrite and aluminum hydroxide (Karamalidis and Dzombak, 2010). It was confirmed that these models could successfully represent the experimental results of neutralization, co-precipitation and adsorption processes, such as fluoride uptake by aluminum hydroxide (Haraguchi et al., 2011) and zinc uptake by aluminum hydroxide (Tokoro et al., 2015). To accomplish the present objective, an As(V) and a Hg(II) removal treatment models were constructed based on chemical equilibrium calculations and surface complexation with ferrihydrite. In addition, our constructed models were validated for neutralization experiments using simulated waste water that measured the residual concentration of all metal ions in the solution.

MATERIALS AND METHODS

Standards and Reagents

All chemicals and solutions used were analytical graded and purchased from Kanto Chemicals Inc., Japan and Wako Pure Chemical Industries, Japan. The As(V), Hg(II) and Fe(III) solutions were prepared from $Na_2HAsO_4 \cdot 7H_2O$, $Hg(NO_3)_2 \cdot H_2O$ and $Fe(NO_3)_3 \cdot 9H_2O$, respectively. The simulated waste water, containing Hg(II), As(V), Zn(II), Fe(III) and sulfate, was prepared from $HgSO_4$, $Na_2HAsO_4 \cdot 7H_2O$, $ZnSO_4 \cdot 7H_2O$, $Fe_2(SO_4)_3$ and H_2SO_4 solution, respectively. For the co-precipitation and adsorption experiments, the pH and ionic strength were adjusted by the addition of 0.1 M HNO_3, KOH and KNO_3. The ionic strength was fixed at 0.05. All experiments were conducted at 25 ^{0}C.

Co-precipitation Experiments

The co-precipitation experiments involved the formation of ferrihydrite in the presence of As(V) or Hg(II) ions. The Fe(III) and As(V) or Hg(II) solutions were initially combined at pH 2 to maintain the ionic states of Fe(III), As(V) and Hg(II). The pH and ionic strength were then adjusted to the target level and maintained for 1 h with agitation. The pH was kept within ± 0.5 of the target value by the addition of a few drops of 0.1 M KOH during agitation. After agitation, the suspension was filtered through a 0.1μm membrane filter, and the filtrate was analyzed by ICP-AES (Seiko Instruments, SPS-7800, Japan) to determine the concentration of Fe(III) and As(V). Cold vapor atomic absorption spectrophotometry (Japan Instruments, RA-3, Japan) was used to measure the concentration of Hg(II). The initial Hg(II) concentration was fixed at 10 or 25 mg dm^{-3} (0.05 or 0.12 mmol dm^{-3}), the initial As(V) concentration was fixed at 10, 20 or 30 mg dm^{-3} (0.13, 0.27 or 0.40 mmol dm^{-3}), and the initial Fe(III) concentration was varied to give initial Hg/Fe or As/Fe molar ratios ranging from 3 to 0.01 or 5 to 0.125.

Adsorption Experiments

The adsorption experiments were conducted using the same conditions as for the co-precipitation experiments, except that the ferrihydrite suspensions and As(V) or Hg(II) solutions were prepared separately at twice the initial concentration of As(V), Hg(II) and Fe(III). The pH values and ionic strengths of both solutions were already adjusted to the target levels, and equivalent amounts were then combined and agitated for 1 hour. The pH was kept within $\pm$0.5 of the target value by the addition of a few drops of 0.1 M KOH during agitation. After agitation, the Fe(III), As(V) and Hg(II) concentrations in filtrate were analyzed using the procedures described above. In all the adsorption experiments, freshly precipitated ferrihydrite was used to avoid changes caused by aging.

X-ray Diffraction (XRD) Analysis

The filtered residues from the co-precipitation and simple adsorption experiments were analyzed using XRD (RINT Ultima III, Rigaku, Inc., Japan). For the XRD analyses, the initial Fe(III) concentration was adjusted to 10 mg dm^{-3} and the initial As(V) or Hg(II) concentrations were varied according to the targeted As/Fe or Hg/Fe molar ratio. The filtered residues from the co-precipitation and adsorption experiments were freeze-dried at -45 ^{0}C and 10 Pa for at least 24 hours to avoid crystallization or mineralogical transformation.

Powder XRD patterns were obtained using a copper target (Cu-Kα), a crystal graphite monochromator, and a scintillation detector. The X-ray source was operated at 40 kV and 30 mA with step scanning from 2 θ values of 2-80° , in increments of 0.02. A crystal sample holder was used and the patterns were not corrected for background diffraction.

X-ray Adsorption Fine Structure (XAFS) Analysis

The X-ray adsorption fine structure (XAFS) analysis embraces two types of analysis. One is the X-ray Absorption Near-Edge Structure (XANES), and the other is the Extended X-ray Absorption Fine Structure (EXAFS). In this study, both XANES and EXAFS analysis were performed at the beam line BL5S1 in the Aichi Synchrotron Radiation Center, Nagoya, Japan. The K edge of Fe was scanned in the range of 6600 to 8300 eV to obtain both XANES and EXAFS. The ionization chamber was filled with He (70%) + N_2 (30%) gas to monitor the incident beam, whereas N_2 (75%) + Ar (25%) gas was used for the transmission beam. Moreover, the L3 edge of Hg was scanned in the range of 11900 to 13400 eV to obtain both XANES and EXAFS. The ionization chamber was filled with He (85%) + Ar (15%) gas to monitor the incident beam, whereas Ar (100%) gas was used for the transmission beam.

The samples obtained from the co-precipitation and simple adsorption experiments were freeze-dried at -45 ^{0}C and 10 Pa for at least 24 h to avoid crystallization or mineralogical transformation. Then, for the transmission method, they were blended with boron nitride (BN) powder and pressed under 20 kN to create the XAFS sample in tablet form. Ferrihydrite, HgO (red) and HgO (yellow) were synthesized as reference materials to be compared with the forms of Fe(III) or Hg(II) in the samples after the co-precipitation and simple adsorption experiments.

Neutralization Experiments with Simulated Waste water

Two litres of simulated waste water containing As(V), Hg(II), Zn(II), Na(I), Fe(III) and sulfate were prepared for the neutralization experiments, and Table 1 shows the chemical composition.

Table 1 – Chemical composition of simulated waste water (mg dm^{-3})

pH	Fe	Na	Hg	As	Zn	SO_4
2.40	143.4	86.6	0.127	0.40	405.4	2010

In the neutralization experiment, the pH was adjusted to a target level of 1 by adding a drop of 5 N NaOH. In this study a strong neutralizer was used because the objective of this test was to validate the constructed model, and it was necessary to avoid dilution by adding a large amount of a weak neutralizer. At each pH, the pH was strictly controlled for 1 hour with agitation using a magnetic stirrer, and 10 mL solution samples were obtained using a syringe. The concentrations of all the cations and anions in the supernatant solution, filtered using a 0.1 μm membrane, were analyzed by ICP-MS (7700x, Agilent Technology, Santa Clara, America) or ion chromatography (IC) (ICS-90, Dionex, Japan).

MODELLING

Surface Complexation with Ferrihydrite

In this study, the diffuse layer model (DLM) proposed by Dzombak and Morel (1990) was used for the calculation of surface complex formation on ferrihydrite. The surface complexation model for As(V) on ferrihydrite was constructed using their proposed parameters. It was confirmed that the parameters obtained in the present study were in excellent agreement with their proposed values. The surface complexation model for Hg(II) on ferrihydrite was constructed using the value proposed by Tiffreau et al. (1995) and Bonissel-Gissinger et al. (1999). The site density and specific surface area of ferrihydrite were set to 2.31 nm^{-1} and 600 m^2/g, respectively, as in Dzombak and Morel's model. In this study, the exchange capacity of ferrihydrite was calculated to be 0.205 mol mol-Fe^{-1} by Equation 1.

$$E = (N_S \cdot S_A \cdot M \cdot 10^{18})/N_a \quad (1)$$

Where E is the exchange capacity (mol mol-Fe^{-1}), N_S is the site density (nm^{-2}), S_A is the specific surface area (m^2g^{-1}), M is the molecular weight of ferrihydrite (g mol^{-1}), and N_a is the Avogadro constant (mol^{-1}). The surface complexation constants for H^+, OH^-, As(V), Hg(II), Zn(II) and sulfate considered in the model are shown in Table 2.

Table 2 – Surface complexation constants considered in the model for various species on ferrihydrite

Reactions	Log K^{int}	Reference
$\equiv FeOH^0 + H^+ \leftrightarrow \equiv FeOH_2^+$	7.29	Dzombak & Morel, 1990
$\equiv FeOH^0 \leftrightarrow \equiv FeO^- + H^+$	-8.93	Dzombak & Morel, 1990
$\equiv FeOH^0 + H_3AsO_4 \leftrightarrow \equiv FeH_2AsO_4 + H_2O$	8.61	Dzombak & Morel, 1990
$\equiv FeOH^0 + H_3AsO_4 \leftrightarrow \equiv FeHAsO_4^- + H_2O + H^+$	2.81	Dzombak & Morel, 1990
$\equiv FeOH^0 + H_3AsO_4 \leftrightarrow \equiv FeOHAsO_4^{-3} + 3H^+$	-10.12	Dzombak & Morel, 1990
$\equiv FeOH^0 + Hg^{2+} \leftrightarrow \equiv FeOHg^+ + H^+$	4.9	Tiffreau et al., 1995
$\equiv FeOH^0 + Hg^{2+} + H_2O \leftrightarrow \equiv FeOHgOH + 2H^+$	-0.9	Bonissel-gissinger et al., 1999
$\equiv Fe_WOH^0 + Zn^{2+} \leftrightarrow \equiv Fe_WOZn^+ + H^+$	-1.99	Dzombak & Morel, 1990
$\equiv Fe_SOH^0 + Zn^{2+} \leftrightarrow \equiv Fe_SOZn^+ + H^+$	0.99	Dzombak & Morel, 1990
$\equiv FeOH^0 + SO_4^{2-} \leftrightarrow \equiv FeOSO_4^- + H_2O$	7.78	Dzombak & Morel, 1990
$\equiv FeOH^0 + SO_4^{2-} \leftrightarrow \equiv FeOHSO_4^{2-}$	0.79	Dzombak & Morel, 1990

※[≡] shows that this reaction occurs only near the surface

Surface Complexation Model for As(V) with Ferrihydrite

The surface complexation model for As(V) on ferrihydrite was constructed previously by Haraguchi et al. (2013). They investigated the As(V) removal mechanism by a method where the co-precipitation and adsorption experiments were conducted separately at pH 5 and 7. Their results were evaluated using isotherms, XRD analysis and XAFS analysis. From their experimental results, it was confirmed that As(V) removal with ferrihydrite mainly involved surface complexation for both the co-precipitation and adsorption processes. In the co-precipitation process, however, surface precipitation was observed when the initial As/Fe molar ratio was greater than 0.5. The same results and trends were identified from our experimental results. So, their constructed model was used in this study.

Simulated Waste Water Treatment Model

Chemical equilibrium calculations were conducted using PHREEQC 3.0 with Minteq.v4 and phreeqc database (Parkhurst & Appelo, 1999). The elements in the simulated waste water, except Hg(II), were considered in the calculation, and the pH was calculated considering an ion balance in the Minteq.v4 and phreeqc databases. The values of the equilibrium constants for Hg(II) proposed by Martel and Smith (1976, 1982, 1989) are shown in Table 3. The solid phases presented in Table 4 were considered in the calculation.

Table 3 – Equilibrium constants related Hg(II) used in this study

Reactions	Log K^{int}
$Hg^{2+} + H_2O \leftrightarrow HgOH^{+} + H^{+}$	-3.40
$Hg^{2+} + 2H_2O \leftrightarrow Hg(OH)_2 + 2H^{+}$	-6.20
$Hg^{2+} + 3H_2O \leftrightarrow Hg(OH)_3^{-} + 3H^{+}$	-21.10
$Hg^{2+} + 4H_2O \leftrightarrow Hg(OH)_4^{2-} + 4H^{+}$	-36.80
$2Hg^{2+} + H_2O \leftrightarrow Hg_2(OH)_3^{+} + H^{+}$	-3.30
$3Hg^{2+} + 3H_2O \leftrightarrow Hg_3(OH)_3^{3+} + 3H^{+}$	-6.40

Table 4 – Solid phases considered in the model (Parkhurst & Appelo, 1999)

Solid phases	Chemical formula	Log K^{int}
Ferrihydrite	$Fe(OH)_3 + 3H^{+} \leftrightarrow Fe^{3+} + 3H_2O$	3.191
$Zn(OH)_2$	$Zn(OH)_2 + 2H^{+} \leftrightarrow Zn^{2+} + 2H_2O$	11.5
$Zn_4(OH)_6SO_4$	$Zn_4(OH)_6SO_4 + 6H^{+} \leftrightarrow 4Zn^{2+} + SO_4^{2-} + 6H_2O$	28.4

RESULTS AND DISCUSSION

Surface Complexation Model for Hg(II) on Ferrihydrite

Hg(II) adsorption(Ad)/co-precipitation(Co) experiments with ferrihydrite were conducted at pH 4. As a result, it was confirmed that the sorption isotherms obtained by both adsorption and co-precipitation processes were not saturated, as for BET type sorption, Figure 1. The XRD patterns of both co-precipitates and an adsorption sample changed when the initial molar ratio of Hg/Fe was near 0.125 (Figure 2). That is, the peak position changed from 34° characteristic of ferrihydrite to about 25°. Figure 3 shows the K edge of Fe and the L3 edge of Hg XANES results. The L3 edge of the Hg XANES pattern of the co-precipitates changed when the initial molar ratio Hg/Fe was near 0.5. In other words, the peak position changed from 12283 eV for low initial Hg/Fe molar ratios to about 12281 eV. The HgO (red) and HgO (yellow) reference samples have a peak at 12281 eV. However, all peak positions of the high initial Hg/Fe molar ratio samples did not match the peaks of HgO (red) or HgO (yellow). Therefore, these results suggested that when the initial Hg/Fe molar ratio was high, a material having a structure similar to HgO was formed as a surface precipitate. From above results, the main mechanism of Hg(II) removal should be surface complexation of Hg(II) with poorly crystalline ferrihydrite at initial Hg/Fe molar ratios less than 0.125, whereas surface precipitation should be involved at higher molar ratios. Accordingly, the surface complexation model for Hg(II) removal mechanism in the present study was constructed. Figure 4 shows the comparison between the experimental and calculated results for Hg(II). From these results, it is clear that our constructed model can successfully represent the experimental results.

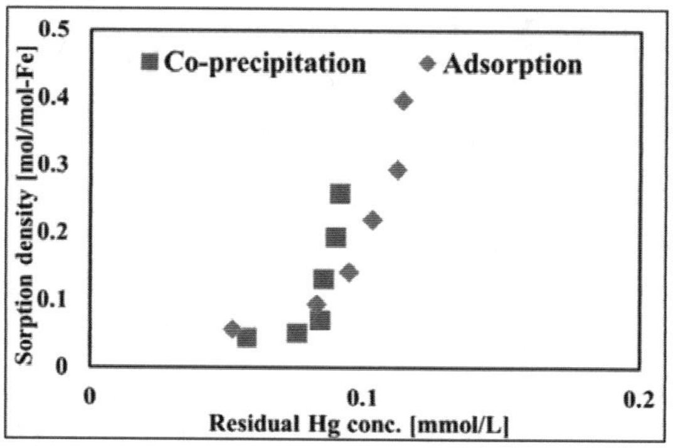

Figure 1- Sorption isotherm of Hg(II) co-precipitation and adsorption with ferrihydrite at pH 4

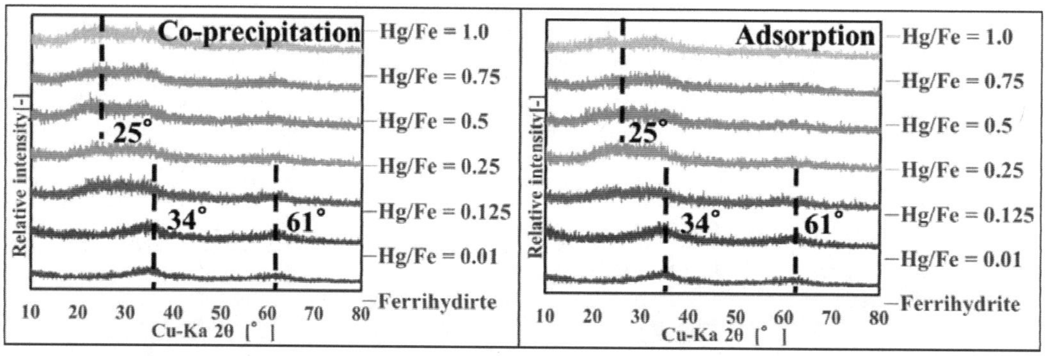

Figure 2 - XRD pattern of Hg(II) co-precipitation and adsorption with changing initial Hg/Fe molar ratio

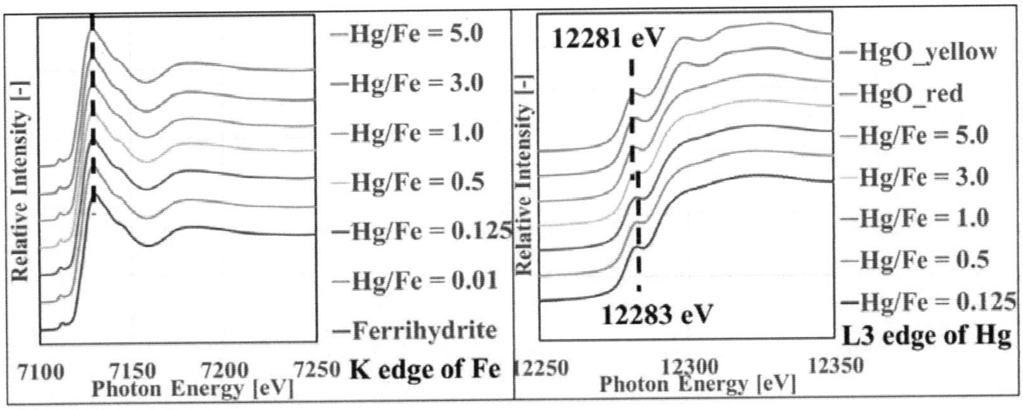

Figure 3 - K edge of Fe and L3 edge of Hg XANES patterns for Hg(II) co-precipitation and adsorption with changing initial Hg/Fe molar ratios

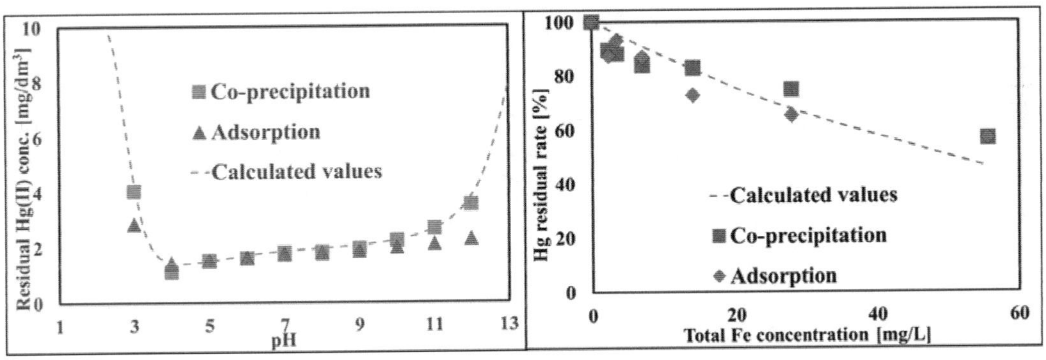

Figure 4 - Validation of the model by comparison with the experimental results

Validation of the Model Using Simulated Waste water

The above mentioned models were combined and validated for the neutralization experiments using simulated waste water. The chemical composition of the simulated waste water is presented in Table 1. Comparisons between the experimental and calculated results for the Fe, Zn, As, Hg and SO_4^{2-} concentrations are shown in Figure 5. The ion balance results are shown in Figure 6. In our constructed model, the chemical equilibria of the solid phases (Table 4) and Hg(II) (Table 3), as well as surface complexation on ferrihydrite (Table 2), were considered.

These results suggested that the concentration of all the ions in the simulated waste water during neutralization, as well as the ion balance, could be successfully represented by our constructed model over almost the entire pH range. However, the results for the ion balance and residual As(V) concentration could not be represented in acidic region because of schwertmannite ($Fe_8O_8(OH)_{8-2x}(SO_4)_x \cdot nH_2O$, $1<x<1.75$) formation. Schwertmannite has the capacity to remove some anions containing As(V) from the water (Bigham, J. M et al., 1996), but schwertmannite was not considered in our constructed model. Thus, our constructed model could not successfully represent the ion balance and residual As(V) concentration in the acidic region.

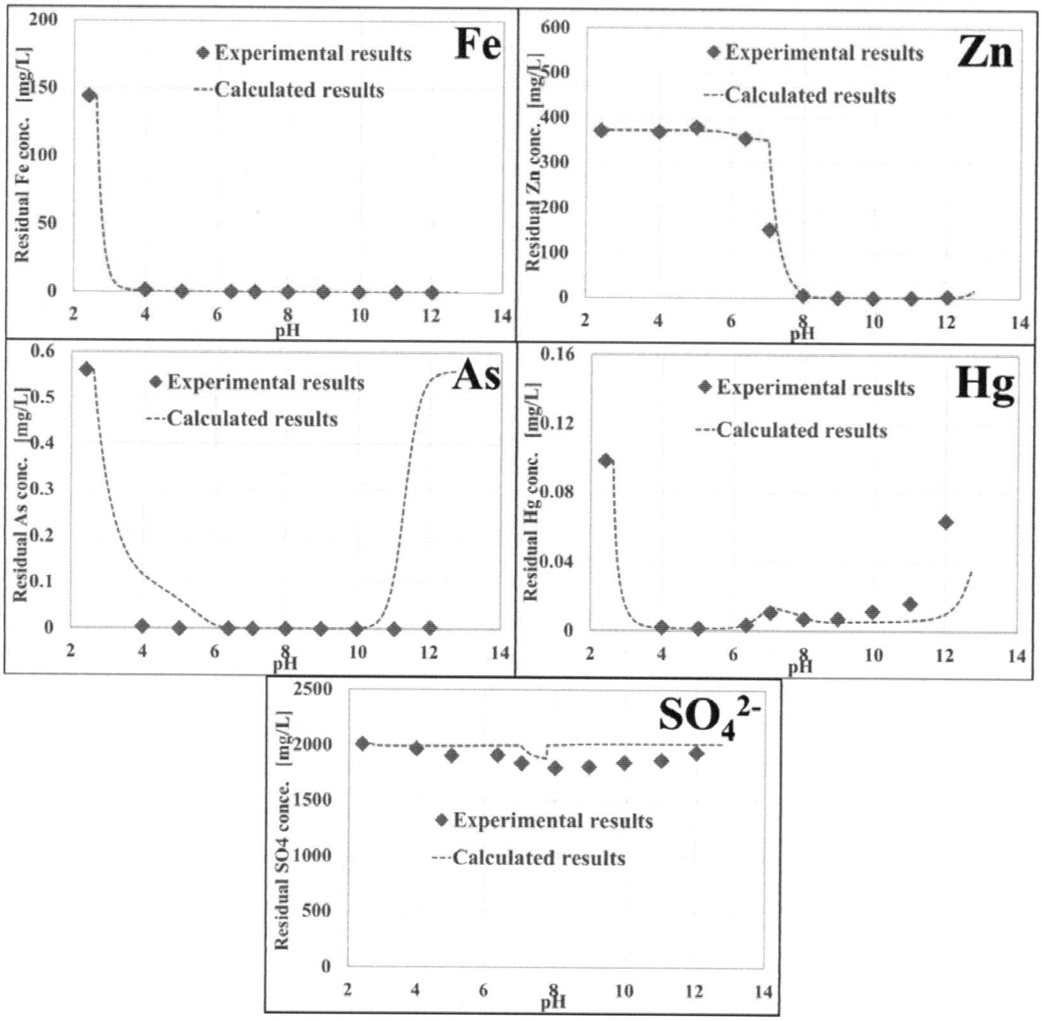

Figure 5 - Comparison of the experimental and calculated results for the simulated waste water neutralization experiments

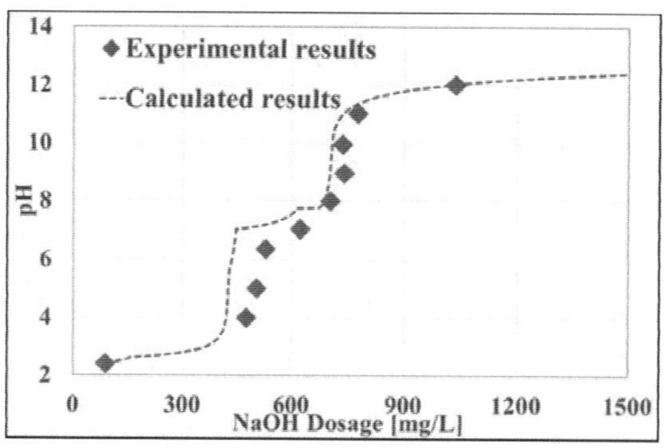

Figure 6 – Ion balance results for the neutralization treatment

CONCLUSIONS

The As(V) and Hg(II) removal mechanisms involved in water treatment using ferrihydrite precipitation were experimentally investigated by comparing the co-precipitation and adsorption processes. A quantitative model was constructed for the optimum design of As(V) and Hg(II) treatment processes involving ferrihydrite precipitation.

To investigate the As(V) removal mechanisms, the co-precipitation and adsorption experiments were separately conducted at pH 5 and 7, and the results were compared using isotherms and XRD analysis. It was confirmed that As(V) removal with ferrihydrite mainly involved surface complexation for both the co-precipitation and adsorption processes. However, for only the co-precipitation processes, surface precipitation was observed when the initial As/Fe molar ratio was greater than 0.5.

To investigate the Hg(II) removal mechanisms, the co-precipitation and adsorption experiments were separately conducted at pH 4, and the results were compared by using isotherms, XRD analysis and XAFS analysis. It was confirmed that Hg(II) removal with ferrihydrite mainly involved surface complexation for both the co-precipitation and adsorption processes. However, in both processes, surface precipitation was also observed when the initial Hg/Fe molar ratio was greater than 0.125 – 0.5.

A quantitative model was constructed based on a surface complexation model. The exchange capacity of ferrihydrite was set at 0.205 mol/mol-Fe by reference to the value proposed by Dzombak and Morel. The surface complexation coefficients for H^+, OH^-, As(V), Hg(II), Zn(II) and SO_4^{2-} adsorption on ferrihydrite were determined by Dzombak and Morel, Tiffreau et al. and Bonnissel-Gissinger et al. The constructed model was validated by the neutralization experimental results using simulated waste water containing As(V), Zn(II) and Hg(II) as toxic elements, and Fe(III), Na(I) and sulfate. As a result, the residual concentration of all the ions contained in the simulated waste water could be successfully represented by the constructed model.

REFERENCES

Bigham, J. M., Schwertmann, U., Traina, S. J., Winland, R. L., & Wolf, M. (1996). Schwertmannite and the chemical modeling of iron in acid sulfate waters. *Geochimica et Cosmochimica Acta*, 60(12), 2111–2121.

Bonnissel-gissinger, P., Alnot, M., Lickes, J., Ehrhardt, J., & Behra, P. (1999). Modeling the adsorption of mercury (II) on (hydr) oxides II : α-FeOOH (Goethite) and amorphous silica. *Journal of Colloid and Interface Science,* 322, 313–322.

Chen, J., Liu, R., Sun, W., & Qiu, G. (2009). Effect of mineral processing waste water on flotation of sulfide minerals. *Transactions of Nonferrous Metals Society of China*, 19(2), 454–457.

Dold, B. (2008). Sustainability in metal mining: from exploration, over processing to mine waste management. *Reviews in Environmental Science and Biotechnology*, *7*(4 SPEC. ISS.), 275–285.

Dzombak, D.A., & Morel, F.M.M. (1990). Surface complexation modeling: Hydrous ferric oxide. John Wiley & Sons, New York City, NY, USA.

Haraguchi, D., Tokoro, C., & Owada, O. (2011). Comparison between aluminum hydroxide coprecipitation and adsorption for removal of fluoride from waste water. *Journal of MMIJ*, 127, 26 – 31.

Haraguchi, D., Tokoro, D., Owada, S., & Izawa, S. (2013). Quantitative modeling of As(V) waste water treatment by ferrihydrite coprecipitation using surface complexation and surface precipitation model. *Journal of MMIJ*, 129, 158 – 164.

Japan Oil, Gas and Metals National Corporation (JOGMEC) (2003). Sansei kouhaisui mondai no taisho doukou nitsuite, *JOGMEC Current Topics* [in Japanese]. Retrieved from http://mric.jogmec.go.jp/public/current/03_03.html.

Karamalidis, A.K., & Dzombak, D.A. (2010). Surface complexation modeling: Gibbsite. John Wiley & Sons, New York City, NY, USA.

Liang, H.C., & Thomson, B. M. (2008). Minerals and mine drainage. *Water Environ. Res.*, 80, 1481-1509.

Parkhurst, D.L. & Appelo, C.A.J. (1999). User's guide to phreeqc (version2) - A computer program for speciation, batch-reaction, one-dimensional transport, and inverse geochemical calculations, water-resources investigations report 99-4259, US Geological Survey, Denver, Colorado. Retrieved from http://pubs.usgs.gov/wri/1999/4259/report.pdf.

Smith, R.M. & Martell, A.E. (1976). Critical stability constants. Vol.4. Plenum Press, New York City, NY, USA.

Smith, R.M. & Martell, A.E. (1982). Critical stability constants. Vol.5 First Supplement. Plenum Press, New York City, NY, USA.

Smith, R.M. & Martell, A.E. (1989). Critical stability constants. Vol.6 Second Supplement. Plenum Press, New York City, NY, USA.

Tokoro, C., Sakakibara, T., & Suzuki, S. (2015). Mechanism investigation and surface complexation modeling of zinc sorption on aluminum hydroxide in adsorption/coprecipitation processes. *Chemical Engineering Journal,* 279, 86–92.

Tiffreau, C., Lützenkirchen, J., & Behra, P. (1995). Modeling the adsorption of mercury(II) on (hydr)oxides: I. Amorphous iron oxide and α-quartz. *Journal of Colloid and Interface Science.* 172, 82–93.

Iron and Arsenic Management During SX/IX/EW Processes

IRON IN ARSENIC REMOVAL - FROM TRADITONAL ARSENIC PRECIPITATION TO NOVEL ELECTROCHEMICAL PROCESSES

M. Martikainen[1], T. van der Meer[1], N. Isomäki[2], A. Mäkinen[2], M. Arvola[2]

[1]*Outotec Finland*
Rauhalanpuisto 9
P.O. Box 1000
02230 Espoo, Finland
*(*Corresponding author: mika.martikainen@outotec.com)*

[2]*Outotec Research Center*
Kuparitie 10, P.O.Box 69
28101 Pori, Finland

ABSTRACT

Arsenic is a compound that has no significant market value. However, it needs to be removed from ores, concentrates and other process streams because of environmental or technical processing concerns. A wide range of arsenic containing minerals exists in nature, and in many instances these minerals are also associated with the ones that have commercial value. For example, sulfide copper ores are normally closely associated with minerals like enargite, Cu_3AsS_4 and tennantite $Cu_{12}As_4S_{13}$. After mineral processing these arsenic sources can report to the waste waters and these effluents need to be treated before discharge or recycling. Ferric arsenate precipitation is a well-established method for stabilizing arsenic in effluents. Electrochemical arsenic removal utilizes the same adsorption/precipitation ideology as ferric arsenate precipitation. Electrochemical arsenic removal, even though it is a simple process in terms of equipment, is a complex system involving chemical and physical mechanisms taking place simultaneously to remove dissolved and/or solid arsenic components from wastewaters. The whole process can be described by the following three steps: 1) electrolytic oxidation at a sacrificial anode (dissolving electrode metal, iron for example), 2) chemical reaction between dissolved iron and arsenic to form a solid precipitate and 3) aggregation of the particles to form flocs, which can be removed by known solid/liquid separation methods. Electrochemical arsenic removal is a very efficient way to remove arsenic from different waters to very low residual concentrations.

KEYWORDS

Arsenic, electrochemical oxidation, sacrificial iron anode, Outotec, iron-arsenic reaction

INTRODUCTION

Arsenic Removal

Many hydrometallurgical processes produce arsenic containing effluents. Depending on the process that releases the arsenic from the minerals and state in which the arsenic occurs in the processed minerals, the arsenic can be present in the trivalent (As^{3+}) or pentavalent form (As^{5+}). In metallurgical applications, arsenic is present in aqueous solution typically in the acidic form, as H_3AsO_3 or H_3AsO_4. Arsenic is an exception among most soluble metals, because it occurs as oxyanions in hydrometallurgical processes and not as a cation. For this reason, arsenic removal requires a separate consideration in effluent treatment.

The most typical methods for removing arsenic from effluents in the mining and metallurgical industry include the following key processes:

- Calcium based precipitation ($Ca_3(AsO_4)_2$ or $Ca_3(AsO_3)_2$)
- Basic ferric arsenate precipitation ($FeAsO_4$ * $xFe(OH)_3$)
- Scorodite precipitation ($FeAsO_4$ *$2H_2O$)
- Arsenic precipitation as sulfides (As_2S_3)
- Adsorption
- Coagulation
- Electrocoagulation
- Ion exchange
- Membrane separation
- Biological precipitation

Electrocoagulation

Electrocoagulation in the treatment of water has been employed for more than 100 years, as the first electrocoagulation treatment plant was built in London 1889 (Moreno, Cocke, Gomez, Markovsky, Parga, Peterson & Garcia, 2009). In 1909, in the United States, J.T. Harries received a patent for wastewater treatment by electrolysis with sacrificial aluminum and iron anodes (Vik, Carlson, Eikun, & Gjessing, 1984). An "Electronic Coagulator", which electrochemically dissolved aluminum ions into solution and reacted them with the hydroxyl ions formed at the cathode to form aluminum hydroxides was used in the 1940s for water treatment. Electrocoagulation (EC), is a process where reagents/coagulants are generated in situ by the electrochemical dissolution of electrodes. Metallic cation formation occurs at the anode while, typically, hydrogen gas (H_2) is formed at the cathode (Hansen, Nuñez, Raboy, Schippacasse, & Grandon, 2007).

In electrocoagulation processes the dissolved metallic cations are further reacted to different metal hydroxide species. These metal hydroxides are effective coagulants as they provide a highly active surface for different adsorption processes (Ali, Gupta, Khan, & Asim, 2012). The sacrificial metal anodes are used continuously to produce cations. Coagulation/chemical reactions occur when the cations are combined with the negative particles/elements in the water. The mechanisms of contaminant removal from wastewater streams are either chemical, by precipitation, or physical by means of physical or chemical attachment to colloidal materials generated by electrode erosion (Liu, Zhao, & Qu, 2010). The formed solid particles are typically removed from the aqueous suspension by filtration and sedimentation.

Electrocoagulation is a fast and easily controllable method for the removal of dissolved elements, and especially, for the removal of trace metals. With no additive chemicals, sludge formation in EC is significantly lower than in chemical precipitation processes. EC based processes can reduce sludge volumes by 75-80%, and at the same time, the capability to remove contaminants is much higher compared with chemical precipitation (Davis, 2011). Less sludge means less residual material for treatment or disposal, and therefore, lower disposal costs. Several heavy metals can be decreased by 95 to 99% of their original aqueous concentrations (Fu & Wang, 2011).

The characteristics of the electrochemically formed floc are different compared with the ones generated by chemical coagulation/precipitation. Typically, electrocoagulated flocs bind less water and are more shear resistant. The sludge produced by electrocoagulation is more readily filtered than the sludge produced by chemical coagulation. The metal oxides/hydroxides formed in the sludge are typically capable of passing most environmental leachability tests; therefore, they are often considered to be stable (Davis, 2011; Mickley, 2004).

In addition to its heavy metal removal abilities, electrocoagulation removes numerous other substances from aqueous solutions. Organic materials such as bacteria, cysts, viruses, pesticides, as well as other organic compounds, can be removed. Textile, pulp and paper, mining and oil industry effluents can be treated with electrocoagulation (Davis, 2011; Mickley, 2004).

Electrocoagulation has a minimal setup time, and the electrochemical processes in the cell are controlled electrically. The absence of moving parts decreases the maintenance costs (Mollah et al., 2004). The electrocoagulation technology has proven to work at a variety of temperatures and over a wide pH range (Kuokkanen et al., 2013; Mickley, 2004). Electrocoagulation can be considered as a pre-treatment method for distillation, crystallization and membrane processes (Davis, 2011).

Because no chemicals are used in the electrocoagulation approach, there is no secondary contamination of the treated water, as is the case when coagulants or reagents are utilized (Mollah et al., 2001). Ali et al. (2012a) observed that the analyzed post-treatment water did not include any additional components due to the EC treatment process; there was, however, a slight increase in pH. No dissolved electrode material, aluminum or iron, was detected in the treated water. Therefore, no indication of water quality degradation due to the EC process could be noticed (Ali et al., 2012a).

A known drawback of electrocoagulation is passivation of the electrodes. This can be avoided or decreased by electrode material selections, by using alternative conductive electrolytes, or by periodically reversing the polarity of the DC current. Depending on the concentration of the impurities and the flow rate of the waste water, electrocoagulation can use a large amount of electricity, and this may lead to high operating costs (Mollah et al., 2004).

There are several factors affecting the operation of electrocoagulation. The electrode material selection defines the coagulating agent in the electrocoagulation process. Iron and aluminum electrodes are successfully used. However, different electrode materials can be used in EC systems, for example, inert coated titanium cathodes or a combination of aluminum and iron electrodes (Chen, 2004). Current density directly determines the dissolution of the metal and bubble generation rates; it also strongly influences mass transfer at the electrodes (Liu et al., 2010; Schweitzer, 2010). Higher conductivity of the electrolyte solution affects the current efficiency positively, leading to increased removal efficiency (Liu et al., 2010). The pH affects the conductivity of the solution, electrode dissolution rate, speciation of the hydroxides and zeta-potential of the colloidal particles. Retention time affects the coagulant generation in the EC system, and mass transport has an evident role in the chemistry of aqueous solutions. Temperature affects hydroxide formation during EC treatment as well as the reaction rates and conductivity (Vepsäläinen. 2012).

Electrocoagulation in the Treatment of Arsenic Containing Wastewaters

Arsenic is sensitive to potential differences and is able to form oxyanions. In a dissolved state, arsenic is capable of binding with one or several hydrogen ions, forming a deprotonation series of arsenates and arsenites (Sullivan, Tyrer, Cheeseman & Graham, 2010). According to Lakshmipathiraj et al. (2010), the surface of the iron oxy-hydroxide (FeOOH) is negatively charged above pH 7.8 in aqueous solutions. The presence of arsenic in the form of trivalent or pentavalent ions shifts the isoelectric point towards the acidic region. This indicates the possible interaction of the arsenic species with the iron oxide precipitate. The strong adsorption properties of hydrous metal oxides/hydroxides, such as ferric or aluminum hydroxides, have been observed in electrocoagulation related studies (Emamjomeh & Sivakumar, 2009). Layered Fe(II)-Fe(III) hydroxides have an internal surface area, and therefore, these particles have both reactivity and a specific surface for interacting with aqueous arsenic species (Moreno, Cocke, Gomez, Markovsky, Parga, Peterson & Garcia, 2009). Other than ferric hydroxides, ferrihydrite and goethite have also been found to be strong arsenic adsorbing species (Jain, Raven & Loeppert, 1999, Kumar et al. 2004).

Lakshmipathiraj et al. (2010) proposed that the following interactions between iron oxy-hydroxide and aqueous arsenic species take place under different conditions:

$$FeOOH + H_2AsO_4^{1-} \rightarrow FeOHAsO_4^{1-} + H_2O$$

$$FeOOH + H_2AsO_4^{1-} \rightarrow FeOH_2AsO_4 + OH^{1-}$$

$$3FeOOH + HAsO_4^{2-} \rightarrow (FeO)_3AsO_4^{1-} + H_2O + 2\ OH^{1-}$$

$$Fe(OH)_2 + H_2AsO_4 \rightarrow FeAsO_4 + 2\ H_2O$$

$$FeOOH + H_3AsO_3 \rightarrow FeOH_2AsO_3 + H_2O$$

$$FeOOH + H_2AsO_3^{1-} \rightarrow FeOHAsO_3^{1-} + H_2O$$

$$FeOOH + HAsO_3^{2-} \rightarrow FeOAsO_3^{2-} + H_2O$$

The chemistry of arsenic removal by electrocoagulation is similar to that of arsenic removal by chemical precipitation. The same principles apply: the chemical reagent (for example iron salt) is mixed with the polluted water, and after a certain amount of time the precipitated particles need to be removed from the solution by settling, filtration and/or other processes. In electrocoagulation, the coagulant/reagent is formed with the help of the applied electric current and the sacrificial anodes, whereas in chemical precipitation, the reagent is added to the solution as an separate chemical, for example as ferric chloride or aluminum sulfate salts (Bilici Baskan & Pala, 2010). Kumar et al. (2004) reported that electrocoagulation based arsenic removal processes had not only a very high As(V) removal efficiency but also a high As(III) removal efficiency. They concluded that the reason lies in the electrochemical mechanism itself. In the EC process, the simultaneous oxidation of As(III) to As(V) occurs and the removal of the oxidized species by adsorption/complexation on the metal hydroxides generated in the same process takes place.

Arsenic removal by electrocoagulation has been extensively studied: from ground waters, drinking waters, copper refining industry effluents and abandoned mining area waters. Arsenic has been present in both the As(III) and As(V) forms in these solutions, and the arsenic concentration can be reduced at least by 99% from its initial concentrations of 5 to 100 ppm, depending on the treatment time and current density (Lakshmipathiraj et al., 2010). These results are supported by the studies of Hansen & Ottosen (2010) and Ali et al. (2012), who observed arsenic removal rates above 99% from initial arsenic concentrations of 5 ppm. The removal rate depends not only on the operational parameters but also on the arsenic concentration of the feed solution (Ali et al., 2012).

EXPERIMENTAL

Experimental Set Up

Arsenic removal from three different industrial waste waters was tested with continuous electrocoagulation (EC). The EC reactor was equipped with carbon steel anodes and cathodes. Both the current and voltage were automatically controlled so that the target current density was kept constant during the test run. There was basically no variation of the current density from the set point in any of the test runs during the test periods. Retention time in the EC reactor was calculated based on the empty space of the EC reactor and the flow rate of the waste water entering the EC reactor. Pumps were calibrated each time the flow rate was changed, as for example when the retention time was varied.

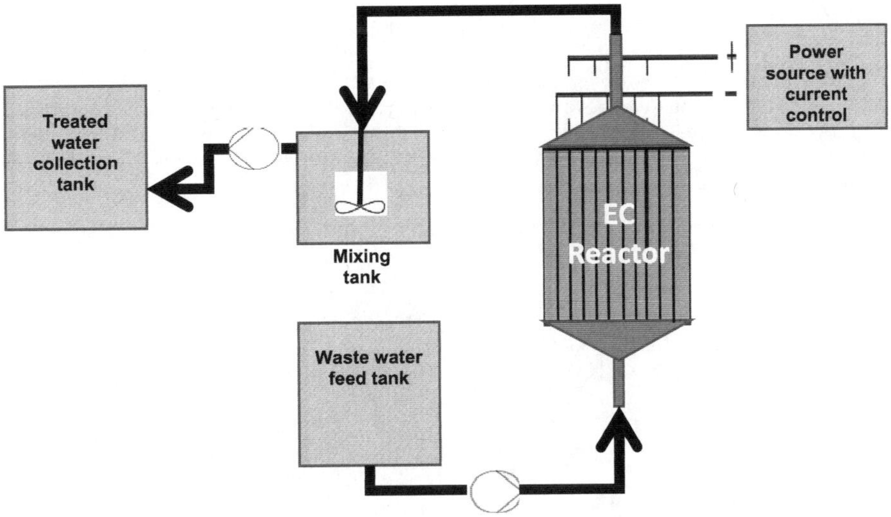

Figure 1 – Simplified flowsheet of the continuous EC experiments

Arsenic Removal from Gypsum Saturated Industrial Waste Water by Electrocoagulation

Arsenic removal by electrocoagulation from industrial gypsum saturated waste water (Ca 749 mg/L and SO_4 1832 mg/L) was tested at three different current densities and three different retention times. The waste water had an initial total As concentration of 8.53 mg/L and a pH of 7.8. Arsenic was in the form of As(III). Results of the continuous EC testes are shown in the Figure 2. Electrocoagulation works efficiently in removing As(III) form the waste water even without any pre-oxidation of As(III) to As(V).

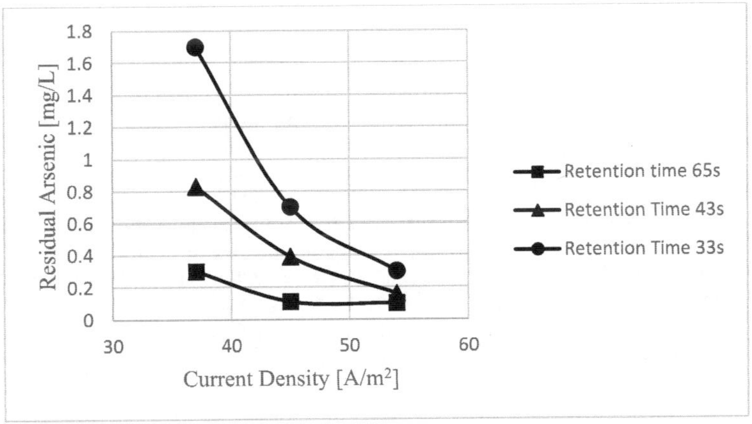

Figure 2 – Arsenic removal from the gypsum saturated waste water with continuous electrocoagulation. Initial arsenic concentration was 8.53 mg/L in all of the tests

Arsenic Removal from Tailings Water by Electrocoagulation

Arsenic removal by electrocoagulation of tailings water containing 8.58 mg/L total arsenic was tested at three different current densities and three different retention times. Results of the continuous EC tests are shown in the Figure 3. The results are similar to those obtained with the gypsum saturated waste water. Arsenic removal by electrocoagulation is efficient and the required purification can be easily controlled by the current density and retention time.

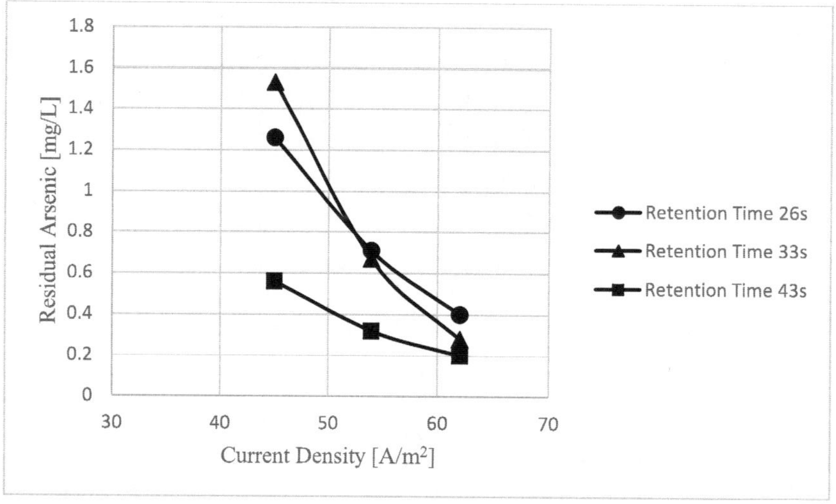

Figure 3 – Arsenic removal from tailings water with continuous electrocoagulation. Initial arsenic concentration was 8.58 mg/L in all the tests

Arsenic Removal from Recycled Mineral Industry Process Water

Waste water was collected from a mineral processing operation where the waste water is recycled and reused as process water. The arsenic content of the process water increases due to the constant reuse of the water. Arsenic removal by electrocoagulation from recycled process water containing 1.61 mg/L total arsenic was tested in two sets of experiments. The first tests were done with a constant retention time of 43 seconds, and the current density was varied between 37 A/m^2 and 54 A/m^2 (Figure 4). The second set of tests was done using a constant current density of 45 A/m^2, and the retention time was varied between 26 seconds and 43 seconds (Figure 5). It is evident that electrocoagulation can treat even low concentration arsenic waters and reach very low residual arsenic concentrations, if needed.

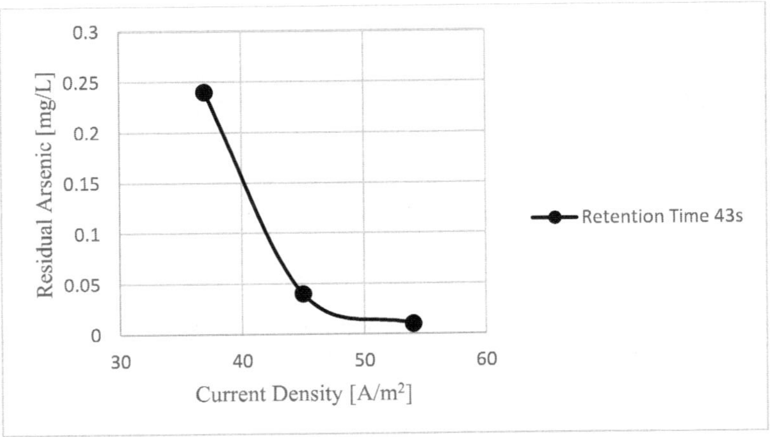

Figure 4 – Arsenic removal from recycled water from mineral processing with continuous electrocoagulation. Initial arsenic concentration was 1.61 mg/L in all the tests

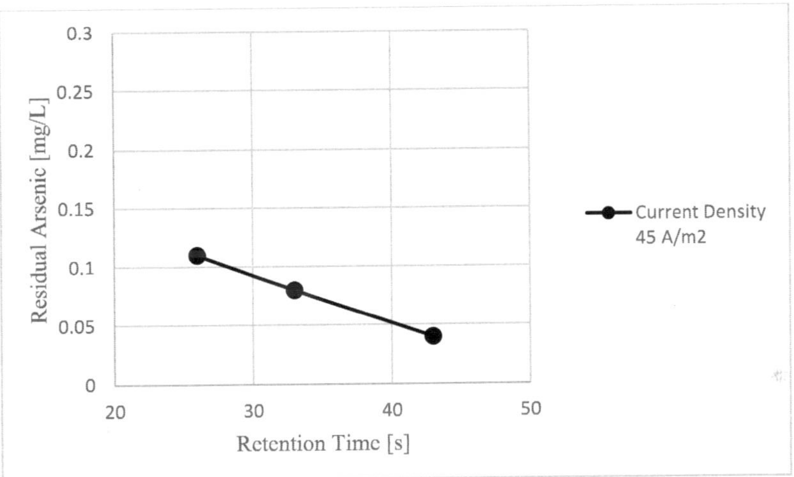

Figure 5 – Arsenic removal from recycled water from mineral processing with continuous electrocoagulation. Initial arsenic concentration was 1.61 mg/L in all the tests

CONCLUSIONS

Tests done using three different industrial waste waters showed that electrocoagulation is an effective way to remove arsenic to very low residual levels. Residual arsenic concentrations can be easily controlled by the retention time and current density applied during electrocoagulation. An additional benefit of electrocoagulation is that pre-treatments, like oxidation of As(III) to As(V) before arsenic removal, can be omitted. Electrocoagulation, even it is an old technology, has now gained a renaissance as recent technological developments in reactor design, power sources, etc. have made it more cost efficient (both CAPEX & OPEX wise) than ever before. Also, the simple control philosophy, easy automation and a minimum need for labor have made it an interesting alternative in different waste water treatment applications.

REFERENCES

Ali, I., Gupta, V.K., Khan, T.A., & Asim, M. (2012). Removal of arsenate from aqueous solution by electro-coagulation method using Al-Fe electrodes. *International Journal of Electrochemical Science,* 7(3), 1898-1907.

Chen, G. (2004). Electrochemical technologies in wastewater treatment. *Separation and Purification Technology,* 38(1), 11-41.

Chen, X., Chen, G., & Yue, P.L. (2000). Separation of pollutants from restaurant wastewater by electrocoagulation. *Separation and Purification Technology,* 19(1–2), 65-76.

Emamjomeh, M.M., & Sivakumar, M., (2009). Review of pollutants removed by electrocoagulation and electrocoagulation/flotation processes. *Journal of Environmental Management,* 90(5), 1663-1679.

Fu, F., & Wang, Q. (2011). Removal of heavy metal ions from wastewaters: A review. *Journal of Environmental Management,* 92(3), 407-418.

Hansen, H.K., Nuñez, P., Raboy, D., Schippacasse, I., & Grandon, R. (2007). Electrocoagulation in wastewater containing arsenic: Comparing different process designs. *Electrochimica Acta,* 52(10), 3464-3470.

Hansen, H.K., & Ottosen, L.M. (2010). Removal of arsenic from wastewaters by airlift electrocoagulation: Part 3: Copper smelter wastewater treatment. *Separation Science Technology,* 45(9), 1326-1330.

Hasson, D., Lumelsky, V., Greenberg, G., Pinhas, Y., & Semiat, R. (2008). Development of the electrochemical scale removal technique for desalination applications. *Desalination,* 230(1–3), 329-342.

Holt, P., Barton, G., Wark, M., & Mitchell, C. (2002). A quantative comparison between chemical dosing and electrocoagulation. *Colloids and Surfaces A: Physicochemical and Engineering Aspects,* 211, 233-248.

Jain, A., Raven, K.P., & Loeppert, R.H., (1999). Arsenite and arsenate adsorption on ferrihydrite: surface charge reduction and net OH^- release stoichiometry. *Environmental Science & Technology,* 33(8), 1179-1184.

Kumar, P.R., Chaudhari, S., Khilar, K.C., & Mahajan, S.P. (2004). Removal of arsenic from water by electrocoagulation. *Chemosphere,* 55(9), 1245-1252.

Kuokkanen, V., Kuokkanen, T., Rämö, J., & Lassi, U. (2013). Recent applications of electrocoagulation in treatment of water and wastewater - A Review. *Green and Sustainable Chemistry,* 3(2), 89-121.

Lakshmipathiraj, P., Prabhakar, S., & Raju, G.B. (2010). Studies on the electrochemical decontamination of wastewater containing arsenic. *Separation and Purification Technology,* 73(2), 114-121.

Liu, H., Zhao, X., & Qu, J. (2010). Electrocoagulation in water treatment. In: C. Comninellis & G. Chen (Eds.), *Electrochemistry for the environment* (pp. 245-262). New York, NY: Springer Science+Business Media.

Mickley, M. (2004). Pretreatment capabilities and benefits of electrocoagulation. Retrieved from: www.dtic.mil/cgi-bin/GetTRDoc?AD=ADA433998.

Mollah, M.Y.A., Morkovsky, P., Gomes, J.A.G., Kesmez, M., Parga, J., & Cocke, D.L. (2004). Fundamentals, present and future perspectives of electrocoagulation. *Journal of Hazardous Materials,* 114(1–3), 199-210.

Mollah, M.Y.A., Schennach, R., Parga, J.R., & Cocke, D.L. (2001). Electrocoagulation (EC) — science and applications. *Journal of Hazardous Materials,* 84(1), 29-41.

Moreno, H., Cocke, D.L., Gomes, J.A., Morkovsky, P., Parga, J., Peterson, E., & Garcia, C. (2009). Electrochemical reactions for electrocoagulation using iron electrodes. *Industrial & Engineering Chemistry Research,* 48(4), 2275-2282.

Moreno, H., Cocke, D.L., Gomes, J.A., Morkovsky, P., Parga, J., Peterson, E., & Garcia, C. (2007). Electrochemistry behind Electrocoagulation using Iron Electrodes. *ECS Transactions, 6*(9), 1-15

Phutdhawong, W., & Buddhasukh, D. (2010). *Applications of Electrocoagulation: A review.* Saarbrücken, Germany: LAP Lambert Academic Publishing AG & Co. KG.

Schweitzer, P.A. (2010). *Fundamentals of corrosion - mechanisms, causes, and preventative methods.* Boca Raton, FL: CRC Press.

Sullivan, C., Tyrer, M., Cheeseman, C.R., & Graham, N.J.D. (2010). Disposal of water treatment wastes containing arsenic — A review. *Science of the Total Environment,* 408(8), 1770-1778.

Vepsäläinen, M. (2012). *Electrocoagulation in the treatment of industrial waters and wastewaters.* (Doctoral dissertation). Retrieved from http://www.vtt.fi/inf/pdf/science/2012/S19.pdf.

Vik, E.A., Carlson D.A, Eikun, A.S., Gjessing, E.T. (1984). Electrocoagulation of potable water, Water Research, 18 (11), 1355-1360

Vlyssides, A.G., Loizidou, M., Karlis, P.K., Zorpas, A.A., & Papaioannou, D. (1999). Electrochemical oxidation of a textile dye wastewater using a Pt/Ti electrode. *Journal of Hazardous Materials,* 70(1-2), 41-52

CONTINUOUS COUNTERCURRENT ION EXCHANGE (CCIX) USED IN HYDROMETALLURGY - THE INFLUENCE OF IRON

*Filip Rochette and Yvan De Busscher

PuriTech, Ltd.
Iz Dikberd 14 – 10 A
2200 Herentals, Belgium
*(*Corresponding author: filip.rochette@puritech.be)*

ABSTRACT

Extraction of minerals can be achieved by a variety of technologies; one of them is ion exchange. There are several ion exchange configurations and mechanical set-ups available, but only one technology will provide more process performance and flexibility compared to the other. The multiple column Continuous Countercurrent Ion Exchange (CCIX) technology provides the highest form of process flexibility with also the highest mineral recovery feasible through ion exchange. In comparison with a fixed bed system, the PuriTech CCIX system offers the benefits of continuous counter-current contact mode. This way the resin capacity is fully exploited leading to reduction in resin inventory and, at the same time, in chemicals and water consumption, and product dilution. For any set of operating conditions the mass transfer zone has a characteristic length (MTZL). The MTZL can be expressed in duration, column length, or volume, but is the measure of the quantity of resin in the system that is doing "work" at any given moment in the process. The objective of continuous ion exchange processing is to eliminate all resin volume from the process equipment that is not within a mass transfer zone. In most commercial ion exchange processes for the recovery of metals, CCIX engineering can eliminate as much as 20 - 80 % of the ion exchange resin inventory. Consumptions of water and chemicals are typically brought down around 40% or more. For each type of resin the selectivity/affinity towards a particular metal depends on the basic process conditions such as pH, concentrations of other competing metals in solution. By changing the pH in the ion exchange process phase, the selectivity of the resin can be changed. By doing this, a split elution, or extraction of metals separately can be achieved and the effect of other competing metals (cross-contamination) can be reduced. For example the DOW resin XUS43578 is commonly used for the recovery of copper or nickel/cobalt separation.

KEYWORDS

Hydrometallurgy, ion exchange, resin, metal recovery

INTRODUCTION

Extraction of minerals can be achieved by a variety of technologies; one of them is ion exchange. There are several ion exchange configurations and mechanical set-ups available, only one technology will provide more process performance and flexibility compared to the other. The multiple column Continuous Countercurrent Ion Exchange (CCIX) technology provides the highest form of process flexibility with also the highest mineral recovery feasible with ion exchange. In comparison with a fixed bed system, the PuriTech CCIX system offers the benefits of the continuous counter-current contact mode. This way the resin capacity is largely exploited leading to reduction in resin inventory and, at the same time, in chemicals and water consumption, and product dilution.

ION EXCHANGE

Ion exchange has been used for many decades for the purification of water, chemical processes, food/ingredient treatment and for the recovery of valuable elements such as metals. Before explaining the CCIX (Continuous Countercurrent Ion Exchange) process behind metal recovery, a quick introduction to some keyword used in this paper and the technology behind it:

- Continuous: all process zones which are used in an ion exchange process: adsorption + displacement + rinse + elution all work at the same time, our CCIX system can produce continuous 24 h per day, 7 days a week.
- Countercurrent: the process flows are moving in one direction while the ion exchange resin is virtually moving in another direction.
- Ion exchange: exchange of ions between two electrolytes or between an electrolyte solution and a complex. In most cases the term is used to denote the processes of purification, separation, and decontamination of aqueous and other ion-containing solutions with solid polymeric or mineral ion exchangers.

Typical ion exchangers are ion exchange resins (functionalized porous or gel polymer), zeolites, montmorillonite, clay, and soil humus (see Figure 1). Ion exchangers are either cation exchangers that exchange positively charged ions (cations) or anion exchangers that exchange negatively charged ions (anions).

Figure 1 – Ion exchange resin

For any set of operating conditions the Mass Transfer Zone has a characteristic length, i.e. Mass Transfer Zone Length (MTZL) (see Figure 2). The MTZL can be expressed in duration, column length, or volume, but is the measure of the quantity of resin in the system that is doing "work" at any given moment in the process. The MTZL is influenced by the average resin bead size, the feed flow rate, resin bed depth and operating temperature.

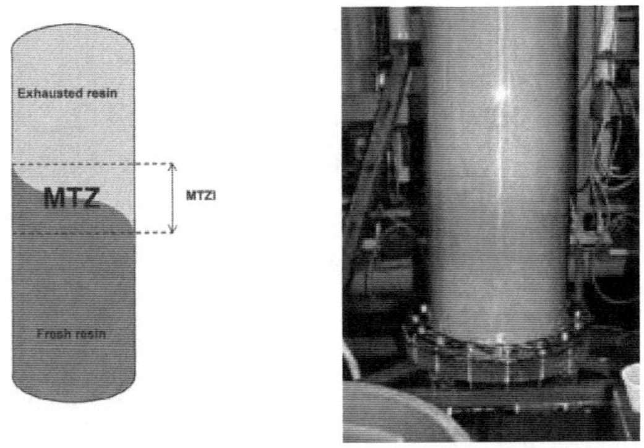

Figure 2 – MTZL visualisation in a resin bed

The objective of continuous ion exchange processing is to eliminate all resin volume from the process equipment that is not within a Mass Transfer Zone. In most commercial ion exchange processes for the recovery of metals, CCIX engineering can eliminate as much as 20 - 80 % of the ion exchange resin inventory. Consumptions of water and chemicals are typically brought down around 40% or more.

METAL SEPARATION OR RECOVERY BY ION EXCHANGE

For each type of resin the selectivity or affinity towards a particular metal depends on the resin bead structure; the basic process conditions such as pH, concentrations of other competing metals in solution. By changing the pH in the ion exchange process phase, the selectivity of the resin can be changed. By doing this, a split elution, extraction of separate metals can become possible and/or the effect of other competing metals can become reduced. There are different types of resins available in the market which have a specify affinity for several metals. The following resins are commonly used in hydrometallurgy applications:

- Carboxylic resins: $Cu^{2+} > Pb^{2+} > Zn^{2+} > Cd^{2+} > Ni^{2+} > Ca^{2+}$
- Phosphonic resins: $Cu^{2+} > Zn^{2+} > Cd^{2+} > Mn^{2+} > Co^{2+} > Ni^{2+} > Ca^{2+}$
- Aminophosphonic resins (at acidic pH): $H^{+} > Fe^{3+} > Pb^{2+} > Cu^{2+} > Zn^{2+}, Al^{3+} > Mg^{2+} \geq Ca^{2+} \geq Cd^{2+} > Ni^{2+} \geq Co^{2+} > Na^{+}$
- Iminodiacetic resins: $Cr^{3+} > In^{3+} > Fe^{3+} > Al^{3+} > Hg^{2+} > UO^{2+} > Cu^{2+} > Pb^{2+} > Ni^{2+} > Cd^{2+} > Zn^{2+} > Co^{2+} > Fe^{2+} > Mn^{2+} > Ca^{2}+ > Mg^{2+}$
- Strong acid resins: $Fe^{3+} > Ca^{2+} > Fe^{2+} > Cu^{2+} > Mg^{2+} > K^{+} > NH_4^{+} > H^{+}$
- Strong base resins: $SO_4^{2-} > CO_3^{2-} > NO_3^{-} > Cl^{-} > HCO_3^{-} > F^{-} > OH^{-}$
- Bis-picolylamine resins: $Cu^{2+} > Ni^{2+} > Fe^{3+} > Zn^{2+} > Co^{2+} > Cd^{2+} > Fe^{2+} > Mn^{2+}$

For example, DOW resin XUS43578 is commonly used for the recovery of copper or the separation of nickel from cobalt solutions. The selectivity chart for this DOW resin XUS43578 is shown in Figure 3.

Figure 3 illustrates that at a pH of < 1, this bis-picolylamine resin will preferably load the copper, secondly nickel and eventually the cobalt. The higher the pH, the more tendency towards ferric precipitation and resin loading, and also an increased potential of resin fouling. Ferrous iron does not load well on the resin below a pH of 2. We can take advantage of this by reducing the iron that has been loaded on the resin. If we apply 1% sodium sulfite that has been pH adjusted to 2, we could scrub the iron from the resin and leaving the nickel or cobalt behind on the resin. In a second step, the nickel or cobalt can be removed from the resin. By changing the pH at the elution zone, the affinity is effected and the release of the metal could be "controlled" using the CCIX technology.

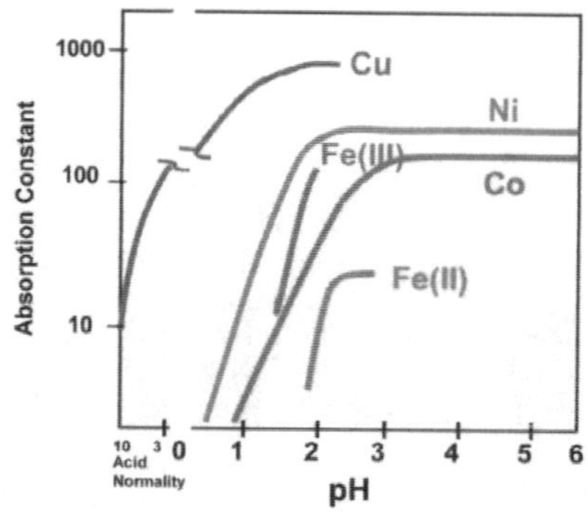

Figure 3 – XUS43578 ions selectivity (DOW technical data sheet)

CCIX SEPARATION TECHNOLOGY

As indicated above, each specific resin has its specific preference or affinity for a specific element. By using multiple ion exchange beds/columns, the Continuous Countercurrent Ion Exchange technology allows to play around with different process conditions. This is shown in Figure 4 with the use of a split elution concept. The pH is changed in between two elution phases and by doing this, the release of each specific metal is dictated to the resin. A process configuration with a split elution is show in Figure 4.

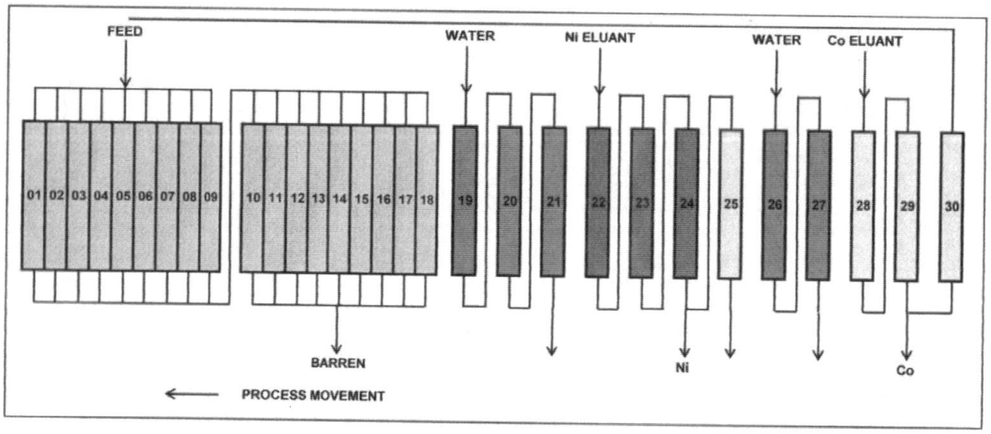

Figure 4 – CCIX process schematic with split elution

Depending on the financial value of the metal which is preferred, a multi pass adsorption could be applied to allow the maximum removal/recovery of this metal as best as possible. A triple adsorption pass will achieve this goal. The process flow will pass first the resin which has been loaded already 2 times. After this first adsorption, the process flow will pass resin which is more active in a second pass. Eventually the process flow will pass the fresh almost virgin resin to remove all remaining elements which have been removed by the 2 earlier passes. A possible process configuration with triple pass adsorption, possible with PuriTech's CCIX technology, is shown in Figure 5.

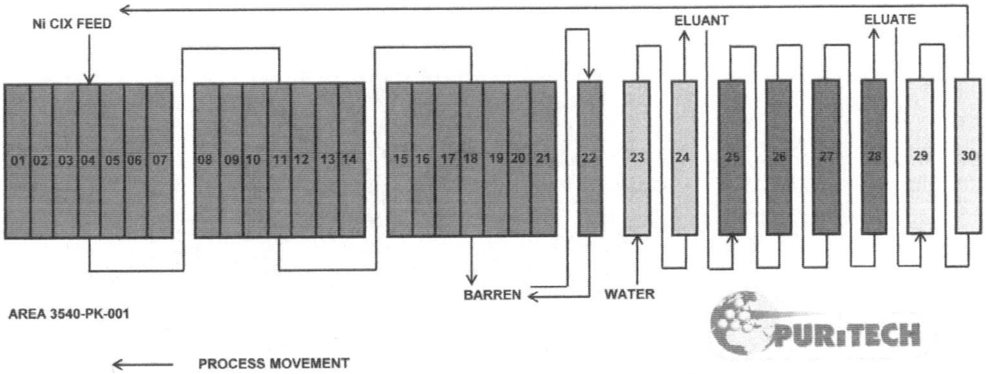

Figure 5 – CCIX process schematic with triple pass adsorption zone

ION-IX™ IN A COBALT PURIFICATION PROCESS

By means of a continuous counter-current ion exchange (CCIX) pilot plant, tests on the client's process liquor have been done (Figure 6). This allows us to develop the process specific to the customer needs and to demonstrate to the highest possible performance of the ion exchange process.

Figure 6 – A pilot installation for cobalt purification

The pilot plant data obtained are used to perform the basic engineering design of the industrial plant. The mass balance and process flow diagram (PFD) is developed during the pilot plant trials. Scale-up from these pilot plant data allows the customer to estimate the CAPEX and OPEX needs for the design of an industrial installation and provides a handy tool to verify the financial feasibility of the project. Afterwards, in collaboration with the client, the P&ID with initial plant lay-out is developed (Figure 7). When equipment lay-out and sizing is finalised, the construction drawings for the vessels, steel structure, piping and electrical drawings are engineered for the local suppliers and contractors.

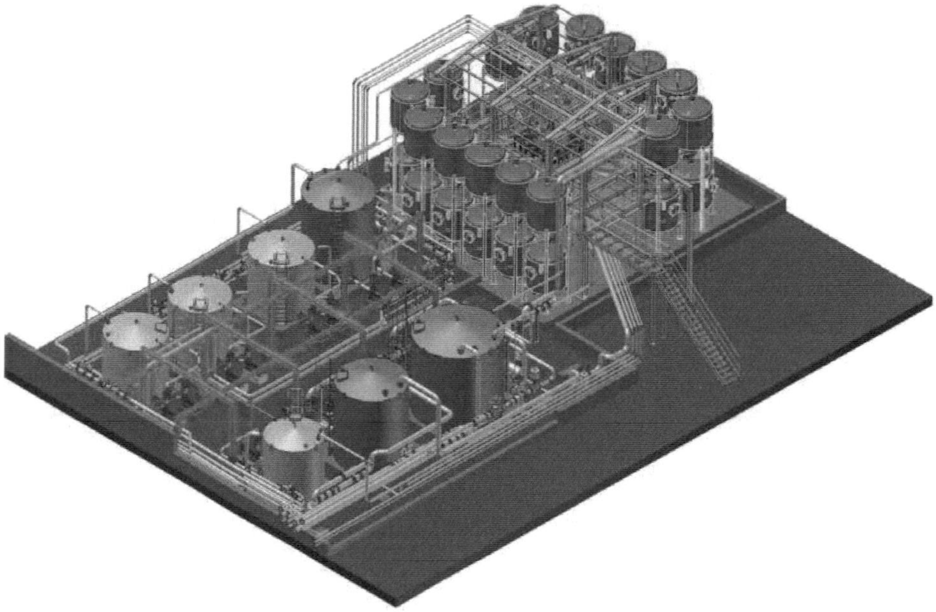

Figure 7 – 3-D dimensional design of a cobalt purification plant

This detailed engineering package is then used to build the plant with local contractors (Figure 8).

Figure 8 – Operational ION-IX™ cobalt purification plant

Analysis of plant operating data shows that the facility was able to meet its performance guarantees.

CONCLUSIONS

Continuous countercurrent ion exchange is a versatile, ion exchange technology that is able to effectively remove elements from a liquor. PuriTech's ION-IX™ is an ingenious mechanical device that allows too not only improve liquid–resin contact efficiency, but also leads to a reduction in plant operating costs and capital expenditure. It provides a wide range of process flexibility to develop high performance ion exchange applications. We have shown above that the ION-IX™ system can easily be tailored to a number of applications in several industries by using different types of resins and flow configurations, and chemicals in the elution zone including salt, acid and alkali. ION-IX™ is a patented and innovative continuous ion-exchange process that can be applied to these applications as well as a number of other separation requirements in various industries.

REFERENCES

DOW Chemicals, technical data sheet for resin XUS43578, Form No. 177-02051-0805, 4 pp.

THE SOLVENT EXTRACTION OF IRON FROM A CHLORIDE LEACH SOLUTION CONTAINING PRECIOUS, RARE AND BASE METALS

*David Dreisinger[1], Niels Verbaan[2], Ron Molnar[3]

[1]*Trimetals Mining Inc.*
580 Hornby Street, Suite 880
Vancouver, Canada V6C 3B6
*(*Corresponding author: david_dreisinger@yahoo.com)*

[2]*SGS Minerals – Lakefield site*
PO Box 4300
185 Concession Street
Lakefield, Canada K0L 2H0

[3]*MetNetH_2O*
1816 Parkwood Circle
Peterborough, Canada K9J 8C2

ABSTRACT

An acidic chloride leach solution was generated from the treatment of an ore containing the following value elements: silver, gold, indium, gallium, copper, zinc and lead. Iron was also co-extracted into solution as a ferric chloride species. Iron extraction was a significant component of the overall acid consumption of the leaching process. The recovery of iron by solvent extraction and water stripping was investigated. Tri-butyl-phosphate (TBP) was very effective at recovering iron from the leachate. The water strip solution could be decomposed to produce ferric oxide (hematite) and hydrochloric acid for recycle via hydrothermal hydrolysis or pyrohydrolysis.

KEYWORDS

Precious, rare, base, metals, iron, solvent extraction, loading, stripping, tri-butyl-phosphate

INTRODUCTION

The Malku Khota deposit is situated in the eastern part of the Bolivian Altiplano at elevations between approximately 3,800 and 4,580 meters above sea level (Figure 1). The project is located 98 km east-southeast of Oruro, and 85 km south of Cochabamba in a relatively remote area.

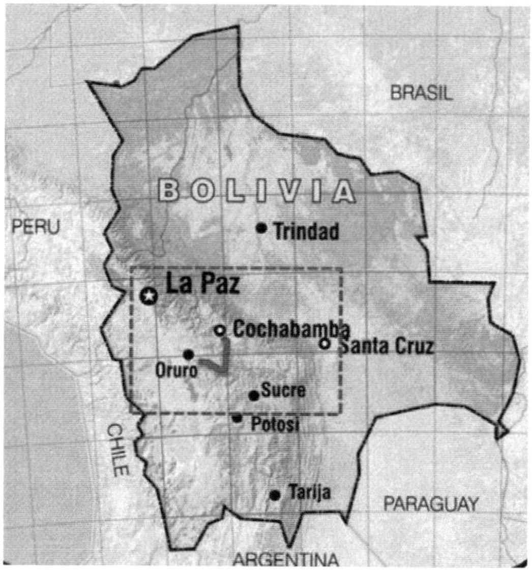

Figure 1 – Project location in Bolivia

The Malku Khota deposit has major values of Ag, In, Ga, Cu, Zn and Pb. Armitage et al. (2011) summarized the measured, indicated and inferred resources for the project as shown below. Small amounts of gold are also found in the deposit.

Table 1 – Resource statement for Malku Khota (Armitage et al., 2011)

Resource		Measured	Indicated	M+I	Inferred
Ag (g/t)	Tonnes	30,989,448	224,001,987	254,991,434	230,013,794
	Grade	33.4	27.3	28.1	18.9
	Ounces	33,319,487	196,960,598	230,280,085	140,027,216
In (g/t)	Grade	6.1	5.8	5.8	4.1
	Tonnes	188	1,293	1,481	935
Ga (g/t)	Grade	4.5	4.3	4.3	4.3
	Tonnes	139	943	1,082	1,001
Cu (%)	Grade	0.02	0.02	0.02	0.02
	Pounds	13,947,823	106,366,881	120,314,704	102,081,329
Pb (%)	Grade	0.07	0.07	0.07	0.07
	Pounds	48,665,367	404,649,086	453,314,453	362,157,607
Zn (%)	Grade	0.02	0.05	0.04	0.05
	Pounds	16,239,090	230,573,723	246,812,812	240,292,377

Notes: The resource cut-off grade of 10 g/t Ag equivalent is based only on the values of Ag (16 USD/oz) and In (550 USD/kg). Estimated metal content does not include any consideration of mining, mineral processing or metallurgical recoveries.

The preliminary economic assessment by Armitage et al. (2011) represented the base case development plan for the Malku Khota project that assumed the treatment of 40,000 t/d of ore using an acid chloride heap leach process. The acid chloride process was successfully tested in order to extract the metals of value from the ore and produce final or intermediate products for sale. An alternative process involving cyanide heap leaching was also examined. As cyanide leaching was not able to recover indium and gallium, the preliminary economic assessment was focused on the acid chloride leach process. The acid-chloride process development has been described in a previous work (Dreisinger et al., 2015).

The simplified flow sheet for the fully integrated process is shown in Figure 2. The process is designed to produce a Au-Ag cement, a Cu cement, an In-Ga precipitate for refining, a lead (PbS) precipitate and a zinc (ZnS) precipitate. However, prior to the recovery of the individual metal products, iron is removed by solvent extraction and water stripped, resulting in a concentrated ferric chloride solution. The ferric chloride solution is hydrolyzed to recycle HCl and produce a hematite by-product.

In this paper the development of the iron solvent extraction process will be described.

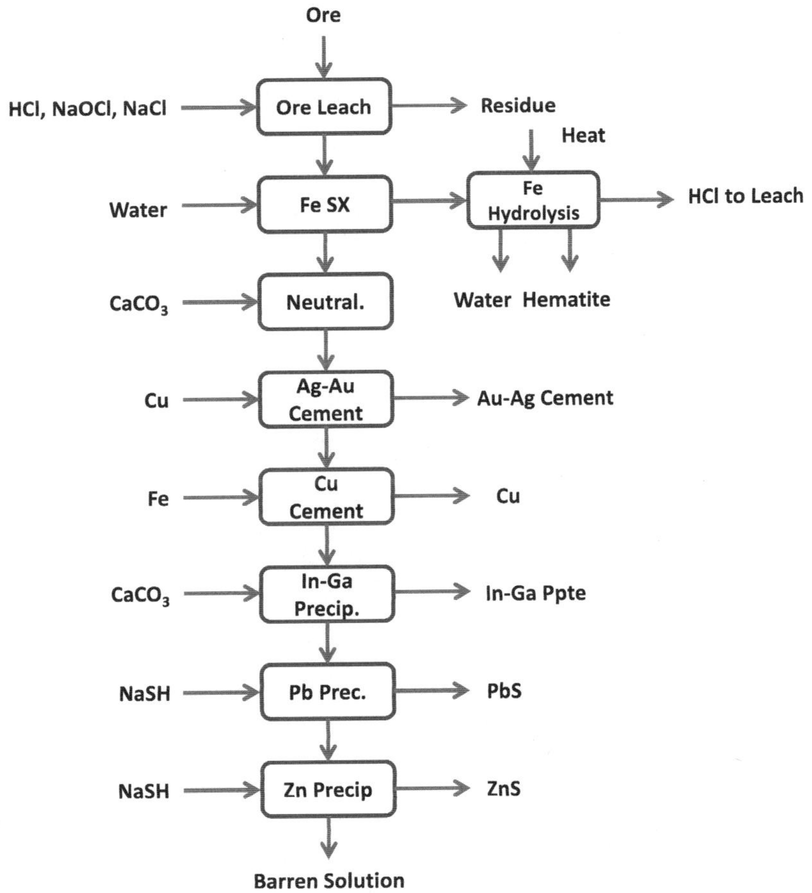

Figure 2 – Simplified process for the acid-chloride leaching process

SOLVENT EXTRACTION OF IRON

The solvent extraction process for iron recovery was designed to be "reagent free". That is, no additional reagent consumption was to be used to minimize the cost. The process involves loading a mixed iron chloride-acid species (e.g., iron chloride-hydrochloric acid species $HFeCl_4$) onto an extractant to form a loaded solvent. The extractant may be, for example, tri-butyl-phosphate (TBP) or tri-alkyl phosphinate (Cyanex 923) or tri-alkyl amine (Alamine 336). These solvating extractants were found to be highly selective in the extraction of the $HFeCl_4$ species from mixed salt solutions. The minor co-extraction of other species may be controlled by scrubbing of the loaded solvent with water or with an acid ferric chloride solution. Finally, water stripping can be used to form a $FeCl_3$ – HCl strip solution.

Reagent Screening

Malku Khota leach liquor was treated via solvent extraction to selectively remove iron and acid. The result was a low-acid, low-iron raffinate containing all of the valuable pay metals and a strip liquor containing high iron and acid concentrations. The feed to the bench SX tests was synthetically generated from chloride salts. Sulphate levels were supplied by addition of sodium sulphate and silver was added as silver nitrate. Three different types of feed solution were prepared:

- High Acid (HA): 80 g/L HCl (5 M Cl)
- Low Acid (LA): 20 g/L HCl (3.8 M Cl)
- Low Acid High Chloride (LAHC): 20 g/L HCl (4.7 M Cl)

The feed composition was based on previous column leach testing of the Malku Khota ore. The composition of this feed liquor is included in Table 2. Various test series were performed, starting with scoping level extraction shake out tests to simulated counter current locked cycle tests.

Table 2 – Solution composition for solvent extraction testing. CL Comp: Column Leach Composite, HA: High Acid, LA: Low Acid, LAHC: Low Acid High Chloride

Element	unit	CL Comp	HA	LA	LAHC
HCl	g/L	37.3	80	20	20.8
In	mg/L	21	24.4	25.5	25.3
Ag	mg/L	62.5	167	130	137
Pb	mg/L	3800	5440	3700	3950
Ga	mg/L	1.1	ND	ND	0.34
Cu	mg/L	220	378	461	457
Zn	mg/L	360	495	549	114
Fe	mg/L	9400	11600	12500	12100
Al	mg/L	680	893	1050	887
Ca	mg/L	270	285	315	306
Na	mg/L	80000	56400	61800	87600
Mg	mg/L	170	206	292	274
As	mg/L	120	ND	ND	ND
Sb	mg/L	170	ND	ND	ND
Cl	mg/L	ND	176600	134500	165800
SO_4	mg/L	ND	2100	2100	2100

Initially, three extractants were selected for the scoping level extraction shake-out tests, and each extractant mixture was contacted with HA and LA solutions at A/O ratios of 1/1 and 2/1.

- 35% TBP in Exxsol D80 diluent
- 30% Cyanex 923 in Exxsol D80 diluent
- 21% Alamine 336 in Orfom SX80 diluent

The following observations and conclusions were made:

- Highest extraction of iron was achieved by Cyanex 923;
- Extraction of iron was better using the high iron feed, which can be attributed to the higher total chloride concentration;
- Greatest degree of selectivity against Ag, In and base metals was obtained with TBP;
- Alamine 336 was the least selective.

Based on these findings, it was decided to conduct full 5-contact isotherms with TBP and Cyanex 923. To increase loading capacity in the extraction isotherm tests, the organic reagent concentration was increased to 100%.

Loading Isotherms

Using the HA and LA feed solutions, four extraction isotherms were generated using 100% TBP and 100% Cyanex 923. From these data the following additional conclusions were made:

- Strong iron extraction from LA and HA feed streams was accomplished using both TBP and Cyanex 923 reagents;
- Acid extraction increased when more organic was available (i.e., low A/O ratios). Generally, Cyanex 923 extracted more acid than TBP;
- Using TBP, Ag and In extractions increased when more organic was available. Co-extraction of Cu, Ag, In, Zn and Pb was low at high A/O phase ratios. Any co-extracted pay metals could probably be scrubbed (crowded) off using a concentrated ferric chloride liquor;
- Using Cyanex 923, In/Zn/Pb co-extraction was high (even at high loadings of iron). This indicated that scrubbing off indium with a concentrated ferric solution was probably not going to succeed. Cu and Ag co-extractions were also high, but high Fe loadings appeared to be able to scrub off these elements.

Based on these findings, further work was carried out with the TBP reagent. It was noted that the extraction of Fe and selectivity against pay metals increased when the high acid feed was used. However, parallel work indicated that the leaching performance was not affected by changing to a low acid lixiviant. Hence, there was a strong preference to treat LA type liquors. Consequently a third feed liquor was prepared that contained a low acid (20 g/L HCl) concentration, but the total Cl molarity was kept equally high (4.7 M Cl) as in the HA feed liquor. The results of this test series are presented in Figure 3.

It can be seen that the highest organic loading of 47 g/L Fe is achieved with the HA feed liquor and the lowest organic loading of 34 g/L Fe with the LA feed liquor. The LAHC liquor gives intermediate results (39 g/L Fe). The McCabe Thiele diagram was constructed for the LAHC feed liquor and showed that two extraction stages are required for an A/O ratio of 2.8. The multi-element analysis data pay metal co-extraction was low when the Fe loading and extraction were high.

The scrubbing of the loaded TBP solution was also studied to minimize losses of the pay metals. The loaded organic composition is shown in Table 3 (composite from load testing).

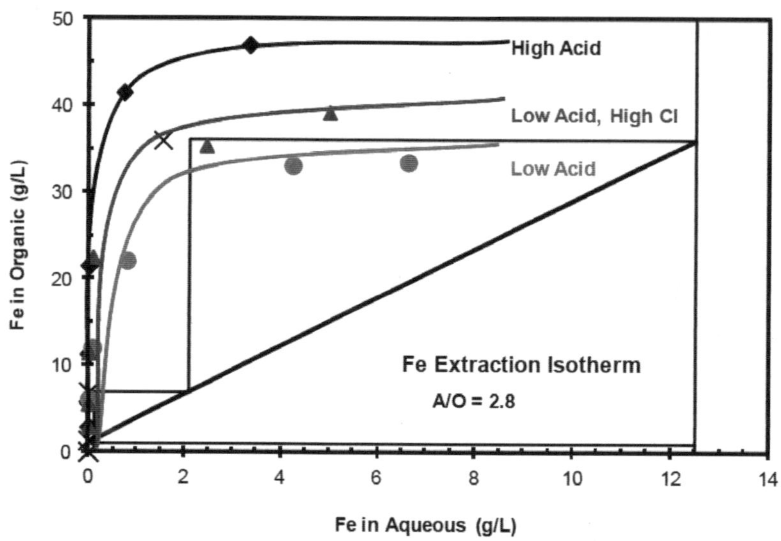

Figure 3 – Iron extraction isotherm with 100% TBP and HA, LA and LAHC feed solutions

Table 3 – Loaded organic composition (100% TBP)

	Fe	HCl	Cu	Ag	In	Zn	Ca	Mg	Na	Al	Pb
mg/L in loaded organic	45000	20236	0.6	0.5	0.5	16.9	354	3	135	0.5	8
% extracted	~100	17.3	0.0	0.4	0.6	0.9	30.7	0.4	0.0	0.0	0.0

The best scrub solution was determined to contain 83 g/L of Fe and 20 g/L of HCl. The scrub results showed:

- Indium is scrubbed off with each scrub feed liquor;
- Scrubbing of other metals increased with increasing Fe and HCl concentration;
- Calcium was the only metal not extensively scrubbed and will likely report to the Fe strip liquor.

A bulk scrubbing test with scrub feed liquor (83 g/L Fe, 25 g/L HCl) was conducted. In this test, 1 L of loaded organic was contacted with 1 L of scrub feed. The scrubbed organic was used in a strip isotherm test series.

A strip isotherm was generated using scrubbed organic and deionized water. The following O/A phase ratios were included in the isotherm: 6/1, 4/1, 2/1, 1/1, 1/2 and 1/4. The isotherm and the associated constructed McCabe Thiele diagram are shown in Figure 4. The data show that at an O/A ratio of 1.1 four stripping stages are required. A maximum strip liquor of 56 g/L Fe was produced.

Tables 4 and 5 show the results of the locked cycle testing of iron extraction using 100% TBP as the organic phase. Loading was performed with three countercurrent stages at an A/O ratio of 3.65. Scrubbing was performed with one stage at an A/O ratio of 2. The scrub solution containing iron (Table 4) was reused for a number of cycles before being refreshed. Stripping was performed with four countercurrent stages at an O/A ratio of 1.2. The loading, scrubbing and stripping operations were performed as sequential locked cycle contacts.

The deportment analysis (Table 5) showed that the iron solvent extraction process resulted in 99.9% removal of iron from the PLS with negligible loss of the pay metals. Gold and gallium were not analyzed in solution but are expected to deport to the raffinate.

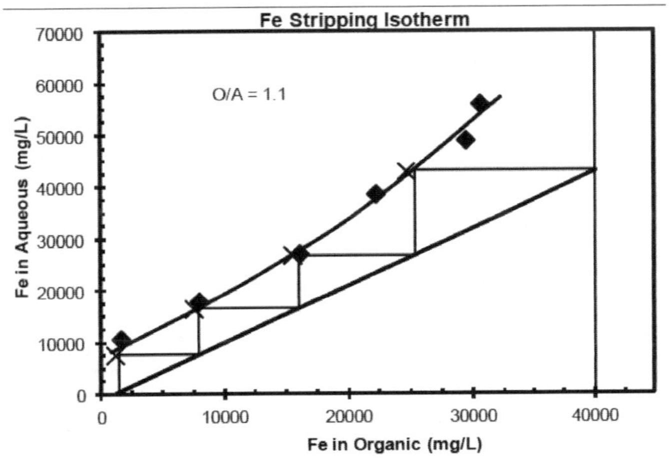

Figure 4 – Iron stripping isotherm with Fe-loaded 100% TBP and deionized water

Table 4 – Solvent extraction of PLS solution with 100% TBP solution

Stream	Chemical Analysis (mg/L)								
	In	Ag	Fe	Al	Ca	Cu	Na	Pb	Zn
PLS	22.7	75.3	13,300	959	259	409	95,900	5,940	523
Raffinate	23.5	98.5	9.7	921	170	366	110,000	6,430	672
Scrub Feed			88,100						
Scrub Product	1.83	0.47	79,300	16	116	7	2620	55	35
Strip Solution	0.18	<0.03	48,300	<8	369	<1	104	<20	<20
Loaded Organic			41,700		148	<0.1	235		15
Scrubbed Organic			50,600		188	<0.1	14		1
Stripped Organic			4,260		<2	<0.1	<2		0.1

Table 5 – Deportment of the various elements through the solvent extraction process

Stream	Deportment (%)								
	In	Ag	Fe	Al	Ca	Cu	Na	Pb	Zn
Raffinate	99.4	100	0.1	99.7	63.4	99.8	99.9	99.9	99
Scrub Liquor	0.4	0	23.8	0.1	2.1	0.1	0.1	0	0.2
Strip Liquor	0.2	0	76.2	0.2	34.5	0.1	0	<0.1	<0.7

Iron Chloride Hydrolysis

The hydrolysis of the strip solution allows the recovery of HCl and the formation of synthetic hematite.

$$2FeCl_3 + 2H_2O = Fe_2O_3 + 6HCl$$

There are a variety of engineering solutions for iron hydrolysis, including spray roasting, pyrohydrolysis or heating to decomposition (hydrothermal hydrolysis). Ideally, the engineered solution would result in (1) a filterable hematite product, (2) a dilute acid solution for recycle to stripping and (3) a concentrated acid solution for recycle to the lixiviant makeup circuit for ore leaching. Dry et al (2011) reported on the development of a thermal hydrolysis process fed by ferric chloride.

The testing of iron chloride hydrolysis was performed using three sequential (but related) experiments. A synthetic solvent extraction strip liquor containing 30 g/L Fe (as $FeCl_3$) and 25 g/L HCl was subjected to a boil test to determine the optimum temperature for hydrothermal iron hydrolysis. The boil test had to be divided into three phases due to the dilute nature of the feed solution. At the start of the second and third phases a synthetic liquid was prepared simulating the composition of the preceding final boiled down product.

At the start of the test (Boil 1), around 3 L of feed liquor was boiled down to 0.88 L at 106°C. The resulting filtrate contained 102 g/L Fe and 81 g/L HCl. The condensate (2.2 L) contained 4.5 g/L HCl, which is equivalent to around 13% of the original quantity of acid.

During the second test (Boil 2), around 3 L of synthetic Boil 1 filtrate was prepared at 98 g/L Fe and 81 g/L HCl. This was boiled down to 0.88 L at 124°C. The resulting filtrate was calculated to contain 329 g/L Fe and 111 g/L HCl. The condensate (2.0 L) contained 72 g/L HCl, which is equivalent to around 59% of the original quantity of acid.

During the third test (Boil 3), around 3 L of synthetic Boil 2 filtrate was prepared at 396 g/L Fe and 111 g/L HCl. This was boiled down to 1.9 L at 187°C. The resulting filtrate was calculated to contain 597 g/L Fe and negligible HCl. The condensate (1.0 L) contained 365 g/L HCl, which is equivalent to around 100% of the original quantity of acid.

The results of the three boil tests were combined and are shown graphically in Figure 5. The point where more than 100% of the original acid has been recovered into the condensate is defined as the temperature where iron hydrolysis started. The graph shows that this point was reached at a temperature just below 180°C. The relationship between the acid concentration of the condensate versus the test temperature is presented in Figure 6; the graph shows that a sharp increase in acid concentration is observed between 120 and 150°C.

Further (continuous) test work is required to confirm the proposed operating conditions. Since the hydrolysis test was carried out before the counter-current locked cycle test was performed (which showed that a 50 g/L Fe strip liquor could be produced), it is also recommended to conduct further hydrolysis tests with such strip liquors. However, the initial indications were favourable with respect to heating the strip solution to precipitate hematite. Progressive heating of the ferric chloride solutions showed that the solutions were very concentrated (over 500 g/L Fe) and had a very high boiling point (>180°C). Nearly complete recovery of the acid to condensate occurred.

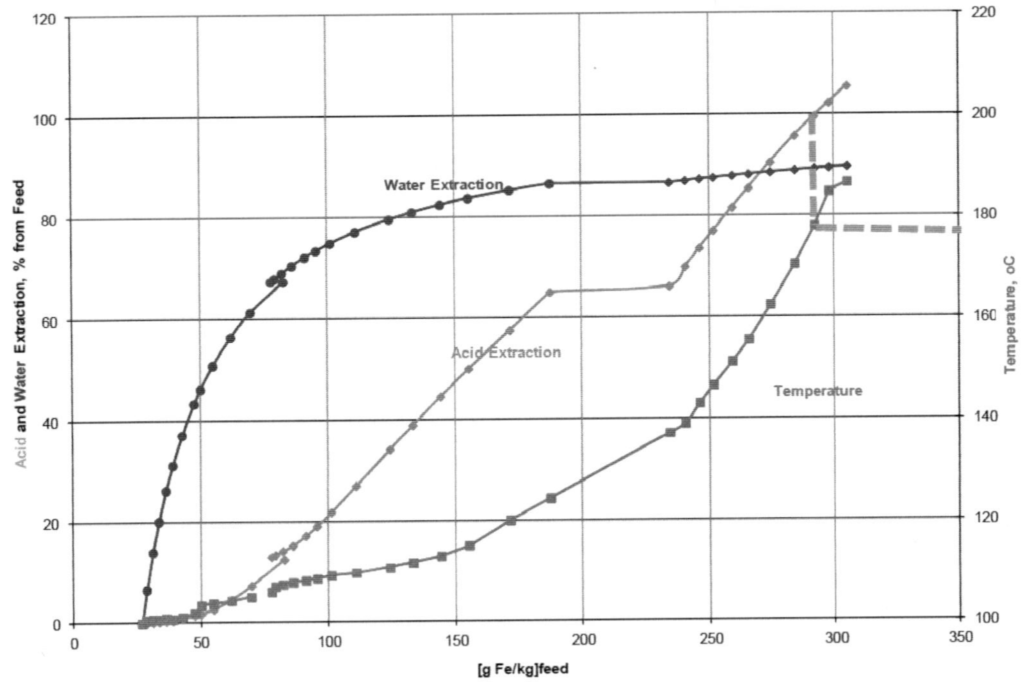

Figure 5 – Combined iron hydrolysis test results for boil tests 1-3

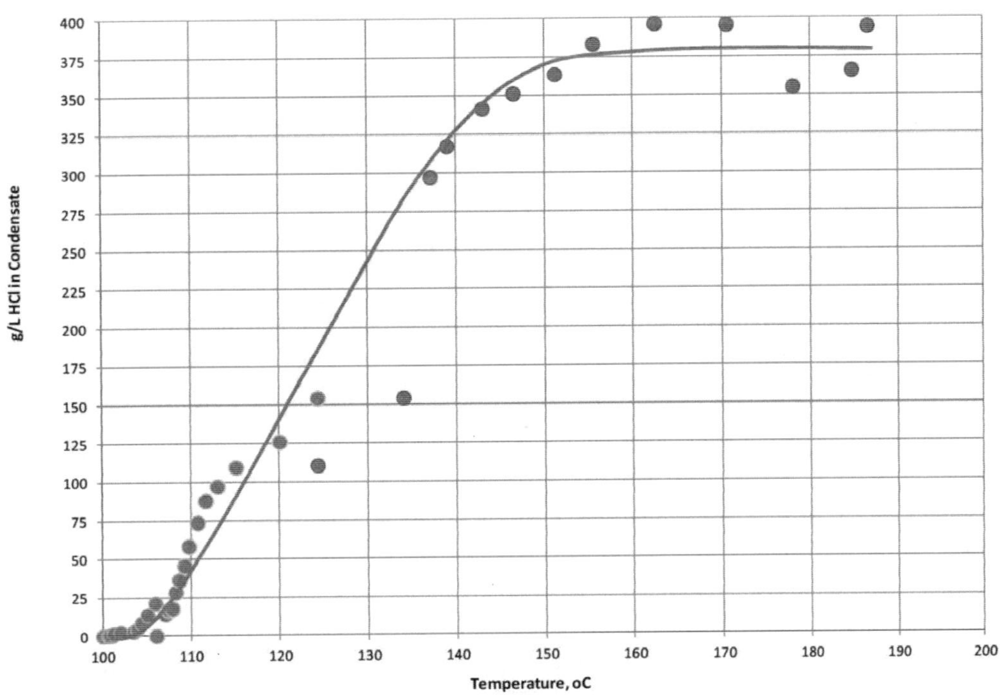

Figure 6 – Acid concentration in the concentrate for boil tests 1-3

CONCLUSIONS

An acid chloride leaching process has been developed to unlock the unique potential of the Malku Khota deposit in Bolivia which hosts significant values in precious, rare and base metals. A novel process (Dreisinger et al, 2013 and 2015) for treatment of the leachate was developed that included the removal of iron and acid by solvent extraction. The stripped iron solution was hydrolyzed to precipitate the iron as hematite and regenerate a concentrated HCl solution for recycle to the leaching stage. The process also includes the production of separate products of value including:

- Ag-Ag cement
- Cu cement
- In-Ga hydroxide precipitate
- PbS precipitate
- ZnS precipitate

The solvent extraction design basis was developed through the screening of three reagents (TBP, Cyanex 923 and Alamine 336), and the development of loading isotherms, scrubbing isotherms and strip isotherms. A batch simulation of the solvent extraction process (repeated batch contact of solutions to simulate a steady state solvent extraction process) was conducted to predict the overall circuit performance. The recovery of iron was excellent and the loss of pay metals was negligible.

The hydrolytic decomposition of the ferric chloride – hydrochloric acid solution was demonstrated by the sequential boiling down of a simulated strip solution to produce a concentrated ferric solution and an acid condensate.

The Malku Khota project is no longer available for development by TriMetals Mining Inc. However, the acid chloride leach process (including iron solvent extraction and acid regeneration) could be used for treatment of alternate ore deposits. The company is actively searching for a suitable deposit to take advantage of the unique characteristics of the process for extracting value from complex deposits containing precious, rare and base metals.

ACKNOWLEDGEMENTS

TriMetals Mining is acknowledged for supporting this work and permitting its publication.

REFERENCES

Armitage, A., Desautels, P., Zurowski, G., Pennstrom, W., Malbran, F., & Fitch, R. (2011). Preliminary Economic Assessment Update. May 10, 2011. (www.sedar.com)

Dreisinger, D., Fitch, R., & Verbaan, C. (2013). Method for recovering indium, silver, gold and rare, precious and base metals from complex oxide and sulfide ores. US Patent 8,585,991 B2, Nov 19, 2013.

Dreisinger, D.B., Verbaan, N., Forstner, C., & Fitch, R. (2015). Hydrometallurgical extraction of precious, rare and base metals using an oxidizing acid chloride heap leach. Proceedings of Rare Metal Technology 2015, Eds: Neelameggham, N.R., Alam, S., Oosterhof, H., Jha, A., Dreisinger, D., Wang, S., Warrendale, PA: TMS, pp. 87-100.

Dry, M., Verbaan, N., Bourricaudy, E., & Moran, M. (2011). Development of a novel high-chloride circuit for the Starfield Resources' Ferguson Lake Project. EPD Congress, TMS.

REMOVAL OF IRON FROM SIMULATED COPPER LEACH SOLUTION USING A HYDROXY-OXIME CHELATING RESIN

Y. Zhang[1], *Q. Liu[2] and L. Li[3]

[1]*Memorial University of Newfoundland*
St. John's, Newfoundland and Labrador, Canada

[2]*University of Alberta*
Edmonton, Canada
*(*Corresponding author: qi.liu@ualberta.ca)*

[3]*Changsha Research Institute of Mining and Metallurgy*
Changsha, China

ABSTRACT

Iron removal is a common procedure in hydrometallurgy of non-ferrous metallic ores. Chemical precipitation and solvent extraction are currently the main methods employed for separation of iron from copper leach solution. However, chemical precipitation methods have the problem of producing large amounts of iron precipitates, and/or requiring high temperature and high pressure conditions for precipitation. The precipitates are difficult to separate from the solution. Emulsification is a general problem for solvent extraction, which causes loss of extractants and contamination of the electrolyte. A common problem for solvent extraction and chelating ion exchange processes is the difficult iron stripping process. Therefore, finding an efficient and environmentally friendly method for iron removal is of great significance. In this paper, new processes of using ion exchange resins to remove iron from copper leach solution have been investigated. Salicylic acid, amino carboxylic acid, amino phosphonic acid and hydroxy-oxime resins were employed to adsorb iron from a simulated copper leach solution containing 40 g/L of Cu^{2+} and 36 g/L of Fe^{3+}. It was found that a hydroxy-oxime resin, named Z-Fe, has excellent adsorption selectivity on Fe^{3+}. Desorption behavior of the resin was studied by two step elution using dilute H_2SO_4 and oxalic acid solution, respectively. After optimizing the adsorption-elution processes, the extent of removal of iron from the simulated copper leach solution using resin Z-Fe was 90%, while copper recovery could reach 100% in closed circuit processing.

KEYWORDS

Iron removal, ion exchange, copper leach solution, hydroxy-oxime resin, oxalic acid, elution

INTRODUCTION

As one of the most abundant elements in the Earth's crust, iron always coexists with nonferrous metals in their ore bodies. The removal of iron is a necessary process in the production of pure nonferrous metals. Copper leach liquors must be purified to remove iron and other impurities before being sent to Cu electrowinning. Chemical precipitation and solvent extraction (SX) are currently the main commercial processes for iron removal in the metallurgy of nonferrous metals.

Initial attempts of iron removal were to precipitate Fe as hydroxides, which are gelatinous, and poorly filterable (Tainton and Leyson, 1924). It is well known that solid–liquid separation and washing processes are improved if the precipitate is dense with a crystalline character and has reasonably large particle size. By using hot, dilute solutions to prevent supersaturation which results in high nucleation rate and precipitation of small particles, crystalline precipitate of coarse particle size typically forms.

The jarosite process was the first iron removal method that realized the production of a filterable iron residue in commercial practice, and is still the most widely used process in nonferrous metal industry today. The first patents of jarosite processes were reported by Asturiana de Zinc (Asturiana de Zinc, 1964), Norzinc (Det Norske Zinkkompani, 1965) and Electrolytic Zinc Company of Australasia (Electrolytic Zinc Company of Australasia, 1965). Ferric ion precipitates slowly from weak acidic sulfate solutions in the form of crystalized $AFe_3(SO_4)_2(OH)_6$ under high temperatures, as illustrated below:

$$3Fe^{3+} + A^+ + 2SO_4^{2-} + 6H_2O = AFe_3(SO_4)_2(OH)_6 + 6H^+ \quad (1)$$

A^+ could be a cation of K^+, Na^+, or NH_4^+. In Jinchuan Nonferrous Metal Company, the iron concentration in copper leach liquor was lowered from 16 g/L to less than 0.05 g/L at pH 1.4~1.6 and temperature above 95℃ using a jarosite process (Jarosite Process Testing Group of Jinchuan Metallurgy Plant, 1980). One advantage of jarosite processes is the removal of SO_4^{2-} ion and other impurities from the leach solution, and their disadvantages are elevated temperature operation to form well crystallized ferric vitriols for fast filtration and the production of large amount of dregs.

Goethite processes remove iron from acidic leaching solutions under relatively low temperature compared with jarosite processes in the form of goethite α-FeOOH by means of the Vieille-Montagne Process (Societe de la Vieille Montagne, 1968). To prevent the formation of $Fe(OH)_3$ colloid, Fe(III) concentration is usually required to be less than 1 g/L, which means the leach liquors should be diluted.

Another major iron removal process by precipitation is the hematite process as practiced at the Iijima Electrolytic Zinc Plant in Japan (Tsunoda et al., 1973). First iron in the solution is reduced to the ferrous state, and is then oxidized by oxygen under high temperature and pressure (1.8-2.0 Mpa) to precipitate the iron as hematite, as shown below:

$$4FeSO_4 + O_2 + 4H_2O = 2Fe_2O_3 + 4H_2SO_4 \quad (2)$$

The advantage of the hematite process is that it produces a small amount of iron residue with 50%-60% Fe content, which can be sold as steel making material or cement additive.

Other iron precipitation processes, such as the paragoethite process (Loan et al., 2006), Zincor process (Claassensa et al., 2002) and phosphate process (Yang et al., 2012), were also explored by researchers.

Although precipitation processes are most widely employed for iron removal, the disposal of solid iron residues represents an environmental drawback. Solvent extraction and ion exchange processes are increasingly being used for iron control (Flett & Monhemius, 1996).

The phosphate group has a strong affinity and selectivity for Fe(III). In the past decades, bis(2-ethylhexyl)phosphoric acid (D2EHPA) (Sato et al., 1985; Biswas et al., 1998) and tributylphosphate (TBP) (Miralles et al., 1992; Sahu et al., 2000; Sarangi et al., 2007; Liu et al., 2014) were used as extractants to remove Fe(III). As a mixture of four trialkyl-phosphine oxides, Cyanex 923 was used to extract Fe(III) from acidic chloride solutions in xylene (Saji et al., 1998).

Ligands containing an oxime group and pyridine ring are popular in coordination chemistry. The majority of the metal complexes of these ligands have been prepared in the last decades. The donor atoms of the 2-pyridyl oximes in metal complexes are the nitrogen atoms of the oxime and the pyridyl groups. The extractive Fe(III) removal from chloride solutions in the presence of Cu(II) and Zn(II) with 2- and 4-pyridyl ketoximes was reported by Parus et al. (Parus et al., 2010).

The volatile diluents like kerosene used in SX exhibit disadvantages such as high volatility and toxicity (Kislik, 2012), which cause evaporation and pollution, respectively. Ionic liquids were introduced as diluents due to their excellent properties such as chemical stability, water immiscibility and negligible vapor pressure which would avoid solvent losses providing a total recyclability. Furthermore, it was found that ionic liquids are advantageous over traditional organic diluents as they exhibit better extractability (Yoon et al., 2010). The selective removal of iron from industrial and synthetic solutions containing mainly Cu(II)/Fe(III) using the ionic liquid 1-butyl-3-methylimidazolium bis(trifluoromethylsulfonyl)imide, [bmim][Tf2N] as the diluent and 1,1,1-trifluoro-2,4-pentanedione (TFA) as the extractant was investigated by E. Quijada-Maldonado et al. (Quijada-Maldonado et al., 2016). However, some researchers think that the hydrolysis of ionic liquids is not always green (Swatloski et al., 2003).

Similar to solvent extraction, a specific/chelating ion exchange resin is generally used for iron removal from copper electrowinning solutions. Chelating ion exchange resins have reactive groups binding with metal ions through the formation of complexes. It is recognized that commercial chelating ion exchange resins used for iron removal are structurally iminodiacetic, picolylamine, sulphonated phosphonic, or aminophosphonic, which have strong affinity for the ferric ion. Due to the strong acidic nature of copper electrowinning solutions, only the phosphonic resins are suitable for this application (Shaw et al., 2006). The phosphonic group likely acts as the specific chelating group for Fe(III) ion in these resins (Chiarizia et al., 1997). Adsorption of ferric ion onto these resins is an ion exchange with hydrogen ions, and for the case of a sulfate based electrolyte can be described by the following equation, where R represents the chemistry of the resin not involved in ion exchange:

$$Fe_2(SO_4)_3 + 6(HR) \rightarrow 2(FeR_3) + 3H_2SO_4 \quad (3)$$

The Diphonix resin is a sulfonated diphosphonic resin used for iron removal. One of the disadvantages of Diphonix resin for iron removal lies in the difficulty associated with stripping iron from the loaded resin. One elution process developed for this resin involved reducing the bound ferric iron to ferrous iron using a solution containing sulphurous acid and copper ions, created by bubbling SO_2 gas through copper electrolyte (Gula et al., 1996). Aqueous diphosphonic acid was tested to be effective in eluting iron from the loaded Diphonix resin (Chiarizia et al., 1997).

As aforementioned, chemical precipitation and solvent extraction are currently the main methods employed for iron removal from copper leach solutions. However, chemical precipitation methods have the problem of producing large amounts of iron precipitates, and/or requiring high temperature conditions for precipitation. The precipitates are difficult to separate from the solutions. As to solvent extraction, emulsification is a general problem, and causes loss of extractants and contamination of the electrolyte. A common problem for solvent extraction and chelating ion exchange processes is the difficulties in iron stripping owing to the strong bonding between chelating groups and ferric ion (Liu et al., 2014; Chiarizia et al., 1997; Gula et al., 1996). Therefore, finding an efficient and environmentally friendly method for iron removal is of great significance. In this paper, new processes of using ion exchange resins with excellent elution properties to remove iron from a synthetic copper leach solution have been investigated.

EXPERIMENTAL

According to the chemical composition of the copper leach liquor from a copper mine in Southern China, simulated leach solution containing 40.0 g/L of Cu and 36.0 g/L of Fe was prepared with analytic reagents $CuSO_4 \cdot 5H_2O$ and $Fe_2(SO_4)_3 \cdot 6H_2O$ for iron removal experiments. Sulfuric acid or sodium hydroxide solution was used to adjust solution pH when needed.

In a resin adsorption test, 25 mL simulated copper leach solution and certain amount of resin were added into a 250 mL conical flask with a rubber stopper, and vibrated using a constant temperature water bath shaker at certain resin dosage, solution pH and temperature. As the adsorption process finished at pre-set time, the resin and solution were separated immediately by filtration. Then the concentrations of Cu and Fe in the processed solution were analyzed. And the adsorption extent η was calculated using following formula:

$$\eta = \frac{C_0 - C}{C_0} \text{ x } 100\% \tag{4}$$

where C_0 and C are the initial and ending concentrations of Cu or Fe in adsorption, respectively. The adsorption amount of Cu or Fe on resin was also determined by metal concentration change after adsorption.

In an elution test, loaded resin from adsorption was mixed with certain volume of rinse solution in a 250 mL conical flask with a rubber stopper, and vibrated using a shaker for pre-set time.

RESULTS AND DISCUSSION

Adsorption Tests

The major metal ions in the simulated copper leach solution are Cu^{2+} and Fe^{3+}. To selectively adsorb Fe^{3+} from the solution, the resin employed should have group(s) with special bonding power to it. However, there is no effective theory for selecting or designing resins with specific adsorption capability to Fe^{3+} ion by now. From the knowledges of inorganic and analytical chemistry, it is known that Fe^{3+} ion has strong bonding preference with OH^- and PO_4^{3-} ions. And from the exploration of extractants and ion exchange resins, the phosphonic group and oxime group may have specific chelating power for Fe(III) ion (Chiarizia et al., 1997; Parus et al., 2010). During our study, a large number of resins were selected for exploratory tests. The adsorption of Fe(III) from simulated copper leach solution using salicylic acid resin SA with functional groups of -OH and -COOH, amino carboxylic acid resin D851 with functional groups of $-NH_2$ and -COOH, amino phosphonic acid resins D405 and LSC-500 with functional groups of -OH and $-PO(OH)_2$, and hydroxy-oxime resin Z-Fe with functional groups of -OH and =C=N-OH were studied in this paper.

Adsorption Properties of Different Resins for Fe(III) and Cu(II) Ions

The adsorption properties of different resins for Fe and Cu from simulated copper leach solution containing 40 g/L of Cu^{2+} and 36 g/L of Fe^{3+} at pH 1.4 and temperature 20°C with adsorption time of 60 minutes are illustrated in Figure 1 and Figure 2, respectively. Both Fe and Cu adsorption extents increase with resin dosage for all resins, among which hydroxy-oxime resin Z-Fe has highest Fe adsorption extent while lowest Cu adsorption extent. Therefore, resin Z-Fe has the most selective adsorption for Fe from copper leach solution compared with other resins. Fe adsorption extent could be more than 91%, while Cu adsorption extent lower than 34% as Z-Fe dosage was 1.2 g/mL.

Resin Z-Fe was chosen for further study of selective removal of Fe from copper leach solution.

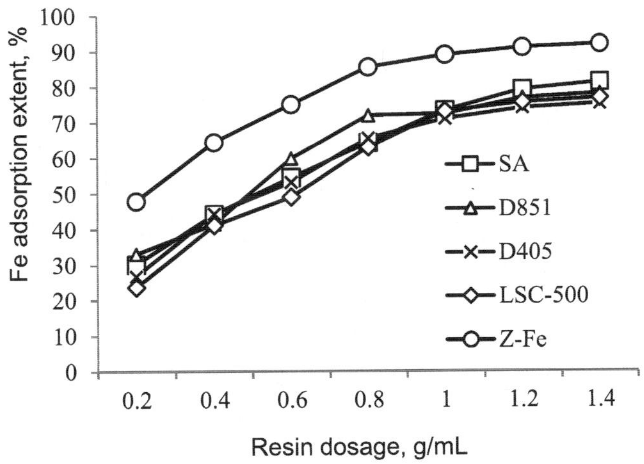

Figure 1 – Fe adsorption properties of different resins

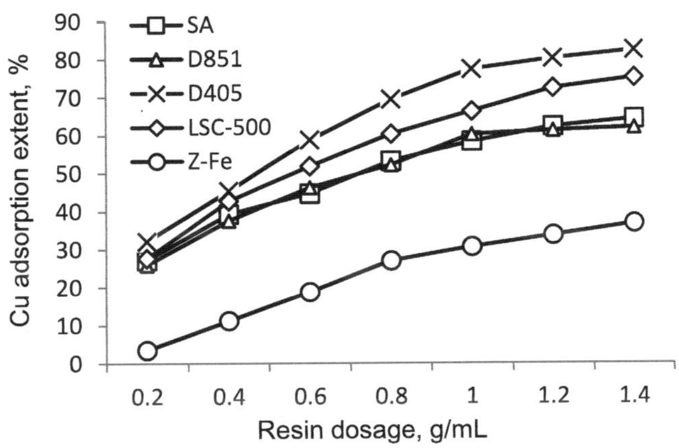

Figure 2 – Cu adsorption properties of different resins

Effect of Solution pH on Adsorption Behavior of Resin Z-Fe

Solution pH usually has remarkable effect on adsorption property of a resin. Figure 3 demonstrates the influence of solution pH on adsorption property of resin Z-Fe under experimental condition of resin dosage 1.2 g/mL, temperature 21°C, and adsorption time 60 minutes. It can be seen that the extent of adsorption of Fe increases from 71% to 92% as pH goes up from 0.6 to 1.9, while the extent of adsorption of Cu is less sensitive to pH.

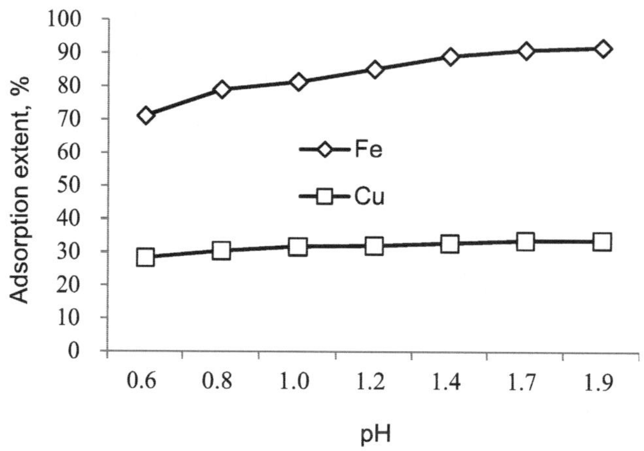

Figure 3 – Effect of solution pH on adsorption property of resin Z-Fe

Effect of Temperature on Adsorption Behavior of Resin Z-Fe

Under the condition of resin dosage 1.2 g/mL, solution pH 1.4, and adsorption time 60 minutes, the influence of temperature on adsorption behavior of resin Z-Fe was studied, as shown in Figure 4. The adsorption extents of both Fe and Cu decreased with temperature rise. It can be deduced that the metal adsorption on resin Z-Fe is an exothermic process.

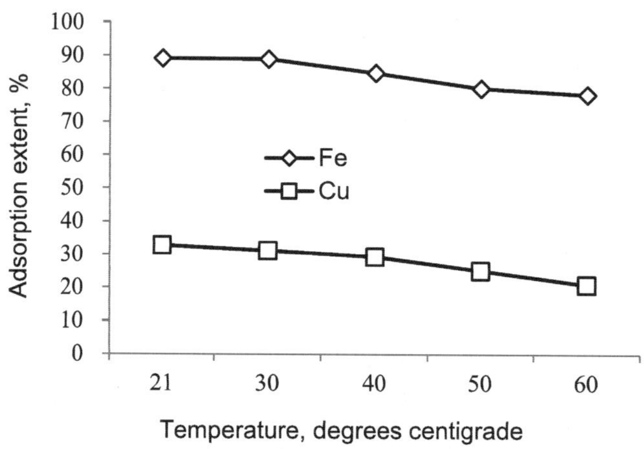

Figure 4 – Effect of Temperature on adsorption property of resin Z-Fe

Effect of Adsorption Time on Adsorption Behavior of Resin Z-Fe

Experiments were carried out under condition of resin dosage 1.2g/mL, pH 1.4, and temperature 20°C. As shown in Figure 5, the adsorption extents of Fe and Cu increase with adsorption time. As adsorption time is longer than 40 minutes, the adsorption extents level off. More than 89% of Fe and about 33% of Cu could be extracted from the simulated copper leach solution by resin Z-Fe. Remarkable adsorption preference to Fe over Cu was obtained. The major bonding atom of resin Z-Fe with Fe(III) may be the nitrogen atom in =C=N-OH oxime group (Parus et al., 2010).

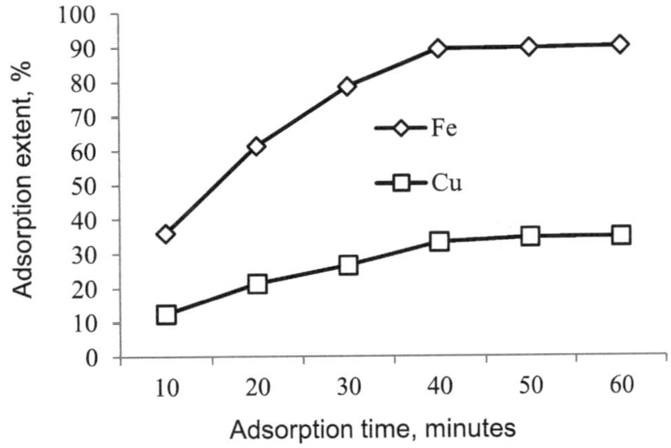

Figure 5 – Effect of adsorption time on adsorption property of resin Z-Fe

Elution Tests

First Stage Elution – Stripping of Copper

To improve the selectivity of the adsorption process using resin Z-Fe, a two stage elution process was employed. In the first elution stage, 10 grams of loaded resin Z-Fe with 11 mg/g Cu and 27 mg/g Fe obtained from the adsorption step was rinsed with 0.05 mol/L sulfuric acid solution. 10 mL sulfuric acid solution was used for rinsing 10 minutes each time. According to the Fe or Cu concentration in rinsed solution, the desorption extent of Fe or Cu was calculated. As shown in Figure 6, the desorption extents of Fe and Cu increase with the volume of rinse solution. As the volume of rinse solution was 40 mL, i.e., 4 times of rinsing, Cu was completely washed out of the resin, while the desorption extent of Fe was 40%.

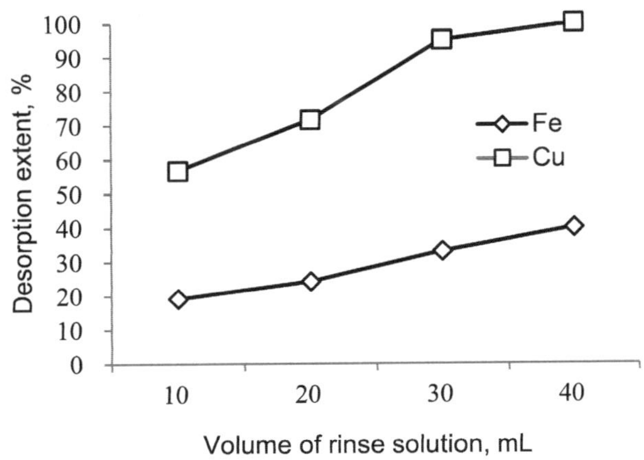

Figure 6 – First stage elution with dilute sulfuric acid solution

Second Stage Elution – Stripping of Iron

To strip off the Fe remaining on the resin after first stage elution and recycle the resin, second stage elution with different complexing agents was explored. 10 grams of resin Z-Fe loaded with 16 mg/g Fe from first stage elution was rinsed in one time using 40 mL different rinsing solutions with a mass concentration of 10% for 40 minutes. As shown in Figure 7, a highest desorption extent of 95% for iron was obtained with oxalic acid solution.

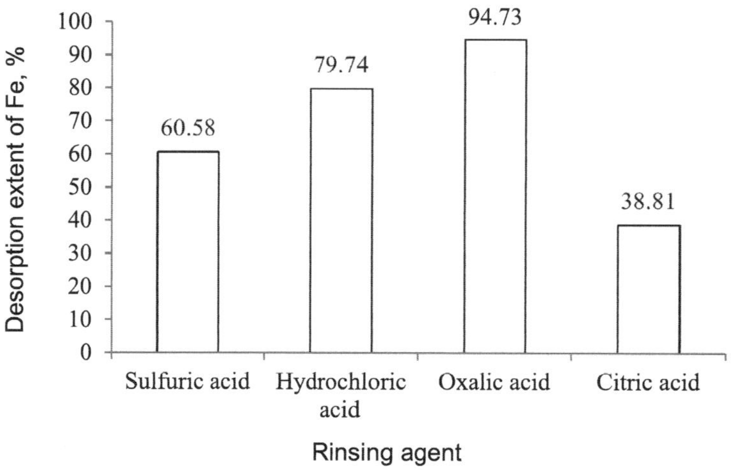

Figure 7 – Second stage desorption using different rinsing agents

Usually, multi-time rinsing is more efficient and complete than one time rinsing for a certain volume of rinse solution. 10 grams of resin Z-Fe loaded with 16 mg/g Fe from first stage desorption was rinsed with 10 mL 10% oxalic acid solution for 10 minutes each time. As shown in Figure 8, the desorption extent of Fe increases with the volume of oxalic acid solution. As the volume of rinse solution was 50 mL, i.e., 5 times of rinsing, Fe was completely stripped from the resin. Therefore, resin Z-Fe has excellent elution property while being rinsed with oxalic acid solution.

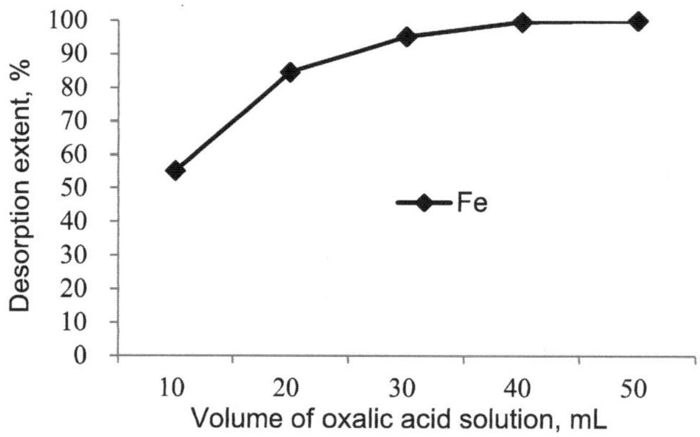

Figure 8 – Second stage rinsing with oxalic acid solution

Removal of Fe from Cu Leach Solution using the Closed Circuit Adsorption-Elution Process

From the above study, it was found that about 33% of Cu in simulated leach solution was adsorbed by resin Z-Fe, and could be completely stripped from the loaded resin with around 40% of Fe desorption extent using 0.05 mol/L H_2SO_4 solution. The rinsed solution contained about 33% of the Cu of the leach solution, and could not be sent to a copper electrowinning system due to high iron content. To recover the copper in the first stage elution, the rinsed solution, which contained about 8.3 /L Cu and 8.1 g/L Fe, can be returned to the Fe adsorption system, as illustrated in Figure 9. A closed circuit process by returning rinsed solution from first stage elution to the Fe adsorption system, which resulted in the decrease of Cu and Fe concentrations, was investigated by adopting the explored conditions listed in Table 1. Six batches of 25 mL simulated copper leach solution were processed, Cu recovery of 100% and Fe removal extent of 90% were obtained.

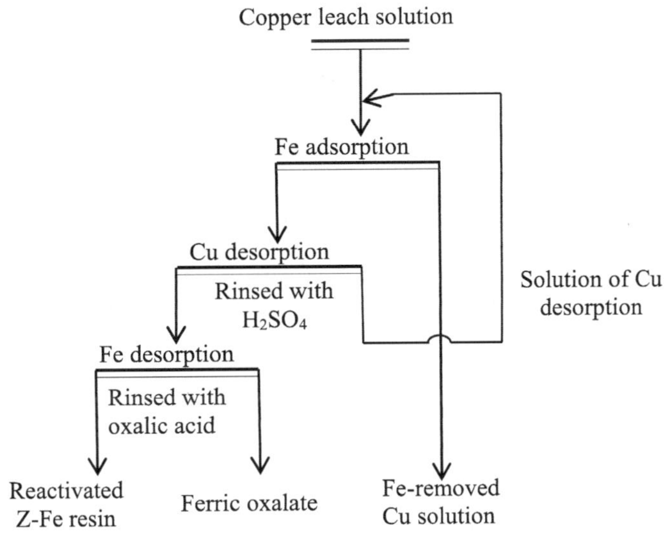

Figure 9 – Flow chart for separation of Fe from copper leach solution using closed circuit process

Table 1 – Optimized condition for closed circuit Fe removal process

Operation	Experimental conditions
Fe adsorption	Resin dosage 1.4 g/mL for original copper leach solution, solution pH 1.4, adsorbing at ambient temperature for 70 minutes
Cu desorption	Rinsed with 0.05 mol/L sulfuric acid solution for 4 times. 10 mL sulfuric acid solution was used for rinsing 8 minutes each time.
Fe desorption	Rinsed with 10 mL 10% oxalic acid solution for 5 times. 10 mL oxalic acid solution was used for rinsing 10 minutes each time.

CONCLUSIONS

New processes of using ion exchange resins to remove Fe(III) from a synthetic copper leach solution have been investigated. Salicylic acid resin SA, amino carboxylic acid resin D851, amino phosphonic acid resins D405 and LSC-500, and hydroxy-oxime resin Z-Fe were employed to adsorb iron from the solution containing 40 g/L of Cu^{2+} and 36 g/L of Fe^{3+}. It was found that hydroxy-oxime resin Z-Fe has the most selective adsorption behavior for Fe from copper leach solution compared with other resins. Iron adsorption extent could be more than 91%, while copper adsorption extent lower than 34% under the condition of resin Z-Fe dosage 1.2 g/mL, pH 1.4, temperature 20°C and adsorption time 60 minutes.

The adsorption extents of both Fe and Cu decrease with temperature rise. It can be deduced that the metal ion adsorption by resin Z-Fe is an exothermic process.

Cu adsorbed in the iron removal procedure could be completely stripped from the resin Z-Fe using dilute sulfuric acid solution. And oxalic acid is efficient for stripping Fe(III) from loaded resin Z-Fe. Therefore, resin Z-Fe has excellent elution properties.

Closed circuit processing by returning rinsed solution from the first stage elution to the Fe adsorption system demonstrated that 100% Cu recovery and 90% Fe removal extent were obtained.

Excellent elution properties and adsorption preference to Fe over Cu will make using hydroxy-oxime resin Z-Fe for iron removal from copper leach solution a promising process.

REFERENCES

Aminian, H., & Bazin, C. (2000). Solvent extraction equilibria in copper (II)-iron (III)-LIX984 system. *Minerals Engineering, 13* (6), 667–672.

Asturiana de Zinc (1964). *Spanish Patent No. 304,601.*

Biswas, R. K., & Begum, D. A. (1998) Solvent extraction of Fe^{3+} from chloride solution by D2EHPA in kerosene. *Hydrometallurgy, 50*(2), 153-168.

Chiarizia, R., Horwitz, E. P., Alexandratos, S. D., & Gula, M. J. (1997). Diphonix resin: a review of its properties and applications. *Separation Science and Technology, 32* (1–4), 1–35.

Claassena, J. O., Meyerb E. H. O., Rennieb J., & Sandenbergh, R. F. (2002). Iron precipitation from zinc-rich solutions: defining the Zincor Process. *Hydrometallurgy, 67* (1), 87–108.

Det Norske Zinkkompani (1965). *Norwegian Patent No. 108,047.*

Electrolytic Zinc of Australasia (1970). *Australian Patent No. 401,724.*

Fernandez-Olmo I., Ortiz A., Urtiaga A., & Ortiz I. (2008). Selective iron removal from spent passivation baths by ion exchange. *Journal of Chemical Technology and Biotechnology, 83* (12), 1616–1622.

Flett, D. S., & Monhemius, A. J. (1996). Solvent extraction for iron control in hydrometallurgy. In A. E. Dutrizac & G. B. Harris (Eds.), *Proc. Second Intl. Symposium on Iron Control in Hydrometallurgy. Canadian Institute of Mining, Metallurgy & Petroleum* (pp. 331–356). Montreal, Canada.

Gula, M. J., Dreisinger, D. B., & Horwitz, E. P. (1996). *United States Patent No. 5,582,737.* Washington, DC: U.S. Patent and Trademark Office.

Jarosite Process Testing Group of Jinchuan Metallurgy Plant (1980). Iron Control in Copper Matte Processing Using Jarosite Process. *Jinchuan Science and Technology,* (3), 99-101.

Kislik, V. S. (2012). In V. S. Kislik (Ed.), *Solvent Extraction* (pp. 451–481). Amsterdam, Netherlands: Elsevier.

Liu, Y., Nam, S. H., Lee, M. (2014). Stripping of Fe(III) from the Loaded Mixture of D2EHPA and TBP with Sulfuric Acid Containing Reducing Agents. *Bulletin of the Korean Chemical Society*, 35 (7), 2109-2113.

Loan, M., Newman, O. M. G., Cooper, R. M. G., Farrow, J. B., & Parkinson, G. M. (2006). Defining the Paragoethite process for iron removal in zinc hydrometallurgy. *Hydrometallurgy, 81* (2), 104-129.

McKevitt B., & Dreisinger D. (2009). A comparison of various ion exchange resins for the removal of ferric ions from copper electrowinning electrolyte solutions part I: Electrolytes containing no other impurities. *Hydrometallurgy, 98*(1), 116–121.

Miralles, N., Sastre, A. M., Figuerola, E., & Martinez, M. (1992). Solvent extraction of iron(III) by bis(2,4,4-trimethylpentyl)phosphinic acid: Experimental equilibrium study. *Hydrometallurgy, 31* (1), 1-12.

Paiva, A.P., & Costa, M.C. (2005). Application of N,N'- tetrasubstituented malonamides to the recovery of iron(III) from chloride solutions. *Hydrometallurgy, 77* (1), 103-108.

Parus, A., Wieszczycka, K., & Olszanowski, A. (2011). Solvent Extraction of Iron(III) From Chloride Solutions in the Presence of Copper(II) and Zinc(II) Using Hydrophobic Pyridyl Ketoximes. *Separation Science and Technology, 46*(1), 87-93.

Quijada-Maldonado, E., Romero, J., and Osorio, I. (2016). Selective removal of iron(III) from synthetic copper(II) pregnant leach solutions using [bmim][Tf2N] as diluent and TFA as extracting agent. *Hydrometallurgy, 159*(1), 54–59.

Sahu, K. K., & Das, R. P. (2000). Mixed solvent systems for the extraction and stripping of iron(III) from concentrated acid chloride solutions. *Metallurgical and Materials Transactions B, 31*(6), 1169-1174.

Saji, J., Rao, T. P., Iyer, C. S. P., & Reddy, M. L. P. (1998). Extraction of iron(III) from acidic chloride solutions by Cyanex 923, *Hydrometallurgy, 49* (3), 289–296.

Santos, M. S. C. S., & Paiva, A. P. (2014). Iron(III) extraction from chloride media by N,N′-tetrasubstituted malonamides: an interfacial study. *Journal of Colloid and Interface Science, 413*(1), 78–85.

Sarangi K., Parhi, P. K., Padhan, E., Palai, A. K., Nathsarma, K. C., Park, K. H. (2007). Separation of iron(III), copper(II) and zinc(II) from a mixed sulphate/chloride solution using TBP, LIX 84I and Cyanex 923. *Separation and Purification Technology, 55* (1), 44–49.

Sato, T., Nakamura, T., & Ikeno, M. (1985). The extraction of iron(III) from aqueous acid solutions by di(2-ethylhexyl)phosphoric acid. *Hydrometallurgy, 15* (2), 209-217.

Shaw, R., Vance, S., Illescas, J., Dreisinger, D., & Wassink, B. (2006). Ion exchange for iron impurity control in the base metal industry. In J. E. Dutrizac & P. A. Riveros (Eds.), *Proceedings of the Third International Symposium on Iron Control in Hydrometallurgy* (pp. 757–769). Montreal, Canadian Institute of Mining, Metallurgy and Petroleum, Canada.

Societe de la Vieille Montagne (1968). *Belgian Patent No. 724,214.*

Swatloski, R. P., Holbrey, J. D., & Rogers, R. D. (2003). Ionic liquids are not always green: hydrolysis of 1-butyl-3-methylimidazolium hexafluorophosphate. *Green Chemistry, 5*(4), 361–363.

Tainton, U. C., & Leyson, L. T. (1924). Electrolytic zinc from complex ores. *Transactions of the American Institute of Mining, Metallurgical, and Petroleum Engineers*, (LXX), 486–522.

Tsunoda, S., Maeshiro, I., Ewi, M., & Sekine, K. (1973). *The construction and operation of the Iijima electrolytic zinc plant.* Presented at AIME (Paper No. A73-65), Warrendale, PA, USA.

Xue, S. S., Gula, M. J., Harvey, J. T., & Horwitz, E.P. (2001). Control of iron in copper electrolyte streams with a new monophosphonic / sulphonic acid resin. *Minerals & Metallurgical Processing, 18* (PART 3), 133–137.

Yang, Q., Liu, J., Jiang, B., & Liu, H. (2012). Technique of removing iron by phosphate in high iron-low copper solution. *Journal of Central South University: Science and Technology, 43* (10), 3763-3768.

Yoon, S. J., Lee, J. G., Tajima, H., Yamasaki, A., Kiyono, F., Nakazato, T., & Tao, H. (2010). Extraction of lanthanide ions from aqueous solution by bis(2-ethylhexyl)phosphoric acid with room-temperature ionic liquids. *Journal of Industrial and Engineering Chemistry, 16* (3), 350–354.

IRON MANAGEMENT IN ELECTROWINNING PROCESSES

* Q. Ricoux, A. Maihatchi, F. Goettmann

Extracthive,
Bat 426, Centre de Marcoule
30207 Bagnols-sur-Cèze,
France
*(*Corresponding author: Quentin.ricoux@extracthive.eu)*

ABSTRACT

Electrodeposition processes are widely used in the metallurgical industry to produce, extract and refine metals from various feed materials. However, raw materials used to provide non-ferrous metals often contain high amounts of iron. In the presence of iron, while metals are plated at the cathode, electrowinning processes produce corrosive and highly oxidative ferric ions at the anode that are harmful for the process sustainability. In fact, these ferric ions cause dramatic current efficiency loss, induce high maintenance expenditures to limit the corrosion of devices and it results in the formation of impure and dendritic deposits on the cathode. Consequently, the management of iron in electrodeposition process became a major issue in non-ferrous metals production. Actually, numerous methods have been developed to limit the ferric ion concentration in the electrolyte or, at least, to limit its harmful effect. Many methods aim to limit the iron(III) concentration in the electrolyte, generally by coupling electrowinning to extraction process such as solvent extraction (SX) (Bender & Emmerich, 2015), ion exchange (IX) (Gula, Dreisinger, & Horwitz, 1996) or cementation (Kust, 1979) processes to remove ferric ions. Other routes consist of (i) using the ferric ions oxidant properties to leach raw materials, either directly in the electrolytic cell (Fanta & Tributsch, 1975) or through a secondary circuit (Cain, 1950), or (ii) reducing ferric ions through periodic reversal of direct current to enhance the formation of Fe^{2+} (Loutfy & Bharucha, 1980). On the other side of the spectrum, papers discuss methods to valorize the iron from by-products of the non-ferrous metal industry, that generates vast quantities of iron containing wastes (A. Agrawal, Sahu, & Pandey, 2004). Many studies on iron electroplating and electrolytic iron production from ores have been carried out (Cain, 1950; Mostad, Rolseth, & Thonstad, 2008), but very few have been conducted on waste or by-products. Therefore, we propose to build on the previous literature review to develop a process for the production of electrolytic iron from by-products.

KEYWORDS

Iron, non-ferrous metals production, valorization, electrowinning

INTRODUCTION

Non-ferrous metals and their alloys are used in a large variety of applications. Aluminum and titanium are used for their strength, light weight and corrosion resistant properties in structural applications. Copper has been used in various applications requiring high thermal and electrical conductivity, as kitchen products and electrical wires. Nickel and zinc are commonly used as plating materials; nickel offers a wide range of working temperature and zinc is used for steel plating. Gold, silver, and platinum group metals are used in electrical/electronic applications, jewelry and chemistry as catalysts.

Non-ferrous metals are usually obtained from sulphide, carbonate, and silicate ores, which can contain iron as a major impurity (Reuter et al., 2013). Their extraction processes focus on the production of pure metals and therefore on their separation from undesirables, especially iron. In the case of hydrometallurgical production routes, a first step of metals leaching from ores is typically followed by metals separation through precipitation, solvent extraction or ion exchange processes and in numerous cases electrodeposition to produce pure metals. The presence of soluble iron during non-ferrous metals electrowinning significantly reduces the energy efficiency and affects the quality of the deposited metal. In order to address these issues, many methods have been developed to limit the iron concentration in electrolytic cells and to limit the effect of residual iron on current efficiencies and metal deposits.

The iron, which is separated during non-ferrous metals production as iron-rich sludge is mostly landfilled and/or reemployed in low added-value application, despite the fact that they contain significant amounts of both toxic and valuable metals (Al-Abed, Jegadeesan, Purandare, & Allen, 2007; C. Klauber, Gräfe, & Power, 2011; Craig Klauber, Gräfe, & Power, 2009). Major consideration is now given to the stabilization and the valorization of by-products in the metallurgical industry (Herck, 2000) and developments in extractive metallurgy emphasize the reprocessing of metals from landfilled by-products (Herck, 2000). Recent research often focuses on the recovery of valuable metals from these wastes, even though their selective extraction is generally hindered by very low valuable metal contents (Musina, Bocokic, Lavric, & van Zutphen, 2014; Ricoux et al., 2015). On the opposite, only a few studies focus on processes to first extract iron in order to access more valuable non-ferrous metals which were previously too diluted (because of the iron species) to be extracted (Barnett & Mezner, 2001).

The first part of this review article focuses on non-ferrous electrodeposition processes from raw materials containing iron species and the description of methods to limit their deleterious effects in electrolytic cells. The second part deals with a way to valorize iron rich landfilled by-products from the production of non-ferrous metals, especially red mud, by production of electrolytic iron and valuable metals.

EFFECT OF IRON IN NON-FERROUS METALS ELECTRODEPOSITION

Electrodeposition processes aim at plating metallic cations from electrolytes onto a substrate as a result of an electric field. A typical electrolytic cell (Figure 1) involves cathodes and anodes (respectively negative and positive electrodes) immersed in a conductive solution containing the metal to be plated (called electrolyte). A power source supplies a current to the anode, oxidizing species from the electrolyte and/or the electrode, while dissolved metal ions from the bath are reduced onto the cathode. Energy consumption is a critical aspect of electrowinning processes and it is determined by three different factors:

- The cell voltage, which is the difference of potential between the cathode(s) and anode(s), depends on the configuration of the cell, the electrodes, the electrolyte conductivity, the pH and the temperature.
- The applied current, which determines the rate of metal deposition, is generally fixed by upstream processes. The deposition rate increases with the current up to a limiting value, above that point electrons are consumed by side reactions such as the evolution of hydrogen.
- The current efficiency, which represents the ratio of electrons that are really used to plate metal, is calculated as the ratio of the 'theoretical current needed to produce the deposit / the real supplied one'. The current efficiency depends on possible side reactions at anodes and cathodes, such as short circuiting and hydrogen evolution.

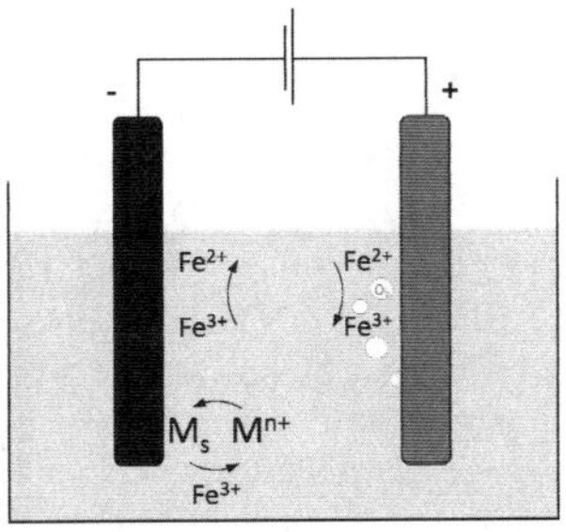

Figure 1 – Electrodeposition cell

Whereas hydrogen evolution depends on operating parameters, cell design and electrolyte composition; unexpected co-deposition and short-circuiting are mainly due to the presence of impurities. In non-ferrous metals production, iron is one of the main sources of current efficiency losses (via a Fe^{2+}/Fe^{3+} loop). Iron is associated with non-ferrous metals in many raw materials (as gangue material in ores and finely associated in alloys), and it is generally difficult to separate completely by standard beneficiation approaches. Its presence, even at trace levels, can affect electrodeposition process efficiency in many ways:

- Unexpected deposition of iron together with non-ferrous metals decreases the purity and alters the microstructure of the deposit, with major impacts on the stress behavior of metals.
- Due to cyclic oxidation and reduction of ferrous/ferric ions, the presence of traces of iron induces additional electric energy consumption. Potentials applied at the cathode in non-ferrous metals electrodeposition (e.g. $E^0 = 0.34$ V for copper deposition) are often enough to reduce ferric ions to ferrous, while the oxygen evolution at the anode generates a sufficiently high potential to oxidize the ferrous ions back to ferric. These phenomena cause short-circuits and decrease the operated current efficiency (Figure 1).
- Ferric ions produced at the anode are corrosive and attack deposited metals, thereby affecting the deposition rate and the surface properties. The partial re-dissolution of the deposit slows the apparent kinetic of deposition down, hinders the deposition of smooth layers, and causes the occlusion of impurities that increase the stress of the deposit.

These consequences are well illustrated in copper and zinc electrodeposition. Das et al. report an increase in current consumption when the Fe(III) concentration is raised beyond about 1.0 g/L during copper electroplating (Das & Gopala Krishna, 1996). Ehsani et al. reported a decrease in current efficiency from 96% to 88% when the ferric concentration increases from 0 to 2.5 g/L during electrowinning processes (Ehsani, Yazıcı, Deveci, & Erdemir, 2012). More generally, Bender et al. evaluated that presence of iron in the electrolytic cell is responsible for a drop of current efficiency of about 3-5% for each gram of iron per liter of solution (Bender & Emmerich, 2015).

Similar behaviors were reported in zinc electroplating but with a higher sensitivity to iron concentration. Saba et al. report that current efficiency can decrease from 96% to 82% and 52% when the concentration of iron increased from 0 to 50 and 150 mg/L, respectively (Saba & Elsherief, 2000).

To avoid these dramatic effects of ferric ions on electrochemical processes, iron is generally removed upstream of electrodeposition processes and methods were developed to limit the effects of ferric ions during the extraction and/or purification process. The most common methods for limiting the concentration of iron species upstream of electrolytic cells are low-cost and efficient processes such as cementation and precipitation.

IRON MANAGEMENT IN NON-FERROUS METALS PRODUCTION

Methods Based on Iron Species Extraction

Recovering zinc from waste products such as electrolytic zinc plants residues and brass mill residues generally involves an electrowinning step. To achieve suited purity and favorable economic, electrolytes have to contain residual concentrations of iron lower than 20 mg/L (Kust, 1979). The precipitation of iron and zinc hydroxides occurs at different pH ranges (pH 2-3 for Fe and pH 7-8 for Zn), thereby allowing extraction by addition of zinc-containing materials (Menendez & Fernandez, 1976). When effluents contain high amounts of valuable metals, coupling precipitation and cementation allows extraction of metals with a higher selectivity. For example, Kust et al. developed a process to produce zinc metal from brass dust materials and valorize all metals present in the leach liquor (Kust, 1979). After the dissolution of raw materials in sulphuric acid (pH 1-2), iron and copper are selectively extracted by controlled neutralization and addition of zinc containing materials. Copper present at high concentration precipitates first as a hydroxide at pH 2.5-3. Zinc addition to the deoxygenated liquor induces the co-precipitation of nickel and cobalt while iron (<2 g/L) and zinc are still in solution. Zinc is finally electrowon from the purified electrolyte after the precipitation of iron as either an oxide, hydroxide or jarosite at pH 4.5.

Since copper hydroxide precipitation occurs at a similar pH to that of iron (~pH=4), extraction processes with higher selectivity have to be used to separate copper and iron species upstream of the electrolytic cells. One of the most frequently used processes is leach-solvent extraction–electrowinning, which consists of the following steps: (i) the leaching of the sulphide ores containing copper in sulphuric acid solutions, (ii) the extraction of copper from the leach liquor via solvent extraction and (iii) the copper electroplating from the strip solution (Archana Agrawal et al., 2008). Despite the upstream solvent extraction step, iron traces are still present in the copper electrowinning cells, creating current efficiency losses around 3-5% per gram of iron in electrolyte (Bender & Emmerich, 2015; Gula et al., 1996). Moreover, while the copper is continuously plated, iron is accumulated in the electrolytic solution, dramatically affecting the overall process efficiency.

Many methods were thus developed to avoid iron accumulation in electrolytes and maintain high current efficiencies. A well-known method is the bleeding of a portion of the iron-rich electrolyte and its replacement with an iron-depleted solution. However, the loss of additives (such as cobalt added to prevent lead anode corrosion and cathode contamination) during the bleed of the spent electrolyte represents a major operating expense for copper electrowinning plants that use this method.

To prevent accumulation during the Cu-EW step, iron can be continuously extracted directly from the spent electrolyte. Gula and Lee, for example, proposed methods to remove ferric ions (Fe^{3+}) from spent electrolyte using ion exchange resin containing phosphonate groups (Gula et al., 1996; Lee & Nicol, 2007). Phosphonate groups are known to preferentially interact with trivalent cations rather than mono-valent and di-valent species. From an electrolyte containing divalent copper and cobalt and ferric ions, phosphonate groups selectively remove ferric ions, limiting their accumulation and their deleterious effects. The regeneration of the loaded resin can be achieved by reduction of Fe^{3+} to Fe^{2+} using eluents containing sulfurous acid and catalytic amounts of Cu ions (<7 g/L) (Gula et al., 1996) or in presence of sulphuric acid and titanium(III) (Lee & Nicol, 2007). The continuous extraction of ferric ion by selective adsorption presents significant advantages over bleeding methods. For example, (i) it avoids cobalt losses when lead anodes are used, (ii) it uses compact equipment and (iii) it offers the possibility to valorize the extracted iron. Bender and Emmerich (2015) proposed a similar method, based on the solvent extraction of ferric ions from the spent electrolyte using phosphate mono-alkyl ester. The extracted ferric ions are then stripped from the organic phase using sulphuric acid solution and can be used to leach copper from ores.

Methods Based on Ferric Ions Reduction

Many studies have highlighted the importance of the Fe^{3+}/Fe^{2+} ratio in electrolyte on current efficiency during copper and iron electrodeposition (Cain, 1934; Bender & Emmerich, 2015), recommending to keep the Fe^{3+}/Fe^{2+} ratio lower than 1 in order to avoid short-circuiting and anode corrosion effects, which results in loss of current efficiency.

Loutfy et al. proposed a method to extract copper from leach liquor of low grade copper oxide ore that contains high amounts of iron (10 to 40 g/L). To avoid short-circuit and cathode corrosion, they reduced ferric ions to ferrous directly in the electrolytic baths during the copper electrowinning by periodically reversing of the applied current (Loutfy & Bharucha, 1978). During the forward cycle, they applied a current of 2 A/dm^2 and classical reactions take place normally. During the reverse cycle, the reduction of copper and ferric ions takes place at the inert electrode, but only the copper oxidation occurs at the copper electrode. Indeed, the oxidation of ferrous ions to ferric does not take place at the copper electrode since its potential is not sufficiently high. Consequently, the periodic reversal of the direct current (5 seconds every 180 seconds) enhances the formation of Fe^{2+} that are not detrimental to copper electrowinning and limits the concentration of Fe^{3+} in the electrolytic solution (generally below 4 g/L). Under such conditions, pure metallic copper is generated at the cathode with acceptable overall current efficiency compared to the iron concentration.

The reduction of ferric to ferrous ions in the electrolytic solution can also be achieved by the addition of raw materials that contain the metal to be plated, as minerals or chips from the stripping of the cathode. Fanta et al. described a method consisting of copper sulphide ore addition directly to the electrolyte during copper EW (Fanta & Tributsch, 1975), presenting the double advantage of reducing ferric ions and leaching copper ores. Indeed, iron(III) sulphate reacts with copper sulphides to produce a solution enriched in copper and ferrous ions, leading to a high current efficiency during copper electrowinning.

Although techniques aiming at reducing the impact of iron in non-ferrous metals electrodeposition processes provide satisfying results, most widely used methods are those consisting in the extraction of iron species upstream. The main drawback of such methods lies in the outlet for extracted iron that often ends up landfilled, especially from copper, zinc and lead production (Agrawal et al., 2004).

VALORIZATION OF IRON RESIDUE USING ELECTROWINNING

Red Mud –A Potential Resource?

Metallurgical industries generate vast quantities of solid wastes. As is self-evident from the above considerations, most of the solid wastes produced during non-ferrous metal extraction contain iron and minor amounts of valuable metals (A. Agrawal et al., 2004). Due to the important dilution of non-ferrous metals in iron and its low value, valorization of iron rich by-products is often unsustainable and materials are just dumped. Main environmental issues caused by solid waste dumping are directly linked to the copper, lead, zinc and alumina industries.

Red mud, for example, is an infamous by-product of the Bayer process employed to convert bauxite ore into aluminum oxide. Bauxite ores are composed of around 30 to 50 wt.% alumina species, $Al(OH)_3$, α and γ-AlO(OH) mixed with iron oxides, kaolinite ($Al_2Si_2O_5(OH)_4$) and anatase (TiO_2). In the Bayer process, aluminum compounds are dissolved as sodium aluminates with hot sodium hydroxide. The mixture is then filtered to extract solid residues containing iron oxides, unreacted alumina, titanium oxide and traces of heavy metals from the sodium aluminate-rich phase. The leach liquor is then cooled down, to precipitate alumina which is calcined in rotary kilns to produce aluminum oxide. The residue, mainly composed of iron oxide, is considered as a waste of alumina production. Consequently, for every ton of alumina produced, more than one ton of highly alkaline slurry of fine particles and salts containing iron oxide (up to 60% of the mass of the red mud) is wasted. Red mud disposal methods are dry stacking, lagooning and sometimes sub-marine tailings discharge. Such disposal results in large land area occupation (stored bauxite residue in land-based impoundments is estimated to be over 2.7 billion tons), possibility of leaks and floods with dramatic consequences or release of toxic compounds and the destruction of ecosystems (Power, Gräfe, & Klauber, 2011). Nowadays, the red mud disposal represents the main problem in the alumina production. To reduce the volume of waste to manage, research is performed to find (i) other route of alumina production (Boudreault, Biasotto, Fournier, & Alex, 2012) (ii) uses for dry bauxite residues (Power et al., 2011).

Possible applications reported include in chemistry catalysts and adsorbents, in environmental applications to treat waste water and in metallurgical applications as a source of iron in steel making (Craig Klauber et al., 2009). However, none of these disposals and applications is entirely satisfactory, since none of these applications extract all the value present in the residue.

Another way of valorization would be the production of electrolytic iron coupled with the extraction of residual valuable metals such as aluminum, titanium and rare earth elements (Borra, Pontikes, Binnemans, & Van Gerven, 2015; Liu & Naidu, 2014; Table 1). Iron electrowinning from red mud presents these significant advantages, over other iron extraction processes: (i) to concentrate iron and reduce the environmental footprint of the bauxite residue and (ii) to produce a valuable material that can help balance the cost of the remediation process.

Table 1 – Typical composition of red mud (Borra et al., 2015)

Compound	Wt.%
Fe_2O_3	44.6
Al_2O_3	23.6
CaO	11.2
SiO_2	10.2
TiO_2	5.7
Na_2O	2.5

Many studies have been carried out on electrolytic iron production from ores, but very few have been conducted from by-products as red mud. Therefore, we seek to adapt methods developed to produce electrolytic iron from classical sources to bauxite residues.

Bauxite residues could be electro-extracted in two ways: after leaching of dry bauxite residues in acidic liquors or directly from alkaline streams coming from the Bayer process.

Iron Electroextraction from Dry Bauxite Residue

Bauxite residues from dry stacking first have to be dissolved. Iron oxide leaching is generally operated in sulphuric or hydrochloric acid containing solutions that are used later as electrolyte. Sulphate based electrolytes are less expensive than chloride ones and give smoother deposit when the electrodeposition is operated at a low temperature (Izaki, 2010). On the other hand, chloride based electrolytes allow for the dissolution of higher amounts of iron and faster deposition rates at high temperature (over 85 °C) with operating current densities two times higher than for sulphate baths (Izaki, 2010). The dissolution of iron oxide from red mud leads to extraction efficiencies ranging from 46% to 95% for iron and lower than 55% and 64% for aluminum and titanium dioxide (Uzun & Gülfen, 2007; Agatzini-Leonardou, Oustadakis, Tsakiridis, & Markopoulos, 2008; Liu, Zeng, Zhao, & Li, 2009). The resulting leach liquors contain high amounts of iron, aluminum, sodium, silicon, calcium and low concentrations of titanium, vanadium, chromium, zirconium and rare earth elements (Wang, Pranolo, & Cheng, 2013). Those minor elements could be extracted from the iron-depleted electrolytes after its electroextraction.

Due to the high concentration of iron in the electrolytes, specific electrowinning cells have to be used during iron electrodeposition from acidic baths in order to balance the harmful effects of ferric ions produced at the anode. Electrolytic cells were often equipped with asbestos diaphragms or more recently with ion exchange membrane to separate the anolyte and the catholyte (Figure 2) and to avoid short-circuits and contact between ferric ions, produced in the anolyte, and the cathode which is immersed in the catholyte (Ramage, 1911; Cain, 1934, 1950; Mostad et al., 2008). The ferric ions-rich anolyte is often continuously withdrawn and used to leach the iron containing raw materials, leading to a liquor enriched in ferrous ions (Cain, 1934), which is sent to the catholyte and anolyte compartments.

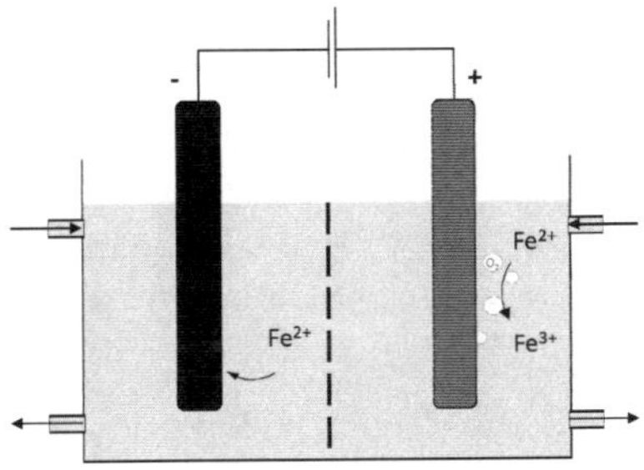

Figure 2 – Electrolytic cell for the production of electrolytic iron

The main impurities in the electrolyte during the iron electroextraction would be aluminum species. However, with a reduction potential lower than the potential applied during iron electrowinning, aluminum would be poorly extracted. The current efficiency and the quality of the deposited metals would be greatly dependent on the operating conditions such as the composition of the electrolytic bath, the concentration of iron, the temperature, the current density and the geometry of the electrolytic cell. Electroplating rates and current efficiency generally increase with the electrolytic metal concentration. Low concentrations result in powdery and brittle deposits, while increasing iron concentration induce homogenous layers, high deposition rates and current efficiency. Consequently, particular attention should be paid to the leaching stage, especially as the leaching of hematite from red mud in sulphuric acid is highly dependent on operating conditions such as the temperature and eventual raw material pretreatment. Design of electrolytic cells can also be optimized for the recovery of metals from dilute solutions. Electrolytic cells with rotating cathodes have especially been developed to prevent current efficiency loss and heterogeneous deposit from rinsing solutions with low metal concentrations (Figure 3) (Recéndiz, León, Nava, & Rivera, 2011; Almazán-Ruiz et al., 2015).

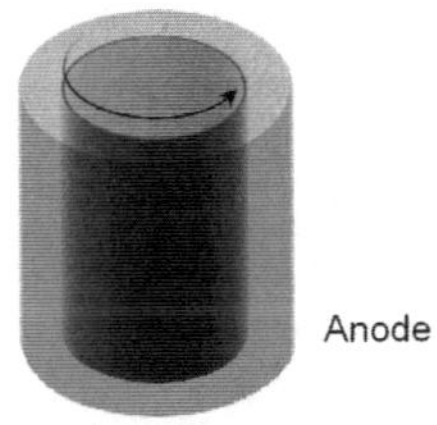

Figure 3 – Electrolytic cell with inner rotating cathode

Although the filtration and drying of the bauxite residues represents significant additional energy costs, electroextraction of iron from dry red mud (Bauxaline) presents some advantages such as the possibility to use acidic electrolyte that are better leaching liquor of iron oxides than alkaline effluents.

Electroextraction of Iron from Alkaline Electrolytes

Another route would be the electrodeposition of iron from alkaline solutions (sodium hydroxide and iron hydroxides) (Estelle, 1918; Allanore, Lavelaine, Valentin, Birat, & Lapicque, 2007, 2008; Yuan, Kongstein, & Haarberg, 2009; Allanore, Lavelaine, Birat, Valentin, & Lapicque, 2010). Iron electrodeposition from alkaline electrolyte is generally operated at temperatures around 110 °C in sodium hydroxide solution (around 50 wt.%). Iron solubility in such alkaline effluents is very low, approximately 1.10^{-3} mol/L. Consequently iron electrodeposition from alkaline baths involves different deposition mechanisms and management issues than from acidic electrolytes where iron solubility is several moles per liter. While iron electrodeposition from acidic bathes is governed by ferrous and ferric ions diffusion, iron deposition from alkaline solutions is mainly due to the direct reduction of iron oxide onto the cathode when particles meet the cathodes as represented below.

$$Fe_2O_{3\,(s)} + 3H_2O + 6e^- \rightarrow 2Fe_{\,(s)} + 6OH^-$$

In this situation, the main issue is not the corrosive effect of ferric ions but hydrogen evolution at the cathode, surface deposit heterogeneity and current efficiency (ranged from 60% to 95%; Yuan et al., 2009). The last one is managed by the transport of iron oxide particles to the cathodes. Consequently, studies on electrolytic cell designs were conducted to optimize the transfer of hematite particles to the cathodes (Allanore et al., 2010). Most efficient electrolytic cells involve outer rotating cathodes and horizontal plate cathodes (Figure 4), where centrifugal and gravitational forces project particles onto the cathode and favor the direct reduction of hematite particles. Experiments carried out in presence of both silica and alumina, impurities present in red mud, showed that it may be possible to obtain pure electrolytic iron directly from electrolytes containing silica and alumina without purification stages (Tang & Haarberg, 2010).

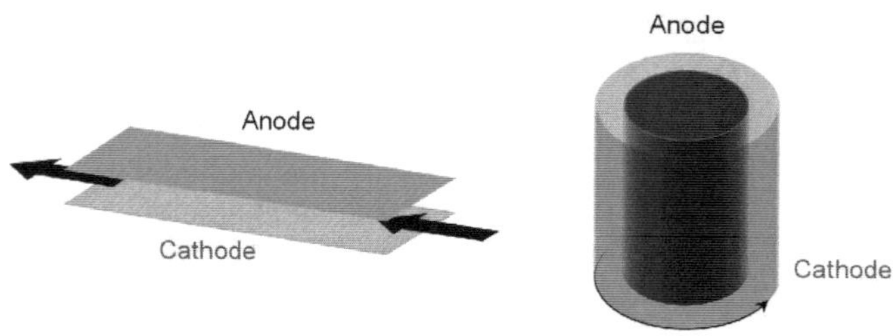

Figure 4 – Electrolytic cell design optimization for the production of iron from alkaline effluents, left- horizontal plate reactor, right- outer rotating cathode (Allanore et al., 2010)

Many studies were carried out on the production of electrolytic iron from acid and alkaline electrolytes and demonstrated the possibility to electro extract iron from complex matrices with interesting purity and efficiency. We plan now to build on these studies to develop a process to reduce the volume of iron-rich waste from the non-ferrous metal production and recover the landfilled iron as electrolytic iron.

CONCLUSIONS

The presence of iron in non-ferrous metals electrowinning can cause dramatic effects on the purity, the morphology of the deposit and especially on the economics of the process through the current efficiency. Indeed, the presence of iron during copper electrodeposition induces a decrease in current efficiency of about 3% to 5% per gram of iron and about 0.3% per milligram iron during zinc electrodeposition. To avoid such energy losses, iron is extracted through cementation and precipitation processes upstream of zinc electrodeposition cells and through solvent extraction and ion exchange resins upstream during copper electrowinning. As a consequence, non-ferrous metals production generate ferrous wastes which are often landfilled.

To limit the volume of landfilled waste, we are developing a technology based on the production of electrolytic iron from these iron-rich from materials. As a part of our project, we focus on the electrowinning of iron from red mud generated during the production of alumina via the Bayer process.

REFERENCES

Agatzini-Leonardou, S., Oustadakis, P., Tsakiridis, P. E., & Markopoulos, C. (2008). Titanium leaching from red mud by diluted sulfuric acid at atmospheric pressure. *Journal of Hazardous Materials*, *157*(2-3), 579–86. http://doi.org/10.1016/j.jhazmat.2008.01.054

Agrawal, A., Manoj, M. K., Kumari, S., Bagchi, D., Kumar, V., & Pandey, B. D. (2008). Extractive separation of copper and nickel from copper bleed stream by solvent extraction route. *Minerals Engineering*, *21*(15), 1126–1130. http://doi.org/10.1016/j.mineng.2008.02.016

Agrawal, A., Sahu, K. K., & Pandey, B. D. (2004). Solid waste management in non-ferrous industries in India. *Resources, Conservation and Recycling*, *42*(2), 99–120. http://doi.org/10.1016/j.resconrec.2003.10.004

Al-Abed, S. R., Jegadeesan, G., Purandare, J., & Allen, D. (2007). Arsenic release from iron rich mineral processing waste: Influence of pH and redox potential. *Chemosphere*, *66*(4), 775–82. http://doi.org/10.1016/j.chemosphere.2006.07.045

Allanore, A., Lavelaine, H., Birat, J. P., Valentin, G., & Lapicque, F. (2010). Experimental investigation of cell design for the electrolysis of iron oxide suspensions in alkaline electrolyte. *Journal of Applied Electrochemistry*, *40*(11), 1957–1966. http://doi.org/10.1007/s10800-010-0172-0

Allanore, A., Lavelaine, H., Valentin, G., Birat, J. P., & Lapicque, F. (2007). Electrodeposition of Metal Iron from Dissolved Species in Alkaline Media. *Journal of The Electrochemical Society*, *154*(12), E187. http://doi.org/10.1149/1.2790285

Allanore, A., Lavelaine, H., Valentin, G., Birat, J. P., & Lapicque, F. (2008). Iron Metal Production by Bulk Electrolysis of Iron Ore Particles in Aqueous Media. *Journal of The Electrochemical Society*, *155*(9), E125. http://doi.org/10.1149/1.2952547

Almazán-Ruiz, F. J., Caballero, F., Cruz-Díaz, M. R., Rivero, E. P., Vazquez-Arenas, J., & González, I. (2015). Nickel recovery from an electroplating rinsing effluent using RCE bench scale and RCE pilot plant reactors: The influence of pH control. *Chemical Engineering Research and Design*, *97*, 18–27. http://doi.org/10.1016/j.cherd.2015.02.022

Barnett, R. J., & Mezner, M. B. (2001, June 19). Process for treating red mud to recover metal values therefrom.

Bender, J., & Emmerich, N. C. (2015, September 10). Methods And Systems For Controlling Impurity Metal Concentration During Metallurgic Processes. United States.

Borra, C. R., Pontikes, Y., Binnemans, K., & Van Gerven, T. (2015). Leaching of rare earths from bauxite residue (red mud). *Minerals Engineering*, *76*, 20–27. http://doi.org/10.1016/j.mineng.2015.01.005

Boudreault, R., Biasotto, F., Fournier, J., & Alex, S. (2012, May 24). Process for extracting aluminum from aluminous ores. Canada.

Cain, J. R. (1934, November 13). Method of making ductile electrolytic iron from sulphide ores. United States.

Cain, J. R. (1950, April 11). Method of making electrolytic iron. United States.

Das, S. C., & Gopala Krishna, P. (1996). Effect of Fe(III) during copper electrowinning at higher current density. *International Journal of Mineral Processing*, *46*(1-2), 91–105. http://doi.org/10.1016/0301-7516(95)00056-9

Ehsani, A., Yazıcı, E. Y., Deveci, H., & Erdemir, F. (2012). The Influence of Impurity Ions on The Electrowinning of Copper from Waste PCBs Leaching Solutions. In *Proceedings of XIIIth International Mineral Processing Symposium-IMPS 2012*. Bodrum. http://doi.org/10.13140/RG.2.1.2701.5200

Estelle, A. T. C. (1918, August 6). Process or electrolytically producing iron. United States.

Fanta, F., & Tributsch, H. (1975, October 28). Electrowinning process for the improved recovery of metal.

Gula, M. J., Dreisinger, D. B., & Horwitz, E. P. (1996, December 10). Ion exchange and regeneration process for separation and removal of iron (III) ions from aqueous sulfuric acid metal ion-containing solutions. United States.

Herck, P. Van. (2000). Zinc and lead removal from blast furnace sludge with a hydrometallurgical process. ... *Science & Technology*.

Izaki, M. (2010). Electrodeposition of Iron and Iron Alloys. In M. Schlesinger & M. Paunovic (Eds.), *Modern Electroplating*. Hoboken, NJ, USA: John Wiley & Sons, Inc.

Klauber, C., Gräfe, M., & Power, G. (2009). Review of Bauxite Residue "Re-use" Options.

Klauber, C., Gräfe, M., & Power, G. (2011). Bauxite residue issues: II. options for residue utilization. *Hydrometallurgy*, *108*(1-2), 11–32. http://doi.org/10.1016/j.hydromet.2011.02.007

Kust, R. N. (1979, April 17). Hydrometallurgical brass dust reclamation. United States.

Lee, M. S., & Nicol, M. J. (2007). Removal of iron from cobalt sulfate solutions by ion exchange with Diphonix resin and enhancement of iron elution with titanium(III). *Hydrometallurgy*, *86*(1-2), 6–12. http://doi.org/10.1016/j.hydromet.2006.10.002

Liu, Y., & Naidu, R. (2014). Hidden values in bauxite residue (red mud): Recovery of metals. *Waste Management*, *34*(12), 2662–2673. http://doi.org/10.1016/j.wasman.2014.09.003

Liu, Z.-R., Zeng, K., Zhao, W., & Li, Y. (2009). Effect of temperature on iron leaching from bauxite residue by sulfuric acid. *Bulletin of Environmental Contamination and Toxicology*, *82*(1), 55–8. http://doi.org/10.1007/s00128-008-9576-5

Loutfy, R. O., & Bharucha, N. R. (1978, November 7). Electrowinning of copper in presence of high concentration of iron.

Loutfy, R. O., & Bharucha, N. R. (1980, May 20). Electrowinning of copper in presence of high concentration of iron. Canada.

Menendez, F. J. S., & Fernandez, V. A. (1976, October 12). Process for recovering zinc from ferrites.

Mostad, E., Rolseth, S., & Thonstad, J. (2008). Electrowinning of iron from sulphate solutions. *Hydrometallurgy*, *90*(2-4), 213–220. http://doi.org/10.1016/j.hydromet.2007.07.014

Musina, A., Bocokic, V., Lavric, V., & van Zutphen, S. (2014). Phosphorus-Based Polymers for Selective Capture of Platinum Group Metals. *Industrial & Engineering Chemistry Research*, *53*(34), 140810102656007. http://doi.org/10.1021/ie502153f

Power, G., Gräfe, M., & Klauber, C. (2011). Bauxite residue issues: I. Current management, disposal and storage practices. *Hydrometallurgy*, *108*(1-2), 33–45. http://doi.org/10.1016/j.hydromet.2011.02.006

Recéndiz, A., León, S., Nava, J. L., & Rivera, F. F. (2011). Mass transport studies at rotating cylinder electrode during zinc removal from dilute solutions. *Electrochimica Acta*, *56*(3), 1455–1459. http://doi.org/10.1016/j.electacta.2010.10.021

Reuter, M., Hudson, C., van Schaik, A., Heiskanen, K., Meskers, C., & Hagelüken, C. (2013). *Metal Recycling: Opportunities, limits, infrastructure, A Report of the Working Group on the Global Metal Flows to the International Resource Panel.* UNEP.

Ricoux, Q., Bocokić, V., Méricq, J. P., Bouyer, D., Zutphen, S. v., & Faur, C. (2015). Selective recovery of palladium using an innovative functional polymer containing phosphine oxide. *Chemical Engineering Journal*, *264*, 772–779. http://doi.org/10.1016/j.cej.2014.11.139

Tang, S. H., & Haarberg, G. M. (2010). Electrowinning of Iron from Alkaline Solution. In *ECS Transactions* (Vol. 28, pp. 309–315). The Electrochemical Society. http://doi.org/10.1149/1.3367923

Uzun, D., & Gülfen, M. (2007). Dissolution kinetics of iron and aluminium from red mud in sulphuric acid solution. *Indian Journal of Chemical Technology*, *14*(3), 263–268.

Wang, W., Pranolo, Y., & Cheng, C. Y. (2013). Recovery of scandium from synthetic red mud leach solutions by solvent extraction with D2EHPA. *Separation and Purification Technology*, *108*, 96–102. http://doi.org/10.1016/j.seppur.2013.02.001

Yuan, B., Kongstein, O. E., & Haarberg, G. M. (2009). Electrowinning of Iron in Aqueous Alkaline Solution Using a Rotating Cathode. *Journal of The Electrochemical Society*, *156*(2), D64. http://doi.org/10.1149/1.3039998

Iron Control During Zinc Processing

TREATMENT OF HIGH IRON CALCINES IN THE ELECTROLYTIC ZINC PLANT OF METALLURGICA MET-MEX PEÑOLES, S.A DE C.V.

*Claudio Rocha Rosales

Metalurgica, Met-Mex Peñoles S.A. C.V.
Electrolytic Zinc Plant
Boulevard Laguna 3200 Poniente, Colonia Metalurgica
Torreon, Coahuila, Mexico
*(*Corresonding author: claudio_rocha@penoles.com.mx)*

ABSTRACT

Peñoles is a mining group with integrated operations for the smelting and refining of non-ferrous metals, and it is also a chemicals producer. Peñoles is the world's foremost producer of refined silver, metallic bismuth and sodium sulphate. It is also the leading Latin American producer of refined gold, lead and zinc. The Met-Mex Peñoles Zinc Plant commenced operations in 1973 with an annual production capacity of 105,000 t/y zinc. Since then, production capacity has increased, and at present, it stands at 240,000 t/y zinc. In this paper, the leaching process and particularly the operation of the jarosite circuit is reviewed, specifying process improvements to treat high iron calcines.

KEYWORDS

Jarosite, zinc, flow sheet, plant description

INTRODUCTION

Twenty years ago the Electrolytic Zinc Plant in Peñoles was treating high quality zinc concentrates with low concentrations of iron (~7.5 wt.%) and silica (1.8 wt.%). Nowadays the quality of zinc concentrates is poor with a typical iron content of around 11 wt.% and silica at 2.0 wt.%. This has created numerous processing problems in our roast-leach-purification- electrolysis sections. In order to meet customer requirements it is necessary to find processing solutions to assure the operational continuity of the plant. These processing solutions are the focal point of this paper.

REVIEW OF THE LEACHING PROCESS

The purpose of the leaching process is to recover the metals, mainly zinc, through a series of extraction stages with sulphuric acid solutions, which progressively increases the zinc dissolution. After the series of acidification stages, the solution, which is rich in iron, is sent to the jarosite circuit to separate zinc from iron (Garcia & Valdez, 1996). The first step of the leaching process is the neutral leach circuit which produces a zinc sulphate solution that is sent to the purification section for the removal of impurities such as cadmium, cobalt and copper. Then this solution is sent to the tank house to feed the electrolytic cells.

Leaching Process

The Leaching Process consists of the following ten stages, which are presented in Figure 1. They are (i) neutral leach, (ii) fourth purification, (iii) acid leach, (iv) hot acid leach, (v) jarosite formation, (vi) jarosite acid washing, (vii) sulphates purge, (viii) lead–silver residue filtration, (ix) jarosite filtration and (x) jarosite neutralization.

Neutral Leach

The purpose of this unit operation is to leach most of the zinc present as ZnO in the calcine. The zinc extraction is in the order of 85%. In this circuit, there is a 320 m^3 reaction tank with a 1 hour residence time in which the acidity is controlled at pH 1 to 1.5. The acid is composed of spent electrolyte, manganese acid solution from the cell house and zinc solution recovered from the purification section.

Fourth Purification

The purpose of this circuit is to recover zinc that was dissolved in the residue treatment stages, because the solution has an average acidity of 15 g/L H_2SO_4 while the average iron concentration ranges from 4 to 8 g/L. This solution is used to leach calcine, and simultaneously, the iron hydrolyzes and precipitates impurities like germanium, arsenic, antimony, indium, selenium and tellurium. In this stage, the reactor tank (320 m^3, 0.7 hours retention time) is controlled at a pH of 4.0 to 4.5.

Acid Leaching

The purpose of this unit operation is to leach the remaining zinc, present as ZnO, which was not leached in the previous stages to achieve a cumulative extraction of about 90%. There are three reactors (each 80 m^3 and 90 minutes retention time) where acidic overflow from the hot acid thickeners is added to achieve a high zinc extraction. The free acidity in the first tank ranges from 10 to 30 g/L H_2SO_4.

Hot Acid Leaching

The purpose of this circuit is to leach the zinc ferrites ($ZnO.Fe_2O_3$), and in this way, to increase the cumulative zinc extraction to 98%. In addition, a lead-silver residue is generated that is filtered and sent to the adjacent lead smelter to recover the lead and silver values. There are 10 reactor tanks (each measuring 80 m^3 and having an 8 hours retention time), where sulphuric acid and spent electrolyte are added to maintain the free acidity in the first tank at 150 to 170 g/L H_2SO_4; steam is added to these tanks to maintain the temperature at 85°C.

Jarosite Formation

This operation eliminates iron from the solution by precipitating jarosite which is filtered and removed from the circuit. During jarosite precipitation, other impurities such as germanium, arsenic, antimony, indium, selenium and tellurium are co-precipitated and are also eliminated from the circuit. This process section consists of 9 tanks, i.e. 3x320 m^3 and 6x90 m^3.

Jarosite Acid Washing

This stage consists of six reactors (1x200 m^3 and 5x80 m^3) which extract the metals present in the calcine that was added as a neutralizing agent in the jarosite formation circuit, thereby increasing the overall recovery of zinc. Spent electrolyte is added to obtain a free acidity of 70 g/L H_2SO_4 and steam is used to maintain a temperature of 92 °C.

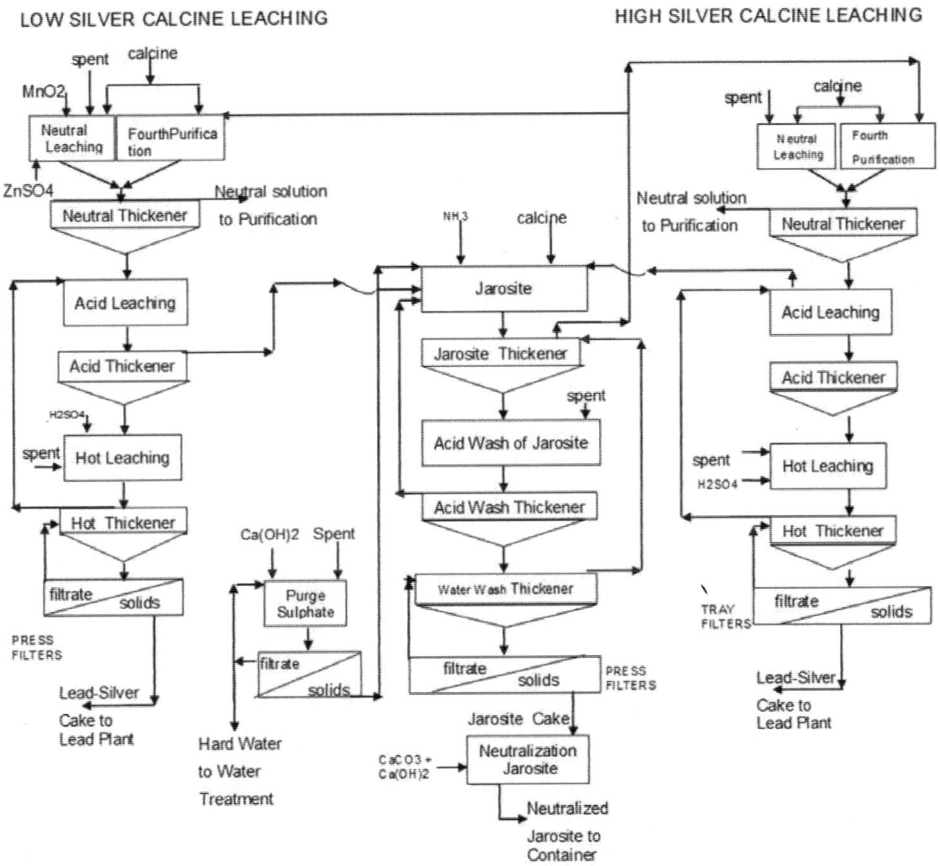

Figure 1 – Leaching circuit flow diagram

Sulphates Purge

The purpose of this circuit is to bleed sulphates from the circuit. However, it also reduces the concentrations of other salts, such as manganese and magnesium sulphates as well as chlorides and fluorides, to acceptable operating levels. The sulphates bleed is important because it allows an increased addition of sulphuric acid to the circuit, and this achieves an increase in the overall zinc recovery. The flow of spent acid fed is 13 m^3/h.

Lead – Silver Residue Filtration

This stage is used to recover the lead and silver contained in the zinc concentrates and it also contributes to the solids balance in the leaching circuit. The filtration of this residue, with adequate washing, also helps to recover the soluble zinc present in the filter cake. There are two filter presses.

Jarosite Filtration

The purpose of this stage is to remove the jarosite to balance the solids and reject impurities from the circuit. It also increases the overall zinc recovery by recovering the soluble zinc in the wash liquor from the filter cakes. Six filter presses are currently in use.

Jarosite Neutralization

The jarosite coming from the filters is neutralized with mixture of $Ca(OH)_2$ and $CaCO_3$ (at a rate of 400 kg per dry metric tonne jarosite) to produce a physically stable material that can be stockpiled. The neutralization process also stabilizes any minor metals content that might be leached from the jarosite residue by rain and subsequently discharged into the water table.

Year after year the quality of our concentrates decreases and the Met-Mex Peñoles Zinc Plant treats concentrates with high contents of total iron, ferrous iron, cobalt and silica to produce a calcine with 11.5-12.5% Fe and 0.018 to 0.030% Co. Table 1 shows the typical analysis of the calcine treated. To treat the high iron calcines, improvements were made to the Metalurgica Met-Mex Peñoles Zinc Plant.

Table 1 - Typical analysis of Peñoles zinc calcine

Element	Average Content (wt. %)	Range (wt. %)
Zn	57	56 - 58
Fe	12	11.5 - 12.5
Pb	1.0	0.8 - 1.20
Cu	1.20	1.00 - 1.40
Fe ++	1	0.8 - 1.2
Cd	0.42	0.4 - 0.44
Insol.	2.0 - 4.0	3
Co	0.24	0.018 - 0.30
Ni	0.004	0.003 - 0.007

PROCESS IMPROVEMENT DESCRIPTIONS

The following process improvements were implemented between 2013 and 2015.

- Increase free acid in Weak Acid Leaching from 10 to 20 g/L H_2SO_4. How? We installed a decanter (D-1) in the underflows from neutral thickeners 32 and 33 to better separate the neutral solution from the solids. The decanter increased the solids density.
- Increase free acid in Hot Acid Leaching from 130 to 160 g/L H_2SO_4. How? We reduced water addition in reagents preparation circuits to increase the spent flow. For example, flocculant water addition was reduced from 6 m^3/h to 2 m^3/h.

- Increase retention time in Jarosite Precipitation from 6 to 7 hours. How? We installed Reactor #6 (320 m^3) at the beginning of the jarosite precipitation stage.
- Better Flow Balance. How? We implemented automatic flow controls to achieve the best performance from the chemical reactions.
- Automatic pH control in the Fourth Purification Reactor. We achieved increased calcine leaching efficiency.
- Automatic pH control in the First Jarosite Reactor. We achieved increased iron precipitation in the jarosite stage.
- Increased temperature in the Jarosite Reactors from 88°C to 92°C. How? We installed Reactor # 6 in jarosite stage with an automatic heating system.
- Increased Jarosite Filtration. How? We moved one silver-lead filter press to jarosite filtration. For clarification, when we increased acidity in the hot acid leach we decreased the amount of lead silver residue while increasing the jarosite production rate.
- Increased thickening surface in Jarosite Acid Wash stage. How? We installed one big 400 m^2 thickener to jarosite acid wash stage. It used to be a 200 m^2 thickener.
- Increased Oxidation. How? We installed Neutral Leaching Reactor # 35 and Jarosite Reactor # 6, both with an oxidation system using oxygen (by Outotec Inc.) to oxidize ferrous iron to the ferric state.

Figure 2 indicates each improvement that was implemented in the different stages of the leaching process to allow the treatment of calcines with higher iron contents. Because of the higher impurity contents, several reactors are now used for jarosite precipitation. The jarosite precipitation circuit is fitted with nine reactor tanks (see Figure 2). The process is initiated when the iron rich solution (15 to 30 g/L Fe) from each Weak Acid Leaching thickener overflow is fed to the first reaction tank #6. Here calcine is added to neutralize the acid. Ammonia (4.0 to 5.0 g/L) is added to reactor #6 to form ammonium jarosite, and the slurry in the reactors is heated with steam to a temperature of 92°C. Additional calcine is added in reactors #7 and #8 to continue lowering the acid concentration (see Table 2).

Table 2 - Typical process variables in the jarosite precipitation circuit

Process Variables	Average Content (g/L)	Range (g/L)
Fe (Reactor 6) inlet	20	15 - 25
Fe (Reactors 28 and 32) outlet	6	4 - 8
Acidity (Reactor 6)	22	15 - 30
NH_3 (Reactor 6)	4.5	4 - 5

The reactions for jarosite formation are:

$$2\ NH_3 + 2\ H_2O \rightarrow 2\ NH_4OH \tag{1}$$

$$3\ Fe_2(SO_4)_3 + 10\ H_2O + 2\ NH_4OH \rightarrow 2\ (NH_4)Fe_3(SO_4)_2(OH)_6 + 5\ H_2SO_4 \tag{2}$$

The jarosite formed is sent to three thickeners for sedimentation. The overflow is a zinc solution containing 4-8 g/L Fe, and this is conveyed to the Fourth Purification stage to recover the zinc and to make use of the iron in solution to precipitate any remaining impurities.

Jarosite from the underflows of the thickeners is sent to the acid wash stage to leach the calcine that was added for acid control in the jarosite circuit. After acid washing, the washed jarosite is sent to thickener No. 35. The overflow of thickener No. 35 is returned to the jarosite precipitation circuit because of its high iron concentration. The underflow is processed in two washing thickeners (No. 34 and No. 134) to lower the soluble metals concentration. The overflow from the washing thickeners is sent to the Jarosite Thickeners to improve their clarity, and the underflow slurry is pumped into six pressure filters. The jarosite is neutralized with a mixture of $Ca(OH)_2$ and $CaCO_3$ to stabilize the heavy metals and to impart the proper physical characteristics to the residue which is transported by truck to storage areas outside the plant. A typical jarosite analysis is shown in Table 3.

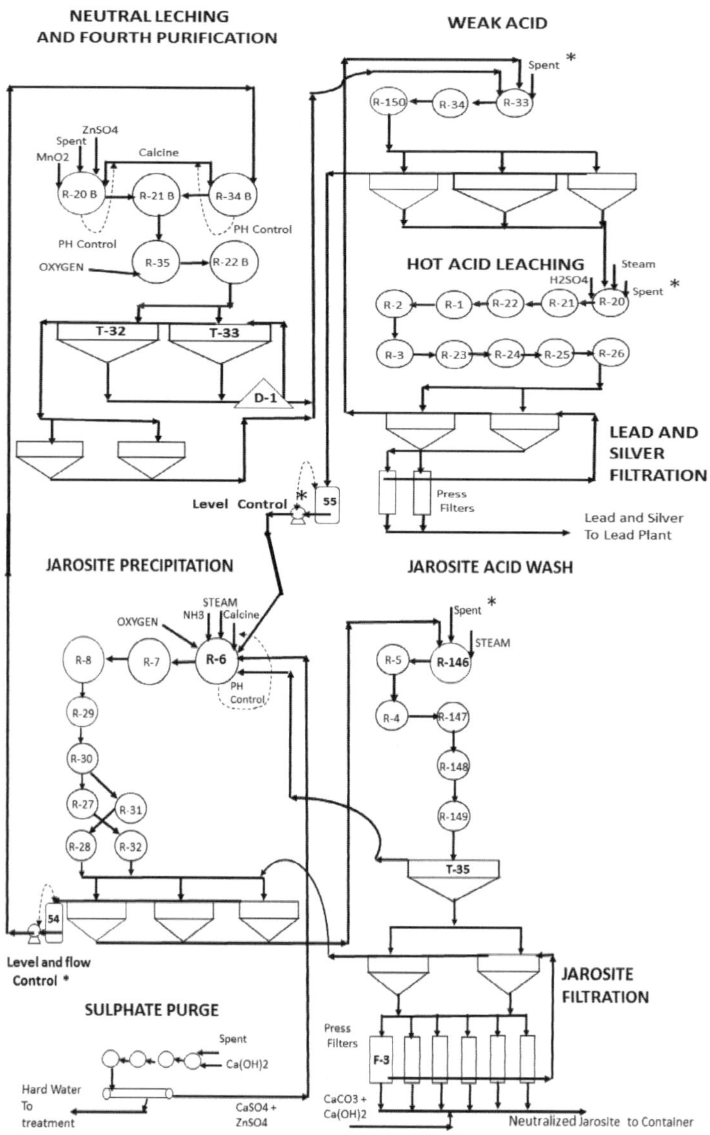

Figure 2 – Low Silver Leaching circuit flow diagram

Table 3 - Typical analysis of the Peñoles jarosite

Element	Average Content (wt.%)	Range (wt.%)
Zn	3.5	3-4
Fe	27	26 - 28
Pb	1.7	1.5 - 2.3
Cu	0.4	0.3 - 0.5
S	13	12 - 14
Cd	0.18	0.16 - 0.20
Insol.	5.5	5.2 - 5.8
As	150	140 - 160
Co	0.007	0.004 - 0.01
Ni	0.004	0.002 - 0.006

REMOVAL OF IMPURITIES DURING JAROSITE PRECIPITATION

The jarosite precipitation stage is very important in the overall leaching process because most of the iron from the solution, as well as a considerable amount of the impurity content, is eliminated in this circuit, as shown in Table 4. The iron precipitated as jarosite also serves as a purge and helps to reduce the recirculation of impurities that have to be removed in the Fourth Purification circuit and in the hot and cold purification stages. Tests in the Peñoles jarosite circuit have shown that the amount of iron removed in the process correlates with the amount of impurities removed (arsenic, antimony, indium, etc.). There is also a direct relationship between the variables that are important for the jarosite formation and the amount of iron precipitated. The correlation between iron removal and impurities elimination has been demonstrated in previous studies (Garcia & Valdez, 1996; Rocha & Ayala, 2006).

Table 4 - Impurities removal in the jarosite circuit

Element	(Reactor 6) inlet (mg/L)	(Reactors 28 - 32) outlet (mg/L)
As	550	150
Sb	41	26
Se	17	12
Te	32	25
Si	879	414
Sn	4.86	3.22
In	48.4	20.36
Cl	355	284
F	46	42
Mn	14530	14180

The most important variables for jarosite precipitation are temperature, acid concentration in the solution and the ammonia concentration. These variables are plotted against the amount of iron precipitated (expressed as g/L from a solution initially containing about 20 g/L) in Figures 3 and 4. For the efficient precipitation of jarosite, these variables must be closely monitored and controlled within specific ranges.

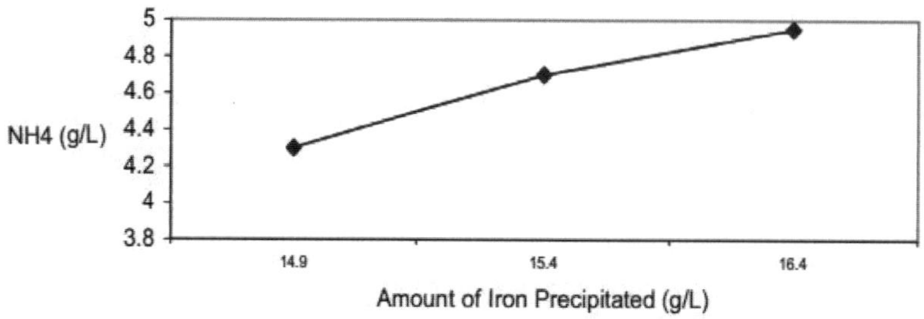

Figure 3 – Amount of iron precipitated (expressed as g/L from a solution initially containing about 20 g/L) as a function of the ammonia concentration of the solution in the Peñoles jarosite circuit

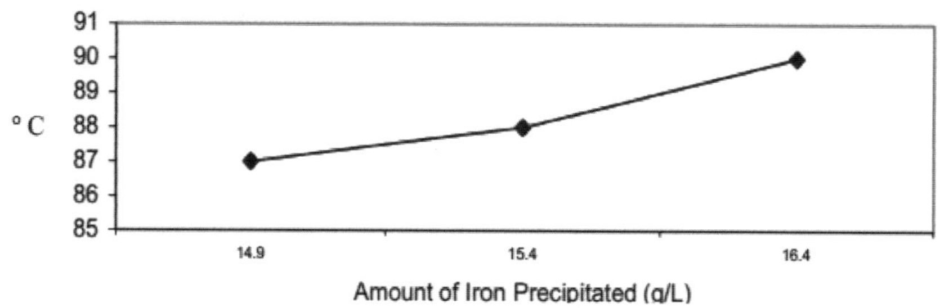

Figure 4 – Amount of iron precipitated (expressed as g/L from a solution initially containing about 20 g/L) as a function of the temperature of the slurry in the Peñoles jarosite circuit

CONCLUSIONS

Year after year and all around the world, the quality of zinc concentrates continues to decrease, and it is becoming difficult to find high quality concentrates. Total iron, ferrous iron, silica, lead, copper and germanium are increasing, and these impurities cause problems in typical roast-leach-electrowin operations. Thus, it is necessary to continue to make improvements to the process to achieve daily production quotas and to meet customer requirements.

REFERENCES

Garcia, C. & and Valdez, C. (1996). Jarosite disposal practices at the Peñoles Zinc Plant, *Iron Control and Disposal,* J.E. Dutrizac and G.B. Harris, Eds., Canadian Institute of Mining, Metallurgy and Petroleum, Montreal, Canada, 643-650.

Rocha, C. & Ayala, J. (2006). Impurity removal in the jarosite precipitation circuit of the electrolytic zinc plant of Met-Mex, Peñoles S.A de C.V. *Iron Control Technologies*, J.E Dutrizac and Patricio Riveros Eds., Canadian Institute of Mining, Metallurgy and Petroleum, Montreal, Canada, 501-512.

RECENT OPERATION AND IMPROVEMENTS OF THE HEMATITE PROCESS AT AKITA ZINC COMPANY, LIMITED

*Manabu Kanno, Akiyoshi Horiuchi, Taro Aichi and Shigeki Sato

DOWA METALS & MINING CO., LTD.
14-1, Sotokanda 4-Chome
Chiyoda-ku, Tokyo 101-0021, Japan
*(*Corresponding author: kannom@dowa.co.jp)*

ABSTRACT

AKITA ZINC Co., Ltd. (Iijima Zinc Refinery), jointly owned by DOWA METALS & MINING Co., Ltd.(DMM), has used the company's own hematite process for iron control since the beginning of operations in 1971. The hematite process consists of reductive leaching with SO_2 gas and iron removal in autoclaves by high temperature hydrolysis. This process enhances not only zinc recovery but also the recovery of gold, silver, copper, lead and rare metals, which is carried out at another DOWA metallurgical plant. All by-products produced in the hematite process are marketed and sold. This paper presents an overview of the hematite process and the recent process improvements made.

KEYWORDS

Zinc, Hematite Process, DOWA, Akita, Iijima

INTRODUCTION

DMM carries out mining and smelting activities for precious metals, PGMs, rare metals and base metals such as copper, lead and zinc. It owns KOSAKA SMELTING & REFINING, producing copper and lead, and the AKITA ZINC Co., Ltd. (AZC) in Akita Prefecture. DOWA is a specialized smelting company which has established a metallurgical complex for copper, lead and zinc, including a recycling business.

AZC was established in 1971 at a production capacity of 72,000 t/y zinc; it is jointly owned by DMM and five other non-ferrous metal companies. AZC has increased its capacity continuously and now produces 210,000 t/y of zinc. Figure 1 shows the changes in zinc production and capacity at AZC.

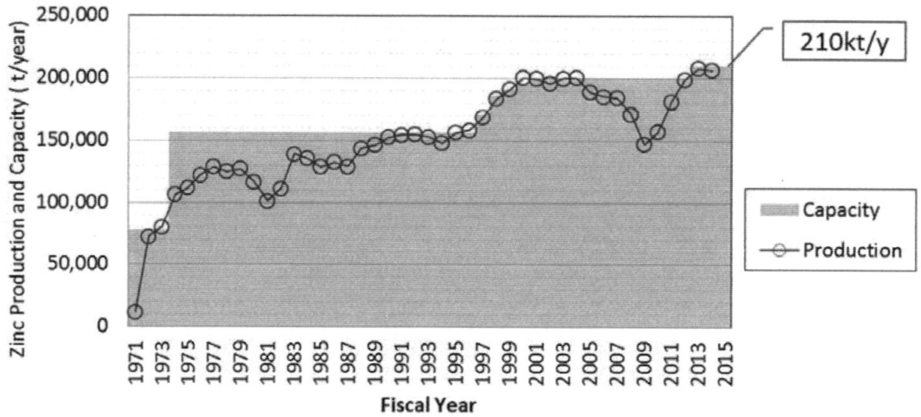

Figure 1 – Annual electrolytic zinc production and capacity at AZC

AZC was established under the following concepts (Yamazaki, 2010).

- Environmentally friendly, non-polluting refinery
- Highly productive refinery requiring the minimum number of operators
- Complete extraction of valuable metals from Japanese black ore
- Highest quality

In order to achieve complete recoveries, AZC has used the company's own hematite process for residue treatment since the start of operations. The iron included in the zinc concentrates is precipitated as zinc leach residue, together with a large number of valuable metals. Therefore, iron control is one of the key steps in the operation. There are several hydrometallurgical methods for zinc leach residue treatment, such as the jarosite process, the goethite process and the hematite process. AZC is the only refinery in the world which presently uses the hematite process. The iron included in the zinc leach residue is separated as hematite, as the name implies. The hematite process is generally perceived to be high in cost and difficult to operate, but we have carried out improvements in order to overcome those challenges. Those actions enabled a recent increase in production. In addition, one of advantages of the hematite process is the high zinc recovery, and it was recently further improved.

OUTLINE OF THE SMELTING AND REFINING PROCESS

Figure 2 shows the schematic flow sheet of AZC. In the next paragraph, details of the hematite process are described. Note that the other unit operations - roasting, leaching, purification, electrolysis - are typical processes.

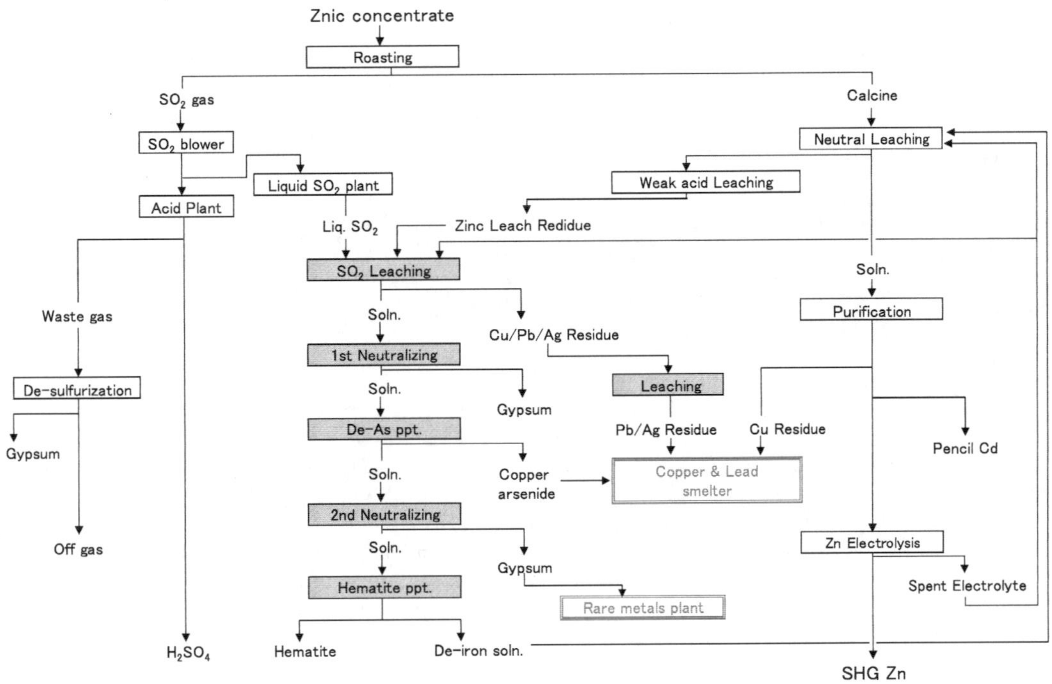

Figure 2 – Schematic flow sheet of the AZC operation

Hematite Process

The iron included in the zinc concentrate converts chemically during the roasting process into insoluble zinc ferrite. Therefore, the major mineralogical component of the zinc leach residue is zinc ferrite. However, lead sulfate and silica are also present. In the hematite process, zinc leach residue is handled as described below and changes occur to form various by-products. Table 1 shows the throughput and composition of those by-products.

Table 1 – Throughput and composition of the by-products

	Throughput	Composition						
		Fe	Zn	Pb	SiO_2	Al_2O_3	Cu	As
	t/M	%	%	%	%	%	%	%
Zinc Leach Residue	11,438	23.1	23.2	6.0	6.6	1.4	0.7	0.4
Pb/Ag Residue	2,947	1.8	3.0	23.2	25.3	-	0.2	-
1st Neutralizing Gypsum	3,009	-	0.2	-	-	-	-	-
Copper arsenide	160	-	0.5	-	-	-	63.2	27.3
2nd Neutralizing Gypsum	815	-	-	-	-	16.0	-	-
Hematite	4,014	54.1	0.6	-	-	-	-	-

SO_2 Leaching

The first process step is the leaching of zinc ferrite by sulfuric acid and SO_2 gas, as shown in Equation 1.

$$ZnFe_2O_4 + 2H_2SO_4 + SO_2 \rightarrow ZnSO_4 + 2FeSO_4 + 2H_2O \quad (1)$$

Figure 3 shows the schematic diagram for the SO_2 leaching process.

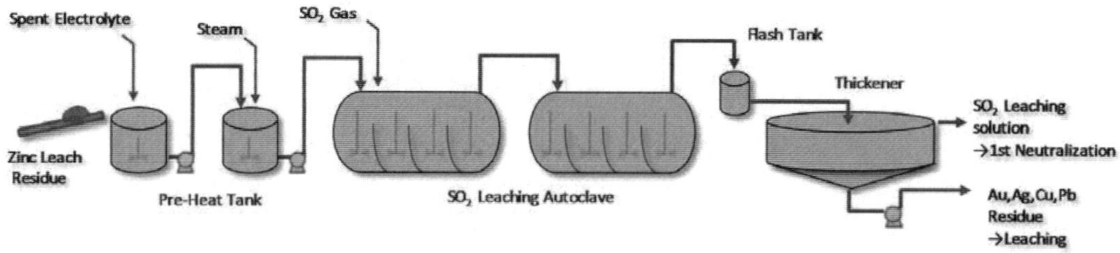

Figure 3 – Schematic diagram for SO_2 leaching

Zinc leach residue is pulped with spent electrolyte, and the slurry is charged into an autoclave for SO_2 pressure leaching. Two 100 m^3 autoclaves made of steel lined with a nickel alloy, each having four compartments, are fed continuously at the rate of 100 m^3/h (Tashita et al., 2014). The temperature of the slurry, charged at 90-95°C, is increased to about 120°C by the exothermic reaction in the autoclaves. The pressure of the vessel is controlled by SO_2 gas at 0.26 MPa in the first autoclave, and at 0.16 MPa in the second autoclave. It is important to use SO_2 gas as the reductant (not zinc concentrate), not only for the recovery of valuable metals but also for the stability of the operation. This occurs because the high zinc concentration of the solution may cause zinc precipitation in the hematite precipitation autoclaves, and kill the hematite precipitation operation. The purity of the SO_2 gas is important too. If the purity is low, discharges of inert gas increase, and maintenance of the temperature becomes difficult. Therefore AZC employs a unique process to produce liquid SO_2 from the off gases of the roasters. The mixing ratio of spent electrolyte to residue is one of the most important parameters to ensure adequate dissolution of the residue and to reduce costs in the down-stream processes. When not enough spent electrolyte is added, dissolution of the residue becomes insufficient. On the other hand, when it is super-abundant, liquid throughput increases and uses excessive power and steam. In the neutralization process, more calcium carbonate will be needed, too. At AZC, the spent electrolyte is mixed with the residue so that the free sulfuric acid concentration of the SO_2 leaching solution is 30 g/L (Arima & Kudo, 2004). The concentration of free sulfuric acid in the spent electrolyte is measured in real time with an auto-analyzer, and the data are used for feed-forward control. A similar measurement of the concentration of free sulfuric acid in the SO_2 leaching solution is done at the same time, and we feed back the data to prepare the mixing ratio of spent electrolyte to residue.

Additional Leaching

In the SO_2 leach autoclave, copper in the zinc leach residue is dissolved as shown in Equation 2. However as shown in Equation 3, a part of the copper re-precipitates in the SO_2 leach autoclave, because the zinc leach residue contains a small amount of zinc sulfide that had not reacted in the roasting process.

$$CuFe_2O_4 + 2H_2SO_4 + SO_2 \rightarrow CuSO_4 + 2FeSO_4 + 2H_2O \quad (2)$$

$$CuSO_4 + ZnS \rightarrow CuS + ZnSO_4 \quad (3)$$

When copper is included in the Pb/Ag residue, it may become a problem in the lead smelter. In addition, the arsenic removal process works well in the high copper concentrations of the feed solution. Therefore, in order to dissolve the copper sulfide in the residue, an additional leaching process was added in 2006 (Nakamura et al., 2007). In the additional leaching process, the residue is pulped with the spent electrolyte again. The copper dissolution reactions shown in Equations 4 and 5 occur with zinc ferrite being added. The source of the zinc ferrite is the zinc leach residue.

$$ZnFe_2O_4 + 4H_2SO_4 \rightarrow ZnSO_4 + Fe_2(SO_4)_3 + 4H_2O \quad (4)$$

$$CuS + Fe_2(SO_4)_3 \rightarrow CuSO_4 + 2Fe_2SO_4 + S \quad (5)$$

The Pb/Ag Residue is treated at KOSAKA SMELTING & REFINING which is located 120 km from AZC.

1st Neutralization Step

The leaching solution containing 30 g/L of free sulfuric acid is neutralized to 5 g/L with calcium carbonate. The gypsum product is sold to the cement industry.

De-Arsenic Process

Copper and arsenic are co-precipitated by zinc dust. The chemistry of the arsenic removal step is shown in Equation 6.

$$6Cu^{2+} + 2As^{3+} + 9Zn \rightarrow 9Zn^{2+} + 2Cu_3As \quad (6)$$

This is the process that was introduced in 2003 to improve the quality of the hematite product (Masuda et al., 2003). The copper-arsenic precipitate is also further treated at KOSAKA SMELTING & REFINING.

2nd Neutralization Step

After the de-arsenic process, the solution is further neutralized to around pH 4.5 using calcium carbonate in the second neutralization stage. At this stage, aluminum, indium and gallium are precipitated as hydroxides, together with the gypsum. Gypsum and the hydroxides are separated by sedimentation. The hydroxides are then further treated to recover indium and gallium at the rare metals plant. The gypsum is recycled back to the 1st neutralization stage.

Hematite Precipitation

The next stage is a process to precipitate the hematite by oxidizing the iron in solution under high temperature conditions, as shown in Equation 7.

$$2FeSO_4 + 0.5O_2 + 2H_2O \rightarrow 2Fe_2O_3 + 2H_2SO_4 \quad (7)$$

Figure 4 shows a schematic diagram of the hematite precipitation process. There are three production lines each having a heat exchanger, a 25 m^3 vertical autoclave, a 100 m^3 horizontal autoclave and a flash tank in series. The iron oxidation reaction mainly occurs in the vertical autoclave, and the hematite crystal growth mainly progresses in the horizontal autoclave. The scale which builds up inside of the vessel and the connecting pipe is cleaned at a frequency of once every 50 days. It is only the vertical autoclave where cleaning is necessary. One production line has two vertical autoclaves in order to create more throughput. The total flow rate is approximately 110 m^3/h. The autoclaves are made of steel lined with 3 mm of titanium to protect against corrosion.

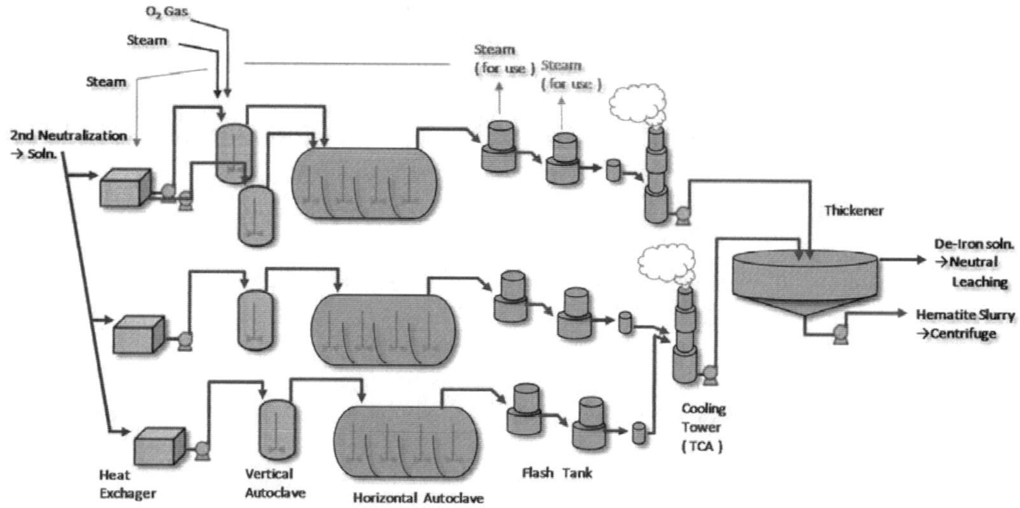

Figure 4 – Schematic diagram of the hematite precipitation circuit

It is necessary to oxidize ferrous to ferric above the boiling point of water, to ensure that iron precipitates as hematite and not jarosite or goethite. That is reason why we use autoclaves. The feed solution temperature is raised to approximately 80˚C by a heat exchanger. The solution is then charged into four vertical autoclaves. In the vertical autoclaves, the temperature is further increased to 185˚C by steam (Arima et al., 2007). Oxygen is concurrently supplied to the autoclave, and the pressure becomes 1.1–1.5 MPa by the saturated steam pressure and the pressure of oxygen. The purity of the O_2 gas is important. If the purity is low, discharges of inert gas will increase and much steam is necessary to maintain the temperature. We use oxygen of 99.6% purity which is made on site in a cryogenic air separation plant. Three flash tanks are necessary in order to efficiently recover the steam that is used in SO_2 leaching, preheating before hematite precipitation and elsewhere. The hematite is filtered with centrifuges, and all is sold to the cement industry. The solution after iron precipitation is recycled back to the neutral leaching stage. The acid generated by the hydrolysis reaction is used for the dissolution of calcine. The return of the de-ironed solution for calcine leaching means that the zinc collected from the zinc leach residue is returned at the same time.

RECENT OPERATION AND IMPROVEMENTS OF THE HEMATITE PROCESS

Figure 5 shows the inventory of zinc leach residue from 2004 to 2007. Changes in the composition of the zinc concentrate and many operational problems that mainly happened in the acid plant caused the increase at that time. It hurt our cash flow. It also caused operational difficulties because it reduced storage space for by-products and zinc concentrate.

Improvement of the Processing Capacity

At first, we re-evaluated the capacity of each facility as far as residue treatment was concerned. Figure 6 shows the result. The load to each facility would vary with the composition of the zinc leach residue. For example, the load to the hematite precipitation process was high when there was excessive iron in the circuit. When there was much lead and silica, the load to the filter press for Pb/Ag residue filtration increased. In this way, the capacity required changes. As shown in Figure 6, there are several facilities which required improvement. We made modification in these bottleneck facilities, as shown in Table 2.

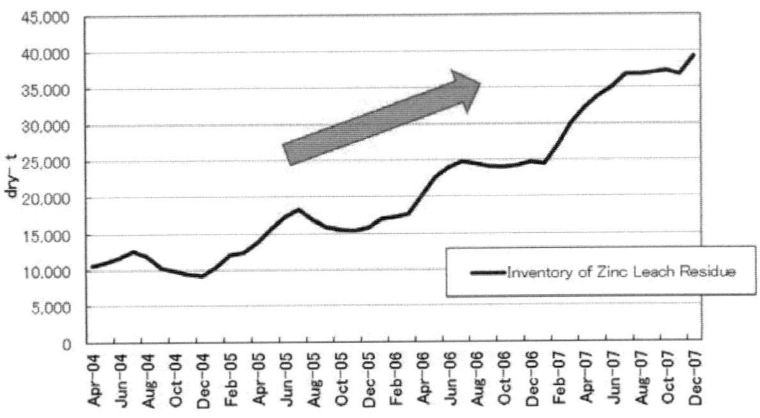

Figure 5 – Inventory of zinc leach residue from 2004 to 2007

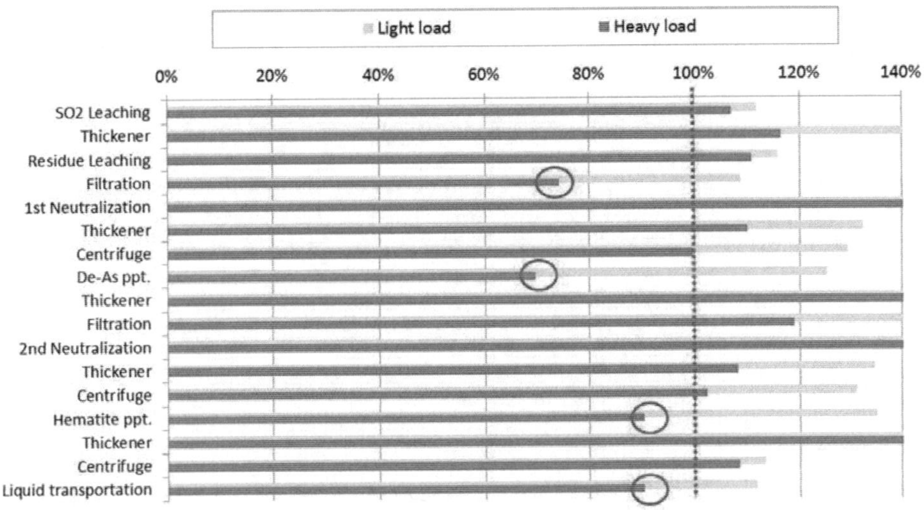

Figure 6 – Evaluation of the processing capacity

Table 2 – Improvement in each facility

Modification in each facility

Facility	Contents of modification
Pb/Ag Residue filtration	Addition of a Filter Press (3 → 4 units)
De-As precipitation	Enlargement of Table feeder
Hematite precipitation	Improvement of pressure control at Flash Tank
Liquid transportation	Increasing capacity of pumps & Extension of piping
SO_2 Leaching	Replaced (steel with homogeneous lead lining and acid resistant brick → steel and nickel alloy lining)

Let us discuss one example of these improvements, i.e. pressure control at the flash tank. Figure 7 shows the schematic diagram of the flash tank at the hematite precipitation stage. At the facility, high temperature slurry discharged from the autoclave is fed to this vessel. Its pressure falls because steam is being released, and its temperature falls. A part of the steam is recovered and is used as a heat source in the plant. In the first flash tank, the pressure is set at 0.3 MPa. When the pressure rises further, steam is released to the atmosphere via TCA. In the second flash tank, the pressure is set at 0.1 MPa. The slurry moves to the next flash tank by the difference in pressure. It was found that the pressure in first flash tank occasionally became too low to push enough slurry out to the second flash tank. Because it is important to stabilize the pressure difference in order to maintain a high discharge rate from an autoclave, the cause needed to be identified. Through detailed observation with electrodes installed, it was concluded that the variation in steam use at other parts of the plant caused the variation of pressure in the first flash tank. Therefore, we installed a control valve to regulate the use of steam. Figure 8 shows the change of the pressure of the flash tank before and after the improvement. As the result, it was possible to increase the charge rate further.

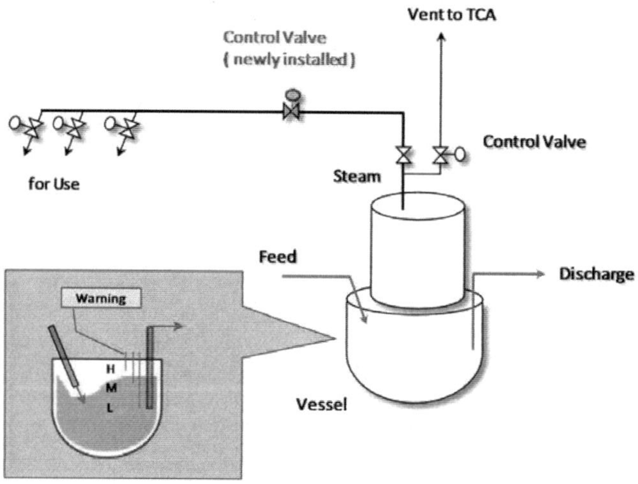

Figure 7 – Schematic diagram of the flash tank

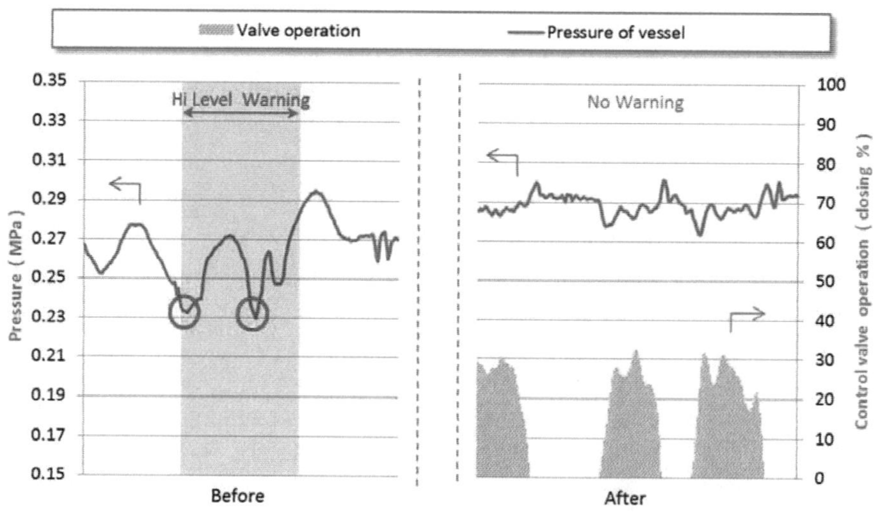

Figure 8 – Pressure of the flash tank before and after the change

Improvement of Process Water Management

A large amount of solution always circulates in the hydrometallurgical process systems of AZC. Figure 9 shows the schematic diagram of the plant water balance. The total water volume is approximately 30,000 m^3. Water enters the system along with additives and wash water used for the by-products. The hematite process requires a lot of water to suspend the calcium carbonate or flocculants, and for washing the Pb/Ag residue, the gypsum and the hematite. This water is also collected in the circuit, as is the steam for heating. On the other hand, water exits the plant by evaporation, mostly in the electrolysis process. Furthermore, we do "bleed off" solutions. Input balances output in the long term. However, it does not balance in the short term. In the winter season, the quantity of water evaporation decreases and it is always a challenge to keep the water volume under control. The zinc recovery is sometimes sacrificed because of reduced wash water availability. In extreme but rare cases, zinc leach residue treatment must be decreased to limit the water input.

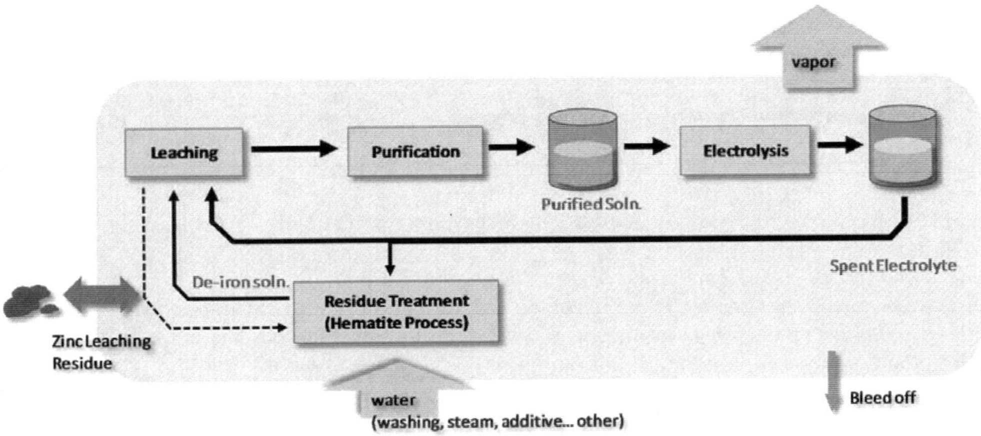

Figure 9 – Schematic diagram of the water balance

There is a reason why we must keep the water volume under control. In AZC, the electrolysis process runs fully during the period of low electric power rates. In the high rate zone, it runs at minimum load. On the other hand, the other processes always operate at the same rate. Therefore, the volumes of the purification solution and the spent solution increased and decreased in a pre-determined manner during a week, as shown in Figure 10. If the quantity of the water is too great (or too little), such operations become impossible.

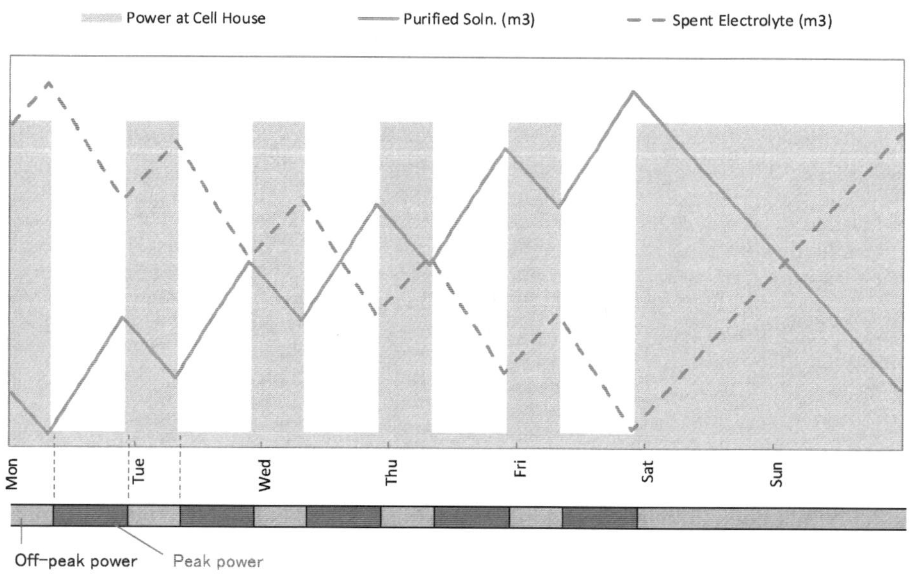

Figure 10 – Variation of the quantity of electricity and quantity of liquid during a typical one week period

Recently, a detailed monitoring practice was developed in order to quickly find the cause of any water increase such as spillage or failure of control valves. The structure of that management is shown in Figure 11. With this practice, everyone from an operator to the plant manager will understand the situation and take proper actions (Kanno et al., 2011).

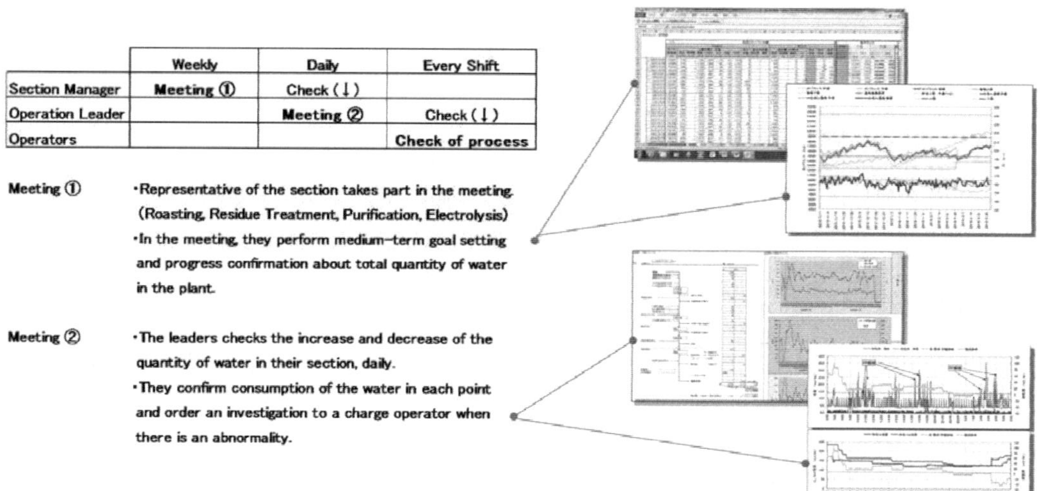

Figure 11 – Structure of the process water management system

As a result of the modifications to the facilities and improvements in process water management, more zinc leach residue can be treated. Figure 12 shows the changes in the processing capacity and the inventory of the zinc leach residue.

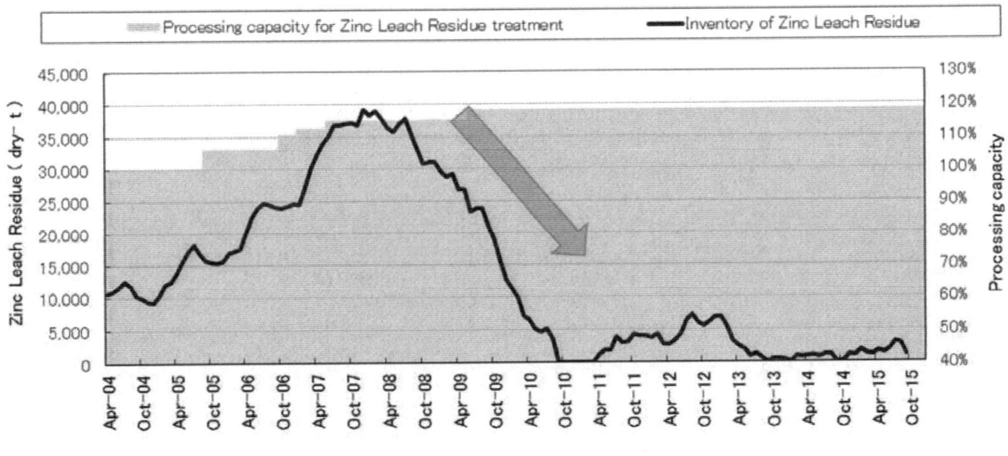

Figure 12 – Changes in the processing capacity and the inventory of the zinc leach residue

Improved Zinc Recovery

Better water management also contributes to better zinc recovery as shown in Figure 13. After keeping the inventory of the zinc leach residue at the proper level, there was room to use more water for washing the Pb/Ag residue, in order to recover more zinc.

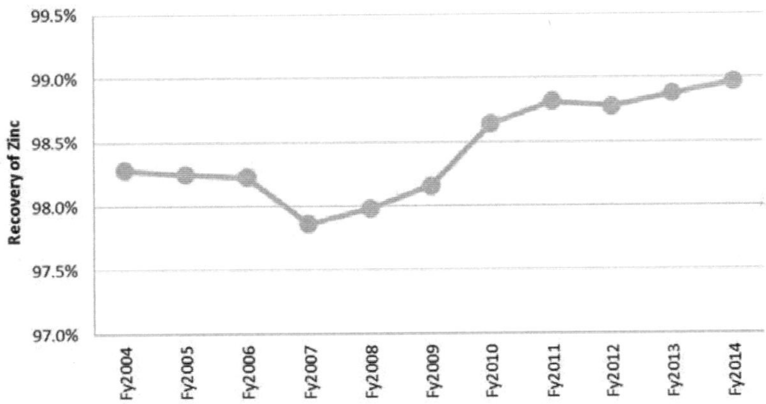

Figure 13 – Zinc recovery at AZC

In summary, in this paper de-bottlenecking of the hematite process and improvements in process water management were discussed. As a result of those improvements, increased throughput in the hematite process and better zinc recovery were achieved.

REFERENCES

Arima, H. & Kudo, Y. (2004). Autoclave application for zinc leach residue treatment by Akita Zinc Co., Ltd. *Pressure Hydrometallurgy Meeting*, M.J. Collins and V.G. Papangelakis, Eds., CIM, Banff, Alberta, Canada, 949-963.

Arima, H., Aichi, T., Kudo, Y., Saruta, K., Kanno, M., & Togashi, R. (2006). Recent improvements in the hematite precipitation process at the Akita Zinc Company. *Iron Control 2006*, J.E. Dutrizac and P.A. Riveros, Eds.., CIM, Montreal, Quebec, Canada, 123-134.

Kanno, M., Aoki, D., Usami, Y., & Saruta, K. (2011). Cleaning out the stockpile of zinc residue at Iijima Refinery. *Journal of MMIJ*, Vol.127, No.6, 362-365.

Masuda, H., Sato, S., Kudo, Y., & Shibachi, Y. (2003). Installation of arsenic removal in the hematite process. *Yazawa International Symposium*, F. Kongoli, K. Itagaki, C. Yamauchi and H.Y. Sohn, Eds., The Minerals, Metals and Materials Society, Warrendale, PA, 99-109.

Nakamura, H., Arima, H., Aichi, T., Kanno, M., Saruta, K., & Aoki, D. (2007). Improvement in quality and throughput of Pb/Ag residue. *Journal of MMIJ*, Vol.123, No.4-5, 236-239.

Tashita, H., Horiuchi, A., Usami, Y., & Kanno, M. (2014). Installation and operation of a new autoclave for leaching zinc residue. *Journal of MMIJ*, Vol.130, No.6, 331-334.

Yamazaki, N. (2010). The historical steps of 40 years in DOWA'S zinc smelting business and the role that the zinc smelting industry in Japan should have. *Lead-Zinc 2010*, CIM, Vancouver, Canada, 57-67.

PROCESS FOR THE PRECIPITATION OF PURE JAROSITE

*B. Saxén, M. Lahtinen, T. Ritasalo and M. Latva-Kokko

Outotec (Finland) Oy
Rauhalanpuisto 9, PO Box 1000
FI-02231 Espoo, Finland
*(*Corresponding author: bjorn.saxen@outotec.com)*

ABSTRACT

The conventional jarosite precipitation process with neutralization by zinc calcine is widely utilized in the zinc industry. However, it has a significant drawback of causing losses of valuable metals like zinc, lead and silver to the jarosite. This comes mainly from the content of these metals in the calcine used for neutralization. As the zinc plants today want to maximize metal recoveries and minimize the amount of residues, this matter has become more and more relevant for zinc producers. Outotec developed and patented in early 2000 an improved version of the jarosite precipitation process in which the need for neutralization can be minimized, thus the metal losses to the jarosite also can be minimized. At the same time, the iron content of the jarosite increases and the amount of produced jarosite is smaller. The main feature of this process is to start the precipitation from ferrous iron after having reduced all the iron by zinc concentrate in a separate step before jarosite precipitation. Iron is then oxidised back to the ferric form simultaneously with jarosite precipitation. By this method, the formation of sulphuric acid is only half that compared to the conventional method and the need for calcine for neutralization can be avoided. High mixing and oxygen dispersion capability are needed in this jarosite precipitation process. Such capabilities are found in the Outotec OKTOP® reactor series. High performance reactors are not only needed to generate high quality precipitates, but also to maximize oxygen utilization; i.e., to minimize operation costs. Outotec's latest development in this area is the OKTOP® plant product in which not only the reactors, but also all the accessories are designed, manufactured and delivered by Outotec. One responsible supplier ensures efficient and fast delivery and can provide mechanical and performance guarantees for the entire plant unit.

KEYWORDS

Jarosite, zinc, precipitation, mixing, ferrous, oxidation, dispersion

INTRODUCTION

An iron precipitation process is needed in the hydrometallurgical production of zinc as, in the leaching of zinc calcine or zinc concentrate, iron in the raw materials is mostly leached together with zinc. Two kinds of iron precipitation process are generally today at zinc plants. The conventional precipitation process of jarosite with neutralization by zinc calcine is most widely utilized in the zinc industry. The other process, goethite precipitation, is operated at a limited number of zinc plants.

In spite of its popularity, jarosite precipitation has a significant drawback in that it causes losses of valuable metals like zinc, lead and silver to the iron residue. This comes mainly from the content of these metals in the calcine. As the zinc plants today want to maximize metal recoveries and minimize the amount of residue, this matter has become more and more relevant for zinc producers. Jarosite precipitation, however, is a very well proven technology and it has certain benefits such as the possibility to achieve high Zn recoveries, the ability to capture some impurities like arsenic, halides and some heavy metals, and it helps to control the sulphate balance (Beaulieu, Gagné, Nasmyth, Cooper & Inostroza, 2006). It is not easily replaced by goethite precipitation or any other iron control method. Thus, optimisation or further development work on the jarosite precipitation process would be well justified and motivated.

One solution presented in this work is a process which Outotec developed and patented in early 2000. It improves the jarosite precipitation process by eliminating the need for neutralization with calcine, thus minimizing metal losses to the jarosite. At the same time, the iron content of the jarosite increases and the amount of produced jarosite is smaller. Here, we present this process with laboratory test results and mass balance calculations. In addition, we discuss the requirements of reactor and process design for optimal operation.

JAROSITE PRECIPITATION

The traditional process for precipitating jarosite starts with iron in the ferric (Fe^{3+}) form according to reaction (1)

$$3\ Fe^{3+} + Na^{+} + 2\ SO_4^{2-} + 6\ H_2O = NaFe_3(SO4)_2(OH)_6 + 6\ H^{+} \qquad (1)$$

From the reaction equation it can be seen that, for each mole of iron precipitated, there are two moles of hydrogen ions (H^+) formed. As jarosite formation occurs favourably only in a certain acidity range, typically 10-20 g/l H_2SO_4, there is a need to neutralize the produced sulphuric acid to enable efficient jarosite precipitation. The normal neutralising agent is zinc calcine, where ZnO is the main component consuming acid and it neutralizes the solution. However, the calcine also contains a number of components which are not leached, but will contaminate the precipitated jarosite. Especially, zinc ferrite ($ZnOFe_2O_3$) in the calcine is problematic, as this will not leach under the conditions used for jarosite precipitation and will thus cause zinc losses. In practice, to achieve high zinc recoveries, there has to be a separate "acid wash" stage after jarosite precipitation where the zinc ferrite is leached at high acidity and high temperature. This brings significant additional cost to the process. In addition to the acid wash stage, there is also needed process stage for washing and/or neutralizing the acidity from the acid washed jarosite before final deposition of the residue. Also, because they are not soluble in sulphuric acid, silver and lead in the calcine used for neutralisation will be lost to the jarosite residue.

There have been developments of the jarosite process beginning soon after the introduction of the process in the 1960's (Ismael and Carvalho, 2003). Regardless of this, the jarosite process is still operated mostly as it was originally developed. Perhaps this occurs mainly because no really major improvements have been found and, on the other hand, because of the reluctance of the zinc producers to make any major modifications to the existing operations. However, there is one jarosite process modification developed by the former Outokumpu, which has not been acknowledged for some reason, even though it offers significant opportunities to improve the performance and economics of the jarosite process.

This process for precipitation of pure jarosite that was patented by Outokumpu (Fugleberg, 2002) - now a technology owned and further developed by Outotec - has the following main features differing from traditional jarosite precipitation:

1. Iron is brought to the ferrous (Fe^{2+}) form by reduction with zinc concentrate prior to jarosite precipitation
2. In the jarosite precipitation stage, oxidation is carried out to transform iron to ferric form (Fe^{3+}) to enable precipitation.

When the oxidation of iron is carried out simultaneously with jarosite precipitation, the reaction is

$$3\ Fe^{2+} + Na^{+} + 2\ SO_4^{2-} + 0.75\ O_2 + 4.5\ H_2O = NaFe_3(SO_4)_2(OH)_6 + 3\ H^{+} \quad (2)$$

In reaction (2), the formation of sulphuric acid is only half that needed in the traditional process, reaction (1). This significantly reduces the need for neutralisation and enables jarosite precipitation without zinc calcine feed additions, thus leading to a pure jarosite obtained without any separate acid washing stage. Simultaneously, a higher silver and lead recovery is obtained as no silver and lead from a neutralizing calcine will be lost in the jarosite. An additional benefit is the possibility for the recovery of indium, germanium and gallium from the solution where iron is in ferrous form, as will be explained later.

The process for the precipitation of pure jarosite is suitable for implementation in most zinc plants, and the modifications needed are not very big. A process flow sheet is shown later in this paper.

EXPERIMENTAL

We conducted test work on a laboratory scale to further verify the concept from early 2000's and to set the basis for the design of the process for pure jarosite precipitation.

Pre-treatment

To produce a solution suitable for precipitation of pure jarosite, the iron containing solution from high acid leaching is treated in a reduction step and a subsequent neutralization step, as explained later in the flow sheet section. Although known technology, we verified these unit operations by tests at the same time as we produced authentic solutions for the jarosite precipitation tests. Some results of the reduction and neutralization tests are shown in Figure 1. It is seen that when iron is reduced to ferrous state, the acidity can be neutralized down to 5 g/l H_2SO_4 without significant precipitation.

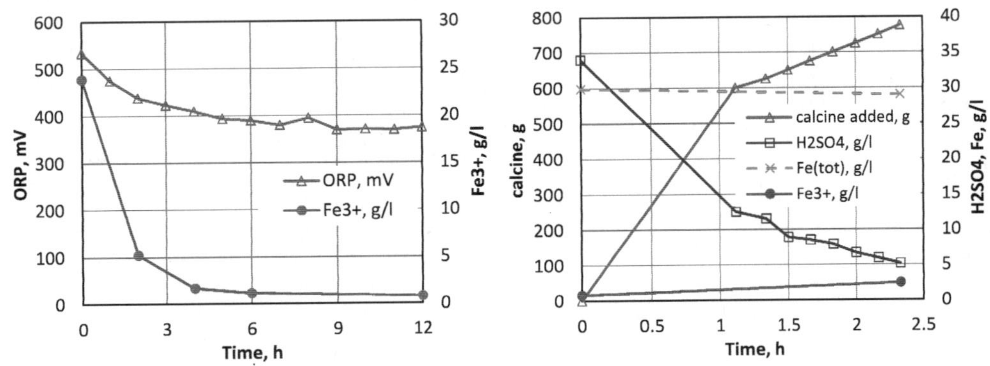

Figure 1 – Left graph: Results from a reduction test where zinc concentrate was added (at time 0 h) to an acidic solution with high ferric iron content. Right graph: Results from a neutralisation test with calcine

Recovery of In, Ge and Ga

Before jarosite precipitation, indium gallium and germanium can be recovered from solution. Because iron is mainly in the ferrous state, neutralisation and cementation can be used for precipitating In, Ge and Ga (Salminen, Riihimäki & Ruonala, 2012) without precipitating significant amounts of iron. We show the results of one such test in Figure 2. It is shown that copper and part of the nickel also precipitate, but these metals can be recovered during the subsequent treatment of the precipitate.

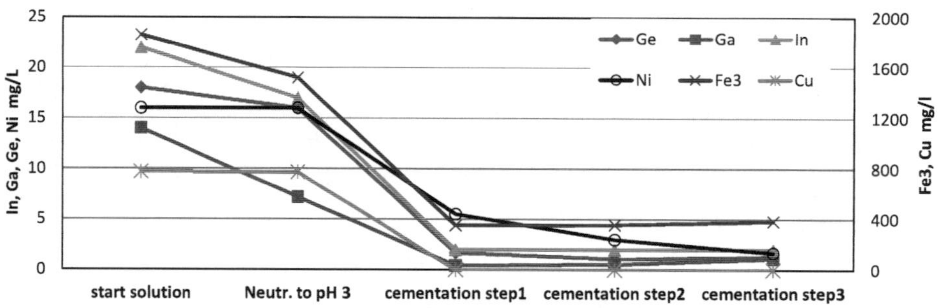

Figure 2 – In, Ga and Ge precipitation by neutralization and cementation with zinc dust

Jarosite Precipitation

We carried out precipitation tests for pure jarosite using the solution produced by the reduction and neutralization tests described above. Sodium sulphate was added to ensure the formation of Na jarosite. Starting solution compositions are given in Table 1. Oxygen gas was fed below the mixing device especially designed for gas dispersion in a three-phase system and the temperature was controlled to 95 °C by heating. Jarosite seed were added at the beginning of the test. The first seeds were produced by a special precipitation procedure with a high Fe^{3+} concentration; later the seed originated from the product of a former test.

The basic concept for pure jarosite precipitation was verified with two tests, which were similar except that in one test all iron was initially in the ferric form and in the other test the iron was initially in ferrous state but was oxidised to ferric form during the test. No neutralisation agent was added during the tests, with the result that the precipitated jarosite was equally clean in both cases. However, as can be seen from Figure 3, precipitation in the test with ferric iron stopped at a high iron concentration where less than half of the iron was precipitated. With iron initially in the ferrous form and with oxidation during the test, the final iron concentration was much lower and approximately two-thirds of the iron precipitated.

More details on the precipitation of pure jarosite are shown in the test results presented in Figure 4. It is demonstrated that even without neutralization (left graph), over 20 g/L iron can be precipitated. The precipitation can be further improved by intensified oxidation and neutralisation with sodium hydroxide (right graph). The use of NaOH is justified because it is needed to compensate for the sodium precipitated with the jarosite.

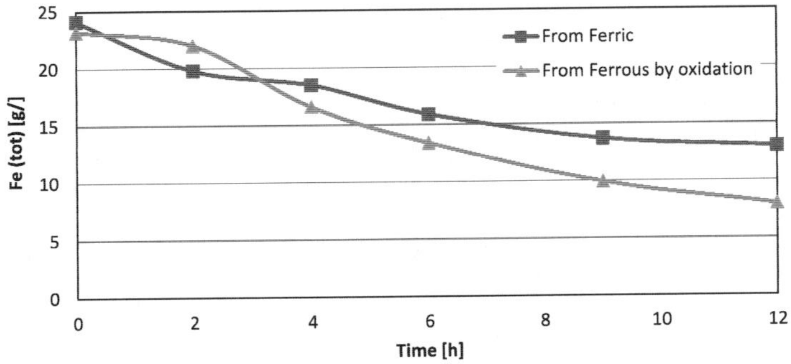

Figure 3 – Jarosite precipitation results from two tests, one starting with a solution having all iron in the ferric form and the other starting with a solution having iron in the ferrous form and with oxidation during the test

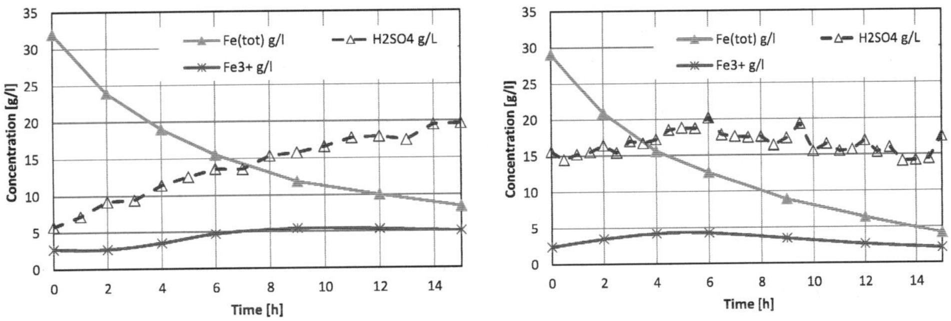

Figure 4 – Results from two jarosite precipitation tests with simultaneous oxidation. Left graph: without neutralisation, right graph: Intensified oxidation and NaOH addition for acidity control

The results shown in Figure 4 indicate that the oxidation of ferrous iron to the ferric iron is rather slow, and that the oxidation efficiency is essential for the overall performance of the process. Thus, high performance oxidation reactor technology is needed for the successful industrial application of this technology. One factor affecting the oxidation is the mixing efficiency, as higher mixing disperses the oxygen into smaller bubbles having a larger total surface area. In Figure 5, a comparison of the oxidation rates obtained in two jarosite precipitation tests with different mixing speeds is shown.

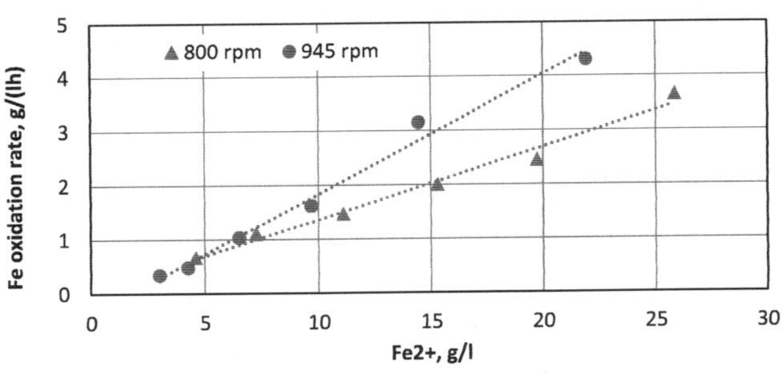

Figure 5 – Effect of mixing efficiency on the iron oxidation rate

Compositions of the precipitated jarosite from two tests are shown in Table 1. The tests differ in the starting solution composition as some As, Al and Cd was added to the starting solution of the second test. The quality of the precipitated jarosite is good, and the zinc content is low. Practically no lead or silver are present in the jarosite as these elements are not dissolved in the sulphate solution. Arsenic and aluminum partly precipitate with the jarosite, as expected. This is even desired as it provides an outlet for these compounds from the zinc plant. XRD analyses of the precipitated jarosites detected Na-jarosite as the only major compound.

Table 1 – Compositions of starting solutions and pure jarosite precipitates from the two tests

			Zn	Fe	Na	Ca	Mn	Mg	Al	Cd	As
Low	Start solution	g/L	130	33	5.8	410	3.2	11	0.48	0.08	0.52
As, Al, Cd	Precipitated jarosite	%	0.22	33.1	3.8	0.02		0.01	0.09	<0.01	0.21
High	Start solution	g/L	140	31	3.6	0.44	3.2	10.5	1.6	0.65	1.2
As, Al, Cd	Precipitated jarosite	%	0.49	32.5	3.8	0.01	0.007	0.028	0.32	<0.01	0.58

REACTOR DESIGN

Precipitation of pure jarosite requires oxidation of ferrous iron to the ferric state. This oxidation is made with oxygen gas. Since the solubility of oxygen in concentrated sulphate solutions at high temperatures is very low, the reaction rate is at least partially controlled by the rate of oxygen transfer from the gas to the aqueous phase. Among all the factors that affect the rate of oxygen mass transfer, it has been stated that the most important is the reactor configuration and geometry (Filippou, Cheng & Demopoulos, 2000). Thus, reactor design plays a very important role in this process. High oxygen feed capacity with efficient gas dispersion is required. Reactors should produce high quality precipitates and obtain high oxygen utilization efficiencies in order to minimize operating costs.

The volumetric mass transfer coefficient k_La can be used as a quantitative measure of the gas to liquid mass transfer capability of a reactor. It is basically a function of mixing power intensity (P/V) and superficial gas velocity v_s. The general equation for the estimation of k_La is shown in Equation 3. The coefficients A, B, and C are reactor and application specific values.

$$k_La = A\,(P/V)^B\,v_s^{\,C} \qquad (3)$$

In order to achieve high oxygen mass transfer rate with high oxygen utilization efficiency, the mixing power efficiency must be significantly high. This presents some special requirements for agitator design in large scale industrial reactors. In addition, the presence of gas affects the performance of the impeller and its ability to suspend solids. The Outotec OKTOP®2100 impeller, shown in Figure 6, is specially designed for three phase processes where gas dispersion is the limiting factor for the overall reaction rate. Its novel design does not contain any welded joints of the impeller blades or their fastenings; these facts make it superior to welded structures in terms of corrosion resistance and mechanical strength. Welded joints can be considered as the weakest link in an impeller because of their decreased corrosion resistance and vulnerability to fatigue failure. The design also enables easy and quick replacement of worn-out blades by new ones and, as a result, the maintenance costs can be decreased.

Figure 6 – OKTOP®2100 impeller with non-welded design

OKTOP® Reactor Plant

Common practice in hydrometallurgical reactor design is to fulfill the industry specific standards whereas the process specific requirements are not the main focus. The split scope delivery with many suppliers rarely allows development of a functional unit providing the best metallurgical performance and functional design to secure high availability with safe and easy operation and maintenance. Common issues in the design and delivery phases of a project are long response times, unclear battery limits, exceeded costs and lead times, case by case variations in quality and solutions that are not always as feasible as those estimated. Outotec's approach to overcome these typical issues is to supply the reactor train as a complete, highly predefined, but still tailor-made, functional unit that is integrated into the process design and automation. The Outotec OKTOP® reactor plant unit concept utilizes proprietary OKTOP® agitators, which are chosen so that the specific reactor requirements of the hydrometallurgical application are fulfilled. Product delivery with standardized plant modules and one responsible supplier ensures efficient and fast delivery and can provide mechanical and performance guarantees for the entire plant unit. (Latva-Kokko, Hirsi, Lindgren & Ritasalo, 2014).

Safety and maintenance features are embedded in the OKTOP® reactor plant unit design. Agitator support with a service platform and a gravity flow launder with isolation valves that allow any reactor to be bypassed for periodic maintenance are shown in Figure 7. In addition, agitators can be removed and installed as a one-piece assembly and there is clear and safe accessibility to all equipment and accessories. Heat transfer is made with effective dual purpose baffles that are especially suitable for slurries prone to scaling, because they can be separately lifted from the reactor for cleaning.

Figure 7 – OKTOP® reactor plant unit. Agitator service center on top left and OKTOP® launder on top right

FLOW SHEET AND COMPARISON WITH TRADITIONAL PROCESS

Figure 8 shows the results of our preliminary flow sheet and mass balance calculations for the pure jarosite precipitation process as applied at a plant producing 100,000 t/a zinc.

The process steps for the neutral leach, acid leach and high acid leach as well as for solution purification and electrowinning represent basic modern zinc plant technology. The reduction and neutralisation steps are "new" in the sense that these are not needed for traditional jarosite precipitation. However, such steps are used at other plants, such as those using goethite precipitation, and thus represent known technology. The sulphur cake from the reduction stage can be fed to the roasters to recover valuables. Indium, gallium and germanium can be recovered as by-products when their content in the raw material is sufficiently high. These advantages are facilitated by the flow sheet having iron in a reduced (ferrous) form in the solution after neutralisation, as described in the experimental section. The precipitate containing In, Ga and Ge can be further treated by leaching and solvent extraction to recover those metals separately.

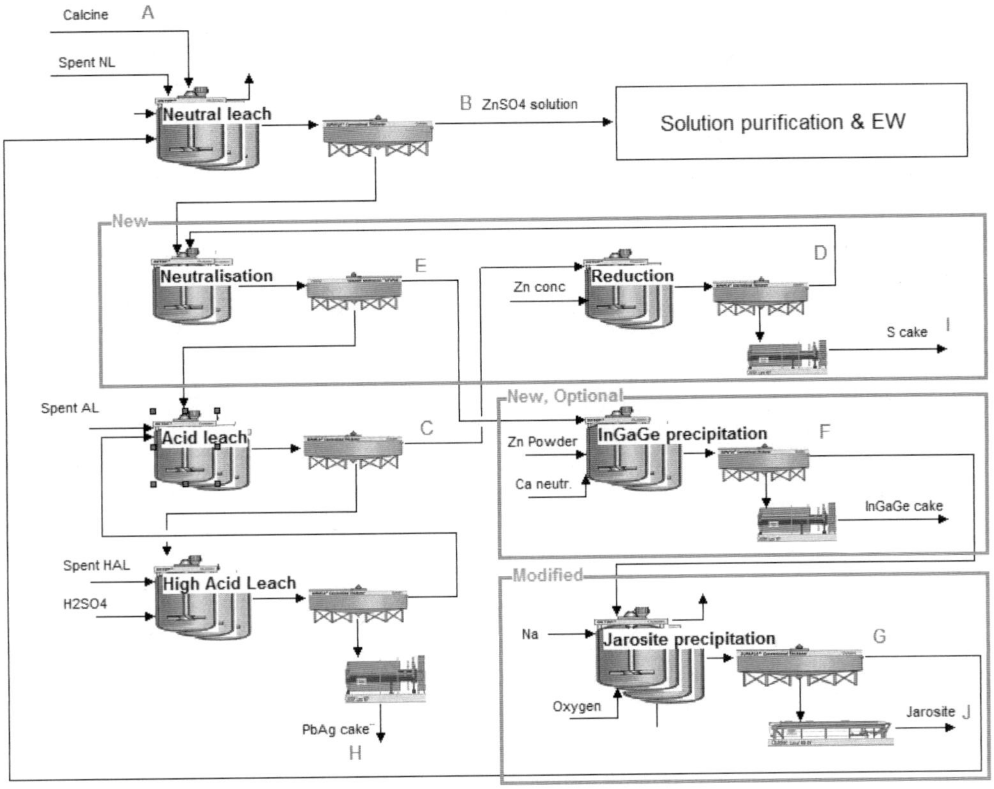

Stream	Content	Solids, t/h	Zn, %	Fe, %	Pb, %	Ag, %	Solution, m3/h	Zn, g/l	H2SO4, g/l	Fe2+, g/l	Fe3+, g/l
A	Calcine	23.0	56.0	10.2	2.7	0.012					
B	NL overflow						130	150	0	0.0	0.0
C	AL overflow						99	94	40	7.9	25.2
D	Red. overflow						98	105	39	33.2	0.5
E	Neutr. overflow						104	133	5	24.7	1.0
F	InGaGe overflow						104	137	1	24.7	0.4
G	Jarosite overflow						95	128	14	3.0	3.0
H	PbAg cake	2.5	5.1	7.5	25.0	0.110					
I	S cake	1.3	10.9	2.8	5.9	0.017					
J	Jarosite	5.8	0.5	33.4	0.0	0.000					

Figure 8 – Flow sheet and preliminary mass balance results. Changes in the flow sheet compared to the traditional process are indicated with green boxes

The core unit operation, jarosite precipitation, differs from the traditional technology mainly in the need for oxidation using oxygen gas and efficient mixing in the precipitation reactors. On the other hand, no calcine feed is needed in the jarosite precipitation step and no acid wash treatment is needed for the jarosite precipitate as the zinc content is low.

The jarosite from a traditional precipitation process typically contains several percent zinc, between 4 and 6% according to Beaulieu et al. (2006). An acid treatment stage for the precipitated jarosite can be utilized to leach zinc ferrites from the calcine and recover part of the contained zinc. However, the losses of lead and silver cannot be avoided in the traditional process. Typically around 20% of the total calcine is used for neutralization in traditional jarosite precipitation; thus, 20% of both the lead and silver in the raw material is lost to the jarosite residue. This implies that a plant producing 100,000 t/a Zn from a concentrate containing 50% Zn, 2% Pb and 100 ppm Ag loses approx. 800 t/a Pb and 4 t/a Ag with the traditional jarosite process.

CONCLUSIONS

Outotec developed and patented in the early 2000s an improved version of the jarosite precipitation process. Recent test work has now verified the processing concept which has the following advantages over the conventional process:

- Need for neutralization can be minimized; i.e., no calcine is needed, thus also the metal losses to the jarosite can be minimized.
- The iron content of the jarosite increases and the amount of produced jarosite is smaller.
- Higher silver and lead recoveries are possible, as no silver and lead from calcine will be lost to the jarosite.
- An additional benefit is the possibilities for the recovery of indium, germanium and gallium from the solution where iron is in ferrous form.

The main feature of this process is the ability to precipitate iron by starting the precipitation from ferrous iron solutions. This is done by reducing all the iron to the ferrous form by zinc concentrate in a separate step before jarosite precipitation. The formation of sulphuric acid in the new process, with simultaneous oxidation of the iron for jarosite precipitation, is only half of that formed in the traditional process.

In order to achieve high oxygen mass transfer rates with high oxygen utilization the mixing efficiency must be significantly high. This presents some special requirements for agitator design in large scale industrial reactors. Outotec can deliver an OKTOP® reactor plant unit equipped with agitators specially designed for this three phase process where gas dispersion is critical. It is also possible to modify an existing jarosite process and to change the operation from the conventional process to the new improved jarosite process.

REFERENCES

Beaulieu R., Gagné, G., Nasmyth M., Cooper, G., & Inostroza, C., (2006). Iron control and management in the zinc industry. In J.E. Dutrizac and P.A. Riveros (Eds.), Iron Control Technologies (pp.45-55). The Metallurgical Society of CIM, 2006.

Filippou, D., Cheng, T. & Demopoulos, G.P. (2000). Gas-liquid oxygen mass-transfer; from fundamentals to applications in hydrometallurgical systems. *Mineral Processing and Extractive Metallurgy Review, 20*, 447-502.

Fugleberg, S. (2002). International patent publication WO 02/46481, WIPO.

Ismael, M.R.C., & Carvalho, J.M.R. (2003). Iron recovery from sulfate leach liquors in zinc hydrometallurgy. Minerals Engineering, 16, 31-39

Latva-Kokko, M., Hirsi. T., Lindgren, M., & Ritasalo, T. (2014). Influence of reactor design to process performance in hydrometallurgical applications. *Hydrometallurgy 2014 – Volume II, Proceedings of the 7th International Symposium*, Canadian Institute of Mining, Metallurgy and Petroleum, 3-13

Salminen J., Riihimäki T. & Ruonala M (2012). International patent publication WO 2012/049361, WIPO.

THE OPTIMIZATION OF AN INDUSTRIAL AMMONIUM JAROSITE PRODUCTION CIRCUIT

*Dr. Corby G. Anderson[1], Camille M. Fleuriault[1] and Dr. Larry G. Twidwell[2]

[1]*Kroll Institute for Extractive Metallurgy*
George S. Ansell Department of Metallurgical and Materials Engineering
Colorado School of Mines
1500 Illinois Street
Golden, Colorado 80401 USA
*(*Corresponding author: cganders@mines.edu)*

[2]*The Center for Advanced Mineral and Metallurgical Processing*
Montana Tech
1300 West Park Street
Butte, Montana 59701 USA
ltwidwell@mtech.edu

ABSTRACT

The selective precipitation of iron as ammonium jarosite is a well-established industrial technology in base metals hydrometallurgy. It is a key technology in minimizing iron tenors which would be disruptive to downstream processing. It also provides a more readily filterable iron solid to better facilitate materials handling. This paper elucidates the fundamentals behind industrial iron precipitation. Then, the details of the successful optimization of an industrial ammonium jarosite circuit are presented.

KEYWORDS

Iron, jarosite, copper, nickel, ammonia, optimization

HYDROMETALLURGICAL IRON REMOVAL FUNDAMENTALS

Iron is the fourth most abundant element in the Earth's crust (5% by weight) and is often found in complex ores such as sulphide-bearing formations. Iron represents an issue though during leaching, a technology which is widely used for the treatment of this type of ore:

- Iron has the particularity of forming hydroxides when leached under certain conditions. Iron hydroxides form a viscous gel, which is difficult to filter and traps valuable elements such as silver.
- Iron is solubilized along with other metals and interferes with the subsequent extraction steps.

As a result, purification of the pregnant leach solutions is required to achieve economical extraction of base metals. In the 1960s, metallurgists developed iron precipitation processes for solution purification (Habashi, 1999). Nowadays, precipitation of iron bearing species, such as goethite, hematite or jarosite, has become one of the most common methods used in the industry to purify leach liquors. To be suitable, the precipitate has to be readily filterable and thus must not be too fine of particle size. Precipitation can take place in autoclaves at medium to high temperatures, under an oxygen or air overpressure (Tourre, 1984, Fleuriault 2016, Fleuriault Anderson Shuey, 2016)).

None of the precipitates from the goethite and jarosite processes is suitable for direct commercialization. Further refining is needed. In the production of hematite, ferrite materials can be considered as marketable products if they are pure enough. Table 1 summarizes and compares the characteristics of the three industrial iron precipitation processes, as presently used in the zinc industry.

Fundamentals of Iron Hydroxysulphate Phase Formation during Iron Precipitation Processes

During sulphide leaching processes, such as pressure oxidation, iron forms ferrous sulphate, which is oxidized to ferric sulphate. Hydrolysis of ferric sulphate then forms an iron precipitate. At temperatures >185°C in an oxidative environment, several iron species can form, but most often, hematite is the desirable product. The conditions for hematite formation are stringent and sometimes difficult to control. It is likely that other iron solids (called iron hydroxysulphates) precipitate along with the hematite (Fleming, 2010).

Iron hydroxysulphates are unwanted constituents in the hematite residue for several reasons:

- Poorer settling and filtration properties
- Relatively unstable compounds which may represent an environmental risk if stockpiled
- Trapping of precious metal values in their crystal structure resulting in lower PGM recovery
- In the case of gold cyanidation circuits, a high concentration of iron hydroxysulphates in the autoclave residue means high lime consumption and difficulties maintaining pH values >10 during the cyanidation step. There is a slight risk of forming toxic HCN.

Several publications have shown that the total amount of free acid in solution determines which iron species preferentially forms (Cheng et al., 2004; Fleming, 2010; Tourre, 984; Umetsu et al., 1977). Depending on the leaching conditions, the limiting acid concentration for the formation of one or the other iron species varies. Because hematite is preferred over iron hydroxysulphates in the final product, the acid concentration is closely monitored during autoclave processing.

Table 1 – Comparison of the goethite, jarosite and hematite processes as used in the zinc industry (Gupta et al., 1990; Gathje, 2006)

Parameter	Goethite	Jarosite	Hematite
Compound formed	FeO.OH	$MFe_3[(SO_4)_2(OH)_6]$ with M=K, Na, NH_4	Fe_2O_3
Temperature (°C)	70-90	90-100	>185
pH	2-3.5	1.5	Up to 2% H_2SO_4
Anion present	Any	SO_4^{2-}	SO_4^{2-}
Cation added	None	Na^+, K^+, NH_4^+	None
Cationic impurities	Medium	Low	Low
Anionic impurities	Medium	High Medium	
Filterability	Very good	Very good	Very good
Fe left in solution (g/L)	<0.05	1-5	3
Metal recovery %	Zn 96 % Cu 90 % Ag 85 %	Zn 96 % Cu 90 % Ag 60-65%	Zn 98.2 % Cu 98.2 % Ag 98.2 %
Residue composition wt.%	Fe 40-45 % Zn 5-10 % S 2.5-5 %	Fe 25-30 % Zn 4-6 % S 10-12 %	Fe 50-60 % Zn 0.5-1 % S 2-3%
Moisture wt.%	50	50	10
Amount produced/t concentrate	0.25	0.40	0.18
Zn loss in t/t slab	0.025	0.025	0.002

IRON HYDROXYSULPHATE PRECIPITATE CHARACTERIZATION

The nomenclature of iron hydroxysulphate species varies depending on the authors. These compounds can also be referred to as ferric hydroxysulphate, basic iron sulphate or iron hydroxyl sulphate. Early work by Posnjak and Merwin (1922) reported many ferric sulphate salts but many of them are not crystalline species. Basic ferric sulphate, or BFS, is metastable and often composed of mixtures. As a result, such precipitates are quite difficult to characterize and few data are available. Posnjak and Merwin identified three series of BFIS. However, at the time, the authors did not exclude the possibility that other hydrated species also exist. The three series were subdivided and organized by the ratio Fe_2O_3 to SO_3. The following Table 2 presents the BFS already identified at the time of this study.

Table 2 – BFS compounds classified by their Fe_2O_3:SO_3 ratio (Posnjak, 1922)

Fe_2O_3 : SO_3 ratio	Formula	Crystal	Color	Name
3:4	$3Fe_2O_3.4SO_3.9H_2O$	Rhombohedral	Light-deep yellow	Carphosiderite[1]
	$Fe_2O_3.2SO_3.H_2O$	Orthorhombic	Orange-yellow	
	$Fe_2O_3.2SO_3.5H_2O$	Monoclinic	Light yellow	
1:2	$2Fe_2O_3.5SO_3.17H_2O$	Orthorhombic	Light-bright yellow	Copiapite
	$Fe_2O_3.2SO_3.7H_2O$		Yellowish	Amarantite
	$Fe_2O_3.2SO_3.8H_2O$		Yellowish	Castanite
2:5	$Fe_2O_3.2SO_3.10H_2O$		Yellowish	Fibroferrite

[1]Carphosiderite is a discredited mineral name and nowadays corresponds to hydronium jarosite.

In the late 1970s, the iron hydroxysulphate minerals were divided in two series (Rossman, 1976 and Scordari, 1981):

- Compounds with the general formula $Fe(OH)SO_4.nH_2O$, with the following members: basic ferric sulphate $Fe(OH)SO_4$, butlerite/parabutlerite (they are polymorphs) $Fe(OH)SO_4.2H_2O$ and fibroferrite: $Fe(OH)SO_4.5H_2O$. According to several publications, the non-hydrated member of the BFS series is synthetic and has been obtained only in the laboratory.
- Jarosites. The main jarosite species are presented later in this paper.

Lazaroff et al. (1982) also make the distinction between crystalline jarosites and amorphous ferric hydroxysulphates, but refer to this last category as BFS. In order to simplify the nomenclature and avoid any confusion within the next sections, we will distinguish BFS ($Fe(OH)SO_4$ and its hydrated species) from the jarosite compounds. We will consider BFS and jarosite as iron hydroxysulphates. The following Table 3 presents the most common BFS encountered in high temperature systems, using the most recent nomenclature.

Table 3 – Selected hydroxysulphates and oxy-hydroxysulfates of iron (adapted from Bigham, 2000)

Name	Formula
Copiapite	$Fe^{II}Fe_4^{III}(SO_4)_6(OH)_2.2H_2O$
Fibroferrite	$Fe^{III}(SO_4)(OH).5H_2O$
Amarantite	$Fe^{III}(SO_4)(OH).3H_2O$
Butlerite	$Fe^{III}(SO_4)(OH).2H_2O$
Basic Ferric Sulphate	$Fe^{III}(SO_4)(OH)$
-	$Fe_4^{III}(SO_4)(OH)_{10}$
Schwertmannite	$Fe_8^{III}O_8(OH)_6(SO_4)\cdot nH_2O$

JAROSITE PRECIPITATE CHARACTERIZATION

Jarosites are compounds of the general formula: $MFe_3(SO_4)_2(OH)_6$, where M can be H_3O^+, Na, K, NH_4^+, Pb, Ag, Tl, Hg, Rb (Tourre, 1984). Silver-jarosite and potassium-jarosite are the two most stable compounds of the family. Other extensive substitutions occur for Fe^{3+}, SO_4^{2-}, OH^-. As for BFS, not all jarosites are naturally occurring. Out of nine species synthetized, only seven occur as minerals. The first mineral to be identified was potassium jarosite, in Andalusia, Spain in 1852. Table 4 lists all known jarosite species.

Table 4 - Chemical and mineral names of jarosites (Dutrizac, 2000)

Formula	Chemical Name	Mineral Name
$H_3OFe_3(SO_4)_2(OH)_6$	Hydronium Jarosite	Hydronium Jarosite
$NaFe_3(SO_4)_2(OH)_6$	Sodium Jarosite	Natrojarosite
$KFe_3(SO_4)_2(OH)_6$	Potassium Jarosite	Jarosite
$RbFe_3(SO_4)_2(OH)_6$	Rubidium Jarosite	None
$AgFe_3(SO_4)_2(OH)_6$	Silver Jarosite	Argentojarosite
$NH_4Fe_3(SO_4)_2(OH)_6$	Ammonium Jarosite	Ammoniojarosite
$TlFe_3(SO_4)_2(OH)_6$	Thallium Jarosite	Dorallcharite
$Pb_{1/2}Fe_3(SO_4)_2(OH)_6$	Lead Jarosite	Plumbojarosite
$Hg_{1/2}Fe_3(SO_4)_2(OH)_6$	Mercury Jarosite	None

IRON HYDROXYSULPHATES COMPARISON

A spectroscopic analysis of the products of ferric sulphate hydrolysis was performed in 1996, with emphasis made on the amorphous species, which are compared to jarosite (Burgina et al., 1996). While jarosite dissolves very slowly, BFS are reactive in water. This difference is explained by the atomic arrangement of each species: jarosites are trimers, and amorphous BFS are tetramers (Figure 1). According to Burgina and al., the reactivity of BFS is caused by the small separation between neighboring iron atoms, short Fe-O distance and strong hydrogen bonds (creating a large number of highly acid centers). This would promote the oxidation of H_2O in O_2 and thus explain BFS reactivity in water.

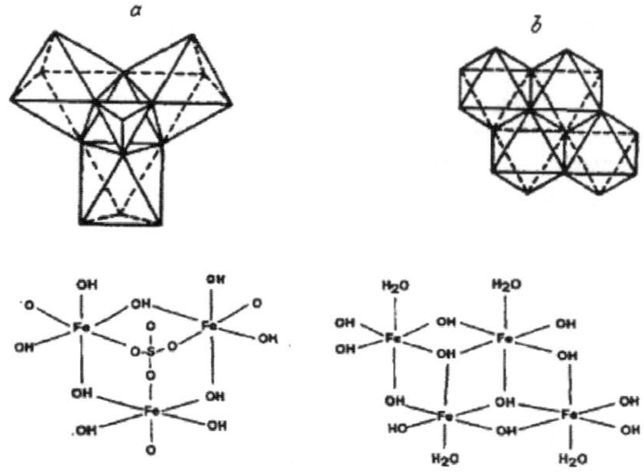

Figure 1 – Molecular structures of jarosite (trimer) and BFS (tetramer) (Burgina et al., 1996)

BFS are also reactive in the atmosphere (Posnjak et al., 1922), decomposing either to ferric hydroxide and gypsum (alkaline conditions) or ferric sulphate (acidic conditions). BFS can be formed by precipitation, hydrolysis or evaporation but none of these processes yielded a product that fully achieved equilibrium; thereby, mixtures were produced. In the case of pressure leaching, the main factors influencing BIS formation are acidity, temperature and iron concentration. BFS readily precipitate at high acidity (>20g/L) and lower temperatures (<200°C) (Fleming, 2010). As mentioned before, jarosite formation is favored by high acidity as well as well as the presence of some cations in relatively high concentrations (Na^{+}, NH4^{+}, K^{+}, Ag^{+} or Pb^{+}). Additionally, BFS precipitation seems to be promoted by increasing ferric concentration. At 225°C, within the stability region of hematite, Dutrizac and Chen (2012) have observed that BFS becomes the predominant phase if the initial [Fe^{3+}] concentration exceeds 22.3g/L (Dutrizac and Chen, 2012). Based on Posnjak and Merwin's work, Tourre (1984) produced a 3-D model of the hematite/sulphate system, as shown in Figures 2 and 3.

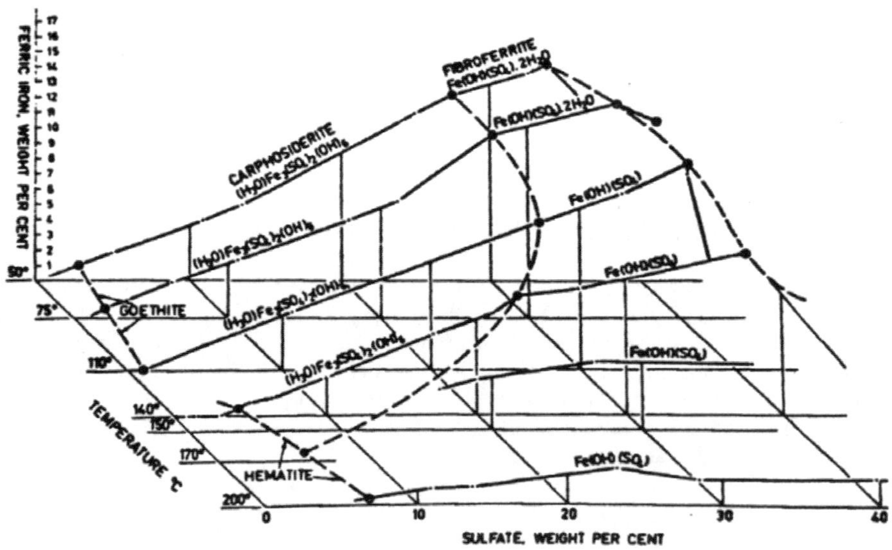

Figure 2 – The $Fe_2(SO_4)_3$-H_2SO_4-H_2O system polytherm at 50 to 200°C and 0 to 40% SO_4 (Tourre, 1984)

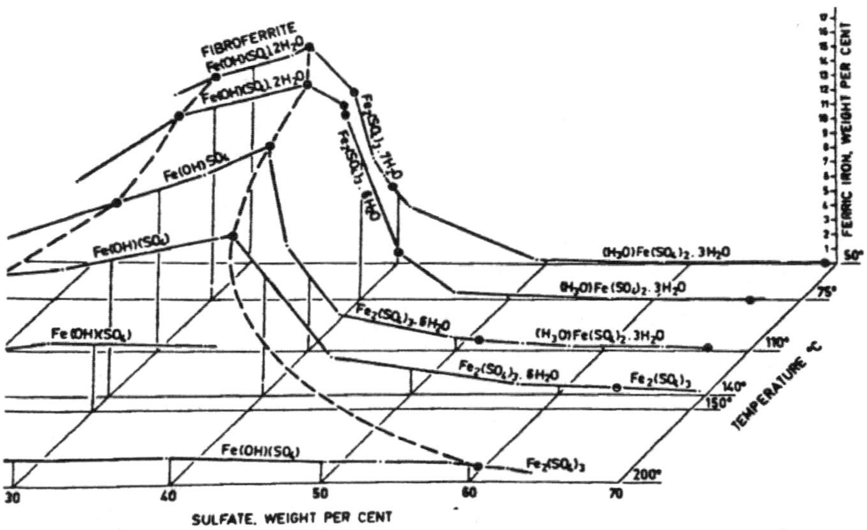

Figure 3 - The $Fe_2(SO_4)_3$-H_2SO_4-H_2O system polytherm at 50 to 200°C and 30 to 70% SO_4 (Tourre, 1984)

These observations were confirmed in 1982 (Lazaroff et al., 1982). In the simple Fe-O-S system, hydronium jarosite has a limited area of stability, as it is one of the least stable jarosite species. If additional ions such as Na^+, K^+ or NH_4^+ were present (which is extremely likely in a real industrial situation), jarosites would certainly replace BFS. In 1971, Babcan (1971) defined the area of stability of iron hydroxysulphates in the Fe-O-S system (see Figure 4). In more recent studies, BFS have been proven to be also stable at conditions which would normally lead to hematite formation (Dutrizac & Chen., 2012; Voigt et al., 1986). If the oxidation conditions were maintained for a longer time, the most stable phase would eventually form. As a result, from a thermodynamic point of view, iron hydroxysulphates are metastable.

Several hematite solubility studies have been conducted in order to identify the optimum conditions for iron oxide precipitation. At equilibrium and room temperature, Umetsu et al. (1977) proposed a linear relation between ferric concentration in g/L and free acidity:

$$\text{Log}\,[\text{Fe(III)}]_{\text{total}} = a*\text{Log}\,[H_2SO_4]_{\text{free}} - b \quad (1)$$

Where a and b are coefficients which depend on temperature and the presence of other metal sulphates. This linear model has been experimentally confirmed and modeled (Figure 5) (Papangelakis et al., 1994) Investigations on the hydrolysis of iron sulphate solution by the same authors highlighted the same linear relationship between the ferric ion and free acid concentrations, but up to a certain point. The solid phase equilibrium curve is actually made of two straight lines of different slopes (Figure 6.). At lower free acidity, the curve describes the hematite equilibrium, and at higher acidity, the curve describes the BFS equilibrium.

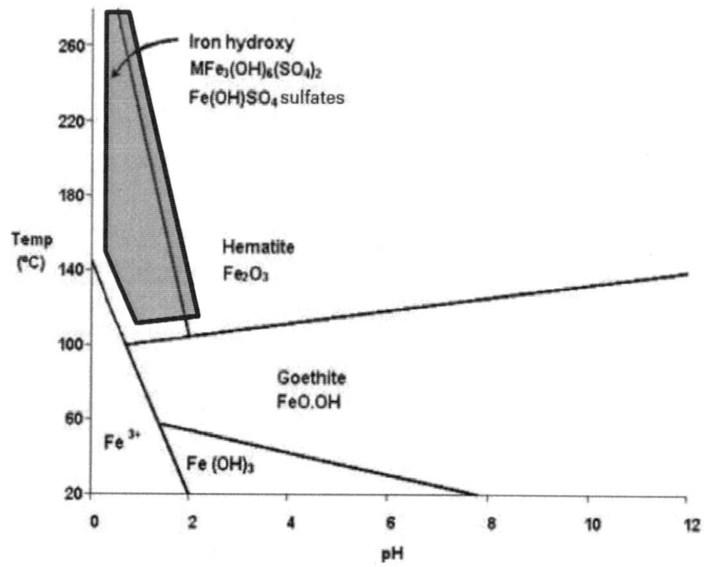

Figure 4 – Areas of stability of the various compounds in the Fe-S-O system (modified after Babcan (1971)

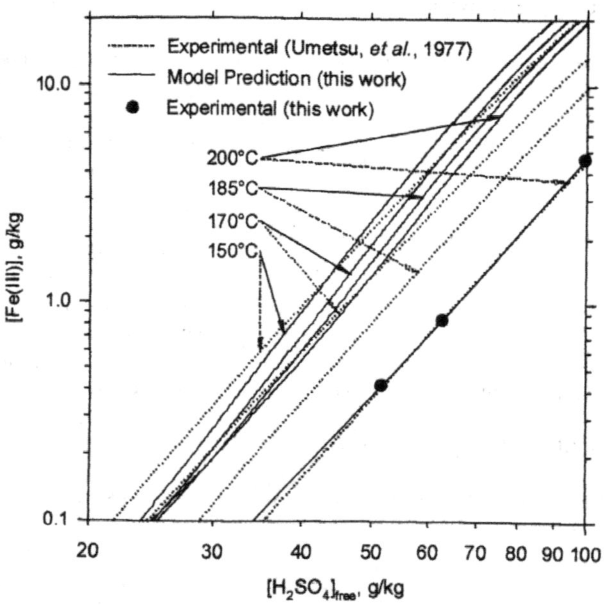

Figure 5 – Hematite solubility – A comparison between model predictions and the experimental data (Papangelakis et al., 1994)

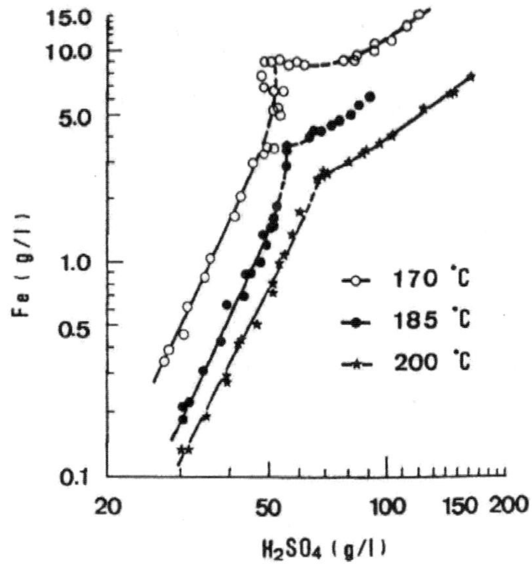

Figure 6 – Relationship between the concentrations of ferric ion and free sulfuric acid in the absence of other metal sulphates (Tozawa & Sasaki, 1986)

Sulphide ores contain many different metals which can be leached along with iron. Common sulphates in leaching solutions are $CuSO_4$, $ZnSO_4$ and Na_2SO_4. Therefore, assessing their impact on hematite or BFS precipitation is important. Figure 7 shows that the hydrolysis of ferric iron is favored at high temperature and low pH (Umetsu et al., 1977). It can also be seen that for some conditions, only iron precipitates and the other ions stay in solution. Hydrolysis is thus a very efficient way to selectively remove iron from solution.

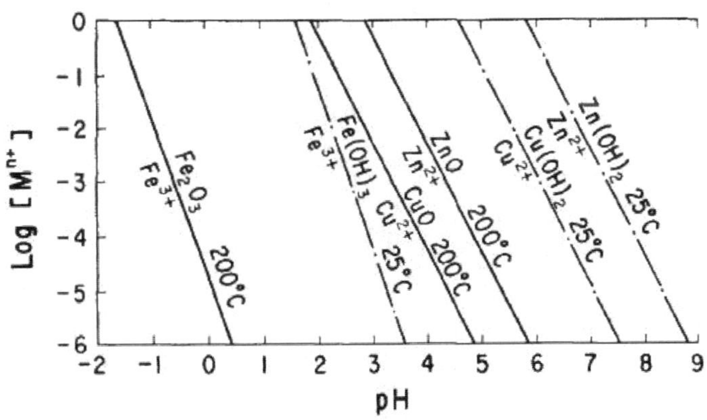

Figure 7 – Relationship between metal ion concentrations and pH at 25°C and 200°C (Umetsu et al., 1977)

OPTIMIZATION OF AN INDUSTRIAL HYDROMETALLURGICAL JAROSITE CIRCUIT

An applied study was undertaken to optimize the performance of the Stillwater Base Metals Refinery jarosite precipitation circuit. The goal was to find the multivariate operating parameters to substantially reduce residual iron in solution, consistently to below 10 ppm. The Stillwater Base Metals Refinery (BMR) process consists of matte grinding, atmospheric leaching, pressure leaching, PGM concentrate separation and iron precipitation (Newman & Makwana, 1997). First, the converter matte from the smelter is ground batch-wise in a tower mill to yield a 70 wt.% solids slurry in water, with 80% of the solids passing 74 microns. The ground matte is leached with the recycled acidic pressure leach solution and oxygen in a series of five cascading agitated tanks. Some of the nickel and iron are extracted from the matte, while some of the copper present in the pressure leach solution is precipitated according to the following typical reactions:

$$Ni^0 + H_2SO_4 + 0.5\ O_2 \rightarrow NiSO_4 + H_2O \quad (2)$$

$$Ni_3S_2 + H_2SO_4 + 0.5\ O_2 \rightarrow NiSO_4 + 2\ NiS + H_2O \quad (3)$$

$$Ni_3S_2 + 2\ CuSO_4 \rightarrow Cu_2S + 2\ O + NiS \quad (4)$$

$$Fe^0 + H_2SO_4 + 0.5\ O_2 \rightarrow FeSO_4 + H_2O \quad (5)$$

The dissolved iron remains ėssentially in the ferrous state under the prevailing low pH (2.0 to 2.2) conditions of the atmospheric leach process. Any PGMs present in the feed solution are also co-precipitated with the copper. The residue from the atmospheric leach process is separated by thickening and is then leached further at elevated temperature and oxygen pressure. The atmospheric leach residue consists essentially of millerite (NiS), digenite ($Cu_{1.8}S$), djurleite ($Cu_{1.96}S$); iron is in the form of magnetite and hydrated ferric oxide. The principal reactions in the pressure leach process are of the type shown below.

$$NiS + 2\ O_2 \rightarrow NiSO_4 \quad (6)$$

$$Cu_2S + H_2SO_4 + 2.5\ O_2 \rightarrow 2\ CuSO_4 + H_2O \quad (7)$$

$$Fe_2O_3 + 3\ H_2SO_4 \rightarrow Fe_2(SO_4)_3 + 3\ H_2O \quad (8)$$

Magnetite, if present, does not dissolve at the relatively low acid concentrations (20 to 25 g/L) and temperatures (130 to 135°C) prevailing in the pressure leach process.

The atmospheric leach thickener overflow solution is polish-filtered and then treated to precipitate most of the iron as ammonium jarosite, which is filtered and returned to the smelter. The iron precipitation process is necessary to meet the requirements of the nickel-copper sulphate solution customer. The iron precipitation process also acts as a backstop to precipitate or catch any soluble and insoluble PGMs that may be present in the atmospheric leach solution. The iron is precipitated as ammonium jarosite according to the following reactions.

$$2\ FeSO_4 + 0.5\ O_2 + H_2SO_4 \rightarrow Fe_2(SO_4)_3 + 3\ H_2O \quad (9)$$

$$3\ Fe_2(SO_4)_3 + 2\ NH_3 + 12\ H_2O \rightarrow 2\ NH_4Fe_3(OH)_6(SO_4)_2 + 5\ H_2SO_4 \quad (10)$$

The plant produces a final readily refined bullion concentrate of about 60% PGM content. As well, by-product nickel sulphate and copper cathode metal are now produced from this zero discharge hydrometallurgical facility.

To carry out this study, bulk representative samples of BMR plant process solutions were provided for laboratory testing. Then, the pertinent controllable plant variables of pH, temperature and reaction time were tested using STAT EASE Design of Experimentation methodology. Because of client budget constraints and time, STAT-EASE statistical software was used to minimize the number of experiments required to identify a statistical representative experimental set. This commercially licensed program based on the fundamentals of design of experiments provides highly efficient:

1. Two-level factorial screening studies so that the vital factors which affect process can be identified
2. Response surface methods to find ideal process settings and achieve optimal performance
3. Mixture design techniques to discover optimal formulations

A complete discourse on the computer software and statistical design of experimentation fundamentals for chemical processing is beyond the scope of this paper, but pertinent references are provided for the reader's convenience (Anderson & Whitcomb 2000; Box & Draper 1987; Box 2000; Cochran 1957; Cornell 1990; Hinkelmann & Kempthorne 1994; Kempthorne 1975; Lorenzen and Anderson 1993; Meyer and Napier-Munn 1999; Montgomery 2001; Napier-Munn and Meyer 1999; Neter 1990; Pazman 1986; STAT EASE 2002;,Wu and Hamada 2000; Wayne, Shari, Pat and Mark. 2009). The full factorial design matrix used for this study is illustrated in Table 5.

The results of testing were modeled with an adjusted R-squared fit value of 0.9929 or, in real terms, near perfection. The resultant values and those predicated by the fitted model are shown in Table 6. Below equation 11 illustrates the final statistical model equation in terms of actual plant factors.

$$\frac{1}{(\text{Solution [Fe]})} = -5456.14+1861.19*pH+24.54*T+22.11*t - 8.35*pH*T - 4.92*pH*t - 0.038*T*t \quad (11)$$

(With concentration [Fe] in g/L, temperature T in F, time t in hrs)

Table 5 – STAT EASE full design matrix for jarosite precipitation optimization

Std	Run	pH	Temp, F	Time, Hrs
4	1	3.5	210	4
7	2	2.9	210	8
6	3	3.5	170	8
2	4	3.5	170	4
1	5	2.9	170	4
3	6	2.9	210	4
8	7	3.5	210	8
5	8	2.9	170	8

Table 6 – STAT EASE model fit of jarosite precipitation optimization diagnostic case statistics

Standard Order	Actual Value	Predicted Value	Residual	Leverage
1	3.65	0.42	3.23	0.875
2	250.00	253.23	-3.23	0.875
3	3.73	6.96	-3.23	0.875
4	62.50	59.27	3.23	0.875
5	2.55	5.78	-3.23	0.875
6	250.00	246.77	3.23	0.875
7	9.43	6.21	3.23	0.875
8	43.48	46.71	-3.23	0.875

Figures 8 is a graphical illustration with time held constant at four hours of the results for the iron concentration as a function of temperature and the pH. In essence, the testing and STAT EASE modeling indicated that plant iron concentrations consistently below 10 ppm would be achieved by increasing the pH to 3.5 and lowering the operating temperature to 77°C while keeping the plant residence time the same. In retrospect, this is consistent with the use of ammonia based hydrometallurgical systems. These recommendations were made, implemented in the plant and proved successful.

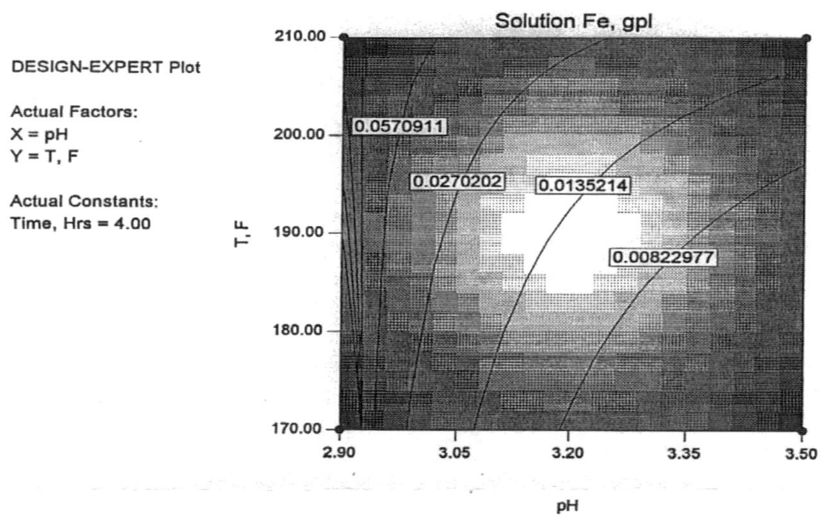

Figure 8 – Graphical illustration of the residual iron concentration at constant time as a function of temperature and pH

SUMMARY

This paper has outlined the fundamental aspects of hydrometallurgical iron precipitation. Then, these principles, coupled with statistical design of experimentation, were successfully applied to optimize iron removal as ammonium jarosite an operating industrial hydrometallurgical facility.

REFERENCES

Anderson, C.G. and Fayram, T.S, (2013). The use of design of experimentation software in applied flotation testing. *Journal of Powder Metallurgy and Mining,*

Anderson, M.J. & Whitcomb, P.J., (2000) *DOE Simplified: Practical Tools for Effective Experimentation*, ISBN: J-56327-225-3, Published May 2000, Productivity, Inc.

Babcan, J., (1971). Synthesis of jarosite $KFe_3(SO_4)_2(OH)_6$. *Geol. Zb*, *22*(2), 299-304.

Bigham, J.M. & Nordstrom, D.K. (2000). Iron and aluminum hydroxysulphates from acid sulphate waters. *Reviews in Mineralogy and Geochemistry*, *40*(1), 351-403.

Box, G.E.P., & Draper, N.R., (1987). *Empirical Model Building and Response Surfaces* ISBN: 0-471-81033-9, Published January 1987, Wiley-Interscience.

Box, G.E.P., (2000). *Box on Quality and Discovery: with Design, Control, and Robustness*, Published 2000, Wiley, New York.

Burgina, E.B., Kustova, G.N., Nikitenko, S.G., Kochubei, D.I., & Elizarova, G.L. (1996). Comparative study of the structures of some basic iron (III) sulphates. *Journal of Structural Chemistry*, *37*(2), 240-246.

Cheng, T.C. & Demopoulos, G.P. (2004). Hydrolysis of ferric sulphate in the presence of zinc sulphate at 200 C: precipitation kinetics and product characterization. *Industrial & engineering chemistry research*, *43*(20), 6299-6308.

Cochran, W.G., (1957). *Experimental Designs*, Published 1957, Wiley, New York.

Cornell, J., (1990). *Experiments with Mixtures: Designs, Models, and the Analysis of Mixture Data*, 2nd Ed., ISBN: 0-471-52221-X, Published April 1990, John Wiley and Sons.

Dutrizac, J.E. & Jambor, J.L., (2000). Jarosites and their application in hydrometallurgy. *Reviews in Mineralogy and Geochemistry*, *40*(1), 405-452.

Dutrizac, J.E., & Chen, T.T., (2012). Behavior of various impurities during the precipitation of hematite from ferric sulphate solutions at 225° C. In *TT Chen Honorary Symposium on Hydrometallurgy, Electrometallurgy and Materials Characterization*, (pp. 489-507). John Wiley & Sons, Inc.

Fleming, C.A., (2010). Basic iron sulphate–a potential killer in the processing of refractory gold concentrates by pressure oxidation. *Minerals & Metallurgical Processing Journal*, *27*(2), 81-88.

Fleuriault, C.M. (2016). Iron phase control during pressure leaching at elevated temperature, MSc Thesis, Kroll Institute for Extractive Metallurgy, Colorado School of Mines, Golden, Colorado

Fleuriault, C.M., Anderson, C. G. & Shuey, S. (2016) Iron phase control during pressure leaching at elevated temperature, Manuscript in publication.

Gathje, J.C., (2006). Iron reduction in copper leach liquors using controlled precipitation of sulphate species during high temperature pressure oxidation of base metal ores. *Iron Control Technologies*, CIM, Montreal, 2006, 231-246.

Gupta, C.K., and Mukherjee, T.K., (1990). *Hydrometallurgy in Extraction Processes*. Vol. 2. CRC Press, 196-200.

Habashi, F., (1999). *Textbook of Hydrometallurgy. Métallurgie Extractive Québec.*

Hinkelmann, K. & Kempthorne, O., (Editors), (1994). *Design and Analysis of Experiments: Introduction to Experimental Design*, ISBN: 0471551783, Published March 1994, John Wiley & Sons.

Kempthorne, O., (1975). *The Design and Analysis of Experiments*, Published 1975, Robert E. Krieger Publishing Co., Huntington, New York.

Lazaroff, N., Sigal, W., & Wasserman, A., (1982). Iron oxidation and precipitation of ferric hydroxysulphates by resting Thiobacillus ferrooxidans cells. *Applied and Environmental Microbiology*, *43*(4), 924-938.

Lorenzen, T.J. and Anderson, V.L., (1993). *Design of Experiments: A No-Name Approach*, ISBN: 0-8247-9077-4, Published 1993, M. Dekker, New York.

Meyer D. & Napier-Munn T.J., (1999). *Optimal Experiments for Time Dependent Mineral Processes*, Australian and New Zealand Journal of Statistics, 1999.

Montgomery, D.C, (2001). *Design and Analysis of Experiments*, 5th Ed., ISBN: 0-471-31649-0, Published 2001, John Wiley and Sons.

Napier-Munn T.J. & Meyer D.H., (1999). A modified paired T-test for the analysis of plant trials with auto correlated in time, *Minerals Engineering*, 12(9), 1093-1109.

Neter, J., (1990). *Applied Linear Statistical Models: Regression, Analysis of Variance, and Experimental Designs*, Published 1990, Irwin, Homewood, Illinois.

Newman, L. and Makwana, M., (1997). Commissioning of the Stillwater Mining Company Base Metals Refinery, *Hydrometallurgy and Refining of Nickel and Cobalt*, Proceedings of the 27th Annual Hydrometallurgical Meeting, The Metallurgical Society of CIM, Sudbury, Ontario, August 1997.

Papangelakis, V.G., Blakey, B.C., & Liao, H. (1994). Hematite solubility in sulphate process solutions. In *Hydrometallurgy'94* (pp. 159-175). Springer, The Netherlands.

Pazman, A. (1986). *Foundations of Optimum Experimental Design*, Published 1986, D. Reidel, Boston.

Posnjak, E., & Merwin, H.E. (1922). The system Fe_2O_3-SO_3-H_2O. *Journal of the American Chemical Society*, *44*(9), 1965-1994.

Rossman, G.R. (1976). Spectroscopic and magnetic studies of ferric iron hydroxy sulphates: the series Fe (OH) $SO_4 \cdot nH_2O$ and the jarosites. *American Mineralogist*, 61(5-6), 398-404.

Scordari, F. (1981). Fibroferrite: A mineral with a {Fe (OH)(H_2O) $2SO_4$} spiral chain and its relationship to Fe (OH) SO_4, butlerite and parabutlerite. *Tschermaks mineralogische und petrographische Mitteilungen*, *28*(1), 17-29.

STAT-EASE (2002). Design-Expert Reference Manual, STAT-EASE Corporation, Minneapolis, Minnesota, 2002.

Tourre, J-M. (1984). A study of the conditions of formation of silver-containing jarosites in the acid pressure leaching of pyritic ores. PhD Diss. Colorado School of Mines.

Tozawa, K., & Sasaki, K. (1986). Effect of coexisting sulphates on precipitation of ferric oxide from ferric sulphate solutions at elevated temperatures. *Iron Control in Hydrometallurgy*, Ellis Horwood, Chichester, England, 454-476.

Turk, D.J., (2000). *Stillwater Mining Company Nye Concentrator Operation*, SME Annual Meeting, Preprint 00-10, Salt Lake City, Utah, February, 2000.

Umetsu, V., Tozawa, K., & Sasaki, K.I. (1977). The hydrolysis of ferric sulphate solutions at elevated temperatures. *Canadian Metallurgical Quarterly*, *16*(1), 111-117.

Van Santen, R.A. (1984). The Ostwald step rule. *The Journal of Physical Chemistry*, *88*(24), 5768-5769.

Voigt, B., & Göbler, A. (1986). Formation of pure haematite by hydrolysis of iron (III) salt solutions under hydrothermal conditions. *Crystal Research and Technology*, *21*(9), 1177-1183.

Wayne, Shari, Pat and Mark". (2009). *Handbook for Experimenters. Stat-Ease 08.1.*

Wu, C.J.F. & Hamada, M. (2000). *Experiments: Planning, Analysis, and Parameter Design Optimization*, ISBN: 0-471-25511-4, Published 2000, Wiley.

AN ENVIRONMENTALY FRIENDLY PROCESS FOR THE FULL RECOVERY OF ZINC AND OTHER VALUABLE METALS FROM ROASTED ZINC CONCENTRATES

*Francisco Tamargo

TAM 5 S.L.
Costa Brava 24, P4 – 5B
28034 Madrid, Spain
*(*Corresponding author: tamzinco@telefonica.net)*

ABSTRACT

This paper refers to a new process for the full recovery (above 98%) of zinc and other valuable metals (Pb, Ag, Au) accompanying zinc in its concentrates while iron is separated as a marketable product free of any harmful impurities and suitable for sale to the cement industry. The process has been especially designed to achieve the following goals: maximizing the recovery of zinc and those valuable metals that are present in most zinc concentrates (lead, silver, gold); allowing the recovery of other valuable metals such as indium or germanium that could be present in some zinc concentrates; avoiding or minimizing the need for residue disposal in safe areas as it is the common practice nowadays in the zinc industry. Finally, the process achieves the previously related goals at a competitive cost, in line with the existing more efficient operations while using conventional equipment working under atmospheric conditions. The new process can be easily implemented in those smelters using the jarosite or goethite processes as well as in those smelters using direct leaching as a part of the processing facilities. A study has already been conducted on the use of jarosite in the cement manufacturing process, in the raw mix preparation as well as for a set controller. It shows that use of clean jarosite in the manufacture of ordinary Portland cement can result in improved quality of the clinker. Jarosite showed compatibility to influence the burning characteristics of the cement raw mixes. The presence of SO_3 bearing mineral phases such as natrojarosite showed the suitability of clean jarosite for use as a set controller in cement production.

KEYWORDS

New process, full recovery, zinc, iron, valuable metals, clean jarosite, cement manufacturing

INTRODUCTION

Iron is by far the most abundant impurity accompanying zinc in its concentrates, and consequently, iron residue disposal is a major environmental concern for many zinc smelters in the world. This is especially true during the last 30 years when environmental restrictions became more and more strict.

Much research work has been carried out during the last 30 years to try to find a solution for this important issue, but with little success. Only the jarofix process represents a partial solution to this problem as jarofix is stored nearby the smelters with certain limitations as a non-contaminant residue. However, it does not avoid the need to dispose of an important amount of residues every year, representing an environmental liability. Meanwhile, the number of countries limiting or banning the disposal of iron residues generated by the zinc smelters is growing, so that the solution to this problem is becoming an urgent matter right now. At the same time, the lack of appropriate available land for iron disposal is also a major concern in many cases.

On the other hand, the recovery of those valuable metals present in zinc concentrates is an important target that zinc smelters need to achieve in order to help them to be profitable.

Whatever new technology is advanced to displace the existing ones needs to provide a positive response to these important aspects of zinc practice: eliminating or mitigating the need for residue disposal, maximizing the recovery of zinc and other valuable metals present in the zinc concentrates and, finally, achieving these targets at a competitive cost in line with the current processes in use.

The process developed and patented by Tamargo (2011) and shown in this presentation provides a positive answer to the above queries as it allows the generation of a clean iron residue which can be sold to the cement industry. It also facilitates the full recovery (above 98%) of zinc, lead and silver at the same time that makes possible the recovery of indium or germanium when present in the concentrates; and makes these achievements at a competitive operating cost, in line with the jarosite + jarofix process (Tamargo, 2011).

PROCESS DESCRIPTION

The new process (the Tamzinco Process) is able to make the following achievements at a competitive cost: a very high recovery of all valuable metals present in the zinc concentrates, the generation of clean residues able to be reutilized in other industrial processes, and an efficient management of the energy resources generated by the process itself (Tamargo & Bhatter, 2015). The general process flowsheet of the Tamzinco Process is shown in Figure 1.

The process is based on Fe(III) first being reduced to Fe(II) in the solution resulting from the acid leaching stages. It is later oxidized to Fe(III) after the acidity of the solution has been neutralized with calcine, before iron precipitation as jarosite. At the same time, the iron content of the solution is adjusted, before oxidation and its precipitation as jarosite, by diluting it with zinc sulphate solution low in iron. Thus, during iron oxidation and jarosite precipitation, the working acidity is sufficiently low for the iron to precipitate as jarosite in an efficient manner. This prevents excessive iron recirculation through the leaching stages, as this could alter the nature and proper operation of the indicated operations, particularly the acid leaching stages during which a high volume of steam is consumed. This procedure makes possible, on the one hand, the neutralization of the acidity accompanying the iron in solution and the removal of impurities from solution before proceeding to jarosite precipitation. The solids generated during neutralization can be recycled to the process or transferred to an indium (In) recovery plant, if a sufficient amount of In is present in the concentrates. On the other hand, iron is precipitated as clean jarosite free of harmful impurities, able to be treated in other industrial processes such as the cement industry.

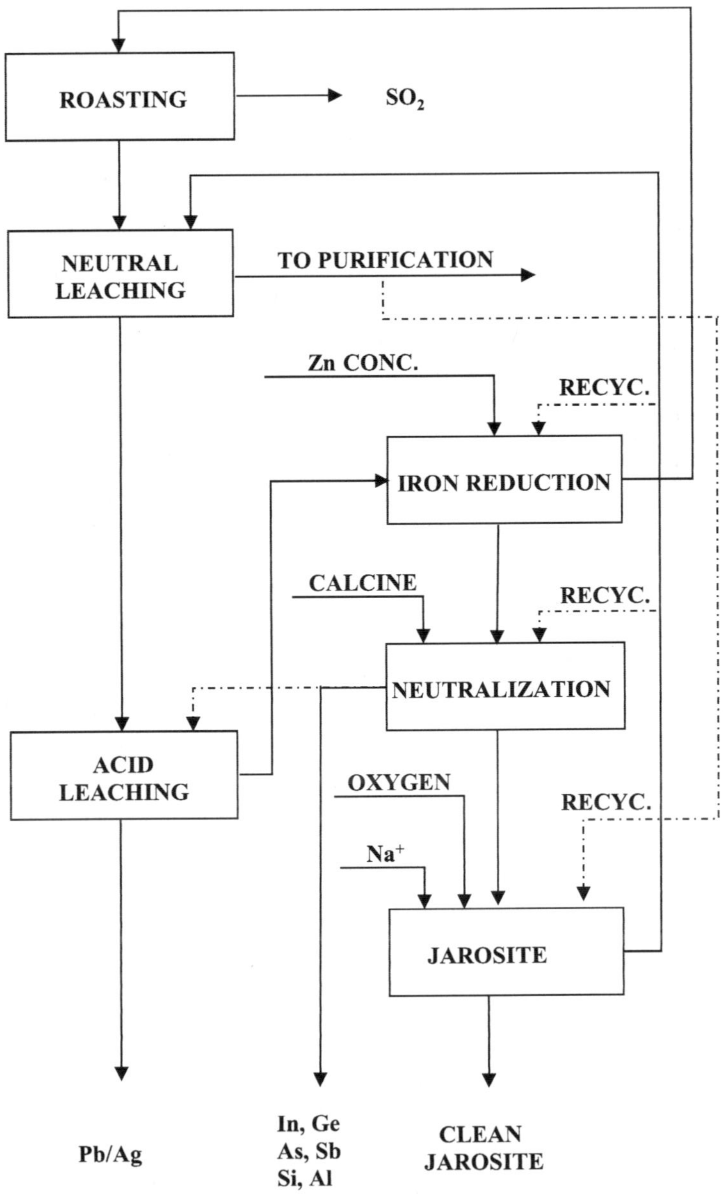

Figure 1 – General process flowsheet

With the exception of the last stage (iron oxidation and jarosite precipitation) the flowsheet is working well nowadays in some zinc smelters; thus, we can conclude that this part of the process is a proven and well known technology. The differences that characterize this new process is the last stage and the way the acidity is controlled in order to achieve efficient iron precipitation as clean jarosite without the need to feed calcine or other neutralizing agents that could affect the quality of the precipitate.

In order to minimize the acid generated during iron precipitation as jarosite it is recommended to add the required amount of sodium as caustic soda instead of sodium sulphate. In this context the main reactions taking place at this stage are:

$$4FeSO_4 + O_2 + 2H_2SO_4 = 2Fe_2(SO_4)_3 + 2H_2O \quad (1)$$

$$3Fe_2(SO_4)_3 + 2Na(OH) + 10H_2O = 2[(SO_4)_2Fe_3(OH)_6]Na + 5H_2SO_4 \quad (2)$$

According to reaction (1) 0.88 g/L of acid are consumed for each 1 g/L of iron oxidized from Fe(II) to Fe(III). According to reaction (2) 1.46 g/L of acid are generated for each 1 g/L of iron precipitated as jarosite.

Taking into account the above reactions (1) and (2) it's clear that if we make use of the possibility of adjusting the Fe(II) content in solution at the inlet of the jarosite stage, then we can fix the final acidity at the end of this stage at ideal values to allow the efficient precipitation of iron as jarosite. As an example, we can contemplate a scenario where 13.5 g/L of Fe(II) are oxidized to Fe(III) while 12 g/L of Fe(III) are precipitated as jarosite; the final acidity resulting from these reactions is 5.65 g/L.

Impurities like As, Sb, In, and Ge are precipitated in the neutralization stage so that the iron precipitated as jarosite is free from those elements as well as from Pb, Cd, or any other impurity that could prevent this product from being reutilized in cement plants.

Impurities Removal

Normal practice at existing zinc smelters is to remove impurities like As, Sb, Ge or In together with the iron residue (jarosite or goethite) and to dispose of these residues in safe ponds (after stabilization in the case of jarosite) because the high load of contained impurities does not allow application of the "dirty" jarosite in other industrial processes.

With the Tamzinco Process, impurities removal can be tailored according to the characteristics of the smelter and the type of concentrates treated, because all those impurities are contained in the neutralization cake. This cake can be sent to an ISF plant or a lead smelter if one of those plants is situated near the zinc smelter. Alternatively, the cake can be processed within zinc smelter to recover its zinc content (around 2% of the total zinc feed to the plant) and produce a much lower amount of residue where the impurities are confined (10% of the residue currently generated).

MAIN FEATURES OF THE TAMZINCO PROCESS

The main achievements achievable when applying the Tamzinco Process are highlighted as follows:

- A very high recovery rate of metals (Zn, Pb, Ag, Au), above 98% even when treating concentrates with high iron contents.
- Possibility to recover metals as indium or germanium when present in the zinc concentrates.
- At least 90 % of the iron is removed as a saleable clean jarosite.
- Operating cost in line with the jarosite + jarofix process.
- Low investment cost. In most the cases, the payback on the investment is below one year.
- Possibility to combine the Tamzinco Process with the Direct Leaching Process.
- Impurities removal can be tailored depending on the characteristics of the smelter.

EXPERIMENTAL WORK

The process block diagram of the leaching and residue treatment operations for the proposed new process are given in Figure 2. It was tested by conducting the experiments in a batch mode for each of the steps. The set up used for conducting these process steps involved glass/stainless steel beakers having capacities of 2-3 litre. The steps were Neutral Leaching, Weak Acid Leaching, Strong Acid Leaching, Acid Reduction and Pre-neutralization.

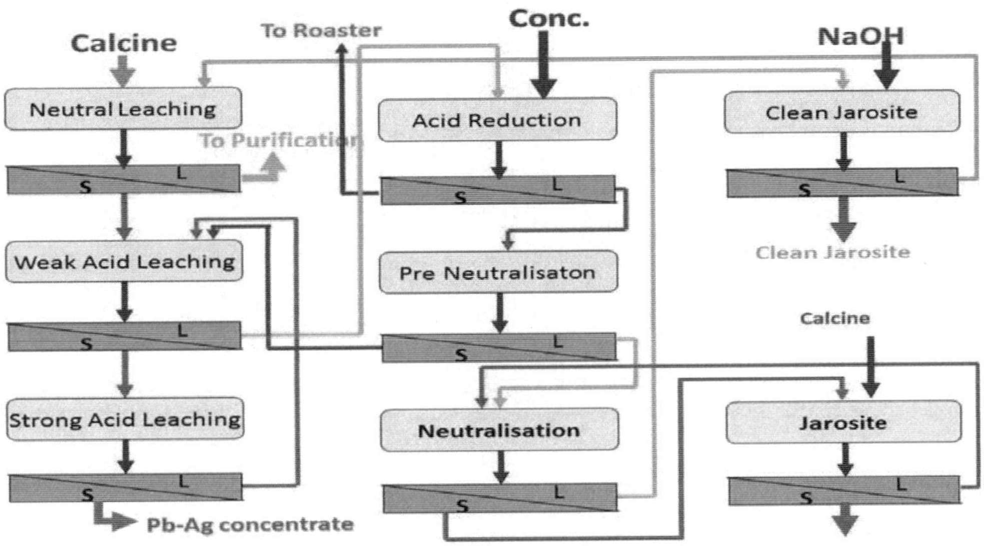

Figure 2 – Proposed process for leaching and separation of the three residues

Neutralization and clean jarosite residue separation were carried out in batches in a beaker stirred using a lab model Remi stirrer. Temperature control was done using a Yokogawa temperature controller, model UT 150. The pH in the neutral leach and neutralization steps was controlled using a Mettler, Model-M400, pH meter. The heating was done using an electrical hot plate. Oxygen sparging, for conversion of ferrous iron to ferric, was done from an oxygen cylinder. Neutral leaching was done using a mixture of spent electrolyte and jarosite overflow solutions. About 480 g of calcine was consumed in the neutral leach. Calcine was also used for the pre-neutralization and neutralization processes; 140 g of calcine was consumed in pre-neutralization and 80 g of calcine was consumed in the neutralization process. The Pb-Ag residue and the clean jarosite residues were analyzed using a Perkin Elmer Inductively Couple Plasma Optical Emission Spectrophotometer (ICP OES), Model Optima 2100DV.

The three residues, Pb-Ag cake, clean jarosite and jarosite, were settled, filtered and washed to remove the soluble salts, dried, weighed and analyzed. The weight of the residues along with the total quantity of calcine used in the neutral leach, pre-neutralization and neutralization stages are given in Table 1, along with the relevant chemical compositions. Based on the Zn content of the total calcine used and the three residues generated, the extraction of Zn was calculated to be 99.2%. It was observed that the total residue generated in the laboratory study was 26 g, of which about 66.7% was clean jarosite along with 23.8% Pb-Ag cake and 9.6% jarosite.

Table 1 - Operating parameters for the various process steps

Sl. No.	Process Stage	Operating Parameter	Operating Value during Experiment
1	Neutral Leach	pH	4.1
2	Weak Acid Leach	Final Acidity	50 g/L
		Temperature	90 °C
3	Strong Acid Leach	Acidity	120 g/L
		Temperature	95 °C
4	Acid Reduction	Acidity	30 g/L
		Temperature	85 °C
5	Pre-Neutralization	Final Acidity	<10 g/L
6	Neutralization	pH	4.2
7	Clean Jarosite	Final Acidity	10 g/L
		Temperature	95 °C

Figure 3 shows the XRD pattern of the clean jarosite. The XRD pattern of clean jarosite shows natrojarosite as the only phase. No other phase was detected.

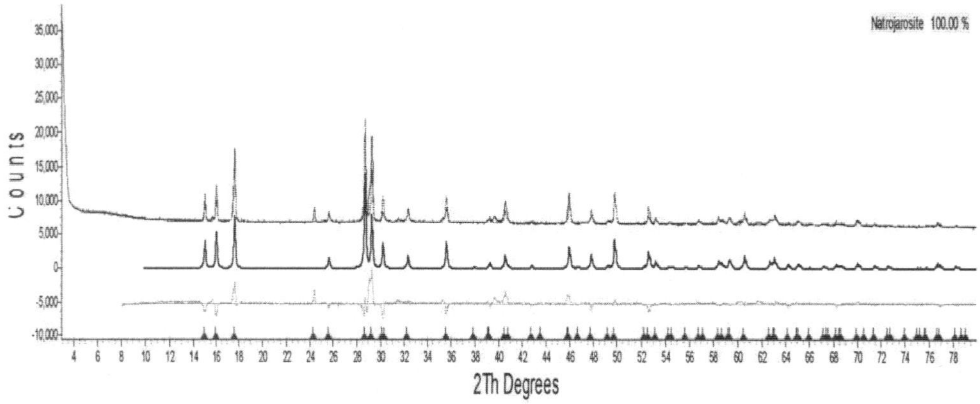

Figure 3 – X-ray diffraction pattern of the clean jarosite

Figure 4 shows pictures of the clean jarosite produced in the laboratory compared to the jarosite produced at the plant. Table 2 gives the chemical composition of both products. The difference in colour of the residues can be observed in the picture. The analytical results show that the clean jarosite contains very low levels of Pb, As, Cd, etc. in comparison to the jarosite that is being generated in the plant. This clearly shows that the clean jarosite is free of hazardous components, and can be classified as non-hazardous.

From the laboratory study it can be observed that the hazardous residue quantity could be reduced considerably. The Pb-Ag cake can be used for the recovery of Pb and Ag. The clean jarosite can find application in the cement manufacturing process.

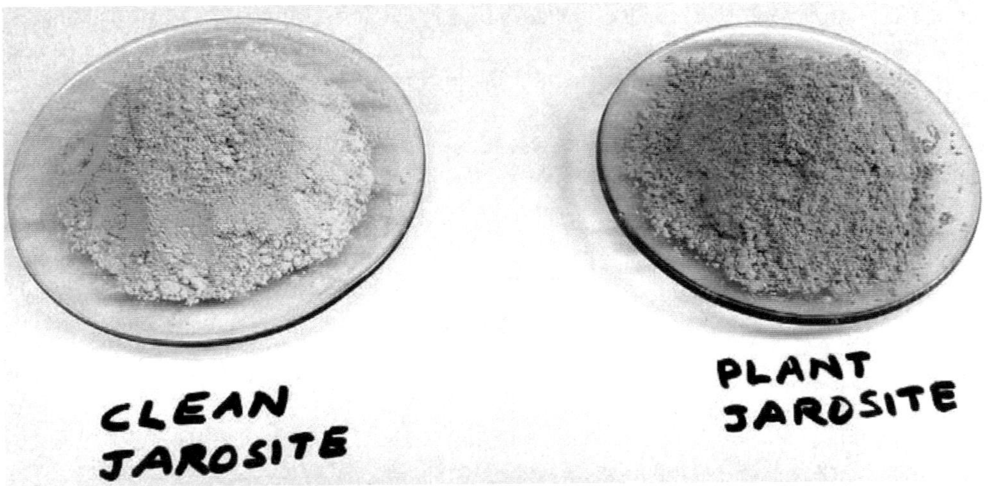

Figure 4 – Photo of the clean jarosite and the typical jarosite from the plant

Table 2 – Chemical composition of jarosite from plant and clean jarosite

Constituents	Clean Jarosite (wt.%)	Jarosite from Plant (wt.%)
Silver (as Ag)	0.0004	0.032
Aluminium (as Al)	0.0043	0.59
Arsenic (as As)	ND	0.17
Calcium (as Ca)	0.012	5.4
Copper (as Cu)	0.0009	0.45
Iron (as Fe)	33.0	21
Magnesium (as Mg)	0.019	0.06
Manganese (as Mn)	0.02	0.07
Sodium (as Na)	2.6	1.5
Lead (as Pb)	<0.1	5.2
Antimony (as Sb)	0.001	0.014
Sulphate (as $SO_4^{=}$)	40.2	38.2
Zinc (as Zn)	0.5	2.7

Studies have already been conducted on the use of the existing plant jarosite in the cement manufacturing process, as a component in the raw mix preparation as well as a set controller (Suresh, Sharma, Kuchya et al., 2012; Agarwal, Ali & Ashwani, 2014; Agarwal, Ali, Ashwani & Suresh, 2014). Those studies show that the use of jarosite in the manufacture of ordinary Portland cement (OPC) can result in improved clinker quality. Jarosite showed compatibility to influence the burning characteristics of cement raw mixes. The presence of sulphate-bearing mineral phases, such as natrojarosite, showed the suitability of jarosite for use as a set controller in cement production. Setting characteristics of a cement sample prepared using the same weight proportion of jarosite in place of mineral gypsum showed longer setting behavior compared to a control cement containing mineral gypsum. Hydration studies of a cement sample containing jarosite showed the development of hydration products similar to those in the control sample. However, jarosite utilization in cement manufacturing has not yet started, possibly because of its hazardous nature. Thus the clean jarosite, without any harmful impurities, will enable its use in the cement manufacturing process.

Implementation Stage of the New Process

BZL initiated the implementation of this new process in its Zinc Smelter at Binanipuram, India. Basic engineering, detailed engineering and preparation of equipment specifications is completed. Ordering of equipment and other service contracts was in progress until disruptions of the operations at the plant site.

CONCLUSIONS

The Tamzinco Process provides an excellent recovery of valuable metals usually present in zinc concentrates (Zn, Pb, Ag, Au) and allows the recovery of other metals like In or Ge. At the same time it produces a clean jarosite able to be sold to the cement industry, thereby avoiding or minimizing the need for residue disposal in safe areas. All of these achievements are realized at a competitive operating cost, with a low capital investment and without major technological risks. It is also posible to implement the Tamzinco Process in those plants using direct leaching as a part of the overall process. Finally, it has to be pointed out that the removal of impurities like As, Sb, Ge, etc. takes place in a separate secondary circuit with a very low interference with the main circuit.

The new process for separating iron as a non-hazardous product was tested in the laboratory by conducting each process step, from neutral leaching to residue separation, in a batch mode. Three residues were generated; viz Pb-Ag cake, clean jarosite and jarosite. The quantity of jarosite residue, which is hazardous, could be reduced to about 10% of the total residue mass. The XRD analysis indicates that the only phase formed in the clean jarosite precipitate is natrojarosite. Analysis of the chemical composition of the clean jarosite showed that As, Pb, Cd, etc. are present in very low concentrations, and hence, it will not be classified as a hazardous product. Studies with the existing plant jarosite show the potential for use in the cement manufacturing process. Hence, the clean jarosite without any harmful impurities will enable its utilization in cement manufacturing.

REFERENCES

Agarwal, S.K., Ali, M.M., & Ashwani, P. (2014). Utilization of by-product jarosite in the manufacture of ordinary portland cement. *Advances in Cement Research*, 26 (1), 41-51.

Agarwal, S.K., Ali, M.M., Ashwani P., & Suresh, K. (2014). Properties and performance of cement prepared using jarosite as mineralizer. *ALIT inform International Analytical Review*, No 3 (35), 4-13.

Suresh, K., Sharma, J.M., Kuchya, M., et.al. (2012). Jarosite waste as a possible alternative to gypsum in portland cement manufacture. *Cement International*, 10 (4), 91-97.

Tamargo, F. (2011). PCT/ES2011/070265 Madrid, Spain: Hydrometallurgical method for recovery of zinc in sulphuric medium starting from sulphidic zinc concentrates with high iron content.

Tamargo, F., & Bhatter, S. (2015). A new process for the full recovery of zinc and other valuable metals from zinc concentrates. *Pb-Zn 2015*, GDMB, Düsseldorf, Germany.

NEW HYDROMETALLURGICAL PROCESS FOR THE RECOVERY OF ZINC AND IRON FROM EAF DUST AS ZINC SULPHATE AND IRON PIGMENT

*T. Boiselle, C. Gielen, J.M. Bier, and P. Henry

HYDROMETAL SA
Rue du parc industriel, 3
4480 Engis, Belgium
*(*Corresponding author: tboiselle@hydrometal.be)*

ABSTRACT

Hydrometal is specialized in the production of zinc sulphate from complex secondary raw materials. Electric Arc Furnace (EAF) dust is one of the more interesting potential sources to produce zinc sulphate. In order to minimize waste quantities generated from EAF dust treatment, our R&D team has developed and patented a new process to recover the Fe contained in the dust as iron oxide product to be used and sold into the pigment market. The process is based on a specific reductive leaching step of the zinc ferrite contained in EAF dust, followed by a purification step to remove impurities (Ni, Cr, Cu...) and finally the controlled precipitation of the iron oxide. The aim of the iron precipitation step is to obtain the right shape and length of crystals in order to guarantee the right color and tinting strength of the pigment. The innovative aspects of the process results in the controlled oxidation of the iron to get the right crystal size and good purity material. The iron product can be valorized as red or yellow pigment often produced by the Penniman-Zoph process.

KEYWORDS

Zinc sulphate, EAF dusts, iron, waste, pigment

INTRODUCTION

A Few Words about HYDROMETAL

Hydrometal SA (Engis, Belgium) is a subsidiary of Jean Goldschmidt International SA (JGI) in Brussels. JGI is mainly involved in metal trading activities (Goldschmidt, 1995). In 1985, JGI acquired the leaching circuit of a Belgian electrolytic zinc producer located in Engis, a few kilometers south of Liège, near the river Meuse and it was decided to name this plant: SA HYDROMETAL. Hydrometal is well connected by water, rail and road from its central location in the industrial region of Liege, in the heart of Europe. The company received in 1997 an additional license to process and recycle toxic and hazardous waste. Figure 1 below shows a picture of the plant.

Figure 1 - Aerial view of Hydrometal plant in Engis, Belgium

Hydrometal's business is focused on the recycling of complex materials containing non-ferrous metals. This recycling concept helps the non-ferrous industry to keep all the valuable metals out of the landfill storage area, which is in line with the Critical Metals Recycling concept (European Commission, 2010). Feed materials for our plant mainly consist of by-products generated by the industrial activities of Hydrometal's suppliers and clients. These feeds arrive in the form of fumes, drosses, slags, sludges, dusts, ashes, cements, filter cakes, spent catalysts, complex alloys, turnings, grindings, manufacturing rejects and alike. Hydrometal treated 55000 wet metric tons of raw material in 2015.

In this paper, we will focus on Electric Arc Furnace Dusts (EAFD) generated by the steel industry. It contains mainly zinc, but it also a high iron content. Hydrometal is specialized in the production of zinc sulphate from complex secondary material. EAFD is one of the more interesting potential sources to produce zinc sulphate. In order to minimize waste quantities generated from EAF dust treatment, our R&D team has developed and patented a new process to recover the iron contained in the EAFD as an iron oxide product to be used in the pigment market.

Zinc is used predominantly in the galvanizing process for steel. The end of life galvanized steel scrap is re-melted in electric arc furnaces (EAFs). During melting most of zinc evaporates. Due to high gas flows, a large amount of iron is carried over and collected together with the condensed zinc in a bag house filter as EAFD. This dust is classified as hazardous waste, because it contains a significant amount of zinc and others leachable metals. Currently, the Waelz kiln process is commercially operated for the intermediate treatment of steelmaking dust worldwide. This process is based on carbothermic reduction and performs a simple extraction process of zinc oxide without valorizing the iron as this is mainly dumped in Europe (Chairaksa-Fujimoto & Nagasaka, 2015).

A specific flow sheet has been developed at Hydrometal to treat the EAFD by removing the undesirable components and separating the valuable products such as zinc sulphate and iron oxides pigment (Figure 2).

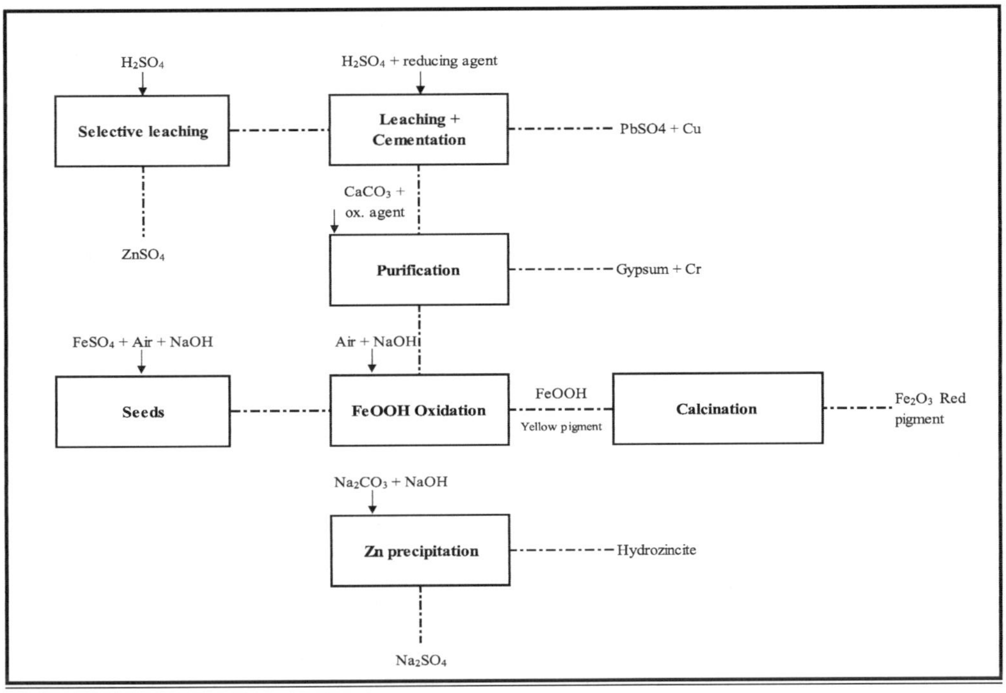

Figure 2 - Simplified flow sheet of the process

Process Description

A typical composition of the EAFD treated at Hydrometal is shown in Table 1.

Table 1 – EAFD composition

Element	Range (wt. %)
Zn	18-30
Fe	10-25
Mn	1-5
Cr	0.5-5
Cu	0.1-0.5
Ni	0-1
Pb	0.5-5
Cl	0.5-5

Leaching and Impurities Removal Steps

The first process step is the selective leaching of zinc with sulfuric acid to produce zinc sulphate which can be sold as is. Zinc contained can also be precipitated as hydrozincite, $Zn_5(CO_3)_2(OH)_6$, and transformed into zinc oxide by our mother company (SILOX).

The remaining cake is treated by a second leaching step operating under severe conditions with a specific reducing agent in order to dissolve the zinc ferrites ($ZnFe_2O_4$). The pregnant leach solution is purified by removing the copper, cadmium, nickel and part of the chromium using a cementation process. A second step of purification is then done in order to remove the remaining chromium still in solution. In this step, a controlled amount of Fe^{2+} is oxidized to Fe^{3+} which allows the co-precipitation of the chromium. The pH of the reaction is controlled by addition of $CaCO_3$ in order to get an easy filterable gypsum residue.

The purified solution, principally an Fe^{2+} sulphate solution plus the zinc sulphate from the zinc ferrites, advances to the iron oxidation stage where iron is oxidized from Fe^{2+} to Fe^{3+} and precipitates as goethite (JGI, 2005).

Valorization of the Iron as Iron Oxide Pigment

The main innovation of this iron precipitation processing route is related to the crystallization step and the growth of the crystals to get the right crystal size and good purity material. The aim of the precipitation is to obtain the right shape and length of crystals, giving the guarantee to get the right color and tinting strength for the pigment. This structure can be obtained by a controlled oxidation of the iron. Commercial quality crystals are achieved by the continuous control and amendment of the ratio between iron, which has just been oxidized, and the available surface area of crystals on which they can be precipitated. That is why oxidized Fe^{3+} species need to be precipitated onto existing nucleating seed crystals.

In order to obtain the right crystallographic structure, seeds of the same structure must be pre-synthetized. In short, a goethite structure is required to grow goethite crystals without any co-precipitation, as the solution contains others metallic elements. For clarification, it is necessary to take into account that precipitation on an isomorph surface (goethite onto goethite) requires less activation energy than a nucleation from species in solution. Therefore, the Fe^{3+} ions will precipitate preferentially onto the surface of the isomorph seeds rather than in the solution.

However, the ratio between the actual oxidised Fe^{3+} species (produced during the oxidation reaction) and the available sites for precipitation (initially provided by the seeds) at any time, but specifically in the early hours of the crystallization, is very critical for growing crystals of appropriate shape and size. Too few seed sites and the Fe^{3+} radicals will combine with each other to form nano-sized crystallites, which themselves will act as seed sites for future precipitation. Consequently, the particle size distribution of the final crystals produced will become broader thereby affecting the pigment quality.

In presence of excess of oxygen, the crystal growth is controlled by the gradual addition of soluble iron. Looking at the pigment quality, we succeeded to determine the optimum rate of reaction.

This optimum ratio and rate will guarantee the right shape and size of the crystals (as defined by Buxbaum & Pfaff, 2005), but it will not guarantee the purity of the iron oxides precipitated. For clarification, the pH needs to be accurately regulated by adding automatically the alkali reagent, which neutralizes the sulphuric acid liberated during precipitation, and maintains this value at a pre-set level depending upon which metals or salts contaminants that one wishes to avoid. Typically the pH must be above 2 to avoid the formation of natrojarosite, as is shown later in the characterization paragraph, and below 4 because of the contaminant metals present in solution. In our specific case, sodium hydroxyde has been used but other reagents can be added such as zinc oxide or ammonia.

Final Zinc Precipitation

The remaining zinc sulphate can be recycled internally or precipitated as hydrozincite and sold to nearby industries for the production of zinc oxide.

Iron Oxides Characterization

Iron oxides are well known for their pigment quality since the Palaeolithic painters. The pigment quality is mainly defined by the particle size and shape, which will determine the colour and tinting strength.

Yellow Iron Oxides

Yellow pigments were defined by Buxbaum & Pfaff (2005), as outlined in Table 2:

Table 2 – Goethite properties

Properties	
Goethite	α-FeOOH
Crystal type	Needle-shaped
Crystal length	300 - 800 nm
Length to width ratio	1.5 to 8

The desired shape and size of the crystals obtained by Hydrometal was examined under the Scanning Electron Microscopic, as illustrated in Figure 3.

Figure 3 - Example of SEM picture showing goethite produced with Hydrometal's process

The pigment's color shown in Figure 4 can be characterized by the CIELAB system.

Figure 4 - Picture of the goethite produced at Hydrometal

CIELAB is the second of two systems adopted in 1976 by "Commission Internationale de l'Eclairage" as models that better showed uniform color spacing in their values. CIELAB is a color system based on the earlier (1942) system of Richard Hunter called L, a, b. As shown in Figure 5 the central vertical axis represents lightness (signified as L*) whose values run from 0 (black) to 100 (white). On each axis the values run from positive to negative. On the a-a' axis, positive values indicate amounts of red while negative values indicate amounts of green. On the b-b' axis, yellow is positive and blue is negative (Agoston, 1979; McLaren, 1976).

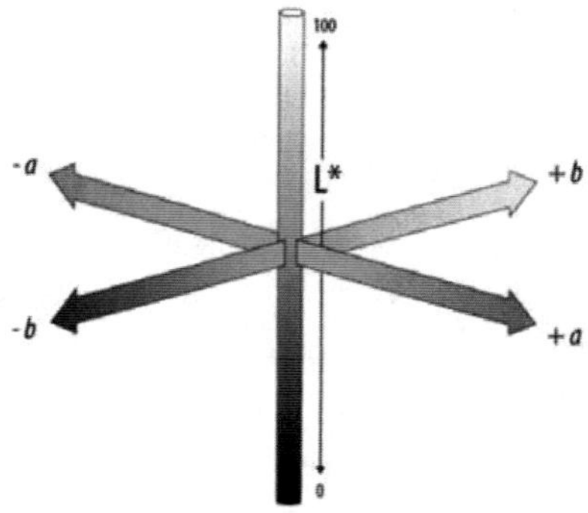

Figure 5 – CIELAB axes drawing

Results from the color evaluation of a typical Hydrometal pigment are presented in Table 3. It was found that the iron oxides produced with this new process are on target with the pigment market.

Table 3 - Example of yellow iron oxide colors produced at Hydrometal

Colour parameters	Hydrometal pigment	Market pigment
L	53	54
A	16	16
B	46	48

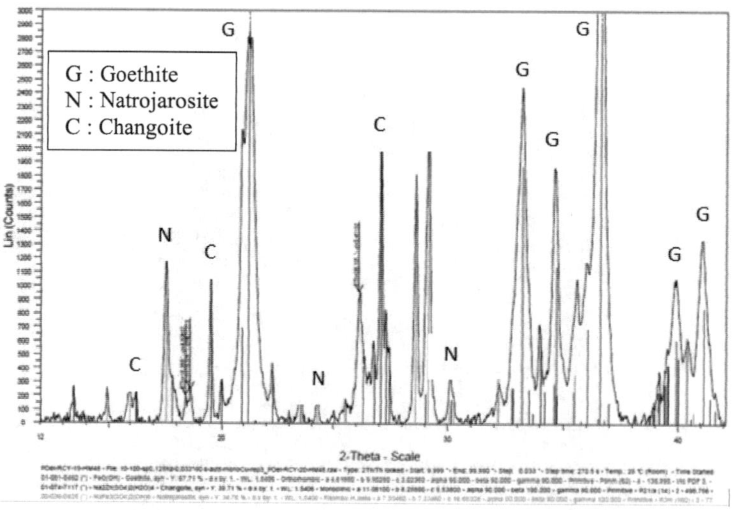

Figure 6 - XRD pattern of goethite contaminated with natrojarosite and changoite

The XRD pattern of goethite produced under not controlled oxidation conditions is presented in Figure 6. The main peaks indicate that the yellow iron oxide contains significant quantities of natrojarosite ($NaFe_3(OH)_6$ $(SO_4)_2$) and even changoite ($Na_2Zn(SO_4)_{2,4}(H_2O)$)). However, when using the Hydrometal's process we succeeded to produce pure goethite, as documented in Figure 7.

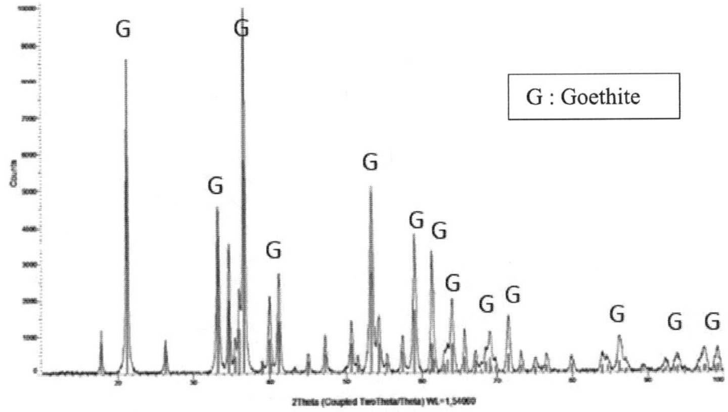

Figure 7 - XRD pattern of "pure" goethite produced at Hydrometal

Table 4 – Hydrometal iron oxide goethite composition

Element	Content (wt. %)
Fe	61
Zn	1
Mn	0.05
Cr	< 0.02
Cu	< 0.001
Ni	< 0.01
Pb	< 0.002
Cd	< 0.001
Cl	< 0.02

The Fe_2O_3 content of our yellow pigments lies between 87 and 88 wt. % corresponding to FeOOH contents of 97 to 98 wt. %.

Red Iron Oxides

When heated, goethite dehydrates and transforms to hematite. Various red shades of hematite can be prepared from the thermal decomposition of the precipitated goethite at temperatures ranging from 600 to 900 °C. Red Iron Oxide pigments are defined by Buxbaum & Pfaff (2005), as shown in Table 5

Table 5 – Hematite properties

Properties	
Hematite	α-Fe2O3
Crystal type	Hexagonal or triangular
Crystal length	100 to 1000 nm

Thanks to Hydrometal's process we succeeded to produce pure hematite from goethite, as demonstrated via XRD in Figure 8.

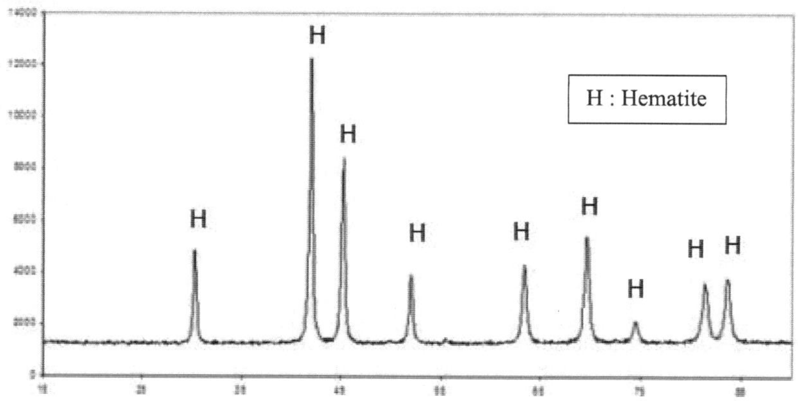

Figure 8 - XRD pattern of pure hematite produced at Hydrometal

Table 6 – Composition of the red iron oxide produced at Hydrometal and shown in Figure 9

Element	Content (wt. %)
Fe	70
Zn	1.5
Mn	0.04
Cr	< 0.002
Cu	< 0.01
Ni	< 0.01
Pb	< 0.002
Cd	< 0.001
Cl	< 0.02

Figure 9 - Example of SEM picture showing hematite produced at 800°C with Hydrometal's process

Results from the color evaluation of a typical Hydrometal pigment are presented in Table 7. For reference, the iron oxides produced are on target with the pigment market specification.

Table 7 – Colors of the red iron oxide produced at Hydrometal as shown in Figure 9

Colour parameters	Hydrometal pigment	Market pigment
L	33	32
A	33	33
B	30	30

The pigments qualities obtained with the new process developed by Hydrometal, obtained from treating hazardous waste materials, are comparable to the most widely used production methods for iron oxide pigments such as Penniman-Zoph and others (Buxbaum & Pfaff, 2005).

SUMMARY

Hydrometal has developed and patented a unique hydrometallurgical process to recover both zinc and the resulting waste containing the iron form EAFD. This becomes the ideal route to treat a variety of materials with low Zn/Fe ratio. Thanks to this approach we valorize the valuable components contained in EAFD and return products such as zinc sulphate and hydrozincite back to the zinc market and iron oxides to be sold as pigment instead of being dumped.

REFERENCES

Agoston, G. A. (1979). *Color theory and its application in art and design.* Heidelberg, Germany: Springer.

Buxbaum, G., & Pfaff, G. (2005). *Industrial Inorganic Pigment.* Germany: Wiley-VCH.

Chairaksa-Fujimoto, R. & Nagasaka, T. (2015). New Approach to Recover Metallic Zinc from EAF Dust by CaO Treatment. GDMB (Ed.), *Proceedings of Lead-Zinc Conference Pb-Zn 2015* (pp. 849-862). Düsseldorf, Germany.

European Commission (2010). *Report of the working group on defining critical raw materials.* Retrieved from http://ec.europa.eu/entreprise/policies/raw-materials/critical/index_en.htm

Goldschmidt, J. (1995) Hydrometallurgical Recovery of Non-Ferrous Metals from Secondary Sources. P.B. Queneau & R.D. Peterson (Eds.), *Third International Symposium on Recycling of metals and engineered materials* (pp. 5-12). Point clear, AL: TMS.

McLaren, K. (1976). The development of the CIE 1976 (L*a*b*) uniform colour-space and colour-difference formula. *Journal of the Society of Dyers and Colourists,* 92, 338-341.

JGI (2005). WO 2001/062991. Le Cheylas, France : European Patent Office.

INFLUENCE OF LOW IRON INPUTS ON IMPURITY CO-PRECIPITATION IN TECK'S IRON REMOVAL PROCESSES AT TRAIL OPERATIONS

M.G. Heximer, B.R. Neufeld, D.G. Reynolds, *S.J. Roach, and W.G. Van Luven

Teck Metals Ltd.
25 Aldridge Ave
PO Box 1000
Trail, British Columbia, Canada, V1R 4L8
*(*Corresponding author: steven.roach@teck.com)*

ABSTRACT

Teck Metals Ltd. operates an integrated zinc-lead complex at Trail, British Columbia, Canada. In the zinc plant, zinc concentrates are roasted to produce calcine, which is then processed in the leaching circuit for zinc recovery. Calcine is first dissolved at pH <2.0. This is followed by a precipitation stage using additional calcine that removes iron as a ferric hydroxide precipitate which is recycled to the lead operations. It is during this pH adjustment that impurities, notably arsenic, antimony and germanium, are co-precipitated with ferric hydroxide which is the primary removal phase for these elements. Recent processing conditions have resulted in lower iron inputs to the leaching circuit and this has resulted in insufficient co-precipitation of key impurities. This paper summarizes the influence of iron concentrations on impurity removal using laboratory and plant scale surveys, and emphasizes key specification limits for the primary iron removal circuit.

KEYWORDS

Iron removal, impurity co-precipitation, germanium, arsenic, antimony

INTRODUCTION

The Sulphide Leach Plant (SLP) at Teck Metals Ltd., Trail Operations produces a zinc sulphate electrolyte from three different zinc bearing streams:

1. Calcine, primarily containing zinc oxide, from the roasters
2. Zinc electrolyte from the Oxide Leach Plant (OLP) is produced by leaching zinc oxide fume from the pyrometallurgical lead smelting operations
3. Zinc slurry from the Zinc Pressure Leach (ZPL) autoclave

The feed sources are blended and introduced to the iron removal process to precipitate iron in the form of a hydrolyzed iron precipitate which is recycled to the lead smelting operation. Along with iron, impurities, namely arsenic, antimony, germanium and silica, also co-precipitate; this process acts as the primary removal step for these elements. Following iron removal, the electrolyte is further processed in the zinc dust purification circuits. Cadmium and copper are removed in the cold stage circuit via zinc dust cementation, followed by cobalt cementation in the hot stage circuit.

In 2015, impurity removal was limited in the iron removal circuit due to lower iron concentrations which resulted from a variety of circumstances, including feed sources and maintenance activities. The lack of impurity co-precipitation manifested itself as issues in the purification circuits; cadmium and cobalt re-dissolved in the cold stage and hot stage circuits, respectively, causing off-specification electrolyte. The purpose of this paper is to provide a synopsis of the issues experienced in 2015 and define the criteria that outline iron removal and the requirements for acceptable impurity co-precipitation.

Iron Removal Process in the Sulphide Leach Plant

A schematic diagram of the iron removal process is shown in Figure 1. The majority of the calcine from the roasting operations is leached at pH 1.8 in the 1st stage leach in a single reactor. The reaction involves rapid zinc oxide dissolution under exothermic conditions as shown in Equation 1.

$$ZnO + H_2SO_4 \rightarrow ZnSO_4 + H_2O \quad (1)$$

The 2nd stage precipitation involves the introduction of additional calcine to raise the pH to 3.8; this precipitates iron in the form of a hydrolyzed iron species as shown in Equation 2. Iron in the ferrous state is converted to the ferric state with the addition of air to the 2nd stage reactors.

$$Fe_2(SO_4)_3 + (3+x)\, H_2O \rightarrow Fe_2O_3 \cdot xH_2O + 3\, H_2SO_4 \quad (2)$$

The 3rd and 4th leach stages fundamentally achieve the same purpose through a total of four reactors having the capability of controlling the pH. By lowering the pH from 3.8 in the 2nd stage to 3.2, the remaining acid soluble zinc in the iron precipitate (unreacted ZnO) is dissolved before subsequent solid-liquid separation in four acid thickeners. The iron precipitate is directed through the acid thickener underflow to the lead smelting operations where it constitutes a primary feed source for the lead circuit. The acid thickener overflow is directed to the neutral leaching circuit.

The purpose of the neutral leaching circuit is to remove trace levels of iron (in the form of ferrous ions) to achieve appropriate electrolyte quality before the downstream zinc dust purification circuits. Excess calcine is introduced in a series of four reactors to raise the pH further (pH ~5.0); air is also introduced to oxidize ferrous ions to the ferric state for precipitation. Solid-liquid separation proceeds in two neutral thickeners that recycle solids containing significant levels of unreacted calcine and impurities back to the 1st stage leach for recovery. The neutral thickener overflow (<10 mg/L Fe) is then processed downstream in the zinc dust purification circuits for the removal of cadmium, copper and cobalt.

Not all of the zinc units in the calcine are in the form of zinc oxide. Zinc ferrite ($ZnFe_2O_4$) and zinc silicates (Zn_2SiO_4) are also present. A Hot Acid Leach circuit (HAL) is in place that leaches a portion of the acid thickener underflow to dissolve zinc ferrites at higher temperatures (97-98 °C) and elevated acid concentrations (60 g/L H_2SO_4) for zinc recovery. Along with the zinc pressure leach circuit, the HAL recovers iron which is a major source of ferric iron for impurity co-precipitation. Zinc in the form of zinc silicates is recovered in the 1st stage leach at pH <1.8 but is not recovered from the calcine introduced into 2nd stage precipitation.

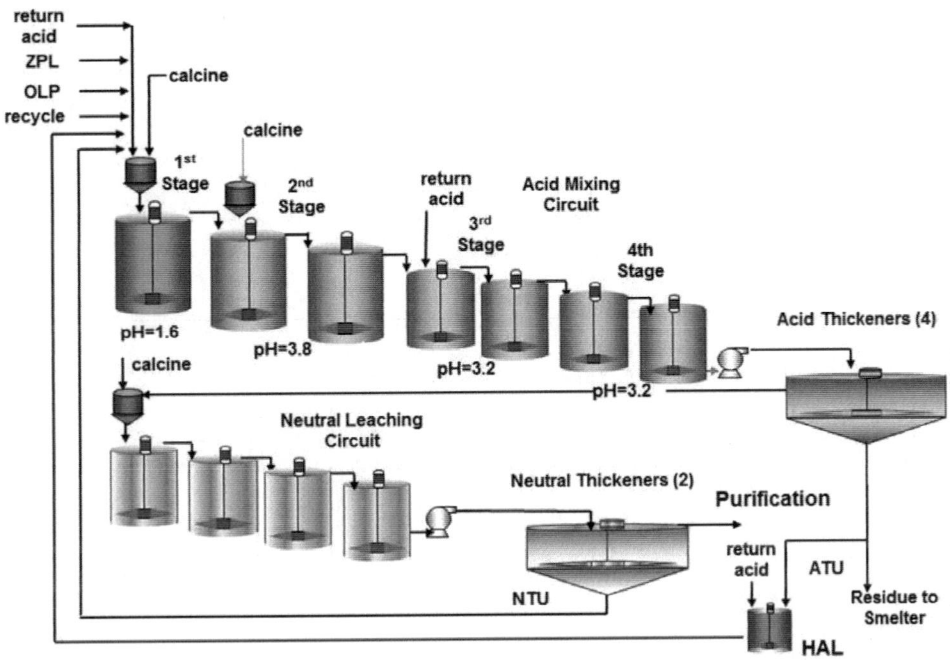

Figure 1 – Overview of the sulphide leaching circuit

The iron present in the precipitate from the 2nd stage will be described as a hydrolyzed ferric iron species for the purpose of this paper. An in-depth review of the speciation of iron precipitates in Trail is beyond the scope of this paper; numerous studies have described compounds produced under similar pH and temperature regimes (Dutrizac, 1979; Casas, Crisóstomo, & Cifuentes, 2005). In the Trail flowsheet, hydrolyzed iron precipitates are more effective for the co-precipitation of impurity elements including arsenic, antimony, germanium and silica compared with other technologies such as the jarosite process. Impurity removal may occur by adsorption or structural incorporation, but for the purposes of this paper, we will refer to the mechanism simply as "co-precipitation with iron".

GERMANIUM BREAKTHROUGH FROM THE IRON REMOVAL CIRCUITS

Low Iron Levels in 2015

Iron levels in 2015 fluctuated due to a variety of circumstances; specifically concentrate feed composition and low iron inputs from a variety of maintenance activities. Figure 2 represents the aqueous iron levels in the 1st stage in the latter half of 2015; the iron concentrations varied between 750 and 2100 mg/L. Much of the variation can be explained from specific maintenance activities in the fall of 2015. The annual shutdown of the zinc pressure leach process limited a principle source of iron to the Sulphide Leach Plant. Other maintenance activities around the HAL circuit also limited the amount of iron normally produced via leaching of zinc ferrites from a portion of the acid thickener underflow.

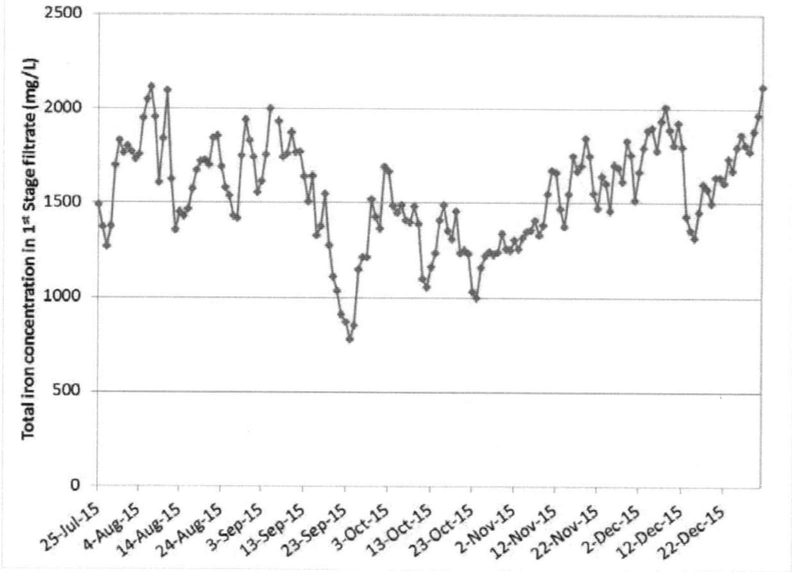

Figure 2 – Total iron concentration in the 1st stage filtrate

During the period of lower iron inputs, the downstream zinc dust purification circuits experienced significant issues. Problems manifested themselves as re-dissolution of cadmium and cobalt in the cold stage and hot stage filtration steps, respectively. The re-dissolution resulted in an exceedance of electrolyte specifications for these elements.

The link between zinc dust purification issues and low iron inputs was not readily identified in the initial stages of the investigation. It was not until the neutral thickener overflow was surveyed for non-routine elements, that germanium breakthrough from the iron removal process emerged as a key factor. Due to analytical constraints, low level germanium in the neutral thickener overflow was not reported routinely. Figure 3 presents plant data which show germanium concentrations increasing in the presence of reduced iron inputs. When germanium in the neutral thickener overflow reached concentrations greater than 1 mg/L, it was speculated that deleterious re-dissolution occurred in the purification circuits. On validation of this hypothesis, supplemental iron was introduced to the circuit via ferrous sulphate reagent addition when the iron concentration of the1st stage filtrate fell below 1200 mg/L.

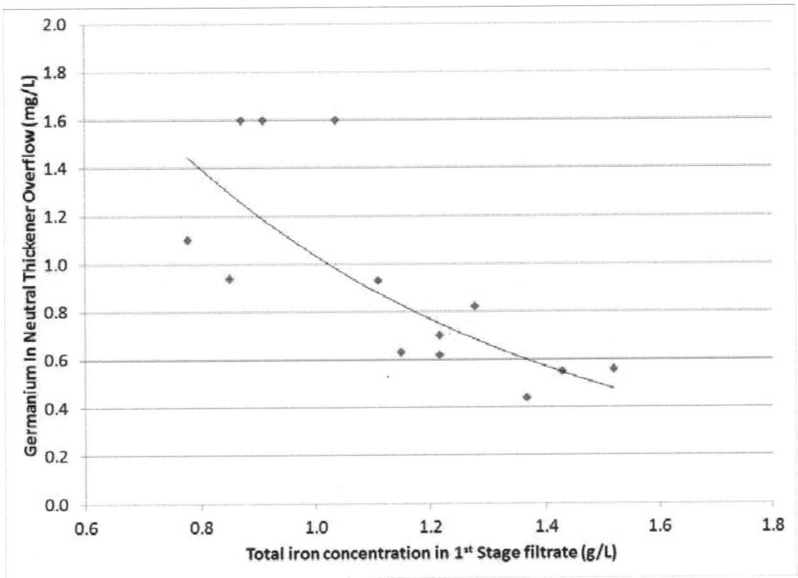

Figure 3 – Germanium in the neutral thickener overflow as a function of total iron concentration in the 1st stage filtrate

Germanium Deportment in the Iron Removal Circuit

As the 1st stage iron concentration decreases, less germanium is removed in the acid leach circuit, resulting in an increased germanium load to the neutral purification circuit. The consequence is an increase in the germanium recycle to the 1st stage leach; this exacerbates the breakthrough of germanium to the zinc dust purification circuits. Figure 4 compares two surveys across the iron purification circuit:

1. A survey conducted on September 24, 2015 with an iron concentration of 870 mg/L in the 1st stage filtrate,
2. A survey conducted on September 30, 2015 with an iron concentration of 1500 mg/L in the 1st stage filtrate.

Germanium concentrations in the neutral thickener overflow were 1.5 and 0.4 mg/L during the September 24th and 30th surveys, respectively. At higher aqueous iron concentrations, the majority of the germanium passes to the lead smelting operations via the acid thickener underflow (60% at 1500 mg/L aqueous iron compared with only 39% at 870 mg/L). The neutral leach circuit takes the brunt of germanium removal when the iron levels are lower in the 1st stage leach filtrate (36% neutral thickener underflow recycle at 1500 mg/L compared with 60% at 870 mg/L). The neutral thickener underflow with elevated germanium content is recycled back to the 1st stage leach. Sustained periods of lower iron concentrations will result in an accumulation of germanium in the iron removal circuit.

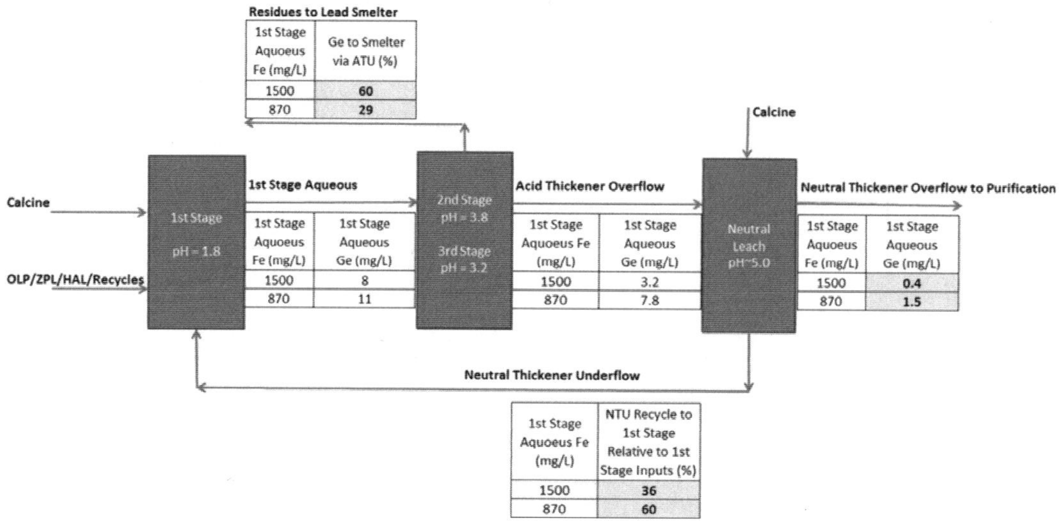

Figure 4 – Germanium deportment evaluated on two separate occasions:
1500 mg/L and 870 mg/L total iron concentration in the 1st stage leach filtrate

Laboratory tests were conducted to determine the impact of germanium levels in the cold and hot stage purification processes. The deleterious impact was substantiated with observations made in the plant. The results of the purification tests are not presented here; instead, the focus is placed on investigating impurity co-precipitation as a function of iron concentration. This was completed via laboratory tests simulating 2nd stage precipitation for the purpose of identifying the co-precipitation limitations for key impurities.

IMPURITY CO-PRECIPITATION TESTS: SIMULATING 2nd STAGE PRECIPIATION

Experimental

A set of laboratory tests was conducted to simulate the 1st stage leach using calcine and return acid from the electrolytic plant to give a pH value of 1.8. Filtrate sample assays provided the total aqueous impurities leached. Additional calcine was then added to raise the pH to 3.8, simulating 2nd stage iron precipitation. The deportment of germanium, antimony, arsenic and silica was inferred by analyzing the filtered product at different levels of ferric sulphate addition. Three separate test groups were conducted:

1. Group A – Baseline calcine leach
2. Group B – Calcine leach with additional germanium
3. Group C – Calcine leach with additional arsenic

Redundant tests were performed within each group to evaluate the repeatability.

Results

The results of the bench test program are presented in Table 1. Data are summarized as impurity concentrations in the pH 1.8 filtrate as well as the concentrations following iron precipitation at pH 3.8. Removal efficiencies are then presented in the last set of columns.

Table 1 – Laboratory test results simulating calcine leaching following by iron precipitation

Group	Concentration Prior to Iron Precipitation (mg/L)					Concentration After Iron Precipitation (mg/L)				Removal Efficiency (%)			
	Fe	Ge	Sb	As	SiO_2	Ge	Sb	As	SiO_2	Ge	Sb	As	SiO_2
A	***720***	4.2	5.0	110	4700	1.2	0.5	0.5	140	71	90	100	97
	1100	4.2	3.0	110	4600	0.48	0.5	0.5	120	89	83	100	97
	1200	4.6	5.0	100	4500	0.62	0.22	1	130	87	96	99	97
	1800	4.6	4.0	110	4500	0.22	0.16	0.5	130	95	96	100	97
B	***670***	48	7.0	100	4300	14	0.42	0.5	130	71	94	100	97
	700	47	7.0	100	4300	16	0.44	1	140	66	94	99	97
	1100	48	5.0	110	4400	6.2	0.31	1	140	87	94	99	97
	1800	48	7.0	110	4400	2.7	0.2	0.5	120	94	97	100	97
C	***780***	4.2	4.0	190	4500	1.4	0.39	21	140	67	90	89	97
	1300	4.3	4.0	200	4600	0.9	0.29	10	130	79	93	95	97
	1300	4.4	4.0	210	4900	0.9	0.17	11	130	80	96	95	97
	2000	4.2	4.0	180	4700	0.5	0.27	5	140	88	93	97	97

Germanium

Test results given in Table 1 reveal that the extent of germanium deportment to the 2nd stage precipitate is a function of the iron concentration in the 1st stage leach, thereby substantiating the plant observations. Figure 5 illustrates this trend and suggests that iron inputs are indeed an appropriate means to ensure aqueous germanium is kept below a critical limit following iron precipitation. In comparing Group A and B tests, Figure 5 also reveals that additional germanium inputs may require additional iron to precipitate the excess Ge. In Group C tests, the additional arsenic resulted in an increased germanium concentration following iron precipitation. This implies that germanium co-precipitation is inhibited to a degree as other impurity concentrations increase.

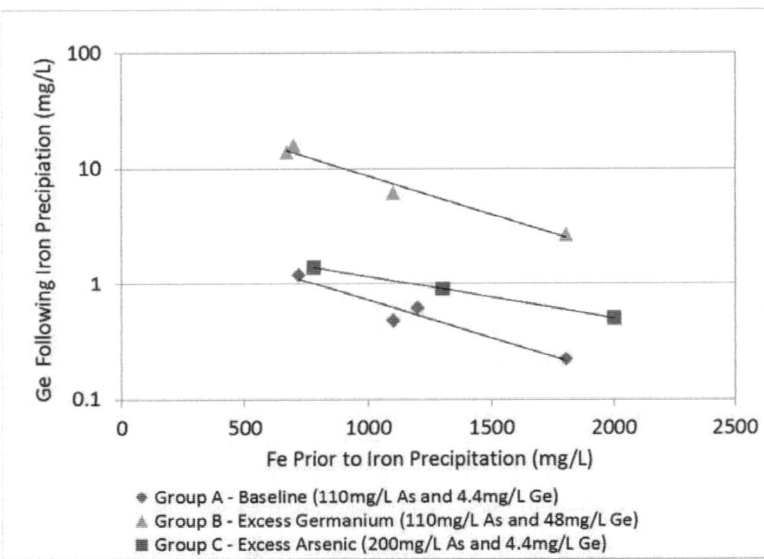

Figure 5 – Aqueous germanium concentration following iron precipitation

Arsenic, Antimony and Silica

In the Group A and B tests, arsenic removal was not influenced by the iron concentration. Group C tests, however, show that more iron was required to remove the additional arsenic. The target iron concentration of the 1st stage leach filtrate may need to be increased for elevated arsenic concentrations (Figure 6).

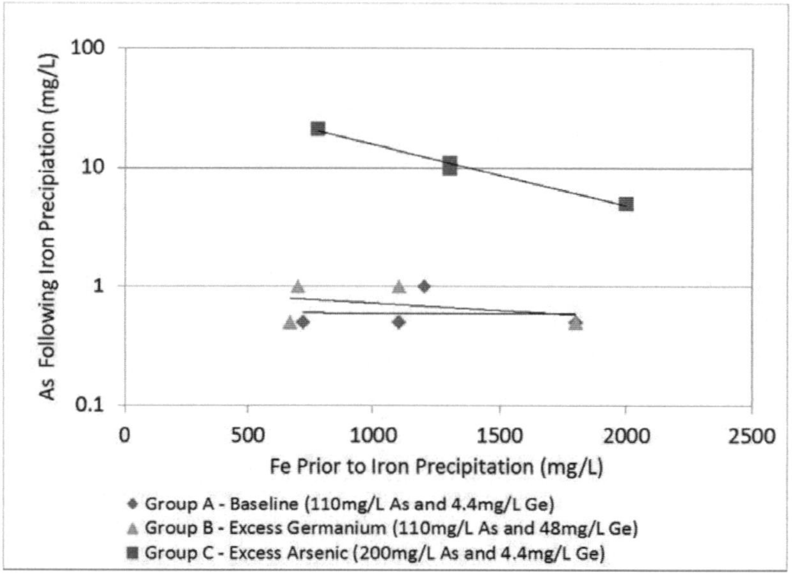

Figure 6 – Aqueous arsenic concentration following iron precipitation

Antimony is also considerably co-precipitated but to a lesser extent compared with arsenic; antimony co-precipitation lacks a clear relationship with the initial iron levels. The results for silica are also summarized; the final concentrations are near the expected solubility limit and illustrate a high removal efficiency (97%) for silica.

CONCLUSIONS

Germanium co-precipiation is more vulnerable to aqueous iron deficiencies in the 1st stage leach compared with other impurities. This is manifested as re-dissolution issues in the zinc dust purification circuits, impacting electrolyte quality. During specific maintenance activities involving key iron input sources such as the zinc pressure leach and hot acid leach circuits, supplemental iron was added to maintain a minimum of 1200 mg/L aqueous iron, to control germanium in the neutral thickener overflow below 1 mg/L. Bench tests confirmed that the sulphide leach plant is more susceptible to the breakthrough of germanium from the iron precipitation circuits operating at lower aqueous iron concentrations, compared to the behavior of arsenic, antimony and silica.

REFERENCES

Casas, J.M., Crisóstomo., & Cifuentes, L. (2005). Speciation of the Fe(II)-Fe(III)-H_2SO_4-H_2O system at 25 and 50°C. *Hydrometallurgy, 80*(4), 254–264. doi: 10.1016/j.hydromet.2005.07.012

Dutrizac, J.E. (1979). The physical chemistry of iron precipitation in the zinc industry. *Lead-Zinc-Tin '80, Conf. Proc.* Eds. J.M. Cigan, T.S. Mackey and T.J O'Keefe, TMS-AIME (pp. 532–563). Warrendale, PA.

IRON AS A LIMITING FACTOR IN THE DIRECT LEACHING OF WAELZ OXIDES

* Juergen Antrekowitsch[1], Stefan Steinlechner[2]

[1]Christian Doppler Laboratory for Optimization and Biomass Utilization in Heavy Metal Recycling
[2]Chair of Nonferrous Metallurgy
Montanuniversitaet Leoben, Franz-Josef-Straße 18, 8700 Leoben, Austria/Europe
*(*Corresponding author: Stefan.Steinlechner@unileoben.ac.at)*

ABSTRACT

Waelz oxide from steel mill dust recycling is usually sold to the primary zinc industry, where it is first charged to the roasters to eliminate the remaining fluorides and chlorides. This limits the overall amount of secondary materials which can be used because of different operating problems in the roasters. Another option, in the case where the halides are low, would be direct leaching with low or medium concentrated sulphuric acid. Unfortunately, because of the carryover of some steel mill dust, iron is also present in the Waelz oxide, reaching contents up to 5 wt.%. Since the objective is to treat the solution more or less directly by electrowinning, with only a few simple preliminary purification steps, iron in the solution is a very important factor. Various trials have been done to investigate the dissolution behavior of the iron contained in the Waelz oxide. Although the zinc extraction should reach a maximum, all other elements, especially iron, should be low. This paper describes the results of our research in this area and offers possible options for Waelz oxide utilization. There is also a focus on how other elements, present in the material, influence the dissolution behaviour of zinc and iron, and whether a loss of zinc is unavoidable, if low iron dissolution is required.

KEYWORDS

Zinc, iron, sulphuric acid, Waelz oxide, electrowinning

INTRODUCTION

One of the most important secondary zinc resources is the zinc oxide originating from steel mill dust recycling. For the treatment of high zinc containing steel mill dusts, the Waelz kiln process is the so called "best available technology" and it dominates the industry, with roughly a 90% market share. The product Waelz oxide contains mostly zinc oxide, but it also contains lead compounds and halides due to evaporation with the zinc. In addition, it contains iron and some slag compounds due to mechanical carry-over. The presence of halides limits the use of the material as an alternative concentrate for primary zinc electrowinning. Such halides can be reduced to a certain extent by washing and/or solvent extraction (see following chapter). Although lead is not troublesome and theoretically represents a second recoverable valuable metal, iron is unwanted and causes additional problems during leaching. Even though Waelz oxide is generally a highly welcome input material in primary zinc metallurgy, due to its high zinc content, it still contains halides, and in the majority of cases, the oxide must first be treated in a roasting process. Different circumstances, like accretion formation because of the lead and halogen compounds as well as unwanted cooling effects in the roaster, limit the amount of Waelz oxide which can be substituted for the primary concentrate. Another, even more limiting factor, is the allowable concentrations of chlorides and fluorides in zinc electrowinning, the last step in the primary zinc processing route (Ruetten, 2011).

Nowadays different optimized methods are available which allow a minimization of the halogen problem, and these methods make it theoretically possible to avoid roasting. In this case, it would make sense to employ a separate weak to medium acid leaching process to generate two products:

1. a zinc containing leach solution of relative high purity
2. a filter cake which contains the lead, silver and other insoluble compounds

Because of the normally high lead content of such filter cakes, they could be sold to the lead industry, while the leach solution is forwarded to purification and electrowinning. However, for such a concept, the purity of the leach solution, the zinc yield and the quality of the filter cake are highly dependent on the iron present in the Waelz oxide.

In the following chapters, the Waelz process and the characterization of the Waelz oxide are described, as this forms the basis for process development targeting direct leaching instead of roasting.

THE WAELZ PROCESS

Due to economic reasons, such as low landfill costs, relatively low zinc and lead metal prices, etc., the Waelz kiln is still the preferred route for processing EAF dust to produce a Waelz oxide and other products, as shown in Table 1. The first step in this process is an agglomeration of the raw materials with reducing agents and slag-forming additives. For the process to be economical, the minimum content of Zn in the feed material should not be less than 18 wt.%. A slight inclination of the kiln causes the material to move through the furnace very slowly while it is heated by the gas which flows countercurrent through the kiln (Ruetten, 2011; Kozlov, 2003).

A minimum temperature of 1,000°C is necessary for the total reduction of zinc and lead- containing residues. A variety of reactions take place in the slowly rotating Waelz kiln, which can be in excess of 100 m long. The reduced Zn, due to its high vapor pressure, is volatilized as Zn gas; it is then combusted to form ZnO powder that is collected in a filter. The oxidation of the zinc liberates heat that is an important part of the heat balance of the kiln. Similarly, other metals such as Pb, Cd and Ag (if present in other residues also fed to the furnace) are also volatilized and collected in the flue dust. Although the Waelz kiln process is the best available technology, improvements in the overall whole process have been introduced in the past few years. An example is the reduction and subsequent re-oxidation in the kiln of the iron which is present in the feed material and later in the Waelz slag. This oxidation reduces the coke and natural gas consumption and increases the throughput and zinc output (Ruetten, 2011; Kozlov, 2003).

The process itself starts in the upper part of the furnace, where the feed material is charged or the Waelz oxide leaves the furnace. Here, the charged material is dried at temperatures up to 150°C. In the following section, volatile matter is evolved from the coke as the temperature increases up to 500°C. At 900°C the sulfates and carbonates are decomposed and halides are volatilized. The main reaction of the zinc starts in the second half of the Waelz kiln. The final step in the kiln is the re-oxidation of the iron at the discharge head to form the slag, at temperatures up to 1120°C. Afterwards the Waelz slag is discharged and is led to a slag granulation plant for cooling. The oxygen supply for the re-oxidation of zinc and iron is delivered by leakage air at the slag discharge head in the lower part of the kiln. Due to the optimization of the process (reduction of iron), there is almost no additional energy necessary for the endothermic chemical reactions in the Waelz kiln (Ruetten, 2011; Kozlov, 2003).

Table 1 - Typical analyses of the products of a Waelz kiln (Meurer, 2000)

	Zn [%]	Pb [%]	Cd [%]	F [%]	C [%]	FeO [%]	Fe(metallic)/Fe	Basicity
Waelz slag	0.2-2.0	0.5-2	<0.01	0.1-0.2	3-8	30-50	80-90	
Basic								1.5-4.0
Acidic								0.2-0.5
Waelz oxide	55-58	7-10	0.1-0.2	0.4-0.7	0.5-1	3-5	-	
Leached oxide	60-68	9-11	0.1-0.3	0.08-0.15	1-1.5	4-7	-	

The high zinc-containing Waelz oxide leaves the furnace in the upper part of the kiln together with the off-gas (N_2, CO_2, H_2O, CO and O_2) and is collected in a dust settling chamber (Kozlov, 2003). However, unwanted compounds, especially halogens present in the EAF dust, are also introduced into the Waelz oxide. Therefore, most Waelz kilns need to have an additional soda-washing step to minimize fluoride and chloride contamination so that utilization of Waelz oxide in primary zinc metallurgy is possible without affecting the electrolysis process. Even though the washing step removes a significant amount of halides, it is generally not possible to substitute more than 15% of the primary zinc concentrate by Waelz oxide, for the above-mentioned reasons (Antrekowitsch & Griessacher, 2013).

CHARACTERIZATION OF WAELZ OXIDE

A typical Waelz oxide from a European producer was characterized to allow an initial evaluation of its behavior in a possible subsequent leaching step. Table 2 gives the chemical analysis of the washed and unwashed Waelz oxide.

Table 2 - Typical analyses of the products of a Waelz kiln (Steinlechner, 2013)

	Washed, %	Unwashed, %	Analysis method
ZnO	87.379	75.803	DIN EN ISO 11885
CuO	0.070	0.056	DIN EN ISO 11885
Pb	5.370	4.100	DIN EN ISO 11885
Cd	0.160	0.130	DIN EN ISO 11885
Fe	2.040	1.630	DIN EN ISO 11885
Ca	1.350	0.930	DIN EN ISO 11885
MgO	0.166	0.116	DIN EN ISO 11885
Al_2O_3	0.132	0.076	DIN EN ISO 11885
MnO	0.284	0.207	DIN EN ISO 11885
K	0.100	4.010	DIN EN ISO 11885
Cl	0.150	7.580	DIN 38405 D1-2
C	0.800	0.260	DIN EN ISO 15350
S	0.050	0.190	DIN EN ISO 15350
F	0.170	0.140	DIN 38405 part1
Na	0.100	2.110	DIN EN ISO 11885

From Table 2, the effect of washing with sodium carbonate can be clearly seen. Most of the chloride is sent to the wash water, while the contents of all the other components increase slightly because of the removal of Cl and other soluble elements.

The final iron content is roughly two percent, and in general, this iron value is seen as low. If this two percent iron combines with zinc oxide to form zinc ferrite and only weak acid leaching is applied afterwards to dissolve zinc, a zinc loss of about 1.3% should be anticipated. If the carry over is not as low and, for example, 5 wt.% iron is present in the Waelz oxide, a zinc loss of more than 3% is more realistic. Zinc ferrite formation will, of course, take place if a roasting process is applied before sulphuric acid leaching. However, when this Waelz oxide is subsequently mixed with iron containing primary concentrate, a hot acid leaching stage, with all its negative aspects, becomes mandatory for conventional zinc electrowinning.

The question is: if the zinc is already bound as zinc ferrite and this cannot be dissolved by a weak acid leaching step, is there an alternative route to avoid the roasting step? In this regard, Figure 1 shows an XRD pattern of the unwashed oxide.

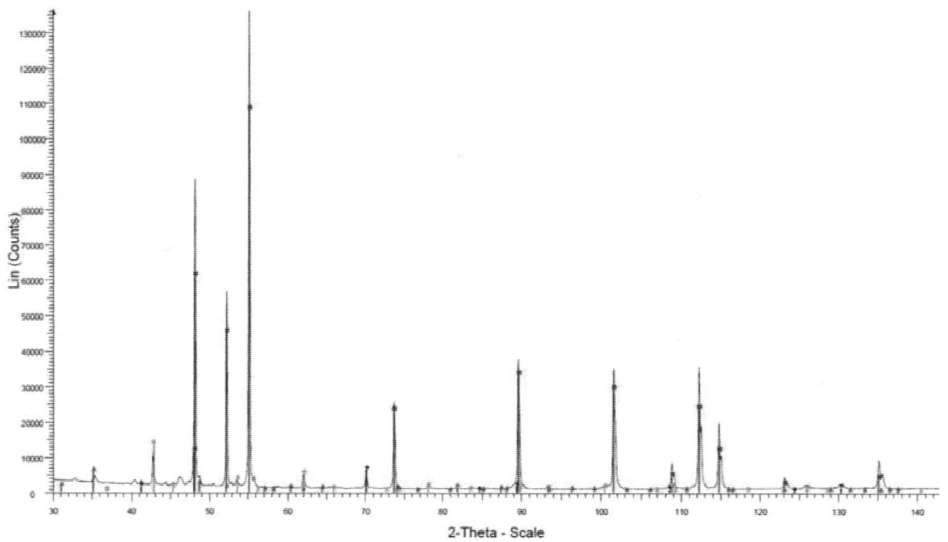

XRD of WOX (■ ZnO, ◆ NaCl, ● KCl, ▲ $ZnPbF_6$, ⌶ $ZnFe_2O_4$)

Figure 1 – X-ray pattern of unwashed Waelz oxide with major detectable components

Even though there are low amounts of iron present, a zinc ferrite phase could not be detected. A quantitative statement cannot be given, but the results give evidence that a certain amount of zinc ferrite must be already present in the Waelz oxide. Furthermore, scanning electron microscopy (SEM) with EDX-based mapping (energy dispersive) was used to detect zinc-iron compounds more clearly. The results are shown in Figure 2. In comparison with the XRD analysis, the mapping gives a much more detailed view of the element correlations and, therefore, of possible compounds in the material. An area where a high number of particles with different morphologies occurred was scanned to ensure a representative result. Iron, zinc, oxygen and other slag forming elements were chosen and their mappings were investigated to identify possible correlations. The first picture of Figure 2 shows the scanned grains and gives an impression of their morphology.

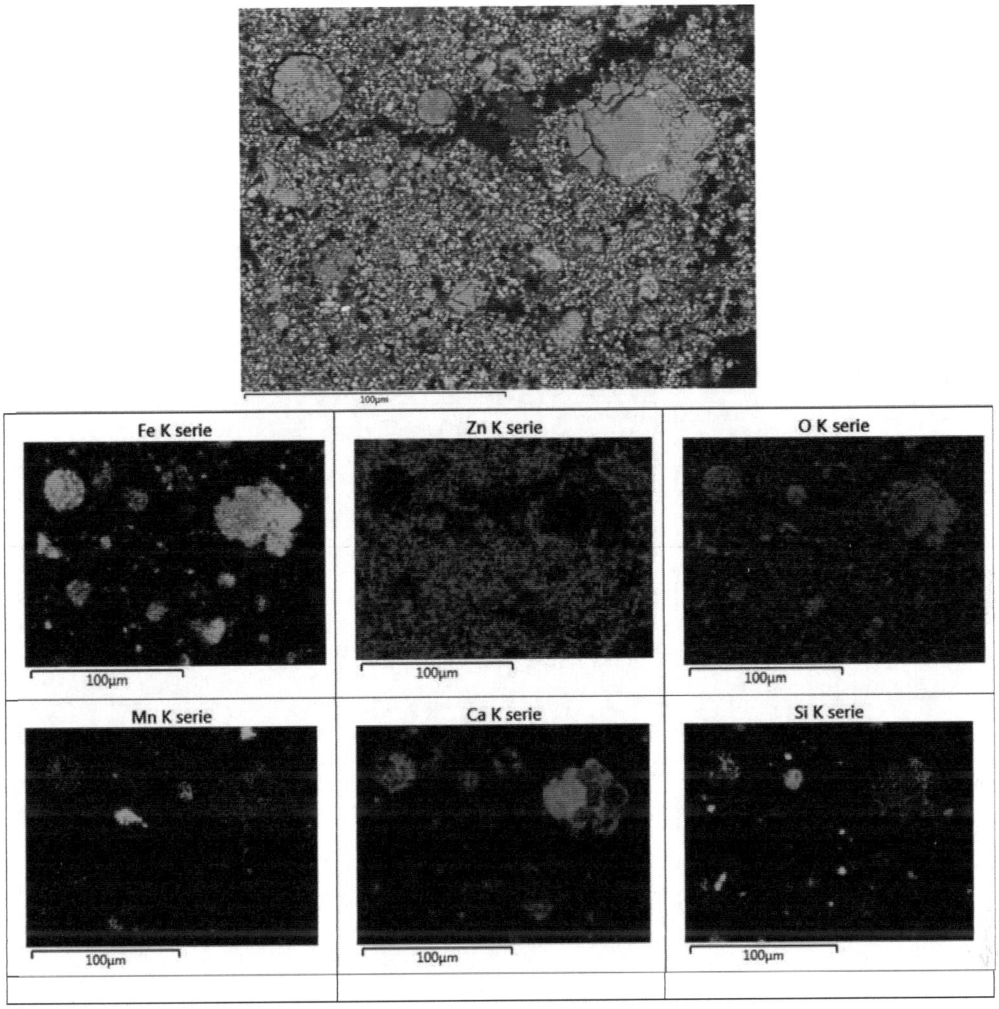

Figure 2 – Scanning electron microscopy and mappings (EDX, energy dispersive) of soda washed Waelz oxide

The mappings for iron and zinc clearly underline that most of the iron is not associated to zinc. Only a few particles show a correlation with zinc and therefore represent the formation of zinc ferrite. The majority of iron is associated with oxygen, and this indicates carried over iron oxides coming from the Waelz kiln. Different forms of appearance are possible:

- hematite (Fe_2O_3): fully oxidized iron, due to the post combustion at the end of the Waelz kiln
- magnetite (Fe_3O_4): partly reduced iron oxide or not fully re-oxidized iron
- wustite (FeO): partly reduced iron oxide or not fully re-oxidized iron (very common in typical Waelz kiln slags)

According to the mappings of manganese, calcium and silicon, most of the iron is associated with the slag particles, and especially with calcium. This is important information because such slag formation may be caused by sintering in the Waelz kiln, and this would influence the solubility of iron in subsequent leaching steps.

In general, bivalent iron is easier to dissolve than trivalent iron. When different slags from the Waelz kiln process where investigated, the bivalent form was by far the dominant occurrence of iron. Because iron finds its way into the Waelz oxide via mechanical carryover, it can be proposed that at the moment of liberation from the slag, iron appears mainly in its bivalent form. However, the upper atmosphere of the Waelz kiln is highly oxidizing at temperatures of 1100-1200 °C. So it is very likely that the iron oxidizes to hematite, if the particles are not embedded in larger grains which could inhibit the access of oxygen to the relevant surface.

The role of slag compounds within the above described processes is difficult to describe. If slag formation takes place because of sintering processes, also under oxidizing conditions, iron could stay in its bivalent forming CaO-SiO_2-Fe_xO-slag phases.

Summarizing the above findings, it is very likely that most of iron is not associated with zinc. Far more iron could be found either in the form of hematite or bound into slag phases in different degrees of oxidation. In both cases, the solubility of iron in a possible subsequent weak acid leach with sulphuric acid should be low. As a consequence, in such a leaching step, zinc losses should be low, and iron contamination of the resulting leach liquor should be low as well. Nevertheless, these theoretical findings must be verified by leaching trials which are described in the following chapter.

LEACHING TRIALS WITH SULPHURIC ACID

The leachability of zinc oxides and iron oxides in sulphuric acid has been investigated in various studies and results have been published several times for different materials. From these studies, the necessary acid concentration for the dissolution of oxides and zinc ferrites is well known, and therefore, there was no need for this topic to be part of this study. The intention of the executed trials was to verify the above assumptions for the specifically investigated Waelz oxides and to apply solutions which are usually available in primary zinc electrowinning.

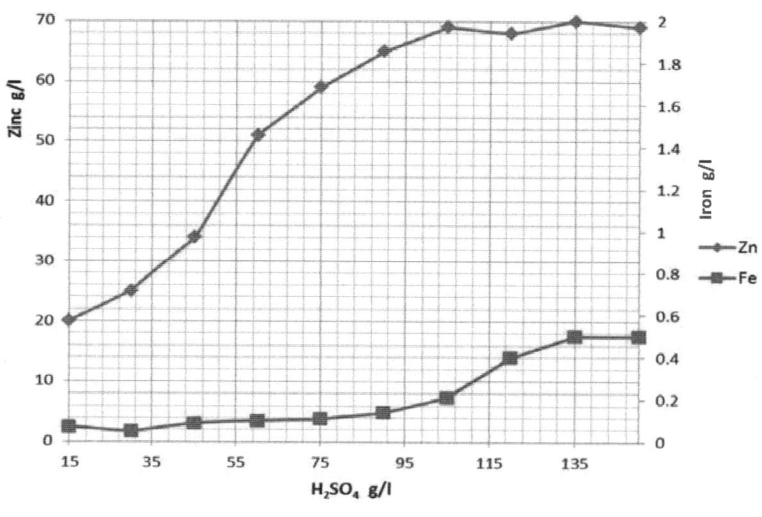

Figure 3 – Iron and zinc concentrations of the leach liquor after leaching

Different sulphuric acid concentrations from 15 g/L to 150 g/L were used to leach 100 g of the above described Waelz oxide. The leaching time was fixed with 60 minutes at a constant temperature of 60°C. A magnetic mixer was used to keep the solids in suspension. The solutions were filtered and the filtrate was analyzed for their iron and zinc values. Figure 3 shows the results of the experiments summarized in one diagram. Each point in the diagram is the average value from three identical trials.

The diagram shows high zinc yields, if the sulphuric acid concentration is 95 g/L or higher. At 95 g/L H_2SO_4, the iron dissolution is still very low. At higher sulphuric acid concentrations, the dissolved iron increases but does not reach values above 0.5 g/L, which corresponds to about 25% dissolution. This comparatively low iron value, even at high acid concentrations, confirms the assumptions in the previous section on the characterization of Waelz oxide; namely, that the iron solubility is low and does not affect the zinc yield in a substantial way. Of course, these results have to be verified for other Waelz oxides because, as described above, the behavior of iron depends on the operating conditions, possible agglomeration or sintering processes and the absolute iron content of the product.

Assuming that the fluoride content can be reduced to lower values, and different options have been studied and developed (Antrekowitsch, 2004), the proposed concept for avoiding the roasting step in primary zinc processing could be realized, as shown in Figure 4.

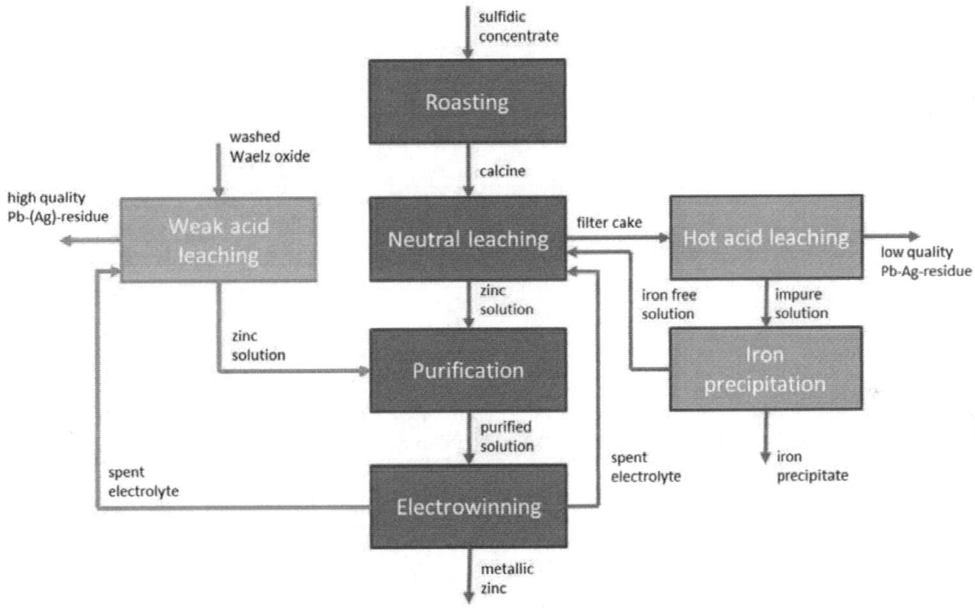

Figure 4 – Hydrometallurgical zinc processing route with separate Waelz oxide treatment

The investigated behavior of the iron compounds and the realization of sufficiently low halide content, either by soda washing or by an additional step during leaching, allows the described processing scheme to exist relatively independent of the conventional hydrometallurgical zinc processing route. The avoidance of the roasting step for washed, high quality Waelz oxides offers the following advantages:

- no limitation on the amount of secondary materials used because of roasting constraints
- additional saleable lead-product (depending on the Waelz kiln feed; silver recovery is also possible)
- higher overall value of the Waelz oxide

- no iron precipitation; iron reports with lead into lead industry and ends up in the slag
- better control of leaching conditions, optimized for the Waelz oxide
- possible additional fluoride removal within the separate leaching step

The different advantages should allow a more economical and far more flexible process, which could even pay for an additional fluoride removal step (if necessary). In such a scenario, a zinc winning process using only secondary zinc oxides, without primary concentrates, is possible (see Figure 5). Such an approach offers the additional advantage of a "zero waste" process, if the low amounts of residue from the purification step are not considered. This approach would lead to a completely new situation for primary zinc producers, similar to the recent introduction of solvent extraction in combination with Waelz kilns (Moats, Cashwell & Staley, 2014).

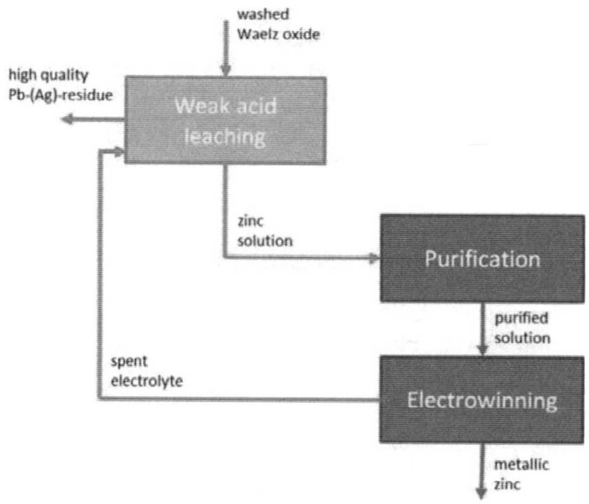

Figure 5 – Simplified concept for zinc production from only secondary input materials

SUMMARY

The behaviour of iron and its compounds in Waelz oxides during sulphuric acid leaching has been investigated. The characterization showed that most of the iron is not associated with zinc in the form of zinc ferrite. Due to the processing conditions in the kiln, the iron in the Waelz oxide is either bound to slag components or occurs in the trivalent form as hematite. Both options support the conclusion that iron is not easy soluble in weak to medium acid leaching solutions. These assumptions were widely verified by leaching trials. The proposed concept where a separate leaching operation, avoiding the roasting step, occurs within a conventional hydrometallurgical primary zinc process offers several advantages. These advantages start with greater flexibility and continue up to better economics due to the production of a marketable lead product. However, such an alternative could only be realized only if a second crucial factor, i.e. sufficiently low halide content, especially fluoride, is guaranteed. Some of the currently produced washed Waelz oxides already fulfil these requirements. Also, new ideas on fluoride removal are continuously being developed and may allow such concepts to be applied to a wider variety of Waelz oxides.

The iron content of the Waelz oxide does not seem to be a limiting factor for the direct leaching of such oxides, as demonstrated in this study. Nevertheless, this case has to be verified for oxides from other Waelz kiln operations.

ACKNOWLEDGEMENTS

The authors want to thank the Christian Doppler Association, the Austrian Research Promotion Agency (FFG) and the Federal Ministry of Science, Research and Economy (BMWFW) for their financial support.

REFERENCES

Antrekowitsch, J. & Griessacher, T. (2009). The current status of recycling of electric arc furnace dusts and perspectives for the future. *Proc.: MicroCad*, Miskolc, 27 – 38.

Antrekowitsch, J. (2004). Aufarbeitung von Stahlwerksstäuben unter besonderer Berücksichtigung der Halogenproblematik, *Dissertation*, Leoben.

Kozlov, P.A. (2003). The Waelz Process. *Ore and metals - publishing house*, Moskau, Russia.

Meurer, U. (2000). Gewinnung von Zinkoxid aus sekundären Rohstoffen - Neue Entwicklungen im Wälzprozess, *Intensivierung metallurgischer Prozesse*, GDMB, Clausthal-Zellerfeld, Germany, 183-196.

Moats, M.S., Cashwell W.J. & Staley, A. (2014). A Review Of The Modified Zincex Process From Skorpion to Horsehead. *Proc. Hydrometallurgy 2014*, Vol II, Victoria, 367-375

Rütten, J. (2011). Various Concepts for the Recycling of EAFD and Dust from Integrated Steel Mills, *3rd Seminar, Networking between Steel and Zinc*, GDMB, Leoben.

Steinlechner, St. (2013). Amelioration and market strategies for zinc oxide with focus on secondary sources, *Dissertation*, Leoben.

EVALUATING THE EFFECT OF SEEDING ON ZINC RECOVERY THROUGH THE ZINCOR IRON REMOVAL CIRCUIT

Diane Taggart and *Jochen Petersen

Department of Chemical Engineering, University of Cape Town,
Private Bag X6, Rondebosch 7701, Cape Town, South Africa
*(*Corresponding author: Jochen.petersen@uct.ac.za)*

ABSTRACT

At an average of 90.3% (for the period of 1 January 2009 to 30 September 2011), Zincor's zinc recovery was well below the industry standard of 96%. The losses through the residues, in particular the iron residue, were the biggest contributors to the overall zinc loss. It was believed that this loss could be decreased with the implementation of seed recycling. A plant trial was run to test the implementation of a seed recycle. The rate of seed addition through recycle was increased over a 4-week period. The presence of seed decreased the insoluble zinc loss by providing additional surface area for iron to precipitate onto, and reducing coating of unleached neutralising agent. The soluble zinc loss was also decreased at low and moderate seed recycle volumes. The improvement was attributed to agglomeration growth of the recycled particles, which improves the dewatering characteristics of the precipitate. In contrast, the soluble zinc loss increased at high seed recycle rates. This result was attributed to particle crowding and a low wash ratio.

KEYWORDS

Zinc processing, iron precipitation, seed recycle

INTRODUCTION

The Zincor refinery, which closed in December 2011, was situated just outside of Springs in the Gauteng Province of South Africa and employed the conventional Roast-Leach-Electrowin process, which accounts for 75 to 80% of the annual primary zinc production in the world. The refinery fed up to 220 000 t/a of zinc sulphide concentrate. The major portion of the raw material was sourced in Southern Africa from the Rosh Pinah and Black Mountain mines. Between 20 000 to 40 000 t/a of material was imported to make up the deficit and to improve the grade of the feed material.

The iron contained in the zinc concentrates, at an average of 6.03% (for the period of 1 January 2009 to 30 September 2011), resulted in the formation of zinc ferrite during roasting, which is insoluble under atmospheric leach conditions. At Zincor, most of the zinc ferrite was dissolved in the residue treatment plant at elevated temperature and high acid strength. The leach liquor from this aggressive leach contained between 15 to 30 g/L of iron, mostly in the ferric form, and was known as the hot iron solution (HIS). The presence of iron in the liquor had an adverse effect on the zinc operation, particularly on the current efficiency in the electrowinning circuit. Therefore the iron needed complete removal.

Iron was removed from the leach liquor through the Zincor Iron Removal Process. This process was similar to the so-called Para-goethite process and employed a unique pH profile with an acid wash in the second reactor. This stage-wise precipitation distinguished the Zincor process from all others.

The Iron Removal process consisted of six 60 m^3 stirred reactors in series with an overflow between each tank. The iron was removed by contacting the HIS with neutralising agent to effect precipitation. Zincor utilised a combination of roaster calcine and oxide fume ('DRC Oxide'), which originated from a copper-cobalt smelting operation in the Democratic Republic of Congo (DRC), for neutralisation. The use of DRC Oxide as a supplement to roaster calcine was implemented during a recovery improvement initiative. The trial resulted in a recovery improvement from 90.3% to 92.6% in the latter part of 2007 (Wyethe et al., 2008). The positive results prompted a number of attempts to increase the DRC Oxide addition in the hope of eventually replacing calcine. However, poor filtration was experienced each time the amount of oxides was significantly more than the calcine. It appeared that the insoluble zinc ferrite complex contained in the calcine acted as a form of iron seed that was essential to achieve acceptable settling and filtration. Since the iron content in the DRC Oxide was very low, excessive use of it depleted the 'seed' necessary for the formation of a good quality precipitate.

The two neutralising agents were fed at a constant rate, while the HIS addition rate was automatically controlled to maintain a slurry pH of 3.2 in the first reactor. Between 85 and 90% of the zinc in the neutralising agents was dissolved and up to 80% of the iron was precipitated in the first reactor (Meyer *et al.*, 1996). An acid wash was implemented in the second reactor by adding more HIS to reduce the pH to 3. The purpose of the acid wash was to dissolve any unleached or entrained zinc-rich neutralising agent and leach any zinc sulphates that may have formed. Additionally, the iron provided by the redissolution of ferrihydrite, together with that contained in the HIS, provided the "cement" required for the precipitation reaction. The temperature was maintained at 65°C. The total residence time was approximately six hours, or one hour per reactor. The slurry from the Iron Removal Stage was thickened and filtered in order to recover the zinc-rich filtrate. The filtrate was returned to the circuit and the cake, known as iron (Fe) residue, was re-pulped with water and sent to the residue disposal site where it was landfilled.

Claassen (2002) proposed that the iron was removed according to Reactions 1, 2 and 3:

$$4Fe_2(SO_4)_3 + ZnO(s) + 13H_2O \rightarrow Fe_8O_8(OH)_6SO_4(s) + ZnSO_4 + 10H_2SO_4 \quad (1)$$

$$3Fe_2(SO_4)_3 + PbSO_4(s) + ZnO(s) + 11H_2O \rightarrow 2Pb_{0.5}Fe_3(SO_4)_2(OH)_6(s) + ZnSO_4 + 5H_2SO_4 \quad (2)$$

$$5Fe_2(SO_4)_3 + ZnO(s) + 23H_2O \rightarrow 2[Fe_5O_7(OH).4H_2O](s) + ZnSO_4 + 14H_2SO_4 \quad (3)$$

One of the biggest restrictions of the conventional iron removal processes is the use of zinc-rich neutralising agents. This was also true of the Zincor Iron Removal Stage. Between 10 and 20% of the insoluble zinc losses was attributed to the presence of insoluble zinc ferrite in the fresh calcine used for neutralisation, while the remainder was attributed to the encapsulation of unleached zinc-rich neutralising agents with iron precipitate (Claassen et al., 2003). The industry standard in terms of zinc recovery is 96% (Claassen et al., 2002). Due to limitations, particularly with the iron removal technology in use, Zincor's maximum achievable zinc recovery was estimated to be 94%. The average recovery achieved during the period of 1 January 2009 to 30 September 2011 was 90.3%, leaving much scope for improvement. At 1.6%, the insoluble zinc loss through the Fe residue was the second biggest contributor to the overall zinc losses. A reduction of 1% in the insoluble zinc in the iron residue was equivalent to an overall zinc recovery improvement of 0.25% (Wyethe et al., 2008). The improvement was to be achieved by focusing on the implementation of seeding.

One way of potentially reducing these losses is to consider alternative neutralising agents that contain no zinc (Claassen, 2002). Claassen (2005) showed that the use of chemically precipitated limestone for neutralisation achieved very good zinc recoveries, but resulted in a large volume of gypsum, with the residue volume being almost double that of the typical iron residue. Claassen (2005) further confirmed that the presence of seed provides additional surface area, so that there are fewer instances of coating of unleached neutralising agent by the precipitate. The presence of seed also promotes both the nucleation and growth processes (Sinclair, 2005). Claassen (2005) found that the presence of even a small amount of seed (of up to 50 g/L) improved the specific filter resistance by up to 80%, due to a reduction in the number of particles, an increase in the overall particle size of the precipitate and a decrease in the surface area. Hove (2008) showed that the presence of seed improved the settling rates of the precipitate. He attributed the improvement in dewatering characteristics of the precipitate to increased size and density of the precipitate particles.

The present study describes the results of a full scale plant trial to implement recycling a portion of the precipitate from the last tank in the train to the first to act as seed for the process. Some laboratory work was carried out prior to the plant trial, and this evaluated alternative neutralising agents as well as the implementation of a seed recycle in a bench-scale mini-plant, mimicking the full-scale operation (Taggart, 2016). While neutralisation with limestone resulted in a big decrease in the insoluble portion of the zinc losses (in the solids), there was a significant increase in the soluble zinc losses (in pore solution). Neutralisation with calcium-based reagents typically results in the formation of a very fine and voluminous precipitate, with poor settling characteristics and high soluble zinc loss through entrainment. DRC Oxide was still identified as the most suitable alternative to calcine. Implementing a seed recycle did not result in the expected decrease in the insoluble zinc loss in the mini-plant. This was attributed to the effective decrease in overall residence time as a result of the seed recycle. Since the laboratory experiments were terminated before reaching steady state, an initial increase was seen in the solids volume in the train as a result of the seed recycle. This had a negative impact on the wash ratio and resulted in an increase in the soluble zinc loss, so that the benefits of the seed recycle were not fully realised.

EXPERIMENTAL PROCEDURE

The plant trial was carried out in the Zincor Iron Removal Stage. Figure 1 is a schematic diagram of the plant operation and indicates the seed recycle, as well as the sampling points.

The plant trial focused on evaluating the impact of a seed recycle together with the combination of roaster calcine and DRC oxide already in use, as switching to limestone would have required a number of modifications on the plant. Seeding was achieved by recycling solids from the pump suction of the thickener underflow via a pipeline into the first reactor. There was a screen box in the pump suction line to remove grit and prevent choking. An automatic valve was installed in this line to regulate the flow rate of the seed recycle to a desired set-point. The line was flushed hourly to prevent solids build-up.

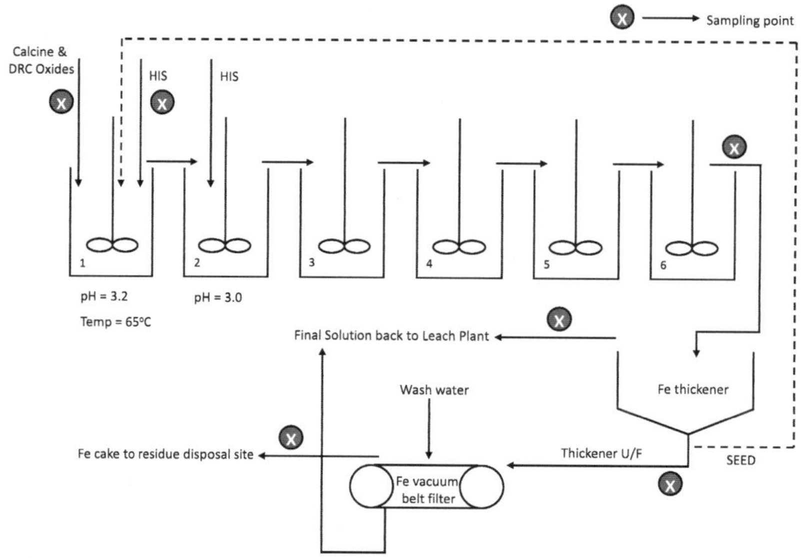

Figure 1 – Experimental set-up for the plant trial indicating the seed recycle and sampling points

The seeding program was run over a period of approximately four weeks. No seed was added during the first week in order to establish a base case. Seeding was implemented at the end of the first week. 'Low', 'medium' and 'high' scenarios were set for the seed recycle volume, with the 'high' scenario (of 6 m^3/h) being equivalent to approximately 50% of the volume of the thickener underflow. The plant was run on each set-point for approximately one week. At the end of the week the seeding set-point would be increased to the next value. No sampling was done for at least 24 hours after a change was made. The test work had to be aborted on the third day of running at 6 m^3/h seed recycle, due to a decrease in the settling and solids build-up in the iron thickener. Zincor operated a three-shift system. Samples of the feed and exit streams were taken on each shift and a composite for the 24-hour period was submitted for chemical, SEM and PSD analyses. The volumes, densities and chemical analyses were used to perform a mass balance. A settling test was also carried out on a thickener feed sample once per shift every day during the plant trial.

RESULTS & DISCUSSION

Outcome of the Mass Balance

The results from the mass balance are shown in Table 1 and Figure 2. Table 1 indicates that the overall zinc recovery shows an improvement over the 4-week period from approximately 90 to 93%. Both Table 1 and Figure 2 show that the soluble loss decreases from 1.4% to a minimum of approximately 0.8% at a seed recycle of 2 m^3/h. The improvement may be attributed to agglomeration growth, which is promoted by the presence of seed (Lewis, 2009; Claassen, 2005). A sharp increase is seen in the soluble zinc loss as the seed recycle is increased to 4 and 6 m^3/h. The maximum soluble loss is between 1.4 and 1.6% zinc. This observation may be attributed to the accompanying increase in the solids loading (from between 15 and 25 m^3 in the first two weeks of the trial to between 30 and 35 m^3 at the higher seed recycle volumes in the latter part of the trial), which results in a decrease in the wash ratio (from approximately 2 to below 1 in the last week of the trial). Wash ratio is an important parameter to minimise zinc losses through entrainment and needs to be adjusted accordingly to accommodate the additional residue volume as a result of the seed recycle. Unfortunately, an increase in wash water addition was often limited by the plant solution inventory.

Table 1 – Summary of plant parameters and mass balance results from the plant trials (Sept to Oct 2011)

PARAMETERS:	27-Sep-11	30-Sep-11	1-Oct-11	2-Oct-11	5-Oct-11	6-Oct-11	7-Oct-11	8-Oct-11
SEED (m^3/hr)	0	0	0	0	1.92	1.92	1.92	1.92
Flowrate (m^3/hr)	53.5	61.7	61.0	59.9	66.2	66.4	53.3	60.8
Temp (°C)	65	63	59	54	54	58	58	61
pH No. 1	3.21	3.22	3.33	3.17	3.12	3.16	3.16	3.22
pH No. 2	2.95	2.91	2.95	3.02	2.94	2.98	3.02	3.12
HIS Fe (g/l)	14.3	12.5	11.0	9.7	5.3	5.4	6.8	5.5
FS Zn (g/l)	20.2	23.9	26.2	26.8	33.2	34.1	20.1	23.5
Wash ratio	2.05	2.33	1.02	2.60	1.39	1.86	1.94	1.61
Residue volume (m^3)	26.79	24.94	26.73	19.75	18.58	14.54	14.57	15.88
RESULTS:	27-Sep-11	30-Sep-11	1-Oct-11	2-Oct-11	5-Oct-11	6-Oct-11	7-Oct-11	8-Oct-11
Zn Recovery (%)	91.27	91.65	90.05	90.04	88.84	89.31	88.25	90.16
Sol Zn losses (%)	1.43	1.30	1.14	0.98	0.90	0.82	0.81	0.92
Insol Zn losses (%)	7.30	7.05	8.81	8.99	10.26	9.87	10.93	8.91
F to residue (%)	94.93	93.68	94.30	94.75	86.85	89.10	87.33	86.92

PARAMETERS:	9-Oct-11	11-Oct-11	12-Oct-11	13-Oct-11	14-Oct-11	15-Oct-11	17-Oct-11	18-Oct-11	19-Oct-11
SEED (m^3/hr)	1.92	3.83	3.83	3.83	3.83	3.83	5.75	5.75	5.75
Flowrate (m^3/hr)	55.7	65.0	63.8	52.8	60.8	62.6	65.2	76.1	64.2
Temp (°C)	64	65	65	62	64	66	65	66	62
pH No. 1	3.26	3.33	3.19	3.35	3.19	3.20	3.12	3.10	3.12
pH No. 2	3.05	3.03	2.95	3.11	2.91	2.93	2.92	2.91	2.99
HIS Fe (g/l)	6.9	11.3	12.6	12.5	15.2	16.6	14.5	13.7	12.9
FS Zn (g/l)	21.2	25.2	22.0	20.6	16.6	20.4	26.2	32.0	26.4
Wash ratio	1.59	1.01	0.96	1.28	0.96	0.90	0.78	0.85	0.18
Residue volume (m^3)	17.35	29.42	30.57	30.64	31.68	32.32	38.20	36.91	36.18
RESULTS:	9-Oct-11	11-Oct-11	12-Oct-11	13-Oct-11	14-Oct-11	15-Oct-11	17-Oct-11	18-Oct-11	19-Oct-11
Zn Recovery (%)	89.13	91.18	91.11	91.36	93.20	93.00	93.13	93.54	91.63
Sol Zn losses (%)	0.93	1.14	1.20	1.13	1.17	1.11	1.54	1.59	1.40
Insol Zn losses (%)	9.94	7.68	7.69	7.51	5.64	5.89	5.33	4.87	6.97
F to residue (%)	89.94	88.76	87.91	87.91	87.76	88.54	89.00	86.59	87.20

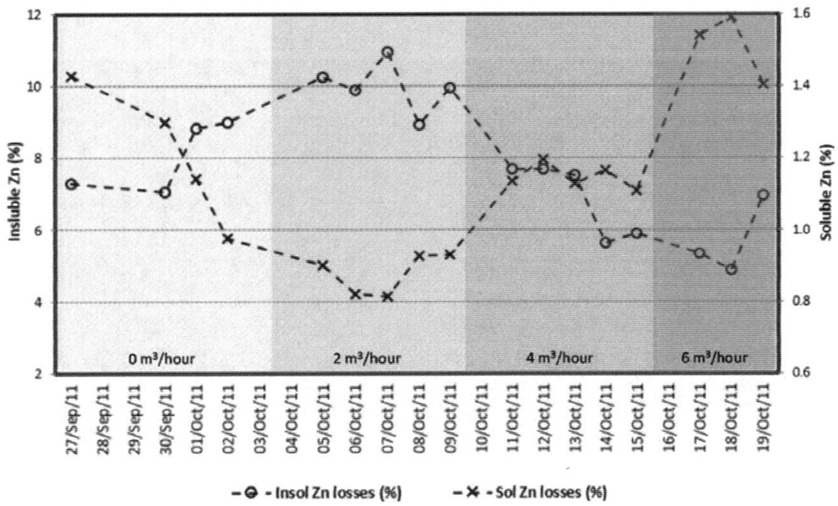

Figure 2 – Plot of soluble and insoluble zinc losses overlapped with seed recycle

Figure 2 shows that during the first half of the plant trial the insoluble zinc losses increase to a maximum of about 11%. This can be explained by the general reduction in both residence time (indicated by an increase in the flow rate from an average of approximately 60 m³/h to almost 70 m³/h as more seed is added) and the operating temperature (which dropped to below 60°C) during this period (refer to Table 1). Haegele *et al.* (1992) found that higher operating temperature improves the efficiency of the precipitation reaction and a longer residence time supports the formation of a better quality residue.

During the second half of the plant trial where the seed recycle is increased to 4 m³/h and then 6 m³/h, the insoluble zinc loss decreases to between 5 and 6%. The reasons may be attributed to the increase in the operating temperature, from a low of approximately 55 °C to approximately 65 °C (refer to Table 1), as well as the presence of seed. The seed recycle rate was increased to between 4 and 6 m³/h in the latter part of the trial. The presence of seed provides additional surface area for precipitation, so that there is less coating of unleached neutralising agent by the precipitate (Claassen, 2005).

Outcome of PSD Analyses, Settling Tests and SEM Analyses

Daily samples from the first, second, fourth and sixth reactors were sent for PSD analyses. Plots of the D_{50} and D_{90} for each of the samples can be seen in Figure 3 and Figure 4.

- In Figure 3 the particles in the second reactor are the largest, whereas those in the fourth and sixth reactors are the smallest. Particles appear to increase in size only to decrease again, and this could indicate attrition.
- Figure 3 also shows some increase in the particle size between the no seed and 2 m³/h seed scenarios, after which there appears to be a decline. While seed addition promotes agglomeration growth (Lewis, 2009; Claassen, 2005), these particles need time to grow (Söhnel and Garside, 1993). It would appear that a seed recycle of more than 2 m³/h is excessive and that the implementation of seed recycle at the expense of residence time may be counter-productive.
- Figure 4 shows the same trend, where there is an increase in particle size between the no seed and 2 m³/h seed scenarios. There appears to be a second increase in particle size with a seed recycle of 6 m³/h, where the biggest particles are found in the last two reactors. This hints at the importance of residence time in agglomeration growth. The presence of seed appears to have the greatest impact in the larger size fractions.

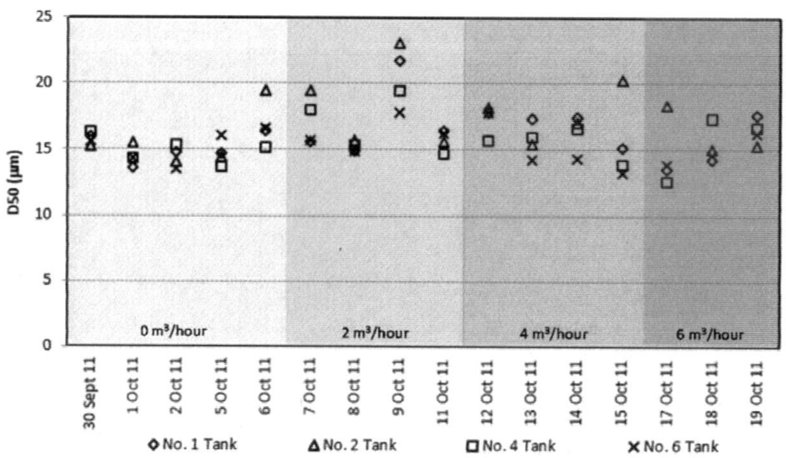

Figure 3 – Plot of 50% passing size for reactors No. 1, 2, 4 and 6

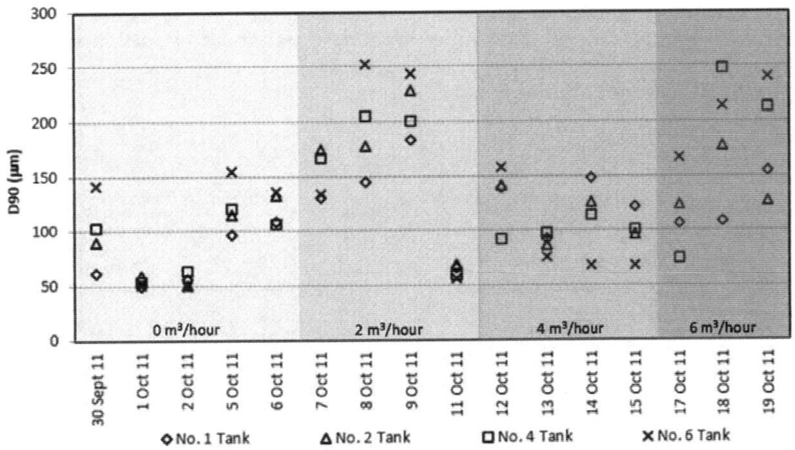

Figure 4 – Plot of 90% passing size for reactors No. 1, 2, 4 and 6

Particles of below 10 µm in size are conventionally classified as fines. This size fraction usually complicates the dewatering and materials handling of a precipitate. Figure5 represents the fraction below 10 µm in size for samples from reactors 1, 2, 4 and 6 for the no seed, 2 m³/h, 4 m³/h and 6 m³/h seed scenarios. The largest portion of fines appear in the fourth and sixth reactors, which may have been the result of attrition. The steady increase in the solids loading across the circuit with an increase in the seed recycle rate may have resulted in particle crowding, which negatively impacts agglomeration growth through dis-agglomeration [Claassen and Sandenbergh, 2006].

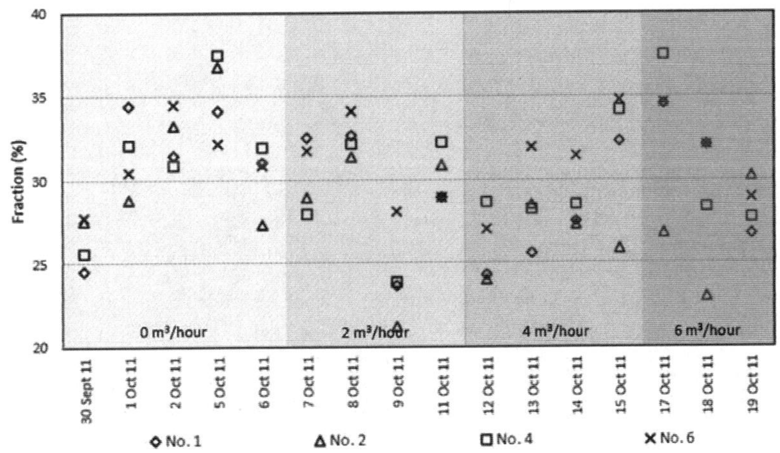

Figure 5 – The fraction of the sample below 10 µm forrReactors No. 1, 2, 4 and 6

Settling tests were performed on thickener feed samples once per shift on every day of the plant trial. The results, shown in Figure 6 show an initial increase in the settling rate between the no seed and 2 m³/h seed recycle scenarios. The PSD results confirmed this outcome, since there was an initial increase in particle size, especially in the coarse fraction, with the implementation of a seed recycle (refer to Figure 4). The settling rate decreases between the 2 m^3/h seed and 4 m^3/h seed scenarios. This may be attributed to particle crowding, since the solids loading showed a gradual increase from 20 to 40 m^3 per day over the period of the plant trial. Particle crowding hinders settling.

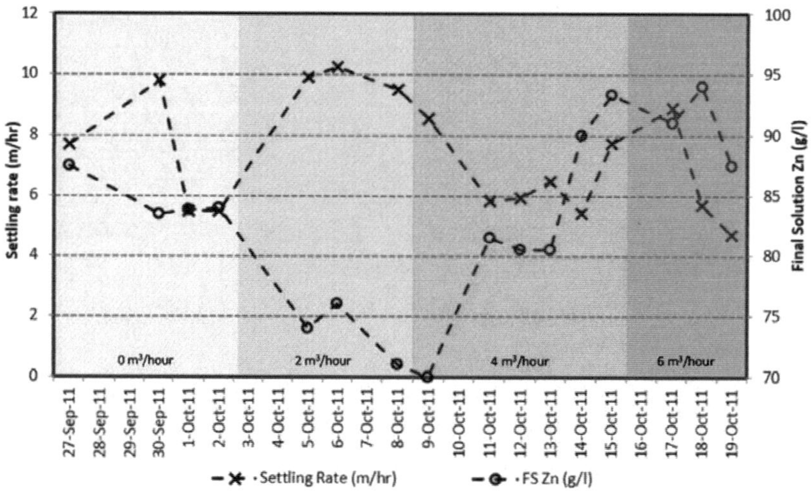

Figure 6 – Plot of settling rate and thickener overflow density overlapped with the seed volume

Figure 6 shows that the settling rates are higher when the solute zinc concentration is lower. Solution viscosity is a function of solute concentration, so that an in increase in solution concentration increases the solution viscosity (Themelis, 1995). High solution viscosity restricts the ease at which particles can move in the liquor, resulting in poor settling and filtration and high soluble zinc losses.

SEM images were taken of selected iron residue samples at no seed, 2 m^3/h, 4 m^3/h and 6 m^3/h seed recycle. The results are shown in Figure 7 and Figure 8.

Figure 7 a.) is an image of a precipitate sample when no seed was being added. One can clearly see that point A, an almost pure oxide particle (attributed to the high Zn and Pb contents), is coated with a thin layer of iron precipitate (indicated by point B). This unleached oxide particle will report directly to the iron residue and will contribute to the insoluble zinc loss.

Figure 7 b.) is an image of a precipitate sample at a seed recycle of 2 m^3/h. Point A is likely a partially leached zinc oxide (due to the high Zn and Pb content) with an iron precipitate coating (indicated by point C).

Figure 8 a.) shows the precipitate at a 4 m^3/h seed recycle. Point B appears to be partially leached neutralizing agent (due to the high Zn content) with an iron precipitate coating (refer to point C). Point D may be lead sulphate. The presence of lead in the iron precipitate is attributed to the presence of lead in the raw materials, especially the zinc oxide.

Figure 8 b.) shows the precipitate at a 6 m^3/h seed recycle. The large particle in the image does not show any evidence of coating. Point A highlights the darker areas in the particle and appears to consist mostly of silica. The lighter areas (as highlighted by point B) consist mostly of Fe, Zn and Pb.

Overall, fewer occurrences of coating were observed as the seed recycle volume increased.

a.)

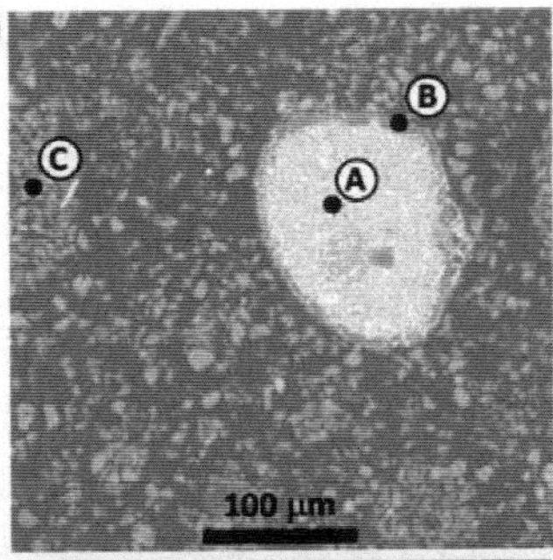

Element	A (Wt %):	B (Wt %):	C (Wt %):
Zn	82.7	23.5	7.4
Fe	2.9	23	52.2
S	1.4	5.2	2.4
Pb	11.2	28.4	12.1
Cu	0.1	9.2	0.9
Al	0.4	6.4	3.8
Mn	0.3	0.3	0.1
Mg	0	0	0.1
Ca	0.1	0.2	2.8
Si	0.9	3.5	10.1
K	0.1	0	0.3
Ba	0	0.3	7.7
Total	100.1	100	99.9

b.)

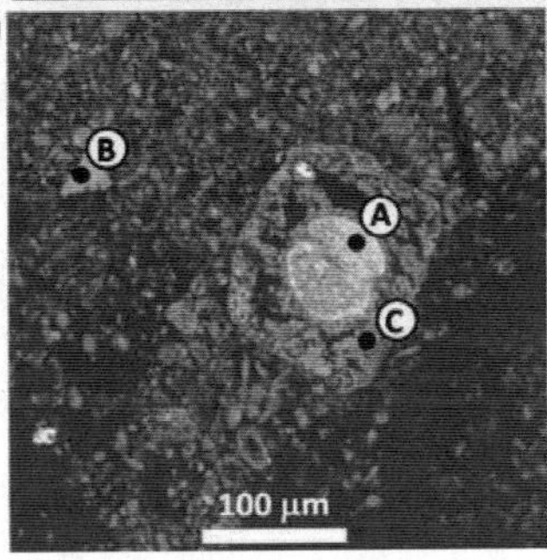

Element	A (Wt %):	B (Wt %):	C (Wt %):
Zn	86.5	31.3	36.4
Fe	6.9	56	33.8
S	0.1	0.2	2.1
Pb	2.6	2.7	8.6
Cu	1.4	2.7	1.8
Al	0.5	0.4	3
Mn	1.4	6	5.8
Mg	0	0.3	0
Ca	0	0.1	0.2
Si	0.5	0.3	5.8
K	0	0	0
Ba	0	0	2.5
Total	99.9	100	100

Figure 7 – SEM image of the iron residue at a.) 0 and b.) 2 m^3/h seed recycle

a.)

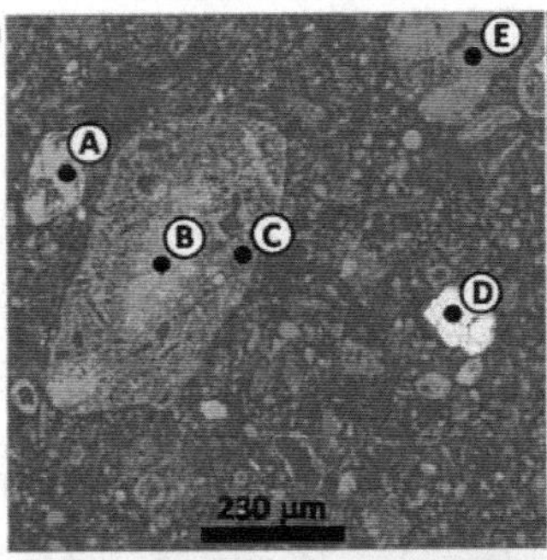

Element	A (Wt %):	B (Wt %):	C (Wt %):	D (Wt %):	E (Wt %):
Zn	33.6	73.7	35.9	1.1	12.6
Fe	51.4	15.8	47.3	2	9.8
S	0.5	4.2	1.5	13.6	11.1
Pb	3.7	1.3	3.9	61.5	37.8
Cu	2.5	2.3	2.6	0.2	1.8
Al	0.1	0.5	2.4	0.1	0.1
Mn	7.2	1.1	3.2	0.4	10.4
Mg	0	0	0	0	0
Ca	0	0.1	0	1.5	13.4
Si	0.4	0.7	2.7	0.2	0.4
K	0	0	0.1	0.1	0
Ba	0.5	0.3	0.4	19.3	2.6
Total	99.9	100	100	100	100

b.)

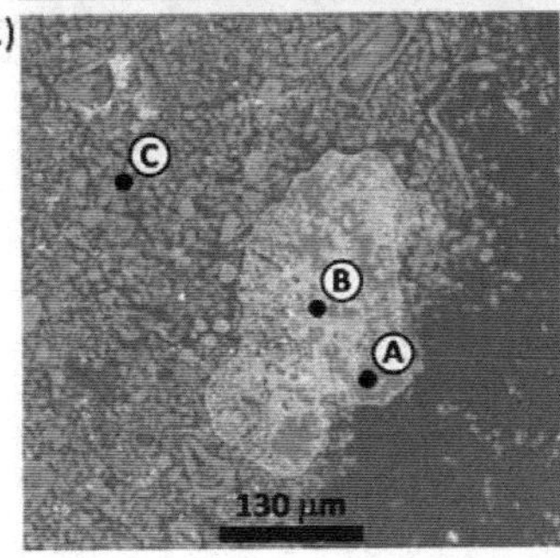

Element	A (Wt %):	B (Wt %):	C (Wt %):
Zn	5	20.7	11.1
Fe	2.6	33.5	60.1
S	0.2	3.8	8.5
Pb	2.8	30.7	10.4
Cu	1.7	1.3	2.1
Al	0.3	0.3	1.8
Mn	0.6	6.1	0.6
Mg	0.3	0	0.1
Ca	0.1	0.3	0.3
Si	86	0.4	3.9
K	0	0	0.7
Ba	0.4	2.8	0.3
Total	100	99.9	99.9

Figure 8 – SEM image of the iron residue at 4 a.) and 6 b.) m^3/h seed recycle

CONCLUSIONS AND RECOMMENDATIONS

The plant trial of implementing a seed recycle in the iron removal stage in Zincor's Springs Refinery showed that the presence of seed provides additional surface area for precipitation to occur, reducing the occurrence of coating of the fresh neutralising agent by iron precipitate, thereby decreasing the insoluble zinc loss. The mass balance and SEM images support this theory.

Seed also promotes agglomerate growth, thereby reducing the soluble zinc loss. The mass balance, PSD and settling results appeared to support this theory. It was seen that an increase in the seed recycle volume resulted in an increase in the residue volume reporting to the thickener and belt filter. This increased the solids loading, negatively impacting on the wash ratio, and resulted in particle crowding and particle breakage. High seed recycle rates also had a negative impact on the residence time. Hence, the implementation of seed recycle should not be done at the expense of wash ratio, settling and residence time. A moderate seed recycle volume of between 2 and 4 m^3/h appears to be ideal under the settings tested.

It would have been preferable to run a second plant trial over a longer period in order to better define the plant parameters and possibly to evaluate the use of limestone for neutralisation. The limestone could have been introduced gradually as a supplement to the roaster calcine, together with a seed recycle. Perhaps the presence of insoluble zinc ferrite (in the roaster calcine), together with a seed recycle, would have addressed some of the issues experienced with the fineness and poor settling characteristics of the precipitate resulting from neutralisation with only limestone. The wash ratio would have to be maximised in order to fully realise the benefits of the seeding.

ACKNOWLEDGEMENTS

This project could not have been completed without the support of a number of Diane's supervisors and colleagues at Zincor. Thank you to Jacolien Wyethe, Rasai Ntsoelengoe, Frikkie de Bruin, Gideon Kriel and the RTP personnel, Glynnis Rawson and the laboratory personnel. Alister MacDonald from Mac One Agencies kindly granted the use of his laboratory facility and equipment, and provided his expertise when it came to the settling tests. A special thanks to Jo Rodrigues for his mentorship and Dr. Johann Claassen for his critical input to this write-up of the study.

REFERENCES

Claassen, J.O. (2002). *Characterization and optimization of the Zincor iron removal process.* M. Eng. Thesis, University of Pretoria, South Africa.

Claassen, J.O., Meyer, E.H.O., Rennie, J., & Sandenbergh, R.F. (2002). Iron removal from zinc-rich process solutions: Defining the Zincor process. *Hydrometallurgy,* 67, 87- 108.

Claassen, J.O., Meyer, E.H.O., Rennie, J., & Sandenbergh, R.F. (2003). Iron precipitation from zinc-rich process solutions: Optimizing the Zincor process. *SAIMM,* 253- 263.

Claassen, J.O., (2005). *Product quality parameters in the reaction crystallization of metastable iron phases from zinc-rich solutions.* PhD Thesis, University of Pretoria, Pretoria, South Africa.

Claassen, J.O. & Sandenbergh, R.F. (2006). Particle growth parameters in the precipitation of metastable iron phases from zinc-rich solutions. *Hydrometallurgy,* 84, 165-174.

Haegele, R. & Smallwood, F. (1992). Optimisation of the iron precipitation process of Zincor's residue treatment plant. *Progress Report 1.* Internal Zincor report.

Hove, M. (2008). *The kinetics and mechanisms of the oxidation and precipitation of iron: The High Density Sludge (HDS) process.* PhD Thesis, University of Cape Town, Cape Town, South Africa.

Lewis, A. (2009). Understanding industrial crystallisation processes (short course). Department of Chemical Engineering, University of Cape Town (unpublished).

Meyer, E.H.O., Howard, G., Haegele, R., & Beck, R.D. (1996). Iron control and removal at the zinc corporation of South Africa. *Proceedings of the Second International Symposium on Iron Control in Hydrometallurgy,* CIM, Montreal, Canada, 163-182.

Sinclair, R.J. (2005). The extractive metallurgy of zinc. *The Australasian Institute of Mining and Metallurgy (AusIMM) Spectrum Series,* Volume 13.

Söhnel, O., & Garside, J. (1992). *Precipitation: Basic principles and industrial applications.* Oxford: Butterworth-Heinemann.

Taggart, D. (2016). *Evaluating the effect of alternative neutralising agent and seeding on the zinc recovery through the Zincor iron removal circuit.* MSc.Thesis, University of Cape Town, South Africa.

Themelis, N., (1995). Transport and Chemical Rate Phenomena. Taylor & Francis.

Wyethe, J.P., Vegter, N.M., & Rodrigues, J.M.S. (2008). Recent recovery improvement initiatives at Zincor: The use of zinc oxide fume and low-ferrite calcine for iron removal to reduce un-dissolved zinc losses. *Southern African Institute for Mining and Metallurgy (SAIMM) - Lead and Zinc Conference 2008*, 57-68.

THE INFLUENCE OF JAROSITE GENERATION ON THE INDIUM LEACH RATE DURING A LEACHING PROCESS OF SECONDARY ZINC OXIDE

CHEN Guo-mu[1], HUANG Meng-yang, *ZENG Peng, YANG Mei-yan, LI Guo-feng, ZHOU Zhou-hua, and WANG Qiu-ying

[1]*Yunnan Yuntong Zinc Co., Ltd.,*
Kunming, China 650120
*(*Corresponding author: peng.zeng@ynz.yunnancopper.com)*

ABSTRACT

Historically, a lot of jarosite was found in the leaching residue when leaching indium from secondary zinc oxide materials. The reasons for the formation of indium bearing jarosite were theoretically analyzed. The influence of controlling the acidity and the addition of reagents on the indium leaching yield during secondary zinc oxide processing was investigated. In order to increase the leaching yield of indium, a measure of controlling the creation and destruction of jarosite was proposed. The production data shows that the leaching yield of indium was increased from 60 to 96%, and the content of indium in leaching residue simultaneously decreased from 900 g/t to 100–200g/t when implementing such controlling measures.

KEYWORDS

Secondary zinc oxide, indium, leaching, acidity, jarosite

INTRODUCTION

Indium does not occur as independent minerals in the earth's crust and is mainly associated with complex sulphide minerals. Specifically some sphalerites can contain as much as 900 g/t of indium. Most indium is recovered from by-products of copper, lead or zinc metallurgy. Therefore, the research on the recovery of by-products in non-ferrous metallurgy is important (Sinclair et al, 2006). The main technologies used for this objective in China include:

- pre-neutralization-reduction-precipitation indium process (Yu et al, 2008),
- countercurrent oxidizing acid leaching process and sulphation roasting-water leaching process (Cao, 2005),
- oxygen pressure acid leaching of zinc concentrate-P204 extraction process (Yang et al, 2006),
- double acid leaching-solvent extraction process on lead dross from a reverbatory furnace flue dust (Liu, 2003),
- an acid leaching-solvent extraction process on smelting ash (Zhou et al, 2005).

Although high pressure and extreme conditions can be used to improve extraction from high iron bearing indium residues (autoclave processing), mild extraction with lower cost and lower environmental impact are preferred.

This paper describes both how the indium extraction using mild conditions in a secondary zinc circuit was quite low and, how through XRD and chemical analysis jarosite formation was found as the main cause. The paper then proceeds to discuss a method for improving the leaching yield and the application of this method in industrial practice.

RAW MATERIAL AND ANALYSIS OF THE CAUSE OF JAROSITE FORMATION

Materials

This experimental program used secondary zinc oxide from a fuming kiln as its raw material. The main components are shown in Table 1. The indium grade is 620 g/t and the iron is as high as 8.0%.

Table 1 – Main components of zinc oxide (%)

Component	Zn	In (g/t)	Fe	S_{total}	$S_{insoluble}$	Pb	Reducibility
Content (%)	50.2	620	8.0	2.98	1.83	8.69	3.50

Process with Secondary Zinc Oxide

The secondary zinc oxide is subjected to a neutral leach, a weak acid leach, and a high acid leach. In the latter the sulphuric acid concentration is about 150 g/L, and an ultra-high acid leach (sulphuric acid concentration is about 200 g/L) is applied, to maximize zinc and indium leaching. Neutralizing and hydrolysis is the method used to recover indium from the low acid leaching liquid. The overall process is shown in Figure 1.

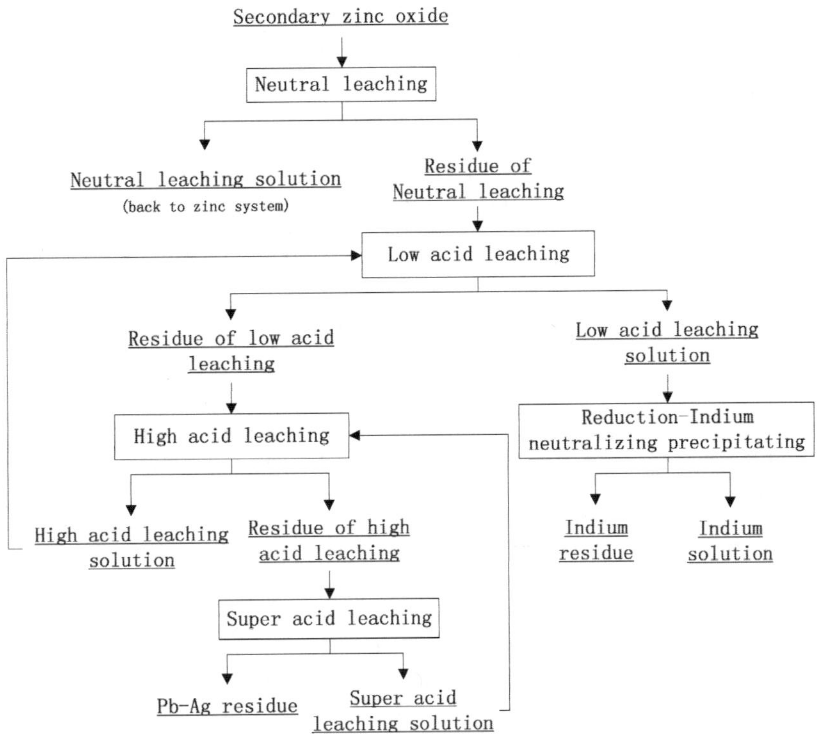

Figure 1 – Zinc oxide dust treatment route

This flow sheet produces a neutral leach overflow which is returned to the main hydrometallurgical zinc flowsheet, a lead silver concentrate, as well as a an indium concentrate (8–10 kg indium per ton of feed) and an iron rich solution which is also returned to the main zinc flow sheet.

The Generation of Jarosite

The average composition of the lead silver residue is shown in Table 2.

Table 2 – Lead silver average composition in one plant

Composition	Zn	In (g/t)	Fe	S	Pb
Content (%)	3.59	929	12.76	16.0	17.60

As Table 2 shows, the indium content is high in the lead-zinc residue, even after triple acid leaching, the indium in residue is still above 900 g/t. The lead silver residue contains a large amount of iron and sulphur; the color of this residue is yellow. The concentration in the leach liquid of ferric iron (Fe^{3+}) is 30 to 40 g/L. This leads to the conclusion that stable jarosite is produced during leaching and at the same time indium is incorporated, which is difficult to leach. The XRD analysis results of three batches of lead silver residue are shown in Figure 2.

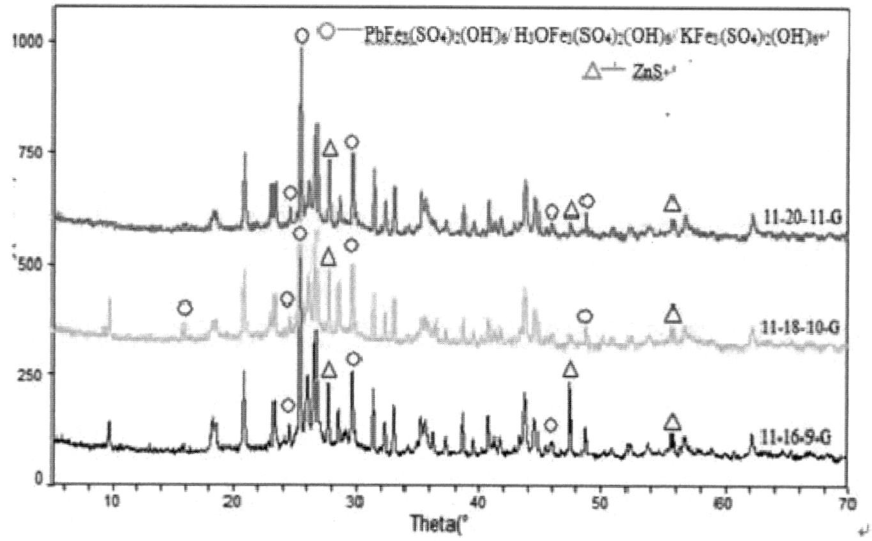

Figure 2 - XRD pattern of the Pb-Ag residue

As seen in Figure 2, the residue contains high zinc sulfide, as well as high lead jarosite, hydronium jarosite and potassium jarosite. The results suggest that jarosite class materials are generated in the lead silver residue, and jarosite partially incorporates indium.

Theoretical Background

Indium and iron seem to be correlated in the lead silver residue, according to the XRD analysis results of lead silver residue,

The mechanism of jarosite formation according to Jiang and Wang (2010) is as follows.

$$3Fe_2(SO_4)_3 + 10H_2O + 2NH_3 \cdot H_2O = (NH_4)_2Fe_6(SO_4)_4(OH)_{12} + 5H_2SO_4 \quad (1)$$

When jarosite is generated, indium can be incorporated by the following reaction:

$$xIn_2(SO4)_3 + (6\text{-}x)Fe_2(SO4)_3 + 10H_2O + 2NH_3{\cdot}H_2O = (NH_4)_2(In_x/Fe_4(SO_4)_4(OH)_{12} + 5H_2SO_4 \quad (2)$$

Because Fe^{3+} and In^{3+} have similar radii (Fe^{3+} 0.1067 nm, In^{3+} 0.1082 nm) in the jarosite process, In^{3+} will substitute part of the Fe^{3+}, and during iron precipitation there will be the formation of jarosite containing indium (Li and Zhou, 2010; J E Dutrizac and Dou Mingmin,1993). In a standard jarosite Fe-removal process, the indium can be recycled. In the secondary zinc oxide leaching process, both indium and iron enter into the residue simultaneously under jarosite formation conditions, and result in high indium in lead silver residue. In Figures 3 and 4 the theoretical conditions under which the jarosite is formed and remains stable have been provided (Zhi Mei. 2003). These figures are provided in support of the reaction mechanism shown above.

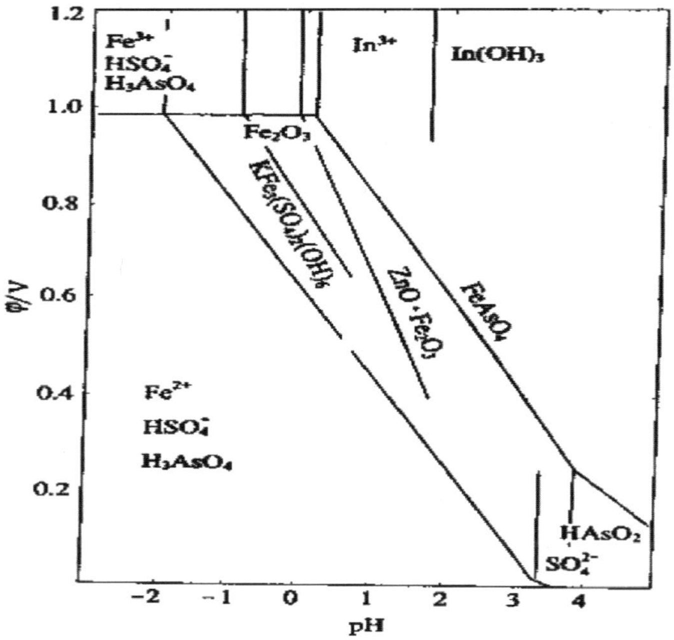

Figure 3 - Eh-pH diagram of Jarosite precipitated iron （t=100 °C，α=1 mol/L）

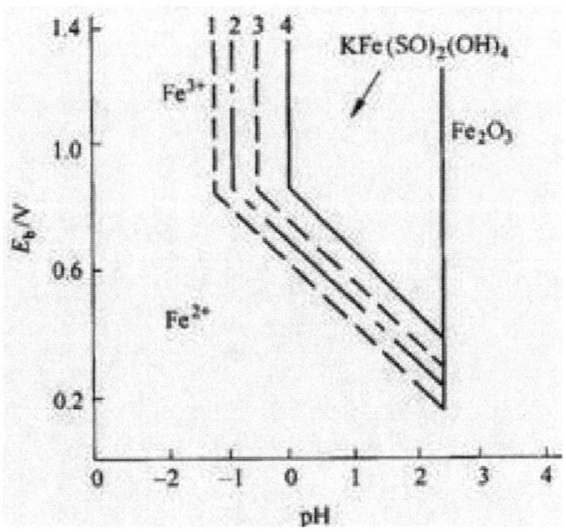

Figure 4 - The stability of jarosite

(Equilibrium concentration in solution（mol/L）1=0.1；2=0.01；3=10^{-3}；4=10^{-4})

THE EFFECTS OF JAROSITE FORMATION ON INDIUM LEACHING RATE

From comparing jarosite formation conditions with the zinc oxide leaching parameters, it became apparent that pH control in the process of acid leaching is the key to influence jarosite production. Tests during the weak acid leaching process, controlling the finish acidity up at 15–20 g/L, intermittent leaching operation, for 30 minutes still formed jarosite that is difficult to leach in the subsequent process and indium leaching still is incomplete with a high indium content in the residue.

Control the Formation of Jarosite

Table 3 – The influence of acidity on leaching rate

No.	Low acid leaching		High acid leaching		Ultra-high acid leaching residue		Indium leaching rate (%)
	T (°C)	Final acid concentration (g/L)	T (°C)	Final acid concentration (g/L)	Zn (%)	In (g/t)	
1	70	5.8	80	105	5.5	550	63.8
2	70	20.6	80	100	5.8	250	85.4

As Table 3 shows, low acid leaching residue after high temperature and high acid leaching is still unable to make the indium leaching complete, so in order to improve the leaching rate of indium, the low acid leaching process should avoid the generation of jarosite.

It is difficult to guarantee the acidity level in the process at a steady value during actual production. Any event of low acid concentration or too high Fe^{3+} concentration; will inevitably lead to jarosite formation. It is possible to recycle indium from the jarosite residue, by roasting at controlled temperature converting it into indium sulfate (Ning and Chen, 1997), but taking this measure will extend the process flow sheet and increase the investment in equipment.

Therefore, a process whereby additives are used to both prevent the formation of jarosite and redissolve indium already incorporated in jarosite was studied. High acid leach lead silver residue test results are shown in Table 4.

Table 4 – Adding reagent in our jarosite test

No.	Additive	High acid leaching		High acid leaching residue		Indium leaching rate (%)	Note
		T (°C)	Final acid concentration (g/L)	Zn (%)	In (g/t)		
1	-	80	108	5.5	554	63.4	none
2	Fe	80	102	5.2	282	83.7	additive
3	ZnS	80	105	5.3	135	90.1	additive

As Table 4 shows, during high acid leaching of residue additives were added, the additives are Fe and ZnS. Both are able to increase the amount of indium leached during high acid leaching. Table 5 shows the effect of reagents Fe and ZnS as a function of dosing.

Table 5 - Fe and ZnS test results

Reagent	Addition % (proportion of leaching residue)	Indium leaching rate (%)
Fe	0	64.2
	5	65.3
	10	84.0
	15	89.4
	20	90.2
ZnS	0	64.2
	5	96.8
	10	96.6
	15	96.5
	20	96.8

As Table 5 shows, reagent Fe at 15%, gives an indium recovery close to 90%; Reagent ZnS at 5%, already gives an indium recovery of 97%. Reagent ZnS is more effective in destroying jarosite.

Production Tests to Improve Leaching Yield of Indium

Effect of Low Acid Leaching Acidity on the Leaching Yield of Indium

The leaching tank was operated such that the addition of leaching solution and residue resulted in an acid content of 30–40g/l. The composition of the resulting lead/silver residue is shown in Table 6.

Table 6 – Low acidic leaching output of lead silver main ingredients

Composition	Zn	In (g/t)	Fe	S	Pb
Content (%)	3–5	200–300	5–8	8–10	25–30

Avoiding Formation of Jarosite

Adding additive Fe during low acid leaching can avoid indium being captured in the jarosite. The lead silver residue composition after reagent Fe addition is shown in Table 7.

Table 7 – The main components of the lead silver residue adding additives Fe

Composition	Zn	In (g/t)	Fe	S	Pb
Content (%)	3–4	100–200	3–5	8–10	25–30

CONCLUSIONS

When leaching secondary zinc oxide to recover indium, the indium leaching yield is low, since part of the indium and iron report to the lead silver residue. Strictly controlling of the acidity in the process of acid leaching, can avoid or reduce the generation of jarosite efficiently, and increase the leaching yield of indium from 63.8 to 85.4%. In the presence of high Fe^{3+} concentration, jarosite generated during low acid leaching is inevitable, by adding proprietary reagents Fe or ZnS to the pulp jarosite in the residue can be destroyed, thereby increasing the indium yield. Reagent Fe should be added at 15%, and reagent ZnS at 5% of residue weight.

REFERENCES

Cao Yingke. Recycling indium from the arsenic ash of copper smelting process. Journal of Hunan Nonferrous Metals, 2005, 21 (1): 5-8.

Dutrizac, J.E. and Dou Mingmin. The Behaviour of Indium during Jarosite Precipitation. International Symposium-World Zinc '93, October 10-13, 1993 in Hobart, Tasmania, 1993: 365-372

Jiang Jibo, Wang Jikun. Research iron vitriol method of sinking indium from acid leaching solution under pressure from zinc sulfide concentration with high indium. Journal of Non-Ferrous Metal Smelting (smelting part), 2010 (5): 5.

Li Hongbin, Zhou Zhihui, et al. Research on the process of sinking jarosite from leaching solution of refining zinc Jarosite. Journal of Hunan Nonferrous Metals, 2010 (4): 27 -28.

Liu Langming. New technology research of extracting indium from lead dross reverberatory furnace soot. Journal of Rare Metal, 2003, 27(1): 112.

Ning Shunming and Chen Zhifei. Recovery of zinc and indium from the jarosite residue. Chinese Journal of Nonferrous Metals, 1997(3): 56-58.

Sinclair, Kooiman, Martin, et al. Geochemistry and mineralogy of indium resources at Mount Pleasant, New Brunswick. Ore Geology Reviews, 2006, 28(1): 123-145.

Yang Zhongping, Qi Xiaozhu, Ma Yajun, et al. Extraction of indium from acid leaching solution under oxygen pressure from zinc concentration. Mineral Resources and Geology. 2006, 20(6): 706-708.

Yu Xiaohua, Li Yonggang, Liu Chunxia, et al. Enrichment indium from leaching solution contained indium. Chinese Journal of Nonferrous Metals, 2008(6), 18(1): 122-125.

Zhou Zhihua, Mo Hongbin, Xu Guorong, et al. Research progress of rare metal indium enrichment and recycling technology. Journal of the Research Progress of Nonferrous Metals, 2005, 57 (1):71-75.

Zhi Mei. Lead and zinc metallurgy. 2003 first edition: 335-336.

Iron Management in Copper Processing Flowsheets

LEACHING CONCENTRATES WITHOUT LEACHING IRON

*M. Maccagni[1], J. Nielsen[2]

[1]Engitec Technologies S.p.A.
Via Borsellino e Falcone, 31
20026 Novate Milanese (MI) – Italy
*(*Corresponding Author: m.maccagni@engitec.com)*

[2]Engitec Technologies USA
813 Marti Drive
Rolla, Missouri 65401, USA

ABSTRACT

Engitec developed the EZINEX® Technology some years ago to recover zinc from oxidized secondary sources. This process works at neutral pH and can take into solution any kind of metal, but iron is rejected and remains in the leach residue. Engitec has also developed the FLUOBORATE System, which is able to recover lead and copper from relevant concentrates. Some iron is leached into solution, but it is controlled through a solution purification bleed treatment. In this paper, we present the two above mentioned technologies. They are very different with regards to the chemical environments and the methods used to control the build-up of iron in solution, but both systems are suitable to treat iron bearing feeds.

KEYWORDS

Fluoboric acid, bleed treatment, sulphates, ammonium chloride, oxides

INTRODUCTION

In the design of hydrometallurgical processes, the most important issues are the water balance, the filtration and the control of impurities. This is particularly important when the process includes an electrowinning unit. Iron has been one of the most problematic elements to remove, especially in acidic environments.

Engitec Technologies has developed several hydrometallurgical processes divided in two different electrolyte lines: the Fluoborate Technology for the recovery of lead and copper from primary and secondary sources and the Ammonium Chloride System based on a chloride electrolyte for the recovery of different non-ferrous metals, among which zinc and lead are the most important. In both cases, iron is present in the feed and needs to be eliminated from the solution. It is not a problem for the quality of the metal produced, but when present in solution, it consumes the basic electrolyte reducing its free concentration which is very important both for both the leaching and electrowinning unit operations. Both process systems are very peculiar, the fluoborate case being much more intriguing.

THE AMMONIUM CHLORIDE SYSTEM

The system is based on an electrolyte composed by a mix of alkali and rare earth chlorides, and working at atmospheric pressure or at high oxygen pressure and temperatures above 70 °C. The system is able to leach most of the metals taking them into solution as amino or chloride complexes. The simplified block diagram for the process is shown in Figure 1.

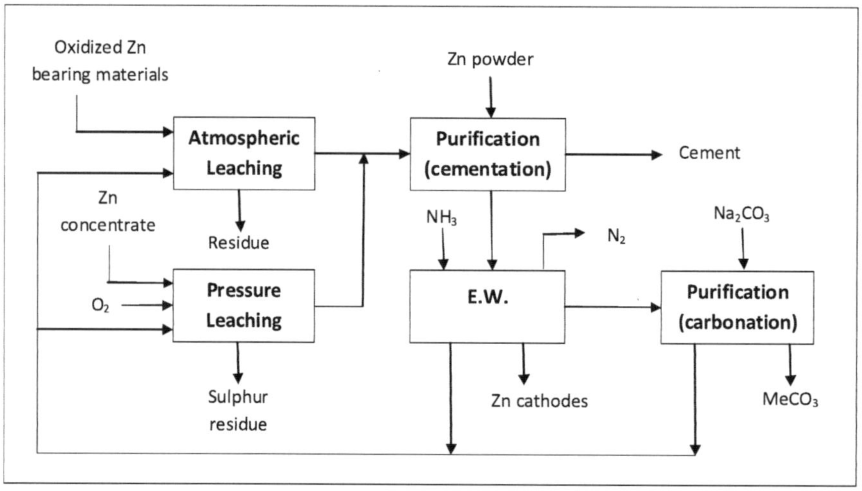

Figure 1 - Ammonium chloride system flow diagram

The atmospheric pressure leaching reactions are the following:

$$Me_2O_X + 2x\ NH_4Cl \rightarrow 2\ Me(NH_3)_xCl_x + x\ H_2O \quad (1)$$

$$Me_2O_X + 2(x+1)\ NH_4Cl \rightarrow 2\ NH_4MeCl_{(x+1)} + 2x\ NH_3 + x\ H_2O \quad [Me = Pb^{2+}, ...] \quad (2)$$

The pressure leaching reaction is the following:

$$Me_2S_X + {}^X/_2\ O_2 + 2x\ NH_4Cl \rightarrow 2\ Me(NH_3)_xCl_x + x\ H_2O + x\ S \quad [Me = Zn, Cu^+, Cu^{2+}, Ag^+, \ldots] \quad (3)$$

$$Me_2S_X + 2x\ O_2 + 2x\ NH_4Cl \rightarrow 2\ Me(NH_3)_xCl_x + x\ H_2SO_4 \quad [Me = Zn, Cu^+, Cu^{2+}, Ag^+, \ldots] \quad (4)$$

$$Me_2S_X + {}^{X}/_{2}\ O_2 + 2(x+1)\ NH_4Cl \rightarrow 2\ NH_4MeCl_{(x+1)} + 2x\ NH_3 + x\ H_2O + x\ S \quad [Me = Pb^{2+}, \ldots] \quad (5)$$

$$Me_2S_X + 2x\ O_2 + 2(x+1)\ NH_4Cl \rightarrow 2\ NH_4MeCl_{(x+1)} + 2x\ NH_3 + x\ H_2SO_4 \quad [Me = Pb^{2+}, \ldots] \quad (6)$$

During the leaching operation, it is possible to dissolve some ferrous salts. However, because of the presence of oxygen in both reactors, the ferrous salts are converted to ferric hydroxide which precipitates, thereby reducing the iron in solution to a very low level (less than 1 mg/L). Sometimes, if the reaction is not complete, some extra oxygen will bring complete oxidation of the iron.

The formation on site of ferric oxide/hydroxide is also very important for the purification of the solution. In fact, the concentrations of As and Sb are also significantly reduced, to about 1 mg/L because of the precipitation of $FeAsO_4$ and $FeSbO_4$. Some other metals are absorbed on the ferric oxide/ferric hydroxide matrix.

As already anticipated, in this case iron control is easy and favourable. In the next case, iron control is more difficult and has been the subject of a patent (Zoppi, 1995).

FLUOBORATE TECHNOLOGY

The Fluoborate Technology is based on the use of fluoboric acid as electrolyte, which is a very good solvent for most of the metals, especially for lead. This technology was developed for the production of lead and copper from primary sources, and sometimes it is also applicable to secondary sources. In fact, the process was initially developed for secondary sources, but the fluoboric acid was very aggressive to metals and it was not possible to find a suitable anode for the anodic oxygen evolution reaction. This problem was overcome by changing the anodic reaction from oxygen evolution to the oxidation of iron from ferrous to ferric state. This was the main reason we started to consider primary sources based on sulphides which require oxidative leaching. This technology can also be used for refining secondary sources, such as lead bullion and copper scrap.

The simplified block diagram for the Fluoborate process is shown in Figure 2.

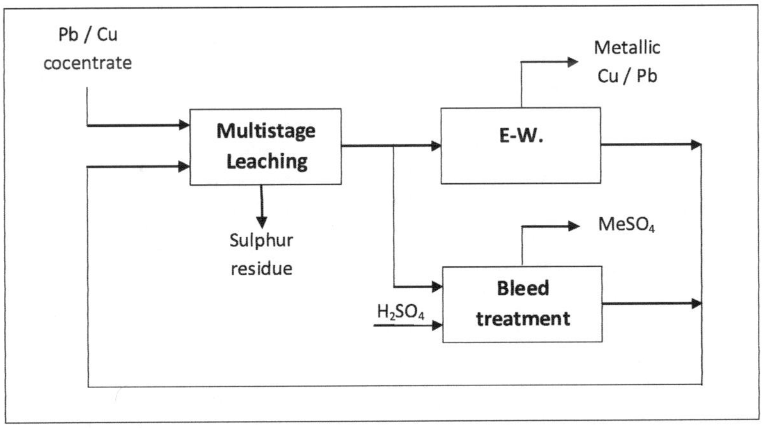

Figure 2 - Fluoborate technology flow diagram

Concentrate leaching in the Fluoborate Technology is performed according to the following reactions:

$$MeS + 2\ Fe(BF_4)_3 \rightarrow Me(BF_4)_2 + 2\ Fe(BF_4)_2 + S^\circ \quad [Me = Pb, Cu] \tag{7}$$

$$MeO + 2\ HBF_4 \rightarrow Me(BF_4)_2 + H_2O \quad [Me = Ca, Mg, Fe, ...] \tag{8}$$

Of course, some other metal sulphides, such as ZnS and Ag_2S, can also be partially leached according to reaction (7). During multistage leaching, and by controlling the leaching parameters, metals more noble than the principal one to be recovered can be precipitated, thereby avoiding the possibility that they will be co-plated.

Iron is among the impurities less noble than the metal to be recovered that is also taken into solution through reaction (8). Iron and the other impurities need to be removed from the solution to regenerate the free acid. There are two reasons for this requirement. Firstly, the free acid helps the kinetics of the leaching reaction, and secondly, the free acid in electrowinning is important for the conductivity of the electrolyte and for the solubility of the anode products. To resolve the problem of impurities, the following bleed treatment was studied and patented (Zoppi, 2001). After some attempts using different techniques, such as ion exchange resins and organic solvents, a process based on the very surprising chemical behaviour of the metallic sulphates in a fluoboric solution was proven in our laboratory. By adding HBF_4 to a saturated solution of a metallic sulphate, the precipitation of the metallic sulphate can be realized with an amount of crystallisation water that is far from known values. Conversely, when some sulphuric acid is added to a concentrated fluoboric acid solution containing dissolved metals, precipitation of metal sulphates takes place. Furthermore, the values for the water of crystallisation are significantly different from those reported in literature. The reaction can be written as follows:

$$Me(BF_4)_2 + H_2SO_4 + x\ H_2O \rightarrow MeSO_4.xH_2O + 2\ HBF_4 \quad [Me = Fe, Zn, Ca, Mg, Cd, \ldots] \tag{9}$$

Through this reaction, the impurities are removed and the fluoboric acid which they consumed in the leaching step is regenerated, thereby maintaining the free acid concentration at the desired level.

Practically, as shown in the conceptual flowsheet in Figure 3, the unit is composed by three main operations:

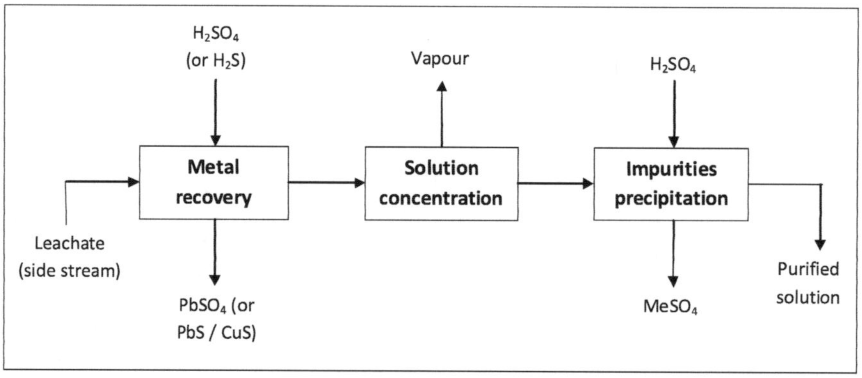

Figure 3 - Bleed treatment flow diagram

Lead / Copper Recovery

Lead and copper must be selectively precipitated to avoid their co-precipitation with the other impurities. Such co-precipitation would represent a significant economic loss for the process. This operation can be done with either H_2S according to the following reaction:

$$Me(BF_4)_2 + H_2S \rightarrow MeS + 2\ HBF_4 \quad [Me = Pb, Cu] \tag{10}$$

or/and with H_2SO_4 according to the following reaction:

$$Me(BF_4)_2 + H_2SO_4 \rightarrow MeSO_4 + 2\ HBF_4 \quad [Me = Pb] \tag{11}$$

Using the first option described, the PbS / CuS produced through reaction (10) can be recycled to the leaching unit, thereby putting the lead / copper back into the process. The second option, much simpler and easier, produces a very pure $PbSO_4$ through reaction (11). This $PbSO_4$ can be used in a secondary lead plant that is present either at the facility or close to the Fluborate Technology plant.

Solution Concentration

The solution is concentrated to reach a fluoborate concentration of about 600 g/L to make the precipitation easy, thereby realizing high precipitation efficiency with the minimum amount of H_2SO_4. In fact, the stabilization of the fluoboric acid requires water, and that water is not available to stabilize the metals in solution which precipitate because of a water deficit.

Sulphates Precipitation

The precipitation of metallic impurities is performed by adding H_2SO_4 according to reaction (9). The solution leaving this unit is ready to be recycled back into the process loop. In fact, under steady state conditions, the leached impurities are precipitated and the relevant HBF_4 is liberated.

CONCLUSIONS

Iron control is one of the most important problems in hydrometallurgy. It usually requires dedicated units and/or separate processes because some iron is commonly leached in each one of these processes. The control of iron in neutral/alkaline environments is usually significantly easier than in acidic media. In acidic environments, as in the case of the traditional zinc electrolytic process, iron control requires a lot of reactants and produces residues which must be disposed a secure area.

In our Ammonium Chloride System, the iron reports to the final residue to be disposed of. Whereas in the Fluoborate Technology, the bleed treatment residue, when metallic values are present, can be easily fed to many existing operations for the recovery of the contained metallic values.

REFERENCES

Zoppi, G. (1995). U.S. Patent No. 5.468.354. Washington, DC: U.S. Patent and Trademark Office

Zoppi, G. (2001). U.S. Patent No. 6.203.769. Washington, DC: U.S. Patent and Trademark Office

AN ALKALINE GLYCINE-BASED PROCESS FOR COPPER RECOVERY AND IRON REJECTION FROM CHALCOPYRITE

*J.J. Eksteen, E.A. Oraby, B. Tanda

Department of Mining Engineering and Metallurgical Engineering,
Western Australian School Mines, Curtin University,
GPO Box U1987, Perth, WA 6845, Australia
*(*Corresponding author: jacques.eksteen@curtin.edu.au)*

ABSTRACT

A process has been developed at Curtin University whereby glycine is used in an alkaline environment to leach gold, silver and copper from their ores, concentrates, process intermediates and wastes. Copper mineralisation investigated includes various oxides, sulfides and native copper. The leach process utilises an oxidising environment such as oxygen. Glycine is a non-toxic, non-volatile, edible, low cost reagent which is produced in large industrial quantities for use in several industries. Glycine and its copper and precious metals complexes are stable over wide pH, Eh and temperature ranges. Other than losses in leach residues, the glycine is recovered and recycled for leaching. Iron dissolution is insignificant in the pregnant leach solution. Iron in pyrite remains unaffected, whereas iron associated with base metal sulfides (e.g. chalcopyrite and bornite) is converted to a readily filterable iron hydroxide. Efficient separation of copper by solvent extraction and conventional electrowinning has been shown to be feasible for a number of organic extractants, whereas gold and silver glycinates adsorb very well onto carbon allowing CIP/CIL/CIC (carbon-in-column) based processes. The process is ideal for ores containing acid-consuming gangue, concentrates with high arsenic or halide contents and for copper-gold ores with significant sulfide mineralisation. The process eliminates many of the problems associated with the processing of copper-gold-silver ores, such as elemental sulfur formation, silica gels, unwanted metal co-dissolution, and jarosite precipitation. The reagents used and process residues are much safer and environmentally benign than conventional process options. The processing technology can be applied to in-situ, heap and vat, or agitated tank leaching at temperatures ranging from ambient to below 100°C at atmospheric pressure and is currently being commercialised. This paper will focus on copper leaching and recovery and iron rejection from a chalcopyrite flotation concentrate.

KEYWORDS

Glycine, copper, silver, gold, cyanide, leaching, solvent extraction, alkali, amino acid

INTRODUCTION

Copper deposits and gold-copper deposits are well known for their mineralogical complexity and geometallurgical variability in both copper and gangue mineralisation. Most deposits are associated with a weathered or oxidised upper zone followed by a zone of supergene enrichment and a hypogene zone as one progresses in depth. This progression has often led to different processing approaches for the various zones and geological domains, in particular (sulfuric) acid leaching of the oxide ores and grinding and flotation of the sulfide ores to produce a concentrate for smelting. Precious metals are often associated with all or many of the zones and domains of the deposit and processing for optimal copper recovery often creates conflicting requirements to optimal recovery of the precious metals. Acid leaching is particularly problematic if the copper minerals are associated with high amounts of acid-consuming gangue and gel-forming silicates. Milling and flotation often becomes uneconomic if the copper grades become too low (less than 0.7% at current prices) and the level of dissemination with associated pyrite becomes fine, making it difficult to prepare a concentrate of sufficient grade at an economic recovery.

We have recently published (Oraby & Eksteen, 2014; Eksteen & Oraby, 2015a) and presented (Eksteen et al., 2015) research that showed that the amino acid, glycine, can be used in alkaline environments to leach gold, silver as well as copper from a broad range of minerals. We have shown that copper can be leached from its oxides, carbonates, hydroxyl-halides and sulfates, native copper, and various sulfides and arseno-sulfides. In particular, our exploratory work indicated the potential for chalcopyrite leaching. The alkaline (pH 7-12) leaching of chalcopyrite offers a technology that can be applied to heap leaching, in-situ leaching and agitated tank leaching. This paper will investigate the glycine based leaching of chalcopyrite flotation concentrates in particular. The leaching of gold, copper, silver, nickel, cobalt, zinc, lead and platinum group metals from materials bearing them (e.g. ores, concentrates, wastes, tailings and process intermediates) with glycine in a neutral to alkaline solution, and the recovery of the metals from solution as well as recovery of the lixiviant is captured in two pending patents (Eksteen & Oraby, 2014; Eksteen & Oraby, 2015b).

Chalcopyrite is a copper iron sulfide mineral refractory to conventional leach processes at room temperature and even mildly elevated temperatures. In a review by Li et al. (2013) they mention that most researchers regard $CuFeS_2$ as having the formal oxidation states of $Cu^+Fe^{3+}(S^{2-})_2$, but that spectroscopic studies of it indicates a high degree of covalency between S and both Fe and Cu. They also indicated that oxidation of chalcopyrite in moist air results in a Fe(III)–O–OH surface layer on top of a Cu rich sulfide layer overlying the bulk chalcopyrite with the formation of Cu(II) and Fe(III) sulfate, and Cu(I)–O on prolonged oxidation. They also noted that Cu_2O and Cu_2S-like species have also been proposed to form on exposure of chalcopyrite to moist air. S_2^{2-}, S_n^{2-} and S^0 form on the chalcopyrite surface upon acid leaching. The latter two of these species along with a jarosite-like species are frequently proposed to result in surface leaching passivation. Chalcopyrite is often associated with gold, or with gold-bearing pyrite. As smelters often put an upper limit of the iron content of concentrates, the pyrite normally has to be removed by flotation. This is problematic for a number of reasons:

- Often the pyrite and chalcopyrite are finely disseminated and in emulsion textures that makes good liberation very difficult and consequently it makes good flotation-based separation to produce a low-pyrite, copper-rich, concentrate and a low-copper, pyrite-rich, tailings difficult.
- Loss of gold/precious metals credits by not being able to recover the gold in the copper concentrate (with gold often deporting to pyrite).
- The remaining copper in a flotation tailings leads to very high cyanide consumption during subsequent cyanidation of the pyrite concentrate and significant generation of weak-acid-dissociable (WAD) cyanides, which requires expensive cyanide detoxification circuits. It also leads to the generation of significant amounts of ferrocyanide and thiocyanate ions from which cyanide recovery is not economic.

For this reason, a process that can leach copper and gold at the same pH and does not dissolve significant amounts of pyrite, or use significant amounts of cyanide with concomitant cyanide-based waste generation, will be highly desirable. The leaching of gold and silver, and their adsorption onto activated carbon has been discussed elsewhere (Oraby & Eksteen, 2015a), so that the emphasis of this paper is on the leaching and metal recovery of copper from chalcopyrite concentrates and the deportment of iron and sulfur in the process.

FLOWSHEET AND PROCESS CHEMISTRY

A process is proposed whereby copper is leached from a concentrate containing chalcopyrite and pyrite at atmospheric pressure and mildly elevated (60°C) temperature, either "as-is", or after a period of pre-oxidation of the concentrate slurry with oxidation. The role of ultrafine grinding of the concentrate will also be discussed. The high level process proposed below consists of the following steps:

i. Reagent make-up;

$$NaOH(aq) + NH_2CH_2COOH(aq) \rightarrow Na^+(aq) + NH_2CH_2COO^-(aq) + H_2O(l) \quad (1)$$

ii. Alkaline Pre-oxidation (partial);

$$2CuFeS_2(s) + 9O_2(g) + 4OH^-(aq) \rightarrow Cu_2O(s) + 2FeO(OH)(s) + 4SO_4^{2-}(aq) + H_2O(l) \quad (2)$$

iii. Leaching;

$$Cu_2O(s) + 4\,NH_2CH_2COOH(aq) + 0.5\,O_2(aq) \rightarrow 2Cu(NH_2CH_2COO)_2(aq) + 2H_2O(l) \quad (3)$$

With background sulfate and goethite FeO(OH) for material that was pre-oxidised. And, for remaining (un-oxidized) chalcopyrite, the following reaction stoichiometry appears to apply:

$$2CuFeS_2(s) + 4\,NH_2CH_2COO^-(aq) + 8.5O_2(aq) + 4OH^-(aq) \rightarrow$$

$$2Cu(NH_2CH_2COO)_2(aq) + 2FeO(OH) + 4SO_4^{2-}(aq) + 1H_2O(l) \quad (4)$$

iv. Solid-liquid separation (SLS) of pregnant leach solution (PLS) from the leach residue;

SLS giving an aqueous solution of $Cu(NH_2CH_2COO)_2(aq)$, $SO_4^{2-}(aq)$, and cationic spectator ions (e.g. Na^+) and a leach residue containing the gangue minerals, unleached chalcopyrite, pyrite and goethite.

v. Copper recovery from solution (through solvent extraction) and glycinate regeneration:

For solvent extraction (SX), copper is removed from solution using LIX 84-I (or similar);

$$2RH(Org) + Cu(NH_2CH_2COO)_2(aq) \rightarrow R_2Cu(Org) + 2NH_2CH_2COOH(aq) \quad (5)$$

and the loaded organic is stripped with a conventional electrowinning (EW) barren solution of copper sulfate in sulfuric acid solution:

$$R_2Cu(Org) + 2H^+(aq) \rightarrow 2RH(Org) + Cu^{2+}(aq) \text{ with } SO_4^{2-} \text{ as spectator ion;} \quad (6)$$

Copper is recovered by electrowinning from the pregnant aqueous solutions with acid regeneration:

$$Cu^{2+}(aq) + 2e^- \rightarrow Cu(s) \text{ at cathode} \quad (7)$$

$$2H_2O(l) \rightarrow O_2(g) + 4H^+(aq) + 2e^- \quad (8)$$

vi. pH re-establishment and precipitation of sulfite/sulfate species or dissolved silicates, phosphates or aluminates by lime addition (only sulfate shown below as predominant species);

$$2NH_2CH_2COOH(aq) + Ca(OH)_2(aq) + SO_4^{2-}(aq) \rightarrow CaSO_4{\cdot}2H_2O(s) + 2NH_2CH_2COO^-(aq) \quad (9)$$

with Na^+ as a typical spectator ion;

vii. Solid-liquid separation of the residue after lime addition (or the precipitate can be maintained as a slurry and recycled back to leach, after which the precipitated gypsum will be mixed in with the leach residue). This step may be eliminated in some cases where the slurry containing the calcium precipitates (such as gypsum) can be recirculated back to the leach stage;

viii. Recycling of the clarified (or slurry containing the calcium precipitates) barren filtrate containing glycine back to the leach stage.

The key uncertainties in the reaction sequence above are the feasibility of the leaching reaction of chalcopyrite with alkaline glycine/glycinate (step ii) and the copper recovery from alkaline glycinate solutions (step iv). These two steps are covered in the research presented in this article. For the other steps there is sufficient verification of their feasibility in the literature. Chalcopyrite leaching in an alkaline glycinate environment will be evaluated under ambient temperature conditions and at elevated temperature conditions. The regeneration of (sodium) glycinate (step vi, reaction 9) is similar to caustic regeneration is dual alkaline circuits where alkali hydroxides (such as caustic soda) are regenerated from sodium sulfate (and/or sulfite) using slaked lime. Industrial examples include the scrubbing of SO_2 and SO_3 from flue gases (flue gas desulfurisation), whereby the resulting sodium sulfite/sulfate solutions after a caustic scrubbing are regenerated using slaked lime (Bezuidenhout et al., 2012).

The proposed high level flowsheet for a possible industrial process is given in Figure 1.

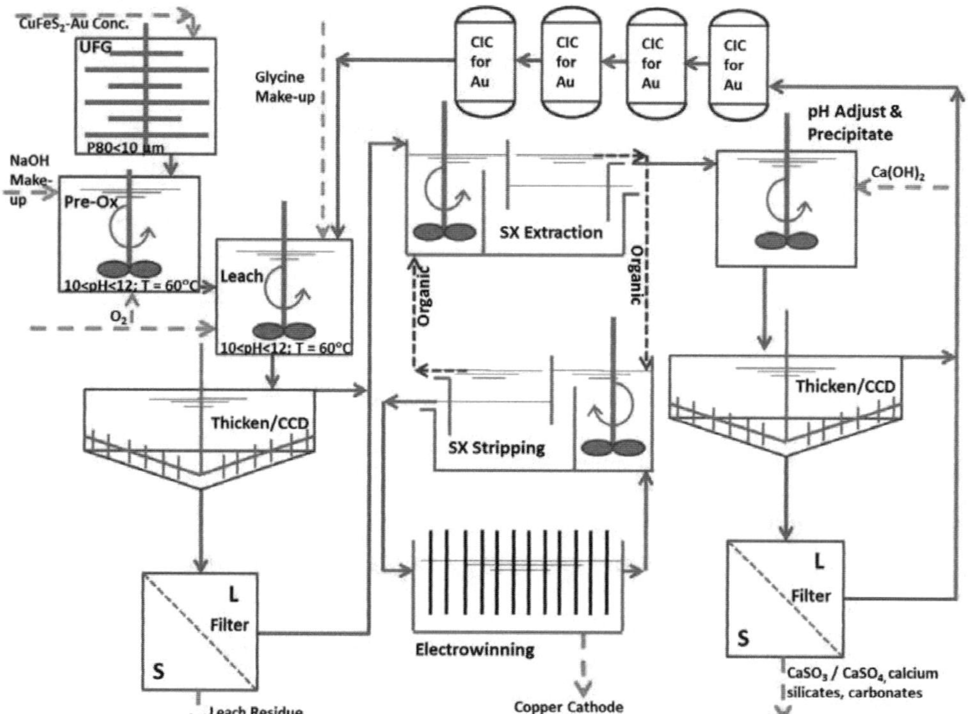

Figure 1 – A high level process flow diagram for the glycine based leaching of copper sulfide ore and concentrates using glycine in an alkaline environment (with concentrate ultrafine grinding, and gold recovery by carbon-in-column included (optional))

The flowsheet in Figure 1 portrays a process which includes the optional steps of pre-oxidation and ultrafine grinding of concentrate, and the recovery of gold by a carbon-in-column process. As the glycine follows the aqueous solution, glycine recovery is directly dependent on the efficiency of solid-liquid separation. Filter wash water would normally be added to wash residues and precipitates and the glycine can be re-concentrated in the wash water through membrane processes, should it be economically justifiable. The leaching and recovery of gold and silver (as glycinates) onto activated carbon have been discussed by Eksteen and Oraby (2015) and Oraby and Eksteen (2015a).

EXPERIMENTAL

Leaching and Iron Rejection

All experiments were carried out using ultrafine chalcopyrite concentrate samples of size 100% passing 10 μm, or "As Is" concentrate, 100% -45 μm. All leaching experiments were conducted using solutions prepared from analytical grade reagents and deionised water. Unless specified, all experiments were conducted using a closed glass reactor. In the atmospheric pre-oxidation tests, an alkaline 550 g slurry (pH 12) containing 50 g chalcopyrite concentrate was oxygenated by adding pure oxygen to the slurry at a flowrate of 75-100 mL/min. Gold leaching was not targeted in the particular test as the leaching of gold and silver has been reported earlier by Eksteen and Oraby (2015b) and Oraby and Eksteen (2015a). The concentrate assay, as analysed by XRF (performed by Bureau Veritas Minerals in Perth, WA), is given in Table 1. Sulfur was analysed by LECO analysis.

Table 1 – Fused disc XRF Assay of the chalcopyrite concentrate (after Loss on Ignition (LOI))

Element (%)	**Si**	**Al**	**Ca**	**Fe**	**Co**	**As**	**K**
Assay (Feed)	3.33	0.17	0.242	27.73	0.032	0.075	0.032
Assay (Leach Residue)	3.66	0.26	19.3	19.2	N/A	N/A	N/A
Element (%)	**Mg**	**Mn**	**Ni**	**Cu**	**Pb**	**Zn**	**S (LECO)**
Assay	1.39	0.023	0.001	24.5	0.136	0.81	31.55
Assay (Leach Residue)	N/A	N/A	BDL	2.06	N/A	N/A	N/A

The pre-oxidation experiments were performed in glass reactor where the slurry stirred at 500 rpm and temperature was maintained at 60 °C under atmospheric pressure for 24 hours, using sodium hydroxide as pH modifier. During the leaching experiments, 500 mL of solution containing 40 g/L glycine at pH 11.0 was heated up to 60 °C oxygenated (0.75 mL/min O_2) and stirred at 500 rpm for 24 hours, using calcium hydroxide to establish the initial pH. This was also the reason for the high calcium content in the leach residue above (where it is present as gypsum). It will be noted that so elements were not assayed (N/A) in the leach residue and this was due to batch submissions on assays at the time when the residues were sent out for assaying (BDL implies below detection limit). Note that the copper recovery cannot be estimated from the residue assay, as the mass change due to a combination of water (or (OH) from goethite) loss and sulfur loss during the LOI and the effect of gypsum precipitation skews the residue assay. In a typical experiment, 5 g of chalcopyrite concentrate was added to the 500 mL of glycine solution. At different times, samples of the leach solution were obtained using a membrane filter. The filtrates were analysed for copper and iron by using atomic absorption spectrophotometry (AAS). The final leach residues were analysed for copper by X-ray Fluorescence (XRF) to calculate the final copper extraction. XRD analyses confirmed that the remaining copper was undissolved chalcopyrite and that the iron converted predominantly to goethite/limonite. Table 2 gives the XRD analysis of the feed and the leach residues. The XRD analysis is not fully quantitative and the % Cu Extraction is best calculated from the assays of the leach solution (by AAS) and the assay of the feed (by XRF). The concentration of Cu in solution was 2.0 g/L after 24 hours for the ultrafine ground material and 1.04 g/L for the "As-received" concentrate (not ground).

Table 2 - Quantitative X-ray Diffraction confirmation of the Chalcopyrite content in the sample

Material	**Chalcopyrite content (Q-XRD), %**
Initial flotation concentrate	73.5
Residue after leaching "as-is" concentrate	39.8
Residue after leaching finely ground of concentrate	6.0

Metal Recovery from Solution

Two approaches were evaluated for metal recovery from solution, through precipitation as covelite, and through solvent extraction. The copper recovery from solution was carried out by either NaSH precipitation or by solvent extraction. During the precipitation test, the required stoichiometric mass of dry NaSH powder was weighed and added to the copper glycinate solution. Copper concentration in solution before and after precipitation was measured using AAS. For the solvent extraction of copper, 10% (v/v) extractants were used from the Mextral product range provided by Hallochem, China. Solvent extraction was performed in conventional separatory flasks. Mextral® 54-100 is a β-diketone in a mixture with high flash point hydrocarbon diluent and Mextral® 84H is a mixture of ketoxime in a high flash point hydrocarbon diluent. DT-100, a pure grade kerosene from Hallochem Group Co. Ltd was used as a diluent for the extractants. No extraction modifiers were used. Copper extraction/ stripping experiments were carried out by mixing required volumes of organic and aqueous phases in a volumetric flask, and then equilibrated by shaking in a thermostatic water bath shaker for the required time and at the established temperature. The mixture was transferred to a separating funnel and after phase disengagement, the raffinate was separated out and the equilibrium pH measured with a field TPS meter. The aqueous solutions were analysed for copper concentration by atomic absorption spectrometry (AAS) using a Varian SpectraAA. The concentration of copper in the organic phase was calculated from the difference between the initial copper concentration in the PLS and its concentration in the raffinate at a fixed organic to aqueous phase ratio. Free glycine concentration was determined using high pressure liquid chromatography (HPLC). Stripping experiments of the copper loaded organic were carried out using 180 g/L aqueous sulphuric acid solution and 30 g/L copper (as sulfate), as is typical of "barren" or return electrolyte from electrowinning tank houses.

RESULTS AND DISCUSSION

Exploratory leaching experiments were conducted using two different grind sizes of chalcopyrite concentrate in solutions containing 40 g/L glycine at pH 11.0 in the presence of oxygen. The particle size distributions of chalcopyrite flotation concentrate which were used in this research were "As Is" 100% -45 μm and ultrafine chalcopyrite of size 100% -10 μm.

Leaching and Iron Rejection

A pre-conditioned/oxidised chalcopyrite concentrate of size 100% - 45 μm was leached at pH 11.0. The copper extraction and pH changes are shown in Figure 2. It can be seen that the rate of initial copper extraction, after pre-oxidation, is quite rapid after which there is a change in rate. The reason of that may attributed to the chalcopyrite that has been partially oxidised and the copper in the oxidised rim is leached fast until reach the surface of chalcopyrite core.

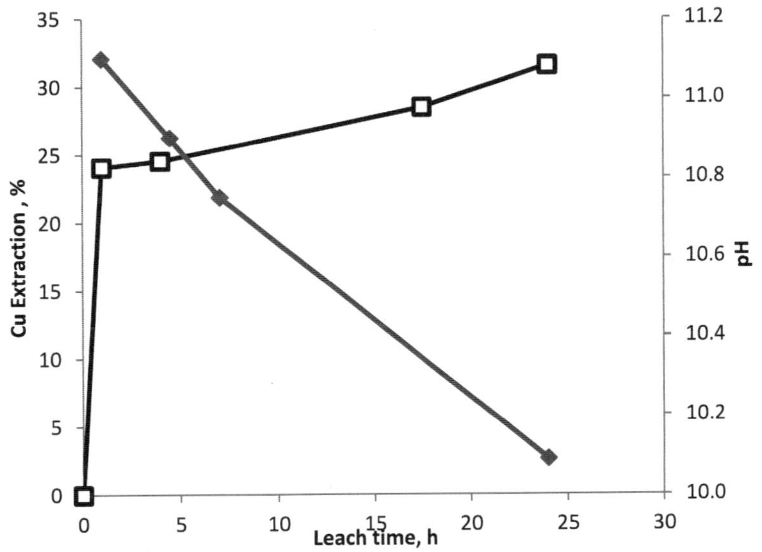

Figure 2 - Copper extraction and pH changes during leaching "as-received" chalcopyrite concentrate (100% -45 μm), after alkaline pre-oxidation

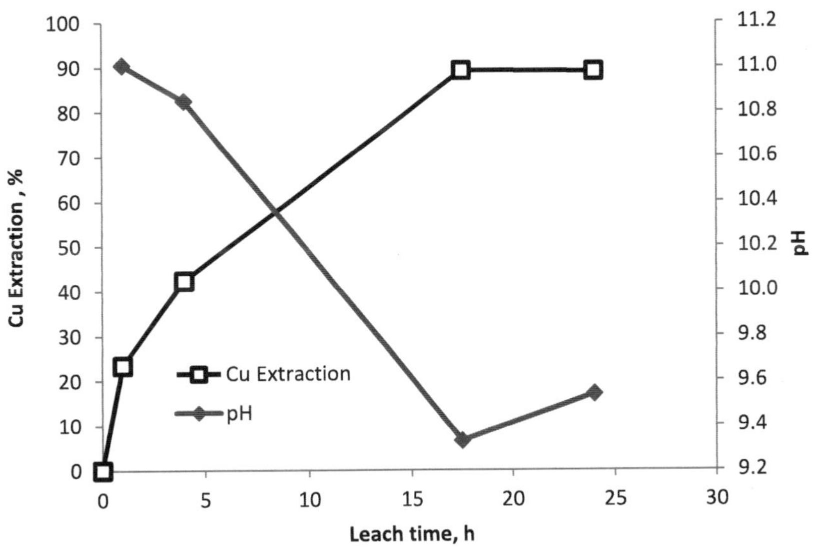

Figure 3 - Copper extraction and pH changes during leaching chalcopyrite concentrate after ultrafine grinding (100% -10 μm), after alkaline pre-oxidation

Copper extraction and pH changes during leaching a pre-conditioned ultra-fine ground chalcopyrite concentrate is shown in Figure 3. It is clear that the copper leach rate is very fast (compared to the "as-received" flotation concentrate) and about 92% of copper was extracted in 17.5 hours of leaching. The iron concentration was less than 13.0 mg/L in the pregnant leach solution after 24 hours. The leach residue was confirmed to be predominantly goethite/limonite and was found to be easily filterable. Iron deportment is therefore controlled through the following mechanisms:

- Pyrite that does not dissolve.
- Iron in chalcopyrite that is oxidised to insoluble Goethite/Limonite.
- Iron dissolution in the leachate is insignificant (typically 13 mg/L or less).

This indicates that iron rejection to the leach residue is quite effective and that jarosites, which are typically associated with acid leaching, are averted.

Metal Recovery from Solution

Precipitation by NaSH Addition

For copper recovery by NaSH precipitation, sulfide ions as NaSH were added to the pregnant liquor at different $Cu:S^{2-}$ molar ratios in order to recover copper from the glycine solution. The copper was successfully recovered up to 99.1% as covelite, CuS at $Cu:S^{2-}$ molar ratio of 1:1 with less than 10 minutes contact time as shown in Table 3.

Table 3 - Copper precipitations at different copper to sulfide molar ratio

Cu to S^{2-} molar ratio	**% Cu precipitated**
1 to 1	99.1
1 to 0.75	77.8

Solvent Extraction

Solvent extraction (SX) experiments show that copper glycinate can be easily extracted from the alkaline aqueous medium using either of the extractants Mextral® 54-100 and Mextral® 84H. Single stage copper extraction into the organic phase of up to 99.7% could be achieved at pH 11.0 as shown in Table 4. The stripping of the organic phase with 180 g/L sulfuric acid solution showed that all the copper can be stripped back into the sulfuric acid aqueous phase. The copper loaded organic phases of 5% Mextral® 84H and 10% Mextral® 54-100 were stripped. The copper stripped from Mextral® 84H was 97.7% at acid concentration of 120g/L while an acid concentration of 50g/L resulted in 100% copper stripping from Mextral 54-100. For the purposes of this research, only aqueous to organic phase ratios of 1:1 are shown. HPLC measurements indicated no glycine transfer to the organic phase during solvent extraction.

Table 4 - Copper deportment during single stage solvent extraction and stripping

Extractant	**% Copper Extracted**	**% Copper Stripped**
Mextral 84H	99.7	97.7
Mextral 54-100	98.6	100

CONCLUSIONS

This research extended earlier research that has showed that numerous copper minerals can be leached in glycine solutions. Due to the importance and predominance of chalcopyrite as copper mineral, this paper focussed on the leaching of copper from chalcopyrite in particular and showed that metal recovery from solution is easily achieved through either solvent extraction or sulfide (NaSH) precipitation. In the case of solvent extraction, complete stripping was achieved in a single stage using an electrolyte composition typical to conventional barren/return electrolyte. It is clear that the chalcopyrite can be leached in an alkaline environment at practical leach rates in the presence of glycine. Chalcopyrite concentrates can either be leached "as-is", or preferentially after fine grinding of the concentrate and prior oxygenation of the pulp, which leads to leaching in excess of 90% in 17 hours. The copper is complexed by the bidentate glycine ligand to form a stable chelate complex. Iron is conveniently rejected to a highly filterable goethite leach residue, whilst pyrite remains essentially unleached. Insignificant iron dissolution occurred and the formation of jarosites, which can lock up heavy metals and precious metals such as silver, is averted. Earlier research has indicated that the presence of copper glycinate significantly enhances leaching of gold under cyanide starved conditions (Oraby and Eksteen, 2015c), indicating that synergies between cyanide and copper glycinates can be exploited to enable gold leaching to occur at the same time as copper leaching and under the same pH conditions. A flowsheet with reagent recycle has been presented and it is clear that the technology enables operations to run flotation circuits at higher mass pulls to produce concentrates at higher copper recoveries (that would normally be unsatisfactory for smelters) which would be totally acceptable for alkaline glycine leaching. The following recommendations can be made with respect to future work, to enhance the leach reaction rate and recovery, compared to the batch tests reported above:

- Optimisation of the oxygen mass transfer during the leach through the use of microbubbles or improved sparging, in order to maximise oxygen utilisation.
- Staged addition of caustic soda during the batch leach to maintain a constant pH, or multiple leach stages in a continuous system with staged addition of pH modifier to maintain constant pH.
- Leach with intermediate surface refreshment, e.g. through a light grind; or rapid agitation with the aim to abrade some surface layers.
- Exploring synergies, such as the "Rapid Oxidative Leach" developed by FL Smidth (Eyzaguirre et al., 2015).
- Continuous removal of the copper glycinate pregnant leach solution, with copper removal, pH readjustment and recirculation of the barren liquor back to leach.
- Evaluating the role of catalytic amounts of cyanide addition on the gold extraction kinetics after the pre-oxidation step and comparing it with gold leach kinetics in the absence of cyanide.

Most of the recommendations from above can be easily explained by considering the main leach reaction (Eq. 4), which explains the formation of the goethite product coating (and need for surface refreshment), and the need to maintain the pH, good O_2 mas transfer (fine bubbles), and the continuous removal of the copper glycinate PLS (following Le Chatelier's Principle), all of which can be executed fairly easily with proper engineering process design. All of these are hard to achieve in a conventional laboratory setup with standard equipment (unless designed as a fit-for-purpose mini-plant).

ACKNOWLEDGEMENTS

The authors would like to acknowledge the financial support provided by Curtin University. The research presented above falls under pending patents, owned by Curtin University. The authors would also like to thank Halochem for the provision of the Mextral range of solvent extractants.

REFERENCES

Bezuidenhout, G.A., Davis, J. Van Beek, B., & Eksteen, J.J. (2012). Operation of a Concentrated Mode Dual-Alkali Scrubber plant at the Lonmin smelter. *Journal of the Southern African Institute for Mining and Metallurgy, 112(7)*, 657-665.

Eksteen, J.J., & Oraby, E. (2014). A process for precious metals recovery, PCT/AU2014/000877.

Eksteen, J.J., & Oraby, E.A. (2015a). The leaching and adsorption of gold using low concentration amino acids and hydrogen peroxide: Effect of catalytic ions, sulphide minerals and amino acid type. *Minerals Engineering, 70*, 36-42.

Eksteen, J.J., & Oraby, E. (2015b). Process for selective recovery of chalcophile group elements. Australian Provisional Patent Application No. 2015900865.

Eksteen, J.J., Oraby, E.A., & Tanda, B.C. (2015). An integrated process for copper extraction from chalcopyrite in alkaline glycinate solutions. *Hydrometallurgy*, In Review.

Eyzaguirre, C., Rocks, S., Klepper, R., Baczek, F., & Chaiko, D. (2015). The FLSmidth® Rapid Oxidative Leach (ROL) Process: A Mechano-Chemical Approach for Rapid Metal Sulfide Dissolution. Hydroprocess 2015. Proceedings of the 7th International Seminar on Process Hydrometallurgy. Eds. T Graber, M.E. Taboada, F. Valenzuela. 22-24 July 2015, Antofagasta, Chile, 11 p.

Li, Y. Kawashima, N., Li, J., Chandra, A.P., & Gerson, A.R. (2013). A review of the structure, and fundamental mechanisms and kinetics of the leaching of chalcopyrite. Advances in Colloid and Interface Science 197–198, pp. 1–32.

Oraby, E.A., & Eksteen, J.J. (2014). The selective leaching of copper from a gold–copper concentrate in glycine solutions. *Hydrometallurgy, 150*, 14-19.

Oraby, E.A., & Eksteen, J.J. (2015a). The leaching and carbon based adsorption behaviour of gold and silver and their alloys in alkaline glycine-peroxide solutions. *Hydrometallurgy, 152*, 199–203.

Oraby, E.A., & Eksteen, J.J. (2015b). Gold leaching in cyanide-starved copper solutions in the presence of glycine. *Hydrometallurgy, 156*, 81-88.

PYRITE'S CRUCIAL ROLE IN COBRE LAS CRUCES ATMOSPHERIC LEACHING PROCESS

*Carlos Frias, Enrique Delgado, Jorge Blanco, and Joaquin Gotor

COBRE LAS CRUCES, S.A.
www.cobrelascruces.com
*(*Corresponding author: carlos.frias@fqml.com)*

ABSTRACT

Since 2009 Cobre las Cruces (CLC) has successfully operated a sustainable mining business near Seville, Spain, applying the most advanced hydrometallurgical technologies to process secondary copper sulphide ores through atmospheric leaching in ferric sulphate media and using oxygen sparging in a cascade of ten reactors. In comparison to conventional copper smelters, the CLC hydrometallurgical process does not generate SO_2 gas, airborne emissions or slags. The copper ore is ground and sent directly to atmospheric leaching reactors; flotation is not required, and consequently, copper extraction is maximized at 93-95%, while producing 72,000 t/y copper cathodes of the highest purity (99.999 wt.% Cu, Grade A). Reserves for the CLC deposit were estimated at 16 million tonnes of ore averaging over 6 wt.% Cu and containing principally a chalcocite mineralization. Other copper bearing species include digenite, covellite, chalcopyrite, bornite and enargite. Pyrite represents over 70 wt.% of the ore. In the CLC leaching process, chalcocite reacts with ferric sulphate at approximately 90 °C and copper is released into solution while ferric ions are reduced to ferrous ions. The ferric sulphate leachant is regenerated continuously by oxygen injection and consumes sulphuric acid. It is remarked that pyrite plays a crucial role to yield high copper leaching rates. When pyrite is oxidized, a small proportion of iron is co-leached and some sulphuric acid is generated. The CLC leaching technology has been designed to allow a certain level of pyrite oxidation, which is positive in many ways: (i) to maintain a Fe/Cu ratio in solution; (ii) to facilitate Fe(II) oxidation; (iii) to produce a high redox potential; and (iv) to counterbalance a bleed solution that removes some iron. Moreover, galvanic interaction between pyrite (cathode) and chalcocite (anode) can substantially increase atmospheric leaching kinetics and efficiency (copper recovery).

KEYWORDS

Copper sulphides, chalcocite, pyrite, atmospheric leaching, pyrite oxidation

INTRODUCTION

The Cobre Las Cruces (CLC) mining company, owned by First Quantum Minerals Ltd, is located approximately 20 km NW of the city of Seville in southern Spain. It comprises of an open pit mine and an integrated on-site hydrometallurgical plant able to produce 72,000 t/y Grade A copper cathodes (Delgado, Frias & Gotor, 2013). Operations started in 2009 and are scheduled to last 11 years.

For a better understanding of the role of iron in the CLC process it is important to know and understand the ore mineralogy. The geological resources are located at a depth of 150 meters and the ore body contains two main ore types, i.e. a massive sulphide ore with 90 wt.% pyrite and a semi-massive ore with 40 wt.% pyrite and over 25 wt.% silica. The copper content in both ore types is rather similar. The main copper bearing minerals include secondary copper sulphides such as chalcocite, digenite, bornite, enargite, and a small proportion of covellite. Other minor species are galena and sphalerite. The main mineral species are shown in Table 1.

Table 1 – Main mineral species in CLC ore

Mineral Species	Average content (%)	Range (%)
Chalcocite, Cu_2S	6.5	6–7
Covellite, CuS	0.9	0.7–1
Chalcopyrite, $CuFeS_2$	0.9	0.5–1.2
Pyrite, FeS_2	68	65–75
Silica, SiO_2	21	15–28
Gangue	2.7	–

The average grade for the whole project (life of mine) is around 6 wt.% copper, and such a high-grade enables direct leaching of the copper ores without a prior concentration stage, such as flotation. The average chemical composition is presented in Table 2.

Table 2 – CLC ore chemical composition

Element	Average content (%)
Cu	6.04
Fe	30.5
Zn	0.20
Pb	0.67
As	0.32
S	36.9

The process plant uses the following main steps, i.e. crushing, grinding, atmospheric leaching, solvent extraction and electrowinning, in order to produce grade A copper cathodes (Delgado, Frias & Del Pino, 2014). The main chemical reactions consist of ferric leaching of copper sulphides and pyrite, ferrous oxidation to ferric with oxygen, and minor impurities being co-precipitated and/or adsorbed into iron oxides precipitates, yielding an extremely efficient purification step.

Iron plays a crucial role throughout the whole CLC hydrometallurgical process, and pyrite oxidation is the key to achieve such a benefit. When pyrite is oxidized, a small proportion of iron is co-leached and some sulphuric acid is generated. The presence of pyrite is rather positive in the CLC leaching process and has the following advantages:

- Release of iron helps to keep a proper Fe/Cu ratio and redox potential in the reaction media.
- Release of acid improves Fe(II) oxidation by oxygen.

- Galvanic interactions between pyrite (cathode) and chalcocite (anode) may substantially increase ferric atmospheric leaching kinetic and efficiency.

This paper presents a review of iron control and management in the CLC hydrometallurgical process plant and its associated processes ranging from leaching reactors to final bleed stream treatment for iron removal. The water balance is also considered.

CLC HYDROMETALLURGICAL PLANT

The hydrometallurgical plant comprises of the following stages:

- Grinding and thickening.
- Leaching.
- Solvent extraction.
- Electrowinning.
- Bleed treatment.

Grinding and Thickening

For copper ore grinding, CLC operates a conventional ball mill running in closed circuit with hydrocyclones to yield ground material 100% passing 150 microns. The cyclone overflow goes to a thickener with the purpose of removing as much water as possible, since additional water would only dilute leachant solution volume.

Leaching

The thickened slurry passes to ferric leaching, which takes place in stirred tanks with oxygen injection operated at atmospheric pressure and 80-90 °C (Figure 1). After leaching, a solid liquid separation step is carried out to separate the leached residue from the pregnant solution bearing mainly copper and iron. The chemical reactions involved in the leaching process are:

$$CaCO_3 + H_2SO_4 + H_2O = CaSO_4.2H_2O + CO_2 \tag{1}$$

$$Cu_2S + 2\ Fe_2(SO_4)_3 = 2\ CuSO_4 + S + 4\ FeSO_4 \tag{2}$$

$$CuS + Fe_2(SO_4)_3 = CuSO_4 + S + 2\ FeSO_4 \tag{3}$$

$$2\ FeSO_4 + H_2SO_4 + \tfrac{1}{2}\ O_2 = Fe_2(SO_4)_3 + H_2O \tag{4}$$

$$FeS_2 + H_2O + 7/2\ O_2 = FeSO_4 + H_2SO_4 \tag{5}$$

The aforementioned reactions are targeted to yield several specific and unique objectives:

- Carbonate neutralization from the gangue minerals in accordance to reaction (1).
- Oxidative copper leaching of secondary copper sulphides by means of ferric sulphate (2, 3). Sulphide ions are oxidized to elemental sulphur and ferric ions are reduced to ferrous. The leaching process is very efficient for secondary copper sulphides yielding 93-95% extraction; however, primary copper sulphides such as chalcopyrite do not leach very well, usually not more than 40%.
- Ferrous to ferric oxidation in heterogeneous media involving aqueous and gas phases (4). Oxygen mixes in the solution and reacts to produce ferric sulphate and water. Net reagents to the system are oxygen and sulphuric acid, since iron is just changing its oxidation state as copper is leached. The aim is to maintain a redox potential around 700 mV (SHE).

- Iron leaching. It is a heterogeneous phase reaction involving oxygen and solids (5). Iron leaching from the pyrite is necessary to some extent to counterbalance the water bleed from the main circuit, which removes iron.

Figure 1 – General overview of atmospheric leaching reactors

The produced pregnant leach solution (PLS) after leaching stage contains the main components shown in Table 3.

Table 3 – Produced PLS typical composition

Component	Range (g/l)
Cu	35-40
Fe	40-45
H_2SO_4	20-25

Solvent Extraction

The aim of the solvent extraction (SX) system is selectively remove copper from PLS by mixing the aqueous phase with an organic phase bearing a chemical reagent (aldoxime in kerosene as diluent) which selectively bonds copper ions. The efficiency of this reaction greatly depends on the pH of the PLS. In the extraction stage, organic and PLS mix in counter-current flows. There are two counter current extraction circuits: a primary circuit for the primary PLS coming from the PLS pond (high copper content) and a secondary circuit for the secondary PLS (low copper content) mainly composed of the leaching tailing filters washing outlet stream. The organic loop is common for both circuits. The loaded organic is washed and then mixed with spent electrolyte in order to strip the copper and send it to the copper tank house.

Electrowinning

In the electrowinning stage, copper is electrodeposited as a massive sheet on the cathode while water is decomposed at the anode and simultaneously generates sulphuric, as shown in next reactions:

Cathode: $CuSO_4 + 2e^- = Cu + SO_4^=$ (6)

Anode: $H_2O - 2e^- = \frac{1}{2} O_2 + 2 H^+$ (7)

Overall: $CuSO_4 + H_2O = Cu + \frac{1}{2} O_2 + H_2SO_4$ (8)

The copper tank house has 144 cells with 84 cathodes into each one. Cathodes are made of stainless steel and anodes of lead-calcium-tin alloy. Harvesting cycles are approximately seven days. Used additives are dextrin and cobalt sulphate aiming to smooth the deposited grains and prevent anode corrosion, and consequently, avoiding cathodes lead polluting. Design current density is over 300 A/m^2.

Bleed Treatment

After copper extraction, the primary raffinate goes back to leaching reactors, and a small portion is bled out to keep the water balance in the main circuit under stable conditions. The bleed solution is sent to a neutralization plant prior to its discharge to the river as a purified effluent. In this stage, milk of lime is added in aerated stirred tanks to precipitate main metals as hydroxides, neutralize sulphuric acid as gypsum, and adsorb very efficiently minor impurities into iron oxides precipitate to meet environmental regulations. The precipitated solids are separated by thickening and filtration while overflow solution goes to a pond and from there to the river as the final purified effluent discharge.

IRON CONTROL AND MANAGEMENT

The CLC hydrometallurgical process is based on atmospheric ferric leaching of secondary copper sulphide ores yielding 93-95% Cu recovery; obviously, iron plays a crucial role to achieve this high process efficiency. In addition to the leaching process, there are other process stages where the iron content has a significant effect as is described below.

Copper Sulphide Leaching

The role of iron in the leaching process is summarized as follows:

- Oxidation of copper sulphide ores in a ferric sulphate media to yield a pregnant leaching solution (PLS) mainly bearing copper and iron, together with other minor soluble species such as As and Zn.
- Oxidation of pyrite to a small extent by means of ferric sulphate, thereby releasing iron and sulphuric acid into the leaching media.

According to our process design, the required iron concentration in the leaching reactors is about 50 g/L with a ferric to ferrous ratio of 1:1 and a redox potential of about 700 mV (SHE). These are the key operational parameters to get high copper extraction levels and plant operators must track and manage them carefully. A useful operating tool is shown in Figure 2, which illustrates how the Fe^{3+}/Cu weight parameter (t/t) varies depending on the total iron content in solution and the ratio Fe^{3+}/Fe^{2+} in the leaching media when assuming a copper feed rate of 10 t/h. The best operating conditions are in the right hand side of the curves.

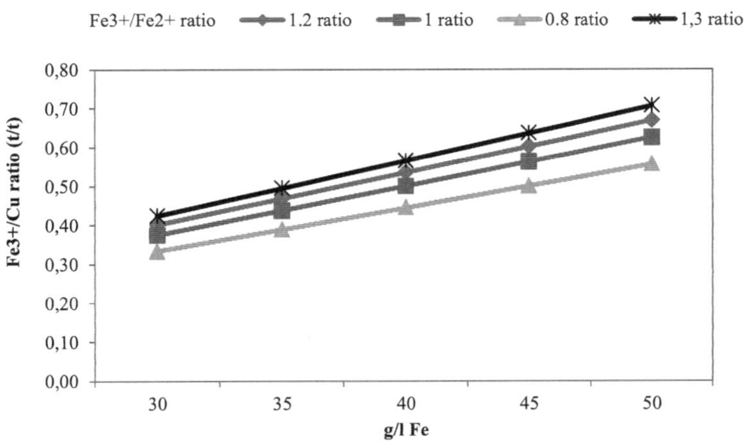

Figure 2 – Variation of Fe^{3+}/Cu parameter in regards total iron in solution and Fe^{3+}/Fe^{2+} ratio

Pyrite Leaching

As the hydrometallurgical plant scheme requires a raffinate bleed stream to fit the water balance, a certain amount of iron is taken out and the leaching process needs to release more iron into solution in order to counterbalance this effect. Pyrite is the major component in the ore feed (68 wt.%), and consequently, it is the main source of iron addition according to reaction (5). Historically, in order to balance the bled out iron the hydrometallurgical process only requires leaching of around 3% of the pyrite, and based upon CLC process and plant operation experience it can be stated that:

- Pyrite leaching behavior fits a sequential kinetic model where copper sulphides leach first and then pyrite is leached.
- Residence time in the reactors cascade is therefore a key factor to achieve both, i.e. efficient leaching of copper and iron to their respective target levels.
- On one hand, it is important to maintain the maximum required residence time for high copper extraction, while on the other hand, additional reaction time is necessary to leach some of the pyrite minerals.

Taking into account the aforementioned philosophy, it is relevant to note that the initial hydrometallurgical plant commissioned in 2009 was designed to run a train of 8 leaching stirred tank reactors with oxygen supply. However, in 2012 and 2013 the original design scheme was modified by adding two larger pre-reactors at the front of the original cascade. The new pre-reactors have increased substantially the overall residence time of the leach, 6-8 additional hours, and that factor has had a very positive impact with respect to copper leaching efficiency, and in particular, to increase iron extraction and improving iron control in the overall leaching process.

Another critical factor affecting the pyrite leaching rate is copper ores mineralogy. When plant feed contains a certain proportion of refractory copper ores, e.g. primary copper ores (chalcopyrite, polymetallic ores) that are difficult to leach in ferric sulphate reaction media, there is less amenable copper to leach, and as a result, more reaction volume is available for pyrite oxidation. This increases the iron concentration in the PLS.

Figure 3 shows a six months operating time window where some refractory copper ore was processed in the CLC hydrometallurgical plant. During October 2014 copper leach recovery decreased substantially, giving more than enough room for pyrite oxidation. Since copper reactions would need less reaction volume, the highest rate of pyrite leaching was observed.

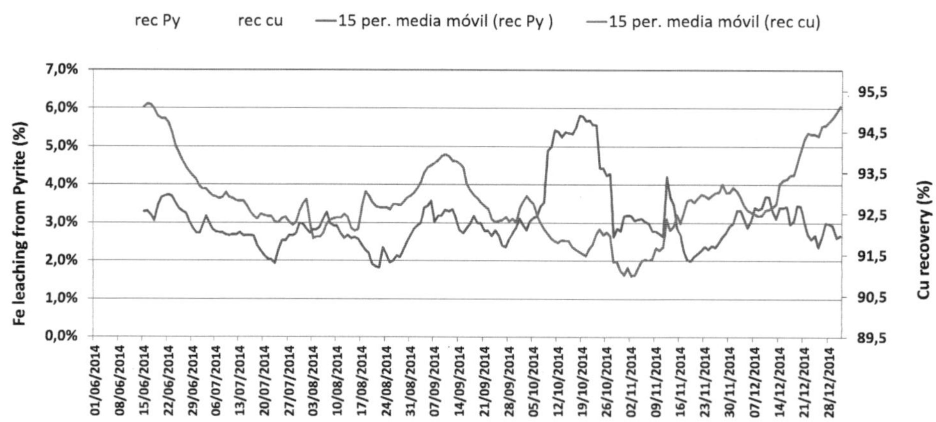

Figure 3 – Effect of primary copper minerals presence in regard to pyrite leaching extent

Iron Oxidation Efficiency in Leaching Reactors

Oxidation of ferrous to ferric takes place by means of oxygen and acid consumption according to reaction (4). Thus, the control of sulphuric acid and oxygen in the leaching reactors is essential.

Sulphuric Acid Control

The copper ores fed to the hydrometallurgical plant do contain a certain amount of carbonate minerals in the gangue, which is neutralized by adding sulphuric acid as shown in reaction (1). Sulphuric acid make up is utilized to drive both reactions (1) and (4), and overall acid consumption is about 35-40 kg per ton of ore. On the other hand, there is some sulphuric acid generation through pyrite oxidation according to reaction (5).

It is important to have some free sulphuric acid in the leaching reactors to facilitate Fe^{+2} oxidation and to avoid any iron precipitation, e.g. jarosite formation. On the other hand, it is desirable to have low acid concentration in the PLS to favor copper solvent extraction. Operators have to keep a proper acid concentration profile in the reactors cascade in line with these considerations.

Oxygen Control

Iron oxidation is a heterogeneous phase reaction in which oxygen gas and ferrous ions in solution react to get a certain redox potential, and all this greatly depends on oxygen solubility and diffusion through the slurry. Higher temperatures improve copper leaching kinetic but reduce oxygen solubility. Thus, oxygen behavior depends on a combination of three main factors, i.e. pulp density, temperature, and agitation.

In order to control the pulp density in the reactors, the leaching circuit has been modified by partial recycling of the PLS to the reactors cascade. Although PLS recycling slightly decreases the residence time, it produces a more dilute pulp, which is positive for pyrite oxidation and the copper leach recovery as shown in Figure 4.

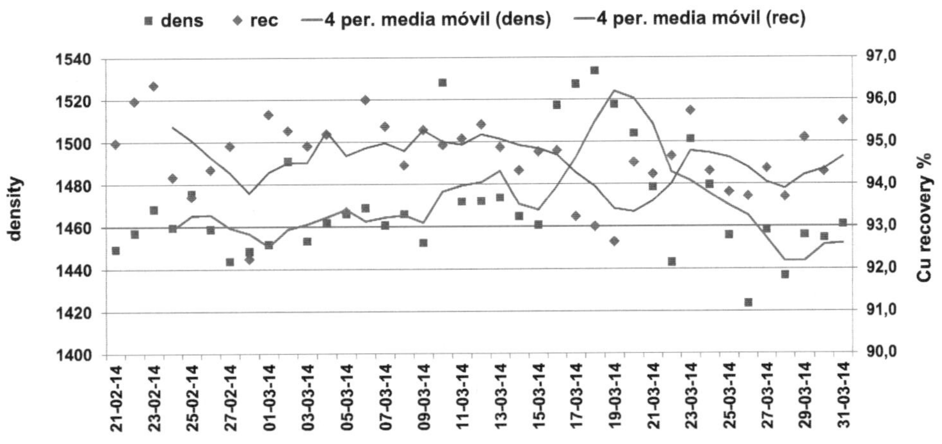

Figure 4 – Effect of pulp density during leaching on the copper recovery rate

Copper sulphide leaching, ferrous oxidation and pyrite leaching are all exothermic reactions, and consequently, the leach reactors are equipped with a cooling system to dissipate reactions enthalpy. CLC operators have developed a specific heat balance to properly control the temperature profile, showing how sulphide oxidation takes place in the beginning of the reactors cascade (first slope: copper sulphide oxidation till reactor 3) and in final reactors cascade (second slope: pyrite oxidation till reactor 8). See Figure 5.

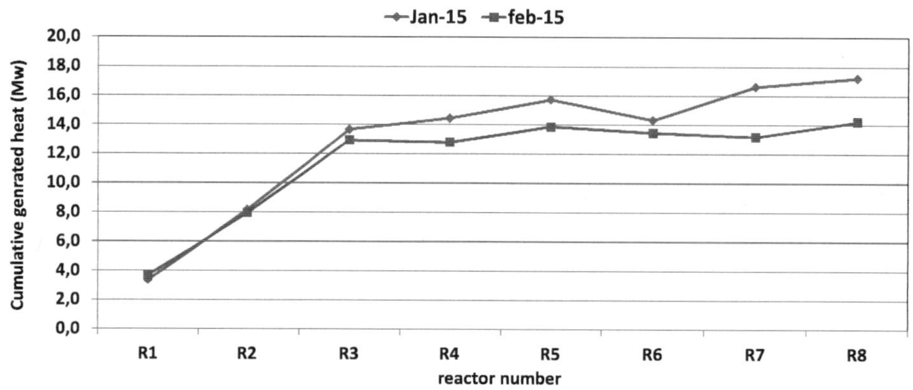

Figure 5 – Effect of pulp density during leaching on the copper recovery rate

In addition, to enhance iron oxidation efficiency, CLC has developed an innovative and low energy agitation system that provides very efficient oxygen transfer resulting in high copper (and also iron) leaching rates. In addition, these new agitators are built of a very resistant material that has doubled its service life in the leaching reactors cascade.

Iron Removal in Bleed Stream Treatment Process

The bleed stream solution from SX raffinate is treated in two steps using a milk of lime and aeration, yielding an extremely efficient process to remove iron and accompanying impurities, and producing a final effluent that fits the most restrictive environmental regulations. Main components of the bleed stream are iron, acid, arsenic and other minor elements.

In the first treatment step, the reaction with lime is performed at pH around 4 obtaining a solid phase composed of gypsum and iron oxides (e.g. goethite) together with ferric arsenate:

$$H_2SO_4 + Ca(OH)_2 = CaSO_4.2H_2O \tag{9}$$

$$Fe_2(SO_4)_3 + 3\ Ca(OH)_2 + 4\ H_2O = 2\ FeOOH + 3\ CaSO_4.2H_2O \tag{10}$$

$$Fe^{3+} + AsO_4^{3-} = FeAsO_4 \tag{11}$$

In the second treatment step, lime is added to reach a pH close to 8.5, aiming to remove metals remaining in solution (Me= Cu, Zn, etc.):

$$MeSO_4 + Ca(OH)_2 + 2\ H_2O = Me(OH)_2 + CaSO_4.2H_2O \tag{12}$$

The neutralized pulp is a blend of gypsum and iron oxides that has the intrinsic capacity to adsorb and collect many impurities, mainly oxyanions (Cornell & Schwertmann, 2003; Sherman & Randall, 2003). Because of iron oxides adsorption property, the final effluent generated after bleed treatment presents maximum purity and fits all required environmental targets for discharge to the river.

Control of iron precipitation depends greatly on the pH and air diffusion. CLC has recently installed a new neutralization reactor provided with hyper-spargers to inject air. This produced satisfactory results; for example, As removal efficiency is > 99.995%, reaching less than 50 ppb, which is the fixed discharge limit.

PYRITE GALVANIC EFFECT

As reported, pyrite plays a positive role in the leach in terms of reaching a specific redox potential and achieving the most effective process performance. There is, however, another crucial benefit of pyrite in the CLC process that has not been mentioned before, i.e. the galvanic effect, consisting of the physical contact between pyrite particles (acting as cathode) and copper sulphide particles (acting as anode) inside the pulp, which catalyzes and enhances the copper sulphide ferric leaching process.

Galvanic interactions occur when particles of pyrite are in contact and are present in a medium which facilitates charge transfer with other conductive metal sulphides that present a lower rest potential, in such a way, that pyrite is able to promote anodic dissolution of those sulphides. Figure 6 (Dreisinger, 2009) shows a galvanic interaction mechanism to leach chalcopyrite ores in ferric sulphate media.

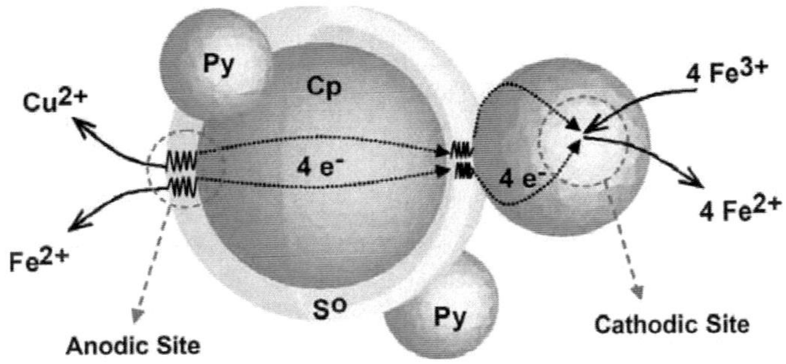

Figure 6 – Galvanic interactions in between pyrite (Py) and chalcopyrite (Cp). (Dreisinger, 2009)

Galvanic effects are well known for a long period of time, and Mehta and Murr (1983) explained such phenomenon in terms of the rest potential of each mineral with respect to the standard hydrogen electrode, which is shown in Table 4.

Table 4 – Rest potentials of some sulphide minerals

Mineral	E_e vs. SHE (mV)
Pyrite, FeS_2	630
Chalcopyrite, $CuFeS_2$	520
Chalcocite, Cu_2S	440
Covellite, CuS	420

Some recent R&D work has been devoted to galvanic interactions between pyrite and secondary copper sulphides (Rivera-Vasquez & Dixon, 2015). In the case of the CLC leaching process, the main sulphides in contact with pyrite are the secondary sulphides chalcocite and covellite, which have a much lower rest potential than pyrite, so galvanic interaction between those species will likely be very intense, favoring fast copper leaching kinetics.

Anyway, it is not easy to propose a specific reaction model or mechanism to describe in detail the pyrite galvanic interactions with chalcocite or covellite inside a so complex physic-chemical system such as the atmospheric leaching reactors running at the CLC hydrometallurgical plant. Probably it is reasonable to assume a model similar to that shown in Figure 6 because real operating conditions at the CLC leaching process fit quite well the established target for such a reaction mechanism:

- Ratio Pyrite/Cu Sulphides: >5
- Temperature: 80-90 °C
- Redox potential: >700 mV (SHE)
- Efficient oxygen transfer and oxidation capacity.

CONCLUSIONS

Iron control in the Cobre Las Cruces hydrometallurgical plant is essential to achieve high process performance to yield 93-95% copper recovery. The main issues in regard to iron control and management can be summarized as follows:

- Iron control comprises of several process stages mainly focused on copper ore leaching, pyrite oxidation/leaching, ferrous oxidation reactions and iron removal in a bleed stream treatment plant.
- Copper and iron leaching correspond to a sequential model, where copper leaches first. Consequently, operating parameters such as residence time are key indicators for iron leaching control.
- Ferrous ions oxidation to ferric in the leaching system depends on oxygen transfer efficiency, and operating parameters such as agitation, temperature, and pulp density have a great influence on that reaction.
- Galvanic interactions between pyrite and secondary copper sulphide species present in the CLC hydrometallurgical plant, such as chalcocite and covellite, likely play a relevant role to enhance copper leaching kinetics.
- Iron control in the final effluent is improved by means of increasing residence time, pH tuning, and oxygen diffusion enhancement.

ACKNOWLEDGEMENTS

The authors would like to express their gratitude to Cobre Las Cruces Company to facilitate the development of this work and authorizing its dissemination.

REFERENCES

Cornell, R.M. & Schwertmann U. (2003). *The iron oxides: Structure, properties, reactions, occurrences and uses.* Second Edition. Wiley-VCH Verlag GmbH & Co. KGaA.

Delgado, E., Frias, C. & Del Pino, M. (2014). Hydrometallurgical processing of copper sulphide ores in Cobre Las Cruces mine. *Proceedings of International Conference on Metallurgy of Non-Ferrous Metals*, November 17-19, 2014, Krakow, Poland.

Delgado, E., Frias, C. & Gotor, J. (2013). Las Cruces copper plant start-up: Beyond design performance through process improvements implementation. *Proceedings of Copper 2013 International Conference* (pp. 25-42), December 1-4, 2013, Santiago, Chile.

Dreisinger, D. (2009). Key note address: Hydrometallurgical process development for complex ores and concentrates. *Journal of South African Institute of Mining and Metallurgy*, 109 (5), 253-271.

Mehta, A.P., Murr, L.E. (1983). Fundamental studies of the contribution of galvanic interaction to acid-bacterial leaching of mixed metal sulfides. *Hydrometallurgy* 9 (1983), 235-256.

Rivera-Vasquez, B.F. & Dixon, D. (2015). Rapid atmospheric leaching of enargite in acidic ferric sulfate media. *Hydrometallurgy* 152 (2015), 149-158.

Sherman, D.M. & Randall, S.R. (2003). Surface complexation of arsenic(V) to iron(III) (hydr)oxides: Structural mechanism from ab initio molecular geometries and EXAFS spectroscopy. *Geochimica et Cosmochimica Acta* 67 (22) (2003), 575-580.

SCORODITE IN THE CESL PROCESS FOR COPPER-ARSENIC CONCENTRATES

*K. Mayhew[1], H. Salomon-De-Friedberg[1], and A. Lossin[2]

[1]*Teck Resources Limited*
12380 Horseshoe Way
Richmond, B.C., Canada
V7A 4Z1
*(*Corresponding author: keith.mayhew@teck.com)*

[2]*Aurubis AG*
Hovestraße 50
Hamburg 20539, Germany

ABSTRACT

The proportion of arsenic-bearing copper concentrates that supply the copper industry is on the increase. The current strategy of miners and smelters to blend feeds to reduce average arsenic levels is rapidly becoming unsustainable as more concentrates with even higher arsenic levels enter the marketplace. In addition, the capability of smelters to manage increasing arsenic levels to more stringent and restrictive environmental standards is limited by available process technologies. CESL Cu-As Technology is a viable technical and economic alternative to traditional smelting. It offers the advantage of simultaneous leaching and precipitation of the arsenic in a single processing step while achieving high copper extractions. The process has been piloted in six separate, fully-integrated pilot campaigns for feeds ranging from ~1% to 12% arsenic in concentration. This paper will focus on the behaviour of arsenic in CESL's leaching/precipitation step and operational factors that influence it, and on the chemical and physical characteristics of the subsequent iron-arsenic residue that represents today's best available technology from a long-term sustainable (capital/operating costs, metal recovery, safety, environmental) perspective. The examples provided will demonstrate copper extractions approaching 99% and arsenic deportment to residue of 99%. Instantaneous settling rates of the residue were measured at 15+ m/hr. Data from subsequent short and long-term stability testing of the residue confidently support the view that the scorodite-bearing residues from the process will be stable and provide an acceptable arsenic management practice. The technology offers increased revenue and environmental certainty for miners, smelters & refiners to support the development of metallurgically challenged copper-arsenic deposits. Furthermore, results will be presented demonstrating the technology's ability to successfully incorporate hazardous arsenic containing smelter by-products such as dust into the process feed and simultaneously extract copper value while fixing the arsenic in a stable residue.

KEYWORDS

Arsenic, scorodite, solid/liquid, copper concentrate, CESL

INTRODUCTION

There are important flowsheet implications for hydrometallurgical processes that treat feeds with appreciable levels of arsenic. Processes that rely on high pH or basic leaching conditions are left with solubilized arsenic, where iron is absent, that must subsequently be fixed in an environmentally sound manner. Processes that operate under acidic conditions have the opportunity to co-precipitate the arsenic along with the iron that is being leached. Since arsenic precipitation with iron removal can be incorporated into a single leaching step, this is seen as advantageous.

Various compounds can be precipitated under hydrothermal conditions (>100°C) when iron and arsenic are found in solution. Swash and Monhemius (1998) reported most of these minerals as shown in Table 1.

Table 1 – Major iron and arsenic mineralogical components precipitated under hydrothermal conditions

Mineral		Ideal Formula
Hematite		Fe_2O_3
Basic Iron Sulphate	(BFS)	$Fe(SO_4)(OH)$
Jarosite		$H_3OFe_3(SO_4)_2(OH)_6$
Scorodite		$FeAsO_4.2H_2O$
Ferric Arsenate sub-Hydrate	(FAsH or Type 1 (Phase 4))	$Fe(AsO_4)(H_2O)_{0.75}$
(Basic) Ferric Arsenate Sulphate	(BFAS or Type 2 (Phase 3))	$Fe_2(AsO_4)(SO_4)(OH).nH_2O$

In an iron only system, hematite is favored at 150 °C when solution acidity is low, jarosite at higher acid levels, while basic iron sulphates tend to dominate at higher temperatures and acidities (Fleming, 2009).

When arsenic is present, scorodite is the dominant phase to precipitate at 150 °C (Gomez et al., 2011) which is the temperature of greatest interest in the present paper. The precipitate is not ideal containing about 1% S in its lattice depending upon initial conditions (Fujita et al., 2012). The finding is consistent with that of other investigators as summarized by Gomez et al. (2008) which is suggestive of minor, partial sulphate substitution for arsenate in the scorodite lattice. The presence of other cations such as copper in solution may cause a short-lived intermediate to form prior to crystallizing to scorodite as suggested by Gomez et al. (2011) which can result in copper and additional sulfur incorporation in the scorodite lattice. The structure of this intermediate approximates that of non-crystalline basic ferric arsenate which is normally found at higher temperature. If conditions are allowed to favor jarosite formation, arsenic incorporation approaches 1% As in sodium jarosite and 2.5% As in the potassium jarosite phase (Dutrizac et al., 1987). However, by limiting ferric levels, scorodite formation is favored. Beudantite ($PbFe_3AsO_4SO_4(OH)_6$), an analogue of jarosite, is believed to form ahead of scorodite (Escobar & Monhemius, 1988) which might make lead in feed undesirable.

The ratio of As:Fe plays an important role in scorodite formation. Not only the rate, but also the extent of arsenic precipitation, decreased with excess iron (Monhemius & Swash, 1998). Any iron in excess of the stoichiometric requirement remained in solution in their testing. As might be expected, higher acid levels led to lower amounts of arsenic precipitated. Free acidity should be kept well below 60 g/L (Swash & Monhemius, 1998). Time did influence the nature of the precipitate as noted above.

In an analysis of a Teck's CESL residue generated when leaching an actual concentrate, XRD analysis failed to detect an arsenic phase perhaps, in part, because the As:Fe weight ratio in feed was only 0.02 (Gomez et al., 2008). Using attenuated total reflectance infrared (ATR-IR) spectroscopy techniques, the investigators concluded that the arsenic had precipitated in a phase resembling basic ferric arsenate sulphate more so than scorodite despite the 150 °C operating temperature.

Subsequent mineralogical analysis of CESL residues produced in the pilot plant demonstrated that the arsenic was precipitated as Type II scorodite ferric arsenate (Bruce et al., 2012).

Only scorodite and basic ferric arsenate sulphate easily pass the standard TCLP test for dissolved arsenic (Gomez et al., 2008) implying the possible existence of two stable forms of iron-arsenic precipitate. Longer term tests, approaching 300 days, indicated continuing stability of these two phases. The phases showed similar leaching characteristics over time, but maintaining a low pH environment was important to both (Gomez et al., 2011). Blending of these types of residues with high pH tailings would therefore appear to be unwise as a long term storage strategy. Swash & Monhemius (1998) pointed out that scorodite, unlike other iron arsenates, is to be found in many climates around the world within a wide range of geological environments, and that this provides compelling evidence for its long term stability.

With CESL Cu-As Technology (operated at 150 °C), the precipitation of arsenic, while dissolving copper from concentrate feed, occurs under conditions that differ somewhat from the idealized studies referred to above. Instantaneous ferric and arsenic solution levels may be lower because leaching and precipitation occur simultaneously. Free acid is also perhaps lower because it is consumed in some of the leaching reactions. Impurities are introduced to the leach solutions at varying concentrations, particularly when operating fully integrated, continuous circuits. Materials providing potential seed surfaces for precipitation can also be varied. The resulting iron-arsenic residues are therefore more complex and their characteristics are the principal topic of this paper.

EXPERIMENTAL

CESL residues were generated in a number of continuous pilot campaigns, as well as in batch bench trials. The pilot flowsheet has been previously described elsewhere (Salomon-de-Friedberg, 2014). Throughput was typically of the order of 100 kg/day of concentrate depending upon its mineralogy which influenced the autoclave heat balance. Closed circuit operation enabled impurity concentrations to reach steady state.

Batch testing was performed in a 2 L titanium Parr autoclave. Concentrates were mixed with acidic, chloride and copper containing feed liquor to a slurry volume of 1.1 L and heated to 150 °C at which point oxygen was introduced. Total vessel pressure was usually kept 14.5 bar. A bleed ensured that oxygen in the gas space was maintained at a minimum of 80% O_2. The reaction was allowed to proceed for a predetermined length of time which was normally 90 minutes.

Slurry discharge samples were filtered and washed with the residue dried at 60 °C overnight prior to assaying. Both solids and filtrate were assayed by ICP-OES (inductively couple plasma optical emission spectrometry). Solids had first been digested in aqua regia. The higher copper levels in filtrates were titrated using recognized methods.

Residues were characterized by the following ways: chemical composition, solid-liquid separation characteristics, XRD patterns, behavior under TCLP conditions and, behaviour in long term stability tests.

RESULTS

Fe & As Precipitation During Leaching

Maximizing arsenic precipitation within the CESL autoclave operating at 150°C is considered advantageous to generate a stable residue. Principal factors influencing the extent of arsenic precipitation were believed to be acidity of the leach filtrate and the arsenic to iron ratio fed to the autoclave as noted by Monhemius & Swash (1999). In the present work, a better correlation was found to exist between the fraction of arsenic dissolved in filtrate and the expression “free acid equivalent”. This is defined as the acid titrated in filtrate plus acid that would have been generated had the soluble iron present precipitated to form acid. The corresponding relationship for a set of a different concentrates (ranging in concentration from 7% to 13% As) is given in Figure 1.

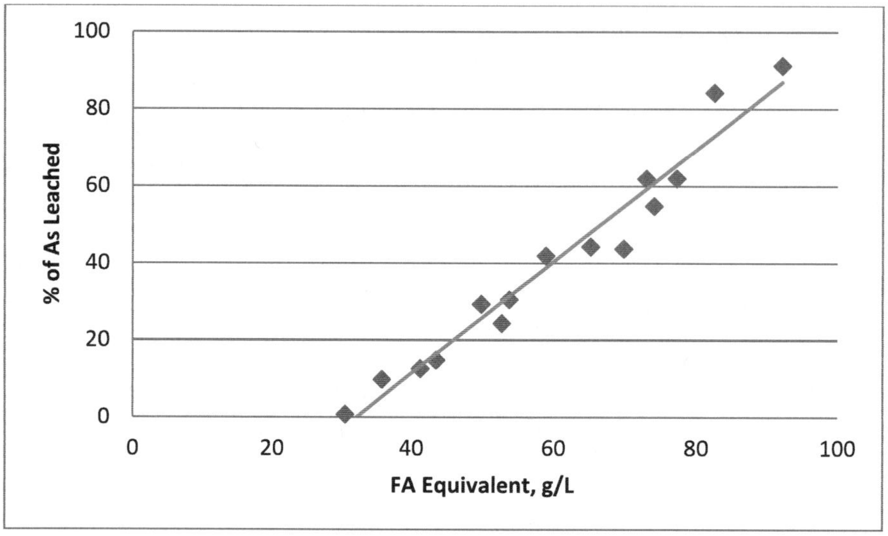

Figure 1 – Arsenic leaching from concentrate versus "Free Acid Equivalent"

This relationship indicates that arsenic in filtrate can be minimized by restricting the acid and iron in feed solution and by limiting acid generating species, particularly pyrite, in the feed concentrate. Minimizing sulphur oxidation in the autoclave is also beneficial from a capital and operating cost perspective.

For a given concentrate, the fraction of solubilized arsenic correlated closely with the final iron in leach filtrate (Figure 2).

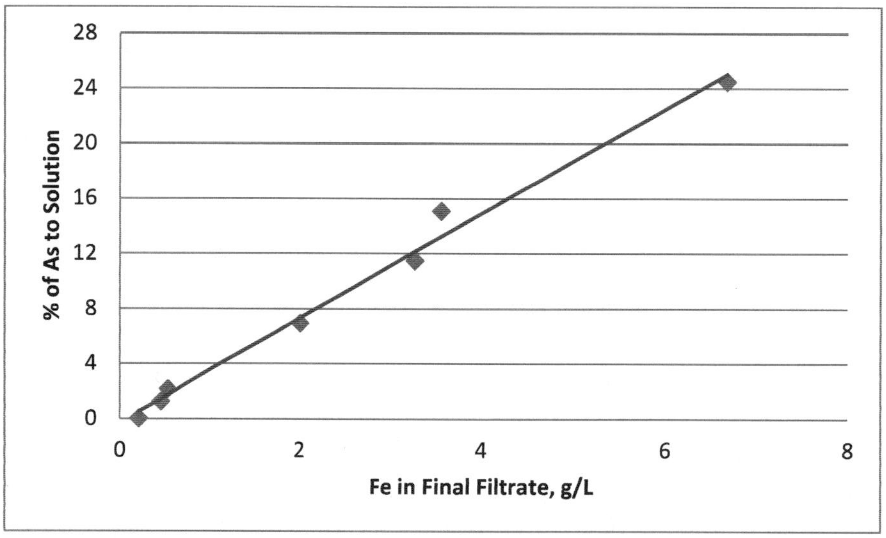

Figure 2 – Dissolved arsenic versus remaining iron in solution

No clear relationship was seen, among a dozen concentrates, between the As:Fe weight ratio in the concentrate and the fraction of arsenic reporting to solution for a given level of final acidity. The type of surfactant used in the leach and the salt content of the leach solution affected arsenic deportment only marginally. Higher oxygen partial pressure and higher temperatures favored more complete oxidation as reflected by the higher final oxidation-reduction values and greater extents of arsenic precipitation.

In continuous pilot runs the relationship between arsenic in autoclave filtrate and free acid equivalent was similar although the slope and intercept were somewhat different (Figure 3). For the five campaigns shown, the data sets all form a similar pattern to that seen in batch testwork despite the wide range of arsenic in feeds (0.7 to 4.4% As) and fluctuating operating conditions across these campaigns.

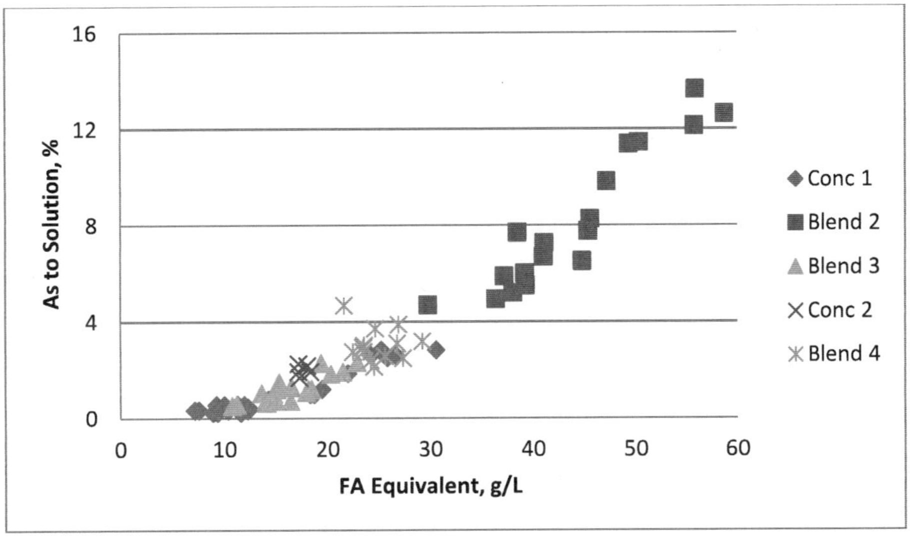

Figure 3 – Continuous operation showing fraction of arsenic left in solution as a function of FA equivalent

Copper Leaching

Batch tests demonstrated copper extraction in some cases approaching 99% for concentrates covering a broad range of arsenic compositions (Table 2). Arsenic in feed did not correlate strongly with copper extraction.

Table 2 – Copper extraction from concentrates with varying arsenic contents

Concentrate	As Content, %	Cu Extraction, %	Cu in Residue, %	Cu as S= in Residue, %
A	0.43	98.5	0.55	0.05
B	0.67	98.3	0.5	0.1
C	0.75	97.7	1.0	0.1
D	0.92	98.9	0.55	0.2
E	2.4	98.7	0.6	0.2
F	3	98.4	1.0	0.1
G	4.8	98.3	1.2	0.2
H	9.7	98.6	1.0	0.6
H-2	10	98.6	1.0	0.7

The unreacted residual copper sulphide component (mostly enargite) in residue varied from one concentrate to another indicating that the reactivity of a mineral for a given set of conditions is not simply a function of its chemical composition. The remaining portion of unrecovered copper is present in an oxidized form that is adsorbed onto the residue surfaces as well as incorporated within the structure of that residue. The copper oxide portion can be variable (0.3 to 1% Cu) and is largely associated with minor amorphous iron phases in the residues (Sahu & Asselin, 2011).

As-Bearing Dust Additions

Copper smelters produce impurity-laden, copper bearing dusts with significant levels of arsenic which must be bled from their circuits to properly manage circuit chemistry. The addition of such dusts to the CESL autoclave offers an opportunity to leach the copper contained in these dusts while precipitating the arsenic in a stable form.

A series of standard bench tests was performed with a blend of smelter dust and concentrate, with the composition of materials as presented in Table 3.

Table 3 – Composition of concentrate and dust fed to bench leach

Material	As, %	Cu, %	Fe, %	S, %
Concentrate	2.19	31	19	35
Dust	10.5	19.3	13.4	8.3

Arsenic bearing dust additions were 0, 10, 20, and 30% by weight of the input concentrate which was kept unchanged at a 100 g/L solids density. With increasing dust charge, the fraction of arsenic reporting to filtrate fell from 1.8% or 0.042 g/L As (no dust addition) to 0.5% or 0.026 g/L As (30% dust addition) despite the steadily increasing arsenic feed grade.

Copper extraction from only the concentrate was 98.5%. Cu extraction from dust was 88 to 89% and was unaffected by the quantity of dust on charge. Consequently, overall Cu extraction slipped slightly with increasing dust levels as shown in Figure 4.

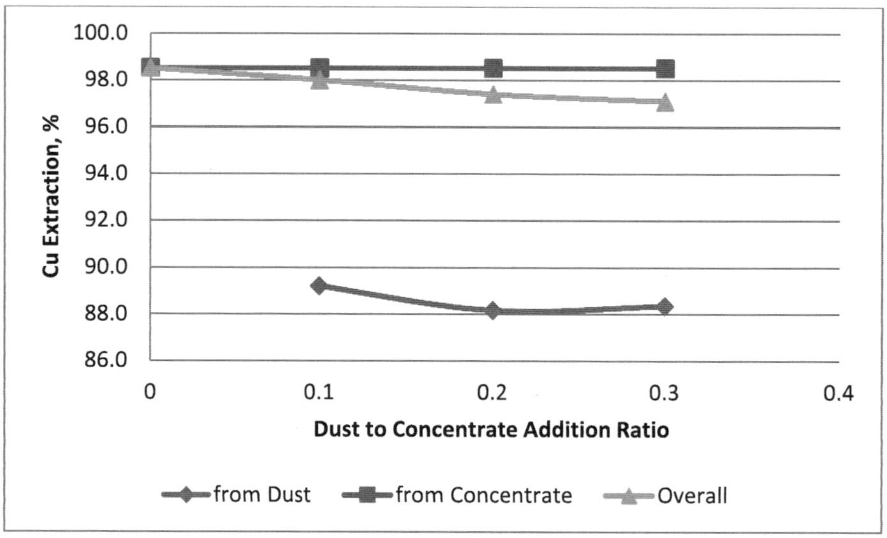

Figure 4 – CESL copper extraction from smelter dust and concentrate blend

The addition of dust to an autoclave operating under standard CESL conditions offers the benefit of increased throughput for the vessel, as it has little impact on the operating heat balance, with no adverse impact on arsenic deportment.

Chemical and Physical Residue Characterization

Mineralogical Characteristics

Mineralogical analysis of residues produced in the continuous, fully integrated CESL pilot plant demonstrated that the arsenic was precipitated as Type II scorodite ferric arsenate, widely accepted as the best available technology for stabilizing arsenic (Ferron & Wang, 2003). A previous paper discussed the different forms of ferric arsenate precipitated depending on autoclave and operating conditions (Bruce et al., 2012).

Since the prior publication, additional quantitative X-ray diffraction analysis has been completed on processing smelter dusts under CESL conditions through bench testing. The results of quantitative phase analysis by Rietveld refinements are given in Table 4. This data represent the relative amounts of crystalline phases normalized to 100%. It is estimated the sample contains, in total, 30% amorphous material.

Table 4 – Quantitative X-ray diffraction analysis of CESL residue produced from smelter dust feed

Mineral	Ideal Formula	Weight, %
Quartz	SiO_2	18
Scorodite	$Fe^{3+}AsO_4.2H_2O$	23
Sulfur	S	13
Hematite	α-Fe_2O_3	3
Maghemite	γ-Fe_2O_3	30
Pyrophyllite	$Al_2Si_4O_{10}(OH)_2$	5
Illite/Muscovite 2M1	~ $K_{0.65}Al_{2.0}Al_{0.65}Si_{3.35}O_{10}(OH)_2$	7
Famatinite	Cu_3SbS_4	1

Physical Characteristics

The formation of residues that exhibit good solid-liquid separation characteristics is a fundamental requirement of a hydrometallurgical process. A previous paper published settling data from residues produced under standard CESL conditions (Defreyne et al., 2006). These residues were principally hematite and elemental sulphur, and contained negligible arsenic.

On CESL residues that contained 16% arsenic with principally a Type II scorodite and elemental sulphur composition, excellent settling characteristics were achieved as shown in Table 5.

Table 5 – Settling results on pilot plant (scorodite) residue

Criteria	CCD
Instantaneous settling rate, m/h	15.5
Unit area, m^2/tpd	0.15
Overflow solids, ppm	50
Underflow solids, wt%	50
Residue P80, μm	46

The scorodite bearing residue exhibits similar settling characteristics as achieved with hematite, which eliminates engineering concerns as compared to poorer settling residues that contain significant jarosite or basic iron hydroxides.

Stability Assessment

Pilot residues, produced during past pilot campaigns, containing up to 16% arsenic were subjected to the Environmental Protection Agency's (EPA) standard hazardous material characterization procedure

(TCLP, Method 113). The consistently low arsenic in the tests (~0.05-0.1 mg/L) versus the EPA limit of 5 mg/L demonstrated that the residues were non-hazardous (Lossin et al., 2015).

Assessment of these non-hazardous residues over a longer time period was equally important. A number of residue samples from previous high arsenic pilot plant campaigns were placed in two-liter bottles with residue at a 20:1 liquid to solid ratio. These are being monitored on an ongoing basis while subjected to mild agitation at room temperature. The equipment used is an orbital shaker (Lab-Line Instruments Orbit Shaker, No. 3590) at 100-110 rpm and 2 cm lateral motion. Results are shown in Figure 5, with residue stability results supporting earlier findings (Gomez et al., 2008) that scorodite and basic ferric arsenate sulphate easily pass the standard TCLP test and longer term geochemical tests, implying the existence of two stable forms of iron-arsenic precipitate.

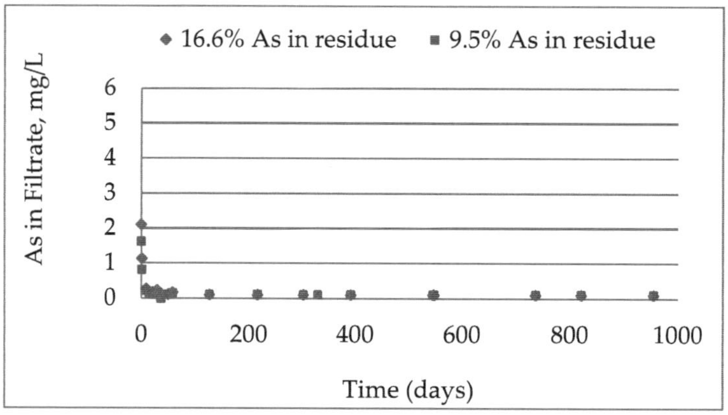

Figure 5 – Long-term residue stability results

INDUSTRIAL APPLICATION

Hydrometallurgical processing of copper concentrates containing elevated levels of arsenic represents the best available option to address environmental implications of produced arsenic residues inherent to the production of copper using complex feed materials. Successful commercialization also requires the technology to meet minimum economic hurdles. The CESL Cu-As Technology has demonstrated its ability to provide a sustainable solution that deals with contained arsenic in an environmentally sound manner, while minimizing the complexity and cost of processing by containing the copper leaching and arsenic precipitation reactions within a single step. Furthermore, settling of the arsenic-containing residue has eliminated engineering concerns, providing further certainty as to the economic viability of the process.

Long-term increasing demand of copper is estimated to require the development of an additional five million annual tonnes of copper capacity from new mine production during the period of 2014 – 2024 (Wood Mackenzie, 2015). Increasing impurity levels in concentrates, e.g. arsenic level, is predicted. Declining head grades combined with higher investment and operational cost will require the reassessment of copper deposits, including those with high arsenic grade, which have been undeveloped so far. China has restricted copper concentrate import at 0.5% arsenic content. The Chinese import ban is an indication of the copper concentrate processing industry's current inability to safely and cost effectively handle high arsenic materials. Typically such high arsenic bearing concentrates are blended with clean concentrates to meet legislative and technical thresholds. The practice of blending is limited by available volume of clean concentrates and the increased costs and logistics of performing the blending. While blending may solve the issue of technical threshold requirements for current copper refining technologies, it does not address the burden that the produced unstable arsenic waste materials place on the environment, nor does it represent a sustainable solution (Drouven, 2015).

Conceptual case studies have been conducted to assess the business case for a commercial CESL Cu-As Technology installation. A copper price of 2 $/lb has been shown as break even for a greenfield mine to metal project with 200 ktpa copper production and 1.5% arsenic content in concentrate (Salomon-de-Friedberg et al., 2014). A further opportunity for improving the economics is the integration of a CESL Cu-As pressure leach operation into an existing copper SX/EW circuit with excess underutilized capacity, as is common for operations with undeveloped hypogene sulphide deposits that underlie late stage copper heap leaching operations. Due to low arsenic concentration in the PLS, the excess sulfuric acid could be utilized for the existing heap leaching operations at the nearby facility, with enhanced economic results.

After 5 years of successful technical development of the CESL Cu-As Technology, Teck and Aurubis have established it as a sustainable technology. Moving forward, they intend to demonstrate the technical and economical capability of this technology in a full scale CESL Cu-As production plant in South America.

CONCLUSIONS

The processing of high arsenic bearing copper concentrates at levels of up to 12% arsenic has been successfully demonstrated in continuous, fully integrated pilot campaigns. Copper extractions in excess of 97% were achieved from concentrates containing chiefly enargite. Arsenic was effectively captured in an iron residue during the same copper leaching step.

Testing at both the bench and pilot scale has demonstrated arsenic deportment to the residue is a function of both the autoclave discharge iron and free acid tenors. Maximizing arsenic deportment to the stable residue is important from both an economic and sustainability perspective. Furthermore, the CESL Cu-As Technology has demonstrated scorodite production when processing smelter dusts which contain iron and arsenic.

Residues from these trials all passed the EPA TCLP regulation that would qualify them as a non-hazardous waste. Long term stability testing over a period exceeding 3 years has shown no tendency for arsenic leaching to increase from residues containing up to 16% arsenic. These tests continue.

Moving forward Teck and Aurubis are evaluating the commercialization of the technically proven CESL Cu-As Technology. The technology provides favorable economics for high-arsenic ores, particularly if the location has a heap leach operation with available SX/EW capacity.

ACKNOWLEDGEMENTS

The authors are grateful for the permission to publish this paper and acknowledge our respective companies, Teck Resources Limited and Aurubis AG. There are many others that deserve recognition for contributing to the guidance, management, operation and support of the work programs, specifically in relation to mineralogy and residue stability.

REFERENCES

Bruce, R., Mayhew, K., Demopoulos, R., & Heidel, A. (2012). Arsenic stability and characterization of cesl process residues. *Conference of Metallurgists Proceedings*, Niagara Falls, ON.

Defreyne J., Grieve W., Jones D.L., & Mayhew K. (2006). The role of iron in the CESL process. *Iron Control Technologies*, Dutrizac & Riveros eds., *CIM*, pp. 205–221.

Drouven, B. (2015, April). *Challenges for modern copper smelters*. CRU 14th World Copper Conference, Santiago, Chile.

Dutrizac J.E., Jambor J.L., & Chen T. T. (1987). The behaviour of arsenic during jarosite precipitation: reactions at 150°C and the mechanism of arsenic precipitation. *Canadian Metallurgical Quarterly, 26*(2), 103–115.

Escobar, Gonzalez V.L., & Monhemius A.J. (1988). The mineralogy of arsenates relating to arsenic impurity control. *Arsenic Metallurgy, Fundamentals and Applications – Proceedings of an International Symposium.* Reddy, Hendrix & Queneau eds., TMS-AIME, Metallurgical Society, 405–418.

Ferron, J. & Wang, Q. (2003). Striving for a Sustainable Mining Industry: Arsenic Control through an Integrated Approach. SGS Lakefield Research Limited. *Mining and the Environment Conference*, Sudbury, ON.

Fleming C.A. (2009). Basic Iron Sulphate – a Potential Killer for Pressure Oxidation Processing of Refractory Gold Concentrates if Not Handled Appropriately. SGS Mineral Services, Technical Paper.

Fujita T., Fujieda S., Shinoda K., & Suzuki S. (2012). Environmental leaching characteristics of scorodite synthesized with ferrous ions. *Hydrometallurgy, 111–112*, 87–102.

Gomez M.A., Becze L., Bluteau M.C., Ke Berre J.F., Culter J.N., Demopoulos G.P. (2008). Autoclave Precipitation and Characterization of Fe(+3)-AsO4-SO4 Phases. *Hydrometallurgy 2008 – Proceedings of the Sixth International Symposium*, pp. 1078–1086.

Gomez M.A., Becze L., Celikin M., & Demopoulos G.P. (2011). The effect of copper on the precipitation of scorodite under hydrothermal conditions: evidence for a hydrated copper containing ferric arsenate sulfate short-lived intermediate. *Journal of Colloid and Interface Science, 360*, 508–518.

Gomez M.A., Becze L., Cutler J.N., Demopoulos G.P. (2011). Hydrothermal reaction chemistry and characterization of ferric arsenate phases precipitated from $Fe_2(SO_4)$-As_2O_5-H_2SO_4 solution. *Hydrometallurgy, 107*, 74–90.

Lossin, A., Mayhew, K., Omaynikova, V., & Salomon-de-Friedberg, H. (2015). Hydrometallurgical considerations in processing arsenic-rich copper concentrates. *Hydroprocess Conference Proceedings,* Antofagasta, Chile.

Monhemius, A.J. & Swash, P.M. (1999). Removing and stabilizing arsenic from copper refining circuits by hydrothermal processing. JOM, September 1999, pp. 30–33.

Sahu, S.K. & Asselin, E. (2011). Characterization of residue generated during medium temperature leaching of chalcopyrite concentrate under CESL conditions. *Hydrometallurgy, 110*(1), 107–114.

Salomon-de-Friedberg, H., Robinson T., Lossin, A., & Omaynikova, V. (2014). Developing copper arsenic resources with CESL technology. *Conference of Metallurgists Proceedings*, Vancouver, BC.

Swash, P.M., Monhemius, A.J. (1998). The Scorodite Process: A technology for the disposal of arsenic in the 21st Century. In Castro S.H., Vergara F., Sanchez M.A. (Eds.). Effluent treatment in the mining industry. Universidad de Concepcion publisher.

Wood Mackenzie (December, 2015). Global copper long-term outlook. *Commodity Market Report.*

THE EFFECT OF IRON REMOVAL FROM RAFFINATE ON THE COPPER BIOLEACHING PROCESS

B. Nazari[1], H. Hani[1], *E. Jorjani[1,2], A. Ghahremaninezhad[2], Z. Manafi[3], F. Hasani[1], A. Riahi[3]

[1] *Department of Mining Engineering, Science and Research Branch, Islamic Azad University, Tehran, Iran*

[2]*The Robert M. Buchan Department of Mining, Queen's University, Goodwin Hall, 25 Union St., Kingston, Ontario K7L 3N6, Canada*
*(*Corresponding author: esmaeil.j@queensu.ca)*

[3] *Hydrometallurgy Research Center, Sarcheshmeh Copper Complex, National Iranian Copper Industries Company, Iran*

ABSTRACT

The presence of a high iron concentration, 9.1 g/L, in the raffinate of the Sarcheshmeh copper heap bioleach process led to the formation of various types of jarosite, $(K,Na,NH_4,H_3O)Fe_3(SO_4)_2(OH)_6$, on the heap mineral surfaces. This study probes iron removal from the raffinate using a bio-oxidation process based on *A. ferrooxidans* bacteria in the presence and absence of jarosite seed. The jarosite seed was biologically synthesized using *A. ferrooxidans* bacteria. The effect of solution pH (1.5 to 2.5) and jarosite seed dosage (0 to 15 wt.%) on iron removal was examined. It was demonstrated that the concentration of iron ions present in the raffinate decreased from 9.1 g/L to 2.5 g/L with the addition of 10 wt.% jarosite seed at pH 2 and 130 rpm mixing rate at 32°C. Three different types of raffinate including (1) native raffinate containing 9.1, 3.4, 5.7 g/L of total, ferric and ferrous iron, respectively, (2) bio-oxidized raffinate in the absence of the jarosite seed containing 8.5 g/L total iron (mainly ferric with negligible ferrous), (3) bio-oxidized raffinate in the presence of jarosite seed containing 2.55 g/L total iron (mainly ferric with negligible ferrous) were used in mesophilic, moderate thermophilic and thermophilic bioleaching processes. The results of the flask tests showed that the highest copper bioleaching recoveries of 9.4, 49.4 and 90.8 % were achieved using raffinate (3) with mesophilic, moderate thermophilic and thermophilic bioleaching processes, respectively.

KEYWORDS

Raffinate, iron removal, jarosite seed, bioleaching, bacteria, copper

INTRODUCTION

The type and concentration of oxidizing agent are two of the main parameters which significantly influence the dissolution kinetics of copper sulphide minerals. An appropriate concentration of ferric iron is usually applied for this purpose (Watling, 2006). Different forms of jarosite precipitates ($(K,Na,NH_4,H_3O)Fe_3(SO_4)_2(OH)_6$) are formed on the surface of the ore particles at a high dosage of ferric ions (Fe^{3+}), and this may result in a decrease in the accessibility of the leaching agent and the bacteria cells to the surface of the minerals (Cordoba et al., 2008). In addition, it is possible that the jarosite precipitates consume certain ions required for bacteria growth (Nazari et al., 2014).

Iron removal is typically implemented in hydrometallurgical processes through the oxidation of ferrous ions (Fe^{2+}) to Fe^{3+}ions, followed by hydrolysis-precipitation (Ritcey, 1986). Various biological and chemical methods were proposed in order to eliminate iron from mining and hydrometallurgical streams (Johnson & Hallberg, 2005; Daoud & Karamanev, 2006; Liu et al., 2007; Gaikwas & Gupta, 2008; Kaksonen et al., 2008; Gramp et al., 2008; Nurmi et al., 2009). Jarosite formation is facilitated in the presence of iron-oxidizing bacteria. Liu et al. (2007) reported that jarosite formation in the 9K medium began when *Acidithiobacillus ferrooxidans* (*A. ferrooxidans*) bacteria underwent the logarithmic growth phase. Gramp et al. (2008) demonstrated that the solubility of $Fe^{3+}$was highly pH-dependent and that it was reduced at a pH above 2. It was illustrated that the amount of jarosite was minimized at a pH range of 1.6-1.7 (Daoud & Karamanev, 2006). Nurmi et al. (2009) conducted a series of experiments using *Leptospirillum ferriphilum* bacteria at 37±2 °C in a fluidized-bed reactor (FBR) in order to control the biological oxidation of Fe^{2+} and precipitate $Fe^{3+}$and sulphate ions from solution.

Dutrizac (1999) stated that the iron precipitation reaction rate was significantly accelerated once jarosite-type compounds such as potassium jarosite, silver jarosite, sodium jarosite, or lead jarosite were used as seed.

A copper heap bioleaching process at a pilot scale has been employed in the Sarcheshmeh copper complex located in Kerman, Iran. One of the main impediments associated with this process is the high concentration of iron (approximately 9 g/L) present in the raffinate which leads to the formation of jarosite precipitates on the surface of the mineral particles. This study aims to biologically synthesize the jarosite seed by means of *A. ferrooxidans* bacteria in order to subsequently minimize the level of iron in the raffinate. Moreover, the impact of solution pH and the dosage of jarosite seed on iron removal from the raffinate will be elucidated. Finally, the effect of iron removal from the raffinate on the mesophilic, moderate thermophilic and thermophilic bioleaching processes will be examined.

MATERIALS AND METHODS

Microorganisms

A mixed culture of bacteria consisting of 40% *A. ferrooxidans,* 40% *A. thiooxidans* and 20% *Leptospirillum ferrooxidans* (*L. ferrooxidans*) was made to carry out the mesophilic bioleaching experiments. Moderately thermophilic and thermophilic bioleaching processes were conducted using *Ferroplasma acidarmanus* and *Acidianus brierleyi* bacteria, respectively. The bacteria were isolated from the acid mine drainage generated by the Sarcheshmeh copper mine operation (Seyedbagheri & Hassani, 2001). Erlenmeyer flasks containing 9K growth media at different pH values were placed in a rotary incubator shaker operating at a rate of 130 rpm and at different temperatures for 7 days (Table 1). Moreover, ferrous sulphate ($FeSO_4{\cdot}7H_2O$) and sulphur as a source of energy were added to the solutions at a predetermined concentration (Table 1). Details of the ingredients present in the 9K growth medium are illustrated in Table 2.

Table 1 –Details of the experimental conditions used for the different bacteria (Donati & Sand, 2007)

Microorganism	*A. ferrooxidans*	*A. thiooxidans*	*L. Ferrooxidans*	*Ferroplasma acidarmanus*	*Acidianus brierleyi*
pH	1.8	2	1.6	1.6	1.6
Temperature (°C)	32	32	45	45	70
$FeSO_4 \cdot 7H_2O$(g/L)	75	-----	45	75	-----
Sulphur (g/L)	-----	10	-----	10	10

Table 2 –Composition of the 9K growth medium (Nemati et al., 1998)

Salt	$(NH_4)_2SO_4$	$MgSO_4 \cdot 7H_2O$	K_2HPO_4	KCl	$Ca(NO_3)_2 \cdot H_2O$
Dosage (g/L)	3	0.5	0.63	0.1	0.014

Copper Ore Sample

A copper ore sample was supplied from the crushed ore stockpile that is used to feed the pilot scale heap bioleaching process in the Sarcheshmeh copper complex. The sample was ground to reduce the particle size (80% passing 108 μm). The X-ray powder diffraction (XRD) pattern of the sample is shown in Figure 1. The results obtained from XRD clearly illustrated that quartz (SiO_2), chlorite ($(Mg,Fe)_6(Si,Al)_4O_{10}(OH)_8$), illite ($(K,H_3O)Al_2Si_3AlO_{10}(OH)_2$), albite ($(Na,Ca)(Al,Si)_4O_8$), muscovite ($KAl_2(AlSi_3O_{10})(F,OH)_2$), pyrite ($FeS_2$) and chalcopyrite ($CuFeS_2$) are the main mineral constituents in the sample.

Microscopic mineralogical studies showed that the copper-bearing minerals are mainly in the form of sulphides, and chalcopyrite is the predominant species (Table 3). Likewise, silicate and alumino-silicates were the most abundant oxide minerals, as depicted in Table 4. The total concentration of copper and iron present in the sample was 0.7 and 7.31 wt.%, respectively.

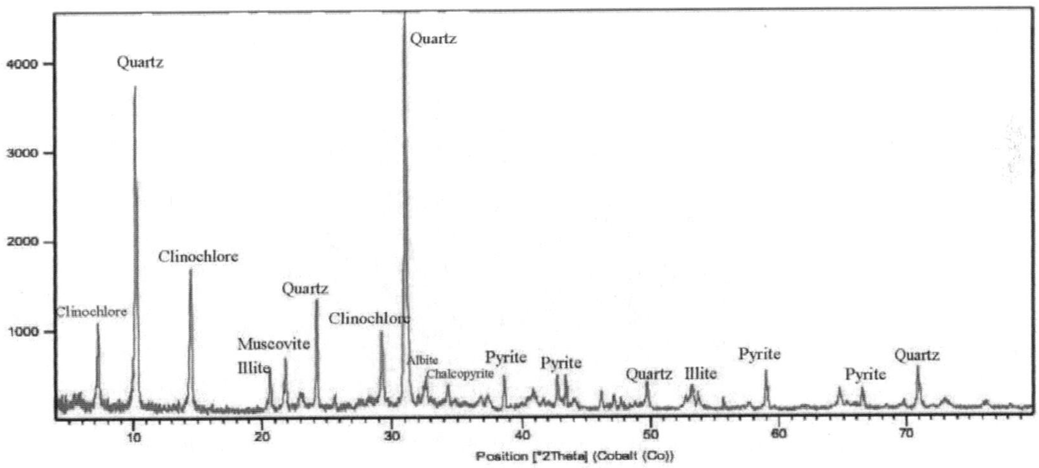

Figure 1 – XRD pattern of the copper ore sample

Table 3 – Main sulphide minerals present in the ore sample

Mineral	Chalcopyrite	Chalcocite	Pyrite	Sphalerite	Molybdenite
Chemical formula	$CuFeS_2$	Cu_2S	FeS_2	ZnS	MoS_2
Concentration (wt.%)	1.9	0.1	13.9	0.1	0.1

Table 4 – Bulk composition of the ore sample

Oxide	Al_2O_3	SiO_2	CaO	K_2O	MgO	TiO_2	CuO
Concentration (wt.%)	17.4	52.9	0.7	4.4	4.3	0.9	Trace

Jarosite Seed Synthesis

Jarosite seed was biologically synthesized in 500 mL Erlenmeyer flasks containing 200 mL of the 9K growth medium, 75 g/L Fe^{2+} and 10% V/V *A. ferrooxidans* bacteria at pH 2. The experiments were systematically performed in the rotary incubator shaker at 130 rpm and 32 °C (Figure 2). The solution was then filtered when the bio-oxidation time reached 13 days. The insoluble particles were dried at room temperature to produce the jarosite seed for iron removal tests (Figure 3).

Figure 2 – Schematic view of test setup for jarosite seed formation using *A. ferrooxidans* bacteria

Figure 3 – Jarosite seed biologically synthesized using *A. Ferrooxidans*

Iron Removal from Raffinate

The iron concentration of the raffinate (which contained Cu: 0.22, Fe: 9.1, Fe^{2+}: 5.7, Fe^{3+}: 3.4 g/L) was decreased biologically in the absence and presence of jarosite seed. A series of iron removal experiments was performed in a 500 mL Erlenmeyer flask containing 200 mL of raffinate. The solution pH was adjusted to 1.5, 2 and 2.5 using sulphuric acid and sodium hydroxide solutions. Jarosite seed was added to the solution at various dosages ranging from 0 to 15 wt.%. The raffinate was then inoculated with 10% (V/V) *A. ferrooxidans* bacteria in the incubator shaker at 130 rpm, 32 °C and for 13 days. Thereafter, the raffinate was filtered to separate the iron precipitates from the solution.

Bioleaching Experiments

The three types of raffinate, shown in Figure 4, were selected in order to carry out the bioleaching experiments:

- Native raffinate (the concentration of iron in solution is ~ 9.1 g/L)
- Bio-oxidized raffinate where iron ions were precipitated from solution in the presence of jarosite seed (the final concentration of iron in solution was ~ 2.5 g/L)
- Bio-oxidized raffinate where iron ions were removed from solution in the absence of jarosite seed (the final concentration of iron in solution was ~ 8.5 g/L)

A mixture of raffinate and copper ore at a liquid:solid weight ratio of 9 was inoculated using 10% (V/V) bacteria in the incubator shaker at 130 rpm for 13 days. Details of the experimental conditions are given in Table 5. No growth medium was added to the raffinate. The insoluble residue was separated from the solution via filtration followed by washing with 5 wt.% sulphuric acid and distilled water to remove the remaining contaminants. The concentration of iron and copper ions present in both the solid residue and solution was quantified using atomic absorption spectroscopy (AAS). Ultraviolet (UV) spectrophotometry was also employed to measure low concentrations of ferrous ions. However, titration using potassium dichromate was used to determine high dosages of ferrous ions.

Figure 4 –Schematic view of native raffinate (NR), bio-oxidized raffinate in the presence of seed (BRPS) and bio-oxidized raffinate in the absence of seed (BRAS)

Table 5 – Details of the batch bioleaching experiments

Microorganism	pH	Temperature (°C)
Mixed mesophilic (40 %*A. Ferrooxidans*, 40 % *A. Thiooxidans*, and *20 %L. Ferrooxidans*)	1.7	32
Moderately thermophilic (*Ferroplasma Acidarmanus*)	1.5	45
Thermophilic (*Acidianus Brierleyi)*	2	70

RESULTS AND DISCUSSIONS

Characterization of Synthetic Seed

XRD pattern of the synthetic seed obviously illustrates the presence of jarosite, hydronium jarosite as well as minor amounts of impurities such as anhydrite and hematite (Figure 5).

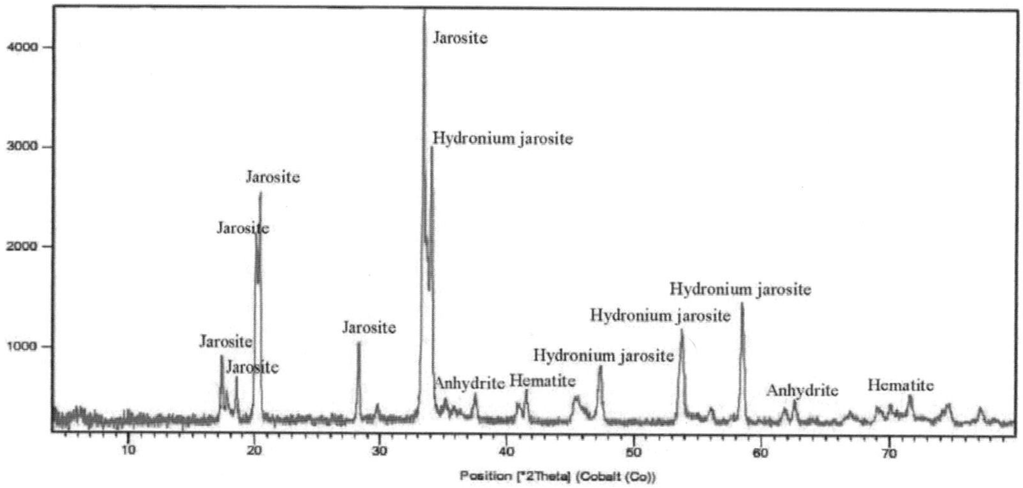

Figure 5 – XRD pattern of the synthetic seed

Biological Removal of Iron from the Raffinate

Effect of Solution pH on Iron Removal

The impact of solution pH on iron removal is shown in Table 6. A rise in solution pH resulted in higher iron removal, until the maximum iron removal (~87%) was achieved at pH 2.5.

Table 6– Effect of the solution pH on iron removal (5% jarosite seed, 32°C and 13 days)

pH	Soluble total iron (g/L)	Soluble ferrous iron(g/L)	Total iron removal (%)
1.3 (pH of native raffinate)	9.1	5.7	0
1.5	9.01	0.01	0.99
2	3.84	0.02	57.8
2.5	1.19	0.01	86.9

Effect of the Solid Seed Fraction

As shown in Table 7, 6.2% of the iron ions were eliminated from solution in the absence of jarosite seed. Increasing the jarosite dosage from 5 to 10 wt.% led to a dramatic increase in iron removal (72%). Whereas, no significant rise in the iron removal was observed with a further seed addition to 15 wt.%. It seems that 10 wt.% seed is sufficient for the nucleation/ growth of the jarosite; so a solution containing 2.55 (g/L) soluble iron was selected as BRPS.

Table 7 –Effect of jarosite seed loading on iron removal from raffinate (pH 2, 32°C and 13 days)

Jarosite seed loading (wt %)	Soluble total iron (g/L)	Soluble ferrous iron (g/L)	Total iron removal (%)
0	8.53	0.01	6.26
5	3.84	0.02	57.8
10	2.55	0.04	71.98
15	2.47	0.01	72.86

Bioleaching using Native and Bio-oxidized Raffinates

The kinetics of copper bioleaching using mesophilic bacteria are illustrated in Figure 6. Increasing the contact time to 7 days led to a sharp increase in copper dissolution for both BRAS and BRPS. However, a plateau was reached for additional increases in the contact time for BRAS. In the case of BRPS, the maximum copper recovery (9.43%) was attained after 13 days. NR showed a different trend for copper recovery compared to that for BRPS and BRAS, and copper dissolution increased to some extent after 7 days. However, a further rise in contact time increased the copper recovery to 3.29% after 13 days. It was also evident that the overall copper dissolution for the three types of raffinate was relatively low due to the refractory nature of chalcopyrite which is the predominant copper mineral in the sample (Table 3). Sterilized (thymol-treated) and not inoculated NR was also applied to examine the Cu dissolution as a result of chemical oxidation. In this case, maximum copper recovery of 2.6 % after 13 days was observed.

Figure 7 shows the impact of contact time on copper dissolution in the presence of the moderately thermophilic bacteria. NR, BRAS and BRPS displayed a similar behaviour and copper dissolution slightly increased to 10% after 10 days. A further increase of the contact time noticeably improved the copper recovery, until the maximum was obtained after 13 days. For sterilized NR and in the absence of bacteria, the Cu chemical dissolution had reached only 10% after 13 days.

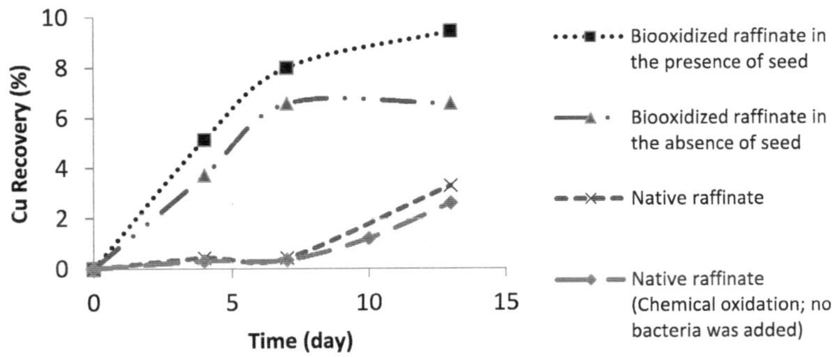

Figure 6 – Effect of contact time on copper recovery using mesophilic bacteria for the bioleaching experiments

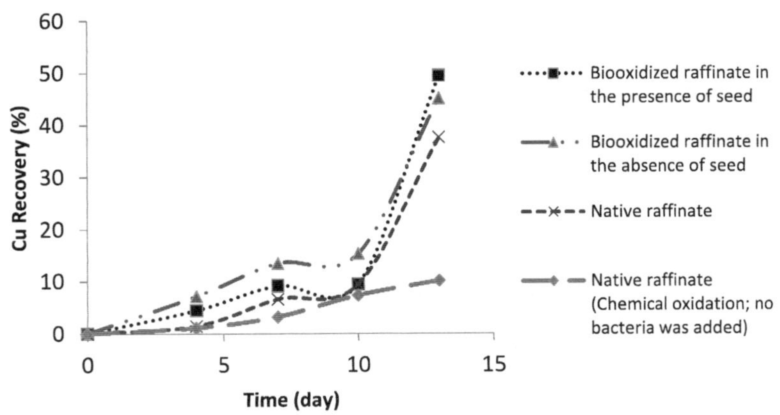

Figure 7 – Effect of contact time on copper recovery using moderately thermophilic bacteria for the bioleaching experiments

Thermophilic bacteria exhibited a significant impact on copper recovery when BRPS was used. As shown in Figure 8, the copper dissolutions were approximately identical for NR, BRAS and BRPS for the first 7 days. An additional increase in the contact time appreciably promoted the copper recovery for BRPS (90.8%) after 13 days. In contrast, copper recovery for BRAS increased when the bioleaching time reached 10 days and then it decreased slightly. The decrease is likely within experimental error. A copper recovery of 16.8 % was achieved for sterilized NR without bacteria inoculation after 13 days.

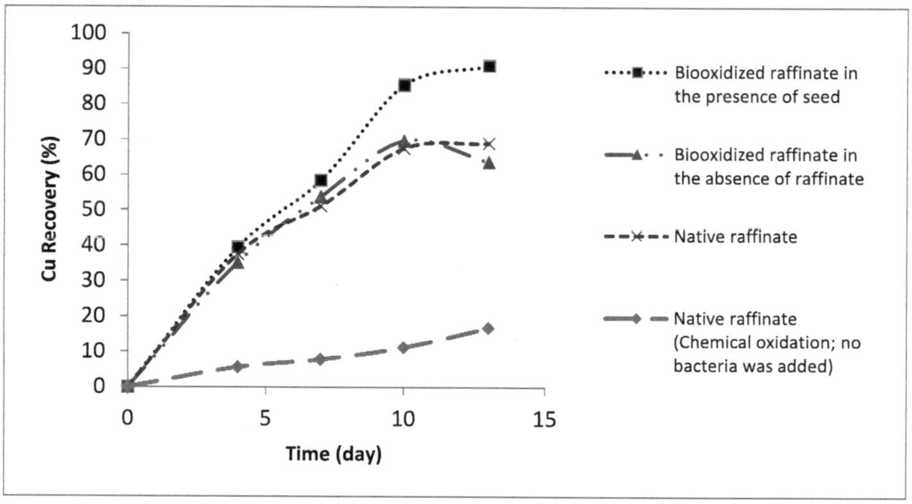

Figure 8 –Effect of contact time on copper recovery using thermophilic bacteria for the bioleaching experiments

CONCLUSIONS

Jarosite seed was biologically synthesized using *A. ferrooxidans* bacteria at pH 2. XRD analysis apparently affirmed that the seed was chiefly composed of jarosite and hydronium jarosite. It was demonstrated that the addition of jarosite seed significantly improved iron removal from the raffinate. Increasing seed dosage to 10% led to approximately 72% iron removal. Likewise, solution pH markedly influenced iron removal and 57.8% iron elimination was achieved at pH 2 and 5% seed dosage. Hence, both a pH of 2 and a jarosite seed loading of 10% were selected as the optimum conditions for iron removal from the raffinate. A series of copper bioleaching experiments was conducted using mesophilic, moderately thermophilic and thermophilic bacteria to investigate the kinetics of copper dissolution in the absence and presence of jarosite seed. The thermophilic bioleaching process exhibited the highest copper recovery of 90.86%, when BRPS was used as the growth medium.

ACKNOWLEDGEMENTS

B.N., H.H., E.J., and F.H would like to acknowledge the R&D division of Sarcheshmeh Copper Complex for funding this project.

REFERENCES

Cordoba, E.M., Munoz, J.A., Blazquez, M.L., Gonzalez, F. Ballester, A. (2008). Leaching of chalcopyrite with ferric ion. Part I: General aspects. *Hydrometallurgy*, 93, 81-87.

Donati, E.R., & Sand, W. (2007). Microbial processing of metal sulphides. Springer, 1-311.

Daoud, J., & Karamanev, D. (2006). Formation of jarosite during Fe^{2+} oxidation by Acidithiobacillus ferrooxidation. *Miner. Eng.*, 19, 960-967.

Dutrizac, J. E. (1999). The effectiveness of jarosite species for precipitating sodium jarosite. *JOM,* 51, 12, 30-32.

Gaikwas, R.W., & Gupta, D.V. (2008). Review on removal of heavy metals from acid mine drainage. *Appl. Ecol. Environ. Res.*, 6 (3), 81–98.

Gramp, J.P.S., Jones, F.S., Bigham, J.M. & Tuovinen, O.H. (2008). Monovalent cation concentrations determine the types of Fe(III) hydroxysulphate precipitates formed in bioleach solutions. *Hydrometallurgy*, 94, 29-33.

Johnson, D.B., & Hallberg, K. B. (2005). Acid mine drainage remediation options: a review. *Sci. Total Environment*, 338, 3–14.

Kaksonen, A.H., Dopson, M., Karnachuk, O., Tuovinen, O.H., & Puhakka, J.A. (2008). Biological iron oxidation and sulphate reduction in the treatment of acid mine drainage at low temperatures. In: Margesin, R., Schinner, F., Marx, J.-C., Gerday, C. (Eds.), Psychro-philes: From Biodiversity to Biotechnology. Springer Verlag, Berlin, 429–454.

Liu, J., Li, B., Zhong, D., Xia, L., &Qiu, G. (2007). Preparation of jarosite by A. ferrooxidans oxidation. J. *Cent South Univ. Technol.*, 623-628.

Nazari, B., Jorjani, E., Hani, H., Manafi, Z., & Riahi, A. (2014). Formation of jarosite and its effect on important ions for Acidithiobacillus ferrooxidans bacteria. *Transactions of Nonferrous Metals Society of China,* 24, 4, 1152-1160.

Nemati, M., Harrison, S.T.L., Hansford, G.S., & Webb, C. (1998). Biological oxidation of ferrous sulphate by Thiobacillus ferrooxidans: a review on the kinetic aspects. *Biochemical Engineering Journal,* 1, 171-190

Nurmi, P., Ozkaya, B., Kaksonen, A.H., Tuovinen, O.H., Riekkola-Vanhanen, M.L., & Puhakka, J.A. (2009). Process for biological oxidation and control of dissolved iron in bioleach liquors. *Process Biochemistry,* 44, 1315-1322.

Ritcey, G.M. (1986). Iron-an overview of its control in solvent extraction of metals. In: Dutrizac, J.E., Monhemius, A.J., (Eds.) Iron Control in Hydrometallurgy. The International Symposium on Iron Control In Hydrometallurgy, Oct 19-22. Toronto: Halsted Press a division of John Wiley and Sons; 247-281.

Seyedbagheri, S.A., & Hassani, H.R. (2001). Isolation and preliminary identification of some iron-and sulphur-oxidizing microorganisms from the Sarcheshmeh copper mine, Ciminelli, S.T., Garcia, O. Jr., Bio hydrometallurgy: Fundamentals, Technology and Sustainable Development. Part A, Bioleaching, Microbiology and Molecular Biology. Amsterdam: Elsevier, 393–396.

Watling, H.R. (2006). The bioleaching of sulphide minerals with emphasis on copper sulphides - A review. *Hydrometallurgy*, 84, 81-108.

Iron Management in the Chloride System

ACID RECOVERY IN THE STEEL AND METALLURGICAL INDUSTRY PROCESS CHARACTERISTICS AND FUEL SAVING OPTIONS

*F. Baerhold and S. Mitterecker

ANDRITZ AG
Eibesbrunnergasse 20
A-1120 Vienna, Austria
*(*Corresponding author: frank.baerhold@andritz.com)*

ABSTRACT

Acid recovery is an important process step in pickling and leaching processes. Pyrohydrolysis in spray roasters and fluidized beds is a well-established technology to recover hydrochloric acid, but the hydrolysis in the liquid phase is more and more considered as an alternative new process route. The characteristic features of these three processes will be discussed along with select industry trends. For example, hydrochloric acid based leaching processes for ilmenite and nickel laterite ores produce far larger volumes of waste acid streams than the steel industry. Scaling-up of the spray roasting technology to the highest possible capacity is necessary in order to become more cost effective. In addition, process routes based on ferric chloride pyrohydrolysis allow the production of super-azeotropic HCl (up to 30 wt.%). Finally, the new ECO*mode* concept optimizes the operation of both pickling lines and the spray roast acid recovery plant. Long term results from industrial applications show that fuel savings in the spray roaster of 25% are easily achieved without additional process steps. An attractive feature is also to upgrade existing installations to higher plant capacities.

KEYWORDS

Acid recovery, hydrochloric acid, pickling, leaching, energy saving,

INTRODUCTION

Pyrohydrolysis is the conversion of metal halides into metal oxides at elevated temperatures in the presence of water vapor. Reaction products are a hydrogen halide gas and a metal oxide. The process was originally developed and introduced into the market by ANDRITZ Ruthner in the 1960's. It has found wide technical application in the field of acid recovery and oxide production. The introduction of this process has revolutionized pickling technology. The use of hydrochloric acid has strongly increased over the last decades, since pyrohydrolysis is available for the total recovery of waste pickle liquors from mild steel pickling, as illustrated in Figure 1.

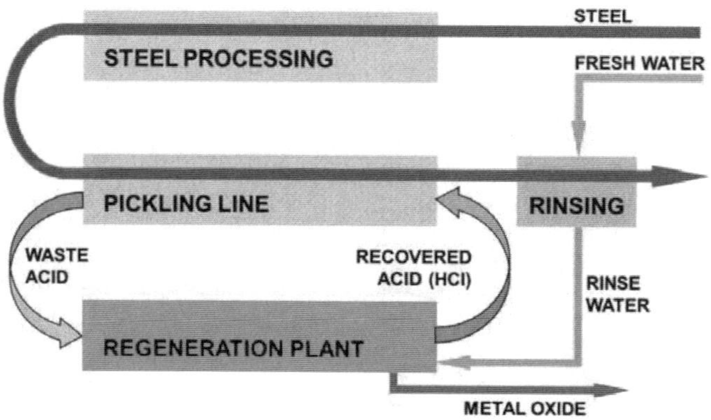

Figure 1 – Closed acid loop in a pickling and acid recovery line

Chloride media are also used increasingly in the metallurgical industry. Oxide production processes for Mg, Co, Ni, and Al have been installed. Processes using hydrochloric acid are known for the leaching of ores as well as for the re-extraction of metals from organic solvents. In both cases the recycling of HCl is of substantial importance to achieve highly economic processes and to reduce waste.

PICKLING/LEACHING PROCESSES

Pickling is defined as the removal of inorganic contaminants from a metal surface with a liquid that results in the chemical dissolution or cracking (removal) of the oxide layers. Pickling is carried out mostly through submersion, flooding or spraying. Additionally the pickling effect may be supported by mechanical or electrolytic processes. The composition, physical structure, and thickness of these layers depend on the preceding treatment of the steel strip. The oxides which build up on the carbon steel surface are converted by pickling to an iron chloride containing solution. The dominant reactions are given below (Steinbach & Baerhold, 2002).

$$Fe_2O_3 + 6\,HCl \Rightarrow 2\,FeCl_3 + 3\,H_2O \quad (1)$$

$$Fe_3O_4 + 8\,HCl \Rightarrow FeCl_2 + 2\,FeCl_3 + 4\,H_2O \quad (2)$$

$$FeO + 2\,HCl \Rightarrow FeCl_2 + H_2O \quad (3)$$

At the same time the free hydrochloric acid attacks the metallic iron and hydrogen is formed, which involves the risk of undesired hydrogen embrittlement.

$$Fe + 2\,HCl \Rightarrow FeCl_2 + H_2 \quad (4)$$

Most of the $FeCl_3$ generated is reduced to $FeCl_2$ by Fe and H_2, which are created during the pickling of the metallic iron.

$$2\ FeCl_3 + H_2 \Rightarrow 2\ FeCl_2 + 2\ HCl \tag{5}$$

$$2\ FeCl_3 + Fe \Rightarrow 3\ FeCl_2 \tag{6}$$

As the chemical dissolution kinetics slow down during the pickling process due to a decreasing free HCl concentration, the solution has to be replaced with fresh liquor. A typical spent pickle liquor stream consists of approximately 190 g/L Cl^-, 120 g/L Fe^{2+} and 1-3 g/L Fe^{3+}.

Treatment of Spent Acid

Due to environmental and economic reasons, disposal through neutralization is only practiced for very small industrial plants. Some treatment processes recover only the free (unbound) acid, while the metal bound halides are discarded. The feasibility of these processes depends on plant capacity and the concentration of free acid, and its profitability is limited to capacities of approximately 50 t/a. Those processes are carried out using technologies like ion exchange, retardation on a resin, partial evaporation and condensation or membrane dialysis (Regel-Rosocka, 2010). In some cases, the iron chloride solution can be sent free of charge to waste water treatment plants. In some cases deep well injection of the waste acid into the ground was common practice (DuPont, USA) or is grandfathered into today's era.

REGENERATION THROUGH PYROHYDROLYSIS

Thermochemical Background

<u>Chlorides</u>

Most metal chlorides and fluorides can be hydrolyzed at elevated temperatures. Some metal chlorides, e.g. lead, zinc, and copper, cannot be hydrolyzed in the pure form, just like most alkaline and alkaline earth metals. In Figure 2 the standard free enthalpies of reaction for the hydrolysis of several metal chlorides are shown, as calculated from Barin & Knacke (1973). For potassium, sodium, calcium, lead, zinc, and copper, the standard free enthalpy of reaction is positive over the whole temperature range shown, thus the reaction will not take place.

The pyrohydrolysis reactions for iron are:

$$2\ FeCl_2 + 2\ H_2O + \tfrac{1}{2}\ O_2 \Rightarrow Fe_2O_3 + 4\ HCl \tag{7}$$

$$2\ FeCl_3 + 3\ H_2O \Rightarrow Fe_2O_3 + 6\ HCl \tag{8}$$

For the design of a pyrohydrolysis process the reactor temperature is the key parameter. In addition, it has to be checked where the melting (or sublimation) points of the chloride and oxide are, and whether the metal chloride has a substantial vapor pressure at the process temperature. Melting compounds lead to unwanted sintering effects. In case of a non-hydrolysable chloride a high vapor pressure results in evaporation of the chloride in the reactor and subsequent condensation at cold spots downstream (e.g. Na, Pb, Ca, K, Zn, Cu). If the vaporized chloride is hydrolysable, the reaction will lead to the formation of very fine dust (e.g. ferric, Ni). This dust cannot be caught in the cyclone and hardly in the venturi scrubber.

The oxidation of HCl to chlorine according to Deacon has a $\Delta G° = 0$ at approximately 600°C. In the temperature range 200 to 600°C the chlorine formation can be substantial, requiring an off gas treatment to meet the emission limits, depending on the excess oxygen available.

$$2\ HCl + \tfrac{1}{2}\ O_2 \Rightarrow Cl_2 + H_2O \tag{9}$$

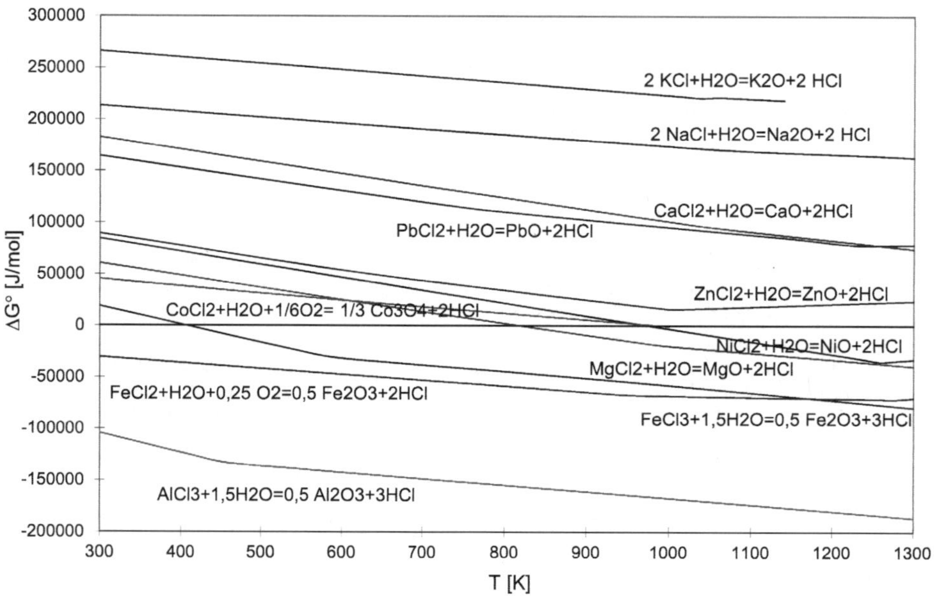

Figure 2– Standard free enthalpy of reaction for some metal chlorides

Mixed Oxides

For the data shown in Figure 2 only the reactions of pure metal chlorides into their corresponding oxides was considered. A very important issue though is the formation of so-called mixed oxides. In the case of processing a solution containing various metal halides, the oxide is a homogeneous crystal lattice comprising of some single oxide components and partially formed mixed oxides. The formation of mixed oxides lowers the free enthalpy of formation of the reaction products, thus helping some components to hydrolyze, which would normally not react in the pure form.

In the case of calcium chloride, for example, the hydrolysis is forced due to the formation of a mixed Ca/Fe/Mg or Ca/Al oxide. Zinc chloride can be hydrolyzed in the presence of iron and magnesium to a Fe/Mg/Zn-mixed oxide (AT 404 031, 1998).

PROCESS OVERVIEW

The pyrohydrolysis process is not only used for the recovery of HCl from spent pickle liquor, but also as a metallurgical process for the production of special metal oxides or ceramic specialties (Jedlicka, 1977). Figure 3 shows the closed hydrochloric acid cycle for the production of pure metal oxides according to the pyrohydrolysis process. The regeneration process can be applied to metal chloride solutions that come from a pickling line, a leaching unit for mineral ore processing (Ni, Co, TiO_2), a purification process for ceramic oxide powder production (Mg, Al), or the production of mixed oxides (Mn/Fe ferrites).

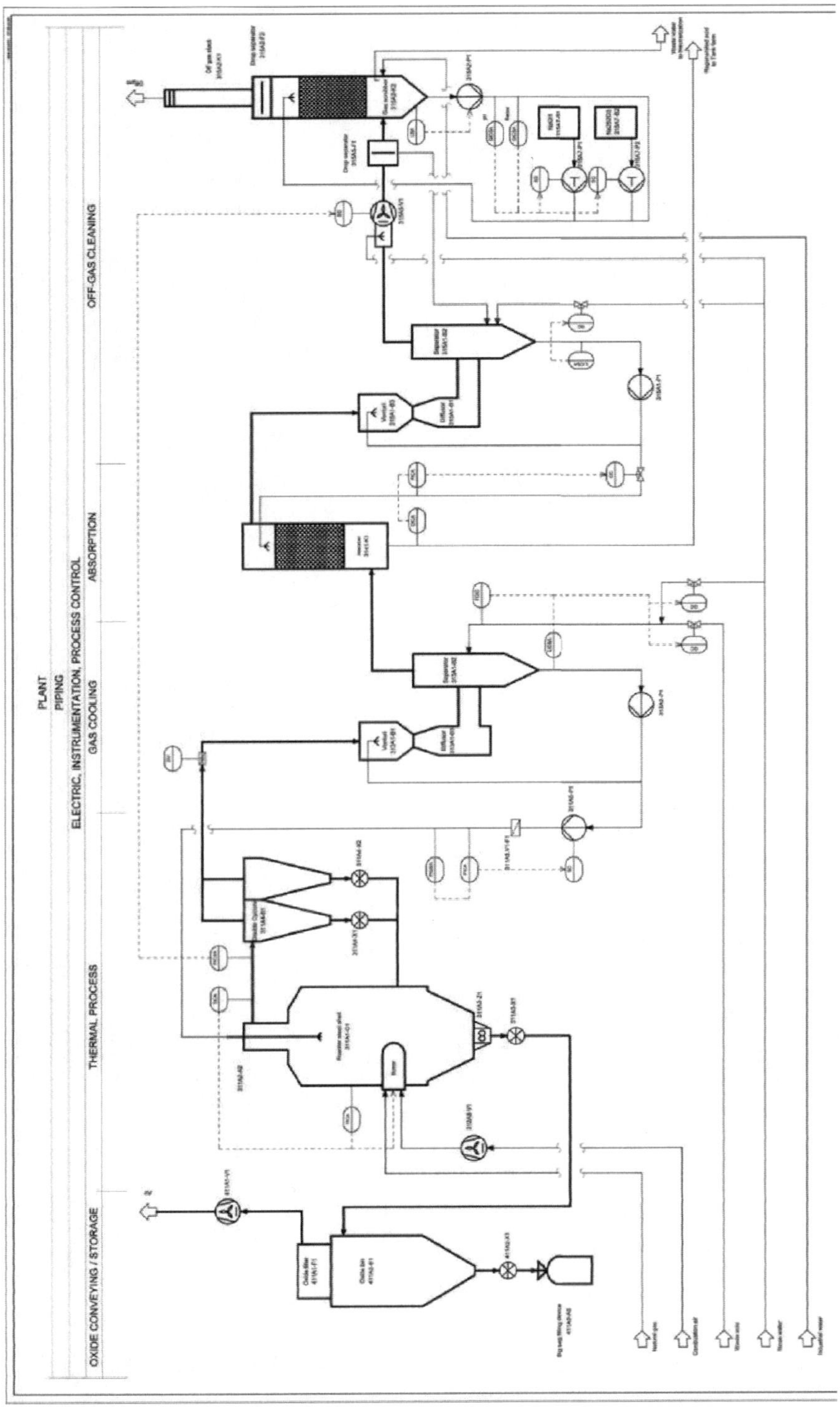

Figure 3 – Flow sheet of the pyrohydrolysis process

The spent acid solution is fed into the preconcentrator (working as a recuperator), where the hot roast gas is cooled by direct heat exchange and the liquid is partially evaporated. The concentrate from the preconcentrator is fed into the reactor, which can be a spray roaster or a fluidized bed. A combustion gas with a certain amount of excess air is let into the reactor and delivers the heat needed for evaporation and the pyrohydrolysis reaction. The off gas leaving the reactor is separated from entrained oxide particles by a cyclone and a venturi scrubber, where the concentrated solution itself serves as a washing solution. The hydrogen chloride vapor produced is converted back into hydrochloric acid in the absorption column fed with a washing liquid coming from a second scrubber. The hydrochloric acid formed can be reused for pickling or leaching. The gas scrubber column absorbs the last HCl residuals from the vapor using fresh water or the wash water from the rinsing section of the pickling line. Only the quantities of hydrochloric acid that are lost due to the presence of thermally non decomposable chlorides and those leaving the process via the off-gas and as residual chlorides in the oxide have to be replaced by fresh hydrochloric acid.

So far approximately 400 pyrohydrolysis process plants (300 spray roasters and 100 fluidized beds) have been built all around the world with a capacity from 0,3 up to 30 m^3/h of spent acid solution.

THE SPRAY ROASTER

The spray roasting reactor mainly consists of an empty cylindrical tower lined with refractory material. The bottom is conical or flat and equipped with a rabble rake to ensure the flow of processed oxide to the outlet. The chloride solution is sprayed through a ceramic (Al_2O_3/SiC) or niobium nozzle assembly into the upper part of the reactor. The droplets descend down in a swirling flow, dry up and are pyrohydrolyzed (see Figures 4 and 5). The dimension of the reactor diameter is related to various factors, such as terminal velocity of the particles, feed concentration, type of metal, fuel used for combustion, desired oxide quality parameters, and plant capacity.

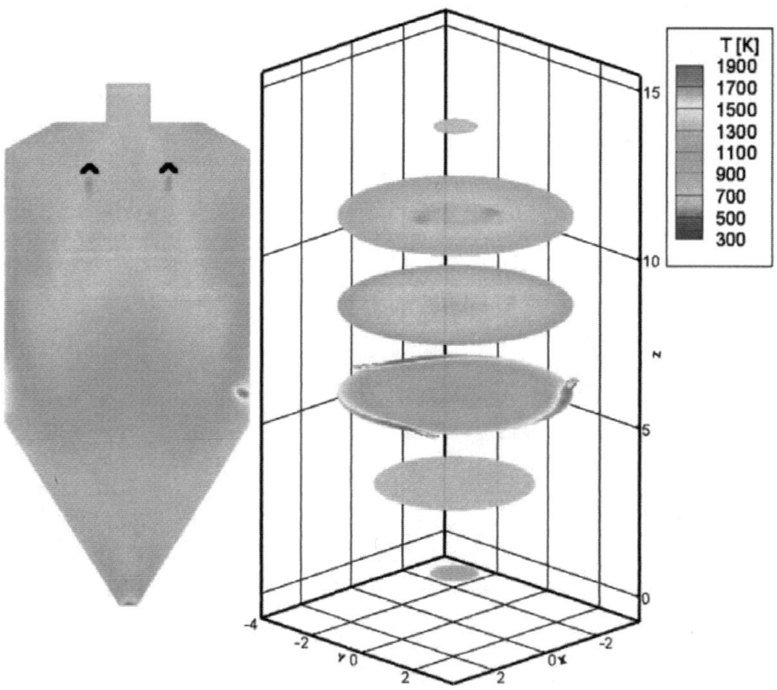

Figure 4 – Typical temperature profile in a spray roast reactor (Schiemann, Wirtz, Scherer & Baerhold, 2013)

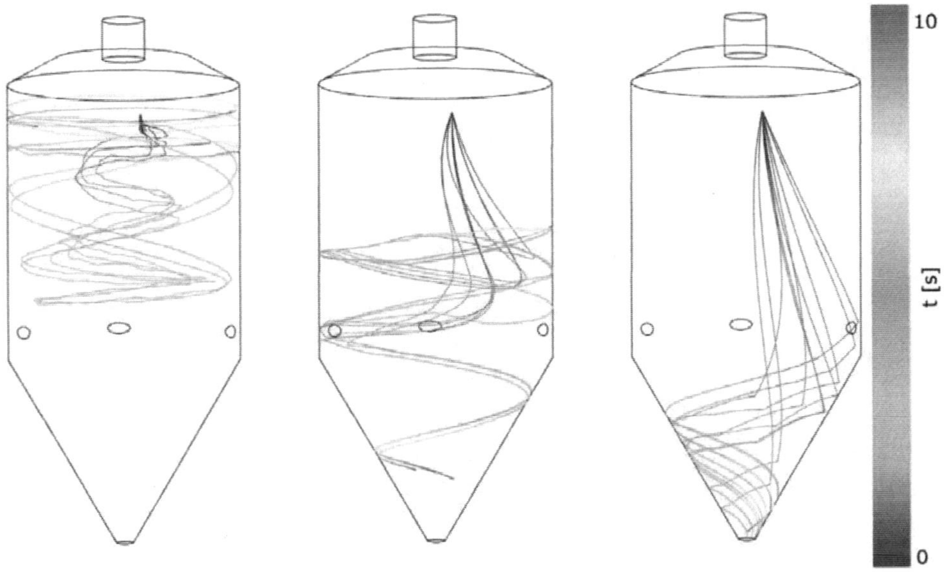

Figure 5 – Particle tracks of 10 representative particles (left to right: 33, 317, and 600 μm). Track length has been limited to 10 s residence time (Schiemann et al., 2013)

On the outside of the reactor one or multiple burner chambers are mounted tangentially. The burners can be operated with gas, liquefied gas or oil, or even coal dust. The hot off-gas from the direct combustion creates a swirling flow pattern. In the reactor very intensive heat yields occur and the temperature profile has a high gradient. The gas outlet temperature is in the range of 350 – 500°C for optimum reaction conditions (mainly depending on the type of feed material and the required oxide product quality) and to avoid condensation of corrosive HCl in the gas duct. A negative pressure must be maintained in the whole plant, so that neither corrosive gases nor oxide powders can escape into the plant environment. The oxide is collected at the bottom, where agglomerates are crushed and continuously removed by a rotary valve which separates the furnace atmosphere from the environment. The oxide is transported by a pneumatic conveying system into an oxide storage bin. From here the oxide powder can be filled into big-bags or directly into trucks for further transportation.

Powders obtained from a ferrous chloride solution by spray roasting pyrohydrolysis show a typical hollow-spherical structure comprising of agglomerated microcrystals (also referred to as primary crystals) with agglomerate diameters ranging from 20 to 500 μm, but usually limited to between 100 and 150 μm. The primary crystals measure between 0.1 and 0.5 μm, depending on the process mode. The oxide powders have a bulk density of 0.2 to 0.8 g/cm³ and a specific surface ranging from 2 to 8 m²/g (BET) for Fe_2O_3 and up to 135 m²/g for other oxides. A post-treatment for the solid product can be added in order to improve the chemical purity or to change the crystallographic structure of the oxide.

THE FLUIDIZED BED

The fluidized bed reactor is more compact compared to a spray roaster (Figure 6). The reactor consists of a brick-lined cylindrical tower. The gas enters the roaster via the bottom, is evenly distributed via nozzle injectors, rise upwards through the bed and the freeboard and leaves the reactor via the top. Fines carried over the bed with the gas stream are retained by increasing reactor cross section in the freeboard and/or separated by a cyclone mounted after the reactor and rejected to the bed. The gas distribution at the roaster bottom consists of multiple tuyeres, where the fuel gas and air are pre-mixed. The combustion takes place directly in the fluidized bed. The fluid bed temperature is in the range of 820–880°C. The combustion off-gas fluidizes the solid particles. The bed diameter has to be designed according to particle diameter and density, residence time of the solids, gas flow (specific energy consumption) and bed height. The solution is fed into the bed via a side-mounted tube made of titanium. No specific injection system is provided, the liquid distributes rapidly when caught by the fluidized solids. Feeding the fluidized bed pyrohydrolyzer with solids or slurries is possible. In some applications, a part of the produced oxide is milled and fed back to improve the granulation process and to control the PSD (Baerhold & Neumann, 1996, Tang-Poy, 1996, Peek, E. M. L., Goedhart, O. & Van Weert, G. 1996).

The pyrohydrolysis reactions take place on the surface of the solids forming the next layer of sintered iron oxide. The roasted particles are discharged via a solid gate or a drain where the solids are screened by a counter-current gas flow. The latter allows a constant diameter of the product. Oxide granulates obtained by fluidized bed pyrohydrolysis are perfectly shaped spheres with an onion-like structure, resulting from the continuous build-up of oxide layers. In case of an iron chloride roaster the typical particle diameters are about 0.5 - 1 mm. The solid density is about 5.4 g/cm³, while the bulk density is around 3.5 g/cm³.

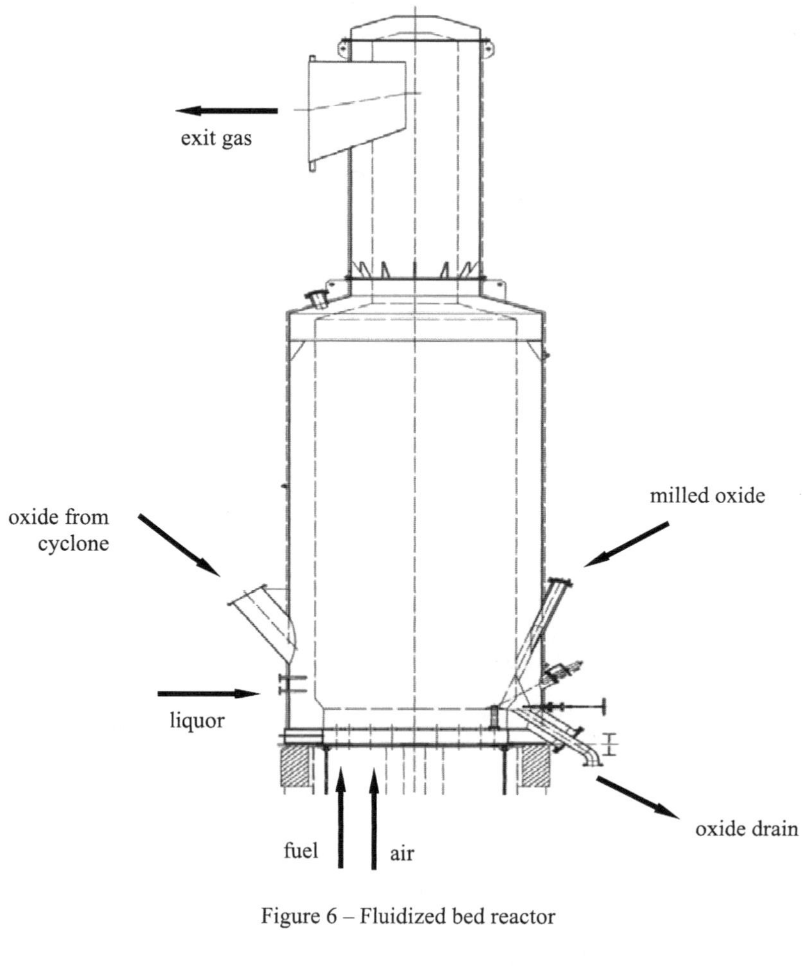

Figure 6 – Fluidized bed reactor

Hydrolytic Distillation

In the 1960s, the PORI Process was developed as a low-temperature alternative to Pyrohydrolysis. It operated briefly in the 1970s (Burtch, 1975). In the last years the basic idea of the PORI Process was further developed by Neomet (Harris & White, 2013; Harris, 2014), McGill University (Becze, Hock. & Demopoulos, 2011), and SMS Siemag (Takahashi, Taki, Weissenbaeck & Vogl, 2009; Weissenbaeck, Nowak, Vogl, & Krenn 2013). In all cases the key process steps to separate iron and to recover the hydrochloric acid are:

- pre-concentration/evaporation of the feed solution to increase the $FeCl_2$ concentration,
- oxidation of Fe(II) to Fe (III) if necessary; this is done at elevated or atmospheric pressure, using oxygen or air; when starting from pure Fe(II) approximately 1/3 of the iron precipitates as hematite in this step; the oxidation reaction is exothermic,
- liquid phase hydrolysis at atmospheric pressure at approximately 180°C; the ferric chloride is converted into gaseous HCl and solid hematite. This reaction is endothermic and mainly controlled by the rate of heat and water input.

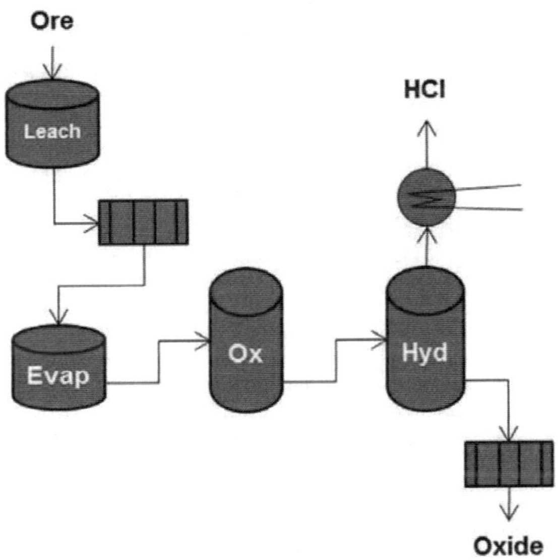

Figure 7 – Hydrolytic distillation

The temperature range for the hydrolysis of the highly concentrated ferric chloride solution is very limited, and is further decreased by the presence of other metals such as manganese. The lower temperature limit is given by the crystallization point, at the upper limit the salt starts to decompose and "dry out". Crystallization is mainly a problem for the filtration step following the hydrolysis. "Dry out" is a problem for the heat input, because the temperature difference at the heat exchanger leads to elevated surface temperatures. A lot of research was done to find an inert matrix solution that expands the temperature range. The filtrate of the hematite separation step is recycled back to the hydrolysis. Components that do not hydrolyze at the actual temperature build up. A bleed stream is necessary to limit the concentration of those components. From the bleed other metal compounds can be separated.

Looking to the aforementioned three process steps as a black box, the input and outputs are the same as for the pyrohydrolysis, thus the energy consumption is almost the same. In case of pyrohydrolysis dilute rinse or wash water can be used for the HCl-absorption and is thereby recycled free of charge. In case of hydrolytic distillation this is not possible. But the process temperatures are on a much lower level, so that relatively low temperature waste heat from other sources can be utilized much easier. One option is, to use the condensation heat of the HCl/H_2O-stream from the hydrolysis to heat the evaporation, which then needs to be operated at reduced pressure (Takahashi et al. 2009).

The most important difference between pyrohydrolysis and hydrolytic distillation is that the latter can be used to selectively separate certain components out of a mixture, whereas pyrohydrolysis produces one mixed oxide product. From the most relevant elements titanium hydrolyses first at relatively low temperatures. The iron hydrolysis follows at a temperature range of 165 – 190°C. Additional heating of the iron-free matrix solution to ~200°C and injecting steam or water causes the base metals, but not magnesium, to hydrolyze and precipitate as basic chlorides, with a generic formula of $Me(OH)_2{*}MeOHCl$, where Me presents for example Cu, Ni or Co (Harris, White, Demopoulos, & Becze, 2008).

The basic metal chlorides can be separated and treated by well-established methods to recover the metals. In a temperature range of 200–230°C magnesium can be precipitated as hydroxy-chloride. Alkaline and earth alkaline metals will not hydrolyze and may be separated via a bleed stream.

ENERGY SAVING OPTIONS

Several attempts have been made to reduce the energy consumption of the spray roasting process. A simple way of process integrated heat recovery is to use the latent heat of the hot reactor off-gas (e.g. after the cyclone) to preheat the combustion air. Unfortunately this is not a technically viable solution, since the dust in the hot off-gas either leads to erosion or to plugging of the heat exchanger.

Pre-evaporation in a multi effect evaporator saves energy, but is limited to the crystallization limit of the metal chloride solution in the pre-concentrator. Due to the extremely corrosive nature of the HCl/metal chloride feed solution investment costs are high.

Other options are to recover heat from off-gas condensation, using e.g. compression and steam generation, heat pumps, or hot water generation (Adham & Lee, 2002). The temperature level of the recovered energy is low or has to be elevated by expensive additional process steps. A disadvantage is also, that a high amount of waste water is created. Energy recovery from the hot oxide is relatively easy to realize, but accounts for only less than 5% of the total fuel requirement.

Another route was published by Frithum (Frithum, Klausner, Szokol & Storek, 2011). The basic idea is to not use the single stage heat recovery from the hot off-gases from either the spray roaster or fluidized bed pyrohydrolysis plant, but to use a multi-effect evaporator to preconcentrate the feed solution up to a concentration close to the saturation limit and then use a rotary kiln for the pyrohydrolysis. Theoretically this way could also lead to a super-azeotropic acid, but corrosion problems in the pre-evaporation and in the rotary kiln and the fact, that this process was never proven on an industrial scale, definitely indicate the limits of this idea.

THE NEW ECO*mode*

As described above, the conventional spray roasting process receives a feed solution from the pickling line with approximately 120 g/L Fe. In the pre-concentrator the feed solution has a concentration of 170–200 g/L Fe. Safe operation of the spray roasting process is possible with much higher concentration. This means, that in the normal mode more water is evaporated than necessary. By reducing the amount of water to be evaporated a lot of energy can be saved (Baerhold, Mitterecker & Stingl, 2015).

The acid recovery process produces a regenerated acid with approximately 200 g/L of free acid and almost no iron. During the pickling process the iron content of the pickling acid increases and the concentration of free acid decreases. In the simplified example described in Figure 8 the waste acid leaving the pickling line contains 120 g/L of Fe and still a free acid of 43.3 g/L.

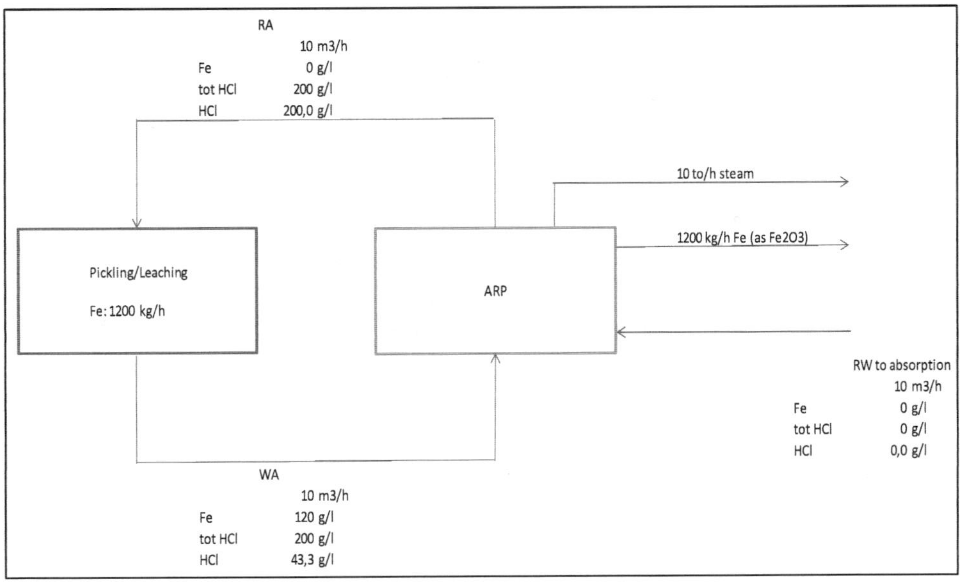

Figure 8 – Standard operating mode of pickling/acid recovery

ECO*mode*

The idea of the ECO*mode* is to shift concentrations and allow acid recovery at higher levels. In the pickling line a fixed amount of scale has to be removed, so a fixed input of Fe into the pickling solution is given. Due to optimized ECO*mode* concentration levels within the system, the acid recovery process is much more energy efficient compared to the conventional concept.

In the ECO*mode* 25% less feed solution has to be fed to the acid regeneration unit to produce the same amount of iron oxide and regenerated acid. The iron separation, which is given by the pickling line, is the same. This results in a reduction of the fuel demand of 25% and a reduction of the electric power consumption (especially for the exhaust/off gas fan). For leaching processes the principles are the same. For each component the HCl-vapor pressure of the system $MeCl_x$-HCl-H_2O has to be considered and the concentration levels need to be optimized.

The system has already been installed in a commercial size pickling/acid recovery plant. Energy efficiency has been proven on a continuous basis for more than a year now. The surface quality of the steel strip leaving the pickling line has always been perfect.

CONCLUSIONS

Pyrohydrolysis via spray roasting and fluidized bed reactors was described. Both systems are still on the market. With the introduction of ECO*mode* the spray roasting process is now more energy efficient. A major issue of HCl-leaching processes in the mineral industry is iron removal. Iron removal and acid recovery via pyrohydrolysis in a spray roaster is a proven and reliable technology, but often consumes a major part of the energy demand of the entire process. The ECO*mode* is ANDRITZ METALS' new milestone in treating spent pickle and waste leaching liquors. 25% energy can be saved by the ECO*mode*.

Hydrolytic distillation is more and more considered as an alternative to pyrohydrolysis. Due to the lower process temperatures compared to pyrohydrolysis internal heat recycling can be applied to reduce the energy consumption. Also external sources of low temperature waste heat may be utilized.

REFERENCES

Adham, K. & Lee, C. (2002), Energy recovery in the metal chloride pyrohydrolysers, in Peek E. & Van Weert G. (Eds.), *Chloride Metallurgy 2002, Vol. II* (pp. 657-672), proceedings of the 32nd annual hydrometallurgy meeting. Montreal, QC, ISBN: 1-894475-28-3

Baerhold, F., Mitterecker, S. & Stingl, A. (2015), Energy savings in the system pickling/acid recovery: the new ECO*mode*, *AISTech Conference*, May 2015, Pittsburgh, US

Baerhold, F. & Neumann, J. (1996), Fluidized bed pyrohydrolyzer for generation of iron oxide, *Proceedings of the iron oxides for hard/soft ferrites conference*, Pittsburgh, 1996

Barin, I., Knacke, O. (1973), Thermochemical properties of inorganic substances, Springer Verlag, Berlin, Heidelberg, New York

Becze, L., Hock, S.J. & Demopoulos, G.P. Precipitation of hematite and recovery of hydrochloric acid from concentrated iron chloride solutions by a novel hydrolytic decomposition process, *Chloride 2011*, TMS EPD Congress 2011, San Diego, CA, March 2011, 529-542.

Burtch, Jack W. (1975). Hydrochloric acid from industrial waste streams - the PORI Process, *CIM Bulletin, Vol. 68,* January 1975, 96-100.

Frithum, G., Klausner, H., Szokol, F. & Storek, F. (2011), Method for retaining or recovering nitric acid and hydrofluoric acid of stainless steel pickling solutions, *International Patent Application, PCT WO2011/130760 A1*

Gamsriegler, D., Braunauer, L. & Rechberger, E. (1998), *Austrian Patent AT 404031*, Verfahren zur Regeneration von HCl-Beizsäuren mit Beimengungen von Zink in einem Sprühröstprozess

Harris, B., White, C., Demopoulos, G. & Becze, L. (2008). Recovery of metal oxides and associated acid from process solutions using water as reagent. *ALTA Copper 2008*, Perth, Australia, June 19-20 2008.

Harris, Bryn & White, Carl (2013). Further development of the chloride process for base and light metals: recent mini plant and first pilot plant data. *ALTA Ni-Co-Cu 2013*, Perth, WA, May 2013.

Harris, G.B. (2014). Making use of chloride chemistry for improved metals extraction processes. *Proceedings of the 7th International Symposium on Hydrometallurgy (HYDRO 2014)*, (Eds. E. Asselin, et al.), Canadian Institute of Mining, Metallurgy and Petroleum, Westmount, Quebec, Canada, pp. 171-184.

Jedlicka, H. (1977), New applications of the spray roasting process in the chloride hydrometallurgy, *Proceedings of the International Symposium of Chloride Hydrometallurgy,* Metallurgie Hoboken and Benelux, Bruxelles, Belgium, September 26-28, 1977, pp. 154 - 180

Kladnig, W. F., Karner, W. (1990), Pyrohydrolysis for the Production of Ceramic Raw Materials, *Ceram. Bulletin*, 69/5, pp. 814 - 817

Peek, E. M. L., Goedhart, O. F. & Van Weert, G. (1996), Process evaluation of steel pickle liquor pyrohydrolysis in a commercial Keramchemie fluid bed reactor, *Proceedings of the Second international symposium on Iron Control in Hydrometallurgy*, Ottawa, Canada, October 1996, pp. 483-504

Regel-Rosocka, Magdalena: A review on methods of regeneration of spent pickling solutions from steel processing, *Journal of Hazardous Materials*, 177 (2010), pp. 57–69

Schiemann, M., Wirtz, S., Scherer, V. & Baerhold, F. (2013), Spray roasting of iron chloride: Numerical modeling of industrial reactors, *Powder Technology*, Vol. 245, pp. 70 - 79

Steinbach, W. & Baerhold, F. (2002), Comparison of spray roasting and fluidized bed granulation for recovery of hydrochloric acid from metallurgical processes through pyrohydrolysis, *in Chloride Metallurgy 2002*, Eds. Peek, E. and Van Weert, G., Canadian Institute of Mining, Metallurgy and Petroleum, Montreal, Canada, pp. 643 - 655

Takahashi, N, Taki, O., Weissenbaeck, H. & Vogl, D. (2009). Processing method for recovering iron oxide and hydrochloric acid. Processing method for recovering iron oxide and hydrochloric acid. *International patent number WO2009/153321 A1,* December, 2009.

Tang Poy, R. (1996), Regeneration of hydrochloric acid from ferrous chloride using the Keramchemie/Lurgi fluidized bed reactor system, *Proceedings of the second international symposium on iron control in hydrometallurgy*, Ottawa, Canada, October 1996, pp. 471-482.

Weissenbaeck, Herbert, Nowak, Benedikt, Vogl, Dieter, & Krenn, Horst (2013). Development of chloride based metal extraction techniques: advancements and setbacks. *ALTA Ni-Co-Cu 2013*, Perth, WA, May 2013.

EVALUATION OF IRON(II) OXIDATION REACTIONS IN CHLORIDE LEACH SOLUTIONS

*C. Dorfling[1], S. Masambi[1], and S.M. Bradshaw [1]

[1]*Department of Process Engineering, Stellenbosch University*
Private Bag X1
Matieland, 7602, South Africa
*(*Corresponding author: dorfling@sun.ac.za)*

ABSTRACT

The presence of iron in base metal leach solutions complicates nickel and copper recovery processes. Leach solutions are commonly purified by precipitating iron as hematite, jarosite or goethite. Fast oxidation of ferrous to ferric is an important step to remove iron by precipitation at low pH values. This study reviews the ferrous oxidation mechanism and reaction rates in chloride media using molecular oxygen. This work contributes to the development of a better understanding of iron precipitation processing steps, which continues to be an important matter in hydrometallurgy. The specific objectives were to investigate (1) the effect of key process variables (acid concentration, copper concentration, and temperature) on the rate and extent of ferrous oxidation by dissolved oxygen and (2) the oxidation mechanism by varying chloride and hydrogen ion concentrations separately. Batch oxidation tests were performed using synthetic chloride solutions containing 45 g/L Fe and 3 g/L Ni at 50 and 80°C, dissolved Cu concentrations (0.3 to 3 g/L) and HCl concentrations (2 to 4.6 M). Regression analysis indicated that the significance of the effect of different variables on the extent of oxidation achieved after five hours decreased in the following order: HCl concentration > Cu concentration > temperature. Increasing the HCl concentration from 4 M to 4.6 M caused the degree of oxidation achieved after two hours to increase from 58.3% to 98.1%. The catalytic effect of dissolved copper on ferrous oxidation was more significant at low acid concentrations. Experimental data suggest that rapid oxidation of ferrous is possible at high HCl concentrations for two main reasons. First, the formation of ferrous chloride complexes allows ferrous oxidation to proceed without the formation of superoxide as an intermediate product (the rate limiting step in dilute acid solutions). Second, the formation of ferric chloride complexes stabilises the oxidation product.

KEYWORDS

Chloride solution, iron precipitation, oxidation

INTRODUCTION

Background

Precipitation remains the preferred method of iron removal for the purification of pregnant leach solutions. In general, precipitation of ferric is preferred to precipitation of ferrous because it can be conducted at a lower pH; this is beneficial because of reduced neutralisation reagent costs and lower valuable metal losses due to co-precipitation. Oxidation of ferrous to ferric is thus an important step in iron precipitation processes used for leach solution purification.

This study focussed specifically on the oxidation of ferrous to ferric present in a chloride leach solution containing nickel and copper as valuable metals by using oxygen gas as oxidising agent, and aimed to review the ferrous oxidation mechanism and reaction rates. The work formed part of a larger project which focussed on the development of a process route for nickel and copper recovery from a leach solution contaminated with iron. The first specific objective was to evaluate the effect of key process variables (acid concentration, dissolved copper concentration, and temperature) on the rate and extent of oxidation. The second objective of the study was to investigate oxidation behaviour at various individual ions' concentrations to better understand the role of chloride and hydrogen ions in the oxidation mechanism.

Oxidation Mechanisms

Oxidation of ferrous by oxygen involves the dissolution of oxygen and the subsequent oxidation of the species in solution with dissolved oxygen; the overall oxidation reaction is shown in Equation 1. The oxidation reaction is usually the rate determining step when performing the oxidation at low pH values in an agitated system (Awakura, Majima, & Iwai, 1986; Bouboukas, Gaunand, & Renon, 1987; Iwai, Majima, & Izaki, 1979; Lamb & Elder, 1931; Posner, 1953; Stumm & Lee, 1961). The slow rate of ferrous oxidation in aqueous solutions has been reported to be due to the formation of superoxide as intermediate product according to Equation 2.

$$4Fe^{2+} + O_2 + 4H^+ \rightarrow 4Fe^{3+} + 2\,H_2O \tag{1}$$

$$Fe^{2+} + O_2 \rightarrow O_2^- + Fe^{3+} \tag{2}$$

For chloride leach solutions specifically, different reaction mechanisms have been proposed to explain the oxidation behaviour. According to Posner (1953), oxidation of ferrous does not proceed via the formation of superoxide as an intermediate species in chloride systems; instead, ferrous chloride complexes (indicated as $HFe^{2+}Cl$) are oxidised directly to ferric according to Equation 3, resulting in faster overall oxidation kinetics. This reaction forms the first step in the modified Haber-Weiss mechanism proposed by Posner (1953). Combining the reaction steps of the modified Haber-Weiss mechanism yields Equation 4 as the overall reaction.

$$HFe^{2+}Cl + O_2 \rightarrow Fe^{3+} + Cl^- + HO_2 \tag{3}$$

$$HFe^{2+}Cl + O_2 + H^+ + 3Fe^{2+} \rightarrow 4Fe^{3+} + Cl^- + 2OH^- \tag{4}$$

Colborn and Nicol (1973) reported that the first step in the modified Haber-Weiss mechanism involves formation and oxidation of a ferrous oxygen complex (indicated as $Fe^{2+}O_2$). The first reaction step and the overall reaction are shown in Equation 5 and Equation 6, respectively. Colborn and Nicol (1973) did also proposed that a degree of chloride complex formation and oxidation contribute to the overall oxidation reaction, depending on the leach solution conditions. In these cases, the first reaction steps are represented by Equation 7 and Equation 8.

$$Fe^{2+}O_2 + Fe^{2+} \rightarrow 2Fe^{3+} + O_2^{2-} \tag{5}$$

$$Fe^{2+}O_2 + 3Fe^{2+} + 2H^+ \rightarrow 4Fe^{3+} + 2OH^- \quad (6)$$

$$FeCl^+O_2 + Fe^{2+} \rightarrow Fe^{3+} + FeCl^{2+} + O_2^{2-} \quad (7)$$

$$FeCl_2O_2 + Fe^{2+} \rightarrow Fe^{3+} + FeCl_2^+ + O_2^{2-} \quad (8)$$

The main difference in the work of Posner (1953) and that of Colborn and Nicol (1973) was the ferrous to hydrochloric acid ratios considered in their experimental work. Colborn and Nicol (1973) investigated the rate expressions for solutions containing 0.48 to 0.97 M ferrous and HCl concentrations ranging between 1 M and 3.9 M; the concentration of ferrous in the solutions investigated by Posner (1953) was an order of magnitude lower than in the case of Colborn and Nicol (1973), while the hydrochloric acid concentration was higher at 8 M.

Oxidation Rate Expressions

A summary of the rate expressions for the oxidation of ferrous in chloride medium by dissolved oxygen that have been proposed in literature is presented in Table 1. Differences in rate expressions are usually ascribed to variations in ferrous concentration, hydrochloric acid concentration, and the presence of catalytic metallic species. In general, the rate expression changes from first order to second order with respect to the ferrous concentration as the ratio of ferrous to hydrochloric acid is increased; the expression at lower ferrous to hydrochloric acid concentration ratios is also a function of the hydrochloric acid concentration, while the hydrochloric acid concentration has an insignificant effect at higher ferrous to acid ratios.

Awakura et al. (1986) reported rapid oxidation at acid concentrations exceeding 4 M but relatively slow oxidation rates at lower acid concentrations. Both cupric and chloride ions were reported to catalyse the ferrous oxidation reaction. Although experiments were not performed to determine which component of the HCl influenced the kinetics of oxidation, it was proposed that the Cl^- ions may have been the species responsible for the increase in the rate of oxidation as a result of its tendency to form complexes with Fe^{2+} and Fe^{3+} compounds. Colborn and Nicol (1973), George (1954), Posner (1953), and Stumm and Lee (1961) have also reported on the catalytic effect of dissolved copper on ferrous oxidation.

Table 1 – Rate expressions for ferrous oxidation by oxygen in chloride solutions

Rate expression	**[Fe^{2+}]**	**[HCl]**	**Reference**
$k[Fe^{2+}]^2P_{O2}$	0.01 – 0.3 M	1 M	McBain, 1901
$k[Fe^{2+}]^2P_{O2}$	0.01 – 0.3 M	1 M	George, 1954
$k[Fe^{2+}]^2P_{O2}$	0.48 – 0.97 M	1 – 3.9 M	Colborn and Nicol, 1973
$k[Fe^{2+}]P_{O2}f\{HCl\}$	0.055 N	8 M	Posner, 1953
$k[Fe^{2+}]P_{O2}\,f\{HCl\}$	0.2 M	6 M	Iwai et al., 1979
$k[Fe^{2+}]P_{O2}\,f\{HCl\}$	0.72 M	2 – 6 M	Awakura et al., 1986

The reaction mechanisms proposed by Posner (1953) and Colborn and Nicol (1973) correlate with the two rate expressions commonly reported in literature and shown in Table 1, assuming that the first step (complex formation and oxidation) is the rate determining step. This observation suggests that oxidation proceeds primarily as a result of oxidation of oxide complexes at high ferrous to acid ratios, but via oxidation of ferrous chloride complexes at low ferrous to acid ratios. This is in agreement with the different experimental conditions used by Posner (1953) and Colborn and Nicol (1973), as discussed previously. The complexity of the expressions proposed for higher acid conditions, which include a relatively complex function of HCl concentration rather than a simple n^{th}-order dependence, confirms that various ferrous chloride complexes with different stoichiometric H+ and Cl^- requirements might be present, and that the oxidation is likely to proceed via a combination of mechanisms rather than a single mechanism.

The range and complexity of iron complexes in chloride solutions have been reported by Apted, Waychunas and Brown (1985), who investigated the iron speciation in chloride solutions for different pH values, ionic strengths, and iron to chloride ratios. It was, for example, reported that ferrous complexes' bond strength and distances depend on the iron to chloride ratio and the acid concentration. Ferric complexes potentially present in acidic chloride solutions were reported to include octahedral $FeCl_6^{3-}$ and $Fe(H_2O)_6^{3+}$, tetrahedral $FeCl_4^-$, mixed octahedral $Fe(H_2O)_{6-x}Cl_x^{(3-x)}$, polymeric tetrahedral $Fe_xCl_{2(x+1)}^{(x-2)}$, and polymeric octahedral $Fe_x(H_2O, Cl)_{4x+2}$. The stability of ferric species was reported to be very sensitive to ferric to chloride ratios.

EXPERIMENTAL

Materials

Synthetic leach solutions containing 1 to 4.6 M HCl, 45 g/L iron, 3 g/L nickel, and 0.3 to 3 g/L copper were prepared using reagent grade chemicals (iron chloride tetrahydrate, nickel chloride hexahydrate, copper chloride, and hydrochloric acid) supplied by Sigma-Aldrich. Sodium chloride used to increase the chloride concentration was procured from Merck. Kimix supplied the phosphoric acid, potassium dichromate, and sodium diphenylamine used for sample analysis by titration.

Experimental Design

The effects of acid concentration, catalytic copper ion concentration, and temperature on the oxidation rate were investigated by considering the following conditions: acid concentrations of 2, 4, and 4.6 M; copper concentrations of 0.3 and 3 g/L; and temperatures of 50 and 80°C. To further elucidate the role of chloride and hydrogen ion concentrations in the oxidation mechanism, additional tests were performed at 80°C with 3 g/L Cu and (a) 1 M HCl, and varying chloride concentrations (2, 3, 4, and 4.6 M), as well as (b) 4.6 M chloride concentration and varying hydrogen ion concentrations (0.65, 1.0, 2.0, 3.0, and 4.6 M).

Equipment

The experimental setup consisted of a 1.6 L closed glass reactor fitted with an overhead mechanical agitation system driving a PTFE coated dual (axial and radial) impeller system. Oxygen required for oxidation was sparged into the synthetic leach solution through a PTFE tube positioned below the bottom impeller. A heat plate with feedback control was used to control the temperature to within 2°C of the setpoint temperature. Sampling was done at fixed time intervals using syringes connected to a sampling tube inserted through the reactor lid.

Procedure and Analysis

The reactor was filled with 800 mL synthetic leach solution, which was agitated at a rate of 600 rpm and heated to the desired temperature. Once the setpoint temperature had been reached, a sample was taken to confirm the initial solution composition. The agitation rate was then increased to 1600 rpm, and oxygen sparging at a rate of 2 L/min was initiated immediately thereafter. The time instance at which oxygen was introduced into the system was considered the start of the experiment, which was run for 5 hours.

Solution samples were taken at the following residence times: 10, 20, 30, 60, 120, 180, 240, and 300 minutes. The ferrous concentration in solution was determined by titration with 0.0084 M potassium dichromate as titrant and sodium diphenylamine as indicator. Phosphoric acid was added to mask ferric in solution.

RESULTS AND DISCUSSION

Effect of HCl Concentration

The test performed at 80°C with 4.6 M HCl and 3 g/L Cu achieved in excess of 98% ferrous oxidation within 2 hours, as shown in Figure 1; this is 38 and 62 percentage points higher than the degree of oxidation achieved after two hours with 4 and 2 M HCl, respectively. Similar results were reported by Awakura et al. (1986), Posner (1953), and Iwai et al. (1975). Published results often report the effect of acid concentration on the oxidation rate in a manner similar to the data shown in Figure 1. However, this is not sufficient to determine the reason for the increase in oxidation rate with an increase in HCl concentration.

The increase in oxidation rate with an increase in acid concentration from 2 M to 4 M could be ascribed to the higher hydrogen ion concentration favouring the forward overall oxidation reactions (e.g. Equations 1, 4 and 6). However, considering the results obtained when changing the chloride concentration at a constant hydrogen ion concentration suggested that the increase in oxidation rate with an increase in acid concentration was exclusively as a result of the increased chloride concentration resulting from the higher HCl concentration. This suggests that formation and oxidation of ferrous chloride complexes contributed significantly to the overall oxidation even at the relatively high ferrous to acid ratio (0.8 M Fe^{2+} and 2 to 4 M HCl) investigated, and will be discussed in more detail in the relevant subsequent section. The increase in oxidation rate when increasing the acid concentration from 4 M to 4.6 M is larger than when increasing the acid concentration from 2 M to 4 M, suggesting that the dominating oxidation mechanism changed as the acid concentration was increased from 4 M to 4.6 M.

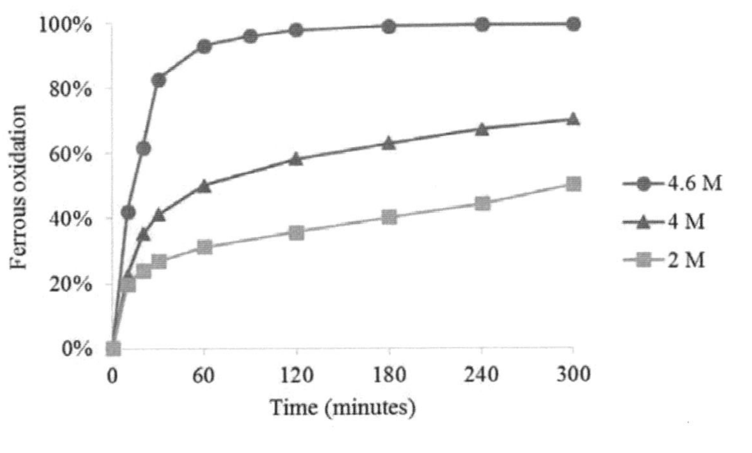

Figure 1 – Effect of acid concentration on the rate of ferrous oxidation at 80°C with 3 g/L Cu

Effect of Temperature

The increase in oxidation rate with an increase in temperature is shown in Figure 2 for tests performed with 3 g/L Cu and 4 M HCl; similar observations were made for other copper and acid concentrations. Despite the decrease in oxygen solubility with an increase in temperature, the oxidation reaction rate increased, which confirms the oxidation was chemical reaction controlled in this study.

Regression analysis with the percentage oxidation achieved after 5 hours as response variable suggested that all investigated variables had a statistically significant effect on the oxidation behavior. The effect of temperature was, however, the least significant followed by the dissolved copper concentration and the HCl concentration, which had the most significant effect.

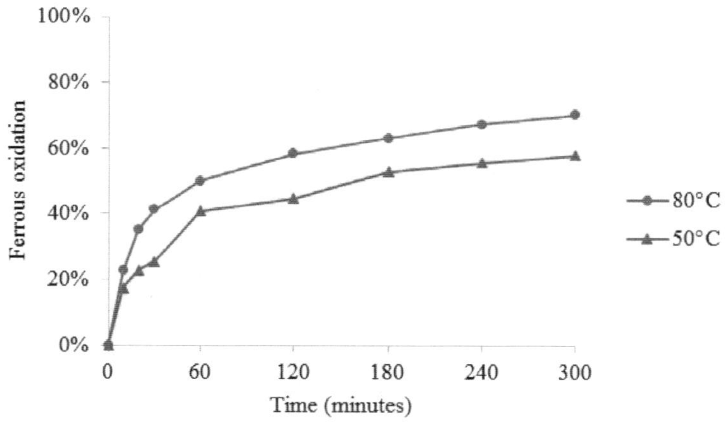

Figure 2 – Effect of temperature on the rate of ferrous oxidation with 3 g/L Cu and 4 M HCl

Effect of Copper Concentration

The catalytic effect of dissolved copper on ferrous oxidation is shown in Figure 3 for tests performed at 80°C and two different acid concentrations. For tests performed with 4.6 M HCl, increasing the dissolved copper concentration from 0.3 g/L to 3 g/L increased the oxidation achieved after two hours from 94.6% to 98.1%. The increase in oxidation rate with an increase in copper concentration was much more significant for the tests conducted with 4 M HCl; the percentage oxidation achieved after two hours increased from 38.2% at 0.3 g/L Cu to 58.3% at 3 g/L Cu.

The catalytic effect of copper is likely to be due to the direct oxidation of ferrous by cupric ions to form ferric and cuprous ions. This oxidation mechanism is electrochemically favourable considering the reduction potentials of the respective half-reactions. The cuprous is subsequently oxidized by dissolved oxygen, which proceeds at a faster rate than ferrous oxidation by dissolved oxygen and hence results in a faster overall ferrous oxidation rate.

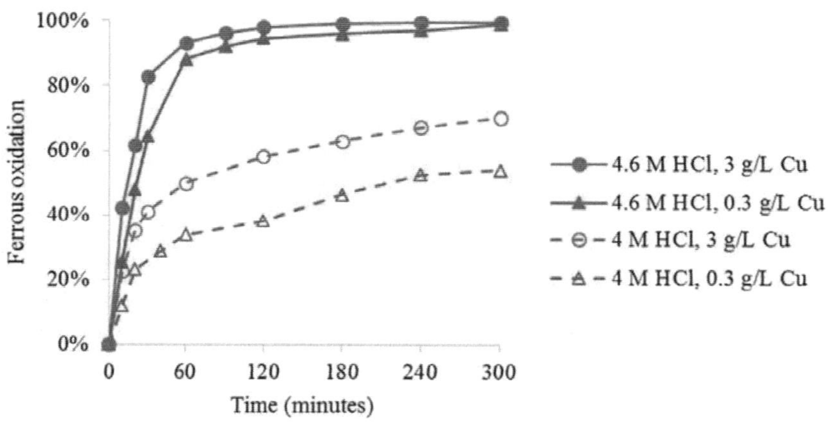

Figure 3 – Effect of dissolved copper concentration on the rate of ferrous oxidation at 80°C and different HCl concentrations

From Figure 3, it is evident that the effect of copper concentration on oxidation rate was more significant at the lower acid concentration. As noted previously, the dominating oxidation mechanism changed as the acid concentration was increased from 4 M to 4.6 M. The different catalytic effects of dissolved copper observed at different acid concentrations provide insight into the relative rates of the rate determining elementary chemical reactions. In the case of 4 M HCl, the direct oxidation of ferrous with cupric was relatively fast compared to the rate of oxidation of ferrous complexes by oxygen, and oxidation by cupric contributed significantly to the overall oxidation. At 4.6 M HCl, ferrous chloride complexes formed more readily. Since ferrous oxidation by cupric played a less important role, it can be concluded that the oxidation of the ferrous chloride complexes by dissolved oxygen occurred relatively fast compared to oxidation by cupric, and dominated the overall ferrous oxidation.

Effect of Cl^- Concentration

The results of the tests performed to evaluate the effect of chloride ion concentration at a constant hydrogen ion concentration of 1 M are shown in Figure 4, which show a clear increase in extent of oxidation with an increase in chloride concentration. The rate of oxidation achieved when using 1 M H^+ and 2 M Cl^- was similar to the results achieved with a 2 M HCl leach solution; similarly, the results obtained with 1 M H^+ and 4 M Cl^- were similar to the results obtained with 4 M HCl. In this chloride concentration range, the oxidation rate was not dependent on the hydrogen ion concentration but rather on the chloride concentration. Considering the possible oxidation mechanisms, these results indicate that oxidation of ferrous chloride complexes already contributed significantly to the overall oxidation at this relatively high iron to acid ratio, since chloride does not participate in the mechanism entailing ferrous oxygen complex formation and oxidation shown in Equations 5 and 6.

An increase in chloride concentration would favour not only the formation of ferrous chloride complexes, and hence the overall ferrous oxidation rate, but also the reduction of ferric (the reverse of the reaction shown in Equation 3). Given the overall increase in oxidation rate, however, it is likely that ferric was stabilised by complexation with chloride preventing the reverse reaction from occurring. Posner (1953) did list the complexation of ferric as important to avoid the ferric from being reduced back to ferrous. Increasing the chloride concentration to 4.6 M at a constant H^+ concentration of 1 M did result in slower oxidation than achieved with 4.6 M HCl; a maximum chloride to hydrogen ion ratio thus exists above which the availability of hydrogen ions starts to limit the oxidation rate.

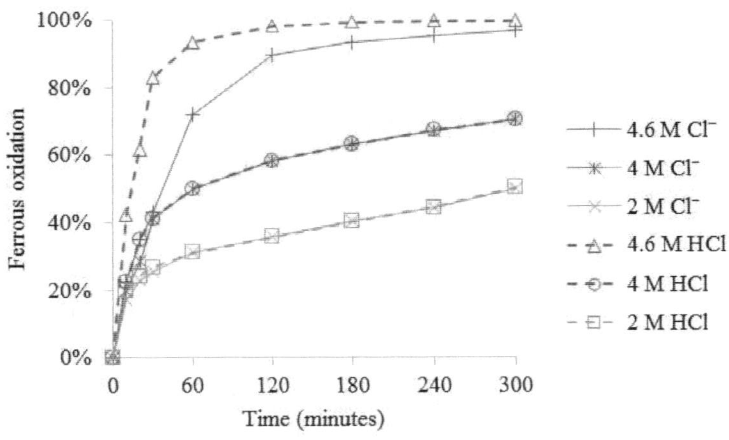

Figure 4 – Effect of chloride concentration on the rate of ferrous oxidation at 80°C with 3g/L Cu and 1 M H^+

Effect of H^+ Concentration

From the results discussed above, it is evident that the hydrogen ion concentration did not affect the ferrous oxidation rate at chloride concentrations of 2 M and 4 M. At a constant chloride concentration of 4.6 M, however, the extent of ferrous oxidation achieved after 2 hours generally increased with an increase in hydrogen ion concentration; the percentage oxidation increased from 82% to 98% with an increase from 0.65 M to 3 M H^+ (as shown in Figure 5).

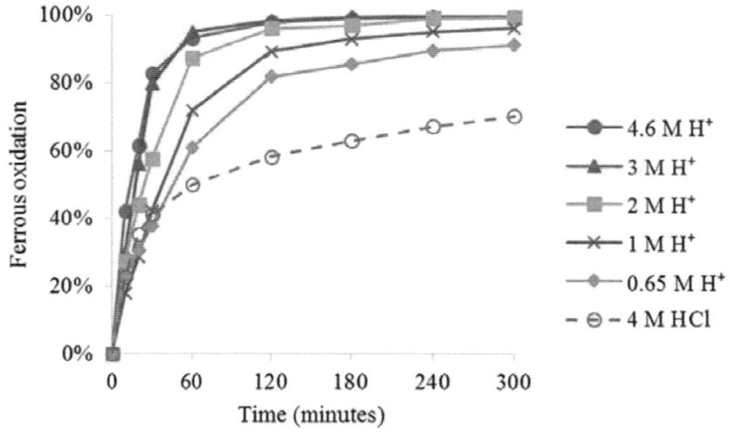

Figure 5 – Effect of H+ concentration on the rate of ferrous oxidation at 80°C with 3 g/L Cu and 4.6 M Cl^-

There was no significant difference in the rate of oxidation achieved when performing the oxidation with 3 M H^+ and 4.6 M Cl^- and when performing the oxidation with 4.6 M HCl; this again suggests that a certain maximum chloride to hydrogen ion concentration ratio exists above which the acid availability becomes the rate determining factor, but below this ratio the acid concentration does not affect oxidation behaviour. This maximum chloride to hydrogen ratio was, however, chloride concentration dependent and this appeared to be the result of the complex speciation in the leach solution.

From Figure 5, it is also clear that the extent of oxidation achieved with 4.6 M Cl^- and 0.65 M H^+ was greater than that achieved with 4 M HCl despite the significantly lower hydrogen ion concentration. The percentage oxidation achieved after 2 hours was 24 percentage points higher for the high chloride and low hydrogen ion concentration conditions than for the 4 M HCl conditions. This result confirms the important role chloride ions play in the ferrous oxidation mechanism, and that the oxidation is more dependent on the chloride concentration than the acid concentration.

CONCLUSIONS

The effects of variations in acid concentration, dissolved copper concentration, and temperature on the rate and extent of ferrous oxidation in synthetic chloride leach solutions were investigated. This was done to facilitate the development of an iron precipitation processing step to be used for impurity removal from an industrial leach solution containing nickel and copper as pay metals. To better understand the role of chloride and hydrogen ions in the oxidation process, the concentrations of these species were also varied independently.

The rate of oxidation increased with an increase in acid concentration, temperature and dissolved copper concentration, with the effect of acid concentration being the most significant. The fastest oxidation was observed for tests performed at 80°C with 3 g/L Cu and 4.6 M HCl; at these conditions more than 98% of ferrous was oxidised within two hours. At the higher acid concentrations, the rate of direct oxidation of ferrous complexes with oxygen is comparable with the rate of ferrous oxidation by cupric. As a result, the catalytic effect of dissolved copper on ferrous oxidation was more significant at lower acid concentrations than at higher acid concentrations.

In the range of 2 – 4 M HCl, the oxidation rate increased with an increase in acid concentration because of the increase in the chloride concentration. Varying the hydrogen ion concentration between 1 and 4 M did not affect the oxidation rate for a specific chloride concentration in this concentration range. It can be concluded that the formation and oxidation of ferrous chloride complexes and the stabilisation of oxidation products through the formation of ferric chloride complexes contributed significantly to the overall oxidation reaction, even at these relatively high iron to hydrochloric acid concentration ratios. The dominating ferrous oxidation mechanism changed when the acid concentration exceeded 4 M, and a maximum chloride to hydrogen ion concentration ratio existed above which the availability of hydrogen ions became the rate limiting step.

REFERENCES

Apted, M.J., Waychunas, G.A, & Brown, G.E. (1985). Structure and specification of iron complexes in aqueous solutions determined by X-ray absorption spectroscopy. *Geochimica et Cosmochimica Acta, 49*, 2081-2089. doi: 10.1016/0016-7037(85)90065-1

Awakura, Y., Majima, H., & Iwai, M. (1986). Oxidation of iron(II) in HCl and H_2SO_4 solutions with dissolved molecular oxygen in the presence and absence of a cupric catalyst. In J.E. Dutrizac & A.J. Monhemius (Eds.), *Iron Control in Hydrometallurgy* (pp. 202-222). Chichester: Ellis Horwood Ltd.

Bouboukas, G., Gaunand, A., & Renon, H. (1987). Oxidation of iron(II) by oxygen in concentrated NaCl solutions: Prediction of stirred gas-liquid reactor performance from homogeneous kinetic data. *Hydrometallurgy, 19*, 25-35. doi: 10.1016/0304-386X(87)90039-9

Colborn, R.P., & Nicol, M.J. (1973). An investigation into the kinetics and mechanism of the oxidation of iron (II) by oxygen in aqueous chloride solutions. *The Journal of the South African Institute of Mining and Metallurgy, 73*(9), 281-289. Retrieved from http://www.saimm.co.za/publications/journal-papers

George, P. (1954). Oxidation of ferrous perchlorate by molecular oxygen, *Journal of the Chemical Society*, 4349-4359. doi: 10.1039/JR9540004349

Iwai, M., Majima, H., & Izaki, T. (1979). A kinetic study on the oxidation of ferrous ion with dissolved molecular oxygen. *Denki Kagaku, 47*, 409-414

Lamb, A.B., & Elder, L.W. (1931). The electromotive activation of oxygen. *Journal of the American Chemical Society, 53*, 137-163. doi: 10.1021/ja01352a019

McBain, J.W. (1901). Oxidation of ferrous solutions by free oxygen. *The Journal of Physical Chemistry, 5*, 623-638. doi: 10.1021/j150036a001

Posner, A.M. (1953). The kinetics of autoxidation of ferrous ions in concentrated HCl solutions. *Transactions of the Faraday Society, 49*, 382-388. doi: 10.1039/TF9534900382

Stumm, W., & Lee, G.F. (1961). Oxygenation of ferrous iron. *Industrial and Engineering Chemistry, 53*, 143-146. doi: 10.1021/ie50614a030

ANDRITZ METALS SPRAY-ROAST ACID REGENERATION PILOT PLANT

*T. Hofbauer[1], F. Baerhold[1], S.K. Mitterecker[1]

[1]Andritz AG
Eibesbrunnergasse 20
1120 Vienna, Austria
*(*Corresponding author: thomas.hofbauer@andritz.com)*

ABSTRACT

Acid regeneration is a very important process step in each acid leach circuit. It is required to recover dissolved metals and HCl from spent leaching acid. Andritz Metals presents its recently commissioned spray roast pilot plant. The facility is now available for testing and demonstrating acid recovery or metal oxide production which is a requirement in several flow sheets under development for the mining and specialized oxides industries. Results from pilot test campaigns are reported here. We are currently focusing on a nickel ore leaching process, which requires both a feasible iron control and acid regeneration step. In addition, we disclose the results of a magnesium ore and aluminium recycling test campaign conducted in our pilot plant facility.

KEYWORDS

Acid Regeneration, Hydrochloric Acid, Spray-Roast, Pilot Plant, Iron control, Oxide production, metal ore, Leaching

INTRODUCTION

Hydrochloric acid regeneration is widely used in the steel industry to regenerate spent pickle liquors into hydrochloric acid at a typical concentration of 18 wt.%. Andritz Metals (formerly known as Ruthner) has built more than 200 acid regeneration plants worldwide (Andritz Metals brochure). As technology leaders Andritz also has vast experience with specialty metal oxide plants, producing metal oxides such as aluminium oxide (Al_2O_3), magnesium oxide (MgO), nickel oxide (NiO), cobalt oxide (Co_3O_4) and mixed oxides from metal chloride solutions (Baerhold F., Lebl A., Starcevic J., 2006), (Jedlicka, H., n.d.), (Ruthner M.J., Krischner H., 1973), (Ruthner Ind. Planungs AG, 1967).

Hydrochloric acid leach circuits have been developed and proven in the industry for many years. Examples of industrial applications are: synthetic rutile production (TiO_2) (Walpole E.A., Winter J.D., 2002), Quebec Iron and Titanium's UGS process (Patoine M.-C., De Mori A., Borowiec K., Coscia C., 2002), magnesium and potassium production (Priemer H., 1973) and Goro Nickel (Vahed A., Colton D.F., Duterque J.P., Karner W., Baerhold F., 2004), (Taylor A., 2013). One of the critical process steps in developing a hydrometallurgical process is acid regeneration, as it is an economic requirement to recover both the dissolved metals and acid (commonly HCl) from the spent leaching solution.

In response to the requests of several customers, Andritz Metals designed and built a spray roast acid regeneration pilot plant to demonstrate the viability of new acid leach processes and specialty metal oxide production, which allows the specific adjustment of practically all relevant process parameters, thus ensuring solid reference data for industrial application and scale up.

ENGINEERING DESIGN

One of the most critical elements of a spray roaster is the spray nozzle. The pilot plant (Figure 1) was designed to use the smallest possible size of spray nozzle with a similar type of spray as used in industrial scale plants. It was decided to install two spray booms, with one spray nozzle each. The roaster can operate with one or two spray nozzles active, which averages a flow rate of 40-70 L/h metal chloride solution. The roaster, the absorption and scrubber columns were designed to meet that flow rate.

a) b)

Figure 1 – Pilot Plant a) front view and b) side view

Another design challenge was to prepare the pilot plant not only for the regeneration of hydrochloric acid from metal chloride solutions, but also for the regeneration of hydrofluoric and nitric (or mixed acid). Some critical parts had to be custom designed for the small scale application in order to be resistant to all these different acids.

The pilot plant was commissioned in March 2014 after a short installation phase. Three examples of test campaigns are disclosed in the experimental section.

Table 1 – Pilot Plant details overview table

Plant Details:	Spray-Roast Acid Regeneration Pilot Plant
Capacity:	40-70 L/h Metal chloride solution
	6-12 kg/h Metal oxide
Total Height:	7,75 m
Regenerated Acids:	Hydrochloric Acid, Hydrofluoric Acid, Nitric Acid
Location:	Vienna, Austria

EXPERIMENTAL

Limonite Leach Liquor (Nickel)

A Ni-limonite flow sheet (Figure 2) had been extensively tested on a laboratory scale. However, the last crucial step – acid regeneration – had to be demonstrated before getting approval and financing for the next project phase. In the process, limonite ore is reduced with coal and leached with hydrochloric acid. The result is a liquor containing nickel chloride and - as a main component - ferrous chloride. A nickel product is extracted using a cementation process. Chemicals used to adjust the pH-value during the cementation step, such as magnesium or calcium hydroxide, accumulate as impurities in the ferrous chloride solution. The objectives of the pilot campaign were to demonstrate the regeneration of the hydrochloric acid, the test production of the iron oxide by-product, and using coal dust as a fuel.

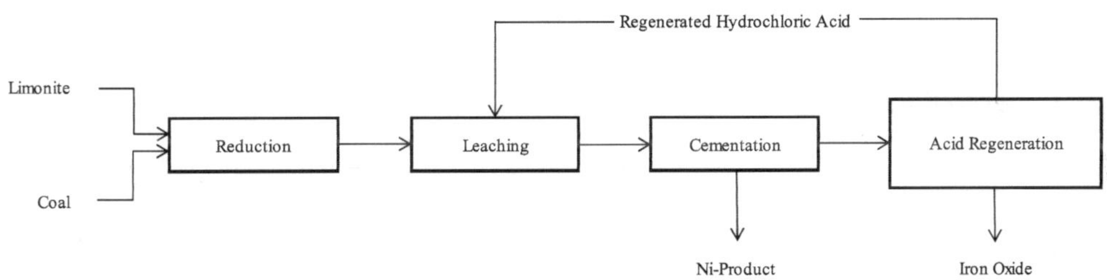

Figure 2 – Ni-Limonite Flow Sheet

Andritz was requested to use coal dust as a fuel for the spray roaster operation because natural gas is not readily available at the remote location, where the industrial limonite treatment facility is being planned. This location in South-East Asia, near the origin of the nickel ore, does have easy access to coal, which is mined in that area.

To the best of our knowledge this pilot plant experiment was the first to use a) an external combustion chamber and b) a pilot coal dust burner for spray roast acid regeneration, as all spray roast plants built so far have used either gaseous (natural gas, coke oven gas, etc.) or liquid fuel (heavy fuel oil) with conventional burners firing directly into the roaster and thus generating the hot air turbulence required for the reaction and residence time in the roaster. A pilot coal dust burner was designed in cooperation with the RWTH Aachen University, Institute of Heat and Mass Transfer (Hillemacher B., Hatzfeld O., Goevert B.,

2015). The pilot burner was positioned on top of an external combustion chamber that created and directed the hot air stream into the spray roaster. The burner itself consisted of a central coal dust inlet, which focused and directed combustion air streams in order to ensure complete and efficient combustion of the coal dust, and an ignition burner operating on natural gas. The ignition burner has to heat up the combustion chamber to the required ignition temperature of the coal dust (600-800 °C). It could be operated with liquid fuel too. At the bottom of the combustion chamber a collection chamber for ash was installed.

The pilot plant was operated using the external coal dust combustion chamber at an average liquid feed rate of 40 L/h for 2 weeks. Oxide samples from the roaster and the cyclone were collected separately and analyzed. In addition, aqueous samples from the absorption and scrubber columns were taken and analyzed. The test successfully demonstrated:

- Spray roaster operation with an external combustion chamber
- Spray roaster operation with a coal dust burner
- Production of sufficient iron oxide material for customer evaluation
- Regeneration of limonite leach liquor into hydrochloric acid

Serpentine Leach Liquor (Magnesium)

Magnesium bearing ores have been leached with hydrochloric acid for many years and several industrial operations have successfully employed this technology (Biangardi S., Marek J., Pawlek F., Ruthner M., 1973). Magnesium chloride is decomposed in the spray roaster at temperatures of 500-600°C to magnesium oxide, which is collected as a white powder. Depending on the intended use, the magnesium oxide can be used directly or is washed and hydrated to magnesium hydroxide. They are valuable products used as refractory material and flame and smoke retardants in plastic compound products.

Ground serpentine ore was leached at 80 °C in a 3 m³ batch reactor. Leach residue consisting mainly of silicate minerals was removed with a filter press, while generating the primary leach liquor containing magnesium chloride, ferrous chloride, aluminium chloride and minor ore specific metal impurities. The leach liquor was purified by oxidation of ferrous to ferric chloride and the subsequent precipitation of ferric and aluminium hydroxide. The precipitate was removed by pressure filtration.

The purified leach liquor – a slightly yellow to colourless liquid – was used as feed solution in the spray roast pilot plant at a feed rate of 40-70 L/h over a period of 3 weeks. The pilot plant operated with two natural gas burners at a reactor top temperature of 580 °C. White magnesium oxide samples from the roaster and the cyclone were collected separately and analyzed. In addition, aqueous samples from the absorption and scrubber columns were taken and analysed.

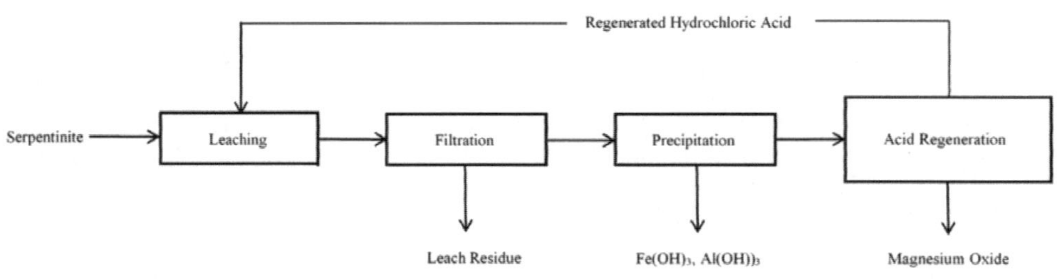

Figure 3 – Mg-Serpentine flow sheet

The magnesium oxide product was then further hydrolysed with water to produce magnesium hydroxide. The test demonstrated successfully:

- The leaching and purification of serpentine ore
- The production of magnesium oxide and magnesium hydroxide material for commercial evaluation
- The regeneration of serpentine leach liquor into hydrochloric acid

Recycled Aluminium Leach Liquor

Another test campaign focused on recycling aluminium from waste packaging material. After removal of paper and fibre material, the packaging waste was treated with hydrochloric acid to dissolve the aluminium coating from the residual plastic material. The resulting aluminium chloride solution was sprayed into the roaster and decomposed at 400-450 °C into aluminium oxide. The test results showed that:

- Aluminium oxide can be produced both in the alpha- and gamma-crystal structure
- Good leaching performance on waste packaging material can be achieved
- A closed loop process is possible with acid regeneration

Table 2 – Reactions in the spray roast reactor

Thermal decomposition reactions in the spray roaster:			
$2\ FeCl_2 + 2\ H_2O + ½\ O_2$	→	$Fe_2O_3 + 4\ HCl$	(1)
$2\ FeCl_3 + 3\ H_2O$	→	$Fe_2O_3 + 6\ HCl$	(2)
$MgCl_2 + H_2O$	→	$MgO + 2\ HCl$	(3)
$2\ AlCl_3 + 3\ H_2O$	→	$Al_2O_3 + 6\ HCl$	(4)
$NiCl_2 + H_2O$	→	$NiO + 2\ HCl$	(5)
$3\ CoCl_2 + 3\ H_2O + ½\ O_2$	→	$Co_3O_4 + 6\ HCl$	(6)

DISCUSSION

The pilot plant enables researchers and industry to demonstrate their new flow sheets and processes and to produce sample oxide material for further testing or market sampling. The results from piloting test work increase the customer's confidence in an engineering scale-up design and reduce the investment risk for the commercial design.

For the first time a spray roast plant has been operated with solid fuel (coal) while using an external combustion chamber, instead of burners installed directly in the roaster. The flexible and modular design of this pilot plant allows for quick adaptions of the plant for each customer specific test campaign.

ACKNOWLEDGEMENTS

The authors want to thank Werner Polak, the installation and commissioning manager and leader of the pilot plant operation team, and all operation team members who have enabled these test campaigns with their efforts, experience and flexibility. We also want to thank Martin Brazda and his team of design engineers who are responsible for the flexible and modular design of the pilot plant.

The design of the pilot coal dust burner was only possible with the experience and support of the RWTH Aachen University, Institute of Heat and Mass Transfer under the leadership of professors R. Kneer, B. Hillemacher, O. Hatzfeld and B. Goevert.

REFERENCES

Andritz Metals brochure – Regeneration systems AM.ARP.01.eng.11.13 retrieved on January 19, 2016, from http://grz.g.andritz.com/c/com2011/00/02/19/21982/1/1/0/-247662697/me-regenerationsystems-brochure_en.pdf

Biangardi S. & Marek J. & Pawlek F. & Ruthner M. (1973). Austrian Patent No. AT328755, *Verfahren zur Gewinnung von Metallen aus ultrabasischen Gesteinen (process to produce metals from ultrabasic rock,* Vienna, Austria: European Patent Office

Baerhold F., Lebl A. and Starcevic J. (2006), *Recycling of spent acids and iron via pyrohydrolysis* in Dutrizac J.E. and Riveros P.A. (Eds.), *Iron Control Technologies* (pp. 789-803), proceedings of the third international symposium on iron control in hydrometallurgy. Montreal, QC, ISBN: 1-894475-64-X

Hillemacher B. & Hatzfeld O. & Goevert B. (2015, December), *Final Report on Design and Commissioning of coal dust burner system for a spray roast pilot plant.* presented at Andritz project meeting, Vienna, Austria

Jedlicka, H. (n.d.), *Herstellung von Magnesiumoxid (+99% MgO) nach dem Ruthner-HCl-Verfahren (production of magnesium oxide (+99% Mg) with the Ruthner-HCl-process).* published by Ruthner Industrieanlagen Aktiengesellschaft, Vienna, Austria, E-1619-2d-182

Patoine M.-C., De Mori A., Borowiec K. and Coscia C. (2002), *Pyrohydrolysis of a Calcium and Magnesium Bearing $FeCl_2$ Leach Liquor* in Peek E. & Van Weert G. (Eds.), *Chloride Metallurgy 2002, Vol. II* (pp. 699-712), proceedings of the 32nd annual hydrometallurgy meeting, Montreal, QC, ISBN: 1-894475-28-3

Priemer H. (1973). Austrian Patent No. AT340365, *Verfahren zur Erzeugung reinen Magnesiumoxids bzw. –oxihydrats (process to produce pure magnesium oxide resp. oxihydrate)* Vienna, Austria: European Patent Office

Ruthner M.J. & Krischner H. (1973). *Production of alkali-free alumina* presented at I.C.S.O.B.A. 3rd Congress International, Nice, France

Ruthner Ind. Planungs. AG (1967). Austrian Patent No. AT273888, *Verfahren zur kontinuierlichen und verlustlosen Herstellung von feinkörnigem Kobaltoxyd (continuous and loss-free process for the production of fine grained cobalt oxide)*, Vienna, Austria: European Patent Office

Taylor A. (2013), *Laterites – still a frontier of nickel process development,* in proceedings of TMS2013 142th Annual Meeting & Exhibition, San Antonio, TX

Vahed A., Colton D.F., Duterque J.P., Karner W. and Baerhold F. (2004), *Development of a fluid bed pyrohydrolysis process for INCO's Goro Nickel project,* in Imrie W.P. & Lane D.M. (Eds.), *International Laterite Nickel Symposium 2004* (pp. 171-191), proceedings of the TMS Annual Meeting. Charlotte, NC, ISBN: 0-87339-550-6

Walpole E.A. and Winter J.D. (2002), *The Austpac ERMS and EARS Processes for the Manufacture of High-Grade Synthetic Rutile by the Hydrochloric Acid Leaching of Ilmenite,* in Peek E. & Van Weert G. (Eds.), *Chloride Metallurgy 2002, Vol. II* (pp. 401-416), proceedings of the 32nd annual hydrometallurgy meeting. Montreal, QC, ISBN: 1-894475-28-3

IRON CHEMISTRY IN LATERITIC SAPROLITE LEACHING WITH CONCENTRATED MAGNESIUM CHLORIDE BRINES

*Douglass Duffy[1], Michael Carlos[1], Vladimiros Papangelakis[1] and Boyd Davis[2]

[1]*Department of Chemical Engineering and Applied Chemistry*
University of Toronto, 200 College Street
Toronto, ON, Canada M5S 3E5
*(*Corresponding author: douglass.duffy@mail.utoronto.ca)*

[2]*Kingston Process Metallurgy*
759 Progress Ave.
Kingston, ON, Canada K7M 6N6

ABSTRACT

The processing of lateritic saprolite in hyper-concentrated magnesium chloride brines offers several potential advantages for the hydrometallurgical production of nickel. An aggressive HCl-$MgCl_2$ leach at atmospheric pressure can significantly reduce magnesium dissolution, and offers the ability to recover HCl and MgO by a less energy intensive process than conventional techniques via the formation of magnesium hydroxychlorides. Inevitably, the high chloride concentration renders iron fully soluble during leaching and thus requires subsequent iron control and disposal steps. Experimental measurements coupled with OLI™ modeling were conducted to investigate the behaviour of iron in concentrated magnesium chloride brines, in terms of chemistry (nature of iron species) and solubility limits. The solubility of ferric chloride in magnesium chloride solutions of concentrations and temperatures relevant to this process was controlled by a previously unreported solid phase: $2.5FeCl_3 \cdot MgCl_2 \cdot 7.5H_2O$. Iron precipitation tests on synthetic leach solutions without seeding indicate a metastable akaganeite precipitate which transitions to hematite upon equilibration. Magnesium chloride solubility in ferric chloride solutions was also investigated to allow the simulation of a ternary phase diagram for the $MgCl_2$-$FeCl_3$-H_2O system, in order to determine process conditions that avoid unwanted chloride precipitation.

KEYWORDS

Ferric chloride, saprolite, nickel, laterite, HCl recovery, magnesium chloride

INTRODUCTION

In the search for more economical processes to treat lateritic saprolite for nickel production, chloride hydrometallurgy has been increasingly investigated. While High Pressure Acid Leaching (HPAL) can frequently economically treat lateritic limonite, the presence of saprolite negatively affects process economics. The higher magnesium content of saprolite increases acid consumption, because magnesium sulphate, unlike iron and aluminum sulphates, does not hydrolyze in HPAL. This soluble magnesium decreases the sulphuric acid strength by an amount greater than the amount of acid stoichiometrically required to leach magnesium, which must also be disposed of or recovered (Jankovic, Papangelakis, & Lvov, 2009). Due to these issues, pyrometallurgy is most commonly employed to treat lateritic saprolite, via the Rotary Kiln-Electric Furnace (RKEF) process; this process uses a rotary kiln to dry and partially reduce nickel and iron oxides before sending the calcine to an electric furnace for smelting to produce a ferronickel product for the stainless steel market. Although the problems encountered during hydrometallurgical processing are avoided and nickel recoveries are high, high capital and energy costs prevent the application of this process to lower grade lateritic saprolite deposits (De Bakker, 2011).

These issues have led to the investigation of an atmospheric chloride process for the treatment of lateritic saprolites, which could also be applicable to lower grade limonite deposits not economically amenable to HPAL (Duffy, Papangelakis, & Davis, 2014). Chloride systems present a number of advantages, but also require consideration of a few limitations. Metal chlorides are generally very soluble, reducing the potential for scale formation and allowing relatively smaller processing equipment through the use of concentrated solutions. Additionally, hydrochloric acid is a more aggressive lixiviant than sulphuric acid, and the hydrogen ion activity can be augmented by further metal chloride additions (Peek, 2011). The increase in proton activity in concentrated chloride systems has repeatedly been referenced as a motivation for the use of chloride hydrometallurgy, and was first proposed by Jansz in 1983 and followed by others (Becze, Hock, & Demopoulos, 2011; De Bakker, 2011; Demopoulos, Li, Becze, Moldoveanu, Cheng, & Harris, 2008; Harris & Magee, 2003; Harris, Magee, Lakshmanan, & Sridhar, 2004; McDonald, & Whittington, 2008). This activity increase allows an equivalent amount of acid to result in a lower pH in a concentrated solution than in a dilute one. The Falconbridge Matte Leach Process pioneered the use of chloride chemistry to recover nickel in the early 1950's (Van Weert, 2002). Recently, considerable research has been focused on concentrated chloride systems, or brines, using calcium, magnesium or sodium as the cation, to leach a variety of oxide and sulphide feeds (McDonald, & Whittington, 2008).

Despite a number of hydrochloric processes being proposed for lateritic saprolite, none has been commercialized. The main issue for these processes has been the need to regenerate and recycle hydrochloric acid, due to the high price of hydrochloric acid compared with sulphuric acid. In spite of a variety of methods being proposed for the regeneration of hydrochloric acid, the common disadvantage is that they are all energy intensive, and, depending on the concentration of the hydrochloric acid produced, can create water balance problems for the process (De Bakker, 2011; McDonald, & Whittington, 2008). The conventional and most mature technique is solution pyrohydrolysis, where a concentrated solution, generally rich in $MgCl_2$ or $FeCl_3$, is fed to a fluidized bed or spray roaster, producing a wet gas stream of $HCl_{(g)}$ and magnesium/iron oxide. Capital and energy costs associated with this method are high, due to the specialized equipment, high temperatures (700 to 900 °C), and large amount of water being evaporated (Demopoulos et al., 2008). One of the most noted attempts at designing a lateritic saprolite chloride flow sheet, the Jaguar process, incorporated HCl recovery via pyrohydrolysis (Harris et al., 2004).

A recently proposed alternative technique, hydrolytic distillation, relies on the elevated boiling point of these solutions to hydrolyze ferric iron to hematite and HCl (Becze et al., 2011; Demopoulos et al., 2008; Harris & White, 2008; Harris, White, & Ballantyne, 2006; Harris, White, Demopoulos, & Becze, 2008; Harris, White, Jansen, & Pursell, 2006). In addition to forming a wet gas stream of $HCl_{(g)}$, this process does not appear to be applicable to solutions of $MgCl_2$, due to scale formation by hydrated magnesium chloride and the higher temperatures required for hydrolysis (>200 °C).

A novel technique for HCl recovery, locking the HCl into a solid compound that can be decomposed later, has recently been investigated. This method relies on the precipitation, separation and subsequent thermal decomposition of magnesium hydroxychlorides (De Bakker, 2011; De Bakker, LaMarre, Papangelakis, & Davis, 2011). Magnesium hydroxychlorides are compounds of the type $xMgO \cdot MgCl_2 \cdot yH_2O$ and form upon addition of MgO to sufficiently concentrated $MgCl_2$ brines at temperatures above 80°C; the stoichiometry of the precipitate is determined by precipitation temperature and $MgCl_2$ concentration. The two forms of magnesium hydroxychlorides of relevance to this process, the "3-form" (x=3, y=11) and "2-form" (x=2, y=6), require solutions of at least 30 wt.% $MgCl_2$. The reactions involved (shown below) are: (1) – precipitation; (2) – dehydration; and (3) - decomposition (De Bakker, 2011; De Bakker et al., 2011; De Bakker, Peacey, & Davis, 2012; Duffy et al., 2014).

$$2MgO\ (s) + MgCl_2\ (aq) + 6H_2O\ (aq) \rightarrow 2MgO \cdot MgCl_2 \cdot 6H_2O\ (s) \quad (1)$$

$$\Delta H^\circ_{rxn,1} = -6.1\ kJ/mol;\ T \geq 90\ °C$$

$$2MgO \cdot MgCl_2 \cdot 6H_2O\ (s) \rightarrow 2MgO \cdot MgCl_2 \cdot H_2O\ (s) + 5H_2O\ (g) \quad (2)$$

$$\Delta H^\circ_{rxn,2} = 310\ kJ/mol;\ T = 230\ °C$$

$$2MgO \cdot MgCl_2 \cdot H_2O\ (s) \rightarrow 3MgO\ (s) + 2HCl\ (g) \quad (3)$$

$$\Delta H^\circ_{rxn,3} = 200\ kJ/mol;\ T = 560\ °C$$

The above HCl recovery route via hydroxychlorides presents a number of advantages compared with other recovery methods, namely energy savings and alleviation of water balance issues. By transferring the chloride content of the brine to the solid phase, the amount of water that must be heated and evaporated to accomplish HCl recovery is greatly reduced. Additionally, the maximum temperature required in this process (560 °C) is lower than that required for pyrohydrolysis (700 to 900 °C). This considerably reduces the amount of energy required for HCl recovery using "2-form" magnesium hydroxychlorides when compared with pyrohydrolysis (De Bakker, 2011). Furthermore, it has been reported that the energy required to treat saprolite using the "2-form" magnesium hydroxychloride recovery based process is slightly less than the energy required for the RKEF process (De Bakker, 2011). As the temperatures for the dehydration and decomposition reactions for magnesium hydroxychlorides differ, water and hydrochloric acid recovery are decoupled, allowing greater process flexibility.

The use of the magnesium hydroxychloride HCl recovery scheme requires that the process operate in concentrated magnesium chloride media, and this presents both limitations and advantages. The leaching step is aggressive and the acidity is augmented by the increase in proton activity in concentrated $MgCl_2$ systems, allowing for high metal recoveries while suppressing magnesium dissolution (Duffy et al., 2014). Iron forms strong chloride complexes however, and this renders iron fully soluble during the leach, mandating a separate iron control stage. A flow sheet for the process as a whole can be seen in Figure 1.

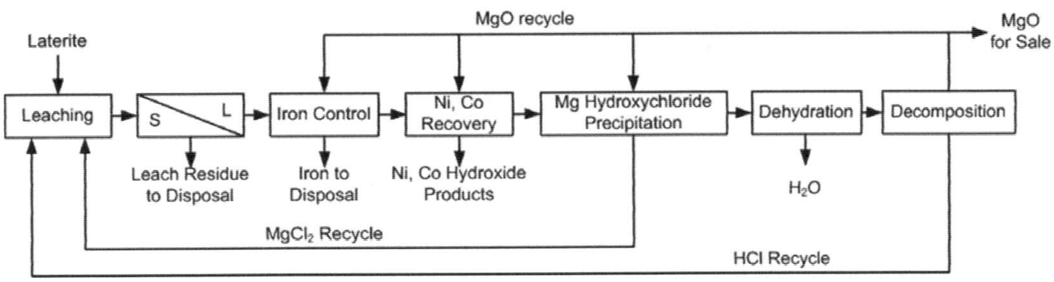

Figure 1 – Flow sheet for lateritic saprolite processing with magnesium hydroxychloride processing (Duffy et al., 2014)

As Fe is the main impurity in lateritic saprolite, iron chemistry and behaviour throughout the process and ramifications for process conditions must be well understood. The objective of this work was to explore the behavior of iron, using both experimentation and simulation through the thermodynamic software OLI^{TM}, and to answer particular questions of relevance to the following processing steps:

- Leaching: what is the solubility of possible solid iron phases and what are the consequences for limits to solids loading;
- Iron control: what is the nature of the solid phase formed?

Furthermore, as the operating window for this process is bound by the solubility of $MgCl_2 \cdot 6H_2O$, the effect of $FeCl_3$ on this solubility must be determined in order to prevent an unwanted drop in $MgCl_2$ concentration. Finally, precipitation of iron as hematite (reaction (4)) leads to an increase in $MgCl_2$ concentration, therefore this effect must be quantified in the context of $MgCl_2 \cdot 6H_2O$ solubility as well.

$$3MgO\ (s) + 2FeCl_3\ (aq) \rightarrow Fe_2O_3(s) + 3\ MgCl_2\ (aq) \quad (4)$$

METHODOLOGY

Analytical Techniques

Aqueous metal concentrations were determined by ICP-OES, while metal content of solids were determined by digestion in aqua regia followed by ICP-OES. Solid phases were also examined using XRD to determine which solid phases were present.

Solubility Experiments

Experiments to determine the solubility of $FeCl_3$ in $MgCl_2$ solutions and of $MgCl_2$ in $FeCl_3$ solutions were undertaken from the dissolution pathway. For the ferric chloride solubility experiments excess $FeCl_3$ was added, while excess $MgCl_2$ was added to the magnesium chloride solubility experiments. Samples were filtered using 0.45 μm syringe filters before being diluted 10-fold using 5 wt.% HCl (solutions containing HCl were diluted with deionized water). Free acid concentration was determined by automated acid-base titration, while iron and magnesium concentrations were determined by complexometric titration (Carlos, 2013; Kosak & Ballozo, 1971; Mendham, Denney, & Barnes, 2000; Zatka, 1975). Solid samples were vacuum filtered without washing (due to extreme solubility in any solvents used, including water, alcohol and acetone) and stored in vacuum desiccators prior to analysis.

Solubility experiments were conducted in Erlenmeyer flasks submerged in a water-polypropylene glycol mixture maintained at the desired temperature using an immersion circulator. The flasks were mixed by magnetic stirrers and sealed using rubber stoppers and parafilm, to minimize losses of water vapor and $HCl_{(g)}$. The rubber stoppers permitted the introduction of a thermometer in each Erlenmeyer flask to monitor the interior temperature in addition to the bath temperature.

Ferric chloride solubility experiments in $MgCl_2$ solutions at 100 °C were conducted in a 1 L jacketed Pyrex® reactor heated using an oil bath maintained at the reaction temperature using a VWR temperature controlled circulator. The interior reaction temperature was monitored using a thermometer and the vessel was stirred using a glass-shafted PTFE impeller (Carlos, 2013).

Iron Control Precipitation Experiments

Iron control precipitation experiments were conducted using the same 1 L reactor set-up previously described in section 2.2. The reactor was loaded with a synthetic leach solution composed of 0.7 m $FeCl_3$, 0.1 m $NiCl_2$, 1.0 m HCl and $MgCl_2$ at 2.0 or 4.5 m, based on data from preliminary leaching experiments (Duffy et al., 2014). The solution was heated to 95 °C and a stoichiometric amount of MgO was added to neutralize the HCl present. Subsequently, 2.0 g of MgO (approximately 10% of the amount required to precipitate all iron) was added every half an hour to effect iron precipitation from solution. Stepwise addition was utilized so as to produce a more crystalline solid, likely more amenable to analysis by XRD. At these intervals a sample was taken and diluted 10-fold prior to filtering using a syringe filter, with the filtrate being analyzed by ICP-OES. This process continued until the stoichiometric amount of MgO required to precipitate all iron from solution had been added. The resulting slurry was vacuum filtered, and the filtrate volume was recorded prior to analysis by ICP-OES. The iron precipitate that was recovered was washed with acetone and filtered again. These solids were vacuum dried at 70 °C prior to analysis by XRD and digestion by aqua regia followed by ICP-OES.

OLI™ Modeling

The OLI™ software (www.olisystems.com) is a series of powerful software applications based on two speciation-based thermodynamic models that enable reliable simulation and estimation of properties and chemical equilibria in electrolyte solutions of interest to hydrometallurgical applications. The two thermodynamic frameworks are the aqueous (AQ) model for dilute solutions and the mixed-solvent-electrolyte (MSE) model for concentrated complex solutions (Wang, Anderko, & Young, 2002; Wang, Springer, Anderko, & Young, 2004). Both models rely on ion interaction parameters to calculate activities of species and new experimental solubility data can be added to extend the range or to improve the performance of the commercial OLI™ database. The solubility data obtained in this work was used to customize the commercial OLI™ MSE database, version 9.0.1, and all simulation results shown here were the product of this custom database (Carlos, 2013).

RESULTS AND DISCUSSION

Iron Behaviour during Leaching

Preliminary leaching tests revealed as expected iron to be very soluble (Duffy et al., 2014). However, the solubility limit of iron chloride must be known in order to avoid chloride losses. Iron solubility will also dictate the maximum solids loading during leaching. A literature review revealed that there is limited information on the $FeCl_3$-$MgCl_2$-H_2O phase diagram at elevated temperatures.

Ferric chloride solubility experiments in water at varying temperatures agreed with reported literature values (Linke & Seidell, 1965). However, all experiments in solutions of $MgCl_2$ above 40 °C produced a solid phase whose XRD spectrum, shown in Figure 2, did not correspond to any ferric or magnesium oxide or chloride (Carlos, 2013).

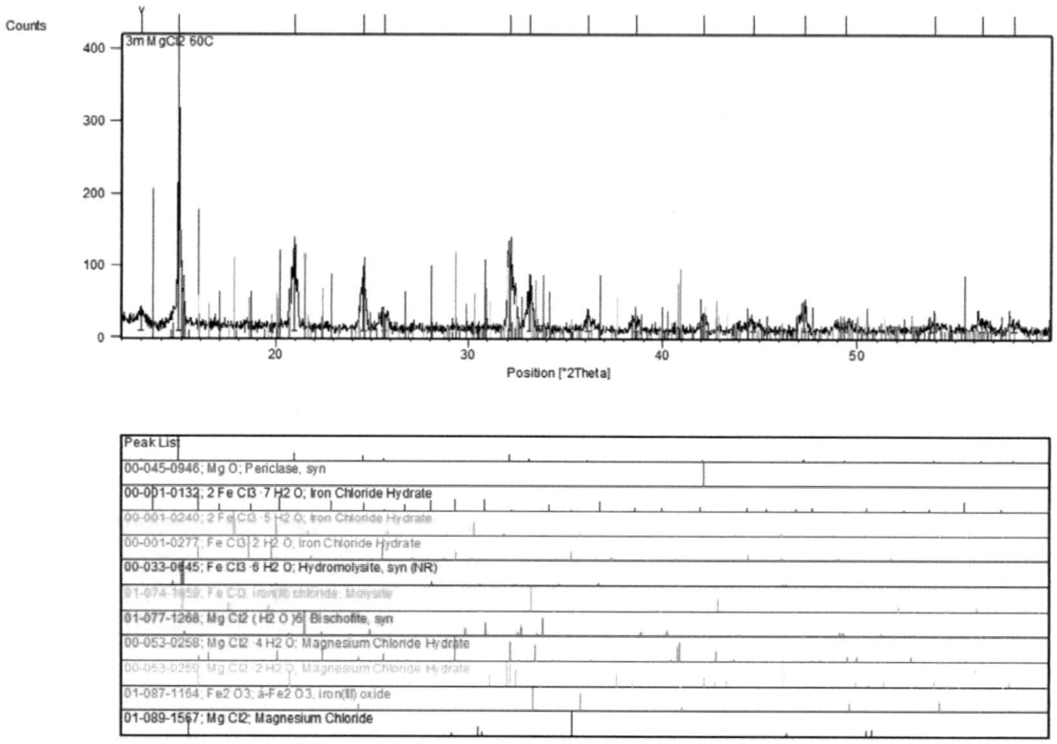

Figure 2 – XRD spectrum of unknown solid phase with best-matched reference patterns (Carlos, 2013)

In order to identify this solid phase, a sample was dissolved in 5 wt.% nitric acid; the resulting solution was analyzed by complexometric titration to determine Mg and Fe concentrations and by AgCl gravimetric analysis to determine the chloride concentration. The unknown solid phase was determined to be $2.5FeCl_3 \cdot MgCl_2 \cdot 7.5H_2O$. Solubility data for this double salt compound at 25 °C was determined through the precipitation pathway by decreasing temperature from 40 °C to 25 °C. This solubility data is shown below in Figure 3, with solid lines denoting the OLI^{TM} modeled solubility of the double salt and dashed lines denoting the OLI^{TM} modeled solubility of the various hydrates of $FeCl_3$.

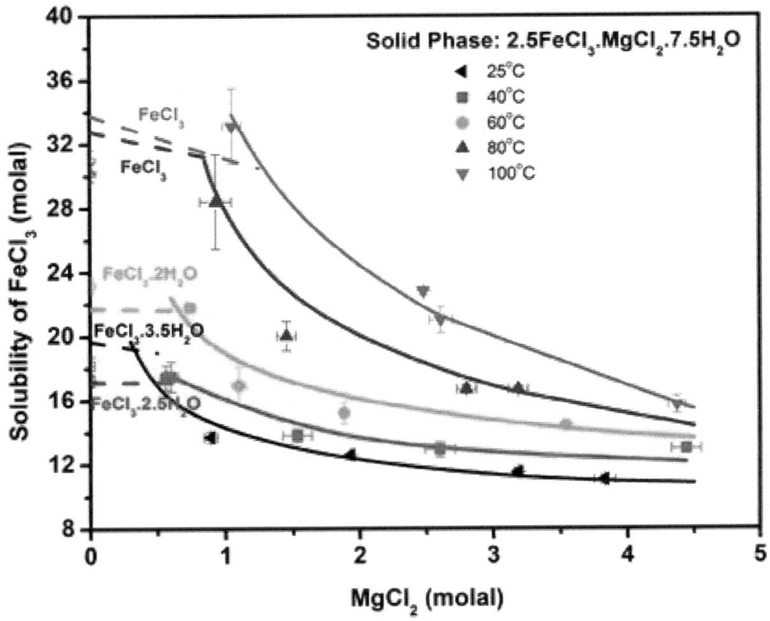

Figure 3 – $FeCl_3$ solubility in $FeCl_3$-$MgCl_2$-H_2O system

Ferric chloride, or one of its hydrates, is the OLI™ modeled solid phase up to 1 m $MgCl_2$ at 100°C, though this decreases to approximately 0.5 m $MgCl_2$ at 25°C, whereas the double salt, $2.5FeCl_3 \cdot MgCl_2 \cdot 7.5H_2O$, is the predicted solid phase at all higher $MgCl_2$ concentrations. The solubility of the double salt increases with temperature and is also influenced by the common ion effect, with its solubility decreasing with increasing $MgCl_2$ concentration. The maximum solids loading during leaching, based on this solubility data, was estimated to be 75 wt.% for 2 m $MgCl_2$ and 60 wt.% for 4.5 m $MgCl_2$ leach solutions, respectively.

Iron Control by Precipitation with Magnesia

Iron control experiments were performed on synthetic leach solutions to investigate the efficacy of iron control by pH adjustment with MgO as well as the phase of iron precipitate formed. Precipitation experiments at 95 °C were performed on synthetic leach solutions of 2 m and 4.5 m $MgCl_2$. The synthetic leach solutions also contained 0.7 m $FeCl_3$, 0.1 m $NiCl_2$ and 1.0 m HCl based on the results of the preliminary leaching experiments (Duffy, 2013; Duffy et al., 2014).

The iron and nickel profiles resulting from these experiments are shown below in Figure 4. Due to the presence of the 1.0 m HCl, approximately 35 % of the total stoichiometric amount of MgO added served to neutralize the acidity of the solution prior to iron precipitation. The stoichiometric concentration lines shown in Figure 4 are plotted as if all MgO added after acid neutralization is used to precipitate iron from solution according to reaction (4). It can be seen that iron, and other trivalent cations (Al and Cr, not shown, at initial concentrations of 0.001 M and 0.0001 M respectively), are precipitated from the synthetic leach solutions without affecting the nickel content. It is possible that these other trivalent cations, present as impurities in the salts used to prepare the synthetic solutions, contribute to the consistent deviation of the actual iron concentration being slightly above the stoichiometric iron concentration.

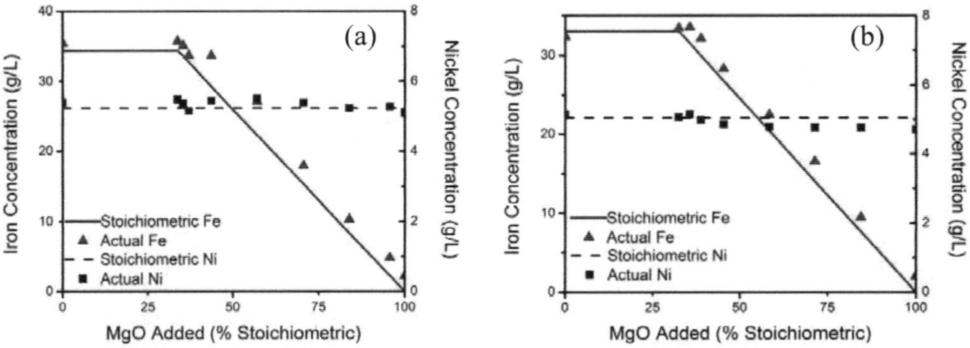

Figure 4 – Observed and stoichiometric concentrations during iron precipitation
(a) 2.0 m $MgCl_2$ (b) 4.5 m $MgCl_2$

The iron precipitates were analyzed by XRD (not shown) and digested in aqua regia in order to determine the composition and phase(s) present. The precipitate was found to contain akaganeite, a ferric oxide-hydroxide/chloride compound, and bischofite, $MgCl_2 \cdot 6H_2O$, with the bischofite peaks more pronounced in precipitates formed in 4.5 m $MgCl_2$ than those formed in 2 m $MgCl_2$. One precipitation experiment was allowed to remain at 95 °C overnight resulting in a phase change from akaganeite to hematite. This is consistent with akaganeite being meta-stable and hematite being the thermodynamically stable product, in agreement with previous studies on the precipitation of hematite from ferric chloride media (Dutrizac & Riveros, 1999). The akaganeite precipitates were found to be approximately 26% Fe, 5 % Mg and 0.3 % Ni, while the hematite precipitate contained 44% Fe, 3% Mg and 0.3% Ni. These concentrations correspond to losses of 9% of Mg and 6% of Ni in the akaganeite precipitates, while in the hematite precipitate these are reduced to 6% of Mg and 4% of Ni. It is expected that the Mg and Ni losses, as well as chloride losses which were not quantified here, can be minimized further through increased retention time or the use of hematite seed, as was observed in the precipitation of hematite from ferric chloride media (Dutrizac & Riveros, 1999).

Effect of Iron on Magnesium Chloride Phase Diagram

During iron removal by precipitation with magnesia, the $MgCl_2$ concentration increases steadily towards the solubility limit of bischofite, $MgCl_2 \cdot 6H_2O$. This could result in precipitation of bischofite and an unwanted loss of chloride content. Solubility experiments were carried out on the $MgCl_2$-$FeCl_3$-H_2O system from 25 to 100 °C to define the $MgCl_2$ solubility limit under varying $FeCl_3$ concentration. $MgCl_2$ solubility experiments in water at varying temperatures agreed with reported literature values, validating the experimental procedures performed here (Linke & Seidell, 1965).

The solubility of $MgCl_2$ was found to decrease with an increase in $FeCl_3$ concentration, consistent with the common ion effect, and to increase with an increase in temperature, as seen below in Figure 5. Once again, solid lines represent the modeled solubility based on the customized OLI™ database. The solid phase was analyzed by XRD and confirmed to be bischofite. The effect of $FeCl_3$ was also seen to decrease with increasing temperature, such that by 100 °C the addition of 1 m $FeCl_3$ reduces the $MgCl_2 \cdot 6H_2O$ solubility by less than 5 %. The deviation of the model (created using both $MgCl_2$ and $FeCl_3$ solubility data) from the observed $MgCl_2$ solubility at 100 °C, less than 5%, is likely due to experimental error.

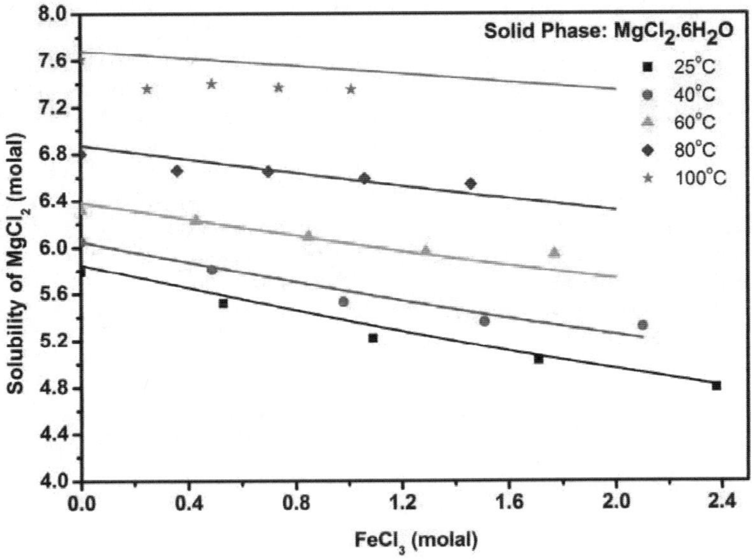

Figure 5 – $MgCl_2 \cdot 6H_2O$ solubility in $FeCl_3$ solutions

Maximum Solids Loading for Process and Ternary $MgCl_2$-$FeCl_3$-H_2O Phase Diagram

In Section 3.1 it was shown that the maximum solids loading during the leaching step was limited by the solubility of $FeCl_3$ to 75 wt.% for 2 m $MgCl_2$ and 60 wt.% for 4.5 m $MgCl_2$ leach solutions. However, in practice the maximum solids loading for the proposed process is much lower, due to the increase in $MgCl_2$ concentration that occurs through the use of MgO as a neutralizing and precipitating agent and the solubility of $MgCl_2$ being limited to approximately 7.7 m at 100 °C.

The largest contributor to the increase in $MgCl_2$ concentration is the iron control step, due to the large amount of iron in solution. However, the increase from the starting $MgCl_2$ brine concentration is also affected by the percent extraction of Mg from the ore, the residual free acid concentration after leaching and the amount of Ni/Co products precipitated. After accounting for all of these factors, the maximum solids loading for the proposed process decreases to approximately 35 wt.% for a 2 m $MgCl_2$ leach at 100 °C, as outlined below in Table 1 (Carlos, 2013). The resulting $MgCl_2$ concentration of 7.5 m remains below the solubility limit of $MgCl_2 \cdot 6H_2O$ at 100 °C (7.7 m) and above the minimum $MgCl_2$ concentration required for the precipitation of both the "3-form" and "2-form" magnesium hydroxychlorides. In theory, one could increase the temperature further to allow for even more concentrated $MgCl_2$ solutions prior to magnesium hydroxychloride precipitation, however the temperature range over which these aqueous solutions are stable at atmospheric pressure becomes steadily smaller and so these options were not considered here.

Table 1 – $MgCl_2$ concentration increase factors

Factor	Change in $MgCl_2$ Concentration (m)
Input $MgCl_2$ Concentration	2
Mg Dissolution from Ore (assuming 50% extraction)	0.41
Free Acid Neutralization (assuming 0.5 m Free Acid)	0.25
Iron Precipitation	4.38
Ni/Co Product Precipitation	0.42
$MgCl_2$ Conc. (Prior to HCl Recovery)	7.46

The solubility data generated during this research was used to customize the standard OLI™ database by regressing the standard thermodynamic properties of the new double salt ($\Delta G°_f$, S°, and C_p) and the mid-range binary interaction parameters between major species in the solution ($FeCl^{2+}$ - Mg^{2+} and $FeCl_2^+$ - Mg^{2+}). Based on the new U of T customized database, a ternary $MgCl_2$-$FeCl_3$-H_2O phase diagram was constructed (Figure 6) for $FeCl_3$ concentrations from 0 to 3 m. Superimposed on this diagram is the progression of $MgCl_2$ concentration with the various processing steps, outlined above with numerical values in Table 1. The same effects of $FeCl_3$ on $MgCl_2$ solubility as seen above in Figure 5 are observed; the increase in the boiling point of the solution with $FeCl_3$ addition is also quantified. It can be observed that the opposing effects of the decrease in $MgCl_2$ solubility and increase in boiling point with $FeCl_3$ addition have a relatively neutral effect on the operating window for the process. As iron precipitation progresses, this ternary phase diagram will revert to the binary $MgCl_2$-H_2O phase diagram (or the 0 m $FeCl_3$ lines as shown here), and through proper consideration of these effects unwanted precipitation of $MgCl_2 \cdot 6H_2O$ can be avoided.

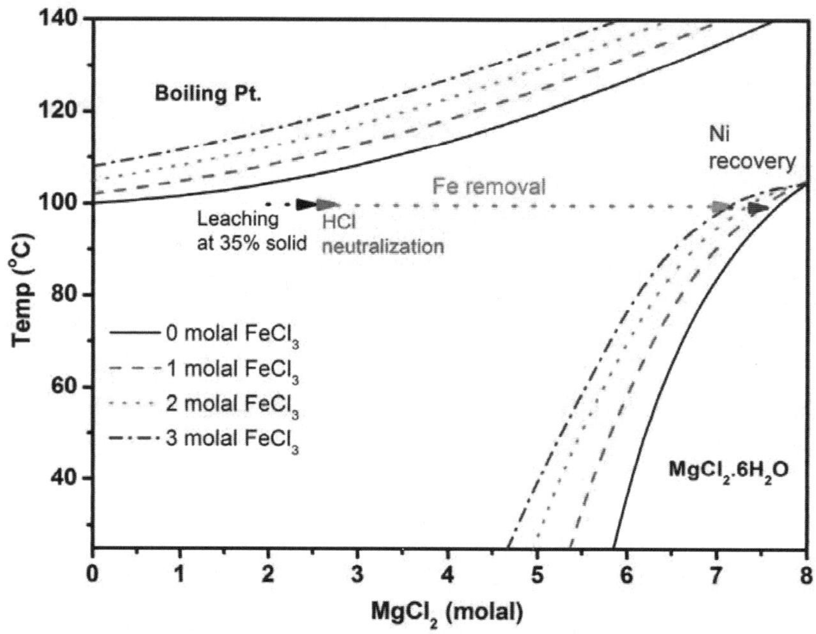

Figure 6 – OLI™ predicted $MgCl_2$-$FeCl_3$-H_2O ternary phase diagram

CONCLUSIONS

The chemistry and solubility limits of iron species in concentrated $MgCl_2$ solutions during the processing of lateritic saprolite have been investigated in this work. The leaching of lateritic saprolite with $MgCl_2$ brines relies on magnesium hydroxychloride precipitation as an alternative HCl regeneration scheme; soluble iron is present in high concentrations during all main steps of the process, from leaching through precipitation and removal, therefore its behaviour must be well understood.

As loss of chloride content during leaching must be avoided, the solubility of $FeCl_3$ in $MgCl_2$ solutions relevant to this process, 0 to 4.5 m $MgCl_2$, was studied for the temperature range 25 - 100 °C. A previously unreported double salt solid phase was observed, analyzed and determined to have the formula $2.5FeCl_3 \cdot MgCl_2 \cdot 7.5H_2O$. However, this double salt presents no practical limit to solids loading during leaching as mass balance considerations regarding the increase in $MgCl_2$ concentration through the proposed processing steps indicate that the solubility of $MgCl_2 \cdot 6H_2O$ is the limiting solubility.

Iron control experiments were conducted on synthetic leach solutions at 95 °C by pH adjustment using MgO addition. Results indicated the ability to maintain soluble Ni in the $MgCl_2$ solution while rejecting iron to the solid phase. The metastable phase akaganeite is formed initially (more kinetically favorable), which via equilibration overnight at 95 °C converts to the thermodynamically stable hematite, consistent with previous hematite precipitation studies in ferric chloride media (Dutrizac & Riveros, 1999). Further study using hematite seed would likely increase the kinetics of this phase transformation and promote the precipitation of hematite directly from the $MgCl_2$ solution.

The effect of $FeCl_3$, on the solubility of $MgCl_2$ from 25 to 100 °C was also studied experimentally in order to identify process conditions that would result in unwanted chloride loss due to magnesium chloride precipitation. The $MgCl_2$ solubility was found to decrease with $FeCl_3$ addition due to the common ion effect, with the effect being more pronounced at lower temperature. These two sets of solubility data were regressed using OLI™ to create a customized database capable of modeling these concentrated solutions of magnesium and ferric chloride. This database improved the modeling of the $FeCl_3$-$MgCl_2$-H_2O system and permitted the creation of a ternary phase diagram to aid selection of process conditions that avoid chloride loss despite the increase in $MgCl_2$ concentration occurring throughout the process. It was determined that leaching with 2 m $MgCl_2$ would allow a maximum solids loading of 35 wt. %, resulting in a $MgCl_2$ concentration of 7.5 m, after neutralization and iron and nickel precipitation, below the solubility limit of 7.7 m $MgCl_2$.

ACKNOWLEDGEMENTS

The authors gratefully acknowledge Vale Canada Ltd. and NSERC for financial assistance.

REFERENCES

Becze, L., Hock, S.J., & Demopoulos, G.P. (2011). Precipitation of hematite and recovery of hydrochloric acid from concentrated iron chloride solutions by a novel hydrolytic decomposition process. *Proceedings of TMS* (pp. 529-538). San Diego, CA.

Carlos, M.K. (2013). *Chemical Modeling of Iron(II)/(III) Solutions in Hydrometallurgy using OLI™* (Master's thesis). Retrieved from TSpace at University of Toronto.

De Bakker, J. (2011). *The recovery of magnesium oxide and hydrogen chloride from magnesium chloride brines and molten salt hydrates* (Doctoral dissertation). Retrieved from QSpace at Queen's University.

De Bakker, J., LaMarre, J., Papangelakis, V., & Davis, B. (2011). HCl leaching and acid regeneration using $MgCl_2$ brines and molten salt hydrates. *Proceedings of TMS* (pp. 521-528). San Diego, CA.

De Bakker, J., Peacey, J. & Davis, B. (2012). Thermal decomposition studies on magnesium hydroxychlorides. *Canadian Metallurgical Quarterly, 51* (4), 419-423.

Demopoulos, G.P., Li, Z., Becze, L., Moldoveanu, G., Cheng, T.C., & Harris, B. (2008). New technologies for HCl regeneration in chloride hydrometallurgy. *World of Metallurgy, 61*, 89-98.

Duffy, D.W. (2013). *Saprolite Leaching and Iron Control in Concentrated Magnesium Chloride Brines* (Master's thesis). Retrieved from TSpace at University of Toronto.

Duffy, D., Papangelakis, V., & Davis, B. (2014). Lateritic saprolite leaching in concentrated magnesium chloride media. *Canadian Metallurgical Quarterly, 53* (4), 432-439.

Dutrizac, J.E., & Riveros, P.A. (1999). The Precipitation of Hematite from Ferric Chloride Media at Atmospheric Pressure. *Metallurgical and Materials Transactions B, 30* (6), 993-1001.

Harris, B. & Magee, J. (2003). Atmospheric chloride leaching: The way forward for nickel laterites. *Proceedings of TMS* (pp. 501-515). San Diego, CA.

Harris, G.B., Magee, T.J., Lakshmanan, V.I., & Sridhar, R. (2004). The Jaguar Nickel Inc. Sechol laterite project atmospheric chloride leach process. *Proceedings of TMS* (pp. 219-241). Charlotte, NC.

Harris, B., & White, C. (2008). Recent developments in the high-strength chloride leaching of base metal sulphide ores. *Proceedings of ALTA* (pp. 1-15). Perth, WA, Australia.

Harris, B., White, C., & Ballantyne, B. (2006). Chloride leaching of polymetallic massive sulphide. *Proceedings of ALTA* (pp. 1-21). Perth, WA, Australia.

Harris, B., White, C., Demopoulos, G., & Becze, L. (2008). Recovery of metal oxides and associated acid from process solutions using water as reagent. *Proceedings of ALTA* (pp. 1-16). Perth, WA, Australia.

Harris, B., White, C., Jansen, M., & Pursell, D. (2006). A new approach to the high concentration chloride leaching of nickel laterites. *Proceedings of ALTA* (pp. 1-20). Perth, WA, Australia.

Jankovic, Z., Papangelakis, V.G., & Lvov, S.N. (2009). Effect of Nickel Sulphate and Magnesium Sulphate on pH of Sulphuric Acid Solutions at Elevated Temperatures. *Journal of Applied Electrochemistry, 39,* 751-759.

Jansz, J.J.C. (1983). Estimation of ionic activities in chloride systems at ambient and elevated temperatures. *Hydrometallurgy, 11*, 13-31.

Kosak, A., & Ballozo, H. (1971). Stufenweise chelatometrische titration von calcium und magnesium mit visueller endpunktsanzeige nehen hoheren eisen(III)-konzentrationen. *Zeitschrift fur Analytische Chemie, 253,* 262-266.

Linke, W.F., & Seidell, A. (1965). *Solubilities: Inorganic and metal-organic compounds* (Vol. II). New Jersey: Von Nostrand Co.

McDonald, R.G., & Whittington, B.I. (2008). Atmospheric acid leaching of nickel laterites review. Part II. Chloride and bio-technologies. *Hydrometallurgy, 91*, 56-69.

Mendham, J., Denney, R.C., & Barnes, J.D. (2000). *Vogel's Quantitative Chemical Analysis.* Essex: Pearson Education Ltd.

Peek, E. (2011). Chloride metallurgy – process technology development. *Proceedings of TMS* (pp. 469-501). San Diego, CA.

Van Weert, G. (2002). Some Observations on the Chloride Based Treatment of Nickel-Copper-Cobalt Mattes. *Proceedings of MetSoc Chloride Metallurgy* (pp. 277-298). Montreal, QC, Canada.

Wang, P., Anderko, A., & Young, R.D. (2002). A Speciation-Based Model for Mixed-Solvent Electrolyte Systems. *Fluid Phase Equilibria, 203*, 141-176.

Wang, P., Springer, R.D., Anderko, A., & Young, R.D. (2004). Modeling Phase Equilibria and Speciation in Mixed-Solvent Electrolyte Systems. *Fluid Phase Equilibria, 222* (223), 11-17.

Zatka, V.J. (1975). The alkalimetric determination of free acid in leach liquors. INCO private communication.

IRON SEPARATION FROM IRON BEARING POLYMETAL CHLORIDE SOLUTIONS BY MEANS OF HYDROLYSIS AND CRYSTALLIZATION

*B. Nowak, M. Oesterreicher, R. Stastny, and H. Weissenbaeck

SMS Siemag Process Technologies GmbH
Daffingerstraße 4,
1030 Vienna, Austria
*(*Corresponding author: benedikt.nowak@sms-siemag.at)*

ABSTRACT

The producers of certain metals like nickel and cobalt increasingly face a shortage of easy-to-mine and easy-to-process ore bodies. Their focus is shifting towards alternative raw materials which in turn demand new, different extraction techniques. One potential route for processing a ferrous ore with a non-ferrous target metal is atmospheric leaching with hydrochloric acid followed by purification of the pregnant leach solution (PLS). For the separation of iron from the PLS, and, in turn, the extraction of the non-ferrous compounds, the following flow sheet is proposed. Pre-concentration of the PLS, oxidation of ferrous (if present) to ferric chloride, hydrolysis of ferric chloride leading to the precipitation of hematite and the evaporation of hydrochloric acid, followed by crystallization of non-hydrolysable metal chlorides by means of sparging hydrogen chloride into the solution. Hydrogen chloride required in the latter step is produced internally by super-azeotropic distillation of regenerated hydrochloric acid. The liquid filtrate from the crystallization step is recycled to hydrolysis. Depending on the specific composition of the PLS, the hematite produced contains only traces of other impurities, whereas the crystallized mix of non-hydrolysable metal chlorides – containing also the target metal in enriched form – contains ferric chloride only in an amount referring to the filter cake moisture. These crystals can then undergo further hydrometallurgical separation and extraction processes. As an example, this flow sheet is proposed for treating nickel laterite ores. Experiments related to the demonstration of crucial assumptions in the flow sheet are presented. Based on those experiments a mass and energy balance of the flow sheet was calculated and is presented here together with a table of the major consumables in order to provide a basis for an operating cost estimate and other process economics.

KEYWORDS

Iron Chloride, Hydrolysis, Nickel Laterite, Hydrochloric acid

INTRODUCTION

Regardless of the mineralogical speciation the separation of iron from any non-ferrous metal is a requirement in various metallurgical processes. Typical examples are the recovery of alumina from bauxite, rare earth elements from their ore bodies, nickel and cobalt from laterites or valuable metals from secondary raw materials. For easy-to mine and easy-to-process raw materials, various well-established processes exist. However, the shortage of these types of ore bodies and the complexity of secondary resources shift the focus to new separation techniques. One potential new processing route is chloride leaching with hydrochloric acid or metal chlorides or mixtures thereof. Without discussing the leaching process itself, most of these processes have in common that at one point or another iron has to be separated from the non-ferrous compounds. If the leaching agent is hydrochloric acid, then this acid has to be recovered for economic and environmental reasons.

For the production of hematite and the recovery of hydrochloric acid from steel pickle liquors, SMS Siemag developed the Hydrothermal Acid Regeneration Plant technology, which consists of these major process steps: (1) pre-concentration of waste acid by means of vacuum evaporation, (2) oxidation of iron from ferrous to ferric by means of autoclave oxidation, and (3) hydrolysis of the ferric chloride yielding hematite and hydrochloric acid vapor (also referred to as hydrolytic distillation). Hematite is filtered from the last process step and the hydrochloric acid vapor is condensed, whereas the latent heat of condensation is used for driving the vacuum evaporation process, which reduces the energy consumption dramatically compared to pyro-hydrolysis processes such as spray or fluidized bed roasting.

Furthermore, the hydrolytic distillation has the advantage that at the proposed operating temperature, which is typically below 200°C, ferric ions hydrolyze almost selectively yielding clean iron oxide and hydrochloric acid vapor: Earlier work has shown that Fe(II), Na, Ca, Pb, Zn (Riveros and Dutrizac, 1997), Co, Ni (Vogl et al., 2012, Zhang et al., 2015), K, Mg, Mn (Nowak et al., 2013) do not hydrolyze. Al and Cr (III) only seem to hydrolyze to a small extent (Nowak et al., 2013). This allows for the selective separation of iron from mixed chloride brines.

However, at conditions where iron hydrolysis yields are almost 100%, the acid vapor concentration is comparably low and a partial vacuum has to be applied (Zhang et al., 2015). Therefore, instead of targeting 100% conversion of $FeCl_3$ to iron oxide, a bleed of the solution has to be sent, after filtration, to an additional treatment step, where non-hydrolysable metal chlorides are removed.

At lower temperatures (<125°C), iron is preferably precipitated as akaganeite, whereas higher temperatures lead to the formation of hematite (Riveros and Dutrizac, 1997). As hematite is thermodynamically more stable, its formation can be observed after longer reaction times (Dutrizac and Riveros, 1999) and it is also formed if hematite seed is used (Riveros and Dutrizac, 1997, Dutrizac and Riveros, 1999). The addition of hematite seed also enhances the filtration behavior of the hematite slurry (Cohen et al., 2005).

Since metal chlorides can be crystallized out of chloride solutions by increasing the concentration of free HCl in the solution (Linke and Seidell, 1965, Linke and Seidell, 1958), sparging of HCl gas into the bleed solution is proposed by Vogl et al., 2012 in order to separate non-hydrolysable metal chlorides. This method has been previously investigated for the refining of aluminum chloride (Edwards et al., 1930, Eisele and Bauer, 1983). However, when comparing literature data provided earlier (Vogl et al., 2012, Nowak et al., 2013) with the ones by other authors (Linke and Seidell, 1965, Linke and Seidell, 1958, Eisele and Bauer, 1983), it became apparent that the presence of ferric chloride decreases the amount of free HCl required for crystallization in a multi-metal chloride system. The multi-metal chloride filter cake precipitated is almost free of ferric chloride (Nowak et al., 2013).

The aim of this work is to present a flow sheet for the separation of iron from a poly-metallic chloride brine. For the treatment of nickel laterite pregnant leach solution, experimental data of hydrolysis and crystallization are presented. With these results, a heat and mass balance is calculated and presented.

FLOW SHEET FOR SEPARATING IRON FROM POLYMETALLIC CHLORIDE BRINES

A flow sheet for the separation of iron from chloride brines is proposed in Figure 1. Steps 1 to 9 listed below are part of the original SMS Siemag Hydrothermal Acid Regeneration Plant technology, which focuses mainly on the removal of iron as hematite. The process steps 10 to 13 have been introduced in order to separate the non-hydrolysable metal chlorides from the bleed required in the hydrolysis step. The steps are:

1. Pre-concentration of the polymetallic solution (1) by means of vacuum evaporation. Recovered heat (30) is used for driving the process. The liquid phase (3) is sent to the oxidation process.
2. Condenser. Vapors (2) from pre-concentration are condensed by means of indirect cooling (31). The condensate (4) is collected in a distillate tank.
3. Distillate tank. This step is used for distributing the distillate internally; the distillate consists mainly of water, e.g. to the hydrolyser (6) or filter cake washing (7). The rest of this distillate (5) can be used in upstream or downstream processes.
4. The preconcentrated brine (3) is oxidized by means of autoclave oxidation with pure oxygen (8). Indirect heat (26) is used for heating the solution to approximately 150°C, and the following reactions take place:

$$FeCl_2 + \frac{1}{4}\,O_2 + HCl \rightarrow FeCl_3 + \frac{1}{2}\,H_2O \qquad (1)$$

$$2\,FeCl_3 + 3\,H_2O \rightarrow Fe_2O_3 + 6\,HCl \qquad (2)$$

   If HCl, which might be contained in the feed solution, is consumed by the first reaction, the second reaction, hydrolysis, will take place yielding HCl which in turn will be consumed by further oxidation.
5. Hydrolyser. The oxidized brine (9) is sent together with internal iron bearing flows such as wash water from hematite washing (11) or the reflux from crystallization (10) to the hydrolyser. In the hydrolyser, the temperature is >170°C and iron hydrolyses with water according to the hydrolysis reaction (2). In order to keep the boiling temperature constant, distillate (6) is fed into the reactor. Since the hydrolysis reaction generates hydrochloric acid vapor (12), indirect heat (27) is required for driving the process. The hydrochloric acid vapor concentration depends on temperature, pressure, and the concentrations of ferric and non-hydrolysable metal chlorides. As non-hydrolysable metal chlorides concentrate in the reactor, a bleed (13), which also contains the hematite made, is required.
6. Recuperator. In the recuperator, hydrolysis vapors (12) are condensed and the condensation heat (3) is used for preconcentrating the polymetallic chloride solution.
7. Cooler. The hydrolysis bleed is cooled prior to solid/liquid separation.
8. Solid liquid separation. The solids (16) are separated from the cooled bleed (13). The filtrate (17) is sent to non-hydrolysable metal chloride separation.
9. Cake washing. Distillate (7) is used for filter cake (16) washing. The wash water (11) is internally re-used; the washed filter cake (18) contains the iron separated from the polymetallic chloride solution.
10. The filtrate is mixed with hydrogen chloride (19). Non-hydrolysable metal chlorides will precipitate and leave the crystallizer as a suspension (20). Cooling (33) is required for removing the heat of dilution and to control the optimum process temperature.
11. Solid/liquid separation. The non-hydrolysable metal chlorides (21) are separated from the remaining ferric chloride/hydrogen chloride (22). This salt filter cake is sent to downstream processing which might be any metal chloride separation process (such as liquid/liquid extraction or precipitation by pH adjustment; depending however on the specific metal chloride mix).
12. Evaporator. As the filtrate (22) is rich in HCl, a concentrated HCl vapor (23) can be removed through heating (29). This vapor is sent directly to acid purification whereas the liquid remainder (10) is sent back to hydrolysis.
13. Acid purification. Condensed preconcentrator vapor [14] and return HCl vapor (23) are sent to a super-azeotropic distillation column. The top-product (19) is used for crystallization; the sump product (24) is the hydrochloric acid product.

Regarding the chloride balance of this process, the hematite made will contain more or less traces of chlorides depending on the filter cake rinsing in step 9. The absolute chloride amount can be further reduced by post-processing such as roasting. As the filter cake (21) is a mixture of solid metal chlorides, a major amount of chloride will end-up there; this chloride recovery is part of downstream processing. Recovered acid (24) will contain mainly the chlorides that have been associated with iron. Free acid contained in (1) will end up in both (5) and (24).

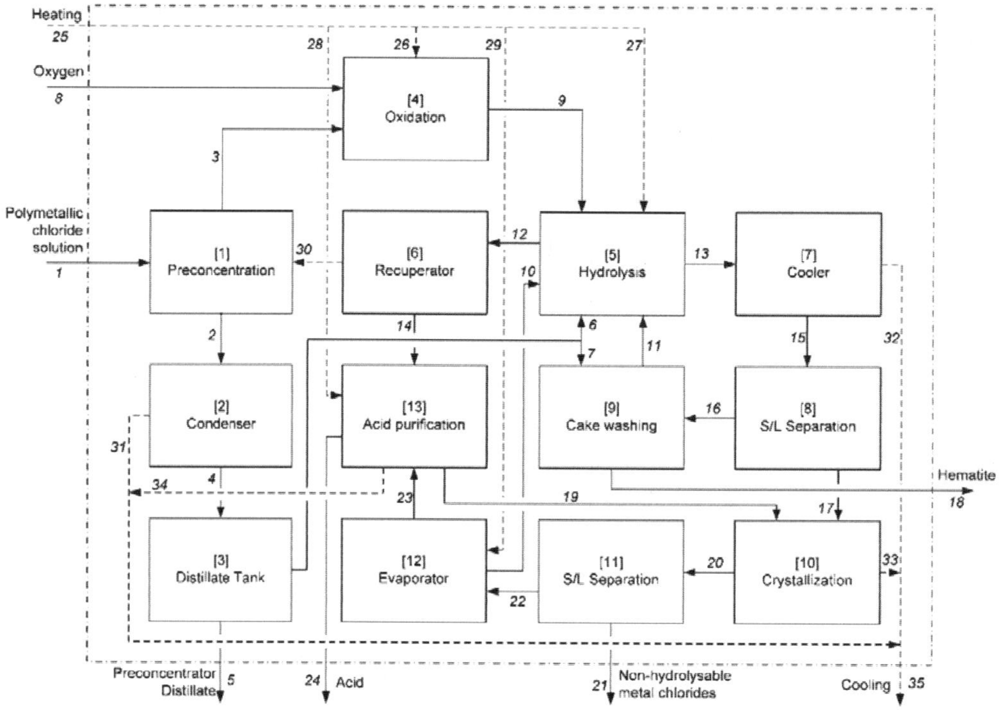

Figure 1 – Flow sheet for separating iron from polymetallic chloride brines

MATERIAL AND METHODS

Chemicals Used

This study was carried out with analytical grade chemicals.

Hydrolysis

For the selective hydrolysis of ferric chloride, a similar experimental setup as in previous studies (Vogl et al., 2012, Nowak et al., 2013) was used (Figure 2). In the beginning of the experiment, a 70 wt.% $FeCl_3$ solution is placed in the reactor and heated to the target temperature, which is 180°C. Then, de-ionized water is fed into the reactor. The distillate is sampled and when the distillate concentration is constant, the feed is switched to a solution containing 30 wt.% $FeCl_3$, 60 wt.% H_2O, and 10 wt.% non-hydrolysable metal chlorides. In order to refine the mass balance presented, the chosen ratios of non-hydrolysable metal chlorides shown in Table 1 is similar to the one from a previous study's mass balance (Nowak et al., 2013). Samples from both the distillate and the reactor's solution are taken every 15-30 minutes. They are analyzed by ICP-OES and titration. At the end of the experiment, the reactor's content is filtered. The filter cake is rinsed with water, dried, and analyzed by Glow Discharge Mass Spectrometry (GDMS) and XRD.

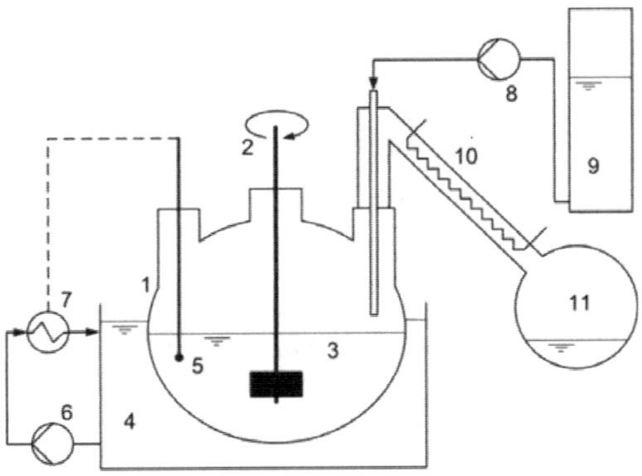

Figure 2 – Setup of hydrolysis experiments. 1 Glass reactor, 2 agitator, 3 reactor's solution, 4 thermo-oil bath, 5 thermometer, 6 thermo-oil circulation pump, 7 thermostat, 8 feed pump, 9 feed reservoir, 10 condenser, 11 distillate flask

Table 1 – Ratio of non-hydrolysable metal chlorides used in the hydrolysis experiments and found in the mass balance

Non-hydrolysable metal chloride	Ratio Experiment (%)	Ratio mass balance (%)
$AlCl_3$	10.9	16.1
$CaCl_2$	6.2	4.5
$CoCl_2$	4.3	2.9
$CrCl_3$	5.4	6.4
$FeCl_2$	0.01	0.01
KCl	4.4	0.6
$MgCl_2$	59.6	59.2
$MnCl_2$	1.8	2.0
NaCl	1.4	1.0
$NiCl_2$	6.1	7.2
Sum	100	100

Crystallization

In order to determine the solubility limits of non-hydrolysable metal chlorides in a ferric chloride solution as a function of the free hydrogen chloride concentration, the experimental setup shown in Figure 3 was used. A synthetic solution containing ferric chloride and non-hydrolysable metal chlorides (with a composition as shown in Table 2) is placed in the glass reactor. Gaseous hydrogen chloride is sparged into the reactor and dissolves in the solution. Samples from the aqueous phase are taken and analyzed by ICP-OES and titration. Solids are continuously separated from the solution by means of filtration and are analyzed as well.

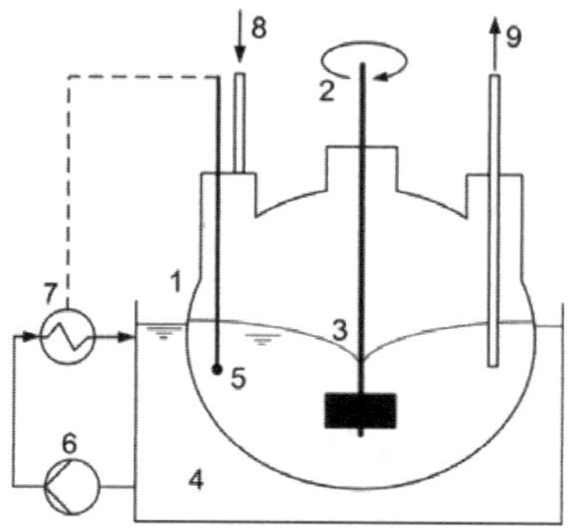

Figure 3 – Setup of crystallization experiments. 1 Glass reactor, 2 agitator, 3 reactor's solution forming a spout, 4 thermo-oil bath, 5 thermometer, 6 thermo-oil circulation pump, 7 thermostat, 8 HCl gas (from bottle), 9 reactor samples

Table 2 – Composition of the synthetic solution used for crystallization experiments

Metal chloride	Concentration (%)
$FeCl_3$	50
$FeCl_2$	0.0
$AlCl_3$	1.3
$CaCl_2$	0.7
$CoCl_2$	0.5
$CrCl_3$	0.6
KCl	0.5
$MgCl_2$	7.0
$MnCl_2$	0.2
NaCl	0.2
$NiCl_2$	0.7
H_2O	rest
Sum	100

Analysis

Chemical Analysis

Metals are generally analyzed by ICP-OES (Inductively coupled plasma optical emission spectroscopy). The device used is a Thermo Scientific iCAP 6000 Series ICP Spectrometer.

Ferrous iron and ferric iron are titrated with $KMnO_4$ and EDTA, respectively. Chloride is analyzed by automatic titration with $AgNO_3$ and chloride-ISE. Hydrogen chloride in distillate is titrated with NaOH. The concentration of free chloride in the crystallization experiment is titrated with NaOH to pH = 2, and, for comparison, it is assumed that it equals the difference between the total chloride concentration determined by titration and the amount of chloride associated with the dissolved metals in the solution.

Oxide Analysis

The phase composition of solid samples is investigated by X-ray diffraction (XRD) using a PANalytical X'Pert PRO powder diffractometer with a Cu(LFF) anode ($\lambda(K_{\alpha1}) = 1.5406$, $\lambda(K_{\alpha2}) = 1.5444$ Å; 40 kV, 40 mA) and a X'Celerator detector (Ni K_β filter). Scans are recorded in the region 2θ = 5-90° within a measuring time of 14 min. For qualitative and quantitative interpretation the databases Powder Diffraction File (PDF; International Centre for Diffraction Data ICDD, Newtown Square, PA/USA, 2001) and the

Inorganic Crystal Structure Database (ICSD, Karlsruhe/Germany, 2007) are used, respectively. The error of quantitative analysis is approximately 3% absolute for main phases (>15%); for minor constituents it is approximately 20% relative.

Trace elements in oxide samples are analyzed by GDMS.

Mass and Energy Balance

For the calculation of the mass and energy balance of the flow sheet shown in Figure 1, the following is assumed:

- 3100 kg/h polymetallic chloride solution with a composition provided in Table 3 enters the flow sheet. The composition is the same as in a previous study (Nowak et al., 2013) and refers to a leaching experiment of Australian nickel laterite ore with hydrochloric acid. The mass flow of 3100 kg/h is the result of leaching 1000 kg/h of ore.
- The preconcentrator [1] operates at 65°C. Pure water is driven off.
- The oxidizer [4] operates at 150°C. Pure oxygen (i.e. 100% O_2) is used for oxidation. 99% of Fe(II) ions are oxidized.
- In the hydrolyser (step 5), one data point from the hydrolysis equilibrium (see later in Figure 6) is used. The operating temperature is 180°C.
- The filter cake moisture (including dissolved compounds) for the solid/liquid separation steps 8 and 9 is 30 wt.% and 15 wt.%, respectively. 500 kg/h wash water is used for washing (stream 7).
- For the crystallization process in step 10, an operating temperature of 40°C and the experimentally derived crystallization limits shown later in Figure 7 (13 wt.% HCl) are used.
- For the solid/liquid separation step 11, the filter cake moisture (including dissolved compounds) is assumed to be 40 wt.%.
- For the acid purification step 13, an operating pressure of 1 MPa is assumed, as this shifts the azeotropic hydrochloric acid concentration to approx. 12 wt.%. The top product (19) contains 99 wt.% HCl and is used for crystallization.
- Heating and cooling demands are estimated via standard enthalpies.

Table 3 – Composition of the polymetallic chloride solution

Metal chloride	Concentration (%)
$FeCl_3$	15
$FeCl_2$	0.15
$AlCl_3$	2.1
$CaCl_2$	0.37
$CoCl_2$	0.06
$CrCl_3$	0.24
KCl	0.03
$MgCl_2$	4.6
$MnCl_2$	0.25
NaCl	0.10
$NiCl_2$	0.88
H_2O	rest
Sum	100

RESULTS AND DISCUSSION

Hydrolysis Experiment

Metal chlorides are continuously fed into the reactor. With the exception of $FeCl_3$, all metal chlorides are being concentrated in the reactor, as shown in Figure 4. At the end of the experiment the solids formed are filtered, rinsed with water, and analyzed with GDMS as shown in Table 4 and XRD, as illustrated in Figure 5.

For the chlorides shown in Figure 4(a) and $MgCl_2$ in Figure 4(b), the slope of the measured concentrations is approximately linear to their respective feed concentration (note that the data point after 100 minutes is obviously an outlier; see also data point at ~4.5 wt.% in Figure 6). However, $AlCl_3$ and $CrCl_3$, shown in Figure 4(c), co-hydrolyze. Although the feed's $AlCl_3$ concentration is almost 80% higher than the $CaCl_2$ concentration (for reference also shown in the diagram), after 200 minutes and after 300 minutes it is only approximately 40% and 10% higher, respectively. The $AlCl_3$ concentration appears to converge to approximately 1.1 wt.%. In the case of $CrCl_3$, the pure $FeCl_3$ solution still contains some traces of $CrCl_3$. The feed's $CrCl_3$ concentration is 5.4 wt.% and therefore between the KCl and $CaCl_2$ concentration. However, in the reactor $CrCl_3$ does not exceed 0.8 wt.%.

The above results are for Al in accordance with other studies (Harris and White, 2011), and for both elements (Al, Cr) with the trace element filter cake analysis, shown in Table 4. Except for Al and Cr, all metals in the oxide are in the ~100 ppm range, which is probably due to insufficient rinsing applied in the experiment when compared to previous work (Nowak et al., 2013). All other metals are found in the filter cake in a similar ratio as in the feed.

Regarding filter cake phase analysis, shown in Figure 5, pure hematite was identified (probably with an amorphous phase present as well). This is in good agreement with other studies (Riveros and Dutrizac, 1997, Vogl et al., 2012). The hematite is formed from $FeCl_3$ hydrolysis according to reaction 2. HCl evaporates as hydrochloric acid and its equilibrium concentration depends on temperature, pressure, and non-hydrolysable metal chlorides. Figure 6 shows the hydrochloric acid concentration, the ferric chloride concentration, and the measured temperature, as a function of the sum of non-hydrolysable elements (Al and Cr aqueous concentrations are added as well). Over the entire experiment, non-hydrolysable metals are concentrated up to almost 16 wt.%. Simultaneously, the $FeCl_3$ concentration decreases from approx. 70 to 50 wt.%. The HCl vapor concentration is up to approximately 12 wt.% while non-hydrolysable metals are always between 20 and 30 wt.%. At high concentrations of non-hydrolysable metals, the lower activity of hydrolysable ions leads also to lower HCl concentrations, which are at the end of experiment as low as 8 wt.%.

Table 4 – GDMS analysis of the oxide produced during hydrolysis
(in ppm, unless indicated differently)

Al	~ 1.1 wt.%
Ca	80
Co	120.0
Cr	~ 2.0 wt.%
K	67
Mg	580
Mn	55
Na	40
Ni	110
Cl	~ 5.8 wt.%

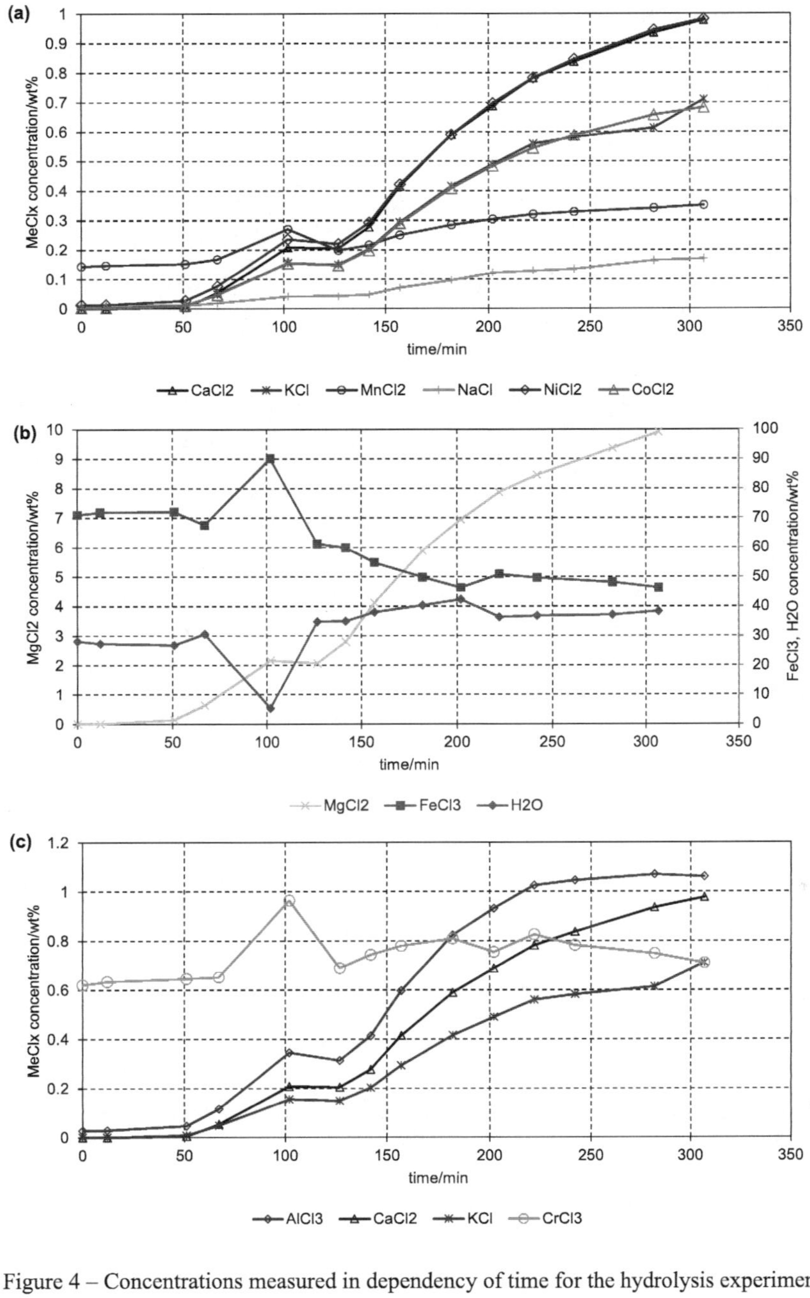

Figure 4 – Concentrations measured in dependency of time for the hydrolysis experiment

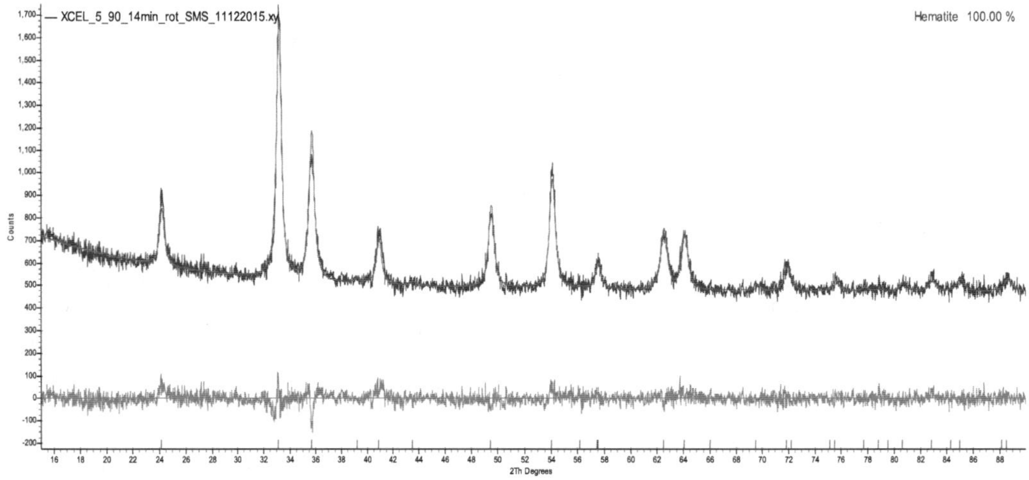

Figure 5 – Hydrolysis filter cake XRD analysis

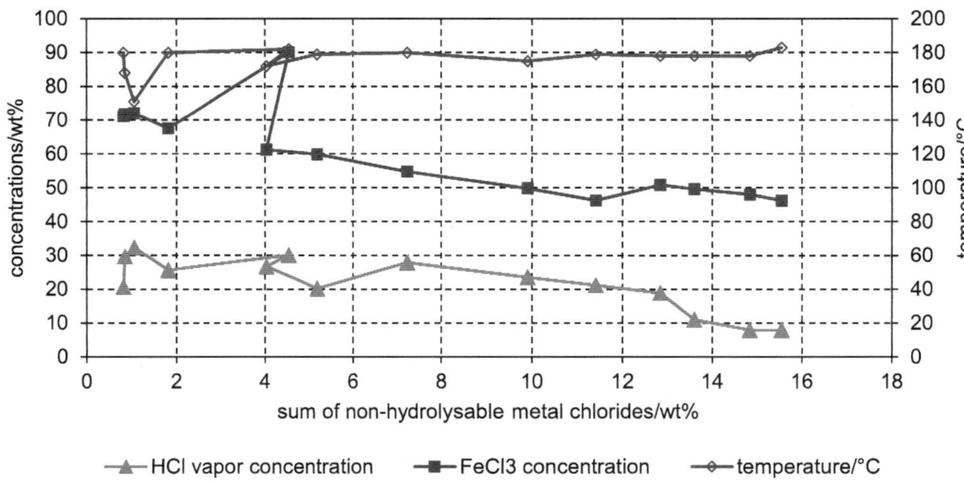

Figure 6 – Hydrochloric acid vapor concentration as a function of the sum of non-hydrolysable metal chlorides

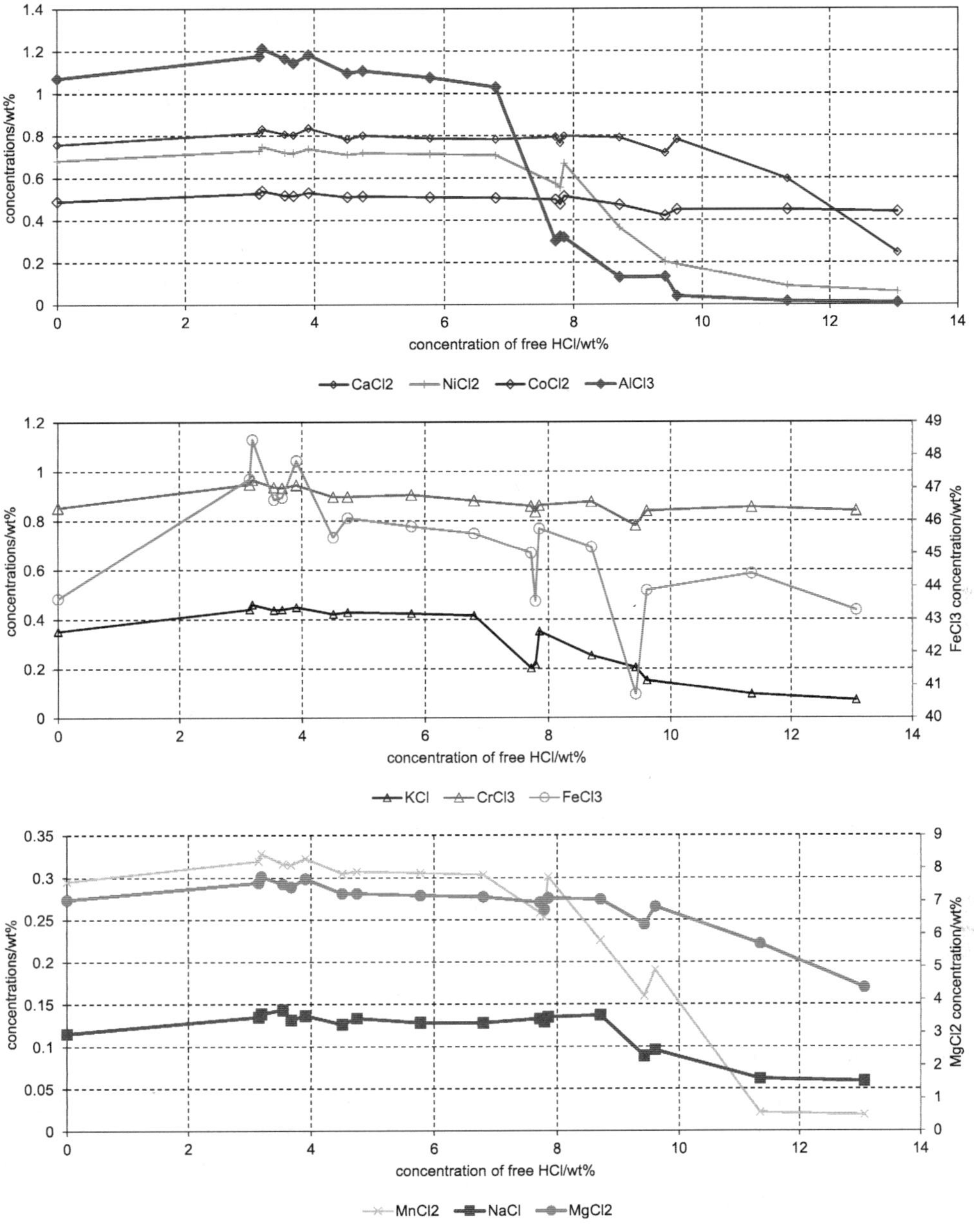

Figure 7 – Crystallization experiment: metal chloride concentrations as a function of free HCl. Note axes

Table 5 – Crystallization experiment filter cake composition; in wt.%.

	Cake 1	Cake 2	Cake 3	Cake 4
$FeCl_3$	31.5	39.0	25.2	10.7
$AlCl_3$	20.3	8.1	0.7	1.5
$CaCl_2$	0.4	0.4	2.1	6.4
$CoCl_2$	0.6	0.7	0.3	0.2
$CrCl_3$	1.1	0.8	0.5	0.3
KCl	2.4	4.8	1.3	0.9
$MgCl_2$	6.3	9.0	27.9	34.4
$MnCl_2$	0.8	1.6	2.9	1.7
NaCl	0.1	1.2	0.7	0.6
$NiCl_2$	3.3	6.0	1.7	1.7
Total $MeCl_x$	66.8	71.7	63.4	58.4

Crystallization Experiment

The aim of the crystallization experiment is to determine the solubility limits of non-hydrolysable metal chlorides as a function of free HCl concentration. Free HCl is measured by two methods previously mentioned. The maximum deviation between free chloride analysis by means of NaOH titration (values shown in diagrams) and difference between total Cl and associated Cl is 3.2 wt.%, which is considered as good agreement.

Regarding the concentration of metal chlorides as a function of HCl concentration at 40°C, shown in Figure 7, at some point, the metal chloride solubility limit is reached and the respective concentration decreases, e.g. for $AlCl_3$ above 7 wt.% HCl and for $NiCl_2$ above 8 wt.%. Note that Fe(II) is always below its limit of detection and therefore not shown in diagrams.

At several points filter cake samples were taken and analyzed, as shown in Table 5. Cake 1 refers to a reactor HCl concentration of roughly 8.7 wt.%, cake 2 to 9.4 wt.% and cake 3 to 11.3 wt.%. Cake 4 was separated from the reactor's solution at the end of the experiment by vacuum filtration. The total metal chlorides yields are below 100%, as the balance refers to either crystal water or specific chloride phase composition.

In filter cake 1-3, the concentration of $FeCl_3$ is comparatively high. This is probably due to the specific filtration method (pressure filtration) used. For cake 4, which was obtained by vacuum filtration, and dried by means of sucking air through the cake, $FeCl_3$ is lower and acceptable. In previous experiments using vacuum filtration, $FeCl_3$ has even been as low as 0.4% (Nowak et al., 2013). Regarding stream 21 from Table 6, which is the filter cake in the heat and mass balance, the chosen moisture of 40 wt.% yields 20 wt.% $FeCl_3$ in the cake, which is in an experimental order of magnitude.

Mass and Heat Balance

The mass and heat balance for the flow sheet shown Figure 1 was calculated. The efficiency of the major process steps hydrolysis (step 5) and crystallization (step 10) determines the total energy balance. Experimental data from the results presented here have been used for the calculations. For the hydrolysis step, the data point at 13 wt.% non-hydrolysable metal chlorides, 50 wt.% $FeCl_3$ concentration and 20 wt.% HCl vapor concentration from Figure 6 was chosen. For the crystallizer, the HCl concentration set-point is 13 wt.%. The concentrations from Figure 7 are assumed to be the crystallization limits (except for $FeCl_3$). The result of the heat and mass balance are presented in Table 6 and Table 7.

Table 6 – Stream table of the mass flows of the flow sheet shown in Figure 1

Stream	Temperature (°C)	Mass flow (kg/h), phase	Composition (in wt.% per phase)
1 Polymetallic chloride solution	40	3100 (l)	76% H_2O, 15% $FeCl_3$, 4.6% $MgCl_2$, 2.1% $AlCl_3$, 0.88% $NiCl_2$
2 Preconcentrator vapor	65	1436 (g)	100% H2O
3 Preconcentrated solution	65	1664 (l)	55% H_2O, 28% $FeCl_3$, 8.5% $MgCl_2$, 3.9% $AlCl_3$, 1.7% $NiCl_2$
4 Condensed vapor	60	1436 (l)	100% H_2O
5 Preconcentrator distillate	60	727 (l)	100% H_2O
6 Water for hydrolyser	60	208 (l)	100% H_2O
7 Filter cake wash water	60	500 (l)	100% H_2O
8 Oxygen	25	0.29 (g)	100% O_2
9 Oxidized solution	150	1664 (l), 0.96 (s)	(l): 55% H_2O, 28% $FeCl_3$, 8.5% $MgCl_2$, 3.9% $AlCl_3$, 1.7% $NiCl_2$; (s): 100% Fe_2O_3
10 Reflux	180	2920 (l)	56% $FeCl_3$, 34% H_2O, 4.9% $MgCl_2$, 3% HCl, 0.9% $CrCl_3$
11 Used wash water	64	539 (l)	93% H_2O, 5.9% $FeCl_3$, 0.9% $MgCl_2$, 0.2%$AlCl_3$, 0.1% $CrCl_3$
12 Hydrolyser vapor	180	1500 (g)	20% HCl
13 Bleed from hydrolyser	180	3676 (l), 156 (s)	(l): 50% $FeCl_3$, 37%H_2O, 7.9% $MgCl_2$, 1.8% $AlCl_3$, 1.0% $CrCl_3$; (s): 100% Fe_2O_3
14 Condensed hydrolyser vapor	80	1500 (l)	20% HCl
15 Cooled bleed	100	3676 (l), 156 (s)	(l): 50% $FeCl_3$, 37%H_2O, 7.9% $MgCl_2$, 1.8% $AlCl_3$, 1.0% $CrCl_3$; (s): 100% Fe_2O_3
16 Unwashed hematite	100	67 (l), 156 (s)	(l): 50% $FeCl_3$, 37%H_2O, 7.9% $MgCl_2$, 1.8% $AlCl_3$, 1.0% $CrCl_3$; (s): 100% Fe_2O_3
17 Filtrate	100	3609(l)	50% $FeCl_3$, 37%H_2O, 7.9% $MgCl_2$, 1.8% $AlCl_3$, 1.0% $CrCl_3$
18 Washed hematite	64	28 (l), 156 (s)	(l): 93% H_2O, 5.9% $FeCl_3$, 0.9% $MgCl_2$, 0.2%$AlCl_3$, 0.1% $CrCl_3$; (g): 100% Fe_2O_3
19 HCl for crystallization	86	476 (g)	99% HCl
20 Suspension after crystallization	40	3606 (l), 479 (s)	(l): 50% $FeCl_3$, 31% H_2O, 13% HCl, 4.4% $MgCl_2$, 0.8% $CrCl_3$; (s): 57% $MgCl_2*6H_2O$, 24% $AlCl_3*6H_2O$, 10% $NiCl_2*6H_2O$, 2.9% $CaCl_2*6H_2O$, 2.6% $MnCl_2*4H_2O$, 1.7% $CrCl_3*6H_2O$, 0.6% NaCl, 0.19% $CoCl_2*6H_2O$, 0.14% KCl
21 Non-hydrolysable metal chlorides	40	319 (l), 479 (s)	(l): 50% $FeCl_3$, 31% H_2O, 13% HCl, 4.4% $MgCl_2$, 0.8% $CrCl_3$; (s): 57% $MgCl_2*6H_2O$, 24% $AlCl_3*6H_2O$, 10% $NiCl_2*6H_2O$, 2.9% $CaCl_2*6H_2O$, 2.6% $MnCl_2*4H_2O$, 1.7% $CrCl_3*6H_2O$, 0.6% NaCl, 0.19% $CoCl_2*6H_2O$, 0.14% KCl
22 Filtrate after crystallization	40	3287 (l)	50% $FeCl_3$, 31% H_2O, 13% HCl, 4.4% $MgCl_2$, 0.8% $CrCl_3$
23 HCl vapor	180	367 (g)	93% HCl
24 Acid purification sump product	183	1391 (l)	12% HCl

Table 7 – Stream table of the enthalpy flows of the flow sheet from Figure 1

Stream	Enthalpy flows (GJ/h)
25 Total heat requirement	6.7
26 Heat for oxidation	0.4
27 Heat for hydrolysis	3.7
28 Heat for acid purification	0.9
29 Heat for evaporation	1.7
30 Recovered heat for preconcentration	3.6
31 Cooling for condensation	3.4
32 Cooling for bleed	0.7
33 Cooling for crystallization	1.1
34 Cooling for acid purification	0.1
35 Total cooling requirement	5.3

Table 8 – Process economics (without upstream or downstream processing)

Item	Unit cost	Consumption per ton of ore	Total cost per ton of ore
Heating (e.g. steam)	3 EUR/GJ	6.7 GJ/t	19 EUR/t
Cooling (e.g. cooling water)	0.05 EUR/m³	63 m³/t	3.15 EUR/t
Oxygen	1 EUR/kg	0.29 kg/t	0.29 EUR/t
Power (electric)	0.1 EUR/kWh	10 kWh/t	1.0 EUR/t
Waste water	1 EUR/m³	0.15 m³/t	0.15 EUR/t
Maintenance			1.0 EUR/t
Total			24.7 EUR/t

6.7 GJ/h thermal energy is required for driving the process. The hydrolysis is the biggest energy consumer. The cooling requirement is in a similar order of magnitude as the total heat demand. When comparing experimental data of the hydrolysis experiment with those actually occurring in the mass balance (Table 1), the agreement is good for the major constituents of non-hydrolysable metal chlorides, like $MgCl_2$. However, for other compounds such as $AlCl_3$ (almost 50% more in balance) or KCl (4.4 versus 0.6% in experiment versus balance) bigger deviations are observed. These deviations are expected to be reduced by running a new set of experiments on both crystallization and hydrolysis. It is not expected that this will affect the total energy consumption dramatically, also as the initial ferric chloride content used for the crystallization experiment is in good agreement with the mass balance (both 50 wt% $FeCl_3$). In the preceding study (Nowak et al., 2013), agreement of experiments and mass balance was not as good, which resulted in an estimated energy consumption of 12 GJ/h and the conclusion to refine experimental concentrations.

In Table 8, the operating costs are summarized for this process, whereas unit costs are estimated for a random location. Waste water, electric power and maintenance are estimated on a similar type of plant basis. In total, approximately 25 EUR (27 USD) per ton of ore are required. Considering the ore's nickel content, the iron separation process costs presented here amount to 1980 EUR per ton of nickel. The nickel sales price is approximately 8000 EUR/t. This is of course only a rough estimate. Besides potential sales value for by-products such as hematite, additional costs such as those for upstream or downstream processing have been neglected here. Also, due to the novelty of the process, capital cost requirements have not been assessed yet.

With respect to the operating costs, heating is the major driver, which depends significantly on the hydrolysis equilibrium. The higher the ratio of non-hydrolysable metal chlorides to ferric chloride, the less bleed, cooling for crystallization, and re-heating is required. For higher HCl vapor concentrations, less water has to be evaporated in the hydrolyser and therefore less energy is required. For example, if the hydrolysis equilibrium was 30 wt.% HCl, 48 wt.% $FeCl_3$, and 15 wt.% non-hydrolysable metal chlorides, the total energy consumption would be approximately 4.5 GJ/h, i.e. only two thirds.

Therefore, future work should focus on the optimization of the hydrolysis equilibrium by means of varying temperature and/or pressure focusing on high HCl vapor concentrations and high non-hydrolysable metal chloride concentrations.

The separation of non-hydrolysable metal chlorides contained in stream 21 is not part of this current study. However, it is proposed to treat this stream – potentially after re-dissolution in water – similarly to existing hydrometallurgical laterite processes, e.g. by solvent extraction followed by HCl recovery from the metal chlorides (Zhu et al., 2012, Fernandes et al., 2013, Shen et al., 2008, Cheng, 2000).

CONCLUSIONS

This paper presented a flow sheet for the separation of iron as hematite from polymetallic chloride brines. A heat and mass balance was calculated based on laboratory experiments of crucial process steps for the purpose of separating iron from a nickel laterite hydrochloric acid pregnant leach solution.

It t could be shown that the hematite produced experimentally is rather pure with aluminum and chromium from the laterite ore being the biggest non-ferrous metal impurity. This is due to co-hydrolysis of these elements in the ferric chloride brine. The major fraction of the non-ferrous (i.e. non-hydrolysable) elements ends up in the crystallization filter cake, which also contains $FeCl_3$ as an impurity, depending on the solid/liquid separation efficiency.

For operating this described process on nickel laterite pregnant leach solution, 6.7 GJ indirect heat addition per ton of ore was calculated. The cooling demand is in a similar order of magnitude. Based on the specific needs for heating, cooling, oxygen, and the costs for power, waste water, and maintenance chosen (for a random location), the operating costs for this iron separation process amount to approximately 25 EUR (27 USD) per ton of ore or 1980 EUR per ton of nickel. The heating cost is almost 80% of the total cost. Thus, future work has to focus on the optimization of the energy demand. Process steps outside the presented flow sheet such as mechanical processing, thermal pre-treatment, or leaching have to be considered as well.

REFERENCES

Cheng, C. Y. (2000). Purification of synthetic laterite leach solution by solvent extraction using D2EHPA. *Hydrometallurgy*, 56 (3): 369–386.

Cohen, B., Shipley, D., Tong, A., Casaroli, S., and Petrie, J. (2005). Precipitation of iron from concentrated chloride solutions: Literature observations, challenges and preliminary experimental results. *Minerals Engineering*, 18: 1344–1347. Selected papers from Bio and Hydrometallurgy 2005, Cape Town, South Africa.

Dutrizac, J. and Riveros, P. (1999). The precipitation of hematite from ferric chloride media at atmospheric pressure. *Metallurgical and Materials Transactions B*, 30B: 993–1001.

Edwards, J. D., Frary, F. C., and Jeffries, Z. (1930). The aluminum industry in two volumes: Aluminum and its production. McGraw-Hill Book Company, Inc., New York.

Eisele, J. A. and Bauer, D. J. (1983). Recovery of alumina from anorthosite by an HCl-F-gas sparging crystallization process. Bureau of Mines Report of Investigations.

Fernandes, A., Afonso, J. C., and Dutra, A. J. B. (2013). Separation of nickel(ii), cobalt(ii) and lanthanides from spent Ni-MH batteries by hydrochloric acid leaching, solvent extraction and precipitation. *Hydrometallurgy*, 133(0): 37–43.

Harris, B. and White, C. (2011). Recent developments in the chloride processing of nickel laterites. In *Proceedings of ALTA 2011*, Perth, Australia.

Linke, W. F. and Seidell, A., editors (1958). *Solubilities of inorganic and metal organic compounds – A compilation of solubility data from the periodic literature*, volume 1. American Chemical Society, 4th edition.

Linke, W. F. and Seidell, A., Editors (1965). *Solubilities of inorganic and metal organic compounds – A compilation of solubility data from the periodic literature*, volume 2. American Chemical Society, 4th edition.

Nowak, B., Vogl, D., and Weissenbaeck, H. (2013). Nickel extraction from laterite ores by leaching with hydrochloric acid. In *Proceedings of ALTA 2013 Nickel-Cobalt-Copper*, Perth, Australia.

Riveros, P. and Dutrizac, J. (1997). The precipitation of hematite from ferric chloride media. *Hydrometallurgy*, 46: 85–104.

Shen, Y., Xue, W., and Niu, W. (2008). Recovery of Co(II) and Ni(II) from hydrochloric acid solution of alloy scrap. *Transactions of Nonferrous Metals Society of China*, 18 (5): 1262–1268.

Vogl, D., Haidinger, C., Nowak, B., and Weissenbaeck, H. (2012). Development of a versatile HCl-leach based metal extraction technology stack. In *Proceedings of ALTA 2012 Nickel/Cobalt/Copper Conference*, Perth, Australia.

Zhang, P., Guo, Q., Wei, G., Qu, J., and Qi, T. (2015). Precipitation of a-Fe_2O_3 and recovery of Ni and Co from synthetic laterite-leaching solutions. *Hydrometallurgy*, 153: 21–29.

Zhu, Z., Zhang, W., Pranolo, Y., and Cheng, C. (2012). Separation and recovery of copper, nickel, cobalt and zinc in chloride solutions by synergistic solvent extraction. *Hydrometallurgy*, 127–128 (0): 1–7.

HYDROLYSIS FOR IRON CONTROL AND ACID REGENERATION IN CHLORIDE HYDROMETALLURGY CIRCUITS

*M.J. Dry[1] and G.B. Harris[2]

[1]*Arithmetek Inc.*
1331 Hetherington Drive
Peterborough, Ontario, Canada K9L 1X4

[2]*NMR360 Inc.*
159-4148A rue Ste-Catherine West
Westmount, Quebec, Canada H3Z 0A2
*(*Corresponding author: mike.dry@arithmetek.com)*

ABSTRACT

This paper presents basic chemistry and operating conditions for the hydrolysis of ferric chloride into solid hematite and gaseous hydrogen chloride, and shows how this unit operation can be applied in chloride hydrometallurgy to regenerate strong hydrochloric acid and reject iron without also rejecting valuable metals such as nickel, copper and cobalt. Two examples are used to illustrate the hydrolysis technology and compare it to existing technology. The first is the regeneration of spent steel pickling acid, comparing hydrolysis to conventional pyrohydrolysis. The second example is the processing of a hypothetical low grade nickel laterite, comparing the novel chloride hydrolysis route to conventional sulphate HPAL technology.

KEYWORDS

Iron control, acid regeneration, hydrolysis, novel technology

INTRODUCTION

Whilst, in many respects, the control and removal of dissolved iron from chloride leach solutions can resemble the manner in which it is done in sulphate systems, chloride chemistry additionally offers the ability to achieve the same objectives in different ways (Harris, 2014). Conventionally, iron is oxidised to its ferric state (if necessary) and then precipitated with a base, usually magnesia where the feed is a laterite and lime where it is a sulphide, since calcium will precipitate sulphate formed during leaching. One method for controlling iron is to form FeOOH, either β-FeOOH (akaganéite), α-FeOOH (goethite), or a mixture thereof (Fillipou and Choi, 2002). However, if akaganéite is formed, the precipitate will contain a significant chloride component (as high as 7% Cl), leading to chloride losses and difficulty in satisfactorily disposing of the residue. Washing chloride from akaganéite is almost impossible due to the propensity of the solids to undergo peptization. The conventional approach requires the addition of a base, but it is based to some extent on controlled supersaturation precipitation (Demopoulos, 2009), which produces a more uniform and crystalline material and is far more attractive than, for example, the turboaeration process proposed by Great Central Mines in their chloride copper process (Raudsepp & Beattie, 1986).

Although there are presently no commercial plants doing so, the recovery of ferric iron by solvent extraction followed by pyrohydrolysis has been proposed. The reagent utilised is generally TBP (tri-butyl phosphate), wherein predominantly the ferric chloride molecule is extracted, and stripping can be effected with water. Another popular reagent is D2EHPA (di-2-ethyl hexyl phosphoric acid), but this requires extremely strong acid (preferably 10-12M HCl) to strip the iron. The main disadvantages of the iron solvent extraction techniques are that the concentration of iron achievable in the strip liquor is limited to a maximum of around 120-140 g/L, and only ferric iron is extracted. This creates appreciable problems for the subsequent pyrohydrolysis step, which much prefers the iron to be in its ferrous (reduced) form. It also creates appreciable water balance and energy consumption issues. The oxidation of ferrous iron is exothermic, whereas the hydrolysis of ferric iron is endothermic, so having the iron in the feed as ferric dictates that significantly more energy is required. The relatively low concentration of iron means that more water has to be evaporated per unit of iron, and since the evaporation of water is a major energy sink, the low concentrations achievable from SX increase this demand.

More recently, a novel, lower temperature method of effecting iron hydrolysis from chloride has been proposed and tested at the miniplant scale (Harris, 2014). This paper presents and reviews this method, and compares it to more traditional processes.

CHEMISTRY

The chemistry of ferric iron hydrolysis is summarized by the following stoichiometric equation:

$$2Fe^{3+} + (9+n)H_2O \leftrightarrow Fe_2O_3\cdot nH_2O + 6H_3O^+$$

The value of the parameter n, the water incorporated into the solid product, depends on the reaction conditions and is generally zero, one or three. Typically, when the solid is $Fe_2O_3\cdot 3H_2O$ (which can also be written as $Fe(OH)_3$, commonly referred to as ferric hydroxide), the precipitate is hard to filter and wash effectively. When the precipitate is $Fe_2O_3\cdot H_2O$ (which can also be written as FeOOH, goethite or akaganéite), it can be filtered and washed more easily, as long as conditions are managed to not cause peptization. Higher temperature conditions, such as are used in the high pressure acid leaching of laterite, for example, result in the iron being precipitated as Fe_2O_3 (hematite or ferric oxide), which is readily filterable, or sometimes as hydro-hematite, which is also $Fe_2O_3\cdot H_2O$, but is analogous to hematite rather than goethite.

Currently, the majority of iron processing in hydrometallurgy (as opposed to steel pickling, for example) is in sulphate systems, but chloride and nitrate systems have been proposed (Dry & Harris, 2012) and processes using them are under development. A significant difference between sulphate and chloride chemistry is that hydrochloric acid is much more volatile than sulphuric acid. That gives chloride systems a feature not found in sulphate systems, i.e. the ability to remove acid by volatilization instead of neutralization.

This enables regeneration and recycle of the hydrochloric acid, or the production of hydrochloric acid as a by-product.

The facet of chloride chemistry that makes it possible to recycle hydrochloric acid arises from the azeotropic nature of the HCl-H_2O system. Hydrochloric acid dissociates in water, as follows:

$$HCl + H_2O \leftrightarrow H_3O^+ + Cl^-$$

In dilute solutions, the HCl is essentially completely dissociated, but in more concentrated solutions it appears as a distinct species. The HCl molecule is more volatile than water, which gives rise to the azeotropic nature of this system. As a dilute solution of hydrochloric acid in water is evaporated, at first the vapor phase contains essentially only water because practically all of the HCl in the liquid phase is dissociated into non-volatile H_3O^+ and Cl^- ions. As the evaporation proceeds and the remaining solution becomes more concentrated in chloride ions, the dissociation equilibrium shifts towards molecular HCl that is more volatile than water. As the concentration of molecular HCl increases in the liquid phase, relatively more of it enters the vapor phase, until the liquid and the vapor contain the same proportions of H_2O and HCl. This is referred to as the azeotropic composition. Figure 1 is a plot of vapor-liquid equilibrium in the HCl-water system. The dotted line shows where the compositions of the vapor and liquid phases are equal and the black dots are measured data from the Kirk-Othmer Encyclopedia of Chemical Technology. The curves were generated using commercially available process simulation software known as Aspen Plus®.

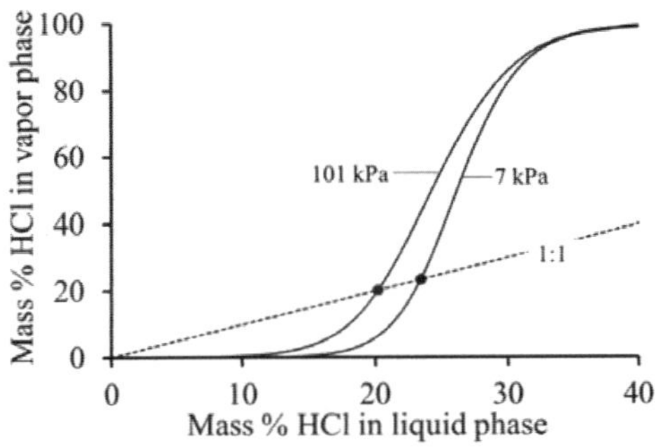

Figure 1 – HCl-H_2O equilibrium

In solutions containing chloride salts as well as hydrochloric acid, the dissolved salts raise the concentration of chloride ions and shift the dissociation reaction of hydrochloric acid towards HCl, increasing the amount of un-dissociated HCl as the concentration of the salt increases. Increasing the un-dissociated HCl in the solution phase also increases the amount of HCl reporting to the vapor phase when the solution is boiled. Figure 2 illustrates this.

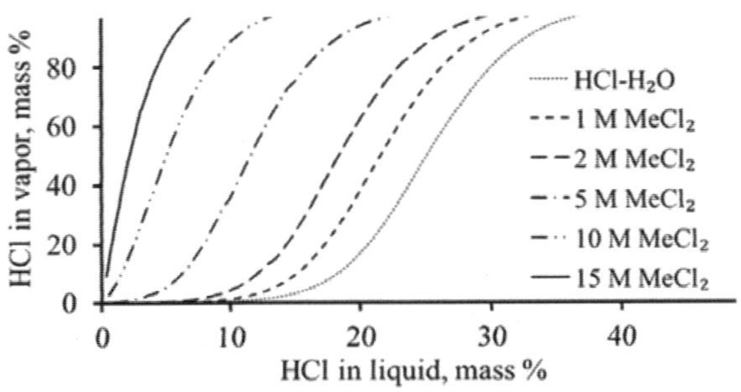

Figure 2 – HCl equilibrium over aqueous HCl-$MeCl_2$

As the concentration of the chloride salt increases in the liquid phase, the HCl in the vapor phase increases substantially for any given level of HCl in the liquid phase. The dissolved salt $MeCl_2$ can be any of a number of chloride salts; this example assumes divalent cations, but monovalent and trivalent chloride salts have the same effect because what matters is the chloride ion concentration.

Dutrizac and Riveros precipitated hematite from solutions of ferric chloride and hydrochloric acid at temperatures between 125°C and 160°C in an autoclave (Riveros & Dutrizac, 1997). Table 1 lists some of the data generated in this work, which was done at the pressures generated by the solutions at the various temperatures, so in that work there would have been little vapor phase and the acid evolved would thus have remained mostly in the liquid phase, limiting the extent of hematite precipitation. The bottom six lines of data in Table 1 are the amounts of hematite precipitated at different levels of free acid in a starting solution containing 0.4M $FeCl_3$. These numbers confirm that increasing free acid in the solution reduces the extent of hematite precipitation. Figure 3 shows the amount of hematite precipitated from a 0.3M solution of ferric chloride versus the precipitation temperature. The symbols are data from the work of Riveros and Dutrizac, and the line is a corresponding prediction generated using an Aspen Plus® process model calibrated to the data.

Table 1 – Data from Riveros and Dutrizac (1997)

Initial conditions			Temperature	Fe_2O_3
HCl	$FeCl_3$	Fe_2O_3	°C	precipitated, %
0.0 M	0.3 M	0 g/L	125	27.6%
0.0 M	0.3 M	0 g/L	130	34.2%
0.0 M	0.3 M	0 g/L	140	43.0%
0.0 M	0.3 M	0 g/L	150	49.3%
0.0 M	0.3 M	0 g/L	160	48.4%
0.5 M	0.4 M	50 g/L	150	-1.6%
0.4 M	0.4 M	50 g/L	150	12.4%
0.3 M	0.4 M	50 g/L	150	22.2%
0.2 M	0.4 M	50 g/L	150	32.1%
0.1 M	0.4 M	50 g/L	150	39.5%
0.0 M	0.4 M	50 g/L	150	44.5%

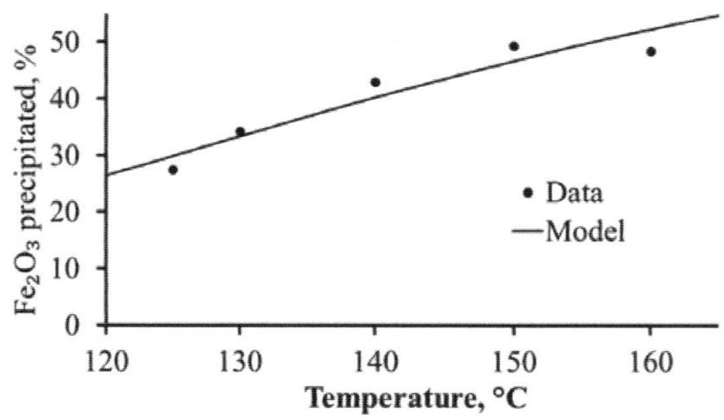

Figure 3 – Hematite precipitation versus temperature, under pressure

Figure 4 is a graph showing the extent of hydrolysis of ferric chloride versus temperature at atmospheric pressure, as calculated using the Aspen Plus® process model calibrated to the data of Riveros and Dutrizac. The dashed line is a prediction assuming that the starting solution is ferric chloride and the solid line is for a system containing ferric chloride plus a divalent chloride salt (chloride matrix). At atmospheric pressure and these temperatures there is a substantial vapor phase, which removes the HCl that is released by the hydrolysis reaction. While the hydrolysis reaction proceeds a little more readily in the first case, in both cases it goes to completion by about 155°C, with all of the ferric chloride converted to solid hematite and a gaseous mixture of steam and HCl. In the development work done to date by Neomet, the standard hydrolysis temperature used has been 180°C to ensure complete conversion of ferric chloride to Fe_2O_3 and HCl. SMS Siemag stated a lower limit of 155°C and a preferred range of 160°C to 200°C for the hydrolysis temperature (Takahashi, et al., 2009). In both cases the acid strength in the vapor phase at complete hydrolysis of the $FeCl_3$ to Fe_2O_3 is close to 37% by mass. This is achieved by manipulating the relative amounts of water and ferric chloride in the feed to the hydrolysis reactor.

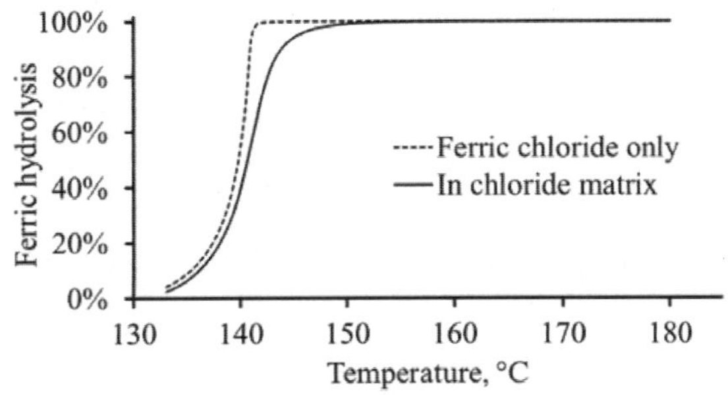

Figure 4 – Hydrolysis at atmospheric pressure

In the absence of a supporting chloride matrix and at complete hydrolysis of the $FeCl_3$, the reactor would have to handle a mainly solid product. In this respect, SMS Siemag stated that it was necessary to maintain the concentration of ferric chloride in the liquid phase at 65% for the reaction to work, making it impossible to precipitate all of the ferric iron in one pass (Takahashi, et al., 2009). In the presence of the supporting matrix and at complete conversion of the ferric chloride to hematite in a single pass, the solid hematite leaves the reactor as a more easily handled slurry, provided sufficient matrix salt is used.

It should be noted that SMS Siemag built a commercial operation based on their patent and employing a ferric chloride-only hydrolysis (Takahashi, 2009). The subsequent failure of this plant was attributed to insurmountable corrosion and fluid handling (notably freezing and scale formation) issues (Weissenbaeck, et al., 2013). Harris and White (2013) pointed out that without the metal chloride matrix, such problems were to be expected, and that with the matrix, operation could proceed quite satisfactorily.

In the established technology known as pyrohydrolysis, the feed solution, preferably containing the iron as ferrous chloride, is oxidized and converted to hematite and steam/HCl in a fluidized bed or a spray roaster. The required energy comes from the combustion of a suitable fuel (e.g. natural gas) at about 400°C to 500°C. Under these conditions the trivalent and most, but not all, divalent metal chloride salts in the feed are converted to solid oxides. The solid product is mainly hematite. Monovalent salts such as sodium chloride are converted to solid chlorides in the product, thus leading to chloride losses, and salts such as nickel and cobalt chloride become oxide impurities in the hematite. Pyrohydrolysis does not separate metals like nickel and cobalt from iron.

A feature of the hydrolysis chemistry is that, while trivalent chloride salts ($FeCl_3$, $AlCl_3$, $CrCl_3$) are converted to their corresponding oxides, divalent chloride salts ($NiCl_2$, $CoCl_2$, $CuCl_2$, etc.) are not hydrolyzed in the temperature range in which the trivalent chlorides are hydrolyzed. Work done by Neomet (Harris and White, 2013) has shown that, after the oxidation and complete hydrolysis of iron, the copper, nickel and cobalt remaining in the molten matrix can be recovered by heating the molten salt further and injecting water or steam, which converts the copper, nickel and cobalt chlorides into solid basic metal chlorides, also releasing gaseous HCl. After that, further heating and addition of water or steam converts magnesium into solid basic magnesium chloride and releases more HCl.

Table 2 lists the relevant heats of reaction for the oxidation and hydrolysis of iron chloride, calculated using the data from the Aspen Plus® databanks. The oxidation of ferrous chloride to ferric chloride is exothermic and the hydrolysis of ferric chloride to gaseous HCl and solid hematite is endothermic. Riveros and Dutrizac reported that the hematite formation reaction appears to be fast, and work done by Neomet indicates that the rate of reaction is essentially controlled only by the rate at which heat can be transferred into the hydrolysis reactor.

Table 2 – Heats of reaction, kJ per mole Fe reacted

$2FeCl_2 + 2HCl + ½O_2 \rightarrow 2FeCl_3 + H_2O$	-85
$6FeCl_2 + 1½O_2 \rightarrow 4FeCl_3 + Fe_2O_3$	-132
$2FeCl_3 + 3H_2O(liq) \rightarrow Fe_2O_3 + 6HCl(vap)$	606
$2FeCl_3 + 3H_2O(vap) \rightarrow Fe_2O_3 + 6HCl(vap)$	485

APPLICATIONS

Two case studies are presented to demonstrate the application of hydrolysis technology, namely regeneration of steel pickling acid and processing a nickel laterite.

Treatment of Spent Steel Pickling Liquor

A convenient illustration of the application of the hydrolysis technology is the regeneration of hydrochloric acid from spent steel pickling acid and rinse water. Table 3 lists typical compositions of these solutions.

Table 3 – Spent steel pickling solutions, mass %

	Spent acid	Rinse water
H_2O	73.9%	98.5%
HCl	3.6%	0.4%
$FeCl_2$	22%	0.8%
$FeCl_3$	0.5%	0.3%
Quantity	2.45 m^3/h	2.04 m^3/h

Oxidation of the ferrous iron can be accomplished in a number of ways. One convenient way is to use oxygen, as air or as commercial oxygen, as the oxidant. If the incoming ferrous chloride feed is evaporated to the appropriate concentration ahead of the oxidation step and commercial oxygen is used, the exothermic oxidation reactions make the oxidation autogenous. The work of Riveros and Dutrizac shows that at and above about 130°C, any iron precipitated in the oxidation is as hematite and that below 120°C the oxidation of ferrous chloride gives akaganéite. Hematite is preferable because akaganéite will lead to chloride contamination of the solids formed in the subsequent hydrolysis step.

Hydrolysis Route

Figure 5 is a diagram of a circuit recovering hydrochloric acid from spent steel pickling acid and rinse water via the hydrolysis of ferric chloride. The incoming spent steel pickling liquor, plus the portion of the rinse liquor not used to dilute the product acid, is first concentrated by evaporation in three stages. In the first stage, the incoming liquor is evaporated under reduced pressure, condensing the acid and steam from the hydrolysis step. In the second stage, the partially concentrated liquor is evaporated further, at reduced pressure, condensing the steam from the third evaporation stage. The liquor from the second evaporation stage is pumped back to atmospheric pressure and evaporated further using an external energy source. The overall extent of evaporation is manipulated to make the subsequent oxidation stage, in which the concentrated pickling liquor is oxidized with oxygen at atmospheric pressure at 130°C, autogenous.

The oxidized slurry is contacted, together with steam, with a circulating flow of 200°C molten chloride hydrate (matrix salt), causing the ferric chloride to decompose fully to HCl and hematite. The steam added is condensate taken from the evaporation sequence and re-evaporated using an external energy source. The amount of steam added to the hydrolysis step is manipulated to make the steam/acid from the hydrolysis step contain 36 percent HCl by mass, which is a concentration that has regularly been achieved by Neomet. The condensate not re-evaporated and sent through the hydrolysis step leaves the circuit as condensate that can return to the pickling line as fresh rinse water. The HCl and residual steam leave the hydrolysis reactor, are condensed in the evaporation sequence and diluted with part of the incoming rinse water, to give the regenerated 18% hydrochloric acid required by the steel pickling line (Andritz technical brochure).

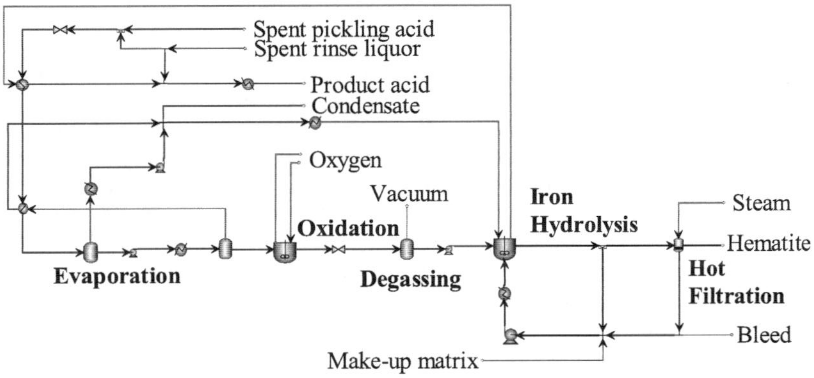

Figure 5 – Oxidation and hydrolysis of spent steel pickling acid

The hematite leaves the hydrolysis reactor as slurry in the molten matrix. Part of this slurry is recirculated to the hydrolysis reactor through a heater using an external energy source to re-heat it to 200°C, to supply the energy needed for the endothermic hydrolysis reaction. The amount recirculated is manipulated to make the temperature in the hydrolysis reactor 180°C. The slurry not recirculated directly is filtered hot and the matrix salt is essentially completely removed by washing the filter cake with steam in a proprietary filter developed by Neomet. The filtrate, less a small bleed to control the accumulation of minor impurities such as manganese, is returned to the circulating slurry and recycled to the hydrolysis reactor. A small amount of matrix salt is added to compensate for loss to the bleed.

This circuit was modelled using Aspen Plus®. Table 4 is a summary of the circuit parameters calculated. A point to note is that the amount of condensate produced, that can return to the pickling line to be used as fresh rinse water, almost exactly matches the volume of spent rinse water from the pickling line.

Table 4 – Summary, treatment of spent pickling liquor

Spent pickling liquor into the circuit	2.45 m^3/h
Spent rinse liquor into the circuit	2.04 m^3/h
Condensate back to the pickling line	2.03 m^3/h
Oxidation temperature	130°C
Oxygen utilisation in oxidation	80%
Hydrolysis temperature	180°C
HCl in steam/HCl ex hydrolysis	36%
HCl in recovered acid	18%
Solids in matrix ex hydrolysis	12%

Table 5 lists the energy inputs into this circuit. It should be noted that the recovered acid is specified as 18% HCl, since this is what steel pickling lines conventionally require, and what is produced by conventional pyrohydrolysis (see below). If a stronger acid product should be required for any reason, it could very easily be made so by not adding as much spent rinse water.

Table 5 – Energy summary, treatment of spent pickling liquor, per kg Fe_2O_3 produced

Recycled energy to evaporation	10.4 MJ
External energy into evaporation	5.8 MJ
External energy into hydrolysis	4.6 MJ
Total external energy demand	10.4 MJ

Pyrohydrolysis

The conventional way of regenerating steel pickling acid is pyrohydrolysis, using fluidized bed or spray roaster technology. Figure 6 illustrates such a circuit, in this case using the spray roaster option (lower energy demand than the fluid bed option) with the same quantities and compositions of spent pickling acid and rinse liquor as for the hydrolysis circuit (Table 3). The spent pickling liquor is pre-heated by heat exchange with the regenerated acid and contacted with hot gas from the spray roaster in a venturi contactor, evaporating some of the water and quenching the hot gas from the spray roaster. The concentrated liquor is then converted to solid hematite and gaseous steam/HCl in the spray roaster, the required energy coming from the combustion of a suitable fuel (methane, in this example) in air. The quenched gas from the venturi is scrubbed with water in an absorber to capture the acid as aqueous hydrochloric acid that returns to the steel pickling line. This circuit was modelled using Aspen Plus®, assuming the roaster temperature to be 450°C and adding enough water to the gas cleaning scrubber to make the regenerated acid contain 18 percent HCl by mass.

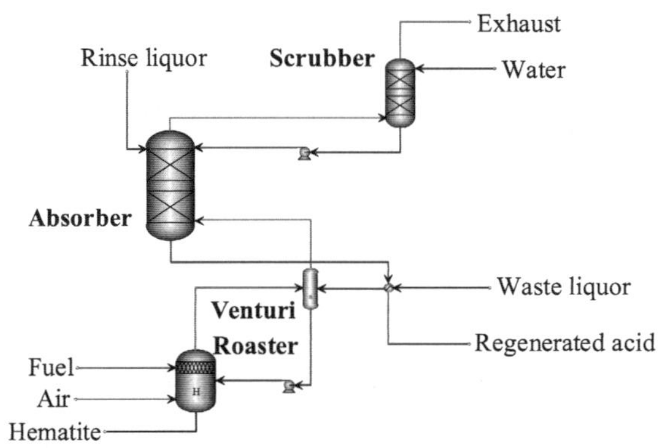

Figure 6 – Pyrohydrolysis of spent steel pickling acid

Table 6 is a summary of the process parameters for the pyrohydrolysis process as modelled. For the same feed liquors, the nominally more complex hydrolysis circuit consumes appreciably less energy. The hydrolysis circuit can also be set up to produce significantly stronger hydrochloric acid than can be produced from the pyrohydrolysis process, which is limited by the water-HCl azeotrope. Also, the hydrolysis circuit regenerates the pickling acid and wash water, whereas the pyrohydrolysis circuit loses more water to the scrubber exhaust than is present in the incoming spent wash liquor.

Table 6 – Summary, pyrohydrolysis of spent steel pickling acid

Pyrohydrolysis temperature	450°C
HCl in recovered acid	18%
Energy input, per kg Fe_2O_3	14.2 MJ

Nickel Laterite

What particularly sets the hydrolysis technology apart from pyrohydrolysis is that the former can be used to reject iron as hematite and separate the iron from more valuable metals, such as nickel and cobalt. This can be conveniently demonstrated using nickel laterite processing, comparing hydrolysis to the more conventional technology, HPAL (High Pressure Acid Leaching). Table 7 shows the composition of a hypothetical laterite that was used in other work comparing various alternative technologies (Dry, 2015).

Table 7 – Laterite composition, mass %

Ni	1.22	MgO	1.54
Co	0.20	MnO	1.51
Al_2O_3	5.61	Na_2O	0.17
Cr_2O_3	3.36	K_2O	0.01
Fe_2O_3	64.31	SiO_2	7.62
CaO	0.06		

<u>Chloride Process</u>

Figure 7 is a diagram of a circuit using chloride technology for processing this laterite (Dry, 2015). It begins with mixing the incoming laterite with recycled wash filtrates and water, and leaching the resulting slurry with strong hydrochloric acid that is sparged into the laterite slurry as a vapor containing steam and hydrogen chloride.

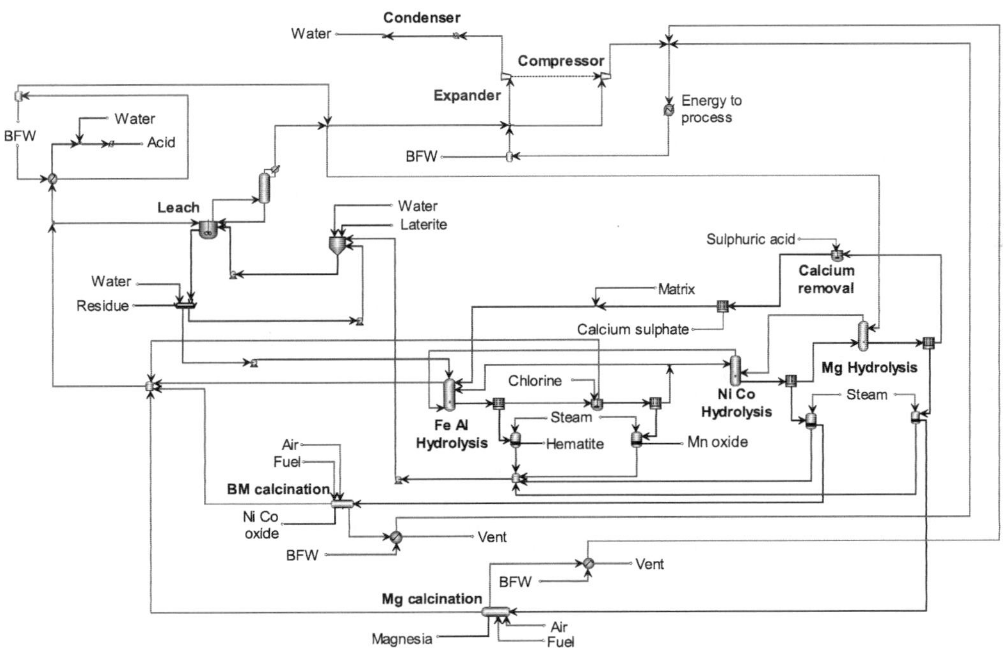

Figure 7 – Laterite processing using hydrolysis

The latent heat in the incoming acid drives the leach temperature up to about 135°C and the incoming steam/HCl does not fully condense but a greater proportion of the HCl does, because it is consumed in the leach reactions. The uncondensed steam/HCl passes to a scrubber column in which the acid is captured by partial condensation of the steam and returned to the leach. The remaining steam is split, part going to the hydrolysis sequence and the balance to an energy recovery step. Adding the acid to the leach as a vapor eliminates the need for an evaporation stage between the leach and the following iron hydrolysis section. It also eliminates the need for cooling in the leach, which greatly simplifies the leach reactors. This method of acid addition was successfully demonstrated in the Magnola plant in Canada.

The leached slurry is filtered and washed with water. The washed filter cake, predominantly benign alumino-silicate gangue, leaves the circuit. The wash filtrate is recycled to the feed mixing tank and the primary filtrate is sent to the iron hydrolysis section, where it is mixed with a circulating molten salt matrix, contacted with steam from the following step and heated to 180°C by indirect heat exchange with condensing high pressure steam. The trivalent (Fe, Al, Cr), chlorides react with the steam, forming solid Fe_2O_3, Al_2O_3 and Cr_2O_3 (referred to collectively as hematite because that is the major oxide) and gaseous hydrogen chloride. The amount of atmospheric-pressure steam going to the hydrolysis reactor is manipulated to give 35 mass percent HCl in the steam/acid leaving the iron hydrolysis step.

The resulting slurry of hematite in the molten matrix is filtered and washed using Neomet's proprietary filter that uses steam to wash the filter cake. The washed hematite leaves the circuit and the wash filtrate is recycled to the leach via the feed mixing tank.

The primary filtrate from the iron hydrolysis step is contacted with gaseous chlorine to oxidize and precipitate manganese as manganese dioxide, which is recovered and washed in the same way as the hematite, then leaves the circuit. This addition of chlorine also acts as chloride make-up in the circuit. The filtrate is split, part returning to the iron hydrolysis stage to build up the level of base metals, and a bleed proceeding to the nickel/cobalt hydrolysis step, where it is contacted again with steam from the subsequent (hotter) magnesium hydrolysis step, which heats it enough to convert the nickel and cobalt to solid basic metal chlorides, releasing the associated chloride ions as gaseous hydrogen chloride. The basic metal chlorides are filtered out and washed in the same way as the hematite. The filtrate is heated further by indirect heat exchange with condensing high pressure steam and contacted with steam from the scrubber in the leach section, causing the magnesium chloride to decompose to gaseous hydrochloric acid and solid basic magnesium chloride that is recovered by filtration and washed in the same way as the hematite.

The basic nickel and cobalt chlorides are calcined to a mixed nickel/cobalt oxide that leaves the circuit as the main product. The basic magnesium chloride is calcined to magnesia that leaves the circuit as a by-product. In these steps the chloride is released as gaseous hydrochloric acid that is recycled to the leach along with the gaseous steam and hydrogen chloride from the hydrolysis section. Calcination of the base metal and magnesium basic metal chlorides is done in externally heated kilns, with the required energy coming from the combustion of natural gas. The hot gas leaving the heating side of each kiln is passed through a heat exchanger in which high pressure steam is raised from boiler feed water, for use as part of the heat source in the hydrolysis section.

Some of the steam leaving the scrubber in the leach section goes to the hydrolysis train and the balance is split, part being expanded through a condensing turbine that drives a compressor in which the other part of the steam is compressed and combined with the steam from the base metal and magnesium calcination section for use as the energy source in the hydrolysis section. The ratio of steam expanded to steam compressed is adjusted to give the required pressure in the compressed steam. The model assumes a gas-fired steam boiler producing high pressure steam to supply the balance of the energy needed in the hydrolysis sequence.

Sulphate Process

The established technology for processing laterite, particularly limonite, is high pressure acid leaching (HPAL). Figure 8 is a diagram of such a circuit that was modelled as part of other work (Dry, 2015). It includes a standard sulphur-burning acid plant that produces the required amount of sulphuric acid and the steam needed for heating the autoclave. The incoming laterite is mixed with recycled process water and pumped through three pre-heating steps in which it is contacted with steam from three flash-down steps after the autoclave. The pre-heated slurry and concentrated sulphuric acid are pumped into the autoclave. Steam from the acid plant is injected into the autoclave to heat the leach to 250°C. After leaching in the autoclave, the pressure of the slurry is reduced to atmospheric in three stages. The steam flashed off is used in the pre-heating sequence.

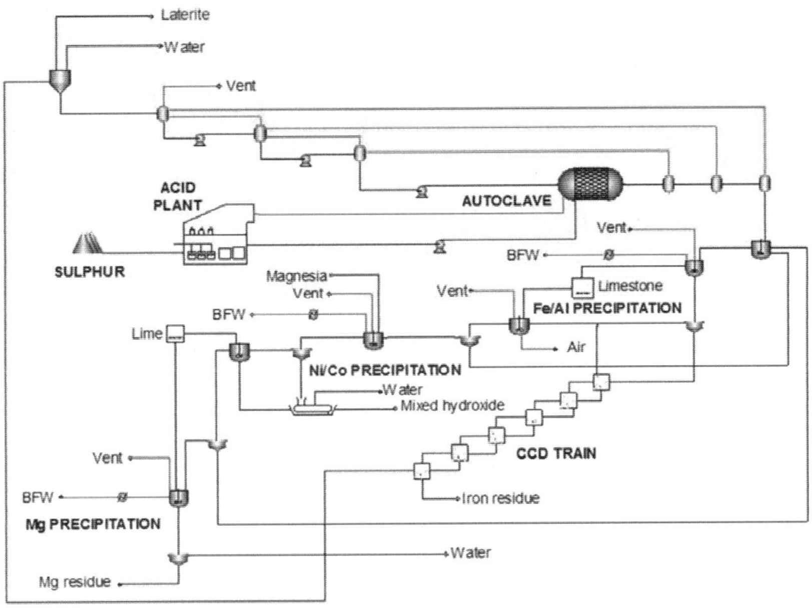

Figure 8 – Laterite HPAL circuit

The leached slurry is first contacted with recycled hydroxide precipitates from downstream to re-dissolve co-precipitated nickel and cobalt and to partially consume the free acid left after the leach. The partly neutralized slurry is neutralized further with limestone to precipitate the bulk of the iron and aluminium while co-precipitating essentially no nickel or cobalt. The resulting iron-aluminium slurry is thickened and the thickener underflow is washed with recycled process water in a six-stage counter-current decantation train. The washed iron-aluminium residue leaves the circuit.

The Fe-Al thickener overflow is combined with the supernatant from the counter-current decantation train and neutralized further with more limestone, to precipitate essentially all of the remaining iron and aluminium. Some nickel and cobalt are co-precipitated in this step, and the underflow from the subsequent thickening step is recycled. The remaining solution is neutralized further with magnesium oxide to precipitate the bulk of the nickel and cobalt as hydroxides. The resulting slurry is thickened, the underflow is filtered and the filter cake is washed with fresh water. The washed filter cake leaves the circuit as the required hydroxide intermediate product. The supernatant and filtrate are combined and neutralized again, this time using lime, to precipitate the remaining nickel and cobalt as hydroxide and gypsum that is thickened, the underflow recycled and the supernatant contacted with lime to precipitate the magnesium and manganese into a hydroxide-gypsum residue that leaves the circuit after thickening. The supernatant from the final thickening step is recycled as process water.

Comparison of the Laterite Processes

The process models for the two laterite circuits were based on thirty thousand tonnes per year of nickel in the feed laterite. Table 8 lists the capital and operating costs that were estimated for each of these circuits from the mass-energy balances generated by the two process models. The methodology used is presented elsewhere (Dry, 2013; Dry, 2015). The fixed operating costs amount to about \$0.09 and \$0.13 per pound of nickel and cobalt, respectively, for the two processes. At the preliminary level of analysis in which these numbers were calculated, the differences in the operating costs are not significant. The variable cost for the HPAL technology depends very strongly on the price of sulphur. The variable cost of the hydrolysis process depends strongly on the cost of energy. The hematite from the hydrolysis technology is essentially pure Fe_2O_3 that could be pelletized and sold into the steelmaking industry at a premium over the price of normal iron ore, or perhaps even into the pigment industry at a higher premium. Samples generated during piloting the oxidation and hydrolysis of spent steel pickling liquor were evaluated and found to possess all the requirements of a high-quality pigment. The magnesium oxide is also essentially pure MgO that could be sold as a byproduct. Assuming that the hematite could be sold for \$40/tonne and the magnesium oxide for \$50/tonne changes the after-credit variable cost appreciably, per unit of nickel and cobalt. The iron residue from the established technology is a waste requiring appropriate disposal.

Table 8 – Economic comparison for laterite processing

Process	Operating costs		Capital cost, \$ million	
	Fixed \$ million/y	Variable \$/lb Ni&Co	Installed process equipment	Mine and infrastructure
Conventional HPAL technology	6.2	2.05	479	666
Hydrolysis, Ni & Co only	8.3	2.24	399	666
Hydrolysis, with Fe_2O_3 & MgO credit	8.3	1.32	399	666

CONCLUSIONS

The hydrolysis technology presented in this paper, while not yet commercially proven, offers a promising new way of rejecting iron and recovering more valuable metals such as nickel and cobalt from circuits using chloride hydrometallurgy. The iron is rejected as pure hematite that could be sold as high quality iron ore or perhaps even as feed to the iron pigment industry. At the very least, the hematite can be impounded with no risk of environmental consequences. Depending on the cost of energy versus the cost of sulphur, and on whether or not the hematite can be sold as a by-product rather than impounded, the costs associated with the hydrolysis route may also be attractive compared to the established technology.

Ongoing pilot work will ultimately show whether or not the hydrolysis technology is a worthy addition to hydrometallurgy's arsenal of processes.

REFERENCES

Demopoulos, G.P. (2009) Aqueous Precipitation and Crystallization for the Production of Particulate Solids with Desired Properties, Hydrometallurgy, Volume 6(3), April 2009, 199-214.

Dry, M. and Harris, B. Nickel and the three mineral acids. *ALTA Nickel-Cobalt-copper conference,* May 2012, Perth, Australia.

Dry, M. Early evaluation of metal extraction projects. *ALTA Nickel-Cobalt-copper Conference,* May 2013, Perth, Australia.

Dry, M. Technical and cost comparison of laterite treatment processes, Part 3. *ALTA Nickel-Cobalt-copper Conference,* May 2015, Perth, Australia, 24-47.

Filippou, D. & Choi, Y. (2002). A contribution to the study of iron removal from chloride leach solutions. *Chloride Metallurgy 2002 Volume 2*, (Eds. E. Peek and G. van Weert), Proceeding of the 32nd Annual CIM Hydrometallurgical Conference, CIM, Montreal, 2002, 729-741.

Harris, Bryn & White, Carl (2013). Further development of the chloride process for base and light metals: recent miniplant and first pilot plant data. *ALTA Ni-Co-Cu 2013*, Perth, WA, May 2013.

Harris, G.B. (2014). Making use of chloride chemistry for improved metals extraction processes. *Proceedings of the 7th International Symposium on Hydrometallurgy (HYDRO 2014)*, (Eds. E. Asselin, et al.), Canadian Institute of Mining, Metallurgy and Petroleum, Westmount, Quebec, Canada, 171-184.

Raudsepp, R. & and Beattie, M.J.V. (1986). Iron Control in Chloride Systems, Iron Control in Hydrometallurgy, (Eds. J.E. Dutrizac and A.J. Monhemius), *Proceedings of 16th Annual CIM Hydrometallurgical Meeting*, Toronto, October 1986, CIM Montreal, 1986, 163-170.

Riveros, P.A. and Dutrizac, J.E. (1997). The precipitation of hematite from ferric chloride media. *Hydrometallurgy 46 (1997)*, 85-104.

Takahashi, N, Taki, O., Weissenbaeck, H. & Vogl, D. (2009). Processing method for recovering iron oxide and hydrochloric acid. Processing method for recovering iron oxide and hydrochloric acid. *International patent number WO2009/153321 A1,* December, 2009.

Weissenbaeck, Herbert, Nowak, Benedikt, Vogl, Dieter, & Krenn, Horst (2013). Development of chloride based metal extraction techniques: advancements and setbacks. *ALTA Ni-Co-Cu 2013*, Perth, WA, May 2013.

The Chemistry of Iron - Some Different Perspectives

THE EVOLUTION OF IRON CONTROL AT RUSTENBURG BASE METAL REFINERS

*L.J. Bryson[1], C.A. Biley[2]

[1]*Anglo American Technical Solutions*
Johannesburg, South Africa, 2091
*(*Corresponding author: Leslie.bryson1@angloamerican.com*)

[2]*Anglo American Platinum*
Rustenburg Base Metal Refiners
Rustenburg, South Africa, 0300

ABSTRACT

Anglo American Platinum's Rustenburg Base Metals Refiners (RBMR) produces Ni and Cu cathode from slow cooled converter matte after Platinum Group Metals (PGMs) are magnetically separated for further processing. In 2010, execution of the RBMR expansion project was commenced to increase the production capacity to 33 ktpa Ni. The expansion, which is currently still in a ramp-up stage, brought about several fundamental changes to the circuit chemistry, including the installation of a dedicated Pressure Iron Removal (PIR) autoclave. This provided an independent outlet for iron from the refinery, which previously deported to the residue from the leach circuit and was toll refined. At the same time as changes in the circuit associated with the RBMR expansion were being made, the plant experienced a significant increase in iron and copper units per nickel unit in the feed. The decoupling of the leaching and iron removal process steps has had significant advantages for the robustness and control of the overall circuit, in particular pertaining to the preparation of a nickel electrolyte of sufficient purity to produce LME grade nickel. This paper details the primary changes to the RBMR circuit chemistry with a particular focus on iron deportment and the benefits derived from the installation of the PIR autoclave.

KEYWORDS

RBMR, base metals, iron precipitation

INTRODUCTION

Anglo American Platinum mines and refines Platinum Group Metals (PGMs) through a complex processing pipeline in its South African operations. Mined ore is first treated through several mineral processing stages, geared to the recovery of base metals and associated PGMs, followed by smelting and refining operations that aim to separate base metals from PGMs and recover the value components to purified, saleable products. The focus of this paper is the Rustenburg Base Metals Refiners (RBMR) complex which receives a PGM-containing converter matte, then physically separates PGMs and base metals, and recovers nickel and copper to cathode by utilising leaching, precipitation and electrowinning processes.

With new mining operations coming online and the continued increase in PGM ore arisings, it was necessary to undertake an expansion project at the Base Metal Refinery (BMR) to increase production capacity from 21 ktpa to 33 ktpa Ni. Complicating the expansion project was the requirement of continued production through all stages of design, construction and implementation and, due to stringent capital constraints, the utilisation of all existing equipment installed during a previous expansion in 1981. Against this backdrop, the expanded BMR flow sheet involved significant changes to the chemistry, especially iron management. These changes in iron management allowed for the separation of iron and copper residues and significantly increased the robustness of the plant to changes in feed mineralogy as well as the potential to increase production duties of the existing equipment.

This paper outlines the evolution of the BMR circuit, highlighting the central role which iron management has played in the ultimate flow sheet configuration. The basic chemistry and short-comings of the pre-expansion circuit are discussed and the developmental journey through piloting and implementation of the expansion project are presented. In addition, key benefits unlocked by the change in circuit configuration hold tremendous value potential for the continued ramp-up of the BMR to full name-plate capacity, and beyond.

PRE-EXPANSION LEACHING CIRCUIT (1981 – 2010)

The feed to the RBMR operation consists of converter matte that is produced at the Waterval Smelter complex. This feed is first milled and treated through a magnetic separation circuit in the Magnetic Concentration Plant (MCP) to produce a PGM-containing magnetic fraction and a non-magnetic sulphide fraction, colloquially referred to as Nickel Copper Matte (NCM). This NCM fraction contains the bulk of base metal sulphides present in the converter matte and is treated in the BMR. The typical composition of NCM in the early 1990's, as published by Hofirek and Kerfoot (1992), is presented in Table 1. In terms of mineralogy, about 90% of the matte mass is comprised of heazlewoodite (Ni_3S_2), djurleite ($Cu_{1.96}S$) and bornite (Cu_5FeS_4). Bornite is the predominant iron bearing phase.

Table 1 - Major element content of NCM ca. 1992

Element	Typical (%)	Range (%)
Ni	43	38 – 45
Cu	29	27 – 32
Fe	1.5	1.0 - 2.0
S	22.5	21 – 24

A simplified block flow diagram of the refinery operation as commissioned in 1981 is depicted in Figure 1. All iron entering this circuit exited with the Secondary Leach Concentrate (Hofirek, 1996). The detailed chemistry of the circuit, which was described by Hofirek and Kerfoot (1992), is not critical to this discussion. However, the basic principle of separation of the Ni, Cu and Fe contained in NCM is discussed in context of the of the original three stage leaching process. These three leaching stages consisted of an atmospheric leach, referred to as Copper Removal, and two pressure leaching stages, referred to as Primary Leach and Secondary Leach. The solid and solution flows in these three stages operated in a counter-current fashion to facilitate the sequential dissolution and separation of Ni, Cu and Fe.

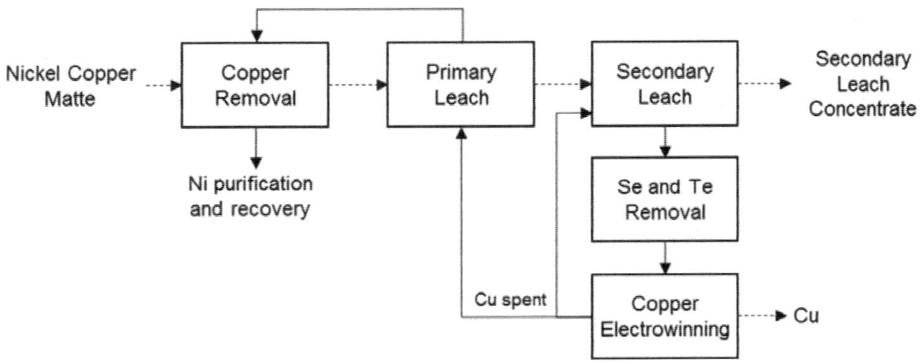

Figure 1 - Simplified block flow diagram of the RBMR leaching circuits commissioned in 1981. Solids flows are shown as dashed lines.

Unlike most matte treatment circuits, the BMR treats a fully sulphidized matte as a result of the MCP removing most of the sulphur-deficient alloy components out of the converter matte. Since only metallic copper, nickel and minor amounts of cobalt sulphate are produced from the refinery, sulphur has to be removed from the circuit separately and, in the case of BMR, sulphur is removed from the nickel electrowinning circuit by separating and crystallising sodium sulphate as a low-value by-product.

Copper Removal

In the first atmospheric leaching stage non-magnetic NCM from the MCP was contacted with solution from the solid-liquid separation stage following the Primary Leach. Ideal feed solution conditions were: pH 1.5, 100 g/L Ni, 5 g/L Cu and 1 g/L Fe. A portion of the comparatively reactive heazlewoodite in NCM was solubilized with oxygen to consume residual acid contained in the Primary Leach liquor such that iron and copper in solution was then precipitated by hydrolysis, mainly as basic sulphates, to levels ideally less than 10 ppm.

Primary Leach

The solid residue from the Copper Removal stage was fed to the Primary Leach along with copper electrowinning spent electrolyte and plant spillages to adjust feed density, resulting in a typical solution composition of 30 g/L Ni, 20 g/L Cu and 50 g/L sulphuric acid. The Primary Leach autoclave was divided into two sections: oxidising and non-oxidising. The oxidising section consisted of two autoclave compartments, each sparged with compressed air operating at 10 bar(g) and 135°C. The oxidising section primarily utilised copper and acid in spent electrolyte to dissolve heazlewoodite to form millerite (NiS), some of which was oxidized to polydymite (Ni_3S_4). Copper in the form of djurleite was oxidised to covellite (CuS) and copper sulphate in solution. Since these oxidation reactions consumed acid, residual acid levels decreased substantially causing copper to precipitate as a basic sulphate which played a vital role in the non-oxidising section of the Primary Leach.

The subsequent two non-oxidising compartments continued to run at temperature, but with no air sparge. The design intention of the non-oxidising compartments was the continued dissolution of millerite and polydymite by metathesis, that is, the exchange of soluble copper ions with nickel in the solid lattice. Some of these metathesis reactions produced acid. Under ideal conditions, this acid generation was balanced by the dissolution of the basic copper sulphates precipitated in the oxidising portion of the Primary Leach. To achieve this, the terminal pH was in the region of 2.3 to 2.7. The intention of this careful pH endpoint control ensured that copper and iron concentrations were in the range of 3-5 g/L Cu and < 1 g/L Fe, which could easily be removed to low levels in Copper Removal, while providing nickel extractions of greater than 75% or, equivalently, a Cu:Ni ratio in residue of greater than 5:1. However, the control of this two stage leach proved to be very difficult. If excess acid was added, a pH of 1.3-1.7 could be achieved which could result in a final solution containing up to 10 g/L Cu and 3 g/L Fe, often resulting in breakthrough of these impurities to the nickel electrowinning circuit. Thus, the preferred operating regime often involved running the Primary Leach in acid deficient mode where pH of greater than 2.7 was targeted. While this resulted in very low residual concentrations of copper and iron in solution, nickel extractions were generally lower than desired due to the less aggressive non-oxidising conditions in the autoclave. Typically, the Cu: Ni ratio in Primary Leach residue was 4:1.

Secondary Leach

Primary Leach residue was fed to the Secondary Leach along with copper electrowinning spent electrolyte and plant spillages to adjust feed density. The objectives of the Secondary Leach were to leach all unleached nickel, copper and iron sulphides and, by careful control of exit acid concentrations, precipitate most of the solubilized iron as hematite or jarosite. This final residue stream would act as the main exit point for iron entering the BMR. However, the product stream also contained un-leached copper sulphides, PGMs not recovered in the MCP and various amphoteric elements such as arsenic, selenium and tellurium. As such, it was dried and exported as a concentrate for further toll refining at a suitable processing facility. Since this concentrate represented a value stream which attracted costs for shipping and toll refining, minimising the bulk mass by producing predominantly hematite was favourable. In addition, hematite was known to be a superior collector of amphoteric elements deleterious to copper electrowinning. However operation of this section, especially with respect to controlling the iron precipitate product, proved difficult to optimize (Hofirek & Kerfoot, 1992).

BMR EXPANSION - PROCESS DEVELOPMENT

In early 2000, Anglo American decided to investigate the expansion of the BMR to an equivalent refined production of 33 ktpa Ni. Due to continued increasing economic pressure associated with capital expenditure the following principles were adopted for the expansion:

- improve overall process safety;
- maximize utilisation of existing BMR assets;
- improve nickel leach circuit control and efficiency;
- reject iron from the leaching circuit in a separate residue in order to reduce costs associated with treatment of the final leach discharge solid stream.

A significant drive underpinning the expansion was the separation of the rejected iron residue from the final copper pressure leach concentrate. The primary reason for this change was an increase in iron content of NCM from typically 1.5% in 1992 to 2.3% in 2006 (Hofirek & Kerfoot 1992, Bryson et al, 2008). A consequence of this for the original circuit configuration is highlighted in Table 2, which shows the total amount of iron residue generated (excluding all unleached copper minerals) based on the NCM iron content and nickel refining capacity on the pre-expansion circuit. Producing over 4 ktpa of residue was considered to be prohibitively costly, and a new strategy for managing iron in the BMR was sought.

Table 2 - Jarosite residue rejection as a function of Fe content of NCM

Ni Refining Capacity (ktpa)	NCM Fe Content (%)	Jarosite (ktpa at 34% Fe)
20	1.5	1.9
20	2.3	2.8
33	2.3	4.4

During process development, two sets of test work were conducted to evaluate circuit configurations at the Anglo American Technical Solutions (then AARL, Anglo American Research Laboratories) in 2005 and later at Sherritt Technologies (then Dynatec) in 2006. Two possibilities existed for the nickel circuit which are presented schematically in Figure 2. Option A involved an oxidising atmospheric leach (NAL) to release the oxidising duty of the original Primary Leach so the autoclaves could be fully utilised for non-oxidising leaching duty (NOX). In this option, the conditions in both leaching stages were to remain acidic enough to maintain iron in solution and allow for a dedicated iron removal stage. Alternatively, option B involved maintaining the existing Primary Leach operation, but control conditions at higher acidities (greater than 10 g/L H_2SO_4) in order to facilitate iron removal by a dedicated stage. For both options, pressure iron removal (PIR) via a medium temperature, oxidative precipitation route was chosen based on similar refining flow sheets (Plasket & Dunn, 1996).

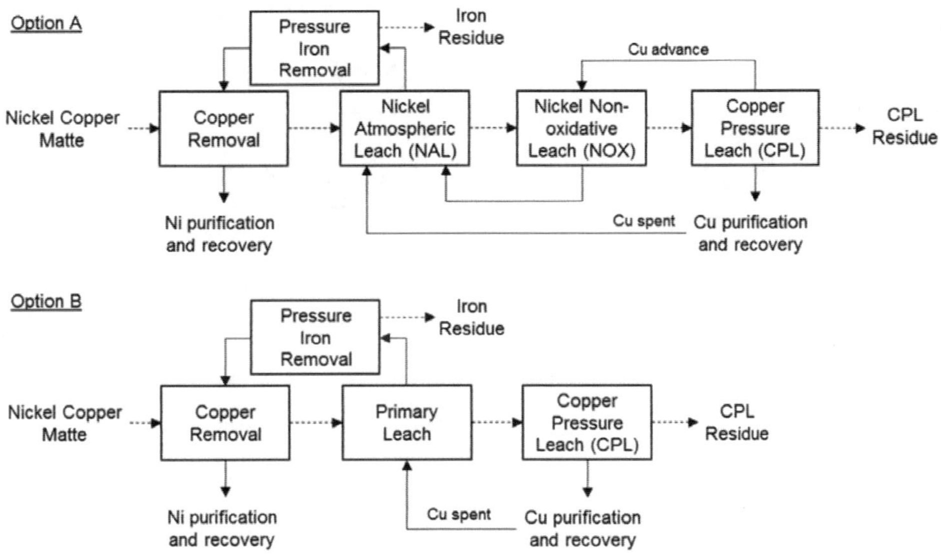

Figure 2 - Simplified block flow diagrams of the two nickel leach circuit considered.

The two leach circuit options were operated under continuous pilot conditions at AARL in 2005. Figure 3 summarises the pertinent results. Option A resulted in a nearly 10-fold reduction in iron content over NCM. These data show a clear distinction between the leach circuit options, highlighting that only the aggressive, non-oxidising NOX conditions, with large amounts of soluble copper, facilitate extensive iron dissolution. Incomplete iron extraction in the new nickel leaching circuit would be problematic as high levels of iron in the Primary or NOX leach residue would be (at least partially) extracted in the copper pressure leach and present difficulties for copper electrowinning (Das & Gopala Krishna, 1996). Moreover, any unleached iron would bulk up the copper residue, with obvious disadvantages.

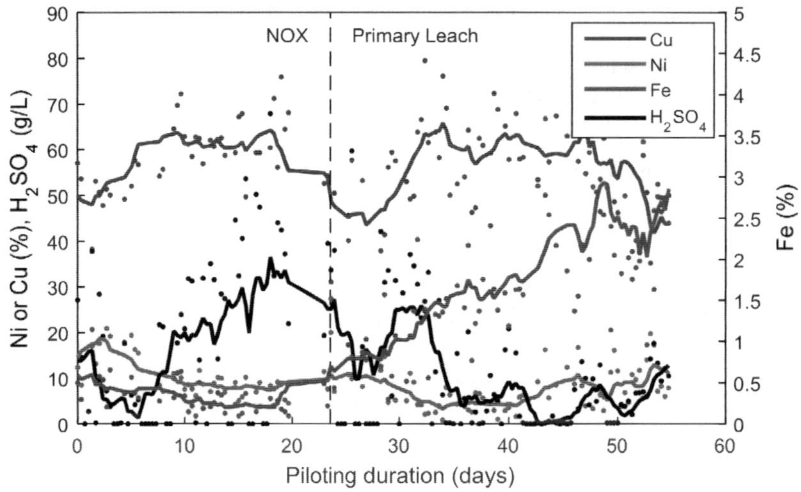

Figure 3 - Solids analyses for Cu, Ni and Fe during piloting of NOX and Primary Leach

On this basis, the preferred circuit for the expanded BMR was Option A, consisting of the atmospheric leach, the NOX pressure leach and the PIR iron rejection configuration. This process, in conjunction with the rest of the leaching circuit, was chosen to pilot continuously at Sherritt Technologies (Bryson et al., 2008). The piloting campaign successfully produced a mass balance for the integrated circuit. The mineralogical analysis of all solid streams confirmed the mechanism underlying the enhanced leaching of iron in the NOX leach. Figure 4 shows the deportment of bornite through the various leaching stages in the Option A configuration and highlights that the NOX leach targets bornite leaching. By comparison, during the piloting campaign the Primary Leach at high acid concentrations produced iron contents in the residue that were similar to the iron content of NCM, suggesting that little or no bornite is leached under these milder conditions.

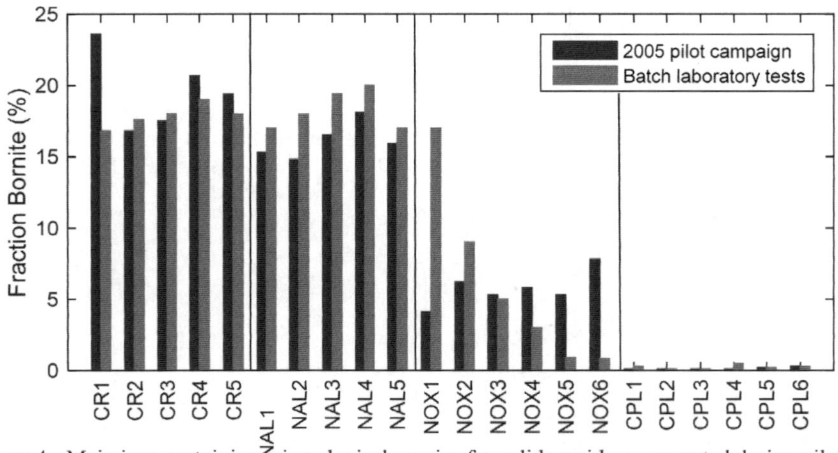

Figure 4 - Main iron containing mineralogical species for solids residues generated during pilot plant campaign (Bryson et al., 2008) and subsequent batch laboratory tests on plant residues (Summerton et al., 2012).

Since NOX conditions are ideal for metathesis, the mineralogical data in Figure 4 confirm that the most effective mechanism of bornite leaching is via the overall stoichiometry presented in Reaction 1.

$$Cu_5FeS_4 + CuSO_4 \rightarrow 2Cu_2S + 2CuS + FeSO_4 \quad (1)$$

Since poor iron leaching was noted during the Primary Leach piloting, which corroborates evidence that minimal bornite leaching occurred in the full scale operational Primary Leach (Hofirek, 1989), it is suggested that little or no metathesis actually occurred in the non-oxidising section. In line with this, anecdotal evidence exists, showing that the Primary Leach was often operated in the acid deficient state to promote the formation of basic sulphates of iron and, especially, copper in the primary leach residue to prevent soluble iron and copper breakthrough to the nickel electrowinning circuit (Hofirek, 2003). This suggested that, in the original circuit, the primary separation mechanism between nickel and copper or iron was via hydrolysis and not metathesis.

BMR EXPANSION – IMPLEMENTATION

Over the period from 2007 to 2010, the original 1981 BMR circuit was reconfigured and modified and commissioning of equipment not central to the fundamental chemistry change began during 2010. Complete commissioning of NAL, NOX and PIR was achieved towards the end of 2011 which materialised the complete leach circuit expansion. To-date, the BMR is still in an on-going ramp-up phase with the name-plate capacity still in excess of current arisings. Detailed accounts of the expanded circuit's leaching chemistry have been previously presented based on piloting results (Bryson et al., 2008) and laboratory leach tests on plant material (Summerton et al., 2012) and only the key factors pertinent to iron management and control are discussed here.

Feed Mineralogy Changes

The realisation that significant changes to feed material occurred between the time of the expansion circuit development and piloting through to installation and commissioning, as highlighted in Table 3, is important to the discussion of the current circuit chemistry and the continued ramp-up of the BMR. The increased iron content in NCM, as compared to the 1992 values presented in Table 1, is primarily a function of a change in converting practice in 2003, in which Pierce Smith Converters operated in batch mode were changed to Top Submerged Lance semi-continuous technology. This conversion was done primarily to reduce environmental emissions. The second significant change is the decrease in the Ni:Cu ratio in NCM, which can be attributed to changes in the upstream mining operations and the partial recycle of copper residues to upstream pyrometallurgical processes.

Table 3 - Major element content of NCM

Element	2003-2006		2011-2016	
	Typical (%)	Range (%)	Typical (%)	Range (%)
Ni	43.5	40 – 44	40.0	33 – 43
Cu	27.6	26 – 30	31.9	24 – 35
Fe	2.2	1.0 – 3.5	2.5	1.5 – 3.9
S	26.4	22 – 29	25.4	22 – 29

The significance of these feed material changes for RBMR is the substantial increase in units of copper and iron per unit nickel from the design of the original circuit. The impact of these differences, particularly in the development of the new circuit, is discussed in the following paragraphs.

Behaviour of Iron

While the analyses for iron in Primary Leach solids were typically not performed, this was commenced just prior to the chemistry changeover and is on-going. As was found during piloting, the iron content of residues dropped significantly in the changeover from Primary Leach to NOX. These results are shown in Figure 5. Importantly, even though it was never the intention of the Primary Leach to extract iron, this marked change in the solids composition is a result of an increased degree of metathesis rather than the lack of hydrolysis at the higher acidity.

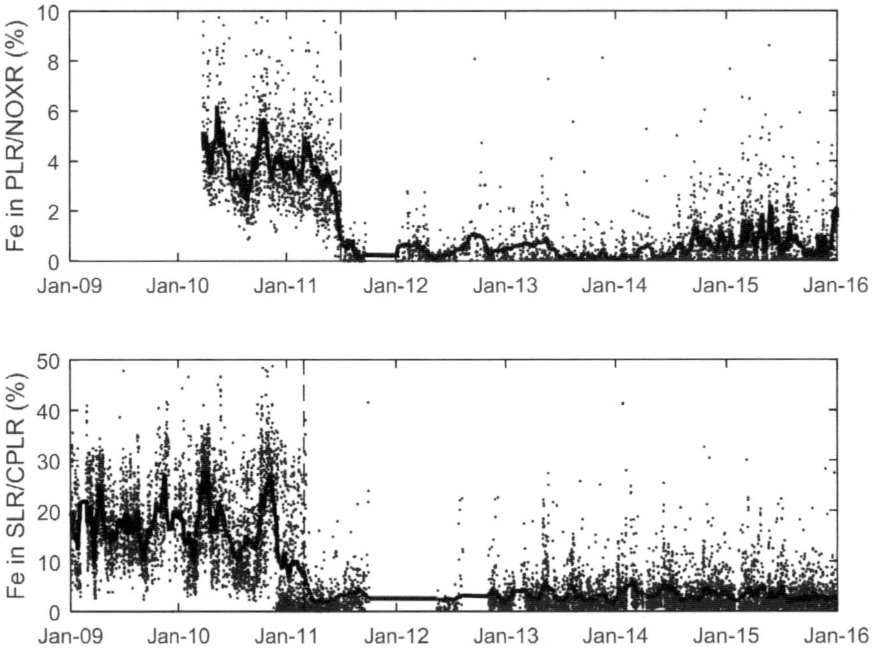

Figure 5 - Iron content of Primary (PLR) and Secondary (SLR) leach residues as well as the NOX and CPL residues after the stage-wise commissioning of the new leaching units. PLR – Primary Leach Residue, SLR – Secondary Leach Residue, NOXR – NOX residue, CPLR – copper pressure leach residue.

Concomitant with the enhanced leaching of iron in the NOX autoclaves, less iron reported to the copper leaching circuit, where commissioning of the CPL autoclaves began a few months prior to NOX. Besides the poor leach efficiencies during the initial CPL commissioning, which depressed the iron content of CPL residue, the clear distinction in operating regimes is apparent from Figure 5. Releasing the requirement of iron precipitation in the Secondary Leach facilitated the operation of CPL at higher acidities. This in turn resulted in significantly decreased mass of residue from CPL, requiring less treatment in the subsequent solid-liquid separation stages. Moreover, ensuring that iron remained largely in the nickel leaching circuit, which includes PIR and Copper Removal to prevent iron breakthrough into the nickel electrowinning circuit, increased the robustness of the overall circuit, particularly with respect to iron breakthrough into copper electrowinning.

Finally, Figure 6 shows the amount of iron removed over the PIR section as the difference in NAL and PIR filtrate solution concentrations. The ability to absorb reasonably large variances in iron and acid concentrations by the adjustment of caustic addition into the PIR autoclave adds significant robustness to the process, and facilitates the optimisation of leaching conditions without the iron removal constraints imposed by the original circuit chemistry.

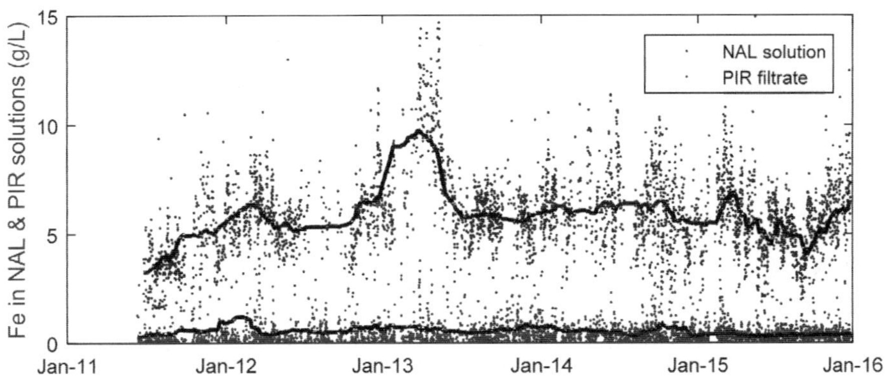

Figure 6 - Iron concentrations in NAL solution and PIR filtrate, highlighting the efficiency of PIR iron removal

Base Metal Leaching

In addition to the new circuit's ability to manage iron more effectively, no less remarkable is its ability to treat significantly lower Ni:Cu ratios in the feed NCM compared to the original circuit. Since both circuits rely on matte leach chemistry for the selective leaching and separation of nickel and copper, an increase in the relative amount of copper versus nickel has a significant effect on the overall balance of the circuit. The increase in relative copper inputs into the circuit (shown in Table 3) ultimately results in increased amounts of acid generated via copper electrowinning. In the original configuration, this would have resulted in a net increase in the sulphuric acid input into the Primary Leach. Since the nickel minerals, i.e., heazlewoodite and millerite, constitute the primary acid sink in the process, acid fed in excess of the neutralising capacity of these minerals would result in a lower terminal pH and the subsequent breakthrough of iron and copper into the nickel electrowinning circuit.

The NAL operation, while operating at atmospheric conditions, has a large capacity to simultaneously consume the majority of acid from copper spent electrolyte in the oxidative leaching of nickel minerals as well as consume, almost entirely, the associated soluble copper by metathesis. Moreover, the high temperature, non-oxidizing conditions and high soluble copper feed (copper advance electrolyte) in the NOX leach, permit the production of a residue containing significantly lower Ni:Cu ratios than the original circuit (Table 4), thereby acting as the key nickel-copper separation stage. This is a direct result of the significant degree of metathesis taking place in the current circuit, which is emphasised in Table 4 by the 170% increase in soluble copper per unit feed being recycled to the nickel leaching circuit as compared to the original circuit. Further confirmation of the degree of metathesis is that, in contrast to the original circuit, the NOX residue contains essentially no basic copper sulphates, but comprises almost entirely of metathesized copper sulphides and some unleached nickel bearing minerals, facilitating a lower Ni:Cu ratio (see Table 4).

The separation of nickel and copper by metathesis as opposed to hydrolysis probably represents the biggest fundamental difference between the old and the new circuits. This has resulted in both improved nickel-copper separation efficiency as well as overall robustness of the circuit. These benefits only materialised as a result of the specific design intent to separate iron removal from the leach stages.

Table 4 – Summary of typical copper and sulphur reagent recycles to the nickel leaching circuit

Typical concentrations		
Copper Spent [H_2SO_4]	100 g/L	
Copper Spent [Cu]	40 g/L	
Copper Advance [H_2SO_4]	15 g/L	
Copper Advance [Cu]	100 g/L	
Circuit parameters	**2002-2003**	**2011-2015**
Recycle fraction as copper spent	100%	15%
Recycle fraction as copper advance	0%	85%
Overall Cu:S in recycle (m/m)	0.76	1.49
Overall Cu in recycle (kg soluble Cu/ton NCM)	140	239
Cu:Ni in Ni leach residue	5:1	7:1

CONCLUSIONS

Iron management and removal in the development of the current BMR circuit has played a central role in the selection of a fundamentally different mechanism of base metal separation and, ultimately, the current flow sheet configuration. The BMR expansion project was undertaken through the austere constraints of maximising the utilisation of existing equipment, minimising capital expenditure and continued production through all stages of development and installation. Nevertheless, the expanded circuit has produced a significantly more robust configuration which essentially stemmed from the design intent to separate iron removal from the leach stages. In addition to the benefits associated with more optimal residue handling, the recent all-time production record with feed material that is significantly different from the original design basis is testament to the inherent flexibility facilitated by the new chemistry. The addition of the NAL and PIR unit operations, while increasing the size and complexity of the overall refinery, has separated key mechanisms of the leach chemistry into different reactors, which has opened up the possibility of significantly improving the performance and efficiency of the various process stages.

As the BMR undergoes continued ramp-up to name-plate capacity, with forecast arisings scheduled to progressively increase into the foreseeable future, the decoupling of iron removal from the various leaching stages has opened up significant opportunities to optimise the various unit operations and their integration with one another.

ACKNOWLEDGEMENTS

The authors wish to acknowledge Anglo American Platinum and Anglo American for permission to publish this work.

This paper is dedicated to the RBMR operation's men and women who have been tasked to deliver an extremely difficult project: from construction through to commissioning and ramp-up on an already operating plant – somewhat like performing open heart surgery while running a marathon.

REFERENCES

Bryson, L.J, Hofirek, Z., Collins, M.J., Berezowsky, R.M.G.S, & Stiksma, J. (2008). New Matte Leaching Developments at Anglo Platinum's Base Metal Refinery. In C.A. Young, P.R. Taylor, C.G. Anderson, & Y. Choi (Eds.), *Hydrometallurgy 2008: Proceedings of the Sixth International Symposium* (pp. 570-579). Phoenix, AZ.

Das, Z. & Gopala Krishna, P. (1996). Effect of Fe (III) during copper electrowinning at higher current densities. *International Journal of Mineral Processing 46(1-2)*, 91-105.

Hofirek, Z. (1989). *Internal Report: Possible causes of poor leaching at the BMR and MC Plant During Period August/September 1988*. Unpublished Manuscript, Technical Development Department, Rustenburg Base Metal Refiners, Rustenburg.

Hofirek, Z & Kerfoot, D.G.E. (1992). The Chemistry of the nickel-copper matte leach and it application to process control and optimisation. *Hydrometallurgy 29(1-3)*, 357-381.

Hofirek, Z. (1996). Iron control in the processing of nickel-copper matte at Rustenburg Base Metal Refiners. In J.E. Dutrizac & G.B. Harris (Eds.), *Second International Symposium On Iron Control In Hydrometallurgy* (pp. 357-381), Montreal, Quebec, Canadian Institute of Mining, Metallurgy and Petroleum .

Hofirek, Z. (2003). *Internal Report: Copper control at BMR.* Unpublished Manuscript, Technical Development Department, Rustenburg Base Metal Refiners, Rustenburg.

Plasket, R.P, & Dunn, G.M. (1996). Iron rejection and impurity removal from nickel leach liquor at Impala Platinum Limited. In J.E. Dutrizac & A.J Monhemius (Eds.), *International Symposium On Iron Control In Hydrometallurgy* (pp. 695-718), Toronto, The Metallurgical Society of CIM, Ellis Horwood Limited.

Summerton, G.C, Craig, D.C, Dinham, P., McCulloch, N. & Dowling, S. (2012). The chemistry and mineralogy of a nickel copper matte leach. In M.J. Collins, D. Filippou, J.R. Harlamovs & E. Peek, (Eds.), *51st Conference of Metallurgists, COM2012, Held in Conjunction with Pressure Hydrometallurgy 2012* (pp.433-456), Niagara Falls.

THE PRESSURE IRON REMOVAL SECTION AT RUSTENBURG BASE METALS REFINERS

*C.A. Biley

Anglo American Platinum
Rustenburg Base Metals Refiners
Rustenburg, South Africa, 0300
*(*Corresponding author: chris.biley@angloamerican.com)*

ABSTRACT

A dedicated iron removal stage was installed at Anglo American Platinum's Rustenburg Base Metals Refiners (RBMR) during an expansion project which began execution in 2010. The Pressure Iron Removal (PIR) process aims to precipitate iron from a nickel leach solution under medium temperature (145 °C) oxidative conditions using caustic for neutralisation. Since commissioning, the process has been robust in removing iron from the refinery but has shown significant variability in the residues produced. These residues, which are typically comprised of ferrihydrite, hematite or natrojarosite, have varying iron contents and different physical properties which have implications for their filtration and ultimate removal from the refinery. Since over 95% of nickel plated at RBMR passes through the PIR section, volumetric constraints caused by poor filtration performance directly influence nickel production and stabilisation of this process remains central to the continued ramp-up of the refinery to full nameplate capacity. This paper details the development of the PIR process and the journey from concept to implementation. Recent laboratory and plant scale tests have highlighted the importance of the initial phase formed in the autoclave for the ultimate residue requiring filtration and how the feed acid concentration is a critical control parameter. On this basis, a new caustic dosing philosophy was implemented which improved pH control in the autoclave. Further optimisation of the PIR process is on-going.

KEYWORDS

Iron precipitation, base metals, hematite, RBMR, refinery, autoclave, nickel

INTRODUCTION

In 2010, an expansion of Rustenburg Base Metal Refiners (RBMR) to increase nickel production from 21 to 33 ktpa began execution. The management of iron was pivotal in the ultimate configuration in the new circuit and one of the primary differences from the original circuit was the addition of a dedicated iron removal stage: Pressure Iron Removal (PIR). A significant drive for this new unit operation was to facilitate the separation of iron from other leach residues. Not only did this result in a significant decrease in the amount of leach residue but, without the constraints associated with iron precipitation during leaching, has permitted more aggressive leaching conditions and, correspondingly, greater throughput and efficiencies.

However, as with many commercial implementations of even mature technologies, the journey from development to implementation has presented challenges for the continued ramp-up of the refinery to full nameplate capacity. Specifically, since all nickel bearing leach solution must pass through the PIR section prior to the nickel purification and recovery circuit, throughput difficulties in the PIR section can directly impact supply of feed to the nickel tank house. Since commissioning, the PIR autoclave has effectively removed iron from the refinery, but has shown significant variation in the solids phases produced in the autoclave. This variation presents difficulties for the filtration of the discharge slurry and represents a key throughput limitation which periodically impacts production. Additionally, the increasing production rates associated with the ramp-up of the refinery have resulted in a significant increase in the required volumetric throughput through the PIR process. This has placed even greater emphasis on improving the stability of the solid phase produced in the autoclave, ultimately targeting an easily filterable residue with high iron content.

This paper presents some recent laboratory and plant scale testwork focussed on understanding the key factors that affect the solid phase variability produced in the PIR autoclave. Findings from this work have initiated a number of on-going optimisation projects around the PIR section that aim to improve the robustness of the process and reduce instances of filtration difficulties.

BACKGROUND

The PIR Process

A simplified block flow diagram of the Base Metal Refinery (BMR) circuit is presented in Figure 1. The PIR section receives solution from the nickel leaching sections that has a typical composition range of 100-120 g/L Ni, 0-5 g/L Cu, 3-9 g/L Fe and 2-15 g/L H_2SO_4 and aims to remove iron from this solution by oxidative precipitation at elevated temperature (145 °C). The iron residue precipitated in the PIR autoclave is filtered from solution and is reverted back to the Anglo Platinum Waterval Smelter where the bulk of the iron contained in the mined ores is rejected from the overall processing pipeline as an environmentally stable slag.

The PIR section consists of a 120 m^3 alloy PIR autoclave which is fed through a set of heat exchangers and discharges into a discharge tank that, in turn, feeds two plate-and-frame filter presses. After filtration of the residue, the PIR filtrate proceeds to Copper Removal, where it is contacted with the plant feed, Nickel Copper Matte (NCM), and oxygen. This causes the pH to rise substantially and iron and copper are precipitated as basic sulfates. These solids are recycled to the nickel leaching circuit while the nickel solution proceeds to the nickel electrowinning circuit for further purification and recovery. With Copper Removal providing an additional barrier to prevent iron breakthrough, the circuit is relatively tolerant to the incomplete removal of iron in the PIR section. However, since over 95% of the nickel plated at RBMR proceeds through the PIR autoclave and filter presses, throughput limitations have a greater detrimental effect to production and can occur more frequently at higher throughputs. The generation of an easily filterable precipitate within the PIR autoclave is thus critical to achieving the required volumetric throughput through the PIR section to balance the circuit.

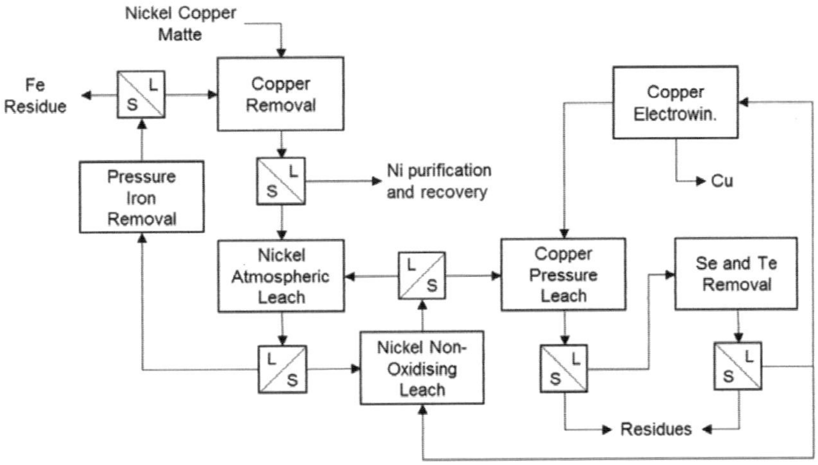

Figure 1 – Simplified block flow diagram of the RBMR leaching circuits

The key reactions involved during iron precipitation in the PIR section are detailed below. In the PIR feed tank, sodium sulfite solution is added to reduce all iron(III) to iron(II) via Reaction 1. This ensures that scaling of the feed heat exchangers by iron precipitation on the hot tube walls is minimised and also ensures that the rate of iron(III) precipitation in the autoclave is controlled (primarily) by the rate of iron(II) oxidation (Reaction 2). The feed solution is pumped into the first compartment of the autoclave and oxygen is supplied into the headspace of the last compartment. Concentrated caustic solution (50 wt.%) is added directly into the autoclave to neutralise acid (Reaction 3) and generate the conditions necessary for the precipitation of iron to form (generally) ferrihydrite, hematite or natrojarosite (Reactions 4 – 6). The target for the PIR section is to maintain solution iron concentration below 500 mg/L which can be easily accommodated in the Copper Removal reactors. Steam is added directly into the autoclave in each compartment to maintain the temperature throughout the autoclave and regular sampling of the fourth compartment for soluble iron, pH and solids is used for process monitoring and control.

$$Fe_2(SO_4)_3 + Na_2SO_3 + H_2O \rightarrow 2FeSO_4 + Na_2SO_4 + H_2SO_4 \quad (1)$$

$$2FeSO_4 + 0.5O_2 + H_2SO_4 \rightarrow Fe_2(SO_4)_3 + H_2O \quad (2)$$

$$H_2SO_4 + 2NaOH \rightarrow Na_2SO_4 + 2H_2O \quad (3)$$

$$9Fe_2(SO_4)_3 + 32H_2O \rightarrow 9Fe_2O_3{\cdot}5H_2O + 27H_2SO_4 \quad (4)$$

$$Fe_2(SO_4)_3 + 3H_2O \rightarrow Fe_2O_3 + 3H_2SO_4 \quad (5)$$

$$3Fe_2(SO_4)_3 + Na_2SO_4 + 12H_2O \rightarrow 2NaFe_3(SO_4)_2(OH)_6 + 6H_2SO_4 \quad (6)$$

The design intent for the PIR section was to target hematite formation in the PIR autoclave, primarily to minimise the mass of solids requiring filtration. However, the commercial implementation of the PIR process, while effective at removing iron from solution, produces a variety of solid phases depending on the conditions in the autoclave.

Iron(III) Precipitation Chemistry

Iron(III) solution chemistry becomes increasingly complex with increasing temperature and pH, where numerous iron(III) hydroxide species exist in solution and ultimately form the building blocks of solid precipitates (Flynn, 1984). Thermodynamic effects dictate the solution phase solubility of iron(III) which is known to rapidly decrease to low levels with increasing pH, i.e., less than 0.06 mg/L at pH 4, 25 °C (Loan et al., 2006). The supersaturation ratio, which is a thermodynamic property, largely dictates the kinetics of precipitation as well as the crystallisation regime; nucleation or growth (Steyl, 2012). Iron(III) precipitation is particularly complex due to the low equilibrium iron(III) solubility at comparably lower pH than other transition metals which is conducive to large supersaturation ratios. Such conditions can result in the nucleation of fine, gelatinous iron precipitates that can be difficult to separate from solution (Flynn, 1984). Thus, in hydrometallurgical operations, it is common practice to control the iron(III) concentrations to low levels by the rate of iron(II) oxidation or by leaching reactions (Ismael & Carvalho, 2003) and to utilise slowly reacting neutralising agents to avoid localised regions of high pH. By careful control of the supersaturation ratio, growth can be favoured over nucleation and larger particles can be produced that can be readily removed from solution.

It is generally accepted that the first solid phase formed during iron precipitation is metastable iron oxyhydroxides (Loan et al., 2002). Evidence has highlighted that iron precipitates commonly referred to as 'amorphous' actually consist of phases with poor long range structure and very fine primary particle sizes of the order of 3-7 nm such as ferrihydrite (approximated as $5Fe_2O_3.9H_2O$) and schwertmannite ($Fe_{16}O_{16}(SO_4)_2(OH)_{12}$) (Jambor & Dutrizac, 1998; Loan et al., 2006). These primary precipitates then transform, probably by a dissolution/recrystallization mechanism, into the stable phases expected under the prevailing conditions. A highly simplified equilibrium stability diagram highlighting the major stability regions of solids in the Fe(III)-S-O system is presented in Figure 2 and should only be used to interpret general trends. At the temperature in the PIR autoclave (145 °C), Figure 2 suggests that the expected phases are hematite or natrojarosite (due to the background of sodium in the process) depending on the pH conditions. However, as will be discussed later, well-structured hematite is rarely produced in the commercial process and ferrihydrite (used in this paper to represent a wide number of potential iron oxyhydroxides), is more common.

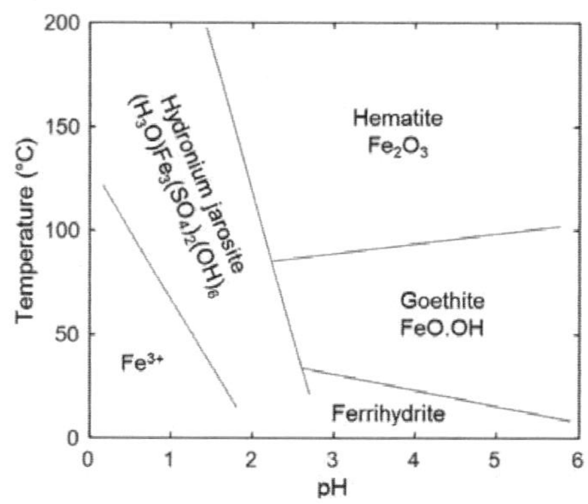

Figure 2 – Simplified stability field diagram of the Fe(III)-S-O system, adapted from Babcan (1971)

DEVELOPMENT OF THE PIR PROCESS

The development of the PIR process during the BMR expansion project was largely based on similar implementations at other commercial refineries (Plasket & Dunn, 1996). The primary objectives were to: (i) oxidise and precipitate the iron from the nickel atmospheric leach solution; and (ii) recover the iron as a residue cake for separate treatment or disposal. The prevailing economic conditions during the expansion project necessitated a strict focus on the minimisation of capital expenditure and thus the maximisation of the use of all existing equipment was imperative.

Although not shown in Figure 1, but discussed in detail elsewhere (Hagemann & Pelser, 2015), the removal of sulfur from the BMR circuit is achieved by neutralising nickel electrowinning spent electrolyte with caustic to produce sodium sulfate as a saleable by-product. Given that the expanded circuit chemistry would make no changes to the method of sulfur removal and all infrastructure for the handling of caustic was already present, the ambitious task of performing iron precipitation with caustic was also specified in the design. Given the complexity involved in removing iron as a relatively minor component (typically less than 3% by mass) from a hydrometallurgical stream using a highly reactive reagent such as caustic, no references could be found in the open literature for similar commercial processes. By comparison, all other nickel refineries utilise more slowly reacting neutralising agents in their iron removal stages, such as NH_3 (Stillwater, Sherritt, Impala) and $NiCO_3$ (Sumitomo, Nikkelverk, Hartley) (Crundwell et al., 2011), or rely on the terminal leach conditions to generate the required conditions for iron removal (Schwarz et al., 1996). A large amount of research has been conducted in the removal of iron from zinc refineries, however this is typically done in the leaching stage where calcine acts as the primary neutralising agent (Ismael & Carvalho, 2003; Loan et al., 2006). Thus, while iron removal via precipitation is common to most, if not all, hydrometallurgical refineries, the PIR process appears to be unique in its use of concentrated caustic for neutralisation.

An integrated piloting campaign of the proposed expansion circuit was performed in 2006 at Sherritt Technologies (then Dynatec) and demonstrated that the PIR process could effectively remove iron from solution under the design conditions and in series with the other leaching operations. The pilot scale PIR process was operated in two, 6.9 litre pot autoclaves operating at 700–600 kPa and 148–151°C respectively and was pressurised with oxygen. Solids analyses of the generated PIR residue highlighted that the primary phase produced during the campaign was probably ferrihydrite or hematite (or a combination), with median iron, sulfur and sodium contents of 52 wt.%, 2.5 wt.% and 0.014 wt.% respectively (Bryson et al., 2008). The solution pH was well controlled between pH 2 and 3 and the acid concentration of the feed, solution reporting from the Nickel Atmospheric Leach (NAL), was relatively stable with a standard deviation of 3.6 g/L. Since the operation of this process proceeded with minimal challenges, detailed mineralogy was unfortunately not carried out on the PIR residues.

The success of the piloting campaign, and the maturity of iron removal technology, provided sufficient confidence in the PIR process for the complete design and manufacture of a new commercial scale autoclave. The PIR autoclave and its ancillary equipment were installed in the BMR during 2010 and commissioning began in early 2011.

PLANT PERFORMANCE AND CHARACTERISATION

Iron Removal Performance

Since commissioning, the PIR autoclave has proven to be highly effective in removing iron from the refinery, highlighted in Figure 3. This is not unexpected since pH and last compartment temperature are typically maintained above 1.5 and 145 °C respectively, where iron(III) is generally unstable in solution. However, also shown in Figure 3, the iron content of the residues produced in the PIR autoclave has a significant variance and typically ranges over 25–55 wt.% iron. Interestingly, the iron content distribution shows bimodality, with components centred at approximately 32 wt.% and 45 wt.%. Typically, higher iron contents are associated with a darker brown–red precipitate which has been confirmed to consist largely of iron oxyhydroxides such as ferrihydrite and can further develop into crystalline hematite. Lower iron contents are attributed to either a more yellow–orange precipitate that contains an abundance of natrojarosite or, in cases of high pH excursions, less structured and often sticky and gelatinous precipitates which are likely also ferrihydrite. Similarly to the PIR piloting campaign, solids produced in higher pH excursions also typically contain higher incorporations of nickel and copper, probably due to their increased surface area and high degree of hydration. Such solids are very problematic for filtration.

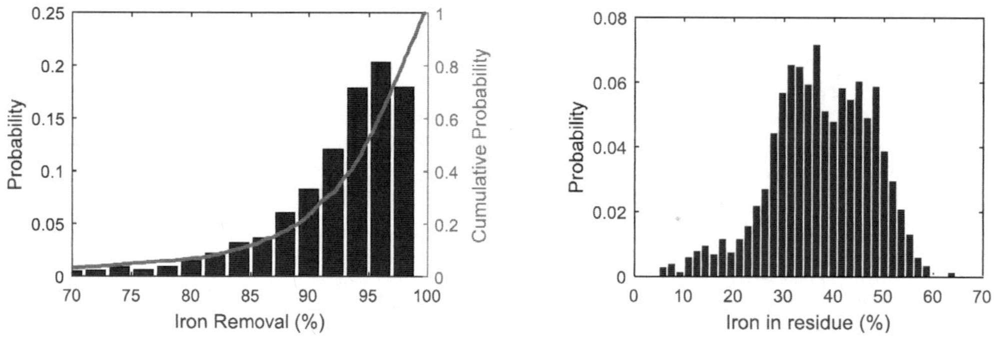

Figure 3 – Probability plots of the instantaneous extent of iron removal over the PIR autoclave based on solution analyses and the corresponding residue iron content over the period 28/06/2011 – 30/12/2015

The fractional removal of iron over the PIR autoclave compartments is not routinely monitored, but several profile samples have highlighted that approximately 50% of the iron is precipitated in the first compartment. This has been observed to be quite constant for a range of first compartment temperatures and feed acid concentrations and suggests, as per the design intent, the controlling rate of iron removal extent is the mass transfer of oxygen into solution.

A major source of variance for the PIR operation is the solution composition reporting from the Nickel Atmospheric Leach (NAL), particularly with respect to acid concentration. The NAL solution acidity is dependent on the leach performance of both the nickel leaching stages as well as the copper leaching and recovery circuit. Nickel Non-oxidative Leach (NOX) solution typically contains 15-30 g/L H_2SO_4 which is generated by some of the acid producing metathesis reactions (Bryson et al., 2008) and is supplied to NAL along with some copper electrowinning spent electrolyte. The oxidative leaching of heazlewoodite (Ni_3S_2) and djurleite ($Cu_{1.96}S$) in NAL constitutes the primary consumption of this supplied acid. Thus, variations in the operation of NAL, NOX or the copper circuit can result in significant variations in the terminal acidity of NAL solution and correspondingly the neutralisation requirements of the PIR section. This presents a key challenge for the accurate control of the pH in the PIR autoclave.

Laboratory Tests

In order to develop a greater understanding into the primary factors affecting the PIR process, a batch laboratory study into the kinetics of iron precipitation under PIR conditions was conducted at Anglo American Technical Solutions. All tests were conducted in a 2 L batch autoclave and all reagents were added up front. All iron was added as analytical grade ferrous sulfate and the autoclave was heated to 150 °C in a nitrogen atmosphere, maintaining the pressure at 600 kPa(g). Tests commenced with the addition of oxygen into the headspace and, after the desired residence time, solids were filtered from the autoclave at temperature. Phase abundances were determined using the chemical analyses which agreed well with several samples which were analysed by detailed mineralogy.

Two significant findings were highlighted from these tests. Firstly, there is a sharp cross-over region between the dominance of natrojarosite or ferrihydrite which was most significantly affected by the initial acid concentration. This is highlighted in Figure 4 which, even for tests conducted at different iron concentrations and temperatures, shows that the most influential variable in the distribution of natrojarosite and hematite is the initial acid concentration. Critically, at lower initial acidities, the acid produced by iron precipitation was not sufficient to stabilise natrojarosite and iron oxyhydroxides were essentially the only phases in the residue. This suggests that, once formed, the conversion from ferrihydrite to natrojarosite does not readily occur under the acidities expected in the PIR autoclave.

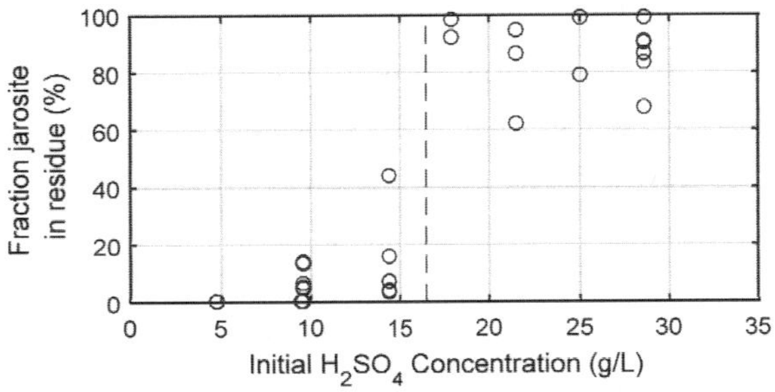

Figure 4 - Summary of batch laboratory tests highlighting the effect of initial acidity on the solid phase produced. Conditions: [Fe] = 6-10 g/L, temp. 120-150 °C, pressure 600 kPa(g), residence time 5-60 min

Secondly, a series of tests was conducted to determine if natrojarosite, once produced, can be converted to ferrihydrite by neutralisation. An initial solution containing 28.6 g/L H_2SO_4 was used to produce essentially pure natrojarosite and, after 20 minutes of reaction, concentrated caustic was added directly into the batch autoclave. After 60 and 120 minutes of total residence time, the residue consisted approximately of 30% and 16% natrojarosite, highlighting how the conversion of natrojarosite to ferrihydrite can be readily achieved by decreasing the solution acidity and allowing sufficient residence time; final acidities in these tests were 5.1 and 8.3 g/L H_2SO_4, respectively.

Effect of Caustic Distribution

Using the laboratory results as a basis, the distribution of caustic into the PIR autoclave compartments was varied and a series of profile samples were taken. The objective of this test was to highlight, on commercial scale, the influence of varying the initial precipitation conditions with minimal risk to production. Samples were analysed for chemical composition by Inductively Coupled Plasma Optical Emission Spectroscopy (ICP-OES) and for mineralogy by Field Emission Gun Scanning Electron Microscopy (FEG-SEM) and powder X-Ray Diffractometry (XRD). The relative abundances of the identified phases were determined using the semi-quantitative Rietveld analysis technique and since XRD cannot detect poorly crystalline material accurately, the amount of this material was estimated by using an internal Si standard (Dinham & Tiddy, 2016).

Sample set A was taken during the dosing of caustic into the first three autoclave compartments, with a relatively low feed pH and generally stable operation. The FEG-SEM images of the precipitates performed under these conditions are shown in Figure 5. The first compartment residue possessed the characteristic platelet structure of natrojarosite which dominates the particle morphology and is confirmed by the XRD results in Table 1, which shows that the residue consisted of approximately 60% natrojarosite. The remaining fraction, also visible in the crevices of the platelets, are finer agglomerates with no apparent structure and are likely ferrihydrite; Energy Dispersive X-ray Spectrometry highlighted such phases to have an iron content typically above 50%. As the material progresses through the autoclave and iron is further removed from the solution, the abundance of natrojarosite decreases while ferrihydrite increases. Additionally, low levels of crystalline hematite (6-8 wt.%) were detected in the last three compartments. The image of the last compartment solids (Figure 5, right) shows significantly larger abundance of less structured ferrihydrite on a sub-structure that still resembles the initial natrojarosite structure.

Table 1 – Solution profiles during plant sampling

	A – NaOH dosing in C#1,2,3 (29/06/2015)					B – NaOH dosing in C#2-3 (25/11/2015)				
	Fe (g/L)	H_2SO_4 (g/L)	C#4 pH	Fe Removal (%)	Fraction Natro-jarosite (%)	Fe (g/L)	H_2SO_4 (g/L)	C#4 pH	Fe Removal (%)	Fraction Natro-jarosite (%)
Feed	6.70	12.7		-	-	7.88	13.9		-	-
C#1	3.37	-		49.7	62	3.66	19.6		53.6	94
C#2	0.49	-	1.73	92.6	40	0.96	9.5	1.63	87.9	88
C#3	0.26	-		96.1	41	0.27	3.6		96.6	85
C#4	0.19	-		97.2	39	0.16	4.1		98.0	78

Figure 5 – PIR Residue sample A (29/06/2015) from the first (left) and last (right) compartments during NaOH dosing into C#1-3. The first and fourth compartments consisted of approximately 62% and 39% natrojarosite by XRD respectively. Optical photos of residues shown as inserts

Figure 6 – PIR Residue sample B (25/11/2015) from the first (left) and last (right) compartments during NaOH dosing into C#2-3. The first and last compartments consisted of approximately 94% and 78% natrojarosite by XRD respectively. Optical photos of residues shown as inserts

On 6 October 2015, the caustic flow into the first compartment was stopped and only fed into compartments 2 and 3. Another profile sample set (Sample B) was taken some time later and the FEG-SEM images of the residues are shown in Figure 6, which were sampled under approximately similar process conditions. In the absence of caustic addition into compartment 1, the yellow precipitate consisted almost entirely of natrojarosite with only very minor ferrihydrite formation noted in the natrojarosite crevices. These are likely the remnants of the initial ferrihydrite produced in the autoclave which had not yet converted to natrojarosite. The last compartment residue (Figure 6, right) shows a greater abundance of the less structured, finer material on the surface of the well-developed natrojarosite substructure which is comparable to sample A, albeit to a lesser extent.

By combining the extents of iron removal with the measured abundance of natrojarosite in these samples, the additional iron removal in compartments 2-4 cannot account for the noted decrease in natrojarosite abundance. This provides evidence for the conversion of natrojarosite to ferrihydrite in the PIR autoclave. On this basis, and the crystalline hematite detected in sample A, it can be inferred that, if allowed sufficient residence time in the neutralised solution at temperature, conversion of the precipitates into hematite would proceed.

Confirmation of the differences noted in samples A and B is highlighted by the routine analyses of the PIR residue, which showed a step change in iron content to 33.5 wt.% (σ = 5.35%) from 40.4 wt.% (σ = 7.89%) during the period of operation where caustic was not added to the first compartment. This further highlights that the initial precipitate generated in the first PIR autoclave compartment is critical to the ultimate residue requiring filtration in the PIR presses. Not unexpectedly, more acidic conditions favour the formation of natrojarosite in the first compartment, even with (partial) neutralisation through the supply of caustic. While these samples provide clear evidence for the conversion of natrojarosite formed in the first compartment the rate of conversion appears to be insufficient to achieve complete conversion to hematite.

Key Implementation Lessons

As suggested above, there are considerable differences in the solid phases produced in the commercial PIR process as compared to the piloting campaign and the potential causes of these can be attributed to several key differences in operation.

Firstly, the caustic solution used for pH control in the pilot autoclave was at a concentration of 50-100 g/L as compared to the concentrated solution readily available in the refinery of 760 g/L (50 wt.%). It is well known that dilute reagents are beneficial to crystallisation processes as lower supersaturation ratios can typically be achieved, which would have contributed to the production of the stable iron residue achieved during the piloting campaign. However, the use of dilute caustic was a source of significant dilution over the pilot autoclave, even with flashing at the discharge, which could not be tolerated in the commercial circuit. Since the nickel purification circuit remained largely unchanged during the expansion project, increased flows due to dilution could not be managed and it was selected to utilise the available concentrated caustic in the commercial PIR installation.

Similarly, due to the elevated operating temperature of the PIR autoclave and the required minimisation of sources of dilution, feed heat exchangers were installed to heat the feed solution to 145 °C to reduce the injection of live steam into the autoclave. Significant difficulties with the heat exchangers were experienced during the initial stages of commissioning which were attributed to the incomplete reduction of iron(III) in the feed tank, as well as insufficient overpressure at the discharge of the heat exchanger which caused flashing and severe scaling. As a result, the PIR feed heat exchangers have operated sporadically since their initial commissioning and, during periods of inoperability, direct steam injection into the autoclave is the only source of heating. While this contributes to dilution, lower temperatures are also observed in the first compartment since the steam lines were not sized for the full heating duty. Consistent with Figure 2, the effect of temperature would contributed to the variation in solid phases produced and increased the impact of pH excursions. Given the evidence for the importance of the initial precipitate, lower temperatures have reduced the region of operation in which stable, easily filterable material with a high iron content are formed.

An additional critical point, relevant to both the use of concentrated caustic solution and the required direct steam addition, is the agitator system selected for the PIR autoclave. During the design of the process this selection was primarily based on the requirement of solid suspension and, since typical solids loading is in the range of 1-2 wt.%, requires only gentle agitation. The specific power input through agitation in the PIR autoclave is approximately 0.36 kW/m^3. This also represents a key difference between the pilot and commercial autoclaves, of which the former would have had approximately an order of magnitude higher specific power input. The combination of more concentrated caustic, direct steam addition and lower specific power input has increased the probability of localised concentration regions in the autoclave and likely significantly contribute to the phase variance in the residue.

Lastly, as mentioned earlier, there are three broad regimes in the type of solid phase produced during typical operation of the plant. While good control of the pH and temperature and low NAL acidities generally produce well developed ferrihydrite or hematite as desired products, pH excursions tend to produce either abundances of natrojarosite or gelatinous solids which are very difficult to filter. Since lower iron contents result in a greater mass of solids requiring filtration, natrojarosite is still detrimental to throughput. However, the solids produced during high pH excursions typically blinds filter cloths and causes throughput to fall rapidly. As such, plant operators prefer to operate at slightly lower pH in the autoclave discharge to prevent such occurrences. While this operational philosophy does safeguard to some extent against periods of significant filtration issues caused by high pH, it does not facilitate production of optimal residue.

CAUSTIC FLOW OPTIMISATION

Given the findings to date, the first step towards stabilising the residue was to optimise the caustic dosing strategy. Since commissioning, caustic dosing has been based on a ratio controller with the autoclave feed flow ratio manually adjusted based on the fourth compartment pH. Since the NAL acid and iron concentration essentially dictate the neutralisation requirement in the PIR autoclave, there was scope to deterministically set the caustic ratio using a feed-forward control strategy. Due to the variability associated with the PIR operational data, a multivariate statistical analysis was used to extract the key relationships among the important process variables. Data were retrieved from the RBMR historian over the period March-June 2015 and the caustic flow ratio was correlated with the NAL acid and iron concentrations using a simple linear model. Maximum correlation of the calculated and actual flow ratios occurred by lagging the NaOH flow ratio by 15 hours, corresponding to the typical solution inventory between the sections, and produced the following expression:

$$(\text{Caustic flow ratio})_{\text{lagged}} = 1.2473 + 0.9433\cdot[H_2SO_4]_{\text{avg.}} + 1.2758\cdot[Fe]_{\text{avg.}} \qquad (7)$$

The regressed parameters in Equation 7 produce caustic flow ratios that lie in between the stoichiometric ratios calculated for ferrihydrite and natrojarosite production, shown in that Table 2. This suggests that typical precipitates over this period comprise of a mixture of ferrihydrite and natrojarosite, which corresponds well with the residue iron content which has a median value of 40 wt. % iron and the mineralogical sample (Sample A in Table 1) which was analysed to constitute about 40% natrojarosite.

Table 2 – Calculated model and stoichiometric caustic flow ratios

	Model	Ferrihydrite/ hematite	Natrojarosite
NAL Fe, g/L		5.65	
NAL H_2SO_4, g/L		7.14	
Caustic:Flow, L/m^3	15.19	18.07	12.82

Implementation of the feed-forward controller commenced in July 2015 and an adjustment of 90% of Equation 7 was made to favor lower caustic flow ratios during the initial testing. To minimise the impact of sampling and analytical errors, 12–hour averages of the NAL iron and acid concentrations were used. Laboratory analyses, updated every 2–4 hours, are validated and entered in the control system by a control room operator. Following this implementation, there was a small but statistically significant shift in the pH distribution, highlighted in Figure 8.

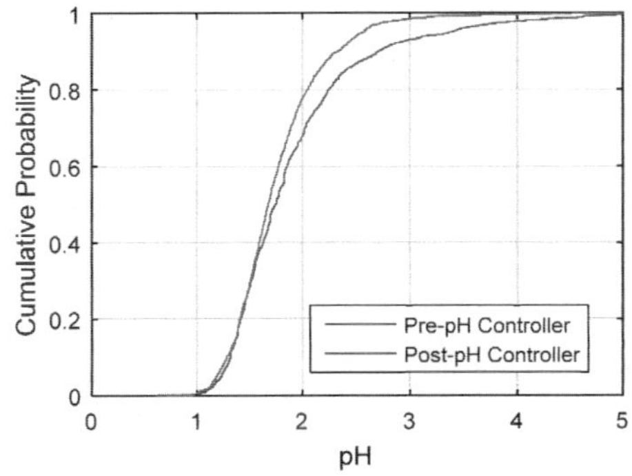

Figure 8 – pH cumulative distribution function highlighting the impact of the feed forward controller on pH. The target pH is 1.75-2

The median pH decreased as expected from the adjustment of Equation 8 but, most importantly, the control strategy produced a significant reduction in high pH excursions and correspondingly the significant production delays associated with colloidal precipitates. A detailed analysis of these events highlighted that, in most instances, delays in the validation of laboratory results were the root cause of the pH excursions. Given the criticality of the initial precipitate highlighted in this work, a natural extension is the control of pH in the PIR feed tank such that most of the free acid can be neutralised before entering the autoclave.

CONCLUSIONS

The PIR operation at Rustenburg Base Metals Refiners represents a unique iron removal process in which concentrated caustic is used to adjust pH, at temperature and pressure. While the process has been shown to be robust in the removal of iron from the circuit, difficulties in the filtration of the iron residue produced in PIR periodically constrains volumetric throughput into the nickel recovery circuit which directly affects production. Typical residues have been quantified by detailed mineralogy as a combination of ferrihydrite, hematite and natrojarosite. The lower iron content of natrojarosite that increases the overall mass requiring filtration, and the occasional formation of gelatinous residues (at higher pH) are both detrimental to filtration and can significantly reduce the throughput through the PIR section. On the basis of laboratory and plant test work, the initial precipitate formed in the PIR autoclave has been shown to be critical to the ultimate residue requiring filtration and is thus significantly dependent on the performance of the upstream leaching operations. An improvement in the pH control strategy based on a feed-forward controller was implemented that was shown to reduce the frequency of high pH excursions. Further development of this controller, extension to control in the PIR feed tank and other optimisation projects on the PIR autoclave are ongoing at RBMR.

The insights uncovered during the installation and commission of the PIR process also highlight that when migrating from pilot scale to commercial scale, even for relatively well developed technologies, often subtle yet critical challenges emerge. With the continued decline of ore grades across many commodities, metallurgical processes are becoming increasingly integrated and complex and their associated challenges are not straightforward and may be more difficult to detect during design stages. Such challenges demand dedicated technical, production and engineering teams to ultimately deliver the objectives laid out on the design desk.

ACKNOWLEDGEMENTS

The author wishes to acknowledge Anglo American Platinum and Anglo American for permission to publish this work and extend thanks to Sibusiso Mdlalose for conducting the laboratory testwork and Petra Dinham and Darren Tiddy for conducting the mineralogical analyses of the PIR residues.

REFERENCES

Babcan, J. (1971). Die synthese van jarosit $KFe_3(SO_4)_2(OH)_6$. *Geologiky Zbornik – Geologica Carpathica,* 22(2). 299-304.

Bryson, L.J., Hofreik, Z., Collins, M.J., Stiksma, J., & Berezowsky, R.M. (2008). New matte leaching developments at Anglo Platinum's base metal refinery. In C.A. Young, C.A. Taylor & Y. Choi (Eds), *Hydrometallurgy 2008, Proceedings of the Sixth International Symposium* (pp. 570–579). Society for Mining, Metallurgy and Exploration, Inc., Littleton, CO, USA.

Crundwell, F. K., Moats, M., & Ramachandran, V. (2011). Extractive metallurgy of nickel, cobalt and platinum group metals. Elsevier.

Dinham, P. & Tiddy, D. (2016). Internal Report: Mineralogical results for PIR autoclave products. Anglo American Technical Solutions.

Flynn, C. M. (1984). Hydrolysis of inorganic iron (III) salts. *Chemical Reviews, 84(1)*, 31-41.

Hagemann, J., & Pelser, M. (2015). Unlocking Rustenburg Base Metals Refiner's sulfur removal section. In *Copper Cobalt Africa 2015 Conference, incorporating the 8th Southern African Base Metals Conference* (pp. 457–468). Livingstone, Zambia, 6-8 July 2015.

Ismael, M. & Carvalho, J. (2003). Iron recovery from sulphate leach liquors in zinc hydrometallurgy. *Minerals Engineering, 16*. 31-39.

Jambor, J. L., & Dutrizac, J. E. (1998). Occurrence and constitution of natural and synthetic ferrihydrite, a widespread iron oxyhydroxide. *Chemical Reviews, 98(7)*, 2549-2586.

Loan, M., Parkinson, G., Newman, M., & Farrow, J. (2002). Iron oxy-hydroxide crystallization in a hydrometallurgical residue. *Journal of crystal growth, 235(1)*, 482-488.

Loan, M., Newman, O.M.G., Cooper, R. M. G., Farrow, J. B., & Parkinson, G. M. (2006). Defining the Paragoethite process for iron removal in zinc hydrometallurgy. *Hydrometallurgy, 81(2)*, 104-129.

Plasket, R.F, Dunn, G.M. (1986) Iron rejection and impurity removal from nickel leach liquor at Impala Platinum Limited. In J.E. Dutrizac & A.J. Monhemius (Eds.), *Iron Control in Hydrometallurgy* (pp. 695–729). Chichester, UK.

Schwarz, J., Channon, W., Dube, M., Makwana, M., Kerfoot, D.G. (1996). Commissioning of the Bindura Pressure Leach Plant. In E.J. Grimsey & I. Neuss (Eds.), *Nickel '96, mineral to market* (pp. 259-264). Kalgoorlie, Western Australia, 27-29 November 1996.

Steyl, J.D.T. (2012). *Simulating the medium temperature chalcopyrite oxidation system in batch and continuous autoclaves* (Doctoral dissertation). Stellenbosch University, Stellenbosch, South Africa.

CHEMICAL MODELLING OF THE $Fe_2(SO_4)_3$–$FeSO_4$–H_2SO_4–H_2O SYSTEM

*C.A. Biley[1] and J.D.T, Steyl[2]

[1]*Anglo American Platinum*
Rustenburg Base Metal Refiners
Rustenburg, South Africa, 0300
*(*Corresponding author: chris.biley@angloamerican.com)*

[2]*Dynamet Insights Pty (Ltd)*
57 Sixth Road, Hyde Park
Johannesburg, South Africa, 2196

ABSTRACT

The chemistry of electrolyte systems containing iron remains generally poorly understood despite their prevalence in many natural and industrial processes, particularly in hydrometallurgy. These systems are typically challenging to study and model due to their affinity for complex formation and precipitation even at low pH. Recent developments in the hydrometallurgical processing of nickel laterites at Anglo American have called for a deeper understanding of the fundamental chemistry of the $Fe_2(SO_4)_3$–$FeSO_4$–H_2SO_4–H_2O system. This paper details the development of an internally consistent, Pitzer ion-interaction model of this system, with explicit recognition of important contact ion pairs (CIPs). The primary application of this model is a thermodynamic framework for the interpretation of kinetic processes occurring during iron(III) reduction with SO_2. Adopting a rigorous thermodynamic basis for the system chemistry facilitates the formulation of simple rate expressions that are capable of inherently capturing complex effects. While the iron(III) reduction case study remains the focus of this work, the developments have more universal applications and the thermodynamic speciation model can be extended into other systems containing acidic ferric sulfate.

KEYWORDS

Ferric sulfate, chemical modelling, Pitzer, activity coefficient, solution speciation

INTRODUCTION

Iron is present in nearly all hydrometallurgical refining circuits typically as both a necessary and unwanted impurity and needs to be removed and disposed in a safe and efficient manner. This is generally performed by precipitation to form a range of solids such as hematite, goethite or jarosite. Numerous studies in the open literature have focussed on the removal of iron from refractory gold sulfide ore (Berezowsky & Weir, 1989; Berezowsky et al., 1991) and zinc pressure leaching (Buban, 1999; Ismael, 2003), and PGM matte sulfide circuits (Dutrizac & Monhemius, 1986). Moreover, since ferric iron is a strong oxidizing agent, it has been used extensively as a leaching agent for sulfide minerals (Dreisinger, 2006). Given the outlook of challenging economic conditions, declining ore grades and increasing levels of impurities, effective iron management is going to become increasingly critical to all hydrometallurgical processes.

Against this background, and considering the widespread importance of acidic iron sulfate systems throughout natural and industrial processes, it is unexpected that the knowledge of the fundamental solution chemistry of these systems contains such significant gaps, particularly at the conditions typically found in hydrometallurgical applications. Existing literature describing the measurement and modelling of solution speciation in these systems have highlighted their significant chemical complexity (Magini & Caminiti, 1977; Magini, 1979; Lee & Tavlarides, 1985; Stipp, 1990; Papangelakis et al., 1994; Gil et al., 1995; Majzlan & Myneni, 2005; Kormanyos et al., 2008; Yue et al., 2014). Partially responsible for the lack of quantitative speciation measurements is the strong tendency of ferric to hydrolyse at even low pH (Flynn, 1984), complicating the laboratory study of such systems. Moreover, the large number of aqueous species that may co-exist simultaneously makes the study of individual species and the modelling of the numerous interactions between these species extremely difficult (Dry & Bryson, 1988; Stipp, 1990; Majzlan & Myeni, 2005; Kobylin et al., 2007).

The ultimate objective of this work is to characterise the kinetics of ferric reduction with sulfur dioxide, specifically applicable to recent developments in atmospheric nickel laterite processing (Smit et al., 2012). To-date, only a handful of studies have experimentally investigated solution speciation in the $Fe_2(SO4)_3$ system (Magini & Caminiti 1977; Magini, 1979; Majzlan & Myneni, 2005) however their data is insufficient to build a reliable framework of the solution chemistry. This paper details the development of an equilibrium solution chemistry framework in the acidic ferric-ferrous sulfate system using Raman spectroscopic speciation measurements (Biley, 2015) as the primary means of speciation quantification. Given the complexity of this system, a pragmatic engineering approach was adopted in cases where insufficient chemical information was available which permitted reliable interpretations of kinetic processes during ferric reduction, which will be detailed in a subsequent paper. A rigorous approach to thermodynamic speciation facilitates an advanced analysis of kinetic processes in which the reaction mechanisms can be investigated and new understanding developed. Moreover, the importance of experimental measurements in the development and validation of chemical models is highlighted. While the experimental conditions favour concentrated solutions below the solution boiling point, the general lack of quantitative solution speciation measurements in this system mean trends can be extended to other systems.

EXPERIMENTAL AND MODELLING DETAILS

Raman Spectroscopy

Raman spectra were measured using a Jobin-Yvon T64000 Raman spectrometer operated in single spectrograph mode with a liquid nitrogen cooled CCD coupled to an Olympus BX40 microscope for sample acquisition. The 514.5 nm line of an argon ion gas laser was used throughout and a 20X long-working-distance (WD = 12.0 mm) objective (Olympus, LMPLFLN20X) was used to focus the laser on the sample, at which approximately 1.5 mW of laser power was applied. Solution samples were prepared from analytical grade chemicals gravimetrically and their composition validated by ICP-OES for metal ions and titration (where applicable) and H^+ titration. A solution of 1.90 mol/kg $(NH_4)_2SO_4$ solution was used as an external standard during all measurements to monitor instrument drift. In all cases, the variation in the measured 981 cm^{-1} band intensity was small in comparison to the variation in the chemical responses of the various solutions and no adjustments were deemed necessary.

Speciation Modelling

Solution phase speciation is described mathematically using mass-action equations where the overall equilibrium constant specifies the ratio in activities of products and reagents. These constants, at practical pressures, are only a function temperature and can be correlated surprisingly accurately using the Density model (Anderson et al., 1991) which exploits the near-constant relationship between lnK and water density with temperature.

$$\ln K(T) = \ln K^{0}_{ref} - (\Delta H^{0}_{ref} \cdot 10^{3}/R_{g}) \cdot (T^{-1} - T_{ref}^{-1}) + \Delta C_{p}{}^{0}_{ref}/(R_{g}T \cdot \partial \alpha / \partial T) \cdot (T^{-1} \cdot \ln(\rho/\rho_{ref}) - (\alpha_{ref}/T) \cdot (T - T_{ref})) \quad (1)$$

The Pitzer ion-interaction model (Pitzer, 1991) has been used extensively to account for solution non-idealities and possesses several key benefits for speciation-type solution modelling. Specifically, the Pitzer ion-interaction approach can be easily extended to include the explicit recognition of contact ion pairs (CIPs) as well as possessing a functional form that is routed in fundamental solution chemistry (Pitzer, 1973) often resulting in reliable interpolation and even extrapolation of thermodynamic data. Recently, Bea et al. (2010) published an efficient matrix-based implementation of the Pitzer model and the main equations for the activity and osmotic coefficients, in matrix notation, are reproduced below:

$$\ln \gamma_{aq} = (\ln \gamma_{DH} + q')\mathbf{z}_2 + \mathbf{q}^{c}\mathbf{z} + (2\mathbf{Q} + Z\mathbf{C})\mathbf{m} + \mathbf{m}^{T}\mathbf{T}\mathbf{m} \quad (2)$$

$$\phi = (2/M)(\ln \gamma'_{DH} + \mathbf{q}^{\phi} + Z\mathbf{q}^{c} + \mathbf{q}^{L} + t) + 1 \quad (3)$$

Where γ_{DH} is the Debye-Hückel limiting slope, $\boldsymbol{\gamma}_{aq}$, $\mathbf{m}$, $\mathbf{z}$, $\mathbf{z}_2$ are vectors of the activity coefficient, molality, species' charge and the square of the charge respectively, $\mathbf{Q}$ and $\mathbf{C}$ are square, symmetric matrices ($N_{aq} \times N_{aq}$) of second ($\beta^{(0)}_{ca}$, $\beta^{(0)}_{ca}$, $L_{cn,an}$, $\varphi_{aa,cc}$) and ternary (C_{ca}) virial interaction coefficients respectively and the tensor, $\mathbf{T}$, includes the higher-order mixing terms ($\psi_{cca,aac}$). In this study, all Pitzer model parameters were, at most, allowed to vary linearly with temperature.

SPECTROSCOPIC MEASUREMENTS

Raman spectra of various solutions in the $Fe_2(SO_4)_3$–$FeSO_4$–H_2SO_4–H_2O systems were measured in order to develop an improved understanding of the dominant species in acidic iron sulfate solutions. While important to the solution model development, the detail of the spectroscopic measurements are outside the scope of this work and are detailed in a separate paper (Biley, 2015). Nevertheless, Figure 1 presents a summary of the dominant Raman bands in this system over the 750-1300 cm^{-1} range. The Raman spectra of unassociated sulfate, $(NH_4)_2SO_4$, and sulfuric acid are well known and are characterised by strong bands at 980 cm^{-1} and 1040/1050 cm^{-1}, which are attributed to the symmetric stretching (ν_1) of the free and associated sulfate units respectively (Rudolph, 1996; Lund Myhre et al., 2003). The introduction of Fe(II) and Fe(III) into the acid-sulfate system results in significantly more complex spectra. In $FeSO_4$ solutions, a high frequency shoulder appears on the 980 cm^{-1} band, clearly visible in Figure 1, which has previously been assigned to the $FeSO_4^0$ CIP (Rudolph et al., 1997). In acidic $Fe_2(SO_4)_3$ solutions, two bands emerge at 1004 cm^{-1} and 1035 cm^{-1} and are assigned to the ν_1-mode of sulfate in the $Fe(SO_4)_2^-$ and $FeSO_4^+$ respectively (Biley, 2015) and are supported by two significant points. Firstly, the larger temperature dependence of the 1004 cm^{-1} band is expected due to the larger entropy term expected by the release of water molecules during $Fe(SO_4)_2^-$ formation from strongly hydrated ferric (and sulfate) ions. Secondly, Sobron et al. (2007) noted a dominant ~1040 cm^{-1} band and a very weak band at 1005 cm^{-1} in iron(III) containing solutions at low SO_4^{2-}:Fe^{3+} ratios of 0.09, emphasising the dominance of $FeSO_4^+$ under these conditions.

Rudolph et al. (1994, 1997, 2001, 2003, 2010) have studied several sulfate systems in a similar manner and, by quantitatively assessing the band intensities associated with the various shifts in the sulfate structure during complex formation, concentrations of inner-sphere sulfate complexes can be extracted. A total of 111 solutions, some containing inert $NaClO_4$ as an internal standard, were analysed by Raman spectroscopy to develop an understanding of the variation in solution speciation with concentration and temperature in the ranges 0-5 mol/kg total sulfate and 25-90°C respectively. All spectra were processed to extract quantitative sulfate species concentrations and, to avoid repetition, species measurements are presented in the following section with the regressed model output.

Validation of the approach was achieved by the excellent agreement between measured and published species distributions for the H_2SO_4 and $FeSO_4$ aqueous systems, presented in Figure 2 and Figure 4. The speciation measurements in the systems containing $Fe_2(SO_4)_3$ provide new speciation data on which an enhanced chemical model can be developed. It must however be stressed that the multiple convoluted bands in the 980-1100 cm^{-1} region make it difficult to extract speciation data from the spectra data. As far as possible the variance in experimental measurements was quantified and is presented by errors bars on the speciation figures in the following section.

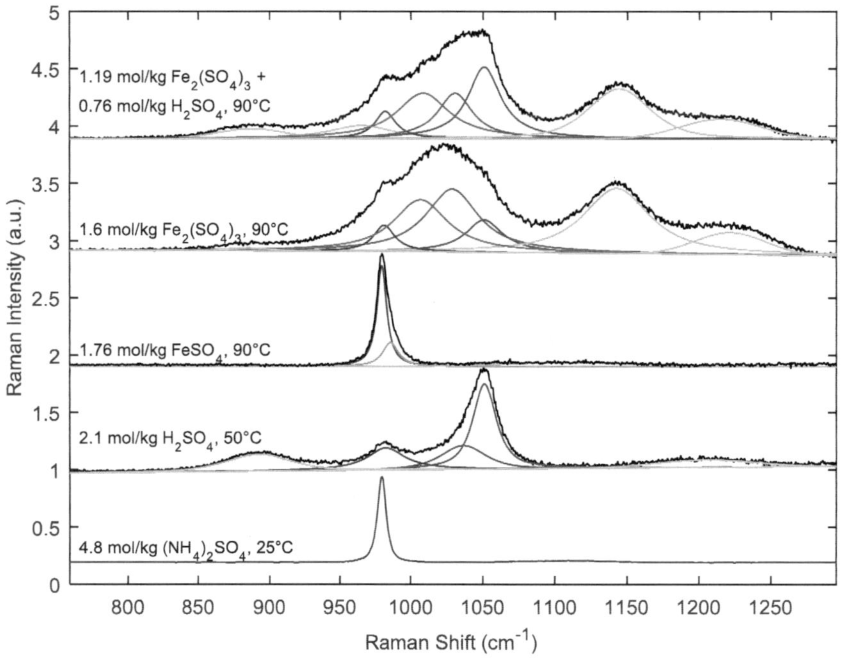

Figure 1 - Summary of major Raman bands in solutions in the $Fe_2(SO_4)_3$–$FeSO_4$–H_2SO_4–H_2O system over the 750–1300 cm^{-1} wavenumber region

CHEMICAL MODELLING

Given the general lack of speciation measurements in this system, a distinct focus for the modelling work was to describe measured speciation and published thermodynamic data using a minimum-parameter, minimum-species approach. Specifically, only those explicit solution species that have been experimentally detected are included. In this approach, and given that the ultimate application of the model is to interpret kinetic phenomena, it is acknowledged that the model will not be capable of explaining all the complex phenomena in these multi-component systems. The primary objective of this work is thus to develop a reasonable framework of the systems' solution chemistry that captures the trends in speciation as a function of concentration and temperature.

H_2SO_4–H_2O System

The first dissociation of H_2SO_4 (Reaction 4) is essentially complete in aqueous solutions (< 40 mol/kg) at room temperature (Clegg et al., 1994) and $H_2SO_4^0{}_{(aq)}$ only constitutes a minor component above 100 °C (Steyl, 2012). However, the second dissociation (Reaction 5) is not complete and must be considered.

This system has been well characterised due to its widespread importance and a variety of reliable thermodynamic data exist, especially at temperatures below 60°C.

$$K'^{0}_{1}: H_2SO_4 = H^+ + HSO_4^- \quad (4)$$

$$K'^{0}_{2}: HSO4^- = H^+ + SO_4^{2-} \quad (5)$$

Thermodynamic data ($\gamma_{\pm}$, φ and α) for this system were taken from the literature (Pitzer, 1977; Holmes & Mesmer, 1992; Clegg et al., 1994; Clegg & Brimblecombe, 1995) and the measured speciation from the Raman measurements (Biley, 2015) were included in the model regression. For the equilibrium constant of the HSO_4^- species, Steyl (2009) highlighted that the empirical model of Dickson et al. (1990) was accurate over a wide temperature range and was adopted in this study (see Table 2). The excellent performance of the Pitzer model with the use of the overall thermodynamic association constant for HSO_4^- highlights that consideration for outer-sphere complexes is not required for this system as the Pitzer model parameters deal with for these effects sufficiently.

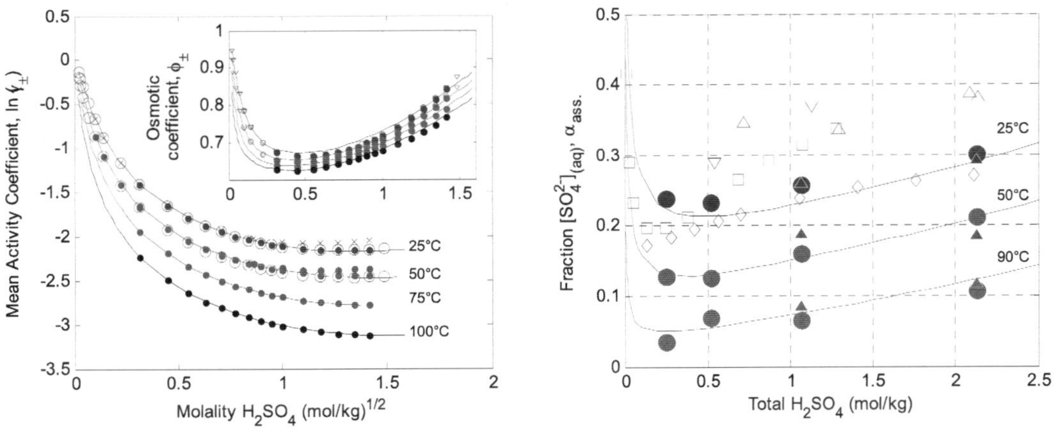

Figure 2 – Mean activity and osmotic coefficients and HSO_4^- association in the H_2SO_4-H_2O system at 25-100° C. Dots: Holmes & Mesmer (1992); open circles: Clegg et al. (1994); crosses: Clegg & Brimblecombe (1995); squares: Lindstrom & Wirth (1969); open triangles: Lund-Myhre et al. (2003); diamonds: Chen & Irish (1972); down triangles: Knopf et al. (2003); filled circles: Biley (2015); filled triangles: Biley (2015) from internal $NaClO_4$ standardised Raman spectra. Solid lines represent the regressed thermodynamic model

$FeSO_4$–H_2SO_4–H_2O System

While some thermodynamic data for the $FeSO_4$ system is available in the open literature, relatively little is known about the stability of the $FeSO_4^0$ $_{(aq)}$ CIP, especially at elevated temperatures and in the presence of acid. In order to model divalent sulfates in metallurgical solutions, it was recently proposed to use the $MgSO_4$ aqueous system as surrogate system for which considerably more thermodynamic data are available (Steyl, 2012). The basis for this approach is the remarkable similarity in the thermodynamic properties, such as overall stability constants and mean activity coefficients, between $MgSO_4$ and other metal-sulfates, such as copper, zinc, iron and nickel sulfate (Steyl, 2009). Where the objectives of the study are engineering inclined, rather than fundamental chemistry, this approach has proven to be extremely useful. Further validity for this approach is provided from the agreement between the association of $MgSO_4^0$ (Hefter, 2006) and that of $FeSO_4^0$ (Rudolph et al., 1997) at 25 °C.

Association constants of CIPs are typically significantly lower than the overall stability constants in systems where outer-sphere interactions are important (Akilan et al., 2006). In a thorough review of divalent metal sulfate systems, and supporting static quantum calculations, Steyl (2012) proposed a value of

$\log_{10}K^0 = 1.5$ for the association of divalent metal sulfates CIPs, as opposed to overall values in the range 2.2 to 2.8 for Mg^{2+}, Fe^{2+}, Cu^{2+}, Zn^{2+} and Ni^{2+} (Martell & Smith, 1976), and this value was adopted in this study.

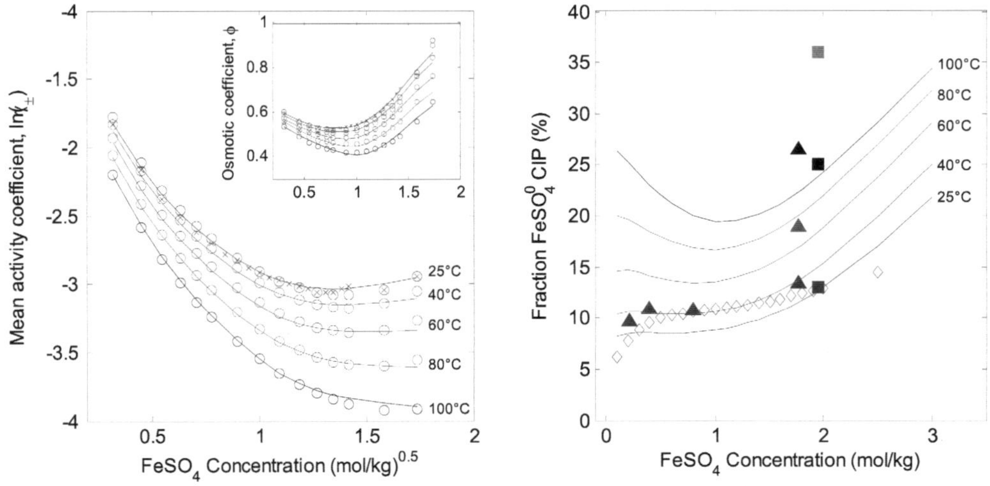

Figure 3 – Mean activity and osmotic coefficients and association of the $FeSO_4^0$ CIP in solutions of $FeSO_4$-H_2O. Crosses: Reardon & Beckie (1987), $FeSO_4$; Circles: Archer & Wood (1985), $MgSO_4$; Squares: Rudolph et al. (1997), $FeSO_4$ (blue: 25°C, black: 90°C, orange: 150°C); Triangles: Biley (2015), $FeSO_4$ (blue: 25°C, green: 50°C, black: 90°C); Diamonds: Hefter (2006), $MgSO_4$, 25°C. Solid lines represent the regressed thermodynamic model

The $FeSO_4$–H_2SO_4 interaction parameters were optimised to the extensive set of solubility data in the $FeSO_4$–H_2O and $FeSO_4$–H_2SO_4–H_2O systems at various temperatures (Linke & Seidell, 1965) as well as the high quality isopiestic study of the $MgSO_4$–H_2SO_4–H_2O system at 25°C (Rard, 1999) via the surrogate approach. Optimised solubility product, K_{sp}, values are not directly comparable to those reported elsewhere due to the use of the neutral CIP species in the solid-liquid equilibrium expressions where n = 1 or 7:

$$FeSO_4{}^0{}_{(aq)} + nH_2O_{(l)} \leftrightarrow FeSO_4{\cdot}nH_2O_{(s)} \tag{6}$$

An additional phase, rozenite ($FeSO_4{\cdot}4H_2O$), is also known to exist near the maximum solubility limit but, given the extra parameters required, inclusion was not deemed valuable. The optimised K_{sp} parameters for the above salts are presented in Table 2. It was furthermore found that the temperature dependence of the $FeSO_4$-H_2SO_4 interaction parameters had little effect and all temperature dependence terms were maintained at zero. The optimised parameters for this system are presented in Table 1 and the resulting calculated solubility surface is presented in Figure 4.

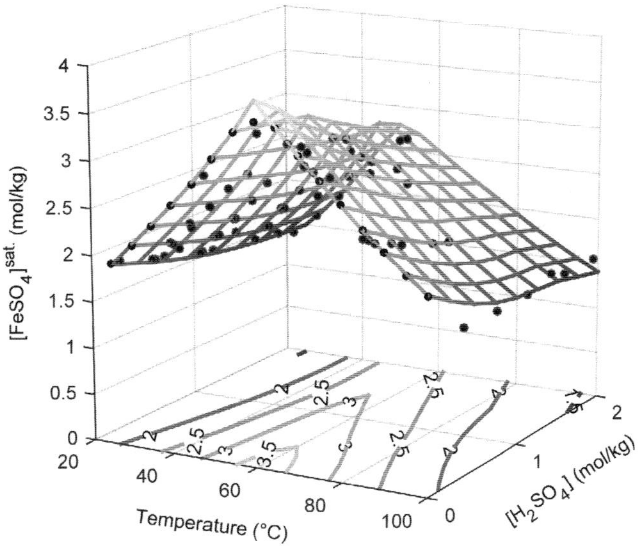

Figure 4 – $FeSO_4$ solubility as a function of temperature and acid concentration. Data points sourced from Linke & Seidell (1965)

$Fe_2(SO_4)_3$–H_2SO_4–H_2O System

Recently, extensive and reliable water activity data of this system have been measured by separate authors up to reasonably high concentrations at 25 and 50°C (Rumyantsev et al., 2004; Velazquez-rivera et al., 2006) and was correlated using an extended Pitzer ion-interaction model (Tosca et al., 2007). The magnitude of the interaction parameters regressed in this study are indicative of a system with a high degree of non-ideality and complex formation, i.e., $\beta^{(1)}_{Fe^{3+},SO_4^{2-}} = 11.968$, $\beta^{(1)}_{Fe^{3+},HSO_4^-} = 7.519$ at 25°C, which are significantly larger than, for example, divalent metal systems (Pitzer, 1991).

In adopting a speciation approach to modelling this system, the possible number of species becomes large and the permutations of interaction parameters rapidly increase. With the Raman measurements suggesting that at least two, previously reported, ferric sulfate CIPs exist in these concentrated solutions, a minimum number of species were selected for inclusion in the model. The presence of significant bisulfate in solutions without added acid suggests that ferric hydrolysis occurred and, in order to replicate the experimental measurements accurately, ferric hydroxyl species needed to be included. Given the uncertainty in the specific hydroxyl species under these conditions, due to very weak Raman scattering of these species, and to avoid unnecessary complication, a surrogate approach for ferric hydrolysis was adopted. For this, only a single $Fe(OH)_n^{3-n}$ species was included in the model and significant improvements in the characterisation of the measured HSO_4^- concentrations were obtained by including only the $Fe(OH)_2^+$ species. This surrogate species was thus considered adequate for modelling the main effects of hydrolysis under the conditions examined. Additional support for this is the noted decreasing stability of the $FeOH^{2+}$ species with increasing sulfate concentration (Lee & Tavlarides, 1985). Necessarily, the $\log_{10} K^0$ and ΔH^0 parameters for the surrogate $Fe(OH)_2^+$ species could not be sourced from literature and were regressed.

The resulting model for the $Fe_2(SO_4)_3$-H_2SO_4-H_2O system thus contains a total of seven solution species. Additional species, reported in the literature (Stipp, 1990; Casas et al., 2005; Papangelakis et al., 1994; Liu et al., 2003), such as $Fe(OH)SO_4^0$ or $FeH(SO_4)_2^0$, were not considered in this framework as there is little direct evidence for their existence and often inferred from the ability to successfully model observed properties or experimental trends such as solubility.

The selection of stability constants for the ferric sulfate CIPs are somewhat arbitrary as no experimental data exist for these quantities and they must be inferred by the overall association constants which have been reported in the literature. Guidance from the Raman results suggests that the majority of Fe(III) in $Fe_2(SO_4)_3$ solutions of moderate concentration are associated as CIPs (Biley, 2015), thus the CIP contribution to the overall association parameter would likely dominate over outer-sphere effects and accepting K_{CIP} to be equal to this would likely not induce considerable error, which would ultimately be absorbed into the regressed model parameters. Additional justification for this approach is the dominant effects of CIPs in charge transfer reactions, making their characterisation the ultimate goal for this study. The ΔH_r^0 and ΔC_p^0 values for the $FeSO_4^+$ species were calculated from data collated by Liu et al. (2003) and for the $Fe(SO_4)_2^-$ species calculated from the Balanced Like Charge Method (Barnes, 1997) using available thermodynamic data for the HSO_4^- and $FeSO_4^+$ species.

The seven species included in the $Fe_2(SO_4)_3$-H_2SO_4-H_2O model can have up to 74 individual Pitzer model parameters, excluding temperature dependent terms. In order to simplify the parameter optimisation, a systematic, manual scan of the most common model parameters (i.e., $\beta^{(0)}$, $\beta^{(1)}$ and C^{φ}) was performed and insensitive parameters were fixed at zero. All interaction parameters involving the $Fe(OH)_2^+$ species were fixed at zero which somewhat validates the surrogate approach, albeit artificially. The regressed model characterisation of the measured solution speciation for this system is presented in Figure 5 for solutions without added H_2SO_4 and in Figure 6 with added H_2SO_4. The solution model effectively captures the overall trends in speciation with temperature and concentration with a total of 9 variable parameters; optimal values of these parameters are presented in Table 1 and Table 2.

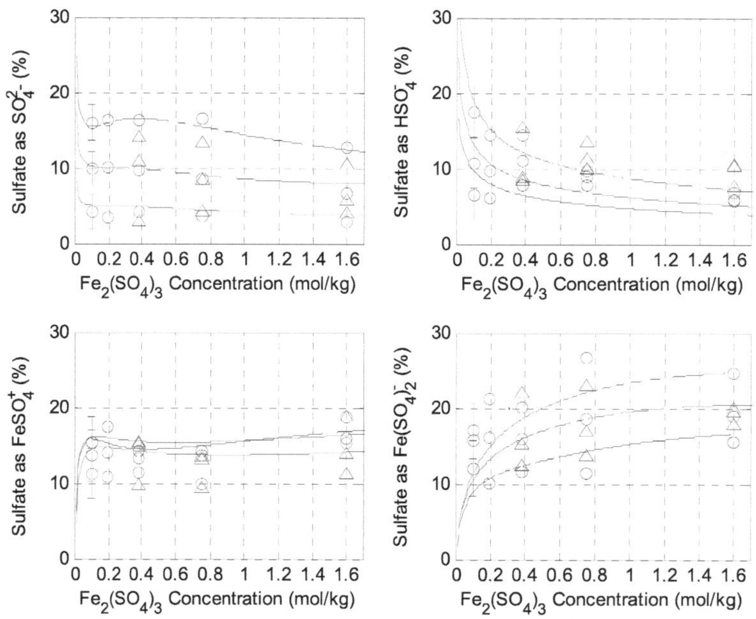

Figure 5 – Speciation in $Fe_2(SO_4)_3$-H_2O solutions. Triangle points represent independent repeated measurements containing 0.5 mol/kg $NaClO_4$ as internal standard

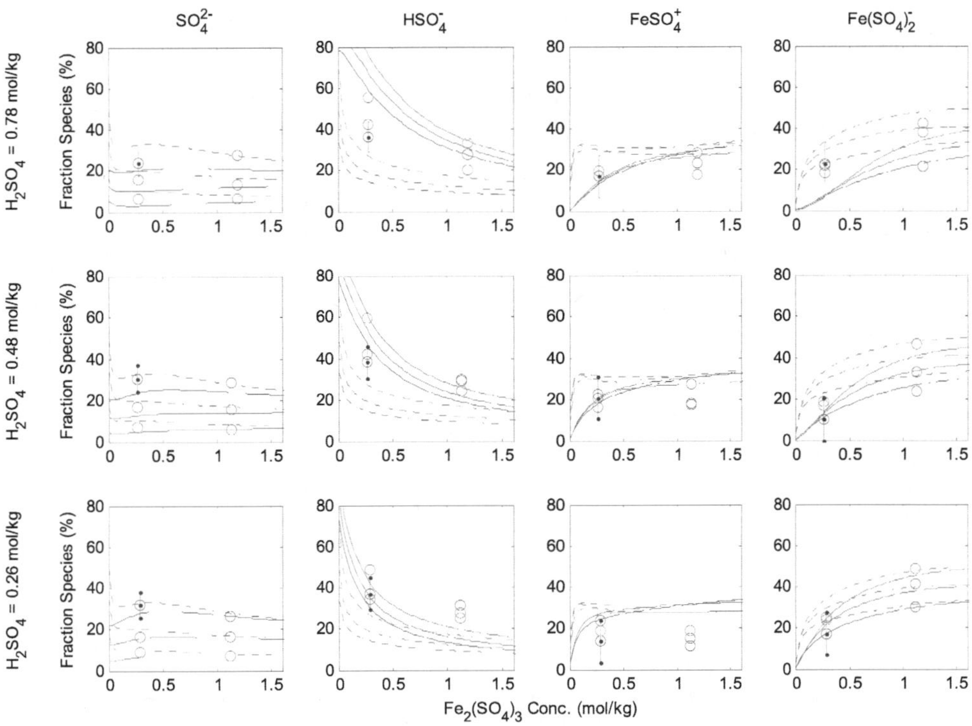

Figure 6 – Speciation in the $Fe_2(SO_4)_3$-H_2SO_4-H_2O system. Dotted lines represent the model regression of the speciation in the absence of acid for comparison

Several important trends are highlighted in the measured speciation data and the modelling thereof. Firstly, at very low concentrations (iron(III) < 0.15 mol/kg) the $FeSO_4^+$ and $Fe(SO_4)_2^-$ species stability decrease significantly, while the sulfate and bisulfate species increases. As the total sulfate concentration decreases, at low acidity, the ferric hydroxyl species become more important. Indeed, at the lower concentrations of $Fe_2(SO_4)_3$, $Fe(OH)_2^+$ is the dominant Fe(III) species in the solution, however, since the model developed in this study was not focused on the low concentration regions and in primarily acidic solutions, the ferric hydroxyl speciation is not well characterised. Ultimately, if speciation data could be extended to more dilute solutions, an improved description of these species may be included in the modelling framework.

A second important point is the fairly constant profile of the $FeSO_4^+$ species, in comparison to the $Fe(SO_4)_2^-$ species, which is observed to increase significantly with temperature and concentration. This buffering effect of the $FeSO_4^+$ species is related to the relatively constant Fe(III):SO_4^{2-} ratio in the solutions investigated and is captured in the model by the $\beta^{(0)}_{Fe^{3+},Fe(SO_4)^{2-}}$ interaction parameter.

Thirdly, from solutions with added acid (Figure 6), it is clear that the main effect of increasing acidity is the significant increase in the stability of the HSO_4^- species and a corresponding decrease in the abundance of the ferric sulfate CIPs, particularly at low concentration. While added acid decreases the Fe(III):SO_4^{2-} ratio, tending to promote CIP formation, the increase in total number of protons in the system has a greater effect on the stability of the HSO_4^- species. The model overestimates the HSO_4^- concentration at higher H_2SO_4:$Fe_2(SO_4)_3$ ratios, which may be due to the presence of additional species containing protons (i.e., $FeHSO_4^+$ or $FeH(SO_4)_2^0$) or simply that the model does not have sufficient flexibility in its minimal-parameter form to account for the changes accurately. However, since the trends in the speciation are considered most important for the engineering modelling objectives of this study, the inclusion of additional model parameters to improve on the fit of the data was not considered valuable.

It is important to note a significant drawback of the solution model developed here is the poor description of the water activity in the $Fe_2(SO_4)_3$-H_2O system which is overestimated considerably. In these concentrated solutions, which contain a number of charged species, including $Fe^{3+}_{(aq)}$, it is expected that significant structuring of the water medium would occur through hydrogen-bonded networks which would significantly affect the water activity. Since such species cannot be easily detected by Raman spectroscopy and no theoretical methods exist for their explicit consideration within a thermodynamic model, no further consideration of these species are made. Given that the description of the solution speciation is the primary objective of this model, it was acknowledged that the model may not have sufficient flexibility to capture the water activity trends, and the addition of species or model parameters to better describe the water activity was not deemed useful.

$Fe_2(SO_4)_3$–$FeSO_4$–H_2SO_4–H_2O System – Model Validation

In addition to the $Fe_2(SO_4)_3$-H_2SO_4-H_2O and $FeSO_4$-H_2O systems analysed by Raman spectroscopy, a set of solutions in the mixed Fe(II)-Fe(III) system were also analysed for validation purposes (Biley, 2015). These solutions consisted of varying amounts of ferric ion at a constant total iron of about 1.6 mol/kg in a background of 0.2 mol/kg H_2SO_4. Species concentrations were extracted in a similar manner to above, however have large uncertainties due to the highly convoluted Raman bands in the mixed system. These data were not incorporated in the parameter regression but only to investigate the predictive capability of the model in a mixed system.

Figure 7 presents the validation data and it is evident that the model reasonably captures the general speciation trends measured in the Raman data. Slight underestimation of the HSO_4^- concentration and slight overestimation of the $Fe(SO_4)_2^-$ concentration is constituted the primary deviations of the model and again are possibly indicative of other poly-cationic species including both H^+ and Fe^{3+}.

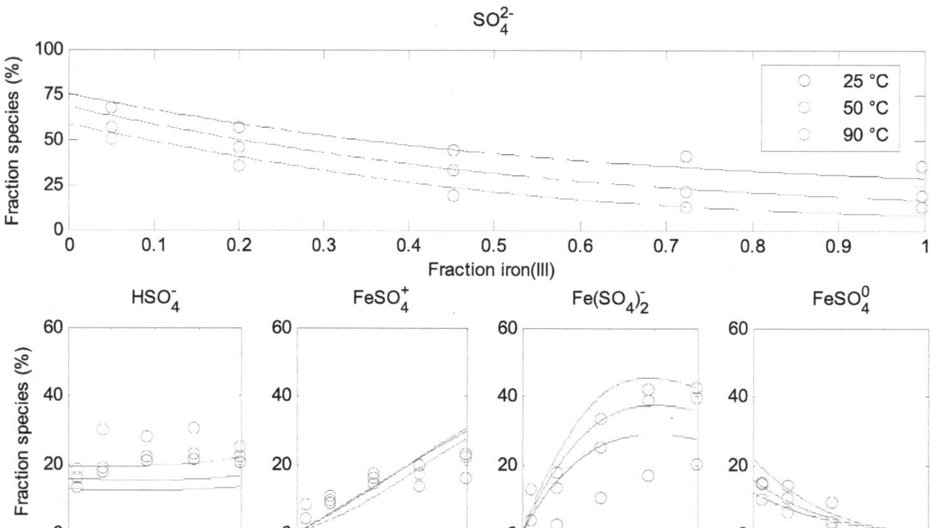

Figure 7 – Sulphate species distribution in the $Fe_2(SO_4)_3$-$FeSO_4$-H_2SO_4-H_2O system. Speciation data not used in the parameter optimisation. $[Fe]_{total}$ = 1.6 mol/kg, $[H_2SO_4]$ = 0.2 mol/kg. Fraction iron(III) is calculated as $[Fe^{3+}]/([Fe^{2+}] + [Fe^{3+}])$

CHEMICAL MODEL PARAMETERS

The overall chemical model of the $Fe_2(SO_4)_3$–$FeSO_4$–H_2SO_4–H_2O is summarised in Table 1 (activity coefficient parameters) and Table 2 (stability constant parameters). The activity coefficient parameters in Table 1 are to be used in the following equation: $X = p_0 + p_1 \cdot (T - 298.15)$, where X is the parameter are temperature and T is the temperature in Kelvin. Parameters p_1, p_2 and p_3 correspond to the lnK0, ΔH_r^0 and ΔC_p^0 for the Density model (Anderson, 1991), presented in Equation 1 and used for the majority of the CIPs in this study. All parameters in Table 1 and only those explicitly stated in Table 2 were regressed in this study. The concentration and temperature regions of applicability for the solution model can be summarised as: [Fe(III)] = 0.05 – 3.2 mol/kg, [Fe(II)] = 0.05 – 3.2 mol/kg, [H_2SO_4] = 0 – 4 mol/kg, Temperature = 25-95°C.

Table 1 – Activity coefficient model parameters

Parameter	Species	p_0	$10^3 \cdot p_1$
$\beta^{(0)}$	H^+, SO_4^{2-}	0.028578	-1.0244
	H^+, HSO_4^-	0.22109	-0.30535
	Fe^{2+}, SO_4^{2-}	0.18995	-2.0163
	Fe^{2+}, HSO_4^-	0.26900	-
	Fe^{3+}, SO_4^{2-}	1.0962	28.677
	Fe^{3+}, HSO_4^-	0.43885	-
	Fe^{3+}, $Fe(SO_4)_2^-$	0.53151	-
$\beta^{(1)}$	H^+, SO_4^{2-}	0.20502	-8.0561
	H^+, HSO_4^-	0.48515	-1.7541
	Fe^{2+}, SO_4^{2-}	3.4302	15.047
	Fe^{2+}, HSO_4^-	0.98649	-
	Fe^{3+}, SO_4^{2-}	6.2122	-
$\beta^{(2)}$	Fe^{2+}, SO_4^{2-}	32	-
C^{φ}	Fe^{2+}, SO_4^{2-}	0.075016	1.3897
	Fe^{2+}, HSO_4^-	0.27526	-
	Fe^{3+}, SO_4^{2-}	-0.18496	-20.8560
L	H^+, $FeSO_4^0$	-0.042489	-
	Fe^{2+}, $FeSO_4^0$	0.035811	-1.3943

Table 2 – Equilibrium constant parameters

Equilibria	p_1	p_2	p_3	p_4	p_5
$H^+ + SO_4^{2-} = HSO_4^-$ (A)	562.7097(1)	-13273.8(1)	-102.5154(1)	0.24775(1)	-1.12E-04(1)
$Fe^{3+} + SO_4^{2-} = FeSO_4^+$ (B)	9.3024(2)	26.11(3)	394.4(3)		
$Fe^{3+} + 2SO_4^{2-} = Fe(SO_4)_2^-$ (B)	12.324(2)	38.872(4)	781.62(4)		
$Fe^{3+} + 2H_2O = Fe(OH)^{2+} + 2H^+$ (B)	-2.8476(5)	15.3258(5)			
$Fe^{2+} + SO_4^{2-} = FeSO_4^0$ (B)	3.4539(6)	15(7)	376.575(5)		
$FeSO_4^0 + 7H_2O = FeSO_4 \cdot 7H_2O_{(s)}$ (C)	23.9438(5)	-41.7097(5)	-11.680(5)		
$FeSO_4^0 + H_2O = FeSO_4 \cdot H_2O_{(s)}$ (C)	7.7552(5)	-24.6195(5)	0.72803(5)		

Extrapolation Equations

A: $\log_{10}K(T) = -(p_1 + p_2/T + p_3 \cdot \ln T + p_4 \cdot T + p_5 \cdot T^2)$

B: $\ln K(T) = p_1 - (p_2 \cdot 10^3/R_g) \cdot (1/T - 1/298) + p_3/(R_g T \cdot (\partial \alpha / \partial T)) \cdot ((1/T) \cdot \ln(\rho/\rho_{298}) - (\alpha_{298}/T) \cdot (T-298))$

C: $\ln K_{sp}(T) = p_1 + (p_2/10^3)T + (p_3 \cdot 10^5)T^{-2}$

References

(1): Dickson et al. (1990)
(2): Stipp (1990)
(3): Liu et al. (2003)
(4): BLCM (Biley, 2015)
(5): Optimised (Biley, 2015)
(6): Steyl (2012)
(7): Rudolph et al. (1997)

MODEL APPLICATION

While having some drawbacks in the accurate quantification of some thermodynamic data, the solution model detailed above provides a rigorous basis on which to interpret kinetic processes in this system; the primary objective of which this work is part. Ultimately, in the $Fe_2(SO_4)_3$–$FeSO_4$–H_2SO_4–SO_2–H_2O reactive system, the kinetics of iron(III) reduction (Reaction 7) can be reduced to a simple reaction involving a ferric sulfite CIP highlighted by the rate law in Equation 8 (Biley, 2015). The effects of temperature, batch or continuous operation and acid and ferric concentrations are captured within the thermodynamic speciation framework and do not need to appear in an "unnatural" rate law containing non-integer exponents.

$$Fe_2(SO_4)_3 + SO_2 + 2H_2O \rightarrow 2FeSO_4 + 2H_2SO_4 \quad (7)$$

$$d[Fe(III)]/dt = -2k[FeSO_3^+] \quad (8)$$

In this system, the competing solution thermodynamic speciation, which takes place much faster than the rate determining step, results in thermodynamic interactions dictating the overall observed rate of reaction. With only a single rate constant, the optimised kinetic framework is capable of accurately modelling iron(III) reduction kinetics over a wide range of concentration, presented in Figure 8 and detailed further in Biley (2015). This further stresses the importance of thermodynamic considerations in the quantification of reaction kinetics, particularly where solution speciation is known to occur.

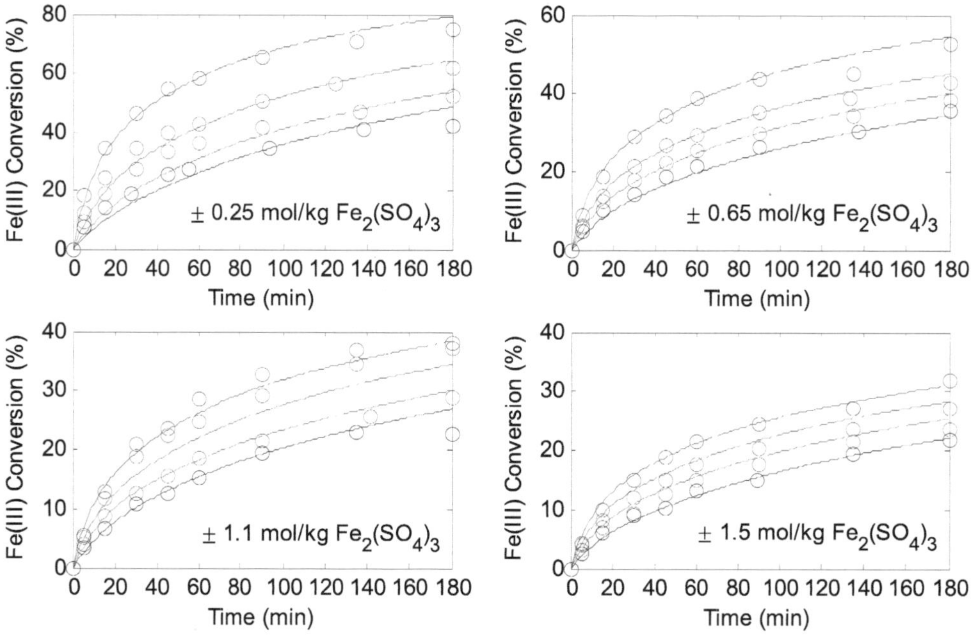

Figure 8 – Model characterisation of atmospheric batch kinetics of iron(III) reduction with SO_2 at 95°C as a function of initial $Fe_2(SO_4)_3$ and H_2SO_4 concentrations (Biley, 2015). H_2SO_4 concentrations: blue: 0.3 mol/kg, green: 0.7 mol/kg, red: 1.1 mol/kg, black: 1.5 mol/kg. Lines represent the kinetic model.

CONCLUSIONS

A solution chemistry model of the $Fe_2(SO_4)_3$–H_2SO_4–H_2O system has been developed using thermodynamic data available in the open literature as well as new speciation measurements in this system. The adopted approach provides a minimum-parameter, minimum-species framework targeted primarily at the description of the main chemical species present in concentrated acidic ferric-ferrous sulfate solutions. The development of this model has highlighted the importance of experimental measurements in the development and validation of solution chemistry models as well as the inclusion of valid, simplifying assumptions and surrogates to facilitate enhanced model robustness. Such robustness provides significant benefits in the modelling of reaction kinetics, for which considerable constraints are imposed by the requirement of solving systems of differential mass-action equations.

REFERENCES

Akilan, C., Rohman, N., Hefter, G., & Buchner, R. (2006). Temperature effects on ion association and hydration in $MgSO_4$ by dielectric spectroscopy. *Chem. Phys. Chem 7(11)*, 2319–2330.

Anderson, G., Castet, S., Schott, J., & Mesmer, R. (1991). The density model for estimation of thermodynamic parameters of reactions at high temperatures and pressures. *Geochimica et Cosmochimica Acta 55(7)*, 1769–1779.

Archer, D.G., & Wood, R.H. (1985). Chemical equilibrium model applied to aqueous magnesium sulfate solutions. *Journal of Solution Chemistry 14(11)*, 757–780.

Barnes, H.L. (1997). *Geochemistry of Hydrothermal Ore Deposits* (3rd ed.). New York. John Wiley & Sons.

Bea, S., Carrera, J., Ayora, C. & Batlle, F. (2010). Modeling of concentrated aqueous solutions: Efficient implementation of Pitzer equations in geochemical and reactive transport models. *Computers & Geosciences 36(4)*, 526–538.

Berezowsky, R. & Weir, D. (1989). Refractory gold: the role of pressure oxidation, in: Gold Forum on Technology and Practices - World Gold, pp. 295–304.

Berezowsky, R., Collins, M.J., Kerfoot, D.G.E. & Torres, N. (1991). The commercial status of pressure leaching technology. *Journal of Minerals, Metals and Materials Society 43(2)*, 9–15.

Biley, C.A. (2015). *Thermodynamic and kinetic modelling of iron(III) reduction with sulfur dioxide gas.* (Ph. D. thesis). Stellenbosch University. Stellenbosch, South Africa.

Buban, K.R., Collins, M.J. & Masters, I.M. (1999). Iron control in zinc pressure leach processes. *Journal of Minerals, Metals and Materials Society 51(12)*, 23–25.

Casas, J., Crisostomo, G. & Cifuentes, L. (2005). Speciation of the Fe(II)-Fe(III)-H_2SO_4-H_2O system at 25 and 50°C. *Hydrometallurgy 80(4)*, 254–264.

Chen, H. & Irish, D. (1971). Raman spectral study of bisulfate-sulfate systems. 2. Constitution, Equilibria, and ultrafast proton transfer in sulfuric acid. *Journal of Physical Chemistry 75(17)*, 2672–2681.

Clegg, S.L. & Brimblecombe, P. (1995). Application of a multicomponent thermodynamic model to activities and thermal properties of 0-40 mol kg^{-1} aqueous sulfuric acid from <200 to 328K. *Journal of Chemical and Engineering Data 40(1)*, 43–64.

Clegg, S.L., Rard, J.A. & Pitzer, K.S. (1994). Thermodynamic properties of 0-6 mol/kg aqueous sulfuric acid from 273.15 to 328.15K. *Journal of the Chemical Society, Faraday Transactions 90(13)*, 1875.

Dickson, A.G., Wesolowski, D.J., Palmer, D.A. & Mesmer, R.E. (1990). Dissociation constant of bisulfate ion in aqueous sodium chloride solutions to 250°C. *The Journal of Physical Chemistry 94(20)*, 7978–7985.

Dreisinger, D. (2006). Copper leaching from primary sulfides: options for biological and chemical extraction of copper. *Hydrometallurgy 83*, 10–20.

Dry, M.J. & Bryson, A.W. (1988). Prediction of redox potential in concentrated iron sulphate solutions. *Hydrometallurgy 21(1)*, 59–72.

Dutrizac, J.E. & Monhemius, A. (1986). Iron control in hydrometallurgy. Toronto, Canada, 19-22 Oct. 1986

Gil, A.F., Salgado, L., Galicia, L., & Gonzalez, I. (1995). Predominance-zone diagrams of Fe(III) and Fe(II) sulfate complexes in acidic media. Voltammetric and spectrophotometric studies. *Tantala 42(3)*, 407–414.

Hefter, G. (2006). When spectroscopy fails - The measurement of ion pairing. *Pure and applied chemistry 78(8)*, 1571–1586.

Holmes, H. & Mesmer, R. (1992). Isopiestic studies of $H_2SO_{4\,(aq)}$ at elevated temperatures: Thermodynamic properties. *The Journal of Chemical Thermodynamics 24(3)*, 317–328.

Ismael, M.R.C., & Carvalho, J.M.R. (2003). Iron recovery from sulphate leach liquors in zinc hydrometallurgy. *Minerals Engineering 16(1)*, 31–39.

Knopf, D.A., Luo, B.P., Krieger, U.K. & Koop, T. (2003). Thermodynamic dissociation constant of the bisulfate ion from Raman and ion interaction modeling studies of aqueous sulfuric acid at low temperatures. *The Journal of Physical Chemistry A 107(21)*, 4322–4332.

Kobylin, P., Kaskiala, T. & Salminen, J. (2007). Modeling of H_2SO_4-$FeSO_4$-H_2O and H_2SO_4- $Fe_2(SO_4)_3$-H_2O systems for metallurgical applications. *Industrial & Engineering Chemistry Research 46(8)*, 2601–2608.

Kormanyos, B., Peintler, G., Nagy, A. & Nagypal, I. (2008). Peculiar kinetics of the complex formation in the iron(III)-sulfate system. *International Journal of Chemical Kinetics 40(3)*, 114–124.

Lee, C.K., & Tavlarides, L.L. (1985) Ionic equilibria in ferric sulfate-sulfuric acid solutions. *Polyhedron 4(1)*, 47–51

Lindstrom, R.E. & Wirth, H.E. (1969). Estimation of the bisulfate ion dissociation in solutions of sulfuric acid and sodium bisulfate. *The Journal of Physical Chemistry 73(1)*, 218–223.

Linke, W.F. & Seidell, A. (1965). *Solubilities: Inorganic and Metal-Organic Compounds; A Compilation of Solubility Data from the Periodical Literature.* (Vol. 2). Washington DC. American Chemical Society.

Liu, H., Papangelakis, V., Alam, M. & Singh, G. (2003). Solubility of hematite in H_2SO_4 solutions at 230-270°C. *Canadian metallurgical quarterly 42(2)*, 199–207.

Lund Myhre, C.E., Christensen, D.H., Nicolaisen, F.M. & Nielsen, C.J. (2003). Spectroscopic study of aqueous H_2SO_4 at different temperatures and compositions: variations in dissociation and optical properties. *The Journal of Physical Chemistry A 107(12)*, 1979–1991.

Magini, M. (1979). Solute structuring in aqueous iron(III) sulphate solutions. Evidence for the formation of iron(III)-sulphate complexes. *Journal of Chemical Physics 70(1)*, 317–324.

Magini, M. & Caminiti, R. (1977). On the structure of highly concentrated iron(III) salt solutions. *Journal of Inorganic and Nuclear Chemistry 39(1)*, 91–94.

Majzlan, J. & Myneni, S.C.B. (2005). Speciation of iron and sulfate in acid waters. Aqueous clusters to mineral precipitates. *Environmental Science & Technology 39(1)*, 188–194.

Martell, A.E. & Smith, R.M. (1976). *Critical stability constants: Inorganic complexes.* Plenum Press.

Papangelakis, V., Blakey, B. & Liao, H., 1994. Hematite solubility in sulphate process solutions, in: Hydrometallurgy '94. Springer, pp. 159–175.

Pitzer, K.S. (1973). Thermodynamics of electrolytes. I. Theoretical basis and general equations. *The Journal of Physical Chemistry 77(2)*, 268–277.

Pitzer, K.S. (1991). *Ion interaction approach: Theory and data correlation. Activity coefficients in electrolyte solutions* (2nd Ed.). Boca Raton, FL. CRC Press.

Rard, J. & Clegg, S. (1999). Isopiestic determination of the osmotic and activity coefficients of zH_2SO_4+(1-z)$MgSO_{4(aq)}$ at T = 298.15 K. II. Results for z = (0.43040, 0.28758, and 0.14399) and analysis with Pitzer's model. *The Journal of Chemical Thermodynamics 31(3)*, 399–429.

Reardon, E.J. & Beckie, R.D. (1987). Modelling chemical equilibria of acid mine-drainage: The $FeSO_4$-H_2SO_4-H_2O system. *Geochimica et Cosmochimica Acta 51(9)*, 2355–2368.

Rudolph, W. (1996). Structure and dissociation of the hydrogen sulphate ion in aqueous solution over a broad temperature range: a Raman study. *Zeitschrift fur Physikalische Chemie 194(1)*, 73–95.

Rudolph, W.W., Brooker, M.H. & Tremaine, P.R. (1997). Raman spectroscopic investigation of aqueous $FeSO_4$ in neutral and acidic solutions from 25°C to 303°C: inner- and outer-sphere complexes. *Journal of Solution Chemistry 26(8)*, 757–777.

Rudolph, W.W., Irmer, G. & Hefter, G.T. (2003). Raman spectroscopic investigation of speciation in $MgSO_{4(aq)}$. *Physical Chemistry Chemical Physics 5(23)*, 5253.

Rudolph, W.W., & Mason, R. (2001). Study of aqueous $Al_2(SO_4)_3$ solution under hydrothermal conditions: sulfate ion pairing, hydrolysis, and formation of hydronium alunite. *Journal of solution chemistry 30(6)*, 527–548.

Rudolph, W.W. (2010). Raman and infrared spectroscopic investigation of speciation in $BeSO_4$(aq). *Journal of solution chemistry 39(7)*, 1039–1059.

Rumyantsev, A.V., Hagemann, S. & Moog, H.C. (2004). Isopiestic investigation of the systems $Fe_2(SO_4)_3$-H_2SO_4-H_2O, $FeCl_3$-H_2O, and Fe(III)-(Na, K, Mg, Ca)Cln-H_2O at 298.15 K. *Zeitschrift für Physikalische Chemie 218(9)*, 1089–1127.

Smit, J.T., Steyl, J.D.T., Pelser, M. (2012). *Process for the recovery of metals such as nickel from an iron-containing ore by leaching with acidic sulfate solution.* EP2462249 A2, Anglo Operations Limited.

Sobron, P., Rull, F., Sobron, F., Sanz, A., Medina, J. & Nielsen, C. (2007). Modeling the physic-chemistry of acid sulfate waters through Raman spectroscopy of the system $FeSO_4$-H_2SO4-H_2O. *Journal of Raman Spectroscopy 38(9)*, 1127–1132.

Steyl, J. (2009). Kinetic modelling of chemical processes in acid solution at T< 200°C. (i) Thermodynamics and speciation in H_2SO_4-Metal(II)SO_4-H_2O systems, in: SAIMM Hydrometallurgy Conference 2009, pp. 401–443.

Steyl, J. (2012). *Simulating the medium temperature chalcopyrite oxidation system in batch and continuous autoclaves.* Ph.D. thesis. Stellenbosch University. Stellenbosch, South Africa.

Stipp, S.L. (1990). Speciation in the iron(II)-iron(III)-sulfate-water system at 25°C and low pH: Sensitivity of an equilibrium model to uncertainties. *Environmental Science & Technology 24(5)*, 699–706.

Tosca, N.J., Smirnov, A. & McLennan, S.M. (2007). Application of the Pitzer ion interaction model to isopiestic data for the $Fe_2(SO_4)_3$-H_2SO_4-H_2O system at 298.15 and 323.15 K. *Geochimica et Cosmochimica Acta 71*, 2680–2698.

Velazquez-Rivera, M., Palmer, D. & Kettler, R. (2006). Isopiestic measurement of the osmotic coefficients of aqueous $xH_2SO_4 + (1-x)Fe_2(SO_4)_3$ solutions at 298.15 and 323.15 K. *Journal of Solution Chemistry 35(12)*, 1699–1730.

Yue, G., Zhao, L., Olvera, O.G. & Asselin, E. (2014). Speciation of the H_2SO_4-$Fe_2(SO_4)_3$-$FeSO_4$-H_2O system and development of an expression to predict the redox potential of the Fe^{3+}/Fe^{2+} couple up to 150°C. *Hydrometallurgy 147-148*, 196–209.

A COST PENALTY ASSOCIATED WITH FERROCYANIDE IN GOLD PROCESSING USING CYANIDE

Chris Fleming*, Mariam Melashvili, Rene Jackman

SGS Canada Inc.
185 Concession Street
Lakefield, Ontario, K0L 2H0
*(*Corresponding author: chris.fleming@sgs.com)*

ABSTRACT

A study was undertaken to develop a flowsheet for treating a high grade silver/gold ore body by cyanidation, CCD and Merrill Crowe precipitation, with tailings detoxification to meet stringent limits for cyanide and thiocyanate concentrations in the final discharge. In one of the flowsheets examined, free cyanide in the final leach liquor was recycled to the mill where it was hoped that it might react beneficially with fresh ore to reduce the cyanide makeup by an equivalent amount. However, laboratory results suggested that most of the cyanide recycled to the mill reacted with iron forming a soluble ferrocyanide complex. This would have had a negative impact on cost, not only because the cyanide recycled to the mill could not be claimed as a credit, but also because the presence of ferrocyanide in the final tailings meant copper sulphate had to be added to the SO_2/air cyanide detoxification process to precipitate the iron cyanide. In an alternative flowsheet that was examined, recycling of cyanide to the mill was abandoned, and this allowed the milled slurry to be pre-aerated in a single mixing tank before cyanidation. During alkaline pre-aeration, any ferrous ions that are generated in the mill are oxidized to ferric ions, which, unlike ferrous ions, are inert in cyanide solution. This significantly reduced cyanide consumption and lowered costs in the detoxification of the tailings. In this paper, the cost implication of these two circuits is compared.

KEYWORDS

Ferrocyanide, ferricyanide, gold leaching, cyanide destruction

INTRODUCTION

Most of the gold that is produced by hydrometallurgical processes is recovered by cyanide leaching. The reason for that lies in the superior thermodynamic stability of the gold cyanide complex and the favorable economics this confers on the process (Fleming, 2014). However, the gold in most of the gold-bearing ores processed is a very minor component, and therefore, the leaching response is often dominated by that of the host minerals.

Many gold ores treated by the cyanidation process contain iron minerals. Cyanide reacts readily with some iron minerals forming ferrocyanide as one of the compounds. Ferrocyanide is stable when solutions are exposed to air and does not retard the dissolution of gold. In fact, it has been claimed that potassium ferrocyanide in the presence of oxygen is a weak solvent for gold (Hedley & Tabachnick, 1958). The main problem with the formation of this compound is the consumption of cyanide. Ferrocyanide contains six molecules of cyanide for each atom of iron $\{Fe(CN)_6^{4-}\}$ so the wasteful consumption of cyanide is greatly increased by its presence. Thiocyanate is also often produced during the reaction of certain iron sulphide minerals with cyanide, and contributes to the wasteful consumption of cyanide. In addition, environmental regulations in certain countries demand that thiocyanate be removed from the tailings, and this adds significantly to the cost of tailings treatment.

A generic gold recovery circuit is based on four main unit operations: (1) crushing and milling the gold- bearing ore to a size in which the gold particles are liberated from the host rock, (2) leaching of gold, which could take place either during or immediately after milling, and uses oxygen as the oxidant and the cyanide ion as the complexing ligand for gold, (3) recovery of gold from the leach solution by carbon adsorption, zinc dust precipitation (Merrill-Crowe), ion exchange, direct electrowinning, etc. and (4) cyanide destruction by SO_2/air, peroxide, hypochlorite, etc. The choice of one method versus another is usually based on a capital and operating cost analysis, and the ability to meet stringent limits for discharge of waste solutions to the natural environment. If the reactive nature of the host minerals with cyanide complicates the processing circuit, then the favorable economics of cyanide leaching of gold could be compromised.

The objective of this study was to establish the processing circuit and operating conditions that achieved the best compromise between the benefits of higher gold and silver recoveries and the cost of higher cyanide consumption when soluble iron sulphide minerals are present. The samples used to produce the test results were composites of high, medium, and low grade ores from the deposit. A number of scenarios were examined to gain some insight into the operating parameters that would be of greatest sensitivity in the comparison. The main variables considered in the test matrix were the ore type (low, mid, and high grade composites) and cyanide concentration in the leach (three levels for each ore type). Two basic flowsheets and two variations on the basic flowsheets (cyanide recovery, cyanide destruction) were compared. It was assumed in each case that cyanide and thiocyanate would have to be destroyed in the final tails solutions, to levels of <5 mg/L and <10 mg/L, respectively, to meet the limits imposed in the country in which the project was being developed.

GENERIC CIRCUIT

The generic circuit involved grinding, cyanide leaching, washing of the leach residue by counter current decantation (CCD), Merrill Crowe (MC) recovery of gold from the pregnant leach liquor (PLS), cyanide and thiocyanate destruction in the tailings. The process that was developed for detoxification of the tailing was a two-step process; first cyanide was destroyed (to <5 mg/L CN_T) by the conventional SO_2/air process, and then thiocyanate was oxidized to (<10 mg/L) using hypochlorite. A simplified block flow diagram is shown in Figure 1. In this flowsheet, part of the MC barren solution is recycled back to the mill in order to reduce new cyanide addition costs and/or reduce cyanide destruction costs.

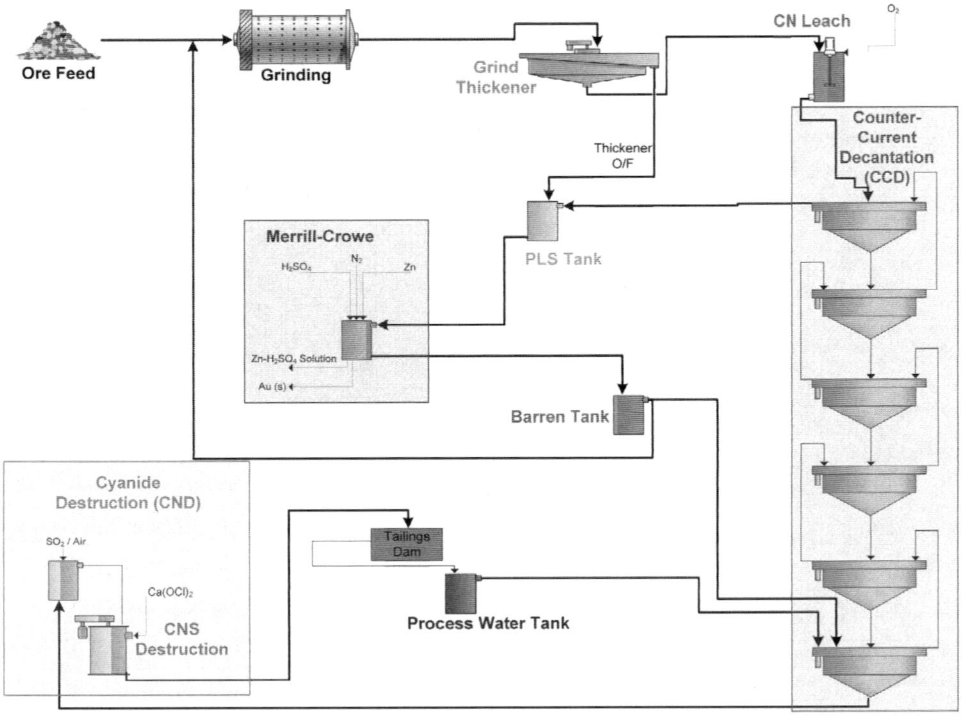

Figure 1 – Generic gold leaching circuit

The water balance for this circuit was based on laboratory data and incorporated the following parameters: 142 t/h ore treatment rate, 174 m^3/h solution flowrate to leach, 45% solids in leach, 120 h leach residence time, six stages of CCD at a 3:1 wash ratio and 521 m^3/h PLS flowrate to MC. The main goal of this exercise was to account for cyanide distribution throughout the circuit shown in Figure 1.

In almost all leach operations, some or all of the cyanide is consumed or lost due to evaporation, to hydrolysis, or to formation of thiocyanate, cyanate, ferrocyanide, copper cyanide and silver cyanide. Cyanide present as silver cyanide is considered to be consumed cyanide since it is converted from the silver cyanide complex to the zinc cyanide complex in the Merrill Crowe process. The concentrations of some of the species in the leach solution are given in Table 1, based on the experimental leach data.

Table 1 – Leach solution composition based on both laboratory data and model calculations for the generic circuit

ORE TYPE	LEACH	LEACH LIQUOR (mg/L)					NaCN CONSUMED (kg/t)
	(mg/L)	SCN	Ag	Fe	Cu	$NaCN_F$	LD
HIGH GRADE	1000	141	235	104	56	267	2.68
	1500	143	247	128	56	420	3.93
	2000	151	251	300	56	524	5.53
MID GRADE	750	79	142	30	54	237	1.91
	1000	83	138	64	56	306	2.52
	1500	81	152	140	56	462	3.79
LOW GRADE	300	93	94	0.2	67	60	0.62
	500	108	112	1.3	70	134	0.96
	750	104	108	22	60	270	1.82

The cyanide consumption figures in the table are based on actual laboratory data (LD) (subtracting the titratable free cyanide remaining in the leach liquor at the end of the 120 h leach period from the total amount of cyanide added during leaching) plus water balances determined by the model calculation.

The calculated distribution of cyanide consumption in various stages of the generic circuit is given in Tables 2 to 4.

Table 2 – Cyanide consumption data for the HIGH GRADE composite - generic circuit

NaCN (kg/t) in Circuit	1000 mg/L NaCN	1500 mg/L NaCN	2000 mg/L NaCN
NaCN makeup (Leaching)	2.35	2.79	4.11
NaCN recovered (Recycle Barren to Grinding)	0.33	1.14	1.42
NaCN makeup total (Recycle consumed in Grinding)	2.68	3.93	5.53
NaCN lost from the circuit (CCD U/F)	0.33	0.51	0.64

Table 3 – Cyanide consumption data for the MID GRADE composite- generic circuit

NaCN (kg/t) in Circuit	750 mg/L NaCN	1000 mg/L NaCN	1500 mg/L NaCN
NaCN makeup (Leaching)	1.27	1.69	2.54
NaCN recovered (Recycle Barren to Grinding)	0.64	0.83	1.25
NaCN makeup total (Recycle consumed in Grinding)	1.91	2.52	3.79
NaCN lost from the circuit (CCD U/F)	0.29	0.37	0.56

Table 4 – Cyanide consumption data for the LOW GRADE composite- generic circuit

NaCN (kg/t) in Circuit	300 mg/L NaCN	500 mg/L NaCN	750 mg/L NaCN
NaCN makeup (Leaching)	0.46	0.60	1.09
NaCN recovered (Recycle Barren to Grinding)	0.16	0.36	0.73
NaCN makeup total (Recycle consumed in Grinding)	0.62	0.96	1.82
NaCN lost from the circuit (CCD U/F)	0.07	0.16	0.33

In Tables 2 to 4, the NaCN makeup includes available free cyanide after leaching plus the cyanide that is lost to tailing from the thickener underflow from the final CCD tank. From these data, the trend of increased consumption of cyanide with higher cyanide concentration in the leach is obvious for all three composites.

In the economic projections for the generic circuit presented in Figure 1, it was assumed that ~40% of the free cyanide in the final leach liquor ($NaCN_F$ in Table 1) would be recycled to the mill, where it would react beneficially with fresh ore. It was therefore assumed (in the initial NaCN makeup calculation) that

NaCN makeup would be reduced by an equivalent amount. However, the tests that were carried out at SGS Lakefield indicated that this is not likely to occur. In one test, NaCN was added into the mill and the solids and the mill water were analyzed for gold and silver. It was found that only 3.5% of the gold and 0% of the silver had leached. Moreover, analysis of the mill water for iron suggested that almost all of the cyanide added to the mill had been converted to ferrocyanide. By the end of the leach process, the iron in solution in this test was 395 mg/L, whereas in an identical test (without cyanide addition to the mill), the final iron in solution after leaching was 231 mg/L. The differential between these two iron values represents 456 mg/L CN (860 mg/L NaCN equivalent) tied up as the ferrocyanide complex, an amount which is more than that available for recycle under any of the experimental conditions depicted in Table 1.

Since the recycled cyanide will almost certainly react with iron and form soluble ferrocyanide in the mill, the total NaCN makeup in this flowsheet will have to incorporate the cyanide recycled to grinding and the "NaCN makeup total". The values in Tables 2 to 4 reflect the increase in cyanide consumption due to this reaction. Furthermore, the higher level of dissolved iron in the tailings means additional copper will have to be added during the SO_2/air process to precipitate ferrocyanide as the $Cu_2Fe(CN)_6$ complex. The requirement for copper is 3.4 mg/L for every 1 mg/L Fe in solution. Because the LOW GRADE ore leach solution already contains 60-70 mg/L Cu and <20 mg/L Fe, it is assumed that no additional copper would be needed in this case. However, if soluble iron accumulates due to cyanide recycle in the mill during MID GRADE and HIGH GRADE ore leaching, it will be necessary to increase the dosage of copper to precipitate the iron.

ALTERNATIVE CIRCUIT

In the first alternative flowsheet examined, recycling of a portion of the MC barren solution to the mill was abandoned and the entire flow of barren solution was sent to the CCD circuit as wash water (rather than the base case of a 60:40 split between CCD and mill recycle). In this case, all of the cyanide in the MC barren solution reports to the CN/SCN detoxification circuit via the underflow from the last CCD thickener plus a bleed stream. This would increase the amount of cyanide that must be destroyed with sodium bisulphite in the generic circuit by a factor of 1.0/0.6 = 1.67.

If cyanide is not recycled to the mill, it is possible to incorporate an alkaline pre-aeration tank into the circuit ahead of the first leach tank, and laboratory testing showed that the iron concentration in solution is lowered by 100-300 mg/L in this way. This has the potential to lower overall cyanide consumption by up to 2 kg/t NaCN, a quantity which is greater than the amount of cyanide that could potentially have been recovered by recycling MC barrens to the mill. During the alkaline pre-aeration stage (pH 10-11), ferrous ions in solution are oxidized to the ferric state, and precipitate as ferric hydroxide. Ferric hydroxide is inert in cyanide solution, and this is the reason that cyanide consumption can be lowered so significantly.

However, laboratory test work on this project showed that the iron that precipitated as ferric hydroxide in the pre-aeration stage and as copper ferrocyanide during the SO_2/air process tended to re-dissolve during hypochlorite treatment of the tailings to destroy thiocyanate. It is speculated that the copper ferrocyanide that was precipitated during the first stage of the detoxification process is oxidized to a soluble ferricyanide complex $\{Fe(CN)_6^{3-}\}$ during the second stage hypochlorite treatment, bringing a significant concentration of cyanide back into solution. This flowsheet, therefore, failed to achieve the cyanide target for discharge of the tailings to the environment.

A third flowsheet was therefore developed. In order to keep iron in solution at as low a level as possible, this flowsheet also incorporated the alkaline pre-aeration step, but the two stage detoxification process (SO_2/air for CN followed by hypochlorite for SCN) was moved from the final tailings slurry to the MC barren solution. A variation on this flowsheet (which was considered as a paper study only) would be to incorporate CN recovery by AVR in place of the SO_2/air destruction of cyanide. In both cases, the final step would still involve thiocyanate destruction with hypochlorite.

It was shown in laboratory tests that the two-step process involving CN destruction with SO_2/air, followed by SCN (plus residual CN) destruction by hypochlorite produced a treated MC barren solution containing <5 mg/L CN_T and <10 mg/L SCN. When this liquor was then used as wash water in a 6 stage CCD circuit (where a wash efficiency of ~99.7% will be achieved), the final tailings from CCD (CCD #6 underflow), also contained <5 mg/L CN and <10 mg/L SCN, and met the environmental target for direct discharge to the tailings dam without the need for further detoxification. This was the flowsheet adopted for the full scale plant when it was built, and a simplified block diagram is shown in Figure 2.

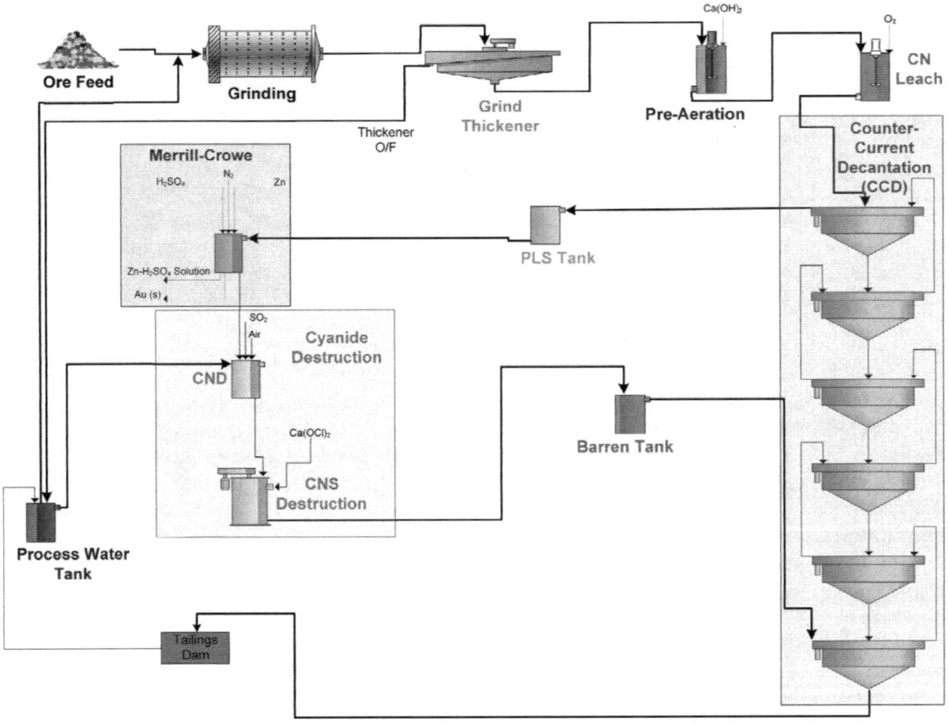

Figure 2 – Alternative gold leaching circuit

The analyzed concentrations of Ag, SCN and Cu in solution in these tests were very similar to those shown in Table 1, but the concentration of iron was significantly lower, because of the incorporation of the pre-aeration stage in this flowsheet. The impact of pre-aeration on the concentration of iron in solution and on cyanide consumption is summarized in Table 5.

Table 5 – Leach solution composition and cyanide consumptions for the alternative circuit

ORE TYPE	NaCN IN LEACH (mg/L)	LEACH LIQUOR, Fe (mg/L) No pre-aeration	LEACH LIQUOR, Fe (mg/L) With pre-aeration	NaCN CONSUMED, LD (kg/t) No pre-aeration	NaCN CONSUMED, LD (kg/t) With pre-aeration	NaCN CONSUMED, LD (kg/t) With pre-aeration & AVR
HIGH GRADE	1000	104	26	2.68	2.10	0.94
	1500	128	32	3.93	2.69	0.92
	2000	300	72	5.53	3.77	1.39
MID GRADE	750	30	6	1.91	1.48	0.66
	1000	64	14	2.52	1.95	0.87
	1500	140	30	3.79	2.94	1.33
LOW GRADE	300	0.2	0.2	0.62	0.62	0.31
	500	1.3	0.6	0.96	0.96	0.42
	750	22	6.0	1.82	1.58	0.66

Calculated consumptions of cyanide are shown in Tables 6 to 8 for a flowsheet that incorporates cyanide recovery from the MC barren solution by AVR.

Table 6 – Cyanide consumption data for the HIGH GRADE composite - alternative circuit

NaCN (kg/t) in Circuit	1000 mg/L NaCN	1500 mg/L NaCN	2000 mg/L NaCN
NaCN makeup (Leaching)	0.94	0.92	1.39
NaCN recoverable by AVR	1.16	1.77	2.38
NaCN makeup (No AVR included)	2.10	2.69	3.77
NaCN lost from the circuit (CCD U/F)	0.00	0.00	0.00

Table 7 – Cyanide consumption data for the MID GRADE Composite - alternative circuit

NaCN (kg/t) in Circuit	750 mg/L NaCN	1000 mg/L NaCN	1500 mg/L NaCN
NaCN makeup (Leaching)	0.66	0.87	1.33
NaCN recoverable by AVR	0.82	1.08	1.61
NaCN makeup (No AVR included)	1.48	1.95	2.94
NaCN lost from the circuit (CCD U/F)	0.00	0.00	0.00

Table 8 – Cyanide consumption data for the LOW GRADE composite - alternative circuit

NaCN (kg/t) in Circuit	300 mg/L NaCN	500 mg/L NaCN	750 mg/L NaCN
NaCN makeup (Leaching)	0.31	0.42	0.66
NaCN recoverable by AVR	0.31	0.54	0.92
NaCN makeup (No AVR included)	0.62	0.96	1.58
NaCN lost from the circuit (CCD U/F)	0.00	0.00	0.00

This alternative circuit design reduces the NaCN makeup values for all three composites. This is mainly due to the incorporation of AVR and the pre-aeration stage, which renders the iron less soluble in cyanide media. In addition, much less reagent ($Ca(OCl)_2$) would be needed to destroy thiocyanate when treating solution (MC barrens) compared to the slurry in the CCD #6 underflow. This happens because there are undoubtedly other oxidizable species in the solid tailings that are not present in the leach solution. This alternative circuit was confirmed to meet the environmental target for direct discharge to the tailings dam without any further requirements.

PROCESS COST COMPARISON

The leaching data indicated that the extraction of gold and silver in all three composites improved with increasing cyanide concentration and the benefit of this improvement has been quantified on a $/t basis using metal price scenarios of 1200 US $/oz for Au and 15 US $/oz for Ag. The residue grades at three levels of cyanide concentration in leaching are shown in Table 9. The residue values were similar at the same free cyanide concentration, regardless of the circuit configuration. The increased value for each ore type is shown relative to the gold and silver recoveries at the lowest cyanide strength.

Table 9 – Gold and silver recoveries and associated costs

ORE TYPE	NaCN IN LEACH (mg/L)	LEACH RESIDUE ASSAY Au (g/t)	LEACH RESIDUE ASSAY Ag (g/t)	INCREASED VALUE ($/t)
	1000	1.44	79.8	0
HIGH GRADE	1500	1.30	67.1	11.5
	2000	1.13	55.3	23.8
	750	1.10	43.4	0
MID GRADE	1000	0.96	33.7	10.1
	1500	0.91	34.7	11.5
	300	0.47	33.2	0
LOW GRADE	500	0.47	24.6	4.1
	750	0.34	15.7	13.5

The improved gold and silver recoveries come at the cost of increased cyanide consumption, as shown by the data in the previous tables. Increasing the cyanide concentration resulted in a large increase in the cyanide consumption and consequently the cost. The large increase in cost is mainly due to the increase in the concentration of dissolved iron in all the flowsheets and to the free cyanide that reports to tailings and is destroyed in the first two flowsheets. The cyanide consumption values for both processing circuits are compared in Table 10. This table also shows the cyanide cost based on a 2.5 US $/kg NaCN reagent price.

Table 10 – Cyanide consumption and the associated cost

ORE TYPE	NaCN IN LEACH (mg/L)	NaCN CONSUMED (kg/t)			NaCN COST ($/t)		
		Generic Circuit	Alternative Circuit	Alternative Circuit (AVR)	Generic Circuit	Alternative Circuit	Alternative Circuit (AVR)
	1000	2.68	2.10	0.94	6.70	5.25	2.35
HIGH GRADE	1500	3.93	2.69	0.92	9.83	6.73	2.30
	2000	5.53	3.77	1.39	13.83	9.43	3.48
	750	1.91	1.48	0.66	4.78	3.70	1.65
MID GRADE	1000	2.52	1.95	0.87	6.30	4.88	2.18
	1500	3.79	2.94	1.33	9.48	7.35	3.33
	300	0.62	0.62	0.31	1.55	1.55	0.78
LOW GRADE	500	0.96	0.96	0.42	2.40	2.40	1.05
	750	1.82	1.58	0.66	4.55	3.95	1.65

The data presented in Table 10 show that the costs related to NaCN purchases are less sensitive to changes in the NaCN concentration in the alternative circuit that incorporates pre-aeration and detoxification of the MC barrens, than in the generic circuit that uses CN/SCN destruction in the CCD #6 thickener underflow. Moreover, this comparison does not include the cost of SO_2 in the cyanide destruction process, or the cost of $Ca(OCl)_2$ in the thiocyanate destruction process, or even the cost of additional copper sulphate that would be required if recycling of cyanide to the mill was contemplated in the generic flowsheet. In all cases, the detoxification-related costs would be substantially higher in the generic flowsheet than in the alternative flowsheet. All of this leads to the conclusion that the operation of the plant can be geared to increasing cyanide concentration to maximize gold and silver recovery if the alternative circuit is selected, whereas the operation of the plant will have to be geared to minimizing cyanide addition if the generic circuit is chosen. Overall, gold and silver recoveries will suffer as a result.

The data in Tables 6 to 8 show that cyanide consumption can be halved by the incorporation of AVR, with significant operating cost benefits (Table 10). Despite this, the process that was incorporated when the full scale commercial plant was built actually employed CN destruction of the MC barrens option rather than AVR. This was based on the much lower capital cost of the SO_2/air process as well as the important consideration that the cyanide destruction process would be much simpler to operate in a remote location.

CONCLUSIONS

A generic flowsheet to recover gold and silver from a high grade deposit by cyanidation was tested. Because of the very high concentration of silver in the ore, Merrill Crowe was selected as the process for recovering the precious metals from the leach liquor. Both cyanide and thiocyanate had to be destroyed in the tailings in the jurisdiction of this deposit, and a two-stage tailings detoxification process was developed. In the first stage, cyanide was oxidized in the well known SO_2/air process, and ferrocyanide in the tailings was precipitated from solution as the cupric ferrocyanide complex, $Cu_2Fe(CN)_6$. Thiocyanate was then oxidized with hypochlorite in the second stage. Detoxification was initially applied to the final tailings slurry, which was generated in a 6 stage CCD circuit in which the leach residue was washed with MC barren solution.

The leaching studies showed that the gold and (particularly) silver recoveries improved as the concentration of free cyanide was increased. However, this improvement came at the cost of increased cyanide consumption and increased cyanide destruction costs. A significant amount of the cyanide added to the leach was consumed by reaction with ferrous ions in the milled slurry, to form the stable ferrocyanide complex. Various flowsheets were examined aimed at reducing the impact of this reaction and lowering cyanide consumption and destruction costs.

Much of the cyanide needed to drive the recovery of gold and silver to higher values remains in the final leach liquor as free cyanide, and in seeking ways of lowering cyanide-related costs, one of the flowsheets examined the impact of recycling a portion of the free cyanide in the MC barren solution to the mill. This was not successful. The cyanide that was recycled to the mill failed to leach additional gold and silver; in fact, it reacted with ferrous ions in the mill to produce much higher levels of soluble ferrocyanide, a process which increased the cost of detoxification of the final tailings.

In a variation of this flowsheet, recycling of cyanide to the mill was abandoned, and this allowed the milled slurry to be pre-aerated before cyanidation. Pre-aeration under alkaline conditions oxidizes ferrous ions to the ferric state; the ferric ions then precipitate as ferric hydroxide. This compound is inert in cyanide solution and cyanide consumption is substantially reduced. The cyanide and thiocyanate in the tailings were again treated in a two stage detoxification process that was applied to the final tailings slurry (CCD #6 thickener underflow). Whilst this flowsheet succeeded in the primary objective of significantly reducing ferrocyanide formation and cyanide consumption, it failed to meet the cyanide target for discharge of the tailings to the environment. This is believed to be due to the oxidation of the precipitated cupric ferrocyanide to the soluble ferricyanide complex, during treatment of the tailings with hypochlorite to destroy thiocyanate.

This led to the development of an alternative flowsheet in which the MC barren solution (rather than the tailings slurry) was treated to reduce the cyanide and thiocyanate concentrations to the required discharge limits. Ferrocyanide formation was again minimized by eliminating the cyanide recycle to the mill and by pre-aerating the milled slurry prior to cyanidation. Once the MC barren solution had been treated by the two-stage detoxification process to remove cyanide and thiocyanate, it was used as wash water in the CCD circuit, and this produced a final tailings slurry that met the required environmental limits for these two species. A paper study was conducted on the potential benefits of recovering and recycling the cyanide in the MC barrens by AVR, rather than destroying it with the SO_2/air process.

The economics looked very favourable, with potential operating cost savings of up to US \$5/t of ore treated. However, the capital cost of AVR would have been substantially higher than that of the SO_2/air process and this, coupled with the fact the SO_2/air process would be much simpler and safer to operate in a remote location, led to the decision to base the full scale commercial plant on cyanide destruction rather than on cyanide recovery. Even with this greatly improved alternative flowsheet, which was adopted for the commercial plant, minimization of ferrocyanide formation in the leach circuit has been critical to achieve the required targets for cyanide (<5 mg/L) and thiocyanate (<10 mg/L) discharge to the environment.

REFERENCES

Fleming, C.A. (2014). SGS Gold Workshop - Cyanidation of Gold and Silver. Chapter 3.

Hedley, N., Tabachnick, H. (1958). Chemistry of cyanidation. Mineral Dressing Notes 23. New York: American Cyanamid Company.

IRON BLEED IN THE MABOUMINE PROCESS

Philippe Ribagnac, *Bruno Courtaud, Jean-Marie Lambert, and Valérie Weigel

Eramet Research, Hydrometallurgy Department
1 avenue Albert Einstein, BP 120, F-78193 Trappes, France
*(*Corresponding author: bruno.courtaud@erametgroup.com)*

ABSTRACT

Maboumié is a polymetallic deposit in Gabon containing niobium (1.2 wt.%), tantalum (0.03 wt.%), rare earth elements (REE) (1.4 wt.%) and uranium (0.03 wt.%) that are mostly carried by pyrochlore minerals. The classical route for niobium recovery is a pyrometallurgical process that consists of pyrochlore flotation, dephosphorization of the niobium concentrate and aluminothermy to obtain ferro-niobium. Pyrochlore concentrates can also be treated by a hydrometallurgical process with a mixture of sulfuric and hydrofluoric acid leaching followed by solvent extraction to separate niobium and tantalum. However, the first process does not allow recovering REE and the second one is not environmentally friendly because of the use of HF. ERAMET is looking for new opportunities on strategic metals market and has recently developed a hydrometallurgical process for the recovery of all these metals based on a sulfuric acid treatment of the ore (Donati, Courtaud, & Weigel, 2014). As for most of the oxidized ore processes, the major impurity is iron (~30 wt.%). Thus, in this hydrometallurgical process, the question arises of the elimination and stabilization of the iron as residue. The iron in solution is ferrous; goethite precipitation seems the most adapted crystalline form. Lab and pilot tests have shown kinetics of precipitation of 6 $g.L^{-1}.h^{-1}$ at pH 4.5. To reduce the salt discharge, sodium-jarosite form is more adapted, but at pH 2, 95°C only 30% of iron is precipitated in 5 hours, compared to 100% in goethite form at the same time and pH 2, 80°C. Different ways of oxidation were tested, air, pure O_2, activated carbon, without drastic effect. The more efficient was ozone use, at room temperature; the kinetics of oxidation is five time faster than the other ways with 30% of sodium removed from the solution.

KEYWORDS

Iron oxidation, goethite, jarosite, hydrometallurgy, ERAMET

INTRODUCTION

ERAMET is a world leader in alloying metals, particularly manganese and nickel, and in high-quality metallurgy. In addition, ERAMET has major research and development projects in new business lines with growth potential, such as titanium dioxide, zircon, lithium, niobium and the rare earths, as well as recycling. The group employs about 14,000 people in 20 countries, through their subsidiaries, like, for example COMILOG in Gabon for manganese, Société Le Nickel in New Caledonia for nickel and Aubert & Duval for alloys in France (ERAMET).

Maboumine, an ERAMET Group subsidiary through COMILOG, is aimed at developing a mining project for the recovery of niobium (1.2 wt.%), uranium (0.03 wt.%) and rare earth elements (REE, 1.4 wt.%) considered as strategic metals from a polymetallic deposit in Mabounié in Gabon (Maboumine). ERAMET has recently developed a hydrometallurgical process to recover those valuable elements. In the upstream part of the process (Figure 1), goethite and crandalite in the non mag ore are first leached with concentrated sulfuric acid so that a part of the REE and the uranium are solubilized into the leaching filtrate (Agin et al., 2012; Donati et al., 2014). During this step, the ORP of the aqueous solution is maintained in order to get iron(II) species in solution and decrease the niobium solubility. Refractory minerals present in the leaching concentrate are then decomposed during pasting with pure sulfuric acid and roasting. Valuable elements are solubilised by leaching the calcined solid with water. Finally, niobium and REE are separated during the thermal precipitation step, which allow obtaining a 10-15 wt.% Nb concentrate whereas REE remain in the aqueous phase. This solid is then purified to obtain a Nb oxide that could be used to produce ferro-niobium (Delvallee et al., 2015; Deblonde et al., 2015a; 2015b; 2016).

The filtrate containing the REE is then recycled to the leach in order to maximize the acid utilization rate. This leads to a Pregnant Leach Solution (PLS) containing up to 2.7 g.L^{-1} REE as well as 90 mg.L^{-1} uranium and 27 mg.L^{-1} scandium. Despite their very low concentrations, uranium and scandium could be economically recovered as by-products by using solvent extraction processes to produce sodium diuranate (SDU) and scandium oxide (Vincec et al., 2014a; Beltrami et al., 2016a).

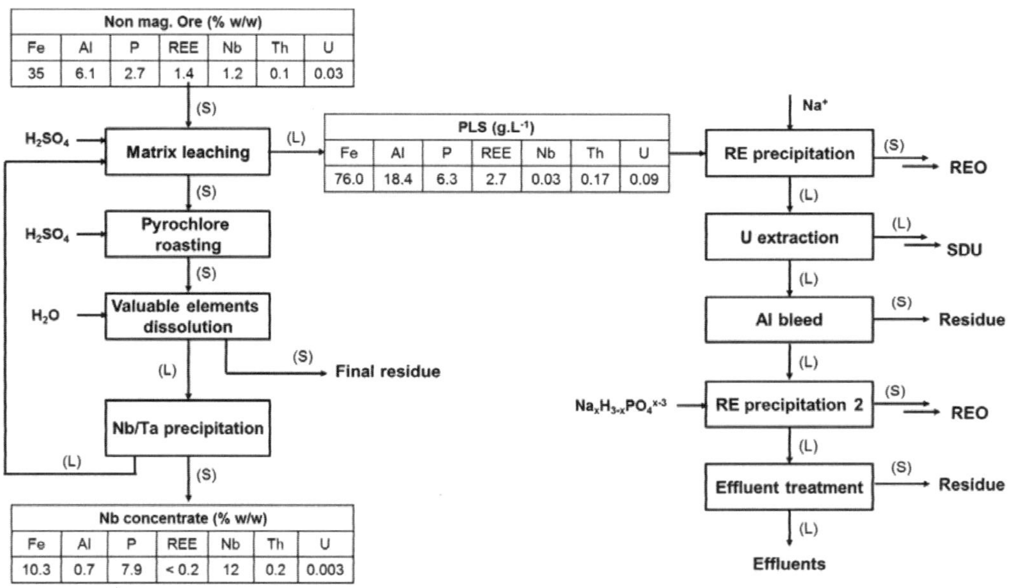

Figure 1 – Simplified flow sheet of the Maboumine hydrometallurgical process main streams

Light and Medium REE from the PLS are easily recovered as REE double salts [$NaREE(SO_4)_2.H_2O$] by the addition of a sodium salt (Beltrami et al., 2016b). After the aluminium bleed, REE remaining in the aqueous solution are precipitated as REE phosphate ($REEPO_4$) by adding phosphate (Vincec et al., 2014b; Beltrami et al., 2015). After the recovery of REE from the PLS, the main stream still contains most of the initial iron and manganese, as well as sodium from the double salts precipitation (Table 1). At this stage, iron is ferrous to avoid its precipitation as phosphate ($FePO_4$) and the pollution of the Heavy REE concentrate. The treatment of this effluent is presented in this paper.

Table 1 – Composition of the PLS after the recovery of REE

Element	Al	REE	U	Th	Ca	Na	Mg	Mn	Fe
$mg.L^{-1}$	< 1	< 1	< 1	< 1	570	5,200	210	2,900	34,300

METHODOLOGY

The Maboumine process development has involved around a hundred people worldwide, with laboratories in France, Australia, Canada, United States, South-Africa and Finland. The Research effort for developing the Maboumine process was organized around ERAMET Research with a forty people team and followed a methodology with several steps, as described hereafter:

- mineralogical study of the ore,
- laboratory tests to define the process steps and the related chemical and physical conditions,
- pilot plant scale operation in order to validate the process in a continuous configuration,
- demonstration plant in order to validate technology options, and to provide data for the scale up to the commercial plant.

Currently, the process has been entirely operated at pilot plant scale in France since the end of 2012 and four patents have been filed on the Maboumine process (Agin et al., 2012; Delvallee et al., 2015; Vincec, 2014a & 2014b).

OXIDATION AND PRECIPITATION OF IRON FROM PLS

Preliminary Experiments

A bibliographic review clearly indicates that two different routes allow iron bleeding: precipitation of jarosite [$NaFe_3(OH)_6(SO_4)_2$] or goethite ($FeOOH$). From our starting solution, these two routes need the oxidation of ferrous iron present in solution to ferric iron. Jarosite presents a major advantage in the co-precipitation of sodium and sulphate, but also a drawback in the pH range that allow its formation. Jarosite indeed appears only in acidic solution ($pH<2$). In more neutralized media, goethite will be formed. In addition, jarosite formation requires temperatures above 95°C (Table 2).

Table 2 – Comparison of jarosite and goethite processes

	Jarosite	Goethite
Advantages	Na and SO_4 bleed	--
Temperature	95°C	80°C
pH	1.5 – 2	2 – 5

Laboratory scale experiments were conducted in order to evaluate the oxidation of iron at two different pH and temperature conditions. Typical experiments were conducted in a 1-liter reactor filed with filtrate coming from previous pilot campaigns. If needed, addition of sodium sulphate was carried out to obtain the composition given in Table 1. The reactor was heated thanks to a hot-plate, insulated and equipped with a reflux condenser. The agitation system consisted of a Rushton impeller, 4 inclined blades and current deflectors. Neutralization was realised using lime.

Firstly, air oxidation was tried combined with a pH regulation. Starting from a solution containing 28 $g.L^{-1}$ of iron(II), a reasonably fast oxidation was reported at pH=4. In five hours at 80°C, the iron(II) concentration fell to 0.25 $g.L^{-1}$. The oxidation speed was thus measured at 5.9 $g.L^{-1}.h^{-1}$. Iron oxidation at pH=2, 95°C and in the same other conditions was slower. The oxidation was not complete and 20 $g.L^{-1}$ iron(II) remained in solution. In these second conditions, the oxidation speed was 1.5 $g.L^{-1}.h^{-1}$ (Figure 2). To be noted in both cases, less than 1 $g.L^{-1}$ of ferric iron was found in solution, meaning a nearly complete precipitation of iron(III) species.

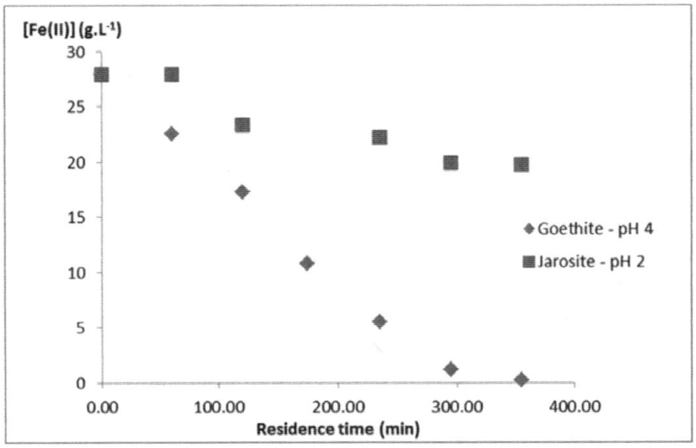

Figure 2 – Iron(II) oxidation in jarosite or goethite conditions

As two parameters varied between the two first experiments, a more complete study was performed to deeply understand the impact of temperature, pH and oxidant flow rate (Figure 3). The following trends appeared:

- Temperature was found as the principal parameter on iron oxidation speed. Reducing the temperature from 95 to 85°C, at pH=2 led to a doubling in iron oxidation speed from 1.5 to 2.8 $g.L^{-1}.h^{-1}$,
- pH appeared also as of prior importance as oxidation occurred two times faster at pH=4 than at pH=2,
- air flow rate finally did not show any influence on the oxidation speed.

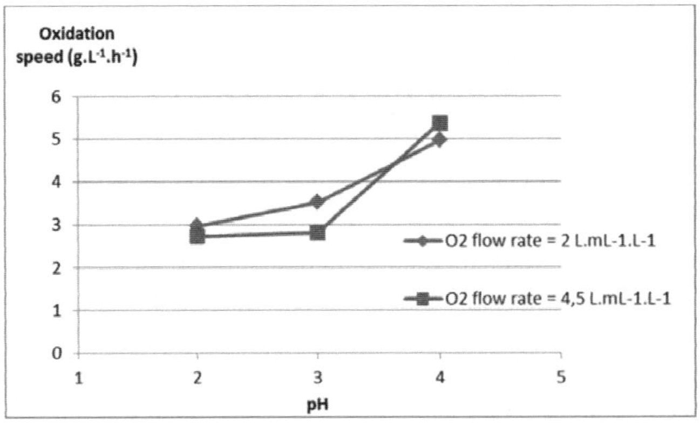

Figure 3 – Iron(II) oxidation by air at different pH and air flow rate

These results are in agreement with what is known for years on the ferrous iron oxidation by air and has been formalized by Goto et al. in 1970 and used in many studies since (Figure 4).

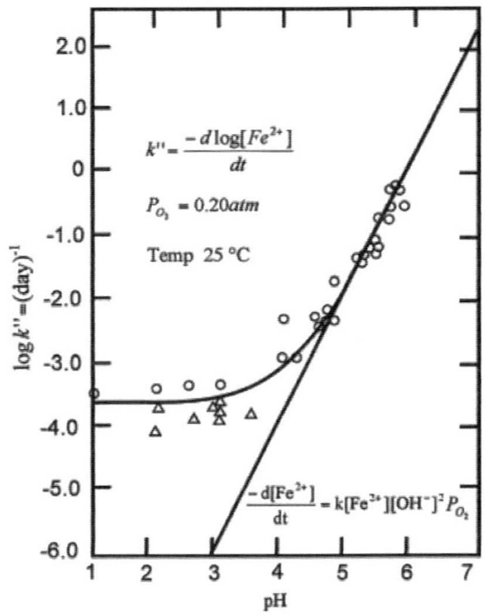

Figure 4 – Oxidation rate of Fe(II) species as a function of pH (Morgan & Lahav, 2007)

Oxidant Comparison

If air oxidation looked promising in the goethite precipitation conditions, bleeding iron as jarosite needed further development to improve the oxidation speed. Tests with two different catalytic systems were tried, activated carbon and SO_2:O_2 mixtures, as well as a more oxidative reagent, ozone.

The first catalyst tested, activated carbon, did not show major oxidation speed improvement, though many studies reported its role in ferrous iron oxidation (Stumm and Lee, 1961; Lowson, 1982). For these tests, 10 g.L^{-1} of coconut charcoal was added. If a slightly faster oxidation from 5.9 to 6.6 g.L^{-1}.h^{-1} was found in the goethite type precipitation (80°C and pH=4), the oxidation in the jarosite conditions was not accelerated. It looked like activated carbon was not efficient in acidic solutions compared to neutral ones (Figure 5).

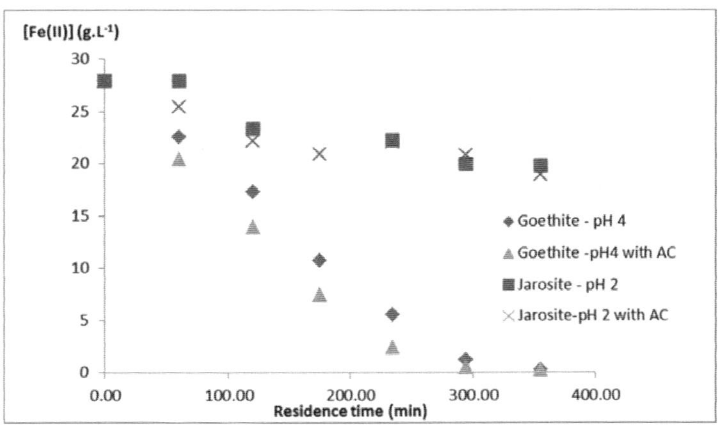

Figure 5 – Effect of activated carbon on iron(II) oxidation by air in jarosite or goethite precipitation conditions

A lot of studies have been carried out on the oxidation with SO_2:O_2 mixtures. To be said first, SO_2 is above all a reductive reagent. It has to be used in very low proportions compared to O_2, instead the mixture becomes reductive. Then, the optimum SO_2:O_2 ratio to use will change with numerous parameters such as the filtrate composition, pH, temperature or gas flow rate. That is to say, results must be each time redefined in the exact industrial conditions (Zhang et al., 2000a, 2000b and 2000c). In the following example, Zhang and Muir optimized the conditions to oxidize up to 5 $g.L^{-1}.h^{-1}$ (=1.5 $mmol.L^{-1}.min^{-1}$) at 75°C and pH=2 (Figure 6).

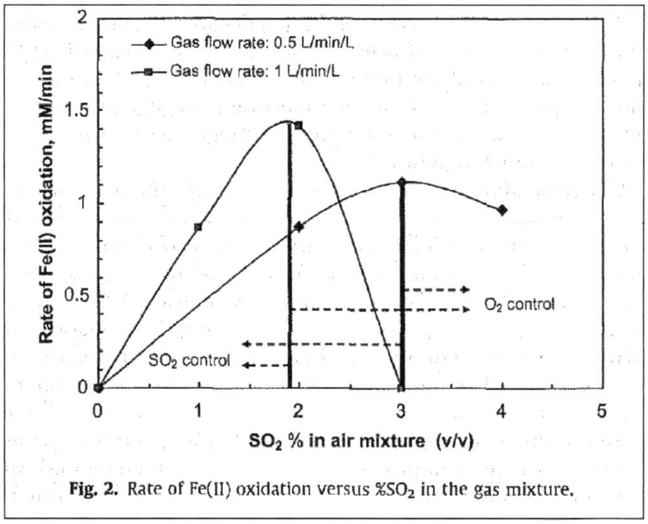

Figure 6 – Fe(II) oxidation speed for different SO_2:O_2 mixtures (Zhang & Muir, 2010)

Tests with different SO_2 ratios were performed for different temperatures at pH=2 and 4. A catalytic effect from SO_2 was observed between tests at 0 and 1% mixture at pH=2, but the effect remained much less than what would be expected. At 3%, the catalytic effect seemed lower for both temperatures tested, confirming that the optimum had been passed. Tests done at pH=4 showed a different pattern with an oxidation speed already reduced with 1% SO_2 compared to air alone, although rapid ferrous iron oxidation has been observed at pH 4 by others using SO_2:O_2 or SO_2:air. No oxidation was seen using a 5% SO_2:O_2 mixture at 85°C. It appears that the highly oxidizing species that are normally formed by mixing SO_2 with oxygen or air in metallurgical processing were not formed in these tests, either because the gas mixing was not optimal, or because species that are known to catalyze the formation of the highly oxidizing species, e.g., specific metal ions, were not present in a sufficient quantity, or because the chosen operating conditions (temperature, agitation) were not in favor of this catalytic reaction (Figure 7).

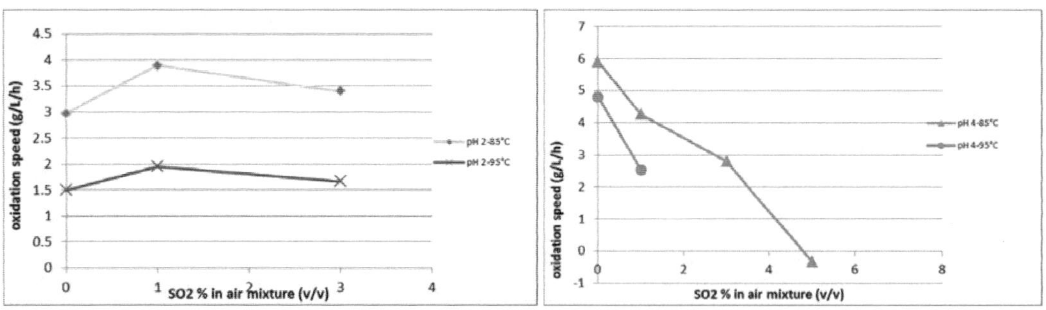

Figure 7 – Fe(II) oxidation speed for SO_2:O_2 mixture at different pH and temperatures

Finally, oxidation with ozone was performed in the jarosite precipitation conditions. This reaction is also well-known (Conocchioli et al., 1965). At 90°C, pH=2, the oxidation and precipitation of an initial solution of 30 $g.L^{-1}$ of iron were quantitatively achieved in 6 h. Final iron concentration was measured at 2.5 $g.L^{-1}$ in the filtrate and the yellowish solid obtained was analysed by XRD as a mixture of natrojarosite and gypsum (Figure 8). ICP analyses confirmed the presence of sodium in a significant amount (2.9 wt.%).

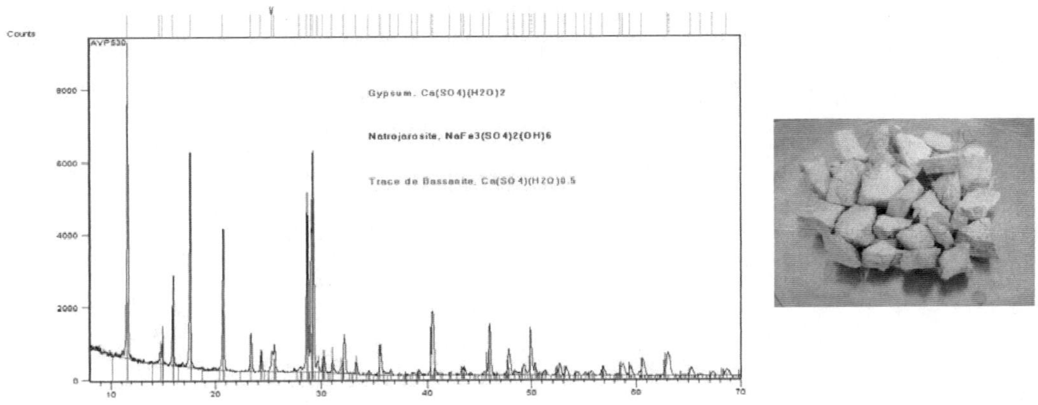

Figure 8 – XRD spectrum and picture of the natrojarosite obtained

Two variations were also conducted with a test at lower pH (1.5) and another at lower temperature (75°C). These two tests also ended with a complete reduction of ferrous iron in 6 h. The only difference was the remaining amount of iron in solution which decreased with the temperature (0.3 $g.L^{-1}$ of iron at 75°C and pH=2).

Pilot Test Results

The iron bleed was tested at pilot scale with a continuous feeding of 35 $L.h^{-1}$ for 3 weeks. The retained process was to bleed iron as goethite. A 10 h residence time was obtained thanks to a cascade of 6 reactors in which air was bubbled. Limestone was added to regulate the pH in reactor 1, 3 and 5. The solid-liquid separation was achieved on a belt filter after a thickening step (Figure 9).

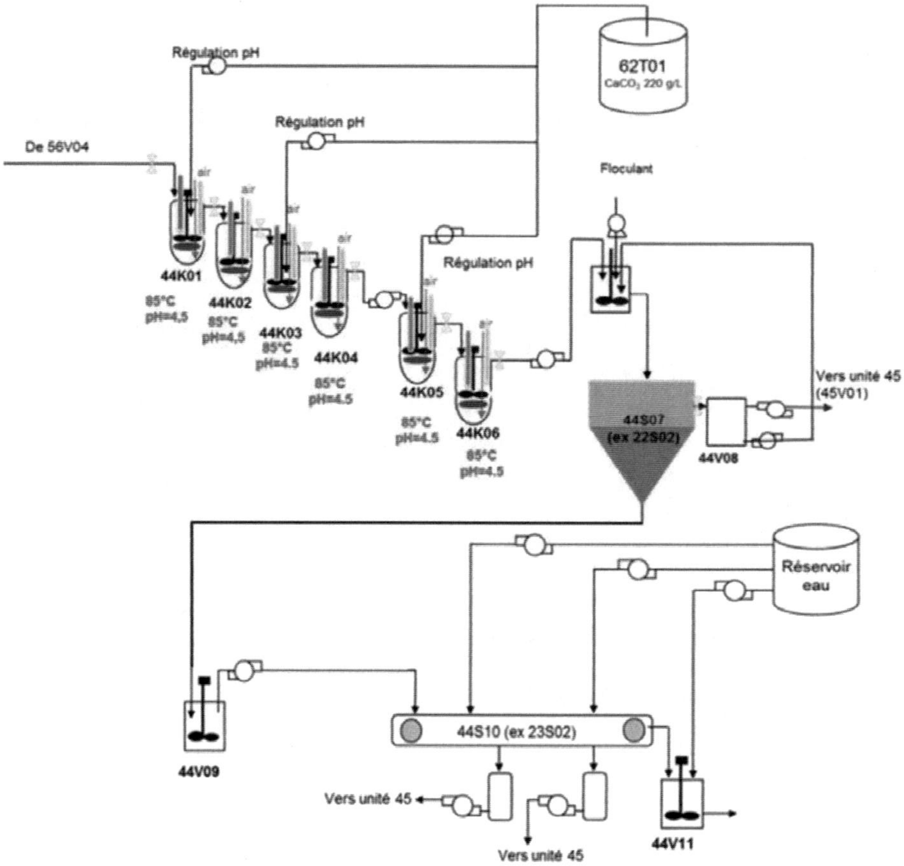

Figure 9 – Flowsheet used during continuous operation

The main results concerning the iron bleed step are summarised below. The feed composition was finely controlled between 30-35 g.L^{-1} Fe(II) and 2.5-3 g.L^{-1} Mn. The unit was started at shift 5, and except for the first three days of stabilization, the step met the objectives of an iron content in the outlet below 0.5 g.L^{-1} and a Fe/Mn in the residue above 100 (Figure 10).

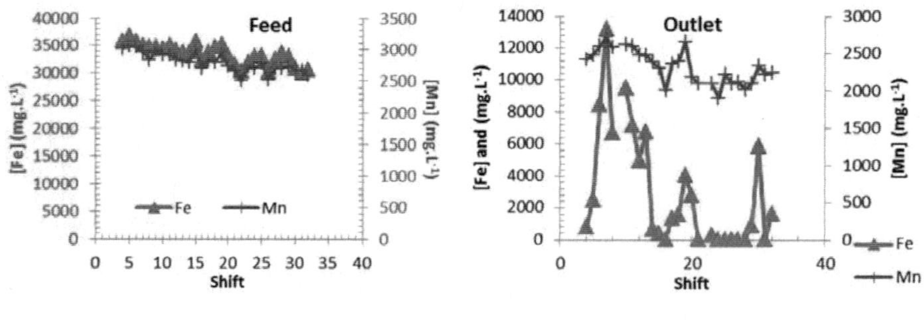

Figure 10 – Fe and Mn analyses for iron bleed feed and outlet during pilot operation (8-hour shift)

Oxidation speed was also measured thanks to analyses on all the different reactors. Laboratory results were confirmed with a precipitation of approximately 6 g.L^{-1}.h^{-1} of iron (Figure 11).

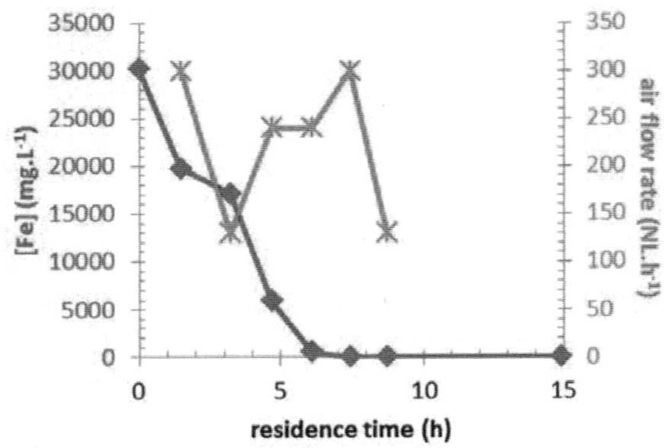

Figure 11 – Iron(II) analyses in the six reactors

CONCLUSIONS

The Maboumine project is aimed at the recovery of valuable elements (niobium, rare earth elements, uranium and scandium) from a Gabonese deposit. The developed process went through an iron(II) scheme. The iron bleed needs thus an oxidation step to produce either jarosite or goethite. Both routes were studied at laboratory scale. At pH=4, air oxidation looked sufficient to oxidise most of the iron in 6 hours which precipitates as goethite. This result was confirmed during a three weeks pilot campaign in continuous operation.

Jarosite precipitation is of more interest as a part of sodium could be co-precipitated. However, conditions needed for jarosite formation are less favourable toward iron oxidation and the iron oxidation speed was reduced at laboratory scale from 6 to less than 2 $g.L^{-1}.h^{-1}$. Three different solutions were tested to improve this oxidation speed. The use of two catalytic systems (activated carbon and SO_2:O_2 mixtures) showed poor benefits. Finally, jarosite was obtained quantitatively after a change of oxidant. The use of ozone indeed enabled a fast precipitation up to 6 $g.L^{-1}.h^{-1}$. Additional works are in progress to optimize jarosite precipitation.

REFERENCES

Agin, J., Durupt, N., Greco, A., Hammy, F., Laroche, G., & Thiry, J. (2012). *U.S. Patent Application 13/978*, 112.

Beltrami, D., Deblonde, G. J.-P., Bélair, S., & Weigel, W. (2015). Recovery of yttrium and lanthanides from sulfate solutions with high concentration of iron and low rare earth content. *Hydrometallurgy*, *157*, 356-362.

Beltrami, D., Moussa Ragueh, D., Bélair, S., & Weigel, W. (2016a). Investigations on the recovery of scandium as a byproduct from Maboumine project. Conference of the XXVIII International Mineral Processing Congress, Quebec City, Canada, September 11-15th 2016.

Beltrami, D., Bélair, S., & Weigel, W. (2016b). Maboumine project: a hydrometallurgical process for the recovery of rare earths elements from a polymetallic ore deposit in Gabon. Conference of the XXVIII International Mineral Processing Congress, Quebec City, Canada, September 11-15th 2016.

Conocchioli, T. J., Hamilton, E. J. Jr., & Sutin, N. (1965). The formation of iron(IV) in the oxidation of iron(II), *J. Am. Chem. Soc.*, *87*, 926-927.

Deblonde, G.J.-P., Chagnes, A., Bélair, S., & Cote, G. (2015a). Solubility of niobium(V) and tantalum(V) under mild alkaline conditions. *Hydrometallurgy*, *156*, 99–106.

Deblonde, G. J.-P., Chagnes A., Weigel V., & Cote G. (2015b). Direct recovery of niobium and tantalum from alkaline solutions using calcium-bearing reagents. *Accepted in Hydrometallurgy.*

Deblonde, G.J.-P., Weigel, W., Bélair, S., & Beltrami, D. (2016). Selective recovery of niobium and tantalum from low-grade concentrates using a simple and fluoride-free process. *Submitted*

Delvallee, F., Lachaize, F., & Weigel, W. (2015). *European Patent WO 2015/004375 A1.*

Donati, L., Courtaud, B., & Weigel, V. (2014). Maboumine process: a promising process for the developing a polymetallic ore deposit – focus on the upstream part of the process. *Hydromet 2014*, Victoria (Canada), June 22-25, 2014.

Eramet: www.eramet.com

Lowson, R.T. (1982). Aqueous Oxidation of Pyrite by Molecular Oxygen. *Chemical Reviews*, *82*(5), 477-485.

Goto, K., Tamura, H., & Nagayama, M. (1970). The mechanism of oxygenation of ferrous ion in neutral solution. *Inorganic Chemistry*, *9*(4), 963-964.

Maboumine: www.maboumine.com

Morgan, B., & Lahav, O. (2007). The effect of pH on the kinetics of spontaneous Fe(II) oxidation by O_2 in aqueous solution - basic principles and a simple heuristic description. *Chemosphere*, *68*(11), 2080-2084.

Stumm, W., & Lee, G.F. (1961). Oxygenation of ferrous iron. *Industrial and Engineering Chemistry*, *53*(2), 143-146.

Vincec, M. (2014a). *European Patent WO 2014/118288 A1.*

Vincec, M. (2014b). *European Patent WO 2014/140492 A1.*

Zhang, W., Singh, P., & Muir, D. (2000a). Iron(II) oxidation by SO_2/O_2 in acidic media: Part I, Kinetics and mechanism", *Hydrometallurgy*, *55*, 229-245.

Zhang, W., Singh, P., & Muir, D. (2000b). Iron(II) oxidation by SO_2/O_2 in acidic media: Part II, effect of Copper", *Hydrometallurgy*, *58*,117-125.

Zhang, W., Singh, P., & Muir, D. (2000c). SO_2/O_2 as an oxidant in Hydrometallurgy", *Minerals Engineering*, *13*(13), 1319-1328.

Zhang, W., & Muir, D. (2010), Oxidation of Fe(II) in a synthetic nickel laterite leach liquor with SO_2/air", *Minerals Engineering*, *23*, 40-44.

AN INITIAL INVESTIGATION INTO THE EFFECT OF Fe (III) CONCENTRATION ON THE PRODUCT CHARACTERISTICS OF FE (III) OXYHYDROXIDE PRECIPITATES DURING LIME PRECIPITATION

Cledwyn Mangunda, Jochen Petersen*, and Alison Lewis

Department of Chemical Engineering
Crystallization and Precipitation Unit
University of Cape Town, Private Bag, Rondebosch 7701, South Africa
*(*Corresponding author: Jochen.petersen@uct.ac.za)*

ABSTRACT

Run-off and waste water streams high in dissolved metal content are generated in all mining operations. The metals pose a threat to life due to their persistent bio-accumulative toxic (PBT) characteristics. Hence, there is a need to treat mine waste water effectively. Lime precipitation is the technology of choice through which Fe, as the major metal species, is removed as Fe (III) oxyhydroxides with other metals co-precipitated and/or adsorbed. Sulphates are removed as gypsum or incorporated into the Fe phases. The resulting sludge has poor dewatering characteristics and is deposited in landfills, with an inherent risk of returning the Fe back into the water cycle through leaching. This study seeks to understand the mechanisms behind the formation of this sludge in terms of product quality. The relationship between process conditions and product characteristics has been investigated through exploring the effect of Fe (III) concentration on the precipitate product characteristics (particle size distribution (PSD), size, morphology and structure) and quality (in terms of filterability, settling ability, etc.) in a simplified system. Neutralisation of acidic synthetic aqueous Fe (III) sulphate streams containing 50, 400 and 800 mg/L Fe(III), respectively, was effected by mixing it in a continuous stirred tank reactor (CSTR) with dissolved calcium hydroxide for pH control at a residence time of 15 minutes. Precipitate samples analysed for morphology and PSD, using nanoSEM and Zetasizer respectively, showed a general decrease in PSD and zeta potential with increases in feed Fe (III) concentration. These results were used to infer the particle size enlargement mechanisms of Fe (III) oxyhydroxide precipitates under a given set of operating conditions. Preliminary product quality results indicate that a high concentration (800 mg/L) leads to the rapid formation of metastable precursors with poor product quality, whilst a low concentration (50 mg/L) leads to formation of thermodynamically more stable precipitates with good product quality. The approach demonstrates a potential route forward in terms of linking operating conditions to the attainment of desirable characteristics rather than specific precipitate phases which are difficult to identify.

KEYWORDS

Mine water, Concentration, Product characteristics, Product quality, Precipitation

INTRODUCTION

Mining surface waters and acid mine drainage (AMD), when discharged into natural water bodies, present a challenge due to their heavy metal and sulphate content (Simate & Ndlovu, 2014), which limit the use of these bodies as water resources. Mining of sulphidic ores produces large quantities of waste material with a high Fe content due to its natural abundance (Pulles, et al., 2014). These waste materials are oxidized, resulting in the release of SO_4^{2-}, Fe (II) and H^+ (source of acidity) into the aqueous phase thereby forming acid surface water and AMD (Jönsson et al., 2006). For example, the total Fe content of AMD from the Grootvlei Mine in South Africa, Vaal Barrage Site 2 and Vaal Barrage Site 3 has been reported as 135 mg/L, 12438 mg/L and 5999 mg/L respectively (Oryx Environmental and Jones & Wagener Consulting Engineers, 2003; Steffen, 1989). Fe (II) can be oxidised further in acidic solution in the presence of air to form Fe (III).

Lime is a low cost, safe and easy-to-handle precipitant (Skousen et al., 1996) that neutralizes the acidity and removes heavy metals from Fe rich streams in order to meet effluent discharge standards and lower liability mitigation (Rao et al., 1996). The treatment of these streams with lime results in the precipitation of a sludge comprising of Fe (III) oxyhydroxides and gypsum. Lime use has its complications, which include: the low solubility of lime at high pH and tendency to armour (Hedin and Watzlaf, 1992), formation of gelatinous sludge which has poor dewatering characteristics, sludge instability necessitating post-neutralisation stabilisation, low solids density resulting in considerable water losses and a large land area required for sludge disposal to landfills (Hove, 2008). The sludge has an inherent risk of leaching the Fe back into the surface water cycle. Fe (III) oxyhydroxide precipitation, and the concomitant mechanisms by which a soluble entity is transformed into a sludge, are not well understood due to the complex chemistry behind their formation (Dutrizac, 1980). Even though a considerable amount of research has been carried out on this subject matter from a thermodynamic standpoint, most work has been predominantly in the domain of earth sciences and in the context of high tenor process solutions at elevated temperatures (Stumm and Sulzberger, 1992; Bigham et al., 1992; Schwertmann et al., 1995). Research from an industrial precipitation view point, with a focus on how operating conditions relate to product characteristics and quality, has been thorough only for iron control in high tenor solutions, but limited in terms of waste water treatment. The two domains of precipitation processes are similar in principle but differ in operating conditions in terms of temperature, reaction time and precipitating agent. Industrial precipitation of Fe from waste waters such as AMD is done at ambient temperature, uses lime and is designed and operated in order to make water treatment as cheap as possible.

The Fe (III) oxyhydroxides precipitate as mixtures (Pedersen, 2006) or polymorphs whose composition is concentration dependent and often determined by the extent of precipitate aging and operating conditions (Reichelt & Bertau, 2015). The exact reaction pathways that lead to these mixtures are difficult to establish and isolate, so that the effect of operating conditions on these pathways has not been concisely understood. The mixture can include several phases (schwertmannite, goethite, hematite and ferrihydrite being a few examples) that possess product characteristics such as high surface area (which renders them as good adsorption sites), varying morphology and crystal structure, multiplicity of oxygen centres and the tendency to entrap liquid. This enables them to act as sinks for dissolved metal cations, which co-precipitate, adsorb or are entrained. The presence of high concentrations of sulphate in the originating waste water introduces further complexity resulting in the formation of further precipitated iron species (especially jarosites) and the formation of gypsum as an additional precipitated phase.

The concentration of Fe (III) in solution determines the supersaturation levels of the system. It also influences the nucleus quality, metastable zone width (MSZW), type and temperature at which precipitates form and the growth mechanisms (Claassen and Sandenbergh, 2006; Voigt and Göbler, 1986; Cheng and Demopoulos, 1997; Dutrizac and Riveros, 1999; Fu et al., 2011). At low concentrations a MSZW exists where particle molecular growth is the preferred method of growth, and this leads to smaller quantities of denser particles of a narrow particle size distribution (Dirksen and Ring, 1991), whilst high concentrations tend to have the opposite effect, dominated by agglomeration (Hove et al., 2008; Claassen and Sandenbergh, 2006). The effect of intermediate concentrations typical of run-off waters (higher than in soils, but lower than in Fe containing process solutions) on mechanisms is not well established, with the potential that these concentrations can lead to an overlap of growth and agglomeration, resulting in mixed product characteristics. Since studies by Grundl and Delwiche (1993) indicate that the growth step is the one which predominantly influences product characteristics, it is possible that the concentration level of Fe (III) results in an interplay of these size enlargement processes. Concentration determines crystal morphology as it impacts the rate at which crystal faces grow, with the slow growing face often determining the morphology (Mullin, 2005). High Fe (III) concentration may result in the formation of small polyhedral crystals which have poor filterability, whereas low concentration may result in bigger, easy to filter spherulitic (sphere-like) crystals (Cornell & Schwertmann, 2003; Fu et al., 2011). Maiti & Devi (2015) found contrasting results with Fe (III) nitrate, as the morphology deviated from spherulites to rods whose lengths and widths increased with decreasing Fe (III) concentration.

The effect of varying Fe (III) concentration on the product quality and characteristics of Fe (III) oxyhydroxide precipitates allows inferring a relationship between these qualities and operating conditions. Hence, it is postulated that it might be a useful approach to develop and populate a properties 'map' for product quality as a function of operating ranges of certain key process parameters. Regions of process conditions thus identified allow the linking of desirable, intermediate or undesirable product characteristics to process operating conditions to be sought or avoided. This will be industrially beneficial in that it relates process conditions to the resulting precipitation product without the complexity of identifying the individual precipitated phases but rather predicting the nature of the precipitate. It is anticipated that this approach will reduce challenges for post-precipitation processes and waste disposal such as dewatering of gelatinous sludge, remobilisation of adsorbed metals in wet waste deposits (landfill sites).

The present study is a preliminary investigation to establish the variation of precipitation product characteristics upon varying certain operating parameters in a bench-scale model system.

MATERIALS AND METHODS

For all the experiments, a synthetic $Fe_2(SO4)_3$ aqueous stream and a $Ca(OH)_2$ reagent grade solution was prepared using Millipore deionised water, with the $Ca(OH)_2$ solution being used for pH adjustment thereby effecting neutralization and precipitation. The Fe (III) concentration was varied at three concentrations of 50 mg/L, 400 mg/L, 800 mg/L based on typical South African AMD concentrations (Oryx Environmental and Jones & Wagener Consulting Engineers, 2003; Steffen, 1989). Residence times (τ) of 15, 30 and 60 minutes were used to investigate attainment of steady state, although the present work focuses on the 15 minutes residence time. A 1.5 litre glass primary precipitation reactor with four equally spaced stainless steel baffles, shown in Figure 1, was used. The reaction proceeded according to equation (1). Precipitate samples were collected at fixed time intervals from the reactor overflow spout, immediately quenched with 0.1M H_2SO_4 before being split up for further analysis. All the experiments were conducted in triplicate in order to establish reproducibility.

$$Fe_2(SO_4)_3+3Ca(OH)_2+6H_2O \rightarrow 2Fe(OH)_3+3CaSO_4{\cdot}2H_2O \quad (1)$$

Atomic Adsorption Spectroscopy (AAS) was used to measure feed concentrations for mass balance computations. A magnetic stirrer mixed the feed solutions continuously whilst being pumped by peristaltic pumps into the reactor. A Rushton turbine stirrer enabled the variation of agitation rates to investigate the effect of mass transfer on kinetics. A Hanna PH212 pH electrode was used to measure the solution pH as the reaction proceeded. Residual Fe^{3+} analysis was done by a Jenway 6400 UV/Vis spectrophotometer and a Nova 60 spectroquant. Particle settling rates were established using 1L Imhoff cones whilst a Zetasizer nano unit was used for PSD and zeta potential. The morphology was analysed using a nanoSEM facility. The reagent containers were rinsed with deionized water prior to addition of solutions. The 1.5L reactor was first be filled with deionized water before reagents were added to it at a flow rate calculated from the set residence time. Filterability was calculated as the volume passing through a 0.22µm Millipore cellulose membrane filter under vacuum for a fixed time. A portion of the precipitated solids was dissolved in order to establish Fe (III) content for mass balance purposes.

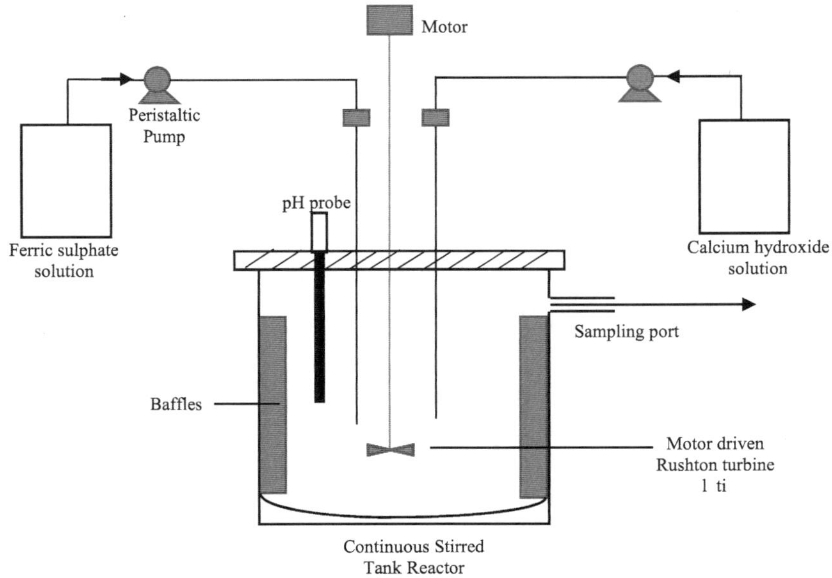

Figure 1 – Experimental setup for continuous precipitation of Fe (III) with lime

RESULTS AND DISCUSSION

Effect of Fe (III) Feed Concentration on Particle Size Distribution

Steady State Verification

The formation of various polymorphs during the precipitation process makes ascertaining steady state conditions for Fe precipitation systems complex. These different morphologies and sizes subsequently affect the agglomeration process and consequently the PSD inside the CSTR. Literature reports a rule of thumb whereby 8 to 15 residence times should be allowed before steady state is attained with respect to consistent PSD (Leubner, 2006). For this study, 12 residence times were assumed to be enough for the system to attain steady state with respect to concentration, PSD and morphology. Figures 2 to 4, which show results for a residence time of 15 minutes with varying Fe (III) concentration, demonstrate the validity of this assumption for the system by showing that the PSD graphs approximate each other increasingly with an increase in the number of residence times. However, they also clearly indicate that steady state was not yet achieved after 12 residence times particularly at low Fe (III) concentration.

The PSDs for the 3 concentrations are monomodal with the peak number % being achieved at 9τ for the 50 mg/L and 400 mg/L concentration. The deviation in this trend to a peak number % being achieved at 11τ for the 800 mg/L case could be as a result of cyclic behaviour typical of systems that have not attained steady state. The cyclic behaviour is due to the effects of irregular mixing which leads to concentration fluctuations and is more pronounced during the transient phase. The average particle size was found to shift positively with an increase in the number of residence times for the 400 mg/L and 800 mg/L cases whereas at 50 mg/L there was a deviation from this pattern. This deviation maybe as a result of breakage and or attrition of the large particles that form at lower Fe (III) concentrations.

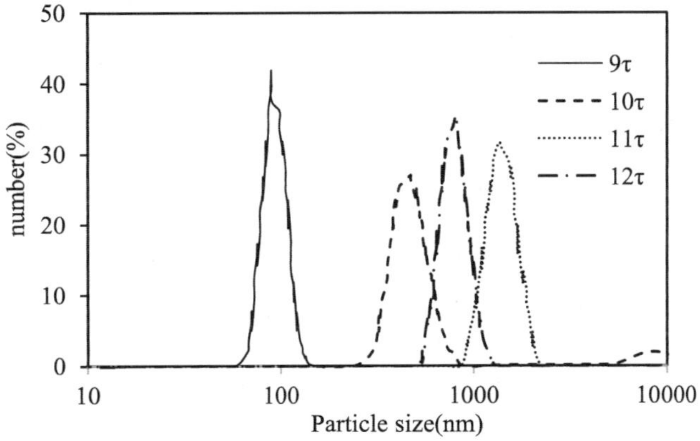

Figure 2 – Particles size distributions measured at 50 mg/L Fe (III)

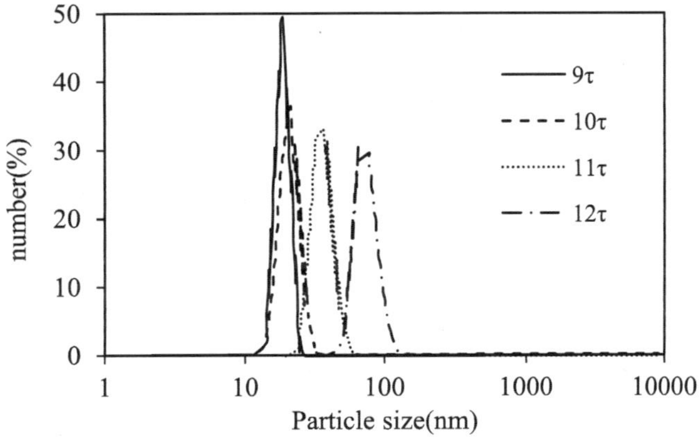

Figure 3 – Particles size distributions measured at 400 mg/L Fe (III)

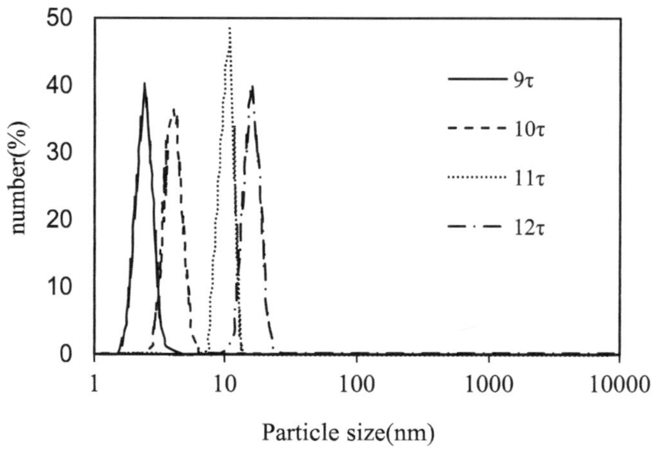

Figure 4 – Particles size distributions measured at 800 mg/L Fe (III)

Figure 5 shows a consolidated plot of the PSD for the 3 concentrations after 12τ for 15 minutes residence time.

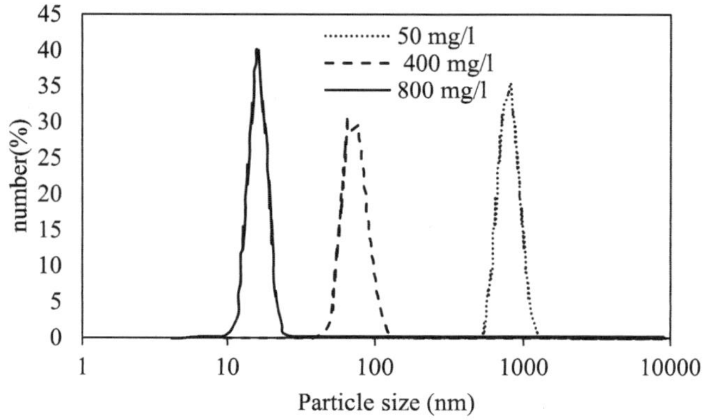

Figure 5 – Particles size distributions measured at 12τ

This very clearly shows that mean particle size decreases with increasing Fe (III) concentration. This is consistent with an increase in fines population due to the dominance of primary homogeneous nucleation arising from high supersaturation (De Yoreo & Vekilov, 2003). The average particle size was greatest for the lowest concentration (50 mg/L) which may indicate size enlargement through growth.

Effect of Fe (III) Feed Concentration on Filtration Rates

The filtration rate was observed to be a function of Fe (III) feed concentration, with the highest filtration rate obtained at 50 mg/L and consequently the lowest obtained at 800 mg/L, as shown in Figure 6. The filtration rates obtained at 400 mg/L and 800 mg/L Fe (III) feed concentration became marginally different as the number of residence times increased—which would correspond to the gradually increasing mean particle size observed in Figure 4. The filtration rates level out beyond 12τ which indicates that 15τ may be a suitable assumption for attaining steady state in the current set-up. Settling rates were also measured and they are presented in Figure 7.

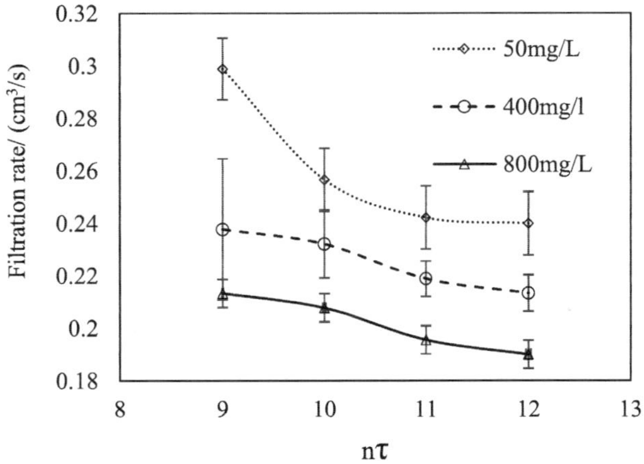

Figure 6 – Filtration rates at different ferric feed concentration

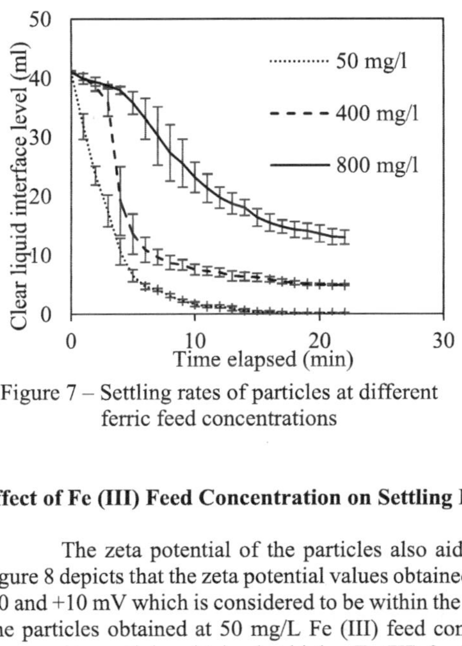

Figure 7 – Settling rates of particles at different ferric feed concentrations

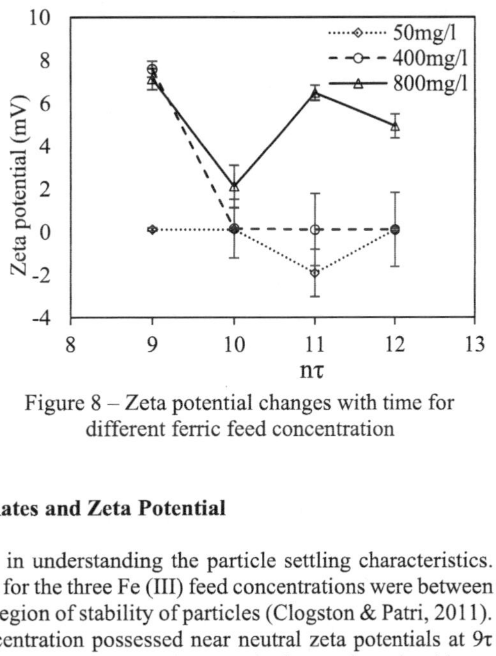

Figure 8 – Zeta potential changes with time for different ferric feed concentration

Effect of Fe (III) Feed Concentration on Settling Rates and Zeta Potential

The zeta potential of the particles also aids in understanding the particle settling characteristics. Figure 8 depicts that the zeta potential values obtained for the three Fe (III) feed concentrations were between -10 and +10 mV which is considered to be within the region of stability of particles (Clogston & Patri, 2011). The particles obtained at 50 mg/L Fe (III) feed concentration possessed near neutral zeta potentials at 9τ compared to particles obtained at higher Fe (III) feed concentration. An increase in the number of residence times resulted in near neutral zeta potentials for the 50 and 400 mg/L Fe (III) feed concentration, whilst the 800 mg/L Fe (III) feed concentration remained high at around 6 mV.

This was corroborated by the observed settling trends which decreased with increasing Fe (III) feed concentration (Figure 7). The settling rate trends observed are supported by the particles sizes obtained since larger particles are expected to settle faster than smaller particles due gravitational forces outweighing the fluid resistance. These results were used to ascertain whether the enhanced settling rates of particles, as Fe (III) concentration increased, was due to agglomeration or other particle size enlargement mechanisms. The average pH for the systems investigated was between 8 and 9. Ideally, negative zeta potential values should be obtained at pH greater than 7. For the higher Fe (III) feed concentrations, this was not consistent with this study, and this could possibly be attributed to the adsorption of the excess Fe(III) onto the surfaces of the resultant small sized precipitates – this merits further investigation.

Effect of Fe (III) Feed Concentration on Morphology

The observed changes in the PSD, with an increase in the number of residence times at a given Fe(III) concentration, suggests that the particles were not primary nucleation particles but rather products of agglomeration. The particles agglomerated especially at the high Fe (III) concentrations which would less support growth. This particle size enlargement mechanism was also observed by Maiti and Devi (2015) and is reported to be the dominant mechanism for precipitation processes (Demopoulos, 2009). The resulting morphology of the particles formed at the 3 different Fe (III) concentrations at a residence time of 15 minutes (taken from the settled sludge) are shown in Figures 9–11 below.

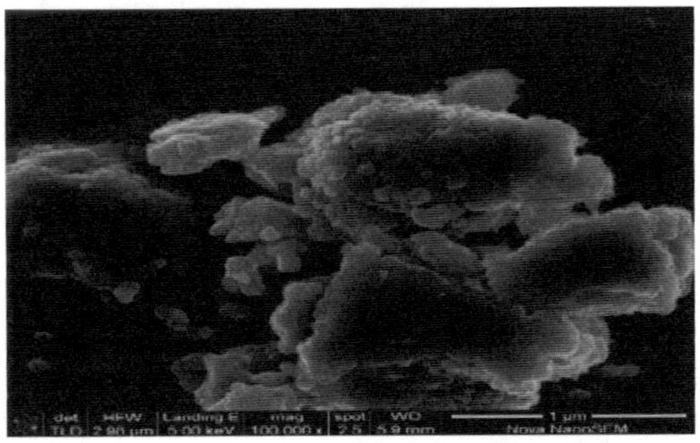

Figure 9 – SEM image obtained at 50 mg/L

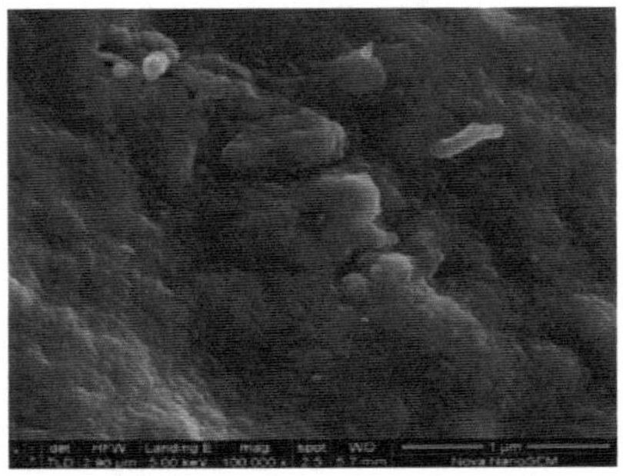

Figure 10 – SEM image obtained at 400 mg/L

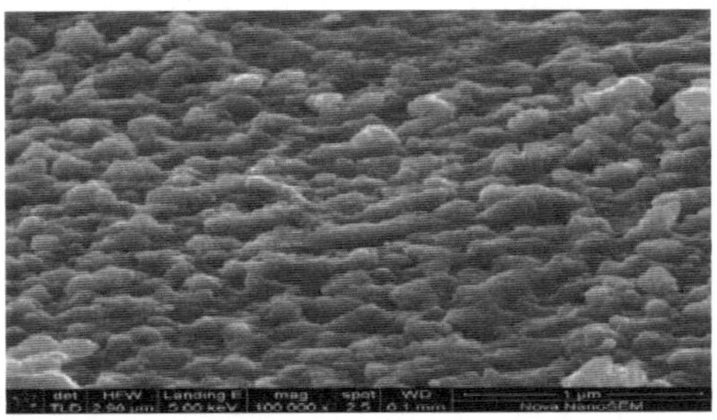

Figure 11 – SEM image obtained at 800 mg/L

These images support the proposed particle size enlargement mechanism as the particles appear to have formed agglomerates. The neutral (less than ±10 mV, Clogston & Patri, 2011) zeta potential values observed in Figure 8 indicate reduced repulsive forces which enhances agglomeration, especially during settling. The agglomerates formed at 50 mg/L appear to be made up of larger spherical particles which settle more easily and are more amenable to filtration. This supports the notion that at such low Fe(III) cocnentrations particle growth is more emphasized before the onset of agglomeration. The smaller particles obtained at 400 and 800 mg/L Fe (III) feed concentration displayed more amorphous characteristics which explains the low filtration and settling rates observed in Figure 6 and Figure 7, respectively.

CONCLUSIONS

Through varying the initial Fe (III) feed concentration, the effect on product characteristics (particle size distribution (PSD), size, morphology and structure) has been studied. The results obtained confirmed that the Fe (III) content in AMD influences the product characteristics of the precipitates. A range of initial Fe (III) concentrations that could yield improved settling ability and a size beyond which such benefits would be realised were established. An increase in mean particle size was observed with decreasing Fe (III) feed concentration. The rapid nature of precipitation processes makes it difficult to ascertain the particle size enlargement mechanism, although results from this study confirmed that it was growth at low Fe (III) concentration and largely agglomeration at the higher values. The neutral zeta potentials for these particles confirmed the small repulsive forces providing little resistance to the agglomeration process. The particles formed from the Fe (III) feed concentration of 50 mg/L at 15 minutes residence time appeared to be discernable individual particles from the SEM images, whereas the particles formed at 400 and 800 mg/L Fe(III) feed concentration, at the same residence time, appeared to be an amorphous network of smaller particles. This explained the better filterability and settling ability of particles obtained at low Fe (III) feed concentration.

Based on these experimental results it can be concluded that low Fe (III) concentrations lead to better product characteristics and quality. These low concentrations support low supersaturation which favours growth over agglomeration. The larger particles possess a smaller specific surface area for adsorption of ions hence the lower zeta potential at low concentrations. The filterability and settling ability of particles can be enhanced by operating at low Fe (III) concentration. The need to study a wider range of intermediate concentrations, residence times and pH among other process conditions, still exists. The results attained from this preliminary study indicate that, given more consistent product characteristics obtained after sufficient time to achieve steady state, valuable information of varying product characteristics can be obtained, with which a precipitate properties map can be drawn, which in turn would allow identification of preferred operating conditions of a waste water treatment facility for a given feed stream

ACKNOWLEDGEMENTS

The authors wish to acknowledge support by the National Research Foundation (NRF) of South Africa under Grant CPRR UID 93556 as well as by the South African Minerals and Metals Research Initiative (SAMMRI) under project S1520.

REFERENCES

Bigham, J.M., Schwertmann, U. & Carlson, L. (1992). Mineralogy of precipitates formed by the biogeochemical oxidation of Fe (II) in mine drainage. In: Biomineralization processes of iron and manganese – modern and ancient environments (Eds HCW Skinner and RW Fitzpatrick), Catena Supplement, 21, 219–232.

Cheng, T.C.M. & Demopoulos, G.P. (1997). In: Mishra, B. (Ed.), EPD Congress 1997. TMS, Warrendale, PA, pp. 599–617.

Claassen J.O. & Sanderbergh R.F. (2006). Particle growth parameters in the precipitation of metastable iron phases from zinc-rich solutions. *Hydrometallurgy,* 84, 165–174.

Clogston, J. & Patri, A. (2011). Characterization of Nanoparticles Intended for Drug Delivery. In: McNeil & S.E, eds. *Methods in Molecular Biology.* New York: Springer.

Cornell, R. M. & Schwertmann, U. (2004). *The Iron Oxides: Structure, Properties, Reactions, Occurrences and Uses.* 2nd ed. s.l.:John Wiley & Sona.

Cornell, R. & Shwertmann, U. (2003). *The Iron Oxides: Structure; Reactivity; Occurences and Uses.* 2nd ed. Weinheim: John Wiley and Sons.

De Yoreo, J. & Vekilov, P. (2003). Principles of Crystal Nucleation and Growth. *Reviews in Mineralogy and Geochemistry*. 54(1), 57–93.

Grundl, T. & Delwiche, J. (1993). Kinetics of ferric oxyhydroxide precipitation. *Journal of Contaminant Hydrology,* 14(1), 71–87.

Demopoulos, G. (2009). Aqueous Precipitation and Crystallisation for the Production of Particulate Solids with Desired Properties. *Hydrometallurgy,* 96, 199–214.

Dirksen, J.A. & Ring, T.A. (1991). Fundamentals of crystallization: Kinetic effects on particle size distribution and morphology. *Chemical Engineering Science*, 46(10), 2389–2427.

Dutrizac, J.E. (1980). The physical chemistry of iron precipitation in the zinc industry, in lead-zinc-tin '80, ed. Cigan J.M., Mackey T.S., and O'Keefe T.J., TMS-AIME, Warrendale, PA, USA, pp. 532–564

Dutrizac, J.E. & Riveros, P.A. (1999). The precipitation of hematite from ferric chloride media at atmospheric pressure. *Metallurgical and Materials Transactions B,* 30B, 992–1001.

Fu, D., Keech, P., Sun, X. & Wren, J. (2011). Iron Oxyhydroxide Nanoparticles Formed by Forced Hydrolysis: Dependence of Phase Compostion on Solution Concentration. *Physical Chemistry Chemical Physics,* 13, 18523–18529.

Garside, J. (1994). *Separation Technology: The Next Ten Years.* 1st ed. Houston: Institution of Chemical Engineers.

Hedin, R. S. & G. R. Watzlaf. (1992). The effects of anoxic limestone drains on mine water chemistry. *Proceedings of the International Land Reclamation and Mine Drainage Conference.* Pittsburgh, 185.

Hove, M. (2008). The kinetics and mechanisms of the oxidation and precipitation of iron: The High Density Sludge (HDS) process. Ph.D. Thesis, University of Cape Town.

Hove, M, van Hille, R.P., & Lewis, A.E. (2008). Mechanisms of Formation of Iron Precipitates from Ferrous Solutions at High and Low pH. *Chemical Engineering Science*, 63, 1626–1635.

Jönsson, J., Jönsson, J. & Lövgren, L. (2006). Precipitation of Secondary Fe(III) Minerals from Acid Mine Drainage. *Applied Geochemistry*. 21, 437–445.

Leubner, I. (2006). Seeded Precipitations in the Continuous Stirred Tank Reactor Crystallizer. *Journal of Dispersion Science and Technology,* 22(4), 373–380.

Maiti, D. & Devi, P. (2015). Selective Formation of iron oxide and oxyhydroxide nanoparticles at room temperature: Crical role of concentration of ferric nitrate. *Materials Chemistry and Physics,* 154, 144–151.

Mullin, J. W. (2005). *Crystallisation and precipitation.* London: John Wiley & Sons.

Oryx Environmental and Jones & Wagener Consulting Engineers (2003). The Grootvlei Proprietary Mines Limited project to reduce and treat the volumes of water pumped from underground, Braamfontein, Oryx Environmental.

Pedersen, H.D. (2006). The transformation of Fe(III) oxides catalysed by Fe^{2+} and the fate of arsenate during transformation and reduction of Fe(III) oxides. Ph.D. Thesis, Technical University of Denmark.

Pulles, W., Heath, R., & Howard, M. (2014). A manual to assess and managethe impact of gold mining operations on the surface water environment.. *WRC,* p. Report No. TT79/96.

Rao, S.R., Leroux, M. & Finch, J. (1996). *Resource recovery from acid mine drainage.* Metals removal from acidic drainage—chemical methods (Part I), MEND Project 3.21.2a, Pointe-Claire, PQ: Noranda Technology Center.

Reichelt, L. & Bertau, M. (2015). Production of ferrihydrite and schwertmannite using a microjet mixer device. *Chemical Engineering Research and Design,* 98, 70–80.

Schwertmann, U., Bigham J.M. & Murad E. (1995). The first occurrence of schwertmannite in a natural stream environment. *European Journal of Mineralogy,* 7, 547–552.

Simate, G. & Ndlovu, S. (2014). Acid mine drainage: challenges and opportunities. *Journal of Environmental Chemical Engineering,* 2, 1785–1803.

Skousen, J. & Hilton, F. (1997). Overview of acid mine drainage. In: *Reclamation of Drastically Disturbed Lands.* West Virginia: West Virginia University.

Steffen, R. (1989). Research on the contribution of mine dumps to the mineral pollution load in the Vaal Barrage, WRC report No 136/1/89, Pretoria, South Africa.

Stumm, W. & Sulzberger, B. (1992). The cycling of iron in natural environments: considerations based on laboratory studies of heterogeneous redox processes. *Geochimica et Cosmochimica Acta,* 56, 3233–3257.

Voigt, B. & Göbler, A. (1986). Formation of pure haematite by hydrolysis of iron (III) salt solutions under hydrothermal conditions. *Crystal Research and Technology,* 21(9), 1177–1183.

MINERALOGICAL CONTROL OF THE FORM OF FERRIC/FERROUS PRECIPITATES FOR EFFECTIVE TREATMENT OF ACID MINE DRAINAGE

Chiharu Tokoro*, Yugo Mitani, Tatsuya Kato, Abdullah Al Mamun

Department of Science and Engineering, Waseda University
*(*Corresponding author: tokoro@waseda.jp)*

ABSTRACT

Acid mine drainage (AMD), which is a major source of mining pollution, often contains ferric and/or ferrous ions together with several toxic elements. The ferric and ferrous ions in AMD precipitate in several mineralogical forms after neutralization, and the ferric/ferrous precipitates become the main component of the sludge generated after AMD treatment. The precipitates are also good absorbents for several toxic elements. To avoid generation of voluminous sludges and the re-dissolution of toxic elements from the sludge, it is very important to generate well-draining and efficiently adsorbing sludges by controlling the mineralogical form of the ferric/ferrous precipitates. Green rust, which is ferric and ferrous hydroxide, is a double layer hydroxide which is able to sorb several toxic elements. The green rust transforms to goethite or lepidocrocite by rapid oxidation, whereas it transforms to magnetite at ambient temperature by slow oxidation. The magnetite is also good absorbent especially for toxic cations and it makes a compact sludge due to its magnetic properties. We investigated the magnetite generation process at ambient temperature in detail. The magnetite generation process can be divided into three steps; (i) oxidation of Fe(II), (ii) Fe(II) surface complexation to Fe(III) hydroxide, and (iii) crystallization of magnetite. More magnetite can form when the oxidation rate is slower than the crystallization rate, and optimum conditions for its formation are at pH 9.5 with mild agitation. Ferrihydrite is the most common precipitate and it is able to sorb many different toxic elements. However, it drains badly and its solid/liquid separation properties are poor. We investigated the removal mechanism of several toxic elements by co-precipitation with ferrihydrite. We confirmed that the removal mechanism by co-precipitation partly involved surface precipitation when the initial ratio of toxic elements to ferric ion was high, although the main mechanism was always surface complexation. When surface precipitation occurred, highly efficient toxic element removal with ferrihydrite could be achieved.

KEYWORDS

Acid mine drainage, ferrihydrite, green rust, magnetite, co-precipitation

INTRODUCTION

Acid mine drainage (AMD) is a global environmental problem (Dold, 2008; Liang & Thomson, 2008). In Japan, there are about 7000 abandoned mines of which 450 generate AMD (Japan Oil, Gas and Metals National Corporation [JOGMEC], 2003). AMD often contains ferric and/or ferrous ions together with several toxic elements. Therefore, we must treat these drainages to decrease the toxic element concentrations under each Japanese effluent standard. However, AMD treatment is costly, and thus, the Japanese government started countermeasure projects in 1971 and has spent more than 100 billion Japanese yen to date (Japanese Ministry of Economy, Trade and Industry, 2006).

The common method to treat AMD in Japan is coagulation-sedimentation using calcium carbonate and/or calcium hydroxide as the neutralizer. AMD often contains ferric and/or ferrous ions which are the source of the ferric/ferrous precipitates. They are very good absorbents for several toxic elements and are also responsible for sludge generation after AMD treatment. Final Japanese disposal sites are limited. Therefore, it is crucial to avoid generation of voluminous sludges, and to stop re-dissolution of toxic elements from the sludge. To accomplish this, there is a need to generate well-drained and efficiently adsorbing sludges by control of the mineralogical form of the ferric/ferrous precipitates.

Green rust is known as Fe(II)-Fe(III) hydroxyl-salt and is able to remove some toxic elements form water (Hayashi et al., 2009). It was reported that green rust has a layered structure similar to brucite-like sheets. These are positively charged owing to the trivalent iron cations, $[Fe(II)_{(1-x)}\text{-}Fe(III)_x(OH)_2]^{x+}$, alternating with interlayers composed of anions, such as Cl^-, CO_3^{2-} or SO_4^{2-}, and water molecules, $[\{x/n\}(a^{n-}) \cdot (m/n)H_2O]^{x-}$ (Bernal et al., 1959). Green rust transforms to magnetite (Fe_3O_4), goethite (α-FeOOH) or lepidocrocite (γ-FeOOH). Among these precipitates, magnetite is also a good absorbent especially for toxic cations and it makes the sludge more compact because of its magnetic properties. Its formation is dependent on the oxidation rate that transforms green rust to magnetite (Tokoro et al., 2006). Thus, we investigated the magnetite generation process at ambient temperature in detail.

Ferrihydrite (commonly and incorrectly identified as simple $Fe(OH)_3$), which is the most common ferric precipitate, is able to sorb many different toxic elements although it drains poorly and its solid/liquid separation properties are challenging. In a previous study, the sorption mechanism of some toxic elements on ferrihydrite was shown to involve not only surface complexation but also surface precipitation (Dzombak & Morel 1990). Thus, studies on the removal mechanism of several toxic elements, especially arsenic (As(V)), during co-precipitation with ferrihydrite have been conducted. To accomplish this, the sorption mechanism on ferrihydrite was investigated experimentally using sorption isotherms, X-ray diffraction (XRD) analysis and X-ray adsorption fine structure (XAFS) analysis. The results were discussed in terms of the co-precipitation and adsorption mechanisms.

MATERIALS AND METHODS

Standards and Reagents

All chemicals and solutions used were made from analytical grade reagents purchased from Kanto Chemicals Inc., Tokyo, Japan and Wako Pure Chemical Industries, Tokyo, Japan. In the oxidation reaction experiments, $FeSO_4 \cdot 7H_2O$ was used as the source for the Fe(II) solution. In the solid phase reaction experiments, FeOOH was prepared from $Fe_2(SO_4)_3 \cdot nH_2O$. In both co-precipitation and adsorption experiments, the As(V) and Fe(III) solutions were prepared from $Na_2HAsO_4 \cdot 7H_2O$ and $Fe(NO_3)_3 \cdot 9H_2O$, respectively. The pH and ionic strength were adjusted by the addition of 0.1 M HNO_3, KOH and KNO_3. The ionic strength was fixed at 0.05. All experiments were conducted at 25 ℃.

Oxidation Reaction Experiments

First, pure water was sparged with 0.5 L/min N_2 gas to remove dissolved oxygen (DO). After the DO was reduced by more than 98%, the target amount of $FeSO_4 \cdot 7H_2O$ was dissolved in the pure water and the experiments started. During the experiments, the pH value was kept constant by the addition of 1 N NaOH using an automatic titrator (Kyotodenshi, APB-410, AT-400), and the pH, DO and oxidation-reduction potential (ORP) were measured.

Solid Phase Reaction Experiments

FeOOH precipitation was conducted using the same conditions as for the oxidation reaction experiments. That is, the DO content of the pure water used for these experiments was kept under 2 % of its saturation value. The target amount of $Fe_2(SO_4)_3 \cdot nH_2O$ was dissolved in the pure water. After this, the pH of the water was maintained at 8.0 by using 6 N NaOH, and ultimately FeOOH precipitated.

$Fe(OH)_2$ formation was carried out using the same conditions as for the oxidation reaction experiments; however, a reductive atmosphere was maintained by the addition of 0.4 L/min N_2 gas during the experiments. After $Fe(OH)_2$ formation, FeOOH suspension was added to react with the $Fe(OH)_2$ suspension. During these experiments, the pH was kept constant using 1 N NaOH.

Co-Precipitation Experiments

The co-precipitation experiments involved the formation of ferrihydrite in the presence of As(V) ions. Fe(III) and As(V) solutions were initially combined at pH 2 to maintain the ionic states of Fe(III) and As(V). The pH and ionic strength were then adjusted to the target levels and were maintained for 1 h with agitation. The pH was kept within ± 0.5 of the target value by the addition of a few drops of 0.1 M KOH during agitation. After agitation, the suspension was filtered through a 0.1 μm membrane filter, and the filtrate was analyzed using ICP-AES (Seiko Instruments, SPS-7800, Chiba, Japan) to determine the concentrations of Fe(III) and As(V). The initial As(V) concentration was fixed at 10, 20 or 30 mg dm^{-3} (0.13, 0.27 or 0.40 mmol dm^{-3}) , and the initial Fe(III) concentration was varied to give initial As/Fe molar ratios from 5 to 0.125.

Adsorption Experiments

Adsorption experiments were conducted using the same conditions as for the co-precipitation experiments, except that the ferrihydrite suspensions and As(V) solutions were prepared separately using twice the initial concentration of As(V) and Fe(III). The pH and ionic strengths of both solutions were adjusted to the target levels, and equivalent volumes of the suspension and solution were then combined and agitated for 1 h. The pH was maintained within ± 0.5 of the target value by the addition of a few drops of 0.1 M KOH during agitation. After agitation, the Fe(III) and As(V) concentrations of the filtrate were analyzed using the procedure described above. In all the adsorption experiments, freshly prepared ferrihydrite was used to avoid changes caused by aging.

X-ray Diffraction (XRD) Analysis

The filter cakes from the co-precipitation and simple adsorption experiments were analyzed using XRD (RINT Ultima III, Rigaku, Inc., Tokyo, Japan). To produce the filter cakes for XRD analysis, the initial Fe(III) concentration was adjusted to 10 mg dm^{-3} and the initial As(V) concentrations were varied according to the target As/Fe molar ratio. The filter cakes from the co-precipitation and adsorption experiments were freeze-dried at -45°C and 10 Pa for at least 24 h to avoid crystallization or mineralogical transformations. Powder XRD patterns were obtained using a copper target (Cu-Kα), a crystal graphite monochromator, and a scintillation detector. The X-ray source was operated at 40 kV and 30 mA with step scanning from 2θ values of 2-80°, in increments of 0.02°. A crystal sample holder was used and the patterns were not corrected for background diffraction.

X-ray Adsorption Fine Structure (XAFS) Analysis

The X-ray absorption near-edge structure (XANES) analysis, which is a kind of XAFS analysis, was performed at the beam line BL-12C in the Photon Factory (PF) of the National Laboratory for High-Energy Physics, Tsukuba, Japan. The K edge of As was scanned in the range of 11535 to 12611 eV to obtain XANES spectra. Ar (100%) gas and N_2 (100%) gas were used for the incident beam and the transmission beam, respectively. Moreover, the K edge of Fe was scanned in the range of 6781 to 7856 eV to obtain XANES spectra. The ionization chamber was filled with N_2 (50%) + Ar (50%) gas to monitor the incident beam, whereas N_2 (100%) gas was used for the transmission beam. The measurement samples obtained from both co-precipitation and simple adsorption processes were freeze-dried at -45°C and 10 Pa at least 24 h to avoid crystallization or mineralogical transformations. Then for the transmission method, the samples were blended with boron nitride (BN) powder and pressed under 20 kN to create thin tablet shaped XAFS samples.

Ferrihydrite and poorly crystalline ferric arsenate were synthesized as reference materials to be compared with the forms of the Fe(III) or As(V) precipitates.

RESULTS AND DISCUSSION

Oxidation Reaction Experiments

Oxidation reaction experiments were conducted under the conditions of 5 mmol dm^{-3} Fe(II) concentration, pH 10 and a stirring rate of 300 rotation per minute (rpm). From the experimental results, it was confirmed that the oxidation reaction was divided into three parts. In the first part, the oxidation reaction of Fe(II) depended on the transport of DO. This was followed by the second part where the transport of DO was preferred to the Fe(II) oxidation reaction because the concentration of Fe(II) had already decreased. Therefore, the concentration of DO increased in this part. In the third part, the concentration of DO increased by the further transport of DO, because the Fe(II) oxidation reaction was completed in this part. It is noteworthy that the concentration of DO over time is represented in equation 1.

$$\frac{dc}{dt} = k_1(C_S - C) - r \quad (1)$$

Where C_S (mg dm^{-3}) represents the saturated DO concentration which is 8.11 (mg dm^{-3}) at 25 degrees, k_1 (min^{-1}) is the overall oxygen transfer capacity coefficient, and r (mg dm^{-3} min^{-1}) is the oxygen consumption rate of the oxidation reaction. In first part of the reaction series, equation 1 could be changed to equation 2 because dc/dt = 0 and c = 0.

$$r = k_1 C_S \quad (2)$$

Equation 2 suggests that the oxidation rate of Fe(II) was determined by k_1 at 25 ℃. That is, we could evaluate the oxidation rate by k_1. k_1 reflects the physical conditions such as stirring rate, aeration rate, etc. As a result, in the experiments, k_1 was not dependent on the pH values.

Solid Phase Reaction Experiments

Lee et al. (1994) reported that Fe_3O_4 was formed by solid/solid reaction when $Fe(OH)_2$ was in contact with FeOOH in the dispersion. This reaction is shown by equation 3 or equations 4 and 5.

$$Fe(OH)_2\ (s) + 2\ FeOOH\ (s) \leftrightarrow Fe_3O_4\ (s) + 2\ H_2O \quad (3)$$

$$Fe(OH)_2\ (s) + 2\ Fe^{3+} + 2\ H_2O \leftrightarrow Fe_3O_4\ (s) + 6\ H^+ \quad (4)$$

$$2\ FeOOH\ (s) + Fe^{2+} \leftrightarrow Fe_3O_4\ (s) + 2\ H^+ \quad (5)$$

From these equations, the formation of Fe_3O_4 is easy under high pH conditions. However, the yield of Fe_3O_4 was greater at pH values above 9.5 and lower at pH values below pH 9.5. The effect is related to the crystallization speed. Tamaura and Buduan (1981) reported that the formation of Fe_3O_4 by solid/solid reaction was dependent on $FeOH^+$ which is a Fe(II) hydrolysis species adsorbed on the surface of FeOOH. In this instance, the formation of Fe_3O_4 by solid/solid reaction was divided into three parts. First part is the hydrolysis reaction. That is $Fe(OH)_2$ was transformed into $FeOH^+$ in suspension. The second part is that $FeOH^+$, which is formed by the hydrolysis reaction, is adsorbed on the surface of FeOOH. The third part is the crystallization of Fe_3O_4. The hydrolysis reaction of $Fe(OH)_2$ is given by equation 6.

$$Fe(OH)_2\ (s) + H^+ \leftrightarrow FeOH^+ + H_2O \quad (6)$$

The adsorption reaction is represented by the surface complexation model (Dzombak & Morel, 1990). Table 1 shows the adsorption reaction considered in this study (Appelo et al., 2002). Among these adsorption reactions, the principal reaction is represented by equation 7 (Zhang, 1992; Liger et al., 1999).

$$\equiv FeOH^0 + FeOH^+ \leftrightarrow \equiv FeOFeOH + H^+ \quad (7)$$

From the Tamaura and Buduan (1981) report, ≡FeOFeOH is gradually changed to Fe_3O_4. This reaction is represented by equation 8.

$$\frac{d[Fe_3O_4\ (s)]}{dt} = k_2[\equiv FeOFeOH] \quad (8)$$

Where $[Fe_3O_4\ (s)]$ is the concentration of Fe_3O_4 (s) and k_2 is the crystallization rate constant from ≡FeOFeOH to Fe_3O_4. Thus, equation 8 can be changed to equation 9.

$$\frac{d[Fe_3O_4\ (s)]}{dt} = k_3[\equiv FeOFeOH] \quad (9)$$

Where k_3 (min^{-1}) is the solid-phase reaction rate constant represented by $k_2E\theta$, where θ (-) is the percentage of ≡FeOFeOH on the surface of FeOOH and E (mol/mol-Fe) is the exchange capacity. In addition, equation 9 can be transformed into equation 10 by using equation 3.

$$\frac{d[Fe_3O_4\ (s)]}{dt} = -2k_3[FeOOH\ (s)] \quad (10)$$

Equation 10 suggests that, over time, changes in the FeOOH concentration depend on the primary reaction at the very beginning of the reaction sequence. Therefore, it is very important to know the k_3 value to evaluate the amount of Fe_3O_4 formation. From the experimental results, it was confirmed that k_3 was affected by the pH value. k_3 value has its highest value at a pH near 9.7.

Table 1 – Equilibrium reactions and constants for adsorption on ferrihydrite

Reactions	Log K^{int}
$\equiv FeOH^0 + H^+ \leftrightarrow \equiv FeOH_2^+$	7.29
$\equiv FeOH^0 \leftrightarrow \equiv FeO^- + H^+$	-8.93
$\equiv Fe_WOH^0 + Fe^{2+} \leftrightarrow \equiv Fe_WOFe^+ + H^+$	-2.98
$\equiv Fe_WOH^0 + Fe^{2+} + H_2O \leftrightarrow \equiv Fe_WOFeOH + 2H^+$	-11.55
$\equiv Fe_SOH^0 + Fe^{2+} \leftrightarrow \equiv Fe_SOFe^+ + H^+$	-0.95
$\equiv FeOH^0 + SO_4^{2-} \leftrightarrow \equiv FeOSO_4^- + H_2O$	7.78
$\equiv FeOH^0 + SO_4^{2-} \leftrightarrow \equiv FeOHSO_4^{2-}$	0.79

Surface Complexation Model For As(V) With Ferrihydrite

As(V) adsorption/co-precipitation experiments with ferrihydrite were conducted at pH 5 and 7. As a result, it was confirmed that the adsorption isotherm was Langmuir type; however, it was BET type for co-precipitation as graphically presented in Figure 1. The XRD pattern of the co-precipitates changed near the initial molar ratio of As/Fe of 0.5 (Figure 2). That is, the peak position shifted from 34° in ferrihydrite to 28° for poorly crystalline $FeAsO_4$. However, the XRD pattern of the adsorption sample did not change when the initial molar ratio of As/Fe changed, as shown in Figure 2. These results indicate that surface precipitation, which generates poorly crystalline $FeAsO_4$, occurred only during the process of co-precipitation. Thus, the mechanism of removal of As(V) on ferrihydrite by co-precipitation was not only surface complexation but also surface precipitation.

Moreover, an attempt to calculate the percentage of surface precipitation in the co-precipitates using XAFS analysis was done. Figure 3 shows the K edge of the As XANES spectra for the As(V) co-precipitates with various initial As/Fe molar ratios at pH 5 and 7. The results showed that, as the initial As/Fe molar ratio increased, the spectra for co-precipitation deviated from the peak position of poorly crystalline $FeAsO_4$. The peak for co-precipitation started changing from 11.88 keV, as the initial As/Fe molar ratio increased. This behaviour is dissimilar to that observed for simple adsorption. These results suggested that both the XRD analysis and the XAFS analysis have similar trends. Furthermore, by comparing both spectra at pH 5 and at pH 7, it was confirmed that surface precipitation occurred more significantly at pH 7 than at pH 5.

Table 2 shows the sorption densities obtained from the As(V) co-precipitation experiments and the surface precipitation rates obtained from the pattern fitting results for the XANES spectra derived from the As(V) co-precipitated ferrihydrite. The curve fitting range was done from 11.87–11.90 keV. From these results, it was also confirmed that the percentage of surface precipitation increased as the initial As/Fe molar ratio increased and that the percentage of surface precipitation at pH 7 was higher than that at pH 5, when the initial As/Fe molar ratio is same.

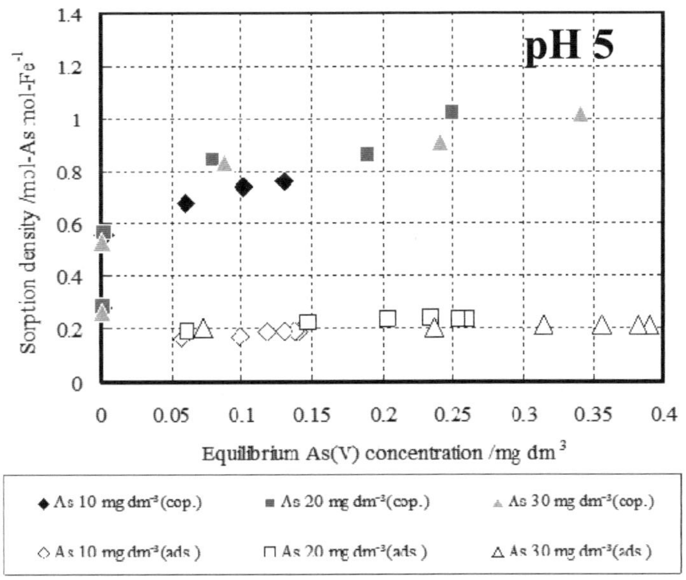

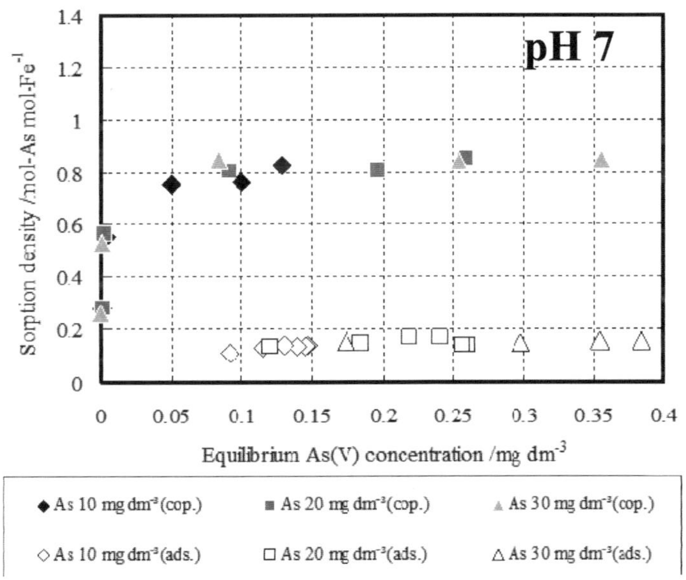

Figure 1 – Sorption isotherm of As(V) co-precipitation and adsorption with ferrihydrite at pH 5 and 7

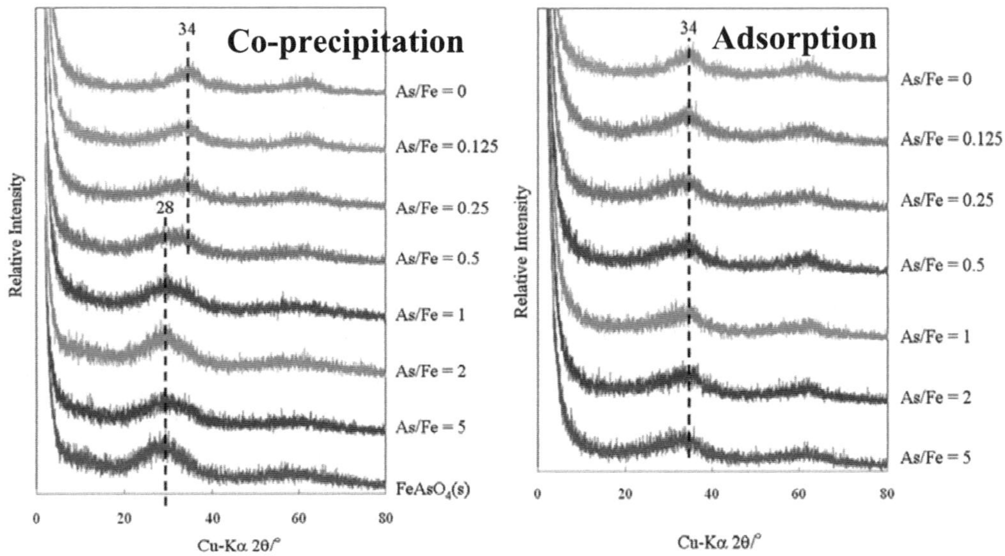

Figure 2 – XRD pattern of As(V) co-precipitation and adsorption with changing initial As/Fe molar ratio

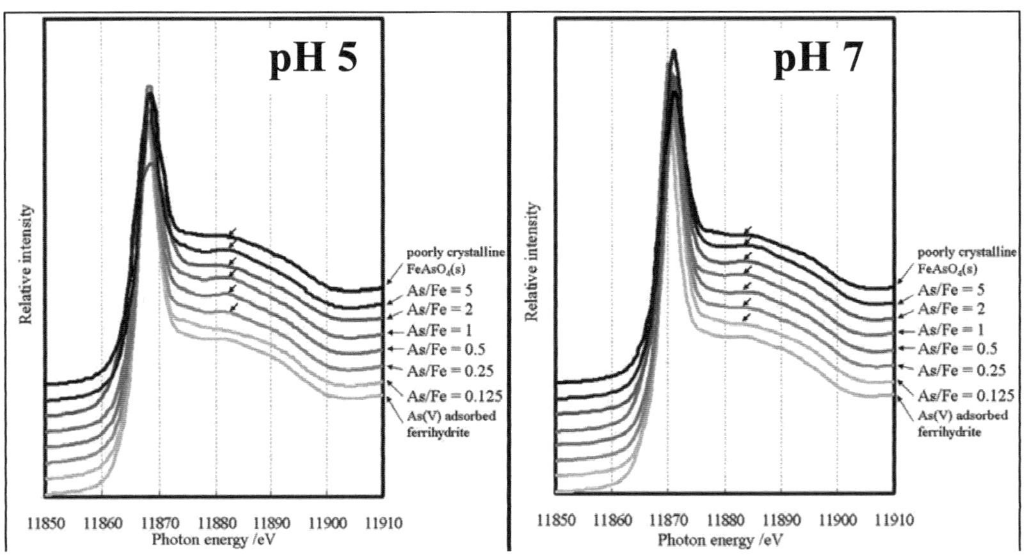

Figure 3 – K edge of the As XANES spectra of the As(V) co-precipitates with changing initial As/Fe molar ratio at pH 5 and 7

Table 2 – Sorption densities obtained from the As(V) co-precipitation experiments and surface precipitation rates obtained from the pattern fitting results for XANES spectra of As(V) co-precipitated ferrihydrite

As/Fe	pH 5		pH 7	
	Sorption density /mol–As mol–Fe^{-1}	Surface precipitation rate /%	Sorption density /mol–As mol–Fe^{-1}	Surface precipitation rate /%
0.125	0.12	35	0.15	56
0.25	0.25	56	0.29	70
0.5	0.40	76	0.45	82
1	0.67	86	0.74	84
2	0.90	90	0.88	92

CONCLUSIONS

The Fe_3O_4 formation processes were evaluated by both the reaction rate of the oxidation reaction and the rate of the solid phase reaction. In the oxidation reaction, it was confirmed that the oxidation rate depended on the overall oxygen transfer capacity coefficient (k_1) because the oxidation reaction was depended on the transport of oxygen. For the solid phase reaction, the equation which represented the change of the FeOOH concentration with time, was confirmed in this study. The solid-phase reaction rate constant was greatly depended on the pH, and the reaction rate constant had its highest value near pH 9.5.

The As(V) removal mechanism during ferrihydrite precipitation was experimentally investigated by comparing the co-precipitation and adsorption processes. The co-precipitation and adsorption experiments were separately conducted at pH 5 and 7. The results were compared using isotherms, XRD analysis and XAFS analysis to investigate the As(V) removal mechanisms. It was confirmed that As(V) removal with ferrihydrite mainly involved surface complexation in both co-precipitation and adsorption processes. Surface precipitation in the co-precipitation processes was observed only when the initial As/Fe molar ratio was greater than 0.5. Moreover, the percentage of surface precipitation in the co-precipitates was calculated by XAFS analysis. As a result, it suggested that the percentage of surface precipitation increases as the initial As/Fe molar ratio increases and that the percentage of surface precipitation at pH 7 was higher than that at pH 5 when the initial As/Fe molar ratio was the same.

REFERENCES

Appelo, C. A. J., Van Der Weiden, M. J. J., Tournassat, C. & Charlet, L. (2002). Surface complexation of ferrous iron and carbonate on ferrihydrite and the mobilization of arsenic. *Environment Science and Technology*, 36, 3096–3103.

Bernal, J. D., Dasgupta, D. R. & Mackay, A. L. (1959). The oxides and hydroxides of iron and their structural inter-relationships. *Clay Miner*, 4, 15–30.

Dold, B. (2008). Sustainability in metal mining: from exploration, over processing to mine waste management. *Review Environmental Science & Biotechnology*, 7, 275–285.

Dzombak, D. A. & Morel, F. M. M. (1990). Surface Complexation modeling: hydrous ferric oxide. John Wiley & Sons, New York City, NY, USA.

Hayashi, H., Kanie, K., Shinoda, K., Muramatsu, A. Suzuki, S. & Sasaki, H. (2009). pH-dependence of selenite removal from liquid phase by reductive Fe(II)-Fe(III) hydroxysulfate compound green rust. *Chemosphere*, 76, 638–643.

Japan Oil, Gas and Metals National Corporation (JOGMEC) (2003). Sansei Kouhaisui Mondai no Taisho Doukou Nitsuite, JOGMEC Current Topics [in Japanese]. Retrieved from internet web site http://mric.jogmec.go.jp/public/current/03_03.html.

Japanese Ministry of Economy, Trade and Industry (2006). Waga Kuni no Kougai Boushi Shisaku Nitsuite. Central Environmental Council [in Japanese]. Retrieved from internet web site http://www.env.go.jp/council/09water/y099-04/mat03.pdf.

Lee, H. S., Kang, N. K. & Oh, J. H. (1994). Magnetite formation of oxidation of $Fe(OH)_2$ suspension. *Journal of MMIJ*, 110, 297–306.

Liger, E., Charlet, L., & Cappellen, L.V. (1999). Surface catalysis of uranium(VI) reduction by iron(II). *Geochim Cosmoshim Acta*, 63, 2939–2955.

Tamaura, Y. & Buduan P.V. (1981). Studies on the oxidation of iron(II) ion during the formation of Fe_3O_4 and α-FeOOH by air oxidation of $Fe(OH)_2$ suspensions. *J. Chem. Soc. Dalton*, 9, 1807–1811.

Tokoro, C., Takao, D. & Sasaki, H. (2006). Quantitative evaluation of ferrite formation at ambient temperature in acid mine drainage containing iron. *Journal of MMIJ*, 122, 155 – 162.

Zhang, Y. (1992). Adsorption of protons, Fe(II) and Al(III) on lepidocrocite (γ-FeOOH). *Colloids and surfaces*, 63, 259–268.

TESTING AND APPLICATION OF IRON PRECIPITATION AS GOETHITE (FEOOH) USING A STRONG AIR FLOW

DAI Xing-zheng[1] and* ZENG Peng[2], LI Yong-xiang[1], HUANG Meng-yang[2], YANG Mei-yan[2], CHEN, Guo-mu[2], LI Guo-feng[2], WANG Hong-liang[2]

[1]*Yunnan Copper Group Limited Liability Company Technical Centre*
Kunming, China 650051
*(*Corresponding author: peng.zeng@ynz.yunnancopper.com)*

[2]*Yunnan Yuntong Zinc Co., Ltd.*
Kunming, China 650120

ABSTRACT

The removal of iron as goethite from the high iron bearing solutions generated in zinc hydrometallurgy using a strong air flow device was investigated. The influence of oxygen flow, copper ion concentration, temperature and reaction time on iron precipitation as goethite was studied. The optimum conditions were: oxygen flow 0.2-0.35 m^3/h, copper ion concentration 0.25-0.5 g/L, temperature 80-90°C, zinc oxide powder neutralizing agent and a reaction time of 5 to 7 hours. Under the optimal conditions, the Fe^{2+} concentration was reduced to 0.025 g/L after iron precipitation, and the goethite residue contained 38.7 wt.% Fe. The SEM and XRD analyses showed that the iron was present as well formed FeOOH crystals in the iron residue, and the amount of iron sulphate was very low. Production practice shows that, when the average iron concentration in the feed solution was 11.61 g/L and 4.08 g/L after precipitation, the weight of iron removed was 1,439 kg/day. When the average iron concentration in the feed solution was 14.51 g/L and 2.73 g/L after precipitation, the weight of iron removed was 2,454 kg/day. This represents a ~70 percent increase in iron removal.

KEYWORDS

Zinc hydrometallurgy, goethite, iron removal, strong airflow

INTRODUCTION

The objective of iron removal is to produce a small volume of iron residue, and our industrial goethite process meets this requirement. The high Fe(II) solution in our reactor is oxidized/hydrolyzed to precipitate goethite and the precipitate is sent to a pyrometallurgical furnace for final residue treatment. Some literature discusses (Deng Yong-gui et al., 2008; Mei guang-gui et al., 2001; Zhong zhu-qian et al., 1980) research on iron hydrolysis; one of the reported rate limiting parameters was the low availability of the oxidant when using air and another was the inhomogeneous air distribution. There are two iron hydrolysis-goethite precipitation methods; the Electrolytic Zinc process (EZ) and the Vieille Montagne process (VM) (Lei Xiang, 2006; Xu guo-qiang, 2007; Zhao Yong et al., 2005). The EZ method hydrolyzes the Fe^{3+} directly to FeOOH. The weakness of the EZ process is the low recovery rate of the associated minor metals. The VM method reduces the iron to the ferrous state and then oxidizes the Fe(II) to Fe(III), with simultaneous hydrolysis/precipitation of the Fe(III) as FeOOH. Before iron hydrolysis, the minor metals, such as indium, could be selectively precipitated, because Fe(II) requires a high pH for hydrolysis. So the VM method has a high recovery rate for the minor metals. The relevant goethite precipitation reaction is:

$$2Fe^{2+} + 1/2O_2 + H_2O + 2ZnO = 2FeOOH + 2Zn^{2+} \quad (1)$$

In this study, the major conditions for iron hydrolysis-goethite precipitation, such as oxygen consumption, copper ion concentration, reaction temperature, reaction time and different neutralizing agents were investigated. Based on these tests, the goal was to identify the optimal conditions for iron hydrolysis in the hydrometallurgical zinc industry and to provide the conditions to improve the iron hydrolysis-goethite precipitation operation at the plant site.

THE TEST MATERIAL ANALYSIS AND THE TESTING DEVICE

Test Material Analysis

Chemical Analysis and Semi-quantitative Analysis

The chemical composition of the Waelz zinc oxide neutralizing agent tested is shown in Table 1. The semi-quantitative X-ray fluorescence analysis of the tested Waelz zinc oxide is shown in Table 2.

Table 1 - Chemical composition of the Waelz zinc oxide

Item	H_2O	Zn	Cd	In	S	Pb	F	Cl
Content (wt.%)	6.50	66.44	1.00	0.095	14.15	5.39	0.089	0.48

From Table 1, the major impurities in the Waelz zinc oxide are cadmium, indium, lead, fluorine and chlorine. The oxide has a high lead content.

Table 1 presents the quantitative chemical analysis results for zinc, lead, iron and sulphur. From the semi-quantitative analyses of Table 2, minor elements such as indium, tin, arsenic, bismuth and aluminum were found.

Table 2 - Semi-quantitative X-ray fluorescence analysis of the Waelz zinc oxide

Atomic number	Element	Content (wt.%)	Atomic number	Element	Content (wt.%)
8	O	18.8	28	Ni	0.004
12	Al	0.055	29	Cu	0.35
13	Si	0.177	30	Zn	66.79
16	S	3.186	33	As	0.757
17	Cl	0.133	38	Sr	0.0158
18	K	0.119	48	Cd	0.71
20	Ca	0.38	49	In	0.12
22	Ti	0.013	50	Sn	0.17
24	Cr	0.007	82	Pb	5.012
25	Mn	0.0765	83	Bi	0.0903
26	Fe	3.078			
27	Co	0.001		Total	100.04

X-ray Diffraction Analysis (XRD) of Waelz Zinc Oxide

Some Waelz zinc oxide was blended and milled to minus 54um. The oxide was dried for two hours at a constant temperature of 120 °C, and then cooled to room temperature. Small samples were then taken for XRD analysis. The result is shown in Figure 1. From the figure, the zinc phases in the Waelz zinc oxide are ZnO, ZnS and $ZnSO_4$; the lead phase is $PbSO_4$.

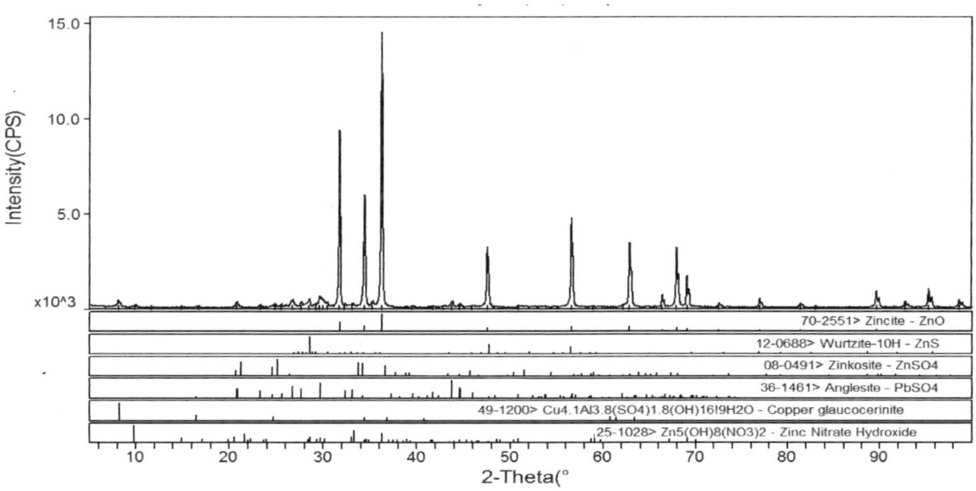

Figure 1 - XRD pattern of the Waelz zinc oxide

Test Equipment

The iron removal test equipment used a special air distribution device. It could disperse the oxygen or air quickly and well. This equipment also had an agitator.

Figure 2 - The testing equipment for iron removal

TESTING RESULTS AND DISCUSSION

A liquid sample before testing analyzed 0.041 g/L Cu(II) and 20.6 g/L Fe(II) at pH 3.0-3.5. For a test, 0.8 L of liquid was added to a 2 L beaker, and the beaker was placed in the iron removal equipment. Another 0.8 L of liquid was added slowly to the beaker, and the contents were stirred at a speed of 400 rpm. The neutralizer was the Waelz zinc oxide (the composition is shown in Table 1). The quantity of the Waelz zinc oxide added was based on the pH control value. Under these experimental conditions, the influence of the oxygen flow, copper ion concentration, temperature and reaction time on the iron precipitation process were researched.

The Influence of the Oxygen Flow on Iron Removal

The copper concentration of the iron test liquid was adjusted to 1 g/L Cu(II). At the 85°C temperature and 7 hour reaction time, the influence of oxygen flow rates of 0.1, 0.15, 0.2, 0.25, 0.3 and 0.35 m^3/h on the extent of iron removal were researched. The test results are shown in Figures 3 to 5.

Figure 3 shows that, when the oxygen flow rate was 0.15 m^3/h, iron removal could be as high as 97.5%. The Fe(II) in solution at the end of the test was 0.65 g/L; based on this value, the calculated oxygen consumption for iron removal was 0.085 m^3/g. When the oxygen flow was 0.2 m^3/h, the iron removal could be as high as 99.1%. The Fe(II) in solution at the end of the test was 0.257 g/L; based on this value, the calculated oxygen consumption for iron removal was 0.085 m^3/g. When the oxygen consumption was higher, the iron removal efficiency was not any better. Based on these tests, the oxygen flow rate was fixed at 0.2 m^3/h.

From Figure 5, the major phases in the precipitate are FeOOH, $Fe_2O_3 \cdot ZnO$, $KFe_3(SO_4)_2(OH)_6$ and SiO_2. The major iron phase was goethite (FeOOH); a minor iron phase was jarosite, $KFe_3(SO_4)_2(OH)_6$. The $Fe_2O_3 \cdot ZnO$ phase came from the Wealz oxide. When the Wealz oxide was used as the neutralizer, the $Fe_2O_3 \cdot ZnO$ did not dissolve, but remained mixed with the residue.

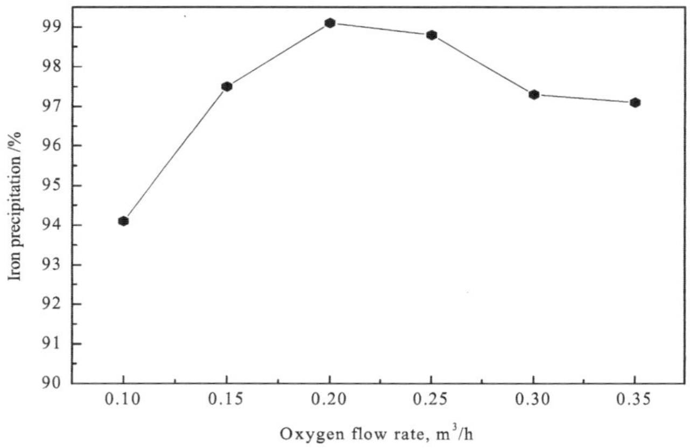

Figure 3 - The influence of the oxygen flow rate on iron removal

Figure 4 - SEM micrograph of the goethite

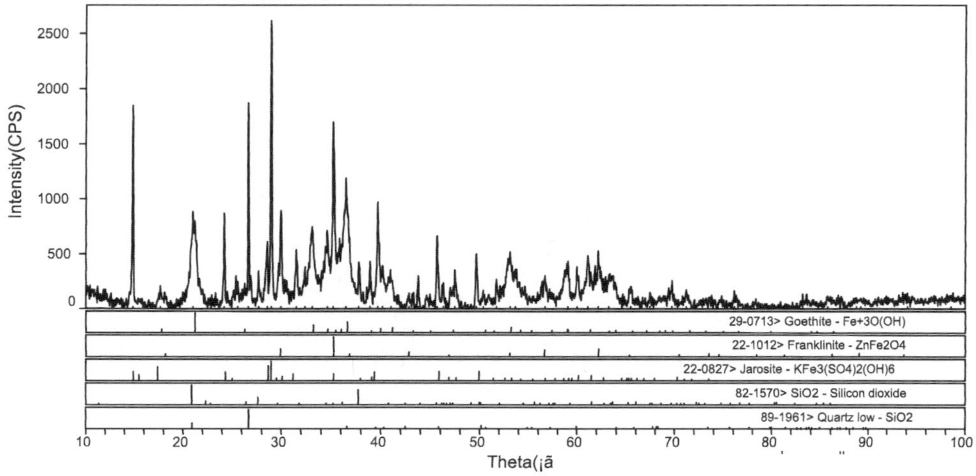

Figure 5 - XRD pattern of the goethite formed during testing

The Influence of the Cu (II) Concentration on the Extent of Iron Removal

The effect of added Cu(II) concentrations of 0.25, 0.5, 0.75, 1.0 and 1.25 g/L on the extent of iron removal were researched under the following test conditions: oxygen flow rate 0.2 m^3/h, temperature 85 °C and reaction time 7 hours. The tests results are shown in Figures 6, 7 and 8.

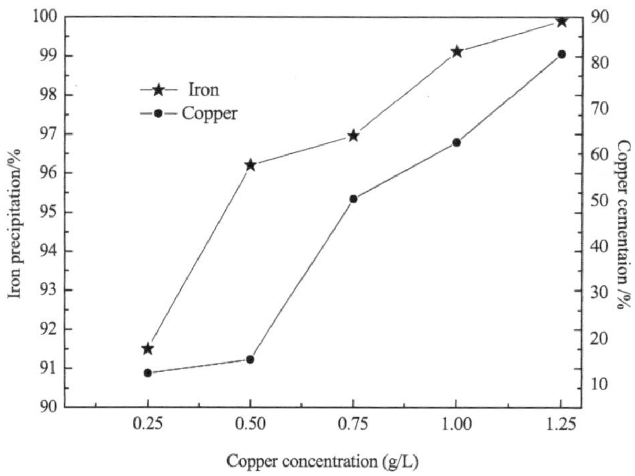

Figure 6 - The influence of the Cu(II) concentration on the amount of iron precipitated

When the copper(II) concentration became higher in the tests, iron removal became faster and was more complete (Figure 6). When the Cu(II) concentration was 0.5 g/L, the efficiency of iron removal was 96.2%; the Fe(II) concentration after testing was 0.87 g/L. Under these conditions, about 15% of the copper deposited with the goethite. The higher Cu(II) concentration resulted in a higher iron removal efficiency, but more copper was precipitated with the goethite. The high copper loss is uneconomical. From this point, the Cu(II) concentration was fixed at 0.5 g/L.

Figure 7 - SEM micrograph of the goethite

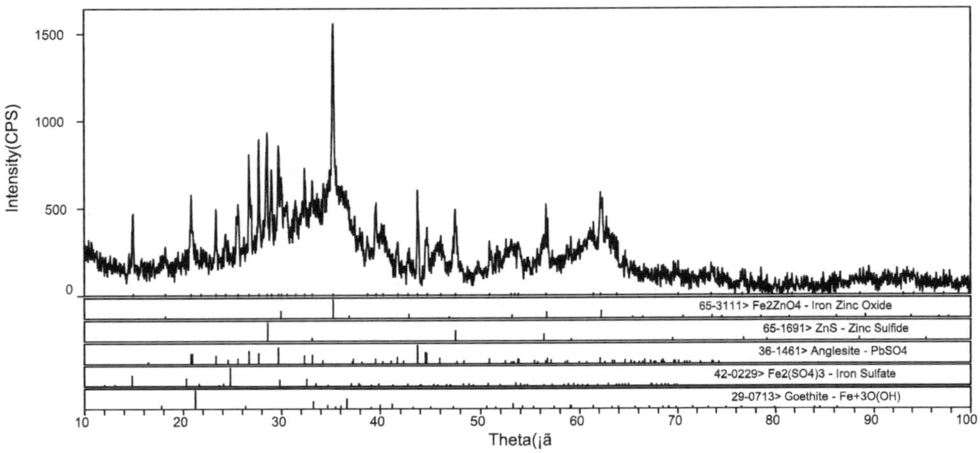

Figure 8 - XRD pattern of the goethite form the test

From the Figure 8, it is apparent that the major phases in the iron precipitate are: FeOOH, $Fe_2O_3 \cdot ZnO$, $Fe_2(SO_4)_3$ and $PbSO_4$. Most of the iron precipitated as FeOOH; $KFe_3(SO_4)_2(OH)_6$ was not detected. The $Fe_2O_3 \cdot ZnO$, $PbSO_4$ and ZnS phases originated from the Waelz oxide.

The Influence of the Reaction Temperature on the Iron Removal Process

The test program examined the influence of the reaction temperature (60 °C, 70 °C, 80 °C and 90 °C) on the extent of iron removal under the following test conditions: oxygen flow of 0.2 m³/h, Cu(II) concentration of 0.5 g/L, and reaction time of 7 hours. The test results are presented in Figure 9.

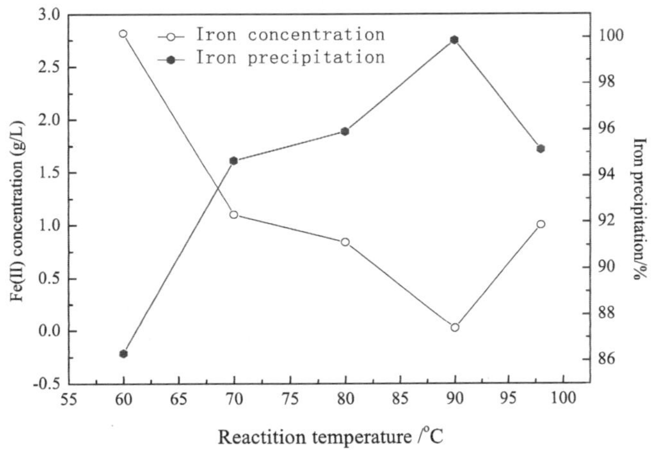

Figure 9 - The influence of the reaction temperature on the extent of iron removal

As shown in Figure 9, when the reaction temperature increased, the iron removal efficiency became higher. When the reaction temperature was 80 °C, the efficiency of iron removal was 95.92%; the Fe(II) concentration after testing was 0.84 g/L. When the reaction temperature was 90 °C, the efficiency of iron removal was 99.87%, and the Fe(II) concentration after the testing was 0.025 g/L. Accordingly, the reaction temperature was fixed between 80 °C and 90 °C.

The influence of the Reaction Time on the Extent of Iron Removal

The test program explored the influence of reaction times of 2.5, 3, 4, 5, 6, 7 and 8 hours on the extent of iron removal, under the following test conditions: oxygen flow 0.2 m^3/h, Cu(II) concentration 0.5 g/L and reaction temperature 85 °C. The test results are given in Figure 10.

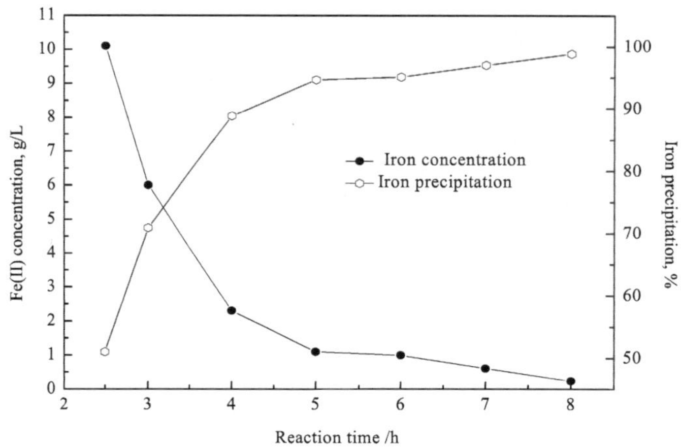

Figure 10 - The influence of the reaction time on the extent of iron removal

When the reaction time increased, the iron removal efficiency became higher, as illustrated in Figure 10. When the reaction time was 5, 6 and 7 hours, the efficiency for iron removal was 94.66%, 95.14% and 97.03%, respectively. The corresponding Fe(II) concentrations after testing were 1.1 g/L, 1 g/L and 0.63 g/L. Accordingly, the reaction time was fixed at 7 hours.

INDUSTRAL APPLICATION

After scale-up of the test device, it was installed in a 35 m^3 reaction tank. The industrial reaction conditions were determined according to the laboratory test results. The results before and after the industrial application of the new test device are shown in Tables 3 and 4.

Before the industrial application of the new device, the Fe(II) concentration was reduced from 11.61 g/L to 4.08 g/L, before and after reaction (Table 3). The iron removal efficiency was 65%, corresponding to 1,439 kg iron per day. After the industrial application, the Fe(II) concentration was reduced from 14.51 g/L to 2.73 g/L before and after reaction (Table 4). The iron removal efficiency was 81%, corresponding to 2,454 kg iron per day. With the new iron removal device for the goethite process and the optimized iron removal conditions, the iron removal efficiency was greatly improved.

Table 3 - The results before the industrial application of the new test device

Date	Fe(II) before reaction (g/L)	Fe(II) after reaction (g/L)	Iron removal efficiency (wt.%)
26-Aug-13	14.8	6.32	57%
27-Aug-13	14.73	4.26	71%
28-Aug-13	13	4.47	66%
29-Aug-13	12.88	3.81	70%
30-Aug-13	9.76	3.65	63%
31-Aug-13	8.01	3.62	55%
01-Sep-13	9.22	3.16	66%
02-Sep-13	9.73	3.51	64%
03-Sep-13	13.69	2.57	81%
04-Sep-13	8.4	2.19	74%
05-Sep-13	13.57	6.07	55%
06-Sep-13	13.42	4.19	69%
07-Sep-13	10.82	4.35	60%
08-Sep-13	10.51	4.9	53%
Average	11.61	4.08	65%

Table 4 - The results after the industrial application of the new test device

Date	Fe(II) before reaction (g/L)	Fe(II) after reaction (g/L)	Iron removal efficiency (wt.%)
26-Aug-14	15.75	2.89	82%
27-Aug-14	12.84	2.33	82%
28-Aug-14	9.1	1.54	83%
29-Aug-14	9.14	1.38	85%
30-Aug-14	15	1.64	89%
31-Aug-14	17.44	2.83	84%
01-Sep-14	15.83	2.64	83%
02-Sep-14	17.95	3.01	83%
03-Sep-14	18.88	2.73	86%
04-Sep-14	16.5	3.35	80%
05-Sep-14	17.66	6.57	63%
06-Sep-14	16.8	2.05	88%
07-Sep-14	12.29	2.41	80%
08-Sep-14	7.9	2.86	64%
Average	14.51	2.73	81%

CONCLUSIONS

In these experiments, the strong air dispersion device gave better iron removal results. The Fe(II) concentration after reaction can be as low as 0.025 g/L in laboratory test work. The optimized parameters were obtained from the iron removal test program. These conditions realized with the Waelz oxide neutralizing agent are: an oxygen flow of 0.2 m^3/h, a Cu(II) concentration of 0.5 g/L, a reaction temperature between 80 °C and 90 °C and a reaction time of 6 to 7 hours. The XRD analyses showed the iron phases in the precipitate to be mostly goethite (FeOOH) together with lesser amounts of potassium jarosite.

REFERENCES

Deng, Yong-gui, Chen, Qi-yuan, Yin, Zhou-lan, & Zhang, Ping-min (2008). Preparation of goethite seeds for removal of ferrous/ferric ions from leaching solution of zinc hydrometallurgy process. *The Chinese Journal of Nonferrous Metallurgy,* 6, 27-31.

Lei, Xian (2006). The technical study for the iron removal by goethite process. *China Nonferrous Metallurgy*, 6, 28-30.

Mei, guang-gui, Wang, de-rui, & Zhou jin-yuan (2001). *Zinc Hydrometallurgy*, Changsha, Central South University Press, 150-165.

Xu, guo-qiang (2007). The iron removal research in the hydrometallurgy $ZnSO_4$ production. *XinJiang Nonferrous Metals (supplement)*, 2007, 67-68.

Zhao, Yong, Jiang, Kai-xi, Wang, De-quan, & Guo Ya-hui (2005). Iron removal from hot acid leaching solution of zinc roasting dust by goethite process. *Nonferrous Metals*, 5, 13-15.

Zhong, zhu-qian, & Mei, zhi (1980). The high zinc calcine hot leaching experiment ($ZnSO_3$ reduced – goethite process). *Nonferrous Metals*, 2, 26-38.

Author Index

Iron Control: Practice and Research

4th International Symposium on Iron Control in Hydrometallurgy
46th Annual Hydrometallurgy Meeting